COLLEGE ALGEBRA

COLLEGE ALGEBRA

MICHAEL SULLIVAN

Chicago State University

DELLEN PUBLISHING COMPANY

San Francisco, California

COLLIER MACMILLAN PUBLISHERS

London

Divisions of Macmillan, Inc.

On the cover: The cover, "Untitled," is an etching executed by Sam Francis in 1986. The work measures $15\frac{1}{2}$ by 13 inches. Francis' work is in many public and private collections throughout the world, including the Museum of Modern Art, the Guggenheim Museum, and the Los Angeles County Museum. His work may also be seen at the Smith Andersen Gallery in Palo Alto and the André Emmerich Gallery in New York.

© Copyright 1987 by Dellen Publishing Company,
a division of Macmillan, Inc.

Printed in the United States of America

Permissions: Dellen Publishing Company
              400 Pacific Avenue
              San Francisco, California 94133

Orders: Dellen Publishing Company
              c/o Macmillan Publishing Company
              Front and Brown Streets
              Riverside, New Jersey 08075

Collier Macmillan Canada, Inc.

Library of Congress Cataloging-in-Publication Data

Sullivan, Michael
  College algebra.

  Includes index.
  1. Algebra. I. Title.
QA154.2.S84   1987   512.9   86-29211
ISBN 0-02-418331-8

Printing: 1 2 3 4 5 6 7 8 9   Year: 6 7 8 9 0

ISBN 0-02-418331-8

C O N T E N T S

Preface   ix

▤ 1 PRELIMINARIES   1

1.1   Real Numbers   2
1.2   The Real Number Line   13
1.3   Integer Exponents   24
1.4   Polynomials   30
1.5   Factoring Polynomials   37
1.6   Rational Expressions   46
1.7   Radicals   55
1.8   Rational Exponents   63
1.9   Complex Numbers   67
1.10 Scientific Notation; Approximations; Calculators   75
1.11 Geometry Topics   79
      Chapter Review   83

▤ 2 EQUATIONS AND INEQUALITIES   87

2.1   Linear Equations   88
2.2   Applications   99
2.3   Quadratic Equations   109
2.4   Quadratic Equations with a Negative Discriminant   125
2.5   Other Types of Equations   128
2.6   Linear Inequalities   136
2.7   Other Inequalities   143
2.8   Equations and Inequalities Involving Absolute Value   151
      Chapter Review   156

▤ 3 GRAPHS   161

3.1   Rectangular Coordinates   162
3.2   Graphs of Equations   172

3.3 The Straight Line   188
3.4 Variation   208
   Chapter Review   216

4 FUNCTIONS AND THEIR GRAPHS   219

4.1 Functions   220
4.2 The Graph of a Function   229
4.3 Graphing Techniques   247
4.4 Operations on Functions; Composite Functions   260
4.5 Inverse Functions   268
   Chapter Review   278

5 POLYNOMIAL AND RATIONAL FUNCTIONS   283

5.1 Quadratic Functions   284
5.2 Polynomial Functions   297
5.3 Dividing Polynomials; Synthetic Division   308
5.4 The Zeros of a Polynomial Function   320
5.5 Approximating the Zeros of a Polynomial Function   332
5.6 Rational Functions   337
5.7 Complex Polynomials; Fundamental Theorem of Algebra   354
   Chapter Review   360

6 EXPONENTIAL AND LOGARITHMIC FUNCTIONS   365

6.1 Exponential Functions   366
6.2 Compound Interest   378
6.3 Logarithms   385
6.4 Logarithmic Functions   397
6.5 Growth and Decay   406
   Chapter Review   413

7 SYSTEMS OF EQUATIONS AND INEQUALITIES   417

7.1 Solutions of Systems of Linear Equations: Substitution   418
7.2 Solutions of Systems of Linear Equations: Elimination   432
7.3 Solutions of Systems of Linear Equations: Matrices   439
7.4 Solutions of Systems of Linear Equations: Determinants   457
7.5 Solutions of Systems of Nonlinear Equations   469
7.6 Solutions of Systems of Linear Inequalities   478
   Chapter Review   485

≡ 8 MISCELLANEOUS TOPICS 491

8.1 Matrix Algebra 492
8.2 Linear Programming 511
8.3 Partial Fraction Decomposition 522
8.4 Vectors 529
       Chapter Review 544

≡ 9 INDUCTION; SEQUENCES 549

9.1 Mathematical Induction 550
9.2 The Binomial Theorem 555
9.3 Sequences 563
9.4 Arithmetic and Geometric Sequences 571
       Chapter Review 581

≡ 10 SETS; COUNTING; PROBABILITY 583

10.1 Sets and Counting 584
10.2 Permutations 591
10.3 Combinations 596
10.4 Probability 600
        Chapter Review 608

≡ 11 THE CONICS 611

11.1 Preliminaries 612
11.2 The Parabola 613
11.3 The Ellipse 623
11.4 The Hyperbola 634
        Chapter Review 647

≡ Tables

Table I   Exponential Functions 649
Table II   Common Logarithms 650
Table III Natural Logarithms 652

Answers A1
Index I1

Textbooks in mathematics must address a variety of interests and concerns if they are to be successful tools for the teacher and the student.

Intent/Purpose

As co-author of a calculus text and a professor at an urban public university, I am aware of the needs of the student who has completed 2 years of high school algebra and intends to go on to study calculus. At the same time, though, the needs of the student who intends to study finite mathematics, business calculus, or discrete mathematics, or who is not planning further study in mathematics, have not been neglected. A conscious effort has been made to serve each constituency through suitable motivation and encouragement.

Contents and Organization

The topics contained in this book were selected and organized with the different types of students who take this course in mind. Thus, throughout the book, a deliberate attempt is made to preview topics and techniques that will be used in later courses.

The first part of the book, Chapters 1–3, provides a detailed presentation of material needed for the later, essential, topics. Chapter 1, in the main, consists of review material. Although complex numbers are introduced in Chapter 1, this section may be postponed at the discretion of the instructor. If the instructor chooses to cover complex numbers later, Section 2.4 (Quadratic Equations with a Negative Discriminant) would also be postponed and then covered with Section 1.9. Whether complex numbers are covered early or late, all exercises dependent upon complex numbers are clearly identified in the directions for the exercise.

The coverage of geometry in Chapter 1 should also be noted. No review would be complete without a discussion of the Pythagorean Theorem and its converse, and the formulas for perimeter and area of a rectangle, area and circumference of a circle, and volume of a sphere.

For some students, Chapters 2 and 3 will represent new material; for others, they will be review. The large majority, of course, fall somewhere in between. Chapter 2 presents an example of careful presentation of topics for the benefit of both instructor and student. In Section 2.3, quadratic equations are treated solely from the real-number point of view; in Section 2.4, quadratic equations with complex-number solutions are discussed. This ensures flexibility of coverage for the instructor and clarity of understanding for the student.

The middle part of the book, Chapters 4–7, covers the essential topics of college algebra. In Chapter 4, special emphasis is placed on functions and the graphs of functions. Graphing is usually done in steps, all of which are illustrated. The graphing techniques introduced in Chapter 4 are utilized and reinforced using the new functions introduced in Chapters 5 and 6.

Because of the need to evaluate polynomial functions, both synthetic division and the nested form of a polynomial are utilized in Chapter 5 as alternatives to substitution. The nested-form technique will be especially appreciated by students of computer science. Precision-type results for finding the zeros of a polynomial, such as the Rational Root Theorem and Descartes' Rule of Signs, are introduced to facilitate graphing of polynomial and rational functions. Approximating results, such as upper and lower bounds on zeros and partition techniques, are discussed in a separate section to provide a clear sense of the difference between numerical and nonnumerical methods.

Chapter 6 treats the exponential and logarithmic functions in the detail necessary and with language consistent for subsequent use in calculus. Applications to both compound interest and growth and decay are given.

Chapter 7 discusses a variety of methods (four in all) for solving systems of linear equations. Sections on systems of nonlinear equations and systems of linear inequalities are also included.

The last part of the book covers topics that may be selected by determining the specific needs of the students. For example, Chapter 8 includes an introduction to matrices and linear programming, topics of use to students intent on taking finite mathematics. Also included is a section on vectors that would be of help to students who will go on to study calculus. Note that partial fraction decomposition is presented in Section 8.3, after systems of equations rather than after rational functions. In this way it provides not only an application for solving systems of linear equations, but also allows for a full and detailed discussion. This should prove useful to students going on to calculus.

Chapter 9 presents mathematical induction, the binomial theorem, and sequences, and Chapter 10 discusses sets, counting, and probability. Each of these chapters contains applications of interest to the student in computer science, business, and mathematics.

The final chapter in the book provides a detailed look at conics. Emphasis is on graphing, analyzing equations, and applications.

Writing Style

The writing style in this book is directed toward the reader; every effort has been taken to be clear, precise, and consistent. Whenever it seemed appropriate, encouragement has been offered; and whenever necessary, warnings have been given.

Applications

Every opportunity was taken to present understandable, realistic applications consistent with the abilities of the student, drawing from such sources as tax rate tables, the *Guinness Book of World Records*, and newspaper articles. For added interest, some of the applied exercises have been adapted from textbooks the students may be using in other courses (for example, economics, chemistry, physics, etc.).

Historical Notes

William Schulz of Northern Arizona University has provided historical context and information in anecdotes that appear as introductory material and at the ends of many sections. In some cases, these comments also include exercises and discussion of comparative techniques.

Examples, Exercises, and Illustrations

The text includes 530 examples and 4000 exercises, of which 500 are word problems. The examples are worked out in appropriate detail, starting with simple, reasonable problems and working gradually up to more complicated ones.

Exercises are numerous, well-balanced, and graduated. They include a number of exercises where the student will need a calculator for their solution; these are clearly marked. A few computer problems also have been included.

Illustrations are abundant, numbering over 450. Full use is made of a second color to help clarify and highlight.

A Word about Format and Design

Each chapter-opening page contains a table of contents for that chapter and an overview—often historical—of the contents.

New terms appear in boldface type where they are defined throughout. The more important definitions are shown in color. New terms always appear in the margin for easy reference and are listed in the Chapter Review at the end of every chapter.

Theorems are set with the word "Theorem" in the margin for easy identification; if it is a named theorem, the name also appears in the margin. When a theorem has a proof given, the word "Proof" appears in the margin to mark the beginning of the proof clearly, and the symbol ≡ is used to indicate the end of the proof.

All important formulas are enclosed by a box and shown in color.

Examples are numbered within each section and identified clearly with the word "Example" in the margin. The solution to each example appears immediately following the example with the word "Solution" in the margin to identify it. The symbol ≡ indicates the end of each solution.

Each section ends with an exercise set. Each chapter ends with a Chapter Review containing a vocabulary list, a selection of fill-in-the-blank questions (to test vocabulary and formulas), and a set of review exercises. Answers are given in the back of the book for all of the odd-numbered exercises.

Supplementary Material

Student's Solutions Manual to accompany College Algebra, by Christopher Lattin, consists of complete step-by-step solutions for the odd-numbered exercises.

Instructor's Solutions Manual to accompany College Algebra is the complement to the above volume. It contains step-by-step solutions to the even-numbered exercises.

DellenTest is an ipsTest which uses computer software to randomly generate tests using an IBM PC.

Transparency Acetates, prepared by Roger Carlsen, consist of approximately 150 transparencies and overlays that duplicate important illustrations in the text. Another 25 are supplementary to those found in the text.

Acknowledgments

Textbooks are written by an author, but evolve from an idea into final form through the efforts of many people. Before initial writing began, a survey was conducted which drew nearly 250 instructor responses. The manuscript then underwent a thorough and lengthy review process, including class-testing at Chicago State University. I would like to thank my colleagues and students at Chicago State, who cooperated and contributed to this text while it was being class-tested.

The following contributors, and I apologize for any omissions, have my deepest thanks and appreciation:

James Africh, Brother Rice High School
Steve Agronsky, Cal Poly State University
Joby Milo Anthony, University of Central Florida
James E. Arnold, University of Wisconsin, Milwaukee
Wilson P. Banks, Illinois State University
Dale R. Bedgood, East Texas State University
William H. Beyer, University of Akron
Trudy Bratten, Grossmont College
Roger Carlsen, Moraine Valley Community College
Duane E. Deal, Ball State University
Vivian Dennis, Eastfield College
Karen R. Dougan, University of Florida
Louise Dyson, Clark College
Don Edmondson, University of Texas, Austin
Christopher Ennis, University of Minnesota
Garret J. Etgen, University of Houston
W. A. Ferguson, University of Illinois, Urbana/Champaign
Merle Friel, Humboldt State University
Richard A. Fritz, Moraine Valley Community College
Wayne Gibson, Rancho Santiago College
Joan Goliday, Santa Fe Community College
Frederic Gooding, Goucher College
James E. Hall, University of Wisconsin, Madison
Brother Herron, Brother Rice High School
Kim Hughes, California State College, San Bernardino
Sandra G. Johnson, St. Cloud State University
Arthur Kaufman, College of Staten Island
Thomas Kearns, North Kentucky University
H. E. Lacey, Texas A & M University
Christopher Lattin, University of Florida
Stanley Lukawecki, Clemson University
Laurence Maher, North Texas State University
James Maxwell, Oklahoma State University, Stillwater
Eldon Miller, University of Mississippi
James Nymann, University of Texas, El Paso
Thomas Radin, San Joaquin Delta College
Ken A. Rager, Metropolitan State College
Stephen Rodi, Austin Community College
Howard L. Rolf, Baylor University
John Sanders, Chicago State University
John Spellman, Southwest Texas State University
Becky Stamper, Western Kentucky University

Neil Stephens, Hinsdale South High School
Tommy Thompson, Brookhaven College
Richard J. Tondra, Iowa State University
Marvel Townsend, University of Florida
Richard G. Vinson, University of Southern Alabama
Carlton Woods, Auburn University
George Zazi, Chicago State University

Recognition and thanks are due particularly to the following individuals for their invaluable assistance in the preparation of this book: Don Dellen, who orchestrated the development of this book and provided clear communication (particular thanks are due Don for his patience, support, and commitment to excellence); Phyllis Niklas, who coordinated the production and kept the schedule intact; Luana Morimoto, Barbara Holland, and Dennis Weiss, who showed such a sincere interest in the development of the book; Dawn Grimmett and Lorraine Bierdz, who struggled with my penmanship while typing the manuscript; Katy Sullivan and Marsha Vihon, who diligently checked my answers to the odd-numbered problems; and my wife Mary and our children, Katy, Mike, Danny, and Colleen, who gave up the dining room table and helped in the preparation of the manuscript in more ways than they'd like to remember.

COLLEGE ALGEBRA

PRELIMINARIES

1.1 Real Numbers
1.2 The Real Number Line
1.3 Integer Exponents
1.4 Polynomials
1.5 Factoring Polynomials
1.6 Rational Expressions
1.7 Radicals
1.8 Rational Exponents
1.9 Complex Numbers
1.10 Scientific Notation; Approximations; Calculators
1.11 Geometry Topics
Chapter Review

The word *algebra* is derived from the Arabic word *al-jabr*. This word is a part of the title of a ninth century work, "Hisâb al-jabr w'al-muqâbalah," written by Mohammed ibn Mûsâ al-Khowârizmî. The word *al-jabr* means "a restoration," a reference to the fact that if a number is added to one side of an equation, then it must also be added to the other side in order to "restore" the equality. The title of the work, freely translated, means "The science of reduction and cancellation." Of course, today, algebra has come to mean a great deal more.

▤ 1.1 REAL NUMBERS

Algebra can be thought of as a generalization of arithmetic in which letters take the place of numbers and symbols are used to indicate the common arithmetical operations of addition, subtraction, multiplication, and division. The laws of algebra are therefore based on the laws of arithmetic.

We will use letters such as x, y, a, b, and c to represent numbers. VARIABLE If the letter represents *any* number, it is called a **variable**; if the letter CONSTANT represents a *particular* number, it is called a **constant**. For example, in the formula for the area of a circle, $A = \pi R^2$, R is a variable representing the radius of the circle, A is a variable representing the area of the circle, and π (the Greek letter pi) is a constant, with a value of approximately 3.14. See Figure 1.

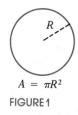

$$A = \pi R^2$$

FIGURE 1

EQUATION   The formula $A = \pi R^2$ is called an **equation**, and the symbol $=$,
EQUAL SIGN   called an **equal sign**, is used to express the idea that the number A on the left of the equal sign is the same as the number π multiplied by R squared on its right. Let's review three important properties of equality; in what follows, a and b represent any number.

REFLEXIVE PROPERTY   1. The **reflexive property** states that a number always equals itself; that is, $a = a$. Although this result seems obvious, it forms the basis for much of what we do in algebra.

SYMMETRIC PROPERTY   2. The **symmetric property** states that, if $a = b$, then $b = a$. This, too, is an often-used property.

PRINCIPLE OF SUBSTITUTION   3. The **principle of substitution** states that, if $a = b$, then we may substitute b for a in any expression containing a.

The familiar operations of addition, subtraction, multiplication, and division of two numbers a and b are symbolized in algebra by

Addition: $a + b$ Subtraction: $a - b$

Multiplication: $a \cdot b$ or ab Division: $\dfrac{a}{b}$ or a/b

Thus, in algebra we avoid using the multiplication sign $\times$ and the division sign $\div$ so familiar in arithmetic. Notice also that, when two expressions are placed next to each other without an operation symbol, it is understood that the expressions are to be multiplied. Thus, $2x$ means 2 times x.

Sets

SET When it is preferred to treat a collection of similar objects as a whole, we use the idea of a **set**. For example, the set of *digits* consists of the collection of numbers 0, 1, 2, 3, 4, 5, 6, 7, 8, and 9. If we use the symbol D to denote the set of digits, then we can write

$$D = \{0, 1, 2, 3, 4, 5, 6, 7, 8, 9\}$$

where the braces { } are used to enclose the objects in the set.

In listing the objects of a set, we do not list an object more than once because the reason for listing the objects is to tell which objects belong to the set. Also, the order in which the objects are listed does not matter. Thus, for example, {2, 3} and {3, 2} both represent the same set.

SUBSET If every object of a set A is also an object of set B, then we say
EQUAL SETS that A is a **subset** of B. If two sets A and B have the same objects, then A **is equal** to B. For example, {1, 2, 3} is a subset of {1, 2, 3, 4, 5}; {1, 2, 3} equals {2, 3, 1}.

Classification of Numbers

It is helpful to classify the various kinds of numbers we deal with in
COUNTING NUMBERS algebra. The **counting numbers** or **natural numbers** are the numbers
NATURAL NUMBERS 1, 2, 3, 4, ... (The three dots, called an **ellipsis**, indicate that the
ELLIPSIS pattern here continues indefinitely.) As their name implies, these numbers are used to count things. For example, there are 26 letters in our alphabet; there are 100 cents in a dollar.

INTEGERS The **integers** or **whole numbers** are the numbers . . . , $-3, -2, -1, 0,$ 1, 2, 3, ...

These numbers prove useful in certain situations. For example, if your checking account has $10 in it and you write a check for $15, you can represent the current balance as $-$5$.

Notice that the counting numbers are included among the integers. Each time we expand a number system, such as from the counting numbers to the integers, we do so in order to be able to handle new, and usually more complicated, problems. Thus, the integers allow us to solve problems requiring both positive and negative counting numbers, such as profit/loss, height above/below sea level, temperature above/below zero, and so on.

But integers alone are not sufficient for *all* problems. For example, they do not answer questions such as "What part of a dollar is 38 cents?" or "What part of a pound is 5 ounces?" To answer such questions, we enlarge our number system to include rational numbers. For example, $\frac{38}{100}$ answers the question "What part of a dollar is 38 cents?" and $\frac{5}{16}$ answers the question "What part of a pound is 5 ounces?"

RATIONAL NUMBER   A **rational number** is a ratio a/b of two integers; the integer a is called
NUMERATOR   the **numerator** and the integer b, which cannot be zero, is called the
DENOMINATOR   **denominator**.

Examples of rational numbers are $\frac{3}{4}, \frac{5}{2}, \frac{0}{4}, -\frac{2}{3}$, and $\frac{100}{3}$. Since $a/1 = a$ for any integer a, it follows that the rational numbers contain the integers as a special case.

Although rational numbers occur frequently enough in applications, numbers that are not rational do also. For example, consider an isosceles right triangle whose legs are each of length 1; the number that equals the length of the hypotenuse, namely $\sqrt{2}$, is not rational. See Figure 2. Also, the number π, referred to earlier in the formula $A = \pi R^2$, is not a rational number.

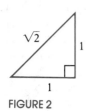

FIGURE 2

IRRATIONAL NUMBERS   Numbers that are not rational are called **irrational**; together, the
REAL NUMBERS   rational numbers and irrational numbers form the **real numbers**.

See Figure 3.

Decimals

DECIMALS   To represent a real number, we can use **decimals**. For example, the rational numbers $\frac{3}{4}, \frac{5}{2}, -\frac{2}{3}$, and $\frac{7}{66}$ are represented as decimals merely by carrying out the division:

$$\frac{3}{4} = 0.75 \qquad \frac{5}{2} = 2.5 \qquad -\frac{2}{3} = -0.666\ldots \qquad \frac{7}{66} = 0.1060606\ldots$$

It can be shown that the decimal associated with a rational number
TERMINATING DECIMALS   is always one of two types: **terminating**, or ending, as in $\frac{3}{4} = 0.75$ and
REPEATING DECIMALS $\frac{5}{2} = 2.5$; or **repeating**, as in $-\frac{2}{3} = -0.666\ldots$, where the 6 repeats, and $\frac{7}{66} = 0.1060606\ldots$, where the block 06 repeats.

Although it may seem that these two types of decimals would be the only types, there are, in fact, infinitely many decimals that neither repeat nor terminate. The decimal $0.12345678910111213\ldots$, in which we write down the positive integers one after the other, for example, will neither repeat (think about it) nor terminate. In fact, every irrational number is represented by a decimal that neither

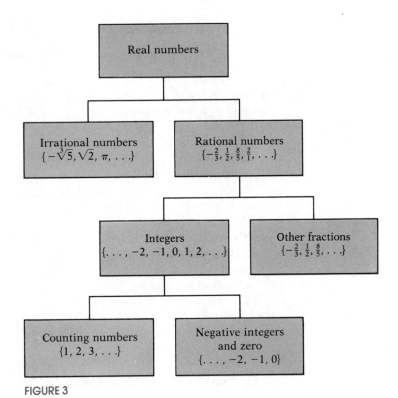

FIGURE 3

repeats nor terminates. For example, the irrational number $\sqrt{2}$ and π have decimals that neither repeat nor terminate:

$$\sqrt{2} = 1.414213\ldots \qquad \pi = 3.14159\ldots$$

Thus, every decimal can be associated with a real number and, conversely, every real number can be represented by a decimal. It is this feature of real numbers that gives them their practicality. In the physical world, many changing quantities like the length of a heated rod, the velocity of a falling object, and so on, are assumed to pass through every possible magnitude from the initial one to the final one as they change. Real numbers as decimals provide a convenient way to measure such quantities as they change.

In practice, it is usually necessary to represent real numbers by approximations. For example, using the symbol $\approx$ (read "approximately equal to"), we can write

$$\sqrt{2} \approx 1.4142 \qquad \pi \approx 3.1416$$

We shall discuss the idea of approximation in more detail in Section 1.10.

Some Conventions

In algebra we agree in advance to certain conventions, three of which are:

1. Given $a \cdot b + c$ or $c + a \cdot b$, we agree to multiply a by b first and then add c. Thus, we have $2 \cdot 4 + 5 = 8 + 5 = 13$ and $3 + 4 \cdot 5 = 3 + 20 = 23$.

2. In arithmetic, mixed numbers are used to mean addition; for example, $2\frac{3}{4}$ means $2 + \frac{3}{4}$. In algebra, use of a mixed number can be confusing because the absence of an operation symbol between two terms is taken to mean multiplication. Thus, we prefer not to use mixed numbers in algebra. The expression $2\frac{3}{4}$ is written instead as 2.75 or as $\frac{11}{4}$.

3. Parentheses are used in algebra to group terms. Thus, $(2 + 3) \cdot 4$ is an instruction to add 2 and 3 and then multiply by 4, so that $(2 + 3) \cdot 4 = 5 \cdot 4 = 20$. When parentheses are nested, the inside parentheses are evaluated first, followed in order by the outer ones. For example,

$$3 - \{2 - [4 + 5 \cdot (8 - 2)]\} = 3 - \{2 - [4 + 5 \cdot 6]\}$$
$$= 3 - \{2 - [4 + 30]\}$$
$$= 3 - \{2 - 34\}$$
$$= 3 - \{-32\}$$
$$= 3 + 32 = 35$$

Inside first —
Second —
Third —

Notice in the above expression that we used braces { } and brackets [ ] to assist us in seeing the relative order of operations. Calculators and computers, needing no such help, will merely nest parentheses. Thus, a computer "sees" the expression

$$3 - \{2 - [4 + 5 \cdot (8 - 2)]\}$$

as

$$3 - (2 - (4 + 5 \cdot (8 - 2)))$$

and will perform the operations from inside to outside.

Properties of Real Numbers

The basic properties of real numbers are given as laws involving addition and multiplication. In the list that follows, a, b, and c represent real numbers:

COMMUTATIVE LAWS

$$a + b = b + a \qquad a \cdot b = b \cdot a \tag{1}$$

The order in which addition or multiplication takes places will not affect the final result.

ASSOCIATIVE LAWS

$$a + (b + c) = (a + b) + c = a + b + c \qquad \text{(2a)}$$
$$a \cdot (b \cdot c) = (a \cdot b) \cdot c = a \cdot b \cdot c \qquad \text{(2b)}$$

The way one adds or multiplies a triple of real numbers will not affect the final result. Thus, expressions such as $2 + 3 + 4$ and $3 \cdot 4 \cdot 5$ present no ambiguity, even though addition and multiplication are performed on one pair of numbers at a time.

DISTRIBUTIVE LAW

$$a \cdot (b + c) = a \cdot b + a \cdot c \qquad \text{(3a)}$$
$$(a + b) \cdot c = a \cdot c + b \cdot c \qquad \text{(3b)}$$

The distributive law is perhaps the most important. With it, we justify such statements as

$$2(x + y) = 2x + 2y \quad \text{and} \quad 3x + 5x = (3 + 5)x = 8x$$

IDENTITY LAWS   The two real numbers 0 and 1 obey the following rules:

$$0 + a = a + 0 = a \qquad a \cdot 1 = 1 \cdot a = a \qquad \text{(4)}$$

INVERSE LAWS
ADDITIVE INVERSE

For each real number a, there is a real number $-a$, called the **additive inverse** of a, having the following property:

$$a + (-a) = -a + a = 0 \qquad \text{(5a)}$$

MULTIPLICATIVE INVERSE

For each *nonzero* real number a, there is a real number $1/a$, called the **multiplicative inverse** of a, having the following property:

$$a \cdot \frac{1}{a} = \frac{1}{a} \cdot a = 1 \qquad \text{(5b)}$$

RECIPROCAL   The multiplicative inverse $1/a$ of a nonzero real number a is also referred to as the **reciprocal** of a.

The additive inverse of a, namely $-a$, is often called the *negative* of a. This practice can be dangerous because use of the term implies that the additive inverse is a negative number, which it may not be. For example, the additive inverse of -3, that is, the real number that, when added to -3, gives 0, is 3, a positive number.

With these five laws for adding and multiplying real numbers, we can now define the operations of subtraction and division.

DIFFERENCE   To subtract b from a, or to find the **difference** $a - b$, means

$$a - b = a + (-b) \qquad (6)$$

RATIO   If b is a nonzero real number, to divide a by b, or to find the **ratio** a/b, means

$$\frac{a}{b} = a \cdot \frac{1}{b} \qquad (7)$$

For example,

$$8 - 5 = 8 + (-5) = 3 \qquad 4 - 9 = 4 + (-9) = -5$$

$$\frac{5}{8} = (5)\left(\frac{1}{8}\right) \qquad\qquad \frac{20}{-5} = (20)\left(-\frac{1}{5}\right) = -4$$

There are additional properties of real numbers that will be useful. Although these can be derived or proven, based on the earlier laws, Equations (1)–(5), we shall not attempt to prove them here. As before, a, b, and c denote real numbers in the properties listed below.

For any number a, the product of a times 0 is always 0; that is,

$$a \cdot 0 = 0 \qquad (8)$$

RULES FOR DIVISION   For a nonzero number a,

$$\frac{0}{a} = 0 \qquad \frac{a}{a} = 1 \qquad a \neq 0 \qquad (9)$$

Division by zero is *not allowed*. One reason is to avoid the following difficulty: $\frac{2}{0} = x$ means to find x such that $0 \cdot x = 2$. But $0 \cdot x$ is always 0, so there can be *no* number x such that $\frac{2}{0} = x$.

RULES OF SIGNS

$$
\begin{aligned}
a(-b) &= -(ab) & (-a)b &= -(ab) \\
(-a)(-b) &= ab & -(-a) &= a \\
\frac{a}{-b} &= \frac{-a}{b} = -\frac{a}{b} & \frac{-a}{-b} &= \frac{a}{b}
\end{aligned}
\qquad (10)
$$

CANCELLATION LAWS   **If c is a nonzero number, then**

$$ac = bc \quad \text{implies} \quad a = b \qquad c \neq 0$$

$$\frac{ac}{bc} = \frac{a}{b} \qquad b \neq 0, c \neq 0$$

(11)

PRODUCT LAW

$$\text{If } ab = 0, \text{ then either } a = 0 \text{ or } b = 0. \tag{12}$$

This important law is used often to draw conclusions. For example, if twice a number x is zero—that is, if $2x = 0$—then the number x must be zero.

ARITHMETIC OF RATIOS

$$\frac{a}{b} + \frac{c}{d} = \frac{ad + bc}{bd} \qquad b \neq 0, d \neq 0 \tag{13}$$

$$\frac{a}{b} \cdot \frac{c}{d} = \frac{ac}{bd} \qquad b \neq 0, d \neq 0 \tag{14}$$

$$\frac{\dfrac{a}{b}}{\dfrac{c}{d}} = \frac{a}{b} \cdot \frac{d}{c} = \frac{ad}{bc} \qquad b \neq 0, c \neq 0, d \neq 0 \tag{15}$$

EXAMPLE 1   Evaluate each expression.

(a) $\dfrac{2}{3} + \dfrac{5}{2}$ (b) $\dfrac{3}{5} - \dfrac{2}{3}$ (c) $\dfrac{8}{3} \cdot \dfrac{15}{4}$ (d) $\dfrac{\frac{3}{5}}{\frac{7}{9}}$

Solution   (a) $\dfrac{2}{3} + \dfrac{5}{2} = \dfrac{2 \cdot 2 + 3 \cdot 5}{3 \cdot 2} = \dfrac{4 + 15}{6} = \dfrac{19}{6}$

By Equation (13)

(b) $\dfrac{3}{5} - \dfrac{2}{3} = \dfrac{3}{5} + \left(-\dfrac{2}{3}\right)$

By Equation (6)

$= \dfrac{3}{5} + \dfrac{-2}{3}$

By Equation (10)

$= \dfrac{3 \cdot 3 + 5 \cdot (-2)}{5 \cdot 3} = \dfrac{9 + (-10)}{15} = \dfrac{-1}{15}$

By Equation (13)

(c) $\dfrac{8}{3} \cdot \dfrac{15}{4} \underset{\uparrow}{=} \dfrac{8 \cdot 15}{3 \cdot 4} = \dfrac{120}{12} = \dfrac{12 \cdot 10}{12 \cdot 1} \underset{\uparrow}{=} \dfrac{10}{1} = 10$

By Equation (14)   By Equation (11)

(d) $\dfrac{\frac{3}{5}}{\frac{7}{9}} \underset{\uparrow}{=} \dfrac{3}{5} \cdot \dfrac{9}{7} \underset{\uparrow}{=} \dfrac{3 \cdot 9}{5 \cdot 7} = \dfrac{27}{35}$

By Equation (14)

By Equation (15)   ≡

LOWEST TERMS   In writing ratios, we shall follow the usual convention and write the ratio in **lowest terms**—that is, write it so that any common divisors of the numerator and the denominator have been removed using the cancellation law, Equation (11). Thus,

$$\frac{90}{24} = \frac{15 \cdot 6}{4 \cdot 6} = \frac{15}{4}$$

$$\frac{24x^2}{18x} = \frac{(4x)(6x)}{3(6x)} = \frac{4x}{3} \qquad x \neq 0$$

≡ HISTORICAL COMMENTS   The real number system, the basic concept of our book, has a history that stretches back at least to the ancient Babylonians (1800 B.C.). It is remarkable how much the ancient Babylonian attitudes resemble our own. As we stated in the text, the fundamental difficulty with irrational numbers is that they cannot be written as ratios or, equivalently, as repeating decimals. Because the Babylonians wrote their numbers in a system based on 60 in the same way we write ours based on 10, it did not upset the Babylonians that π or $\sqrt{2}$ was not a ratio. They would carry as many places for π as the accuracy of the problem demanded, just as today we use

$$\pi \approx 3\tfrac{1}{7} \quad \text{or} \quad \pi \approx 3.1416 \quad \text{or} \quad \pi \approx 3.14159 \quad \text{or}$$

$$\pi \approx 3.14159265358979323846264\overline{3}$$

depending on how accurate we need to be.

Things were very different for the Greeks, whose number system allowed only rational numbers to be expressed. When it was discovered that $\sqrt{2}$ could not be expressed as a ratio, this was regarded as a fundamental flaw in the number concept. So serious was the matter that the Pythagorean Brotherhood (an early mathematical society) is said to have drowned one of its members for revealing this terrible secret. Greek mathematicians then turned away from the number concept, expressing facts about whole numbers in terms of line segments.

In astronomy, however, Babylonian methods, including the Babylonian number system, continued to be used. Simon Stevin (1548–1620), probably using the Babylonian system as a model, invented the decimal system, complete with rules of calculation, in 1585. [Others—for example, al-Kashi of Samarkand (d. 1424)—had made some progress in the same direction.] The decimal system so effectively conceals the difficulties that the need for more logical precision began to be felt only in the early 1800s. Around 1880 Georg Cantor (1845–1918) and Richard Dedekind (1831–1916) gave precise definitions of real numbers. Cantor's definition, though more abstract and precise, has its roots in the decimal, and hence in the Babylonian, numerical system.

Sets and set theory were a spin-off of the research that went into clarifying the foundations of the real number system. Set theory has developed into a large discipline of its own, and many mathematicians regard it as the foundation upon which modern mathematics is built. Cantor's discovery that infinite sets also can be counted, and that there are different sizes of infinite sets, is among the most astounding results of modern mathematics.

≡ HISTORICAL PROBLEMS

The Babylonian number system was based on 60. Thus 2,30 means $2 + \frac{30}{60} = 2.5$, and 4,25,14 means

$$4 + \frac{25}{60} + \frac{14}{60^2} = 4 + \frac{1514}{3600} = 4.42055555 \ldots$$

1. What are the following numbers in Babylonian notation?
   (a) $1\frac{1}{3}$ (b) $2\frac{5}{6}$

2. What are the following Babylonian numbers when written as fractions and as decimals?
   (a) 2,20 (b) 4,52,30 (c) 3,8,29,44   ≡

EXERCISE 1.1

In Problems 1–10 express each statement as an equation involving the indicated variables and constants.

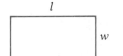

1. The area A of a rectangle is the product of its length l times its width w.

2. The perimeter P of a rectangle is twice the sum of its length l and its width w.

3. The circumference C of a circle is the product of π times its diameter d.

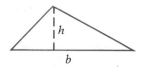

4. The area A of a triangle is one-half the product of its base b times its height h.

5. The area A of an equilateral triangle is $\sqrt{3}/4$ times the square of the length x of one side.

6. The perimeter P of an equilateral triangle is 3 times the length x of one side.

7. The volume V of a sphere is the product of $\frac{4}{3}$ times π times the cube of the radius R.

8. The surface area S of a sphere is the product of 4 times π times the square of the radius R.

9. The volume V of a cube is the cube of the length x of a side.

10. The surface area S of a cube is 6 times the square of the length x of a side.

In Problems 11–22 express each rational number as a decimal, either repeating or terminating.

11. $\dfrac{2}{3}$ 12. $\dfrac{2}{5}$ 13. $\dfrac{1}{8}$ 14. $\dfrac{10}{3}$

15. $-\dfrac{8}{5}$ 16. $-\dfrac{16}{3}$ 17. $\dfrac{1}{9}$ 18. $\dfrac{1}{10}$

19. $\dfrac{4}{25}$ 20. $-\dfrac{5}{6}$ 21. $-\dfrac{3}{7}$ 22. $\dfrac{8}{11}$

In Problems 23–40 evaluate each expression.

23. $-6 + 4 \cdot 2$

24. $8 - 4 \cdot 3$

25. $4 + 5 - 10$

26. $8 - 3 - 2$

27. $4 + \dfrac{1}{3}$

28. $2 - \dfrac{1}{2}$

29. $6 - [3 \cdot 5 + 2 \cdot (3 - 2)]$

30. $2 \cdot [8 - 3(4 + 2)] - 3$

31. $2 \cdot (3 - 5) + 8 \cdot 2 - 1$

32. $1 - (4 \cdot 3 - 2 + 2)$

33. $10 - [6 - 2 \cdot 2 + (8 - 3)] \cdot 2$

34. $2 - 5 \cdot 4 - [6 \cdot (3 - 4)]$

35. $5 - 3 \cdot \dfrac{1}{2}$

36. $18 + 4 \cdot \dfrac{1}{3}$

37. $\dfrac{4}{5} + \dfrac{8}{3}$

38. $\dfrac{2}{5} - \dfrac{4}{3}$

39. $\dfrac{\frac{16}{5} - \frac{1}{2}}{\frac{1}{3}}$

40. $\dfrac{\frac{14}{3} + \frac{1}{2}}{\frac{3}{4}}$

In Problems 41–52 use the distributive law to remove the parentheses.

41. $3(x + 4)$ **42.** $5(2x - 1)$

43. $x(x - 3)$ **44.** $3x(x + 4)$

45. $(x + 2)(x + 4)$ **46.** $(x + 5)(x + 1)$

47. $(x - 2)(x + 1)$ **48.** $(x - 4)(x + 1)$

49. $(x - 8)(x - 2)$ **50.** $(x - 4)(x - 2)$

51. $(x + 2)(x - 2)$ **52.** $(x - 3)(x + 3)$

53. Is subtraction commutative? Support your conclusion.

54. Is subtraction associative? Support your conclusion.

55. Is division commutative? Support your conclusion.

56. Is division associative? Support your conclusion.

57. If $2 = x$, why does $x = 2$?

58. If $x = 5$, why does $x^2 + x = 30$?

≡ 1.2 THE REAL NUMBER LINE

There is a one-to-one correspondence between real numbers and points on a line. That is, every real number corresponds to a point on the line and, conversely, each point on the line has a unique real number associated with it. We establish this correspondence of real numbers with points on a line in the following manner.

ORIGIN   We start with a line that is, for convenience, drawn horizontally. Pick some point on the line and label it O, for **origin**. Then pick another point some fixed distance to the right of O and label it U, as shown in Figure 4.

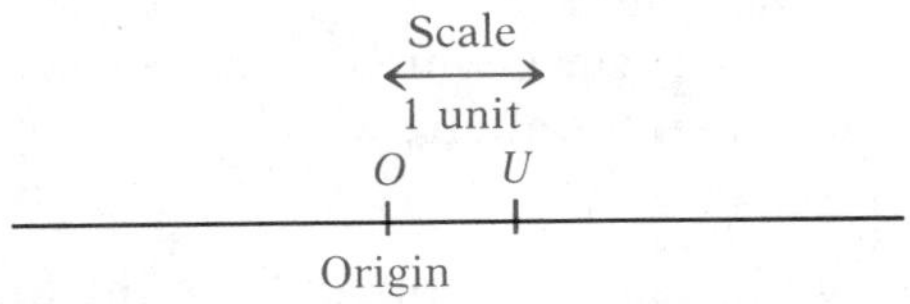

FIGURE 4
Horizontal Line

SCALE   The fixed distance, which may be 1 inch or 1 centimeter, 1 light-year, or any distance, is called the **scale**. We associate the real number 0 with the origin O and the number 1 with the point U. Refer now to Figure 5. The point to the right of U that is twice as far from O as U is associated with the number 2. The point to the right of U that is three times as far from O as U is associated with the number 3. The point midway between O and U is assigned the number 0.5 or $\frac{1}{2}$. Corresponding points to the left of the origin O are assigned the numbers -0.5, -1, -2, -3, and so on, depending on how far they are

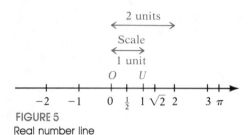

FIGURE 5
Real number line

from O. Notice in Figure 5 that we placed an arrowhead on the right end of the line to indicate the direction in which the assigned numbers increase. Figure 5 also shows the points associated with the irrational numbers $\sqrt{2}$ and π.

COORDINATE   The real number associated with a point P is called the **coordinate**
REAL NUMBER LINE   of P, and a line whose points have been assigned coordinates is called a **real number line**.

A real number line divides the real numbers into three classes (see Figure 6):

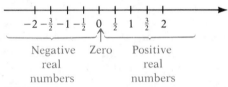

FIGURE 6

NEGATIVE REAL NUMBERS   1. The **negative real numbers**, which are the coordinates of points to the left of the origin O.
ZERO   2. The real number **zero**, which is the coordinate of the origin O.
POSITIVE REAL NUMBERS   3. The **positive real numbers**, which are the coordinates of points to the right of the origin O.

Negative and positive numbers have the following multiplication properties.

1. The product of two positive numbers is a positive number.
2. The product of two negative numbers is a positive number.
3. The product of a positive number times a negative number is a negative number.

Inequalities

Let a and b be two real numbers. If the difference $a - b$ is positive,
GREATER THAN   then we say a is **greater than** b and write $a > b$. Alternatively, if
LESS THAN $a - b$ is positive, we could also say that b is **less than** a and write $b < a$. Thus, $a > b$ and $b < a$ are equivalent statements.

EXAMPLE 1   (a) $3 < 5$ since $5 - 3 = 2$ is positive.

(b) $-6 > -8$ since $-6 - (-8) = -6 + 8 = 2$ is positive.   ≡

On the real number line, if $a > b$, the point with coordinate a is to the right of the point with coordinate b. Based on Figure 6, we conclude that

> $a > 0$ is equivalent to a is positive
>
> $a < 0$ is equivalent to a is negative

GREATER THAN OR EQUAL TO

LESS THAN OR EQUAL TO

If the difference $a - b$ of two real numbers is positive or zero—that is, if $a > b$ or $a = b$—then we say a is **greater than or equal to** b and write $a \geq b$. Alternatively, if $a \geq b$, we can also say that b is **less than or equal to** a and write $b \leq a$.

STRICT INEQUALITY

NONSTRICT INEQUALITY

INEQUALITY SIGNS

Statements of the form $a < b$ or $b > a$ are called **strict inequalities**; statements of the form $a \leq b$ or $b \geq a$ are called **nonstrict inequalities**. The symbols $>$, $<$, $\geq$, and $\leq$ are called **inequality signs**.

If x is a real number and $x \geq 0$, then x is either positive or zero. As a result, we describe the inequality $x \geq 0$ by saying "x is nonnegative."

Properties of Inequalities

In working with inequalities, we will need to know various properties they obey. In the properties that follow, a, b, and c are real numbers.

NONNEGATIVE PROPERTY

For any real number a, we have

> $a^2 \geq 0$

TRANSITIVE PROPERTY

> If $a > b$ and $b > c$, then $a > c$.

ADDITION PROPERTY

> If $a > b$, then $a + c > b + c$.

The addition property states that the sense or direction of an inequality remains unchanged if the same number is added to each side.

MULTIPLICATION PROPERTIES

> (a) If $a > b$ and if $c > 0$, then $ac > bc$.
>
> (b) If $a > b$ and if $c < 0$, then $ac < bc$.

The multiplication properties state that the sense or direction of an inequality *remains the same* if each side is multiplied by a *positive* real number, but the direction is *reversed* if each side is multiplied by a *negative* real number.

$$\text{If } a > 0, \text{ then } \frac{1}{a} > 0.$$

The reciprocal property states that the reciprocal of a positive real number is positive.

  The transitive, addition, multiplication, and reciprocal properties may also be stated as follows:

If $a < b$ and $b < c$, then $a < c$.

If $a < b$, then $a + c < b + c$.

(a) If $a < b$ and if $c > 0$, then $ac < bc$.

(b) If $a < b$ and if $c < 0$, then $ac > bc$.

If $a < 0$, then $\frac{1}{a} < 0$.

Finally, we have the **trichotomy property**, which states that either two numbers are equal or one of them is less than the other. For any pair of numbers a and b,

$$a < b \quad \text{or} \quad a = b \quad \text{or} \quad b < a$$

If $b = 0$, the trichotomy property states that for any real number a,

$$a < 0 \quad \text{or} \quad a = 0 \quad \text{or} \quad a > 0$$

That is, any real number is negative or zero or positive.

EXAMPLE 2   (a) Since $2 < 3$, then $2 + 5 < 3 + 5$ (or $7 < 8$).

(b) Since $1 > -3$, then $1 - 5 > -3 - 5$ (or $-4 > -8$).

(c) Since $4 > 1$, then $4 \cdot 2 > 1 \cdot 2$ (or $8 > 2$).

(d) Since $-5 < -3$, then $-5(-2) > -3(-2)$ (or $10 > 6$).

(e) Since $3 > -1$, then $3(-4) < -1(-4)$ (or $-12 < 4$).   ≡

Graphing Inequalities

We shall find it useful in later work to graph inequalities on a real number line.

EXAMPLE 3   (a) On a real number line, graph all numbers x for which $x > 4$.

(b) On a real number line, graph all numbers x for which $x \leq 5$.

Solution   (a) See Figure 7(a). Notice that we use an unfilled circle to indicate
that the number 4 is not part of the graph.

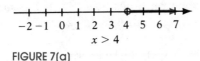

$x > 4$

FIGURE 7(a)

(b) See Figure 7(b). Notice that we use a filled circle to indicate that
the number 5 is part of the graph.

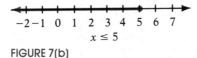

$x \leq 5$

FIGURE 7(b)   ≡

Inequalities are often combined.

EXAMPLE 4   (a) On a real number line, graph all numbers x for which $x > 4$ and
$x < 6$.

(b) On a real number line, graph all numbers x for which $x > 4$ or
$x \leq -1$.

(c) On a real number line, graph all numbers x for which $x > 4$ and
$x > 6$.

(d) On a real number line, graph all numbers x for which $x \geq 4$ or
$x \geq 1$.

(e) On a real number line, graph all numbers x for which $x \leq 4$ and
$x \geq 6$.

Solution   (a) We first graph each inequality separately as illustrated in Figure
8(a). Then it is easy to see that the numbers that belong to *both*
the graph of $x > 4$ *and* the graph of $x < 6$ are the ones graphed
in Figure 8(b). For example, 5 is part of the graph because $5 > 4$
and $5 < 6$; 7 is not on the graph because 7 is not less than 6.

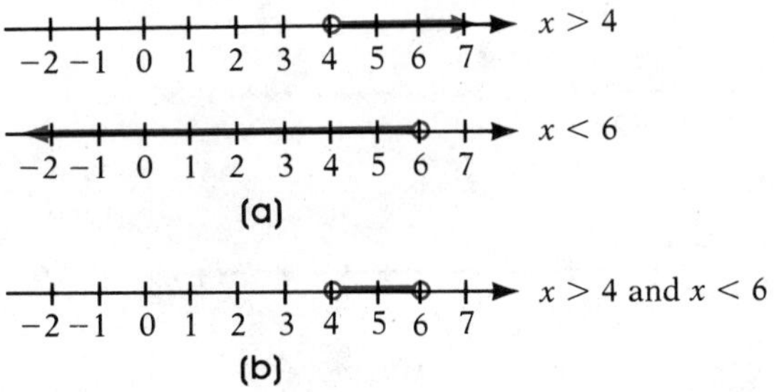

$x > 4$

$x < 6$

(a)

$x > 4$ and $x < 6$

(b)

FIGURE 8

(b) See Figure 9(a), in which each inequality is graphed separately. The numbers that belong to *either* the graph of $x > 4$ *or* the graph of $x \le -1$ are the ones graphed in Figure 9(b). For example, 5 is part of the graph because $5 > 4$; 2 is not on the graph because 2 is not greater than 4 nor is 2 less than or equal to -1.

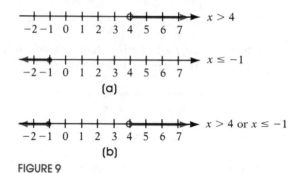

(a)

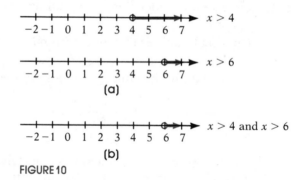

(b)

FIGURE 9

(c) Figure 10 shows the numbers for which *both* $x > 4$ *and* $x > 6$ are true.

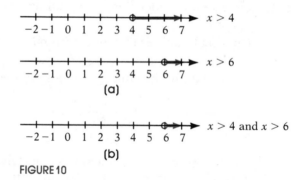

FIGURE 10

(d) Figure 11 shows the numbers for which *either* $x > 4$ *or* $x \ge 1$.

FIGURE 11

(e) Figure 12 illustrates that there are no numbers for which *both* $x \leq 4$ *and* $x \geq 6$ are true.

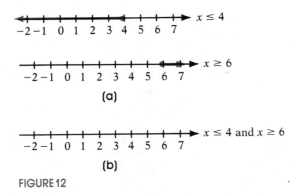

FIGURE 12

The situation in part (a) of Example 4 is one we want to examine further. Suppose a and b are two real numbers and $a < b$. We shall use the notation $a < x < b$ to mean that x is a number between a and b. Thus, the inequality $a < x < b$, called a **chain**, is equivalent to the two inequalities $a < x$ and $x < b$. Similarly, the chain $a \leq x \leq b$ is equivalent to the two inequalities $a \leq x$ and $x \leq b$. The remaining two chains, $a \leq x < b$ and $a < x \leq b$, are defined analogously.

CHAIN

Although it is acceptable to write a chain as $3 \geq x \geq 2$, it is preferable to reverse the inequality symbols and write instead $2 \leq x \leq 3$, so that, as you read from left to right, the values go from smaller to larger. Of course, a statement such as $2 \leq x \leq 1$ makes no sense because there is no number x for which $2 \leq x$ and $x \leq 1$. It makes even less sense to mix inequality symbols, as in $2 \leq x \geq 3$.

The next example will give you some practice for future work involving inequalities.

EXAMPLE 5   Choose a number x that is:

(a) Less than 2   (b) Between 2 and 4   (c) Larger than 4

Use it to determine whether the expression $(x - 2)(x - 4)$ is positive, negative, or zero.

Solution   (a) Any number x less than 2 obeys the inequality $x < 2$. We'll pick the number 1. Then the expression

$$(x - 2)(x - 4) = (1 - 2)(1 - 4) = (-1)(-3) = 3$$

is positive.

(b) Any number x between 2 and 4 obeys the chain $2 < x < 4$. We'll pick $x = 3$. Then the expression

$$(x - 2)(x - 4) = (3 - 2)(3 - 4) = (1)(-1) = -1$$

is negative.

(c) Any number x larger than 4 obeys the inequality $x > 4$. We'll pick $x = 6$. Then the expression

$$(x - 2)(x - 4) = (6 - 2)(6 - 4) = (4)(2) = 8$$

is positive.   ≡

You may wonder what would happen if you picked numbers different from the ones we used in Example 5. Would the answers be different? Try some other choices and see. As it turns out, the answers are always the same. Let's see why.

For any choice of x less than 2, we have $x < 2$. As a result, such an x is also less than 4, so both $x - 2$ and $x - 4$ are negative, making the product $(x - 2)(x - 4)$ always positive for $x < 2$. Similarly, for any choice of x between 2 and 4, we have $2 < x < 4$. Thus, $2 < x$ and $x < 4$ so that $x - 2$ is positive and $x - 4$ is negative, making the product $(x - 2)(x - 4)$ always negative. Finally, for any choice of x greater than 4, we have $x > 4$. But, as a result, such an x is also larger than 2, so that both $x - 2$ and $x - 4$ are positive, making the product $(x - 2)(x - 4)$ always positive. Table 1 summarizes this discussion.

TABLE 1

	$x - 2$	$x - 4$	$(x - 2)(x - 4)$
$x < 2$	Negative	Negative	Positive
$2 < x < 4$	Positive	Negative	Negative
$x > 4$	Positive	Positive	Positive

Intervals

CLOSED INTERVAL

OPEN INTERVAL

HALF-OPEN INTERVAL

Let a and b represent two real numbers with $a < b$. A **closed interval**, denoted by $[a, b]$, consists of all real numbers x for which $a \leq x \leq b$. An **open interval**, denoted by (a, b), consists of all real numbers x for which $a < x < b$. The **half-open** or **half-closed intervals** are $(a, b]$, consisting of all real numbers x for which $a < x \leq b$, and $[a, b)$, consisting of all real numbers x for which $a \leq x < b$.

LEFT ENDPOINT

RIGHT ENDPOINT

In each of these definitions, a is called the **left endpoint** and b the **right endpoint** of the interval. Figure 13 illustrates each type of interval.

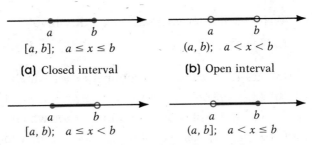

(a) Closed interval   **(b)** Open interval

(c) Half-open (half-closed) intervals

FIGURE 13

PLUS INFINITY
MINUS INFINITY
    The symbol $+\infty$ (read "plus infinity") is not a real number but a notational device used to indicate unboundedness in the positive direction. The symbol $-\infty$ (read "minus infinity") also is not a real number but a notational device used to indicate unboundedness in the negative direction. Using the symbols $+\infty$ and $-\infty$, we can define four other kinds of intervals:

$[a, +\infty)$ consists of all real numbers x for which $x \geq a$

$(a, +\infty)$ consists of all real numbers x for which $x > a$

$(-\infty, a]$ consists of all real numbers x for which $x \leq a$

$(-\infty, a)$ consists of all real numbers x for which $x < a$

Figure 14 illustrates these types of intervals.

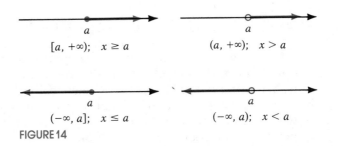

FIGURE 14

EXAMPLE 6   Write each inequality using interval notation.

(a) $1 \leq x \leq 3$ (b) $-4 < x < 0$ (c) $x > 5$ (d) $x \leq 1$

Solution   (a) $1 \leq x \leq 3$ describes all numbers x between 1 and 3, inclusive. In interval notation, we write $[1, 3]$.

(b) In interval notation, $-4 < x < 0$ is written $(-4, 0)$.

(c) $x > 5$ consists of all numbers x greater than 5. In interval notation, we write $(5, +\infty)$.

(d) In interval notation, $x \leq 1$ is written $(-\infty, 1]$.   ≡

EXAMPLE 7   Write each interval as an inequality involving x.

(a) $[1, 4)$ (b) $(2, +\infty)$ (c) $[2, 3]$ (d) $(-\infty, -3]$

Solution   (a) $[1, 4)$ consists of all numbers x for which $1 \leq x < 4$.
(b) $(2, +\infty)$ consists of all numbers x for which $x > 2$.
(c) $[2, 3]$ consists of all numbers x for which $2 \leq x \leq 3$.
(d) $(-\infty, -3]$ consists of all numbers x for which $x \leq -3$.   ≡

Absolute Value

To scientists, the absolute value of a number is the magnitude of that number—that is, the nonnegative value of the number. Thus, the absolute value of 5 is 5; the absolute value of -6 is 6. A more formal definition of absolute value is given below.

ABSOLUTE VALUE   The **absolute value** of a real number x, denoted by the symbol $|x|$, is defined by the rules

$$|x| = x \ \text{ if } x \geq 0 \quad \text{and} \quad |x| = -x \ \text{ if } x < 0$$

For example, since $-4 < 0$, then the second rule must be used to get $|-4| = -(-4) = 4$.

EXAMPLE 8   (a) $|8| = 8$ (b) $|0| = 0$ (c) $|-15| = 15$ ≡

Consider a real number line. The point whose coordinate is -4 is 4 units from the origin. The point whose coordinate is 3 is 3 units from the origin. See Figure 15. Geometrically, then, the absolute value of a number x equals the distance from the origin O to the point whose coordinate is x. Also, $|x - y|$ equals the distance between the points whose coordinates are x and y. For example, the distance between the points whose coordinates are -4 and 3 is

$$|-4 - 3| = |-7| = 7 \text{ units}$$

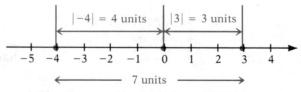

-4 is a distance of $|-4| = 4$ units from the origin
3 is a distance of $|3| = 3$ units from the origin

FIGURE 15

HISTORICAL COMMENT The concept of absolute value has been known for a long time, and various symbols for it have been used, but it was never of great theoretical importance until the reformulation of the basis of calculus in terms of approximation by Weierstrass about 1840. In his lectures, Weierstrass introduced the symbol $|x|$ for the absolute value of x, and the utility of the symbol quickly became obvious. Within 40 years the symbol was in use throughout the world. ■

EXERCISE 1.2 *In Problems 1–10 replace the question mark by $<$, $>$, or $=$, whichever is correct.*

1. $\frac{1}{2}$? 0
2. 5 ? 6
3. -1 ? -2
4. -3 ? $-\frac{5}{2}$

5. π ? 3.14
6. $\sqrt{2}$? 2.14
7. $\frac{1}{2}$? 0.5
8. $\frac{1}{3}$? 0.33

9. $\frac{2}{3}$? 0.67
10. $\frac{1}{4}$? 0.25

11. On a real number line, label the points with coordinates of 0, 1, -1, $\frac{5}{2}$, -2.5, $\frac{3}{4}$, and 0.25.

12. Repeat Problem 11 for the coordinates 0, -2, 2, -1.5, $\frac{3}{2}$, $\frac{1}{3}$, and $\frac{2}{3}$.

In Problems 13–20 write each statement as an inequality.

13. x is positive
14. z is negative
15. x is less than 2
16. y is greater than -5
17. x is less than or equal to 1
18. x is greater than or equal to 2
19. x is less than 5 and x is greater than 2
20. y is less than or equal to 2 and y is greater than 0

In Problems 21–30 graph the numbers x, if any, on a real number line.

21. $x \geq -1$
22. $x < 1$
23. $x \geq 4$ and $x < 6$
24. $x > 3$ and $x \leq 7$
25. $x \leq 0$ or $x < 6$
26. $x > 0$ or $x \geq 5$
27. $x \leq -2$ and $x > 1$
28. $x \geq 4$ and $x < -2$
29. $x \leq -2$ or $x > 1$
30. $x \geq 4$ or $x < -2$

In Problems 31–40 pick a number x that obeys the given inequality and use it to determine whether the given expression is positive, negative, or zero.

31. $3 < x < 5$; $(x - 5)(x - 3)$
32. $1 < x < 3$; $(x - 3)(x - 1)$
33. $-4 < x < 1$; $(x - 1)(x + 4)$
34. $-3 < x < -1$; $(x + 1)(x + 3)$
35. $x > 3$; $(x - 1)(x - 3)$
36. $x > -2$; $(x + 2)(x + 5)$
37. $x < 4$; $(x - 4)(x - 6)$
38. $x < 1$; $(x - 1)(x - 7)$
39. $x > 0$; $x(x + 4)$
40. $x < 0$; $x(x - 3)$

In Problems 41–48 write each inequality using interval notation, and illustrate each inequality using a real number line.

41. $0 \le x \le 4$ **42.** $-1 < x < 5$ **43.** $4 \le x < 6$ **44.** $-2 < x \le 0$

45. $x \ge 2$ **46.** $x \le 2$ **47.** $x < -4$ **48.** $x > 1$

In Problems 49–56 write each interval as an inequality involving x, and illustrate each inequality using a real number line.

49. $[2, 5]$ **50.** $(1, 2)$ **51.** $(-3, -2)$ **52.** $[0, 1)$

53. $[4, +\infty)$ **54.** $(-\infty, 2]$ **55.** $(-\infty, -1)$ **56.** $(-6, +\infty)$

In Problems 57–66 find the value of each expression if $x = 2$ and $y = -3$.

57. $|x + y|$ **58.** $|x - y|$ **59.** $|x| + |y|$ **60.** $|x| - |y|$

61. $\dfrac{|x|}{x}$ **62.** $\dfrac{|y|}{y}$ **63.** $|4x - 5y|$ **64.** $|3x + 2y|$

65. $\big||4x| - |5y|\big|$ **66.** $3|x| + 2|y|$

67. If $a \le b$ and $c > 0$, show that $ac \le bc$. [*Hint:* Since $a \le b$, it follows that $a - b \le 0$. Now multiply each side by c.]

68. If $a \le b$ and $c < 0$, show that $ac \ge bc$.

69. If $a < b$, show that $a < (a + b)/2 < b$. The number $(a + b)/2$ is called the **arithmetic mean** of a and b.

ARITHMETIC MEAN

70. Show that the arithmetic mean of a and b is equidistant from a and b.

▰ 1.3 INTEGER EXPONENTS

Exponents provide a shorthand device for representing repeated multiplications of a real number.

a^n If a is a real number and n is a positive integer, then the symbol a^n means to multiply a times itself n times. That is,

$$a^n = \underbrace{a \cdot a \cdot \cdots \cdot a}_{n \text{ factors}} \tag{1}$$

where it is understood that $a^1 = a$.

In particular, we have

$$a^1 = a$$
$$a^2 = a \cdot a$$
$$a^3 = a \cdot a \cdot a$$

and so on.

BASE
EXPONENT
POWER

In the expression a^n, a is called the **base** and n is called the **exponent** or **power**. We read a^n as "a raised to the power n" or as "a to the nth power." We usually read a^2 as "a squared" and a^3 as "a cubed."

≡ **HISTORICAL COMMENT**

Our method of writing exponents originated with Descartes, although the concept goes back in various forms to the ancient Babylonians. Even after its introduction by Descartes in 1637, the method took a remarkable amount of time to become completely standardized, and, in 1750, expressions like $a^5 + 3aaaa + 2aaa - 4aa + 2a + 1$ remained common. The concept of a rational exponent (Section 1.8) was known by 1400, although inconvenient notation prevented the development of any extensive theory. Wallis, in 1655, was the first to give a fairly complete explanation of negative and rational exponents, and Newton's use of them made them standard in their current form. ≡

EXAMPLE 1

(a) $2^3 = 2 \cdot 2 \cdot 2 = 8$ (b) $5^2 = 5 \cdot 5 = 25$

(c) $10^1 = 10$ (d) $(-2)^4 = (-2)(-2)(-2)(-2) = 16$

(e) $-2^4 = -(2 \cdot 2 \cdot 2 \cdot 2) = -16$

In words, 2 cubed equals 8; 5 squared equals 25; 10 to the first power equals 10; etc. ≡

We define a raised to a negative power as follows:

a^{-n} If n is a positive integer and if a is a nonzero real number, then we define

$$a^{-n} = \frac{1}{a^n} \qquad a \neq 0 \qquad\qquad (2)$$

EXAMPLE 2

(a) $2^{-3} = \frac{1}{2^3} = \frac{1}{8}$ (b) $\left(\frac{1}{5}\right)^{-2} = \frac{1}{(\frac{1}{5})^2} = \frac{1}{\frac{1}{25}} = 25$ ≡

a^0 If a is a nonzero number, we define

$$a^0 = 1 \qquad a \neq 0 \qquad\qquad (3)$$

Notice that we do not allow the base a to be 0 in a^{-n} or in a^0.

Laws of Exponents

There are several general rules that can be used when dealing with exponents. The first one we consider is used when multiplying two expressions that have the same base:

Multiplying

$$3^2 \cdot 3^4 = (3 \cdot 3)(3 \cdot 3 \cdot 3 \cdot 3) = 3 \cdot 3 \cdot 3 \cdot 3 \cdot 3 \cdot 3 = 3^6$$

Same base   2 factors   4 factors   6 factors   Same base

Sum of powers 2 and 4

$a^m \cdot a^n$ In general, if a is any real number and if m and n are positive integers, we have

$$a^m \cdot a^n = \underbrace{(a \cdot a \cdots \cdots a)}_{m \text{ factors}}\underbrace{(a \cdot a \cdots \cdots a)}_{n \text{ factors}} = \underbrace{a \cdot a \cdots \cdots a}_{m+n \text{ factors}} = a^{m+n}$$

Thus, to multiply two expressions having the same base, retain the base and add the exponents. That is,

$$a^m a^n = a^{m+n} \tag{4}$$

It can be shown that Equation (4) is true whether the integers m and n are positive, negative, or zero. Of course, if m or n is zero or negative, then a cannot be 0.

EXAMPLE 3   (a) $2 \cdot 2^4 = 2^5 = 32$ (b) $(-3)^2(-3)^{-3} = (-3)^{-1} = \dfrac{1}{-3}$

(c) $4^{-3} \cdot 4^5 = 4^2 = 16$ (d) $(-2)^{-1}(-2)^{-3} = (-2)^{-4} = \dfrac{1}{(-2)^4} = \dfrac{1}{16}$

Another law of exponents applies when an expression containing a power is itself raised to a power:

$$(2^3)^4 = \underbrace{2^3 \cdot 2^3 \cdot 2^3 \cdot 2^3}_{4 \text{ factors}} = \underbrace{(2 \cdot 2 \cdot 2)}_{3 \text{ factors}}\underbrace{(2 \cdot 2 \cdot 2)}_{3 \text{ factors}}\underbrace{(2 \cdot 2 \cdot 2)}_{3 \text{ factors}}\underbrace{(2 \cdot 2 \cdot 2)}_{3 \text{ factors}} = 2^{12}$$

$$3 \cdot 4 = 12 \text{ factors}$$

$(a^m)^n$ In general, if a is any real number and if m and n are positive integers, we have

$$(a^m)^n = \underbrace{a^m \cdot a^m \cdots \cdots a^m}_{n \text{ factors}} = \underbrace{(a \cdot a \cdots \cdots a)}_{m \text{ factors}}\underbrace{(a \cdot a \cdots \cdots a)}_{m \text{ factors}}\underbrace{(a \cdot a \cdots \cdots a)}_{m \text{ factors}} = a^{mn}$$

$$m \cdot n \text{ factors}$$

Thus, if an expression containing a power is raised to a power, retain the base and multiply the powers. That is,

$$(a^m)^n = a^{mn} \tag{5}$$

It can be shown that Equation (5) is true whether the integers m and n are positive, negative, or zero. Again, if m or n is zero or negative, then a must not be zero.

EXAMPLE 4   (a) $[(-2)^3]^2 = (-2)^6 = 64$ (b) $(3^{-4})^0 = 3^{-4 \cdot 0} = 3^0 = 1$

(c) $(2^{-3})^2 = 2^{-6} = \dfrac{1}{2^6} = \dfrac{1}{64}$ (d) $(5^{-1})^{-2} = 5^{-1(-2)} = 5^2 = 25$ ▤

The next law of exponents involves raising a product to a power:

$$(2 \cdot 5)^3 = (2 \cdot 5)(2 \cdot 5)(2 \cdot 5) = (2 \cdot 2 \cdot 2)(5 \cdot 5 \cdot 5) = 2^3 \cdot 5^3$$

$(a \cdot b)^n$ In general, if a and b are real numbers and if n is a positive integer, we have

$$(a \cdot b)^n = \underbrace{(ab)(ab) \cdots\cdots (ab)}_{n \text{ factors}} = \underbrace{(a \cdot a \cdots\cdots a)}_{n \text{ factors}}\underbrace{(b \cdot b \cdots\cdots b)}_{n \text{ factors}}$$

$$= a^n \cdot b^n$$

Thus, if a product is raised to a power, the result equals the product of each factor raised to that power. That is,

$$(a \cdot b)^n = a^n \cdot b^n \tag{6}$$

It can be shown that Equation (6) is true whether the integer n is positive, negative, or zero. If n is zero or negative, neither a nor b can be zero.

EXAMPLE 5   (a) $(2x)^3 = 2^3 \cdot x^3 = 8x^3$

(b) $(-2x)^0 = -2^0 \cdot x^0 = 1 \cdot 1 = 1$ $x \neq 0$

(c) $(ax)^{-2} = a^{-2}x^{-2} = \dfrac{1}{a^2} \cdot \dfrac{1}{x^2} = \dfrac{1}{a^2x^2}$ $x \neq 0, a \neq 0$

(d) $(ax)^{-2} = \dfrac{1}{(ax)^2} = \dfrac{1}{a^2x^2}$ $x \neq 0, a \neq 0$ ▤

Equations (4) and (6) both involve products. There are two analogous laws for quotients.

a^m/a^n
$(a/b)^n$

If a and b are real numbers and if m and n are integers, then

$$\frac{a^m}{a^n} = a^{m-n} = \frac{1}{a^{n-m}} \qquad a \neq 0$$

$$\left(\frac{a}{b}\right)^n = \frac{a^n}{b^n} \qquad b \neq 0 \tag{7}$$

EXAMPLE 6 (a) $\dfrac{2^6}{2^4} = 2^{6-4} = 2^2 = 4$ (b) $\left(\dfrac{2}{3}\right)^4 = \dfrac{2^4}{3^4} = \dfrac{16}{81}$

(c) $\dfrac{3^{-2}}{3^{-5}} = 3^{-2-(-5)} = 3^3 = 27$

(d) $\left(\dfrac{5}{2}\right)^{-2} = \dfrac{1}{\left(\dfrac{5}{2}\right)^2} = \dfrac{1}{\dfrac{5^2}{2^2}} = \dfrac{1}{\dfrac{25}{4}} = \dfrac{4}{25}$ ▤

You may have observed a shortcut for problems like Example 6(d), namely that

$$\left(\frac{a}{b}\right)^{-n} = \left(\frac{b}{a}\right)^n \qquad a \neq 0, b \neq 0 \tag{8}$$

For example,

$$\left(\frac{2}{3}\right)^{-2} = \left(\frac{3}{2}\right)^2 = \frac{9}{4} \qquad \left(\frac{3}{2}\right)^{-3} = \left(\frac{2}{3}\right)^3 = \frac{8}{27}$$

EXAMPLE 7 Write each expression so that all exponents are positive.

(a) $\dfrac{x^5 y^{-2}}{x^3 y}$, $x \neq 0, y \neq 0$ (b) $\dfrac{xy}{x^{-1} - y^{-1}}$, $x \neq 0, y \neq 0$

Solution (a) $\dfrac{x^5 y^{-2}}{x^3 y} = \dfrac{x^5}{x^3} \cdot \dfrac{y^{-2}}{y} = x^{5-3} \cdot y^{-2-1} = x^2 y^{-3} = x^2 \dfrac{1}{y^3} = \dfrac{x^2}{y^3}$

(b) $\dfrac{xy}{x^{-1} - y^{-1}} = \dfrac{xy}{\dfrac{1}{x} - \dfrac{1}{y}} = \dfrac{xy}{\dfrac{y - x}{xy}} = \dfrac{(xy)(xy)}{y - x} = \dfrac{x^2 y^2}{y - x}$ ▤

Summary

We close this section by summarizing the laws of exponents. In the list that follows, a and b are real numbers and m and n are integers.

Also, we assume that no denominator is zero and that all expressions are defined.

LAWS OF EXPONENTS

$$a^{-n} = \frac{1}{a^n} \qquad\qquad a^0 = 1$$

$$a^m a^n = a^{m+n} \qquad (a^m)^n = a^{mn} \qquad (ab)^n = a^n b^n$$

$$\frac{a^m}{a^n} = a^{m-n} = \frac{1}{a^{n-m}} \qquad \left(\frac{a}{b}\right)^n = \frac{a^n}{b^n}$$

EXERCISE 1.3

In Problems 1–20 simplify each expression.

1. 3^0

2. 3^2

3. 4^{-2} $\;1/16$

4. $(-3)^2$

5. $\left(\frac{2}{3}\right)^2$ $\;4/9$

6. $\left(\frac{-4}{5}\right)^3$

7. $3^0 \cdot 2^{-3}$ $\;1/8$

8. $(-2)^{-3} \cdot 2^0$

9. $2^{-3} + \left(\frac{1}{2}\right)^3$ $\;1/4$

10. $3^{-2} + \frac{1}{3}$

11. $3^{-6} \cdot 3^4$ $\;1/9$

12. $4^{-2} \cdot 4^3$

13. $\frac{8^2}{2^3}$ $\;8$

14. $\frac{4^3}{2^2}$

15. $\left(\frac{2}{3}\right)^{-2}$ $\;9/4$

16. $\left(\frac{3}{2}\right)^{-3}$

17. $\frac{2^3 \cdot 3^2}{2 \cdot 3^{-2}}$ $\;4$

18. $\frac{3^{-2} \cdot 5^3}{3 \cdot 5}$

19. $\left(\frac{9}{2}\right)^{-2}$ $\;4/81$

20. $\left(\frac{6}{5}\right)^{-3}$

In Problems 21–50 simplify each expression so that all exponents are positive. Whenever an exponent is negative or zero, we assume the base does not equal zero.

21. $x^0 y^2$

22. $x^{-1} y$

23. $x^{-2} y$

24. $x^4 y^0$

25. $\frac{x^{-2} y^3}{xy^4}$

26. $\frac{x^{-2} y}{xy^2}$

27. $x^{-1} y^{-1}$

28. $\frac{x^{-2} y^{-3}}{x}$

29. $\frac{x^{-1}}{y^{-1}}$

30. $\left(\frac{2x}{3}\right)^{-1}$

31. $\left(\frac{4x}{5y}\right)^{-2}$

32. $(xy^2)^{-2}$

33. $x^{-1} + y^{-2}$

34. $x^{-1} + y^{-1}$

35. $\frac{x^{-1} y^{-2} z}{x^2 y z^3}$

36. $\frac{3x^{-2} y z^2}{x^4 y^{-3} z}$

37. $\frac{(-2)^3 x^4 (yz)^2}{3^2 xy^3 z^4}$

38. $\frac{4x^{-2}(yz)^{-1}}{(-5)^2 x^4 y^2 z^{-2}}$

39. $\frac{x^{-2}}{x^{-2} + y^{-2}}$

40. $\frac{x^{-1} + y^{-1}}{x^{-1} - y^{-1}}$

41. $\frac{\left(\frac{x}{y}\right)^{-2} \cdot \left(\frac{y}{x}\right)^4}{x^2 y^3}$

42. $\frac{\left(\frac{y}{x}\right)^2}{x^{-2} y}$

43. $\left(\frac{3x^{-1}}{4y^{-1}}\right)^{-2}$

44. $\left(\frac{5x^{-2}}{6y^{-2}}\right)^{-3}$

45. $\frac{(xy^{-1})^{-2}}{xy}$

46. $\frac{(3xy^{-1})^2}{(2x^{-1} y)^3}$

47. $\dfrac{\left(\dfrac{x^2}{y}\right)^3}{\left(\dfrac{x}{y^2}\right)^2}$

48. $\dfrac{\left(\dfrac{2x}{3y^2}\right)^{-2}}{\dfrac{1}{y^3}}$

49. $\left(\dfrac{x}{y^2}\right)^{-2} \cdot (y^2)^{-1}$

50. $\dfrac{(x^2)^{-3}y^3}{(x^3y)^{-2}}$

51. Find the value of the expression $2x^3 - 3x^2 + x - 5$ if $x = 2$. If $x = -1$.

52. Find the value of the expression $3x^3 + 2x^2 - x - 2$ if $x = 1$. If $x = 2$.

▤ 1.4 POLYNOMIALS

We have described algebra as a generalization of arithmetic in which letters are used to represent real numbers. Further, we said that a letter representing *any* real number is a **variable**, and a letter representing a particular real number is a **constant**. From now on, we shall use the letters at the end of the alphabet, such as x, y, and z, to represent variables and use letters at the beginning of the alphabet, such as a, b, and c, to represent constants. Thus, in the expressions $3x + 5$ and $ax + b$, it is understood that x is a variable and that a and b are constants, even though the constants a and b are unspecified. As you will find out, the context usually makes it clear what the intention is when a letter is used. Now we give some basic vocabulary.

MONOMIAL   A **monomial** is the product of a constant times a variable raised to a nonnegative integer power. Thus, a monomial is of the form

$$ax^k$$

COEFFICIENT
DEGREE
where a is a constant, x is a variable, and $k \geq 0$ is an integer. The constant a is also called the **coefficient** of the monomial and k is called its **degree**.

Examples of monomials are:

MONOMIAL	COEFFICIENT	DEGREE	
$6x^2$	6	2	
$-\sqrt{2}x^3$	$-\sqrt{2}$	3	
3	3	0	Since $3 = 3 \cdot 1 = 3x^0$
$-5x$	-5	1	Since $-5x = -5x^1$
x^4	1	4	Since $x^4 = 1 \cdot x^4$

LIKE TERMS   Two monomials ax^k and bx^k with the same degree are called **like terms**. Such monomials when added or subtracted can be combined into a single monomial by using the distributive law. For example,

$$2x^2 + 5x^2 = (2 + 5)x^2 = 7x^2$$
$$8x^3 - 5x^3 = (8 - 5)x^3 = 3x^3$$

BINOMIAL

TRINOMIAL

The sum or difference of two monomials having different degrees is called a **binomial**. The sum or difference of three monomials with three different degrees is called a **trinomial**. For example,

$x^2 - 2$ is a binomial

$x^3 - 3x + 5$ is a trinomial

$2x^2 + 5x^2 + 2 = 7x^2 + 2$ is a binomial

POLYNOMIAL

A **polynomial** is the sum or difference of one or more monomials having different degrees. Thus, a polynomial is an expression of the form

$$a_n x^n + a_{n-1} x^{n-1} + \cdots + a_1 x + a_0 \qquad (1)$$

COEFFICIENTS

where $a_n, a_{n-1}, \ldots, a_1, a_0$ are constants* called the **coefficients** of the polynomial, $n \geq 0$ is an integer, and x is a variable.

TERMS

LEADING COEFFICIENT

DEGREE

ZERO POLYNOMIAL

STANDARD FORM

The monomials that make up a polynomial are called its **terms**. If $a_n \neq 0$, it is called the **leading coefficient** and n is called the **degree** of the polynomial. If all the coefficients are zero, the polynomial is called the **zero polynomial**, which has no degree.

Polynomials are usually written in **standard form**, beginning with the term of highest degree, continuing in descending order according to degree. Examples of polynomials are:

POLYNOMIAL	COEFFICIENTS	DEGREE
$3x^2 - 5 = 3x^2 + 0 \cdot x + (-5)$	$3, 0, -5$	2
$8 - 2x + x^2 = 1 \cdot x^2 - 2x + 8$	$1, -2, 8$	2
$5x + \sqrt{2} = 5x^1 + \sqrt{2}$	$5, \sqrt{2}$	1
$3 = 3 \cdot 1 = 3 \cdot x^0$	3	0
0	0	No degree

Although we have been using x to represent the variable, letters such as y or z are also commonly used. Thus,

$3x^4 - x^2 + 2$ is a polynomial (in x) of degree 4.

$9y^3 - 2y^2 + y - 3$ is a polynomial (in y) of degree 3.

$z^5 + \pi$ is a polynomial (in z) of degree 5.

SUBSCRIPT

*The notation a_n is read "a sub n." The number n is called a **subscript** and should not be confused with an exponent. We use subscripts in order to distinguish one constant from another when a large or undetermined number of constants is required.

Adding and Subtracting Polynomials

Polynomials are added and subtracted by combining their like terms.

EXAMPLE 1   Find the sum of the polynomials

$$8x^3 - 2x^2 + 6x - 2 \quad \text{and} \quad 3x^4 - 2x^3 + x^2 + x$$

Solution   We shall find the sum in two ways.

Horizontal Addition: The idea here is to group the like terms and then combine them.

$$(8x^3 - 2x^2 + 6x - 2) + (3x^4 - 2x^3 + x^2 + x)$$
$$= 3x^4 + (8x^3 - 2x^3) + (-2x^2 + x^2) + (6x + x) - 2$$
$$= 3x^4 + 6x^3 - x^2 + 7x - 2$$

Vertical Addition: The idea here is to vertically line up the like terms in each polynomial and then add the coefficients.

$$
\begin{array}{ccccc}
x^4 & x^3 & x^2 & x^1 & x^0 \\
 & 8x^3 - & 2x^2 + & 6x - & 2 \\
(+) \quad 3x^4 - & 2x^3 + & x^2 + & x & \\
\hline
3x^4 + & 6x^3 - & x^2 + & 7x - & 2
\end{array}
$$ ▀

The difference of two polynomials also can be found in either of the ways outlined above.

EXAMPLE 2   Find the difference $(3x^4 - 4x^3 + 6x^2 - 1) - (2x^4 + 8x^2 + 6x + 5)$.

Solution   *Horizontal Subtraction:*

$$(3x^4 - 4x^3 + 6x^2 - 1) - (2x^4 + 8x^2 + 6x + 5)$$
$$\underset{\uparrow}{=} 3x^4 - 4x^3 + 6x^2 - 1 - 2x^4 - 8x^2 - 6x - 5$$

Be sure to change the sign of each term in the second polynomial.

$$\underset{\uparrow}{=} (3x^4 - 2x^4) + (-4x^3) + (6x^2 - 8x^2) + (-6x) + (-1 - 5)$$

Combine like terms

$$= x^4 - 4x^3 - 2x^2 - 6x - 6$$

Vertical Subtraction:

$$
\begin{array}{ccccc}
x^4 & x^3 & x^2 & x^1 & x^0 \\
3x^4 - & 4x^3 + & 6x^2 & & - 1 \\
(-) \quad 2x^4 & & + 8x^2 + & 6x + & 5 \\
\hline
x^4 - & 4x^3 - & 2x^2 - & 6x - & 6
\end{array}
$$ ▀

The choice of which of these methods to use for adding and subtracting polynomials is entirely up to you. To save space on a printed page, we shall most often use the horizontal format.

Multiplying Polynomials

The product of two monomials ax^n and bx^m is obtained using the laws of exponents. Thus,

$$ax^n \cdot bx^m = abx^{n+m}$$

Products of polynomials are found by repeated use of the distributive law and the laws of exponents. Again, there is a choice of horizontal or vertical format.

EXAMPLE 3 Find the product $(2x + 5)(x^2 - x + 2)$.

Solution *Horizontal Multiplication:*

$$(2x + 5)(x^2 - x + 2) \underset{\uparrow}{=} 2x(x^2 - x + 2) + 5(x^2 - x + 2)$$

Distributive law

$$\underset{\uparrow}{=} 2x \cdot x^2 - 2x \cdot x + 2x \cdot 2 + 5 \cdot x^2 - 5 \cdot x + 5 \cdot 2$$

Distributive law

$$\underset{\uparrow}{=} 2x^3 - 2x^2 + 4x + 5x^2 - 5x + 10$$

Law of exponents

$$\underset{\uparrow}{=} 2x^3 + 3x^2 - x + 10$$

Combine like terms

Vertical Multiplication: The idea here is very much like multiplying a two-digit number by a three-digit number.

$$
\begin{array}{r}
x^2 - x + 2 \\
2x + 5 \\
\hline
2x^3 - 2x^2 + 4x \\
5x^2 - 5x + 10 \\
\hline
2x^3 + 3x^2 - x + 10
\end{array}
$$

	This line is $2x(x^2 - x + 2)$.
(+)	This line is $5(x^2 - x + 2)$.
	The sum of the preceding two lines. ≡

Special Products

SPECIAL PRODUCTS   Certain products, which we call **special products**, occur frequently in algebra. In the list that follows, x, a, b, c, and d are real numbers:

DIFFERENCE OF TWO SQUARES	$(x - a)(x + a) = x^2 - a^2$ (2)
SQUARE OF A BINOMIAL; PERFECT SQUARES	$(x + a)^2 = x^2 + 2ax + a^2$ (3a) $(x - a)^2 = x^2 - 2ax + a^2$ (3b)
MISCELLANEOUS TRINOMIAL	$(x + a)(x + b) = x^2 + (a + b)x + ab$ (4a) $(ax + b)(cx + d) = acx^2 + (ad + bc)x + bd$ (4b)
CUBE OF A BINOMIAL; PERFECT CUBES	$(x + a)^3 = x^3 + 3ax^2 + 3a^2x + a^3$ (5a) $(x - a)^3 = x^3 - 3ax^2 + 3a^2x - a^3$ (5b)
DIFFERENCE OF TWO CUBES	$(x - a)(x^2 + ax + a^2) = x^3 - a^3$ (6)
SUM OF TWO CUBES	$(x + a)(x^2 - ax + a^2) = x^3 + a^3$ (7)

We'll verify Equations (4b) and (5b); derivations of the remaining rules are left as exercises.

$$(ax + b)(cx + d) = ax(cx + d) + b(cx + d) = (acx^2 + adx) + (bcx + bd)$$
$$= acx^2 + (ad + bc)x + bd$$

$$(x - a)^3 = (x - a)(x - a)^2 \underset{\substack{\uparrow \\ (3b)}}{=} (x - a)(x^2 - 2ax + a^2)$$

$$= x(x^2 - 2ax + a^2) - a(x^2 - 2ax + a^2)$$
$$= x^3 - 2ax^2 + a^2x - ax^2 + 2a^2x - a^3$$
$$= x^3 - 3ax^2 + 3a^2x - a^3$$

The formulas in Equations (2)–(7) are used often and should be committed to memory. If you forget one or are unsure of its form, derive it for yourself.

EXAMPLE 4   Find the following products.

(a) $(x - 2)^2$ (b) $(x - 3)(x + 1)$

(c) $(2x + 3)(3x - 5)$ (d) $(x + 2)^3$

(e) $(x - 2)(x^2 + 2x + 4)$

Solution   (a) $(x - 2)^2 \underset{\substack{\uparrow \\ (3b)}}{=} x^2 - 4x + 4$ (b) $(x - 3)(x + 1) \underset{\substack{\uparrow \\ (4a) \\ a = -3, b = 1}}{=} x^2 - 2x - 3$

(c) $(2x + 3)(3x - 5) \underset{\substack{\uparrow \\ (4b) \\ a = 2, b = 3, c = 3, d = -5}}{=} 6x^2 - x - 15$

(d) $(x + 2)^3 = x^3 + 3 \cdot 2x^2 + 3 \cdot 4x + 8 = x^3 + 6x^2 + 12x + 8$
$$\uparrow$$
(5a)

(e) $(x - 2)(x^2 + 2x + 4) = x^3 - 2^3 = x^3 - 8$
$$\uparrow$$
(6)

■

Polynomials in Two or More Variables

POLYNOMIALS IN TWO VARIABLES

COEFFICIENT

DEGREE

Our discussion thus far has involved only polynomials in a single variable. A **polynomial in two variables** x and y is the sum of one or more monomials of the form $ax^n y^m$, where a is a constant—called the **coefficient**—x and y are variables, and n and m are nonnegative integers. The **degree** of the monomial $ax^n y^m$ is $n + m$. The **degree** of a polynomial in two variables x and y is the highest degree of all the monomials with nonzero coefficients that appear.

POLYNOMIALS IN THREE VARIABLES

Polynomials in three variables x, y, and z and polynomials in more than three variables are defined in a similar way. Here are some examples:

$$3x^2 + 2x^3y + 5 \qquad \pi x^3 - y^2 \qquad x^4 + 4x^3y - xy^3 + y^4$$

Two variables, degree is 4   Two variables, degree is 3   Two variables, degree is 4

$$x^2 + y^2 - z^2 + 4 \qquad x^3y^2z \qquad 5x^2 - 4y^2 + z^3y + 2w^2x$$

Three variables, degree is 2   Three variables, degree is 6   Four variables, degree is 4

Adding and multiplying polynomials in two or more variables is handled in the same way as for polynomials in one variable.

EXAMPLE 5

(a) $(x^3 - 3x^2y + 3xy^2 - y^3) - (x^3 - x^2y + y^3)$
$$= (x^3 - x^3) + (-3x^2y + x^2y) + 3xy^2 + (-y^3 - y^3)$$
$$= -2x^2y + 3xy^2 - 2y^3$$

(b) $(xy - 2)(2x^2 - xy + y^2) = xy(2x^2 - xy + y^2) - 2(2x^2 - xy + y^2)$
$$= 2x^3y - x^2y^2 + xy^3 - 4x^2 + 2xy - 2y^2$$

■

EXERCISE 1.4

In Problems 1–10 tell whether the expression is a polynomial. If it is, give its degree.

1. $3x^2 - 5$ 2. $1 - 4x$ 3. 5 4. $-\pi$

5. $3x^2 - \dfrac{5}{x}$ 6. $\dfrac{3}{x} + 2$ 7. $2y^3 - \sqrt{2}$ 8. $10z^2 + z$

9. $xy^2 - 1 + x$ 10. $x^3 - xy + y^3$

In Problems 11–64 perform the indicated operations. Express your answer as a polynomial.

11. $(x^2 + x + 5) + (3x - 2)$

12. $(x^3 - 3x^2 + 2) + (x^2 - x + 1)$

13. $(x^3 - 3x^2 + 5x + 10) - (2x^2 - x + 1)$

14. $(x^2 - 3x - 1) - (x^3 - 2x^2 + x + 5)$

15. $(6x^5 + x^3 + x) + (5x^4 - x^3 + 3x^2)$

16. $(10x^5 - 8x^2) + (3x^3 - 2x^2 + 6)$

17. $3(x^2 - 3x + 1) + 2(3x^2 + x - 4)$

18. $-2(x^2 + x + 1) + 6(-5x^2 - x + 2)$

19. $6(x^3 + x^2 - 3) - 4(2x^3 - 3x^2)$

20. $8(4x^3 - 3x^2 - 1) - 6(4x^3 + 8x - 2)$

21. $(x^2 - x + 2) + (2x^2 - 3x + 5) - (x^2 + 1)$

22. $(x^2 + 1) - (4x^2 + 5) + (x^2 + x - 2)$

23. $9(y^2 - 2y + 1) - 6(1 - y^2)$

24. $8(1 - y^3) + 2(1 + y + y^2 + y^3)$

25. $(x + a)^2 - x^2$

26. $(x - a)^2 - x^2$

27. $(x + a)^3 - x^3$

28. $(x - a)^3 - x^3$

29. $(x + 4)(x - 3)$

30. $(x + 6)(x - 2)$

31. $(x + 8)(2x + 1)$

32. $(2x - 1)(x + 2)$

33. $(-3x + 1)(x + 4)$

34. $(-2x - 1)(x + 1)$

35. $(1 - 4x)(2 - 3x)$

36. $(2 - x)(2 - 3x)$

37. $(x - 1)(x^2 + x + 1)$

38. $(x + 1)(x^2 - x + 1)$

39. $(2x + 3)(x^2 - 2)$

40. $(3x - 4)(x^2 + 1)$

41. $(3x - 5)(2x + 3)$

42. $(2x + 3)(-5x + 1)$

43. $(2x + 5)(x^2 + x + 1)$

44. $(3x - 4)(x^2 - x + 1)$

45. $(3x - 1)(2x^2 - 3x + 2)$

46. $(2x + 3)(3x^2 - 2x + 4)$

47. $(x^2 + x - 1)(x^2 - x + 1)$

48. $(x^2 + 2x + 1)(x^2 - 3x + 4)$

49. $(2x^2 - 3x + 4)(3x^2 - x + 4)$

50. $(3x^2 - 2x + 4)(2x^2 + 4x - 5)$

51. $(x^3 - 3x^2 + 5)(2x^2 - 1)$

52. $(3x^3 - x + 4)(3x^2 - 4x + 1)$

53. $(x + 2)^2(x - 1)$

54. $(x - 3)^2(x + 2)$

55. $(x - 1)^2(2x + 3)$

56. $(x + 5)^2(3x - 1)$

57. $(x - 1)(x - 2)(x - 3)$

58. $(x + 1)(x - 2)(x + 4)$

59. $(x + 1)^3 - (x - 1)^3$

60. $(x + 1)^3 - (x + 2)^3$

61. $(x - 1)^2(x + 1)^2$

62. $(x - 1)^3(x + 1)$

63. $(x^4 - 3x^3 + 2x^2 - x + 1)(x^2 - x + 1)$

64. $(x^4 + 4x^3 - 3x^2 + 2x - 5)(x^2 + 5x - 3)$

In Problems 65–80 perform the indicated operations. Express your answer as a polynomial.

65. $(x + y)(x - 2y)$ **66.** $(x - 2y)(x - y)$

67. $(x^2 + 2x + y^2) + (x + y - 3y^2)$ **68.** $(x^2 - 2xy + y^2) - (x^2 - xy)$

69. $(x - y)^2 - (x + y)^2$ **70.** $(x - y)^2 - (x^2 + y^2)$

71. $(x + 2y)^2 + (x - 3y)^2$ **72.** $(2x - y)^2 - (x - y)^2$

73. $[(x + y)^2 + z^2] + [x^2 + (y + z)^2]$ **74.** $[(x - y)^2 + z^2] + [x^2 + (y - z)^2]$

75. $(x - y)(x^2 + xy + y^2)$ **76.** $(x + y)(x^2 - xy + y^2)$

77. $(2x^2 + xy + y^2)(3x^2 - xy + 2y^2)$ **78.** $(3x^2 - xy + y^2)(2x^2 + xy - 4y^2)$

79. $(x + y + z)(x - y - z)$ **80.** $(x + y - z)(x - y + z)$

81. Explain why the degree of the product of two polynomials equals the sum of their degrees.

82. Explain why the degree of the sum of two polynomials equals the larger of their degrees.

83. Derive Equations (2), (3a), (3b), and (6) (page 34).

84. Derive Equations (4a), (5a), and (7) (page 34).

85. Develop a formula for $(x + a)^4$.

▤ 1.5 FACTORING POLYNOMIALS

Consider the following multiplication problem:

$$(2x + 3)(x - 4) = 2x^2 - 5x - 12$$

FACTORS   The polynomials on the left side are called **factors** of the polynomial on the right side. Expressing a given polynomial as a product of other polynomials—that is, finding the factors of a polynomial—is called **factoring**.

We shall restrict our discussion here to factoring polynomials in one variable, whose coefficients are integers, into products of polynomials in one variable, whose coefficients are integers. That is to say, we are **factoring over the integers**. There will be times, though, when we will want to **factor over the rational numbers** and even **factor over the real numbers**. Factoring over the rational numbers means to write a given polynomial as a product of polynomials whose coefficients are rational numbers. Factoring over the real numbers means to write a given polynomial as a product of polynomials whose coefficients are real numbers. Unless specified otherwise, we will be factoring over the integers.

FACTOR OVER THE INTEGERS
FACTOR OVER THE RATIONAL NUMBERS
FACTOR OVER THE REAL NUMBERS

Any polynomial can be written as the product of 1 times itself or as −1 times its additive inverse. If a polynomial cannot be written as the product of two other polynomials (excluding 1 and −1), then the polynomial is said to be **prime**. When a polynomial has been written as a product consisting only of prime factors, then it is said to be **factored completely**. Examples of prime polynomials are

$$2, 3, 5, x, x + 1, x - 1, 3x + 4, x^2 + 9, x^2 - 2$$

The first thing to look for in a factoring problem is a common monomial factor present in each term of the polynomial. If one is present, use the distributive law to factor it out. For example:

POLYNOMIAL	COMMON MONOMIAL FACTOR	REMAINING FACTOR
$2x + 4$	2	$x + 2$
$3x - 6$	3	$x - 2$
$2x^2 - 4x + 8$	2	$x^2 - 2x + 4$
$8x - 12$	4	$2x - 3$
$x^2 + x$	x	$x + 1$
$x^3 - 3x^2$	x^2	$x - 3$
$6x^2 + 9x$	$3x$	$2x + 3$

Notice that, once all common monomial factors have been removed from a polynomial, the remaining factor is either a prime polynomial of degree one or a polynomial of degree two or higher. (Do you see why?) Thus, we concentrate on techniques for factoring polynomials of degree two or higher that contain no monomial factors.

Special Factors

The list of special products given on page 34, when the equations are read from right to left, provides a list of special factors. For example, Equation (2) states that, if the polynomial is the difference of two squares, $x^2 - a^2$, it can be factored into $(x - a)(x + a)$.

EXAMPLE 1 Factor completely each polynomial, using the equations for special products.

(a) $x^2 - 4$ (b) $x^3 - 1$ (c) $x^3 + 8$ (d) $x^4 - 16$

Solution (a) We notice that $x^2 - 4$ is the difference of two squares, x^2 and 2^2. Thus, using Equation (2), page 34, we find that

$$x^2 - 4 = (x - 2)(x + 2)$$

(b) Equation (6), page 34, demonstrates that the difference of two cubes, $x^3 - a^3$, can be factored as $(x - a)(x^2 + ax + a^2)$. Because $x^3 - 1$ is the difference of two cubes, x^3 and 1^3, we find that

$$x^3 - 1 = (x - 1)(x^2 + x + 1)$$

(c) Equation (7), page 34, demonstrates that the sum of two cubes, $x^3 + a^3$, can be factored as $(x + a)(x^2 - ax + a^2)$. Because $x^3 + 8$ is the sum of two cubes, x^3 and 2^3, we have

$$x^3 + 8 = (x + 2)(x^2 - 2x + 4)$$

(d) Using Equation (2), page 34, for the difference of two squares, $x^4 = (x^2)^2$ and $16 = 4^2$, we have

$$x^4 - 16 = (x^2 - 4)(x^2 + 4)$$

But $x^2 - 4$ is also the difference of two squares. Thus,

$$x^4 - 16 = (x^2 - 4)(x^2 + 4) = (x - 2)(x + 2)(x^2 + 4) \qquad \blacksquare$$

Whenever the first term and third term of a trinomial are positive and are perfect squares, such as x^2, $9x^2$, 1, 4, and so on, check to see whether either of the special products Equations (3a) or (3b), page 34, applies.

EXAMPLE 2   Factor completely each polynomial.

(a) $x^2 + 4x + 4$ (b) $9x^2 - 6x + 1$ (c) $25x^2 + 30x + 9$

Solution   (a) The first term, x^2, and the third term, $4 = 2^2$, are perfect squares. Because the middle term is twice the product of x and 2, we use Equation (3a), page 34, to find that

$$x^2 + 4x + 4 = (x + 2)^2$$

(b) The first term, $9x^2 = (3x)^2$, and the third term, $1 = 1^2$, are perfect squares. Because the middle term is twice the product of $3x$ and 1, we use Equation (3b), page 34, to find that

$$9x^2 - 6x + 1 = (3x - 1)^2$$

(c) The first term, $25x^2 = (5x)^2$, and the third term, $9 = 3^2$, are perfect squares. Because the middle term is twice the product of $5x$ and 3, we use a form of Equation (3a), page 34, to find that

$$25x^2 + 30x + 9 = (5x + 3)^2 \qquad \blacksquare$$

For a trinomial to qualify as a perfect square, both the first term and the third term must be positive and be perfect squares. If this is the case, then check to see whether the middle term is twice the product of the first and third terms.

Factoring Second-Degree Polynomials

Factoring a second-degree polynomial, $Ax^2 + Bx + C$, where A, B, and C are integers, is a matter of skill, experience, and some trial and error. The idea behind factoring $Ax^2 + Bx + C$ is to see whether it can be made equal to the product of two, possibly equal, first-degree polynomials. Thus, we want to see whether

$$Ax^2 + Bx + C = (ax + b)(cx + d)$$

Let's start with second-degree polynomials that have a leading coefficient of 1. Such a polynomial, if it can be factored, must follow the form of the special product

$$(x + a)(x + b) = x^2 + (a + b)x + ab$$

EXAMPLE 3   Factor completely $x^2 + 7x + 12$.

Solution   We take note of the following clues:

1. The first term in $x^2 + 7x + 12$ is x^2, so we start with

$$x^2 + 7x + 12 = (x \quad)(x \quad)$$

2. The third term in $x^2 + 7x + 12$ is $ab = 12$, so the only possibilities for a and b are: $a = 1, b = 12$; $a = 2, b = 6$; $a = 3, b = 4$.

$$x^2 + 7x + 12 = \begin{cases} (x \quad 1)(x \quad 12) \\ (x \quad 2)(x \quad 6) \\ (x \quad 3)(x \quad 4) \end{cases}$$

3. Since the sign of the third term is positive and since the middle (second) term is also positive, the only possibilities are

$$x^2 + 7x + 12 = \begin{cases} (x + 1)(x + 12) = x^2 + 13x + 12 \\ (x + 2)(x + 6) = x^2 + 8x + 12 \\ (x + 3)(x + 4) = x^2 + 7x + 12 \end{cases}$$

4. Of these, only $(x + 3)(x + 4)$ gives the correct middle term $7x$. Thus,

$$x^2 + 7x + 12 = (x + 3)(x + 4) \qquad \equiv$$

EXAMPLE 4   Factor completely each polynomial.

(a) $x^2 - 6x + 8$ (b) $x^2 - x - 12$ (c) $x^2 + 4x - 12$

Solution   (a) The first term, x^2, requires that $x^2 - 6x + 8 = (x \quad)(x \quad)$. The third term, 8, requires the possibilities of 1 and 8, 2 and 4.

$$x^2 - 6x + 8 = \begin{cases} (x \quad 1)(x \quad 8) \\ (x \quad 2)(x \quad 4) \end{cases}$$

The sign of the third term is positive, and the middle term is negative. Thus, the only possible signs are

$$x^2 - 6x + 8 = \begin{cases} (x - 1)(x - 8) = x^2 - 9x + 8 \\ (x - 2)(x - 4) = x^2 - 6x + 8 \end{cases}$$

The middle term we want is $-6x$, so

$$x^2 - 6x + 8 = (x - 2)(x - 4)$$

(b) $x^2 - x - 12 \underset{\underset{\text{First term is } x^2}{\uparrow}}{=} (x \quad)(x \quad) \underset{\underset{\substack{\text{Factors} \\ \text{of } 12}}{\uparrow}}{=} \begin{cases} (x \quad 1)(x \quad 12) \\ (x \quad 2)(x \quad 6) \\ (x \quad 3)(x \quad 4) \end{cases}$

$$\underset{\underset{\substack{\text{Third term is} \\ \text{negative } (-12); \\ \text{the signs must} \\ \text{alternate.}}}{\uparrow}}{=} \begin{cases} (x - 1)(x + 12) = x^2 + 11x - 12 \\ (x + 1)(x - 12) = x^2 - 11x - 12 \\ (x - 2)(x + 6) = x^2 + 4x - 12 \\ (x + 2)(x - 6) = x^2 - 4x - 12 \\ (x - 3)(x + 4) = x^2 + x - 12 \\ (x + 3)(x - 4) = x^2 - x - 12 \end{cases}$$

Thus, $x^2 - x - 12 = (x + 3)(x - 4)$.

(c) We can use the same steps as in part (b) to get

$$x^2 + 4x - 12 = (x - 2)(x + 6)$$ ▬

If none of the possibilities work, the polynomial is prime.

EXAMPLE 5 Show that $x^2 + 9$ is prime.

Solution

$$x^2 + 9 = (x \quad)(x \quad) = \begin{cases} (x \quad 1)(x \quad 9) \\ (x \quad 3)(x \quad 3) \end{cases}$$

$$\underset{\underset{\substack{\text{The third} \\ \text{term is positive } (9); \\ \text{the signs must} \\ \text{be the same.}}}{\uparrow}}{=} \begin{cases} (x + 1)(x + 9) = x^2 + 10x + 9 \\ (x - 1)(x - 9) = x^2 - 10x + 9 \\ (x + 3)(x + 3) = x^2 + 6x + 9 \\ (x - 3)(x - 3) = x^2 - 6x + 9 \end{cases}$$

Since none of the possibilities work, we conclude that $x^2 + 9$ is prime.   ▬

If the leading coefficient is not 1, the same trial-and-error technique works.

EXAMPLE 6 Factor completely each polynomial.

(a) $2x^2 + 5x + 2$ (b) $2x^2 - x - 6$ (c) $6x^2 + 11x - 10$

Solution   (a) $2x^2 + 5x + 2 = (2x \quad)(x \quad) = \begin{cases} (2x \quad 2)(x \quad 1) \\ (2x \quad 1)(x \quad 2) \end{cases}$

$$= \begin{cases} (2x + 2)(x + 1) = 2x^2 + 4x + 2 \\ (2x + 1)(x + 2) = 2x^2 + 5x + 2 \end{cases}$$

Thus, $2x^2 + 5x + 2 = (2x + 1)(x + 2)$.

(b) $2x^2 - x - 6 = (2x \quad)(x \quad) = \begin{cases} (2x \quad 1)(x \quad 6) \\ (2x \quad 6)(x \quad 1) \\ (2x \quad 2)(x \quad 3) \\ (2x \quad 3)(x \quad 2) \end{cases}$

The third term
is negative (-6);
thus, we must
alternate the signs.

$$= \begin{cases} (2x - 1)(x + 6) = 2x^2 + 11x - 6 \\ (2x + 1)(x - 6) = 2x^2 - 11x - 6 \\ (2x - 6)(x + 1) = 2x^2 - 4x - 6 \\ (2x + 6)(x - 1) = 2x^2 + 4x - 6 \\ (2x - 2)(x + 3) = 2x^2 + 4x - 6 \\ (2x + 2)(x - 3) = 2x^2 - 4x - 6 \\ (2x - 3)(x + 2) = 2x^2 + x - 6 \\ (2x + 3)(x - 2) = 2x^2 - x - 6 \end{cases}$$

Thus, $2x^2 - x - 6 = (2x + 3)(x - 2)$.

(c) $6x^2 + 11x - 10 = \begin{cases} (x \quad)(6x \quad) \\ (2x \quad)(3x \quad) \end{cases} = \begin{cases} (x \quad 1)(6x \quad 10) \\ (x \quad 10)(6x \quad 1) \\ (2x \quad 1)(3x \quad 10) \\ (2x \quad 10)(3x \quad 1) \\ (x \quad 2)(6x \quad 5) \\ (x \quad 5)(6x \quad 2) \\ (2x \quad 2)(3x \quad 5) \\ (2x \quad 5)(3x \quad 2) \end{cases}$

x and $6x$
or $2x$ and $3x$ 1 and 10
or 2 and 5

$$= \begin{cases} (x \pm 1)(6x \mp 10) \\ (x \pm 10)(6x \mp 1) \\ (2x \pm 1)(3x \mp 10) \\ (2x \pm 10)(3x \mp 1) \\ (x \pm 2)(6x \mp 5) \\ (x \pm 5)(6x \mp 2) \\ (2x \pm 2)(3x \mp 5) \\ (2x \pm 5)(3x \mp 2) \end{cases}$$

The third term
is negative;
alternate signs.

Now, we want the product to contain the middle term, $11x$. By mentally computing the middle term that results from each of the products in our list, we are led to the final one, namely:

$(2x \pm 5)(3x \mp 2)$ $4x$ and $15x$ can be combined to yield $11x$.

$15x$

$4x$

Now it should be easy to conclude that

$$6x^2 + 11x - 10 = (2x + 5)(3x - 2)$$ ≡

Study the patterns of these examples carefully. Practice will give you the experience needed to use trial and error skillfully and efficiently.

Other Factoring Techniques

Sometimes a common factor occurs, not in every term of the polynomial, but in each of several groups of terms that together make up the polynomial. When this happens, the common factor can be factored out of each group by means of the distributive law. When this technique is used to factor, it is called **factoring by grouping**.

FACTORING BY GROUPING

EXAMPLE 7   Factor completely each polynomial by grouping.

(a) $(x^2 + 2)x + (x^2 + 2) \cdot 3$ (b) $x^3 - 4x^2 + 2x - 8$
(c) $3x^3 + 4x^2 - 6x - 8$

Solution   (a) We see the common term $x^2 + 2$. By applying the distributive law, we have

$$(x^2 + 2)x + (x^2 + 2) \cdot 3 = (x^2 + 2)(x + 3)$$

Since $x^2 + 2$ and $x + 3$ are prime, we are finished.

(b) We need to look carefully at this one to see that $x - 4$ is a common factor of $x^3 - 4x^2$ and $2x - 8$. Once we see this, we group the terms as

$$x^3 - 4x^2 + 2x - 8 = (x^3 - 4x^2) + (2x - 8)$$
$$= (x - 4)x^2 + (x - 4) \cdot 2$$
$$= (x - 4)(x^2 + 2)$$

Since $x - 4$ and $x^2 + 2$ are prime, we are finished.

(c) Here we see that $x^2 - 2$ is a common factor of $3x^3 - 6x$ and $4x^2 - 8$. Hence we group the terms as

$$3x^3 + 4x^2 - 6x - 8 = (3x^3 - 6x) + (4x^2 - 8)$$
$$= (x^2 - 2)(3x) + (x^2 - 2) \cdot 4$$
$$= (x^2 - 2)(3x + 4)$$

Since $x^2 - 2$ and $3x + 4$ are prime (over the integers), we are finished. ≡

Sometimes the addition and subtraction of the appropriate term can be used to factor a polynomial.

EXAMPLE 8   Factor completely the polynomial $x^4 + 7x^2 + 16$.

Solution   If the middle term were $8x^2$ instead of $7x^2$, the polynomial would be the perfect square $x^4 + 8x^2 + 16 = (x^2 + 4)^2$. We shall make it so by adding x^2. But in doing so we must at the same time subtract x^2. Let's see what happens.

$$x^4 + 7x^2 + 16 = x^4 + 8x^2 + 16 - x^2 = (x^4 + 8x^2 + 16) - x^2$$
$$= (x^2 + 4)^2 - x^2 = (x^2 + 4 - x)(x^2 + 4 + x)$$

Difference of squares

$$= (x^2 - x + 4)(x^2 + x + 4)$$

The latter two polynomials are prime, as you can verify by trial-and-error techniques. ≡

Summary

We close the section with a capsule summary.

TYPE OF POLYNOMIAL	METHOD	EXAMPLE
Any polynomial	Look for common monomial factors. (Always do this first!)	$6x^2 + 9x = 3x(2x + 3)$
Binomials— degree two or higher	Check for a special product: Difference of two squares $x^2 - a^2$ Difference of two cubes $x^3 - a^3$ Sum of two cubes $x^3 + a^3$	Examples 1(a), 1(d) Example 1(b) Example 1(c)
Trinomials— degree two	Check for a perfect square $(x \pm a)^2$ Trial and error	Example 2 Examples 3, 4, 5, 6
Three or more terms	Grouping Addition and subtraction of an appropriate term	Example 7 Example 8

EXERCISE 1.5 *In Problems 1–10 factor each polynomial by removing the common monomial factor.*

1. $3x + 6$

2. $7x - 14$

3. $ax^2 + a$

4. $ax - a$

5. $x^3 + x^2 + x$

6. $x^3 - x^2 + x$

7. $2x^2 + 2x + 2$

8. $3x^2 - 3x + 3$

9. $3x^2y - 6xy^2 + 12xy$

10. $60x^2y - 48xy^2 + 72x^3y$

In Problems 11–80 factor completely each polynomial. If the polynomial cannot be factored, say it is prime.

11. $x^2 - 1$

12. $x^2 - 4$

13. $4x^2 - 1$

14. $9x^2 - 1$

15. $x^2 + 7x + 10$

16. $x^2 + 3x - 4$

17. $x^2 - 10x + 21$

18. $x^2 - 4x - 21$

19. $x^2 - 7x - 8$

20. $x^2 - 6x + 5$

21. $x^2 + 2x + 1$

22. $x^2 - 4x + 4$

23. $x^2 + 4x + 4$

24. $x^2 - 2x - 1$

25. $x^2 - 2x - 15$

26. $x^2 - 6x - 14$

27. $3x^2 - 12x - 36$

28. $x^3 + 8x^2 - 20x$

29. $y^4 + 11y^3 + 30y^2$

30. $3y^3 - 18y^2 - 48y$

31. $16x^2 + 8x + 1$

32. $25x^2 + 10x + 1$

33. $4x^2 + 12x + 9$

34. $9x^2 - 12x + 4$

35. $ax^2 - 4a^2x - 45a^3$

36. $bx^2 + 14b^2x + 45b^3$

37. $x^3 - 27$

38. $x^3 + 27$

39. $8x^3 + 27$

40. $27 - 8x^3$

41. $3x^2 + 4x + 1$

42. $4x^2 + 3x - 1$

43. $x^4 - 81$

44. $x^4 - 1$

45. $x^6 - 2x^3 + 1$

46. $x^6 + 2x^3 + 1$

47. $x^7 - x^5$

48. $x^8 - x^5$

49. $2z^2 + 5z + 3$

50. $6z^2 - z - 1$

51. $16x^2 - 24x + 9$

52. $9x^2 + 24x + 16$

53. $16x^2 - 16x - 5$

54. $16x^2 - 11x - 5$

55. $4y^2 + 16y + 15$

56. $9y^2 + 9y - 4$

57. $18x^2 - 9x - 27$

58. $8x^2 - 6x - 2$

59. $8x^2 + 2x + 6$

60. $9x^2 - 3x + 3$

61. $x^2 - x + 4$ *p*

62. $x^2 + 6x + 9$

63. $4x^3 - 10x^2 - 6x$

64. $27x^3 - 9x^2 - 6x$

65. $x^4 + x^2 + 1$

66. $x^4 - 7x^2 + 9$

67. $9x^4 + 8x^2 - 1$

68. $8x^4 + 14x^2 - 4$

69. $x(x + 3) - 6(x + 3)$

70. $5(3x - 7) + x(3x - 7)$

71. $(x + 2)^2 - 5(x + 2)$

72. $(x - 1)^2 - 2(x - 1)$

73. $3(x^2 + 10x + 25) - 2(x + 5)$

74. $7(x^2 - 6x + 9) + 3(x - 3)$

75. $x^3 + 2x^2 - x - 2$

76. $x^3 - 3x^2 - x + 3$

77. $x^4 - x^3 + x - 1$

78. $x^4 + x^3 + x + 1$

79. $x^5 + x^3 + 8x^2 + 8$

80. $x^5 - x^3 + 8x^2 - 8$

1.6 RATIONAL EXPRESSIONS

RATIONAL EXPRESSION

If we form the ratio of two polynomials, the result is called a **rational expression**. Some examples of rational expressions are

(a) $\dfrac{x^3 + 1}{x}$ (b) $\dfrac{3x^2 + x - 2}{x^2 + 5}$ (c) $\dfrac{x}{x^2 - 1}$ (d) $\dfrac{xy^2}{(x - y)^2}$

Expressions (a), (b), and (c) are rational expressions in one variable, whereas (d) is a rational expression in two variables, x and y.

NUMERATOR
DENOMINATOR

REDUCED TO LOWEST TERMS
SIMPLIFIED

Rational expressions are described the same way rational numbers are. Thus, in expression (a) the polynomial $x^3 + 1$ is called the **numerator** and x is called the **denominator**. When the numerator and denominator of a rational expression contain no common factors (except 1 and -1), we say the rational expression is **reduced to lowest terms** or is **simplified**.

A rational expression is reduced to lowest terms or simplified by completely factoring the numerator and the denominator and cancelling any common factors by using the cancellation law

$$\frac{ac}{bc} = \frac{a}{b} \qquad b \neq 0, c \neq 0 \tag{1}$$

We shall follow the common practice of using a slash mark to indicate cancellation. For example,

$$\frac{x^2 - 1}{x^2 - 2x - 3} = \frac{(x - 1)(x + 1)}{(x - 3)(x + 1)} = \frac{x - 1}{x - 3}$$

EXAMPLE 1

Reduce each rational expression to lowest terms.

(a) $\dfrac{x^2 + 4x + 4}{x^2 + 3x + 2}$ (b) $\dfrac{x^3 - 8}{x^3 - 2x^2}$ (c) $\dfrac{8 - 2x}{x^2 - x - 12}$

Solution

(a) $\dfrac{x^2 + 4x + 4}{x^2 + 3x + 2} = \dfrac{(x + 2)(x + 2)}{(x + 2)(x + 1)} = \dfrac{x + 2}{x + 1}$

(b) $\dfrac{x^3 - 8}{x^3 - 2x^2} = \dfrac{(x - 2)(x^2 + 2x + 4)}{x^2(x - 2)} = \dfrac{x^2 + 2x + 4}{x^2}$

(c) $\dfrac{8 - 2x}{x^2 - x - 12} = \dfrac{2(4 - x)}{(x - 4)(x + 3)} = \dfrac{2(-1)(x - 4)}{(x - 4)(x + 3)} = \dfrac{-2}{x + 3}$ ■

Multiplying and Dividing Rational Expressions

The rules for multiplying and dividing rational expressions are the same as the rules for multiplying and dividing rational numbers, namely

$$\frac{a}{b} \cdot \frac{c}{d} = \frac{ac}{bd} \qquad b \neq 0, d \neq 0 \tag{2}$$

$$\frac{\dfrac{a}{b}}{\dfrac{c}{d}} = \frac{a}{b} \cdot \frac{d}{c} = \frac{ad}{bc} \qquad b \neq 0, c \neq 0, d \neq 0 \tag{3}$$

In using Equations (2) and (3) with rational expressions, be sure first to factor each polynomial completely so that common factors can be cancelled. We shall follow the practice of leaving our answers in factored form.

EXAMPLE 2   Perform the indicated operation and simplify the result. Leave your answer in factored form.

(a) $\dfrac{x^2 - 2x + 1}{x^3 + x} \cdot \dfrac{4x^2 + 4}{x^2 + x - 2}$ (b) $\dfrac{\dfrac{x + 3}{x^2 - 4}}{\dfrac{x^2 - x - 12}{x^3 - 8}}$

Solution   (a) $\dfrac{x^2 - 2x + 1}{x^3 + x} \cdot \dfrac{4x^2 + 4}{x^2 + x - 2} = \dfrac{(x - 1)^2}{x(x^2 + 1)} \cdot \dfrac{4(x^2 + 1)}{(x + 2)(x - 1)}$

$$= \frac{(x - 1)^2 (4)(x^2 + 1)}{x(x^2 + 1)(x + 2)(x - 1)} = \frac{4(x - 1)}{x(x + 2)}$$

(b) $\dfrac{\dfrac{x + 3}{x^2 - 4}}{\dfrac{x^2 - x - 12}{x^3 - 8}} = \dfrac{x + 3}{x^2 - 4} \cdot \dfrac{x^3 - 8}{x^2 - x - 12}$

$$= \frac{x + 3}{(x - 2)(x + 2)} \cdot \frac{(x - 2)(x^2 + 2x + 4)}{(x - 4)(x + 3)}$$

$$= \frac{(x + 3)(x - 2)(x^2 + 2x + 4)}{(x - 2)(x + 2)(x - 4)(x + 3)} = \frac{x^2 + 2x + 4}{(x + 2)(x - 4)} \quad ■$$

Note: Slanting the cancellation mark differently for different factors, as in Example 2, is a good practice to follow since it will help in checking for errors.

Adding and Subtracting Rational Expressions

The rules for adding and subtracting rational expressions are the same as the rules for adding and subtracting rational numbers, namely

$$\frac{a}{b} + \frac{c}{d} = \frac{ad + bc}{bd} \qquad b \neq 0, d \neq 0 \tag{4}$$

$$\frac{a}{b} - \frac{c}{d} = \frac{ad - bc}{bd} \qquad b \neq 0, d \neq 0 \tag{5}$$

These are general formulas for adding and subtracting ratios when the denominators b and d have no common factors.

EXAMPLE 3 Perform the indicated operation and simplify the result. Leave your answer in factored form.

(a) $\dfrac{x - 3}{x + 4} + \dfrac{x}{x - 2}$ (b) $\dfrac{x^2}{x^2 - 4} - \dfrac{1}{x}$

Solution (a) $\dfrac{x - 3}{x + 4} + \dfrac{x}{x - 2} = \dfrac{(x - 3)(x - 2) + (x + 4)(x)}{(x + 4)(x - 2)}$

$$= \frac{x^2 - 5x + 6 + x^2 + 4x}{(x + 4)(x - 2)}$$

$$= \frac{2x^2 - x + 6}{(x + 4)(x - 2)}$$

(b) $\dfrac{x^2}{x^2 - 4} - \dfrac{1}{x} = \dfrac{x^2(x) - (x^2 - 4)(1)}{(x^2 - 4)(x)} = \dfrac{x^3 - x^2 + 4}{(x - 2)(x + 2)(x)}$ ▤

If the denominators of two rational expressions to be added or subtracted are equal, we do not use the general rules given in Equations (4) and (5), but use instead the rules

$$\frac{a}{b} + \frac{c}{b} = \frac{a + c}{b} \qquad \frac{a}{b} - \frac{c}{b} = \frac{a - c}{b} \qquad b \neq 0 \tag{6}$$

EXAMPLE 4 Perform the indicated operation and simplify the result. Leave your answer in factored form.

(a) $\dfrac{2x^2 - 4}{2x + 5} + \dfrac{x + 3}{2x + 5}$ (b) $\dfrac{x}{x - 3} - \dfrac{3x + 2}{x - 3}$

Solution (a) $\dfrac{2x^2 - 4}{2x + 5} + \dfrac{x + 3}{2x + 5} = \dfrac{(2x^2 - 4) + (x + 3)}{2x + 5}$

$$= \frac{2x^2 + x - 1}{2x + 5}$$

$$= \frac{(2x - 1)(x + 1)}{2x + 5}$$

(b) $\dfrac{x}{x-3} - \dfrac{3x+2}{x-3} = \dfrac{x-(3x+2)}{x-3} = \dfrac{x-3x-2}{x-3}$

$$= \dfrac{-2x-2}{x-3} = \dfrac{-2(x+1)}{x-3}$$

≡

Least Common Multiple (LCM)

LEAST COMMON
MULTIPLE

If the denominators of two rational expressions to be added or subtracted have common factors, we do not use the general rules given by Equations (4) and (5), but instead use the **least common multiple (LCM)**, namely the polynomial of least degree that contains each polynomial as a factor. We rewrite each ratio using the LCM as the denominator and then use Equation (6). To find the least common multiple of two or more polynomials, first factor completely the polynomials. The LCM is the product of the different prime factors of each polynomial, each factor appearing the greatest number of times it appears in each polynomial. The next example will give you the idea.

EXAMPLE 5   Find the least common multiple of each pair of polynomials.

(a) $x(x-1)^2(x+1)$ and $4(x-1)(x+1)^3$

(b) $2x^2 - 2x - 12$ and $x^3 - 3x^2$

Solution   (a) The polynomials are already factored completely as

$$x(x-1)^2(x+1)   \text{and}   4(x-1)(x+1)^3$$

Start by writing the factors of the left polynomial. (Alternatively you could start with the right one.)

$$x(x-1)^2(x+1)$$

Now look at the right polynomial. Its first factor, 4, does not appear in our list, so we insert it:

$$4x(x-1)^2(x+1)$$

The next factor, $x-1$, is already in our list, so we continue. The final factor is $(x+1)^3$. Since our list has $x+1$ only to the first power, we replace $x+1$ in the list by $(x+1)^3$. The LCM is

$$4x(x-1)^2(x+1)^3$$

Notice that the LCM is, in fact, the polynomial of least degree that contains $x(x-1)^2(x+1)$ and $4(x-1)(x+1)^3$ as factors.

(b) First we factor completely each polynomial:

$$2x^2 - 2x - 12 = 2(x^2 - x - 6) = 2(x-3)(x+2)$$
$$x^3 - 3x^2 = x^2(x-3)$$

We begin by writing the factors of the first polynomial:

$$2(x - 3)(x + 2)$$

Now we look at the factors that appear in the second polynomial. The first factor, x^2, does not appear in our list, so we insert it:

$$2x^2(x - 3)(x + 2)$$

The remaining factor, $x - 3$, is already in our list. Thus, the LCM of $2x^2 - 2x - 12$ and $x^3 - 3x^2$ is $2x^2(x - 3)(x + 2)$. ■

The next example illustrates how the LCM is used for adding and subtracting rational expressions.

EXAMPLE 6 Perform the indicated operations and simplify the result. Leave your answer in factored form.

(a) $\dfrac{x}{x^2 + 3x + 2} + \dfrac{2x - 3}{x^2 - 1}$

(b) $\dfrac{1}{x^2 + x} - \dfrac{x + 4}{x^3 + 1} + \dfrac{3}{x^2}$

Solution (a) First we find the LCM of the denominators:

$$x^2 + 3x + 2 = (x + 2)(x + 1)$$
$$x^2 - 1 = (x - 1)(x + 1)$$

The LCM is $(x + 2)(x + 1)(x - 1)$. Next, we rewrite each rational expression using the LCM as denominator:

$$\frac{x}{x^2 + 3x + 2} = \frac{x}{(x + 2)(x + 1)} \underset{\uparrow}{=} \frac{x(x - 1)}{(x + 2)(x + 1)(x - 1)}$$

Multiply numerator and denominator by $x - 1$ to get LCM in denominator.

$$\frac{2x - 3}{x^2 - 1} = \frac{2x - 3}{(x - 1)(x + 1)} \underset{\uparrow}{=} \frac{(2x - 3)(x + 2)}{(x - 1)(x + 1)(x + 2)}$$

Multiply numerator and denominator by $x + 2$ to get LCM in denominator.

Now we can add by using Equation (6).

$$\frac{x}{x^2 + 3x + 2} + \frac{2x - 3}{x^2 - 1} = \frac{x(x - 1)}{(x + 2)(x + 1)(x - 1)} + \frac{(2x - 3)(x + 2)}{(x + 2)(x + 1)(x - 1)}$$

$$= \frac{(x^2 - x) + (2x^2 + x - 6)}{(x + 2)(x + 1)(x - 1)}$$

$$= \frac{3x^2 - 6}{(x + 2)(x + 1)(x - 1)}$$

$$= \frac{3(x^2 - 2)}{(x + 2)(x + 1)(x - 1)}$$

(b) Again we start by finding the LCM of the denominators:

$$x^2 + x = x(x + 1)$$
$$x^3 + 1 = (x + 1)(x^2 - x + 1)$$
$$x^2 = x^2$$

To get the LCM, we begin by listing the factors of the first polynomial:

$$x(x + 1)$$

Now we work through the factors of the second polynomial. As a result, we need to insert the factor $x^2 - x + 1$ in the list:

$$x(x + 1)(x^2 - x + 1)$$

The factors of the third polynomial require us to insert x^2 in place of x in the list. Thus, the LCM is

$$x^2(x + 1)(x^2 - x + 1)$$

Thus,

$$\frac{1}{x^2 + x} - \frac{x + 4}{x^3 + 1} + \frac{3}{x^2} = \frac{1}{x(x + 1)} - \frac{x + 4}{(x + 1)(x^2 - x + 1)} + \frac{3}{x^2}$$

$$= \frac{x(x^2 - x + 1)}{x^2(x + 1)(x^2 - x + 1)} - \frac{x^2(x + 4)}{x^2(x + 1)(x^2 - x + 1)}$$

$$+ \frac{3(x + 1)(x^2 - x + 1)}{x^2(x + 1)(x^2 - x + 1)}$$

$$= \frac{(x^3 - x^2 + x) - (x^3 + 4x^2) + 3(x^3 + 1)}{x^2(x + 1)(x^2 - x + 1)}$$

$$= \frac{3x^3 - 5x^2 + x + 3}{x^2(x + 1)(x^2 - x + 1)} \qquad \blacksquare$$

If we had not used the LCM technique to add the ratios in Example 6(a) but decided instead to use the general rule of Equation (4), page

48, we would have obtained a more complicated expression, as follows:

$$\frac{x}{x^2 + 3x + 2} + \frac{2x - 3}{x^2 - 1} = \frac{x(x^2 - 1) + (x^2 + 3x + 2)(2x - 3)}{(x^2 + 3x + 2)(x^2 - 1)}$$

$$= \frac{3x^3 + 3x^2 - 6x - 6}{(x^2 + 3x + 2)(x^2 - 1)}$$

$$= \frac{3(x^3 + x^2 - 2x - 2)}{(x^2 + 3x + 2)(x^2 - 1)}$$

Now we are faced with the additional problem of expressing this ratio in lowest terms. It is always best to first look for common factors in the denominators of expressions to be added or subtracted and use the LCM if any common factors are found.

Complex Ratios

COMPLEX RATIO   When sums and/or differences of rational expressions appear as the numerator and/or denominator of a ratio, the ratio is called a **complex ratio**. For example,

$$\frac{1 + \dfrac{1}{x}}{1 - \dfrac{1}{x}} \quad \text{and} \quad \frac{\dfrac{x^2}{x^2 - 4} - 3}{\dfrac{x - 3}{x + 2} - 1}$$

are complex ratios. To simplify a complex ratio, we treat the numerator and denominator separately, perform the indicated operations and simplify, and then simplify the resulting ratio.

EXAMPLE 7   Simplify each complex ratio.

(a) $\dfrac{1 + \dfrac{1}{x}}{1 - \dfrac{1}{x}}$ (b) $\dfrac{\dfrac{x^2}{x^2 - 4} - 3}{\dfrac{x - 3}{x + 2} - 1}$

Solution   (a) $\dfrac{1 + \dfrac{1}{x}}{1 - \dfrac{1}{x}} = \dfrac{\dfrac{x}{x} + \dfrac{1}{x}}{\dfrac{x}{x} - \dfrac{1}{x}} = \dfrac{\dfrac{x + 1}{x}}{\dfrac{x - 1}{x}} = \dfrac{x + 1}{x} \cdot \dfrac{x}{x - 1}$

$$= \frac{(x + 1)x}{x(x - 1)} = \frac{x + 1}{x - 1}$$

(b) $\dfrac{\dfrac{x^2}{x^2-4}-3}{\dfrac{x-3}{x+2}-1} = \dfrac{\dfrac{x^2}{x^2-4}-\dfrac{3(x^2-4)}{x^2-4}}{\dfrac{x-3}{x+2}-\dfrac{x+2}{x+2}} = \dfrac{\dfrac{x^2-3(x^2-4)}{x^2-4}}{\dfrac{(x-3)-(x+2)}{x+2}}$

$$= \dfrac{\dfrac{-2x^2+12}{x^2-4}}{\dfrac{-5}{x+2}} = \dfrac{\dfrac{-2(x^2-6)}{(x-2)(x+2)}}{\dfrac{-5}{x+2}}$$

$$= \dfrac{-2(x^2-6)}{(x-2)(x+2)} \cdot \dfrac{x+2}{-5}$$

$$= \dfrac{-2(x^2-6)(x+2)}{(x-2)(x+2)(-5)} = \dfrac{2(x^2-6)}{5(x-2)} \qquad \blacksquare$$

EXERCISE 1.6 *In Problems 1–20 reduce each rational expression to lowest terms.*

1. $\dfrac{5x+10}{x^2-4}$

2. $\dfrac{4x+8}{12x+24}$

3. $\dfrac{x^2-2x}{3x-6}$

4. $\dfrac{15x^2+24x}{3x^2}$

5. $\dfrac{24x}{12x^2-6x}$

6. $\dfrac{3x-12}{x^2-16}$

7. $\dfrac{y^2-25}{2y-10}$

8. $\dfrac{3y+2}{3y^2+5y+2}$

9. $\dfrac{x^2+4x-5}{x-1}$

10. $\dfrac{x-x^2}{x^2+x-2}$

11. $\dfrac{x^2-4}{x^2+5x+6}$

12. $\dfrac{x^2+x-6}{9-x^2}$

13. $\dfrac{x^2+5x-14}{2-x}$

14. $\dfrac{2x^2+5x-3}{1-2x}$

15. $\dfrac{2x^3-x^2-10x}{x^3-2x^2-8x}$

16. $\dfrac{4x^4+2x^3-6x^2}{4x^4+26x^3+30x^2}$

17. $\dfrac{(x-4)^2-9}{(x+3)^2-16}$

18. $\dfrac{(x+2)^2-8x}{(x-2)^2}$

19. $\dfrac{6x(x-1)-12}{x^3-8-(x-2)^2}$

20. $\dfrac{3(x-2)^2+17(x-2)+10}{2(x-2)^2+7(x-2)-15}$

In Problems 21–34 perform the indicated operation and simplify the result. Leave your answer in factored form.

21. $\dfrac{3x-6}{5x} \cdot \dfrac{x^2-x-6}{x^2-4}$

22. $\dfrac{9x-25}{2x-2} \cdot \dfrac{1-x^2}{6x-10}$

23. $\dfrac{4x^2-1}{x^2-16} \cdot \dfrac{x^2-4x}{2x+1}$

24. $\dfrac{12}{x^2-x} \cdot \dfrac{x^2-1}{4x-2}$

25. $\dfrac{4x-8}{-3x} \cdot \dfrac{12}{12-6x}$

26. $\dfrac{6x-27}{5x} \cdot \dfrac{2}{4x-18}$

27. $\dfrac{x^2-3x-10}{x^2+2x-35} \cdot \dfrac{x^2+4x-21}{x^2+9x+14}$

28. $\dfrac{x^2-x-6}{x^2-4x-5} \cdot \dfrac{x^2-25}{x^2+2x-15}$

29. $\dfrac{\dfrac{2x^2 - x - 28}{3x^2 - x - 2}}{\dfrac{4x^2 + 16x + 7}{3x^2 + 11x + 6}}$

30. $\dfrac{\dfrac{9x^2 + 3x - 2}{12x^2 + 5x - 2}}{\dfrac{9x^2 - 6x + 1}{8x^2 - 10x - 3}}$

31. $\dfrac{\dfrac{8x^2 - 6x + 1}{4x^2 - 1}}{\dfrac{12x^2 + 5x - 2}{6x^2 - x - 2}}$

32. $\dfrac{\dfrac{3x^2 + 2x - 1}{5x^2 - 9x - 2}}{\dfrac{2x^2 - x - 3}{10x^2 - 13x - 3}}$

33. $\dfrac{5x^2 - x}{3x + 2} \cdot \dfrac{2x^2 - x}{2x^2 - x - 1} \cdot \dfrac{10x^2 + 3x - 1}{6x^2 + x - 2}$

34. $\dfrac{x^2 - 7x - 8}{x^2 + 2x - 15} \cdot \dfrac{x^2 + 8x + 12}{x^2 - 6x - 7} \cdot \dfrac{x^2 + 4x - 5}{x^2 + 11x + 18}$

In Problems 35–54 perform the indicated operations and simplify the result. Leave your answer in factored form.

35. $\dfrac{x + 1}{x - 3} + \dfrac{x - 4}{x - 3}$

36. $\dfrac{2x - 5}{3x + 2} + \dfrac{x + 3}{3x + 2}$

37. $\dfrac{3x + 5}{2x - 1} - \dfrac{2x - 4}{2x - 1}$

38. $\dfrac{5x - 4}{3x + 4} - \dfrac{x + 1}{3x + 4}$

39. $\dfrac{\overset{(x-2)}{3}}{x + 1} + \dfrac{4}{x - 2}\;(x+1)$

$3x - 6 + 4x + 4$

$(x+1)(x-2)$

40. $\dfrac{6}{x - 1} + \dfrac{1}{x}$

41. $\dfrac{4}{x - 1} - \dfrac{1}{x + 2}$

$7x - 2$

42. $\dfrac{2}{x + 5} - \dfrac{3}{x - 5}$

43. $\dfrac{x}{x + 1} + \dfrac{2x - 3}{x - 1}$

44. $\dfrac{3x}{x - 4} + \dfrac{2x}{x + 3}$

45. $\dfrac{x - 2}{x + 2} - \dfrac{x + 2}{x - 2}$

46. $\dfrac{2x - 1}{x - 1} - \dfrac{2x + 1}{x + 1}$

47. $\dfrac{x}{x^2 - 4} + \dfrac{1}{x}$

48. $\dfrac{x - 1}{x^3} + \dfrac{1}{x^2 + 1}$

49. $\dfrac{x^3}{(x - 1)^2} - \dfrac{x^2 + 1}{x}$

50. $\dfrac{3x^2}{4} - \dfrac{x^3}{x^2 - 1}$

51. $\dfrac{x}{x + 1} + \dfrac{x - 2}{x - 1} - \dfrac{x + 1}{x - 2}$

52. $\dfrac{3x + 1}{x} + \dfrac{x}{x - 1} - \dfrac{2x}{x + 1}$

53. $\dfrac{1}{x} + \dfrac{1}{x + 1} - \dfrac{1}{x - 1}$

54. $\dfrac{1}{x} - \dfrac{1}{x - 1} - \dfrac{1}{x - 2}$

In Problems 55–62 find the LCM of the given polynomials.

55. $x^2 - 4, \quad x^2 - x - 2$

56. $x^2 - x - 12, \quad x^2 - 8x + 16$

57. $x^3 - x, \quad x^2 - x$

58. $3x^2 - 27, \quad 2x^2 - x - 15$

59. $4x^3 - 4x^2 + x, \quad 2x^3 - x^2, \quad x^3$

60. $x - 3, \quad x^2 + 3x, \quad x^3 - 9x$

61. $x^3 - x, \quad x^3 - 2x^2 + x, \quad x^3 - 1$

62. $x^2 + 4x + 4, \quad x^3 + 2x^2, \quad (x + 2)^3$

In Problems 63–74 perform the indicated operations and simplify the result. Leave your answer in factored form.

63. $\dfrac{x}{x^2 - 7x + 6} - \dfrac{x}{x^2 - 2x - 24}$

64. $\dfrac{x}{x - 3} - \dfrac{x + 1}{x^2 + 5x - 24}$

65. $\dfrac{4}{x^2 - 4} - \dfrac{2}{x^2 + x - 6}$

66. $\dfrac{3}{x - 1} - \dfrac{x - 4}{x^2 - 2x + 1}$

67. $\dfrac{3}{(x - 1)^2(x + 1)} + \dfrac{2}{(x - 1)(x + 1)^2}$

68. $\dfrac{2}{(x + 2)^2(x - 1)} - \dfrac{6}{(x + 2)(x - 1)^2}$

69. $\dfrac{x + 4}{x^2 - x - 2} - \dfrac{2x + 3}{x^2 + 2x - 8}$

70. $\dfrac{2x - 3}{x^2 + 8x + 7} - \dfrac{x - 2}{(x + 1)^2}$

71. $\dfrac{1}{x} - \dfrac{2}{x^2 + x} + \dfrac{3}{x^3 - x^2}$

72. $\dfrac{x}{(x - 1)^2} + \dfrac{2}{x} - \dfrac{x + 1}{x^3 - x^2}$

73. $\dfrac{1}{h}\left(\dfrac{1}{x + h} - \dfrac{1}{x}\right)$

74. $\dfrac{1}{h}\left[\dfrac{1}{(x + h)^2} - \dfrac{1}{x^2}\right]$

In Problems 75–86 perform the indicated operation and simplify the result.

75. $\dfrac{x - \dfrac{1}{x}}{x + \dfrac{1}{x}}$

76. $\dfrac{1 - \dfrac{x}{x + 1}}{2 - \dfrac{x - 1}{x}}$

77. $\dfrac{3 - \dfrac{x^2}{x + 1}}{1 + \dfrac{x}{x^2 - 1}}$

78. $\dfrac{3x - \dfrac{3}{x^2}}{\dfrac{1}{(x - 1)^2} - 1}$

79. $\dfrac{\dfrac{x + 4}{x - 2} - \dfrac{x - 3}{x + 1}}{x + 1}$

80. $\dfrac{\dfrac{x - 2}{x + 1} - \dfrac{x}{x - 2}}{x + 3}$

81. $\dfrac{\dfrac{x - 2}{x + 2} + \dfrac{x - 1}{x + 1}}{\dfrac{x}{x + 1} - \dfrac{2x - 3}{x}}$

82. $\dfrac{\dfrac{2x + 5}{x} - \dfrac{x}{x - 3}}{\dfrac{x^2}{x - 3} - \dfrac{(x + 1)^2}{x + 3}}$

83. $1 - \dfrac{1}{1 - \dfrac{1}{x}}$

84. $1 - \dfrac{1}{1 - \dfrac{1}{1 - x}}$

85. $\dfrac{\dfrac{x + h - 2}{x + h + 2} - \dfrac{x - 2}{x + 2}}{h}$

86. $\dfrac{\dfrac{x + h + 1}{x + h - 1} - \dfrac{x + 1}{x - 1}}{h}$

▦ 1.7 RADICALS

*q*th ROOT OF *a*   Suppose $q \geq 2$ is an integer. A **qth root of a** is a number which, when raised to the power q, equals a.

EXAMPLE 1   (a) A 3rd root of 8 is 2, since $2^3 = 8$.

(b) A 2nd root of 16 is 4, since $4^2 = 16$.

(c) A 3rd root of -64 is -4, since $(-4)^3 = -64$.

(d) A 2nd root of 16 is -4, since $(-4)^2 = 16$.

(e) A 4th root of $\frac{1}{16}$ is $-\frac{1}{2}$, since $(-\frac{1}{2})^4 = \frac{1}{16}$.

(f) A qth root of a is x if $x^q = a$.   ▤

Consider a qth root of a. If q is an odd integer, there is only one real number x for which $x^q = a$. Further, if a is negative, so is the qth root of a; if a is positive, so is the qth root of a. Look back at Examples 1(a) and 1(c).

If q is an even integer and if a is positive, there are two real numbers x for which $x^q = a$: one is positive, the other negative. [Look at Examples 1(b) and 1(d).]

If q is an even integer and if a is negative, there is no real number x for which $x^q = a$. Thus, if q is even and a is negative, there is no qth root for a—it is not defined.

Based on this discussion, we make the following definition:

PRINCIPAL qth ROOT OF a The **principal qth root of a number** a, symbolized by $\sqrt[q]{a}$, is defined as follows:

(a) If a is positive and q is even, then $\sqrt[q]{a}$ is the *positive* qth root of a.

(b) If a is negative and q is even, then $\sqrt[q]{a}$ is not defined.

(c) If q is odd, then $\sqrt[q]{a}$ is the qth root of a.

(d) $\sqrt[q]{0} = 0$.

The symbol $\sqrt[q]{a}$ for the principal qth root of a is sometimes called

RADICAL   a **radical**; the integer q is called the **index**, and a is called the **radicand**.

INDEX   If the index of a radical is 2, we call $\sqrt[2]{a}$ the **square root** of a and omit

RADICAND   the index by writing $\sqrt{a}$. If the index is 3, we call $\sqrt[3]{a}$ the **cube root**

SQUARE ROOT   of a.

CUBE ROOT   Notice that, if it is defined, the principal root of a number is unique. Thus,

$$\sqrt[3]{8} = 2 \qquad \sqrt{64} = 8 \qquad \sqrt[3]{-64} = -4 \qquad \sqrt[4]{\tfrac{1}{16}} = \tfrac{1}{2}$$

PERFECT ROOTS   These are examples of **perfect roots**. Thus, 8 and -64 are perfect cubes, since $8 = 2^3$ and $-64 = (-4)^3$; 64 is a perfect square, since $64 = 8^2$; and $\frac{1}{2}$ is a perfect 4th root of $\frac{1}{16}$, since $\frac{1}{16} = (\frac{1}{2})^4$.

In general, if $q \geq 2$ is a positive integer and a is a real number, we have

$$\sqrt[q]{a^q} = a \qquad \text{if } q \text{ is odd} \qquad \text{(1a)}$$
$$\sqrt[q]{a^q} = |a| \qquad \text{if } q \text{ is even} \qquad \text{(1b)}$$

Notice the need for the absolute value in Equation (1b). If q is even, then a^q is positive no matter whether $a > 0$ or $a < 0$. But if q is even, the principal qth root must be nonnegative. Hence, the reason for using the absolute value—it gives a nonnegative result.

EXAMPLE 2   (a) $\sqrt[3]{4^3} = 4$ (b) $\sqrt[5]{(-3)^5} = -3$

(c) $\sqrt[4]{2^4} = 2$ (d) $\sqrt[4]{(-3)^4} = |-3| = 3$ ▤

Radicals provide a way of representing many irrational real numbers. For example, there is no rational number whose square is 2. Using only decimals, we can at best only approximate the positive number whose square is 2. Using radicals, we can say that $\sqrt{2}$ *is* the positive number whose square is 2.

Properties of Radicals

Let $q \geq 2$ and $p \geq 2$ denote positive integers, and let a and b represent real numbers. Assuming all radicals are defined, we have the following properties:

$$
\begin{array}{ll}
\text{(a) } \sqrt[q]{ab} = \sqrt[q]{a}\sqrt[q]{b} & \text{(b) } \sqrt[q]{\dfrac{a}{b}} = \dfrac{\sqrt[q]{a}}{\sqrt[q]{b}} \\[3mm]
\text{(c) } \sqrt[q]{a^p} = (\sqrt[q]{a})^p & \text{(d) } \sqrt[p]{\sqrt[q]{a}} = \sqrt[pq]{a}
\end{array}
\qquad \text{(2)}
$$

When used in reference to radicals, the direction to "simplify" will mean to remove from the radical any perfect roots that occur as factors. Let's look at some examples of how the rules listed above are applied to simplify radicals.

EXAMPLE 3   Simplify each expression. Assume all variables are positive when they appear.

(a) $\sqrt{32}$ (b) $\sqrt[3]{8x^4}$ (c) $\sqrt[3]{-16x^4y^7}$

(d) $\sqrt{\sqrt[3]{x^8}}$ (e) $\sqrt[3]{\dfrac{8x^5}{27y^2}}$ (f) $\dfrac{\sqrt{x^5y}}{\sqrt{x^3y^3}}$

Solution   (a) $\sqrt{32} \underset{(2a)}{=} \sqrt{16 \cdot 2} = \sqrt{16}\sqrt{2} = 4\sqrt{2}$

16 is a perfect square.

(b) $\sqrt[3]{8x^4} \underset{(2a)}{=} \sqrt[3]{8} \cdot \sqrt[3]{x^4} = 2\sqrt[3]{x^3 \cdot x} = 2\sqrt[3]{x^3}\sqrt[3]{x} \underset{(1a)}{=} 2x\sqrt[3]{x}$

Factor out perfect cube.

(c) $\sqrt[3]{-16x^4y^7} = \sqrt[3]{-16}\sqrt[3]{x^4}\sqrt[3]{y^7} = \sqrt[3]{-8 \cdot 2}\sqrt[3]{x^3 \cdot x}\sqrt[3]{y^6 \cdot y}$

$\qquad = \sqrt[3]{-8} \cdot \sqrt[3]{2}\sqrt[3]{x^3}\sqrt[3]{x}\sqrt[3]{y^6}\sqrt[3]{y}$

$\qquad = -2\sqrt[3]{2}\,x\sqrt[3]{x}\sqrt[3]{(y^2)^3}\sqrt[3]{y}$

$\qquad = -2\sqrt[3]{2}\,x\sqrt[3]{x}\,y^2\sqrt[3]{y}$

$\qquad = -2xy^2\sqrt[3]{2xy}$

(d) $\sqrt{\sqrt[3]{x^8}} \underset{(2d)}{=} \sqrt[6]{x^8} = \sqrt[6]{x^6 \cdot x^2} = \sqrt[6]{x^6} \cdot \sqrt[6]{x^2} = x\sqrt[6]{x^2}$

(e) $\sqrt[3]{\dfrac{8x^5}{27y^2}} = \dfrac{\sqrt[3]{8x^5}}{\sqrt[3]{27y^2}} = \dfrac{\sqrt[3]{8}\sqrt[3]{x^5}}{\sqrt[3]{27}\sqrt[3]{y^2}} = \dfrac{2\sqrt[3]{x^3 \cdot x^2}}{3\sqrt[3]{y^2}}$

$\qquad = \dfrac{2x\sqrt[3]{x^2}}{3\sqrt[3]{y^2}} = \dfrac{2x}{3}\sqrt[3]{\dfrac{x^2}{y^2}}$

(f) $\dfrac{\sqrt{x^5y}}{\sqrt{x^3y^3}} = \sqrt{\dfrac{x^5y}{x^3y^3}} = \sqrt{\dfrac{x^2}{y^2}} = \sqrt{\left(\dfrac{x}{y}\right)^2} = \dfrac{x}{y}$ ▣

Two or more radicals can be combined, provided they have the same index and the same radicand.

EXAMPLE 4   Simplify each expression. Assume all variables are positive when they appear.

(a) $4\sqrt{27} - 8\sqrt{12} + \sqrt{3}$ (b) $\sqrt[3]{8x^4} + \sqrt[3]{-x} + 4\sqrt[3]{27x}$

Solution   (a) $4\sqrt{27} - 8\sqrt{12} + \sqrt{3} = 4\sqrt{9 \cdot 3} - 8\sqrt{4 \cdot 3} + \sqrt{3}$

$\qquad = 4 \cdot \sqrt{9}\sqrt{3} - 8 \cdot \sqrt{4}\sqrt{3} + \sqrt{3}$

$\qquad = 12\sqrt{3} - 16\sqrt{3} + \sqrt{3}$

$\qquad = (12 - 16 + 1)\sqrt{3}$

$\qquad = -3\sqrt{3}$

(b) $\sqrt[3]{8x^4} + \sqrt[3]{-x} + 4\sqrt[3]{27x} = \sqrt[3]{8}\sqrt[3]{x^3 \cdot x} + \sqrt[3]{-1}\sqrt[3]{x} + 4\sqrt[3]{27}\sqrt[3]{x}$

$\qquad = 2\sqrt[3]{x^3}\sqrt[3]{x} + (-1)\sqrt[3]{x} + 12\sqrt[3]{x}$

$\qquad = 2x\sqrt[3]{x} + 11\sqrt[3]{x}$

$\qquad = (2x + 11)\sqrt[3]{x}$ ▣

The special products listed earlier are helpful for finding products involving sums or differences of radicals.

EXAMPLE 5   Perform the indicated operation and simplify the result. Assume all variables are positive when they appear.

(a) $(\sqrt{x} - 3)(\sqrt{x} + 3)$ (b) $(\sqrt{x} + \sqrt{2})^2$
(c) $(\sqrt{x} - \sqrt{3})(\sqrt{x} + \sqrt{2})$

Solution   (a) $(\sqrt{x} - 3)(\sqrt{x} + 3) = (\sqrt{x})^2 - 3^2 = x - 9$
(b) $(\sqrt{x} + \sqrt{2})^2 = (\sqrt{x})^2 + 2\sqrt{x}\sqrt{2} + (\sqrt{2})^2$
$$= x + 2\sqrt{2x} + 2$$
(c) $(\sqrt{x} - \sqrt{3})(\sqrt{x} + \sqrt{2}) = (\sqrt{x})^2 - \sqrt{3}\sqrt{x} + \sqrt{2}\sqrt{x} - \sqrt{6}$
$$= x + (\sqrt{2} - \sqrt{3})\sqrt{x} - \sqrt{6}$$ ≡

Rationalizing

When radicals occur in ratios, it has become common practice to rewrite the ratio so that the denominator contains no radicals. This process is referred to as **rationalizing the denominator**.

RATIONALIZING THE DENOMINATOR

The idea is to find an appropriate expression so that, when it is multiplied by the radical in the denominator, the new denominator that results contains no radicals. For example:

IF RADICAL IS	MULTIPLY BY	TO GET PRODUCT FREE OF RADICALS
$\sqrt{3}$	$\sqrt{3}$	$\sqrt{9} = 3$
$\sqrt[3]{4}$	$\sqrt[3]{2}$	$\sqrt[3]{8} = 2$
$\sqrt{3} + 1$	$\sqrt{3} - 1$	$(\sqrt{3})^2 - 1^2 = 3 - 1 = 2$
$\sqrt{2} - 3$	$\sqrt{2} + 3$	$(\sqrt{2})^2 - 3^2 = 2 - 9 = -7$
$\sqrt{5} - \sqrt{3}$	$\sqrt{5} + \sqrt{3}$	$(\sqrt{5})^2 - (\sqrt{3})^2 = 5 - 3 = 2$
$\sqrt[3]{5} - 1$	$\sqrt[3]{5^2} + \sqrt[3]{5} + 1$	$\sqrt[3]{5^3} - 1 = 5 - 1 = 4$
$\sqrt[3]{2} + 2$	$\sqrt[3]{2^2} - 2\sqrt[3]{2} + 4$	$\sqrt[3]{2^3} + 8 = 2 + 8 = 10$

You are correct if you observed in this list that, after the second type of radical, the special products for differences of squares and sums and differences of cubes are the basis for what to multiply by.

EXAMPLE 6   Rationalize the denominators of each expression. Assume all variables are positive when they appear.

(a) $\dfrac{4}{\sqrt{2}}$ (b) $\dfrac{\sqrt{3}}{\sqrt[3]{2}}$ (c) $\dfrac{5}{\sqrt{3} + 2}$ (d) $\dfrac{\sqrt{x} - 2}{\sqrt{x} + 2}$ (e) $\dfrac{1}{\sqrt[3]{x} - 1}$

Solution   (a) $\dfrac{4}{\sqrt{2}} \uparrow = \dfrac{4}{\sqrt{2}} \cdot \dfrac{\sqrt{2}}{\sqrt{2}} = \dfrac{4\sqrt{2}}{(\sqrt{2})^2} = \dfrac{4\sqrt{2}}{2} = 2\sqrt{2}$

Multiply by $\dfrac{\sqrt{2}}{\sqrt{2}}$

(b) $\dfrac{\sqrt{3}}{\sqrt[3]{2}} \uparrow = \dfrac{\sqrt{3}}{\sqrt[3]{2}} \cdot \dfrac{\sqrt[3]{4}}{\sqrt[3]{4}} = \dfrac{\sqrt{3}\sqrt[3]{4}}{\sqrt[3]{8}} = \dfrac{\sqrt{3}\sqrt[3]{4}}{2}$

Multiply by $\dfrac{\sqrt[3]{4}}{\sqrt[3]{4}}$

(c) $\dfrac{5}{\sqrt{3} + 2} = \dfrac{5}{\sqrt{3} + 2} \cdot \dfrac{\sqrt{3} - 2}{\sqrt{3} - 2} = \dfrac{5(\sqrt{3} - 2)}{(\sqrt{3})^2 - 2^2}$

$= \dfrac{5(\sqrt{3} - 2)}{3 - 4} = -5(\sqrt{3} - 2)$

(d) $\dfrac{\sqrt{x} - 2}{\sqrt{x} + 2} = \dfrac{\sqrt{x} - 2}{\sqrt{x} + 2} \cdot \dfrac{\sqrt{x} - 2}{\sqrt{x} - 2} = \dfrac{(\sqrt{x} - 2)^2}{(\sqrt{x})^2 - 2^2}$

$= \dfrac{(\sqrt{x})^2 - 4\sqrt{x} + 4}{x - 4} = \dfrac{x - 4\sqrt{x} + 4}{x - 4}$

(e) $\dfrac{1}{\sqrt[3]{x} - 1} = \dfrac{1}{\sqrt[3]{x} - 1} \cdot \dfrac{\sqrt[3]{x^2} + \sqrt[3]{x} + 1}{\sqrt[3]{x^2} + \sqrt[3]{x} + 1}$

$= \dfrac{\sqrt[3]{x^2} + \sqrt[3]{x} + 1}{\sqrt[3]{x^3} - 1} = \dfrac{\sqrt[3]{x^2} + \sqrt[3]{x} + 1}{x - 1}$ ≡

In calculus, sometimes the numerator must be rationalized.

EXAMPLE 7   Rationalize the numerator of

$$\dfrac{\sqrt{x} - 2}{\sqrt{x} + 1}$$

Solution   We multiply by $\dfrac{\sqrt{x} + 2}{\sqrt{x} + 2}$:

$$\dfrac{\sqrt{x} - 2}{\sqrt{x} + 1} = \dfrac{\sqrt{x} - 2}{\sqrt{x} + 1} \cdot \dfrac{\sqrt{x} + 2}{\sqrt{x} + 2} = \dfrac{(\sqrt{x})^2 - 2^2}{(\sqrt{x} + 1)(\sqrt{x} + 2)} = \dfrac{x - 4}{x + 3\sqrt{x} + 2}$$

≡

≡ **HISTORICAL COMMENT**   The radical sign, $\sqrt{\ }$, was first used in print by Coss in 1525. It is thought to be the manuscript form of the letter r (for the Latin word *radix* = *root*), although this is not quite conclusively proved. It took a long time for $\sqrt{\ }$ to become the standard symbol for a square root and much longer to standardize $\overset{3}{\sqrt{}}, \overset{4}{\sqrt{}}, \overset{5}{\sqrt{}}$, and so on. The indices of the root were placed in every conceivable position, with

$$\overset{3}{\sqrt{}}8, \quad \sqrt{}\,③\,8, \quad \text{and} \quad \underset{3}{\sqrt{}}8$$

all being variants for $\sqrt[3]{8}$. The notation $\sqrt{}\sqrt{}16$ was popular for $\sqrt[4]{16}$. By the 1700s the index had settled where we now put it.

The bar on top of the present radical symbol, as shown below,

$$\sqrt{a^2 + 2ab + b^2}$$

is the last survivor of the *vinculum*, a bar placed atop an expression to indicate what we would now indicate with parentheses. For example,

$$a\overline{b + c} = a(b + c)$$

≡ **HISTORICAL**
PROBLEMS

1. Christian Dibuadius, a Dane, in 1605 used the early symbols $\sqrt{}$ for square root and $\sqrt{}C$ for cube root. Change the following into modern notation:
   (a) $\sqrt{}C8$ (b) $\sqrt{}C\sqrt{}C512$ (c) $\sqrt{}\sqrt{}C64$ (d) $\sqrt{}C\sqrt{}64$
   (e) $\sqrt{}\sqrt{}81$ (f) $\sqrt{}\sqrt{}256x^8$ (g) $\sqrt{}C - 4 + \sqrt{}8$

2. Oresme (1360) used

   $$\tfrac{1}{2}2^p \text{ for } 2^{1/2} \quad \text{and} \quad \boxed{1\tfrac{1}{2}}\,2^p \text{ for } 2^{1+1/2} = 2^{3/2}$$

   What are:
   (a) $\tfrac{1}{3}8^p$ (b) $\tfrac{1}{4}256^p$ (c) $\boxed{2\tfrac{1}{3}}\,8^p$ (d) $\boxed{-1\tfrac{1}{4}}\,64^p$ ≡

EXERCISE 1.7

In Problems 1–12 evaluate each perfect root. Assume all variables are positive when they appear.

1. $\sqrt{64}$ 2. $\sqrt{81}$ 3. $\sqrt[3]{27}$ 4. $\sqrt[3]{125}$

5. $\sqrt[3]{-1}$ 6. $\sqrt[3]{-8}$ 7. $\sqrt{\dfrac{1}{4}}$ 8. $\sqrt[3]{\dfrac{8}{27}}$

9. $\sqrt{9x^2}$ 10. $\sqrt[3]{27x^6}$ 11. $\sqrt[3]{8(1 + x)^3}$ 12. $\sqrt{4(x + 4)^2}$

In Problems 13–38 simplify each expression. Assume all variables are positive when they appear.

13. $\sqrt{8}$ 14. $\sqrt[4]{32}$ 15. $\sqrt[3]{16x^4}$ 16. $\sqrt{27x^3}$

17. $\sqrt[3]{\sqrt{x^6}}$ 18. $\sqrt{\sqrt{x^6}}$ 19. $\sqrt{\dfrac{32x^3}{9x}}$ 20. $\sqrt[3]{\dfrac{x}{8x^4}}$

21. $\sqrt[4]{x^{12}y^8}$ 22. $\sqrt[5]{x^{10}y^5}$ 23. $\sqrt[4]{\dfrac{x^9 y^7}{xy^3}}$ 24. $\sqrt[3]{\dfrac{3xy^2}{81x^4y^2}}$

25. $\sqrt{36x}$ 26. $\sqrt{9x^5}$ 27. $\sqrt{3x^2}\sqrt{12x}$ 28. $\sqrt{5x}\sqrt{20x^3}$

29. $\dfrac{\sqrt{3xy^3}\sqrt{2x^2y}}{\sqrt{6x^3y^4}}$ 30. $\dfrac{\sqrt[3]{x^2y}\sqrt[3]{125x^3}}{\sqrt[3]{8x^3y^4}}$

31. $(\sqrt{5}\sqrt[3]{9})^2$

32. $(\sqrt[3]{3}\sqrt{10})^4$

33. $\sqrt{\dfrac{2x-3}{2x^4+3x^3}}\sqrt{\dfrac{x}{4x^2-9}}$

34. $\sqrt[3]{\dfrac{x-1}{x^2+2x+1}}\sqrt[3]{\dfrac{(x-1)^2}{x+1}}$

35. $\sqrt{\sqrt[3]{\sqrt[4]{x}}}$

36. $\sqrt[4]{\sqrt[3]{x}}$

37. $\sqrt{\dfrac{x-y}{x+y}}\sqrt{\dfrac{x^2+2y+y^2}{x^2-y^2}}$

38. $\sqrt{\dfrac{x^2+y^2}{xy}}\sqrt{\dfrac{xy(x)}{x^4-y^4}}$

In Problems 39–48 simplify each expression. Assume all variables are positive when they appear.

39. $3\sqrt{2}+2\sqrt{2}-\sqrt{2}$

40. $6\sqrt{5}-\sqrt{5}-4\sqrt{5}$

41. $3\sqrt{2}-\sqrt{18}+2\sqrt{8}$

42. $5\sqrt{3}+2\sqrt{12}-3\sqrt{27}$

43. $\sqrt[3]{16}+5\sqrt[3]{2}-2\sqrt[3]{54}$

44. $9\sqrt[3]{24}-\sqrt[3]{81}$

45. $\sqrt{8x^3}-3\sqrt{50x}+\sqrt{2x^5}$

46. $\sqrt{x^2y}-3x\sqrt{9y}+4\sqrt{25y}$

47. $\sqrt[3]{16x^4y}-3x\sqrt[3]{2xy}+5\sqrt[3]{-2xy^4}$

48. $8xy-\sqrt{25x^2y^2}+\sqrt[3]{8x^3y^3}$

In Problems 49–62 perform the indicated operation and simplify the result. Assume all variables are positive when they appear.

49. $(3\sqrt{6})(2\sqrt{2})$

50. $(5\sqrt{8})(-3\sqrt{3})$

51. $(\sqrt{3}+3)(\sqrt{3}-1)$

52. $(\sqrt{5}-2)(\sqrt{5}+3)$

53. $(3\sqrt{7}+3)(2\sqrt{7}+2)$

54. $(2\sqrt{6}+3)^2$

55. $(\sqrt{x}-1)^2$

56. $(\sqrt{x}+\sqrt{5})^2$

57. $(\sqrt[3]{x}-1)^3$

58. $(\sqrt[3]{x}+\sqrt[3]{2})^3$

59. $(2\sqrt{x}-3\sqrt{y})(2\sqrt{x}+5\sqrt{y})$

60. $(4\sqrt{x}-\sqrt{y})(\sqrt{x}+3\sqrt{y})$

61. $\sqrt{1-x^2}-\dfrac{1}{\sqrt{1-x^2}}$

62. $\sqrt{1-x^2}+\dfrac{x^2}{\sqrt{1-x^2}}$

In Problems 63–76 rationalize the denominator of each expression. Assume all variables are positive when they appear.

63. $\dfrac{1}{\sqrt{2}}$

64. $\dfrac{6}{\sqrt[3]{4}}$

65. $\dfrac{-\sqrt{3}}{\sqrt{5}}$

66. $\dfrac{-\sqrt[3]{3}}{\sqrt{8}}$

67. $\dfrac{\sqrt{3}}{5-\sqrt{2}}$

68. $\dfrac{\sqrt{2}}{\sqrt{7}+2}$

69. $\dfrac{2-\sqrt{5}}{2+3\sqrt{5}}$

70. $\dfrac{\sqrt{3}-1}{2\sqrt{3}+3}$

71. $\dfrac{\sqrt{3}-\sqrt{2}}{\sqrt{3}+\sqrt{2}}$

72. $\dfrac{\sqrt{5}+\sqrt{3}}{\sqrt{5}-\sqrt{3}}$

73. $\dfrac{\sqrt{x+h}-\sqrt{x}}{\sqrt{x+h}+\sqrt{x}}$

74. $\dfrac{\sqrt{x+h}+\sqrt{x-h}}{\sqrt{x+h}-\sqrt{x-h}}$

75. $\dfrac{1}{\sqrt[3]{x}+1}$

76. $\dfrac{1}{\sqrt[3]{x^2}+1}$

77. Rationalize the numerator of

$$\frac{\sqrt{x+h}-\sqrt{x}}{h}$$

78. Rationalize the numerator of

$$\frac{\dfrac{1}{\sqrt{x+h}} - \dfrac{1}{\sqrt{x}}}{h}$$

1.8 RATIONAL EXPONENTS

Our purpose in this section is to give a definition for "a raised to the power p/q," where a is a real number and p/q is a rational number. However, we want the definition we give to ensure that the laws of exponents stated earlier (see page 29) for integer exponents remain true for rational exponents.

We begin by seeking the definition of $a^{1/q}$ where $q \geq 2$ is an integer. If the law of exponents in Equation (5) is to hold, then

$$(a^{1/q})^q = a^{(1/q)q} = a^1 = a$$

Thus, $a^{1/q}$ is a number that, when raised to the power q, is a. But this was the definition we gave of the qth root of a. (See page 55.) This leads us to the following definition:

$a^{1/q}$ If a is a real number and $q \geq 2$ is an integer, then

$$\boxed{a^{1/q} = \sqrt[q]{a} \qquad\qquad (1)}$$

provided $\sqrt[q]{a}$ exists.

EXAMPLE 1 (a) $4^{1/2} = \sqrt{4} = 2$ (b) $(-27)^{1/3} = \sqrt[3]{-27} = -3$
(c) $8^{1/2} = \sqrt{8} = 2\sqrt{2}$ (d) $16^{1/3} = \sqrt[3]{16} = 2\sqrt[3]{2}$ ≡

We now seek a definition for $a^{p/q}$ where p and q are integers containing no common factors (except 1 and -1) and $q \geq 2$. Again we want the definition to obey the laws of exponents stated earlier. Thus, if the law of exponents in Equation (5) is to hold, then

$$a^{p/q} = a^{p(1/q)} = (a^p)^{1/q} \qquad \text{and} \qquad a^{p/q} = a^{(1/q)p} = (a^{1/q})^p$$

$a^{p/q}$ If a is a real number and p and q are integers containing no common factors with $q \geq 2$, then

$$\boxed{a^{p/q} = \sqrt[q]{a^p} = (\sqrt[q]{a})^p \qquad\qquad (2)}$$

provided $\sqrt[q]{a}$ exists.

We have two comments about Equation (2). First, the exponent *p/ q must be in lowest terms and q must be positive.* Second, in actually computing a rational exponent $a^{p/q}$, either $\sqrt[q]{a^p}$ or $(\sqrt[q]{a})^p$ may be used, the choice depending on which one is easier to use.

EXAMPLE 2   (a) $4^{3/2} = (\sqrt{4})^3 = 2^3 = 8$

(b) $(-8)^{4/3} = (\sqrt[3]{-8})^4 = (-2)^4 = 16$

(c) $(32)^{-2/5} = (\sqrt[5]{32})^{-2} = 2^{-2} = \frac{1}{4}$

(d) $4^{6/4} = 4^{3/2} = (\sqrt{4})^3 = 2^3 = 8$ ■

Based on the definition of $a^{p/q}$, no meaning is given to $a^{p/q}$ if a is a negative real number and q is an even integer.

The definitions in Equations (1) and (2) were stated so that the laws of exponents would remain true for rational exponents. For convenience, we list again the laws of exponents.

LAWS OF EXPONENTS   If a and b are real numbers and r and s are rational numbers, then

$$a^r a^s = a^{r+s} \qquad\qquad (a^r)^s = a^{rs}$$

$$(ab)^r = a^r \cdot b^r \qquad\qquad a^{-r} = \frac{1}{a^r}$$

$$\left(\frac{a}{b}\right)^r = \frac{a^r}{b^r} \qquad\qquad \frac{a^r}{a^s} = a^{r-s} = \frac{1}{a^{s-r}}$$

where it is assumed that all expressions used are defined.

Rational exponents sometimes can be used to simplify radicals.

EXAMPLE 3   Simplify each expression. Assume that any variables that appear are positive.

(a) $\sqrt[9]{x^3}$ (b) $\sqrt[3]{4}\sqrt{2}$ (c) $\sqrt{a\sqrt[3]{a}}$

Solution   (a) $\sqrt[9]{x^3} = x^{3/9} = x^{1/3} = \sqrt[3]{x}$

(b) $\sqrt[3]{4}\sqrt{2} = 4^{1/3} \cdot 2^{1/2} = (2^2)^{1/3} \cdot 2^{1/2} = 2^{2/3} \cdot 2^{1/2}$

$= 2^{7/6} = 2 \cdot 2^{1/6} = 2\sqrt[6]{2}$

(c) $\sqrt{a\sqrt[3]{a}} = (a \cdot a^{1/3})^{1/2} = (a^{4/3})^{1/2} = a^{2/3} = \sqrt[3]{a^2}$ ■

EXAMPLE 4   Simplify each expression. Express your answer so that only positive exponents occur. Assume that the variables are restricted so that all expressions are defined.

(a) $\left(\dfrac{2x^{1/3}}{y^{2/3}}\right)^{-3}$ (b) $(x^{2/3}y^{-3/4})(x^{-2}y)^{1/2}$

(c) $\left(\dfrac{9x^2y^{1/3}}{x^{1/3}y}\right)^{1/2}$ (d) $\dfrac{(2x+5)^{1/3}(2x+5)^{-1/2}}{(2x+5)^{-3/4}}$

Solution   (a) $\left(\dfrac{2x^{1/3}}{y^{2/3}}\right)^{-3} \underset{\underset{(8)}{\uparrow}}{=} \left(\dfrac{y^{2/3}}{2x^{1/3}}\right)^3 = \dfrac{(y^{2/3})^3}{(2x^{1/3})^3} = \dfrac{y^2}{2^3(x^{1/3})^3} = \dfrac{y^2}{8x}$

page 28

(b) $(x^{2/3}y^{-3/4})(x^{-2}y)^{1/2} = (x^{2/3}y^{-3/4})[(x^{-2})^{1/2}y^{1/2}]$

$$= x^{2/3}y^{-3/4}x^{-1}y^{1/2}$$

$$= (x^{2/3}x^{-1})(y^{-3/4}y^{1/2})$$

$$= x^{(2/3)-1}y^{(-3/4)+(1/2)}$$

$$= x^{-1/3}y^{-1/4} = \dfrac{1}{x^{1/3}y^{1/4}}$$

(c) $\left(\dfrac{9x^2y^{1/3}}{x^{1/3}y}\right)^{1/2} = \left(\dfrac{9x^{2-(1/3)}}{y^{1-(1/3)}}\right)^{1/2} = \left(\dfrac{9x^{5/3}}{y^{2/3}}\right)^{1/2} = \dfrac{9^{1/2}(x^{5/3})^{1/2}}{(y^{2/3})^{1/2}} = \dfrac{3x^{5/6}}{y^{1/3}}$

(d) $\dfrac{(2x+5)^{1/3}(2x+5)^{-1/2}}{(2x+5)^{-3/4}} = (2x+5)^{(1/3)-(1/2)-(-3/4)}$

$$= (2x+5)^{(4-6+9)/12} = (2x+5)^{7/12} \qquad \blacksquare$$

The next two examples illustrate some algebra you will need to know for certain calculus problems.

EXAMPLE 5   Write each expression as a single ratio in which only positive exponents appear.

(a) $(x^2+1)^{1/2} + x \cdot \dfrac{1}{2}(x^2+1)^{-1/2} \cdot 2x$

(b) $\dfrac{(x^2+4)^{1/2} \cdot 2x - x^2 \cdot \frac{1}{2}(x^2+4)^{-1/2} \cdot 2x}{x^2+4}$

Solution   (a) $(x^2+1)^{1/2} + x \cdot \dfrac{1}{2}(x^2+1)^{-1/2} \cdot 2x = (x^2+1)^{1/2} + \dfrac{x^2}{(x^2+1)^{1/2}}$

$$= \dfrac{(x^2+1)^{1/2}(x^2+1)^{1/2} + x^2}{(x^2+1)^{1/2}}$$

$$= \dfrac{(x^2+1) + x^2}{(x^2+1)^{1/2}}$$

$$= \dfrac{2x^2+1}{(x^2+1)^{1/2}}$$

(b) $\dfrac{(x^2 + 4)^{1/2} \cdot 2x - x^2 \cdot \frac{1}{2}(x^2 + 4)^{-1/2} \cdot 2x}{x^2 + 4}$

$$= \frac{2x(x^2 + 4)^{1/2} - x^3(x^2 + 4)^{-1/2}}{x^2 + 4}$$

$$= \frac{2x(x^2 + 4)^{1/2} - \dfrac{x^3}{(x^2 + 4)^{1/2}}}{x^2 + 4}$$

$$= \frac{\dfrac{2x(x^2 + 4)^{1/2}(x^2 + 4)^{1/2} - x^3}{(x^2 + 4)^{1/2}}}{x^2 + 4}$$

$$= \frac{2x(x^2 + 4) - x^3}{(x^2 + 4)^{1/2}} \cdot \frac{1}{(x^2 + 4)}$$

$$= \frac{2x^3 + 8x - x^3}{(x^2 + 4)^{3/2}}$$

$$= \frac{x^3 + 8x}{(x^2 + 4)^{3/2}} = \frac{x(x^2 + 8)}{(x^2 + 4)^{3/2}} \quad \blacksquare$$

EXAMPLE 6 Factor $4x^{1/3}(2x + 1) + 2x^{4/3}$.

Solution We begin by looking for factors that are common to the two terms. Notice that 2 and $x^{1/3}$ are factors of each term. Thus,

$$4x^{1/3}(2x + 1) + 2x^{4/3} = 2x^{1/3}[2(2x + 1) + x]$$
$$= 2x^{1/3}(5x + 2) \quad \blacksquare$$

EXERCISE 1.8 *In Problems 1–20 simplify each expression.*

1. $8^{2/3}$
2. $4^{3/2}$
3. $(-27)^{1/3}$
4. $(-16)^{3/4}$

5. $16^{3/2}$
6. $64^{3/2}$
7. $9^{-3/2}$
8. $25^{-5/2}$

9. $\left(\dfrac{9}{8}\right)^{3/2}$
10. $\left(\dfrac{27}{8}\right)^{2/3}$
11. $\left(\dfrac{8}{9}\right)^{-3/2}$
12. $\left(\dfrac{8}{27}\right)^{-2/3}$

13. $4^{1.5}$
14. $16^{-1.5}$
15. $\left(\dfrac{1}{4}\right)^{-1.5}$
16. $\left(\dfrac{1}{9}\right)^{1.5}$

17. $(\sqrt{3})^6$
18. $(\sqrt[3]{4})^6$
19. $(\sqrt{2})^{-2}$
20. $(\sqrt[4]{5})^{-8}$

In Problems 21–30 simplify each expression. Express your answer so that only positive exponents occur. Assume that the variables are restricted so that all expressions are defined.

21. $x^{3/4}x^{1/3}x^{-1/2}$
22. $x^{2/3}x^{1/2}x^{-1/4}$
23. $(x^3y^6)^{1/3}$

24. $(x^4y^8)^{3/4}$
25. $(x^2y)^{1/3}(xy^2)^{2/3}$
26. $(xy)^{1/4}(x^2y^2)^{1/2}$

27. $(16x^2y^{-1/3})^{3/4}$
28. $(4x^{-1}y^{1/3})^{3/2}$
29. $\left(\dfrac{x^{2/5}y^{-1/5}}{x^{-1/3}y^{2/3}}\right)^{15}$

30. $\left(\dfrac{x^{1/2}}{y^2}\right)^4\left(\dfrac{y^{1/3}}{x^{-2/3}}\right)^3$

In Problems 31–44 write each expression as a single ratio in which only positive exponents and/or radicals appear.

31. $\dfrac{x}{(1 + x)^{1/2}} + 2(1 + x)^{1/2}$

32. $\dfrac{1 + x}{2x^{1/2}} + x^{1/2}$

33. $2x(x^2 + 1)^{1/2} + x^2 \cdot \dfrac{1}{2}(x^2 + 1)^{-1/2} \cdot 2x$

34. $(x + 1)^{1/3} + x \cdot \dfrac{1}{3}(x + 1)^{-2/3}$

35. $\sqrt{4x + 3} \cdot \dfrac{1}{2\sqrt{x - 5}} + \sqrt{x - 5} \cdot \dfrac{1}{5\sqrt{4x + 3}}$

36. $\dfrac{\sqrt[3]{8x + 1}}{3\sqrt[3]{(x - 2)^2}} + \dfrac{\sqrt[3]{x - 2}}{24\sqrt[3]{(8x + 1)^2}}$

37. $\dfrac{\sqrt{1 + x} - x \cdot \dfrac{1}{2\sqrt{1 + x}}}{1 + x}$

38. $\dfrac{\sqrt{x^2 + 1} - x \cdot \dfrac{2x}{2\sqrt{x^2 + 1}}}{x^2 + 1}$

39. $\dfrac{(x + 4)^{1/2} - 2x(x + 4)^{-1/2}}{x + 4}$

40. $\dfrac{(9 - x^2)^{1/2} + x^2(9 - x^2)^{-1/2}}{9 - x^2}$

41. $\dfrac{\dfrac{x^2}{(x^2 - 1)^{1/2}} - (x^2 - 1)^{1/2}}{x^2}$

42. $\dfrac{(x^2 + 4)^{1/2} - x^2(x^2 + 4)^{-1/2}}{x^2 + 4}$

43. $\dfrac{\dfrac{1 + x^2}{2\sqrt{x}} - 2x\sqrt{x}}{(1 + x^2)^2}$

44. $\dfrac{2x(1 - x^2)^{1/3} + \frac{2}{3}x^3(1 - x^2)^{-2/3}}{(1 - x^2)^{2/3}}$

In Problems 45–52 factor each expression.

45. $(x + 1)^{3/2} + x \cdot \dfrac{3}{2}(x + 1)^{1/2}$

46. $(x^2 + 4)^{4/3} + x \cdot \dfrac{4}{3}(x^2 + 4)^{1/3} \cdot 2x$

47. $6x^{1/2}(x^2 + x) - 8x^{3/2} - 8x^{1/2}$

48. $6x^{1/2}(2x + 3) + x^{3/2} \cdot 8$

49. $3(x^2 + 4)^{4/3} + x \cdot 4(x^2 + 4)^{1/3} \cdot 2x$

50. $2x(3x + 4)^{4/3} + x^2 \cdot 4(3x + 4)^{1/3}$

51. $4(3x + 5)^{1/3}(2x + 3)^{3/2} + 3(3x + 5)^{4/3}(2x + 3)^{1/2}$

52. $6(6x + 1)^{1/3}(4x - 3)^{3/2} + 6(6x + 1)^{4/3}(4x - 3)^{1/2}$

1.9 COMPLEX NUMBERS

One of the properties of a real number is that its square is non-negative. For example, there is no real number x for which

$$x^2 = -1$$

IMAGINARY UNIT To remedy this situation, we invent a number called the **imaginary unit**, which we denote by i, whose square is -1. Thus,

$$i^2 = -1$$

This fact should not surprise you. If our universe were to consist only of integers, there would be no number x for which $2x = 1$. This unfortunate circumstance can be remedied by introducing numbers such as $\frac{1}{2}, \frac{2}{3}$, etc.—the *rational numbers*. If our universe were to consist only of rational numbers, there would be no number x whose square equals 2. That is, there would be no number x for which $x^2 = 2$. To remedy this, we introduce numbers such as $\sqrt{2}$, $\sqrt[3]{5}$, etc.—the *irrational numbers*. The *real numbers*, you will recall, consist of the rational numbers and the irrational numbers. Now, if our universe were to consist only of real numbers, then there would be no number x whose square is -1. To remedy this, we introduce numbers such as i, whose square is -1.

In the progression of ideas advanced above, each time we encountered a situation that was unsuitable, we invented a new number system to remedy the situation. And each new number system contained the earlier number system as a subset. The number system

COMPLEX NUMBER SYSTEM that results from inventing the number i is called the **complex number system**.

COMPLEX NUMBERS **Complex numbers** are numbers of the form

$$a + bi$$

REAL PART
IMAGINARY PART
where a and b are real numbers. The real number a is called the **real part** of the number $a + bi$; the real number b is called the **imaginary part** of $a + bi$.

For example, the complex number $-5 + 6i$ has the real part -5 and the imaginary part 6.

The complex number $a + 0i$ is usually written merely as a. This serves to remind us that the real numbers are just special kinds of complex numbers. The complex number $0 + bi$ is usually written as

PURE IMAGINARY NUMBER
bi. Sometimes the complex number bi is called a **pure imaginary number**. See Figure 16 for an illustration.

Equality, addition, subtraction, and multiplication of complex numbers are defined so as to preserve the familiar rules of algebra for real numbers. Thus, two complex numbers are equal if and only if their real parts are equal and their imaginary parts are equal.

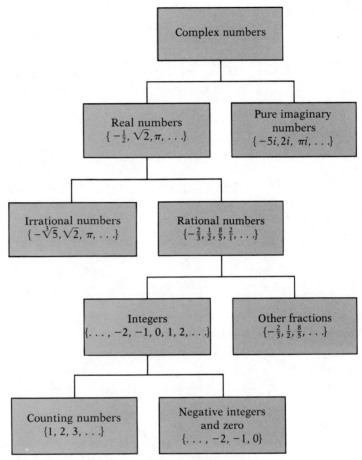

FIGURE 16

That is,

$$a + bi = c + di \quad \text{if and only if} \quad a = c \text{ and } b = d \qquad (1)$$

Two complex numbers are added by forming the complex number whose real part is the sum of the real parts and whose imaginary part is the sum of the imaginary parts. That is,

$$(a + bi) + (c + di) = (a + c) + (b + d)i \qquad (2)$$

To subtract two complex numbers, we follow the rule

$$(a + bi) - (c + di) = (a - c) + (b - d)i \qquad (3)$$

EXAMPLE 1   (a) $(3 + 5i) + (-2 + 3i) = [3 + (-2)] + (5 + 3)i = 1 + 8i$

(b) $(6 + 4i) - (3 + 6i) = (6 - 3) + (4 - 6)i = 3 + (-2)i$ ≣

If the imaginary part of a complex number is negative, such as in the complex number $3 + (-2)i$, we write it instead in the form $3 - 2i$.

Products of complex numbers are calculated as follows.

EXAMPLE 2 $(5 + 3i) \cdot (2 + 7i) \underset{\underset{\text{Distributive law}}{\uparrow}}{=} 5 \cdot (2 + 7i) + 3i(2 + 7i)$

$$\underset{\underset{\text{Distributive law}}{\uparrow}}{=} 10 + 35i + 6i + 21i^2$$

$$\underset{\underset{i^2 = -1}{\uparrow}}{=} 10 + 41i + 21(-1)$$

$$= -11 + 41i$$ ≣

PRODUCT OF COMPLEX NUMBERS   Based on the results of Example 2, we define the **product** of two complex numbers by the formula

$$(a + bi) \cdot (c + di) = (ac - bd) + (ad + bc)i \qquad (4)$$

Don't bother to memorize Formula (4). Instead, whenever it is necessary to multiply two complex numbers, follow the usual rules for multiplying two binomials, as in Example 2, remembering that $i^2 = -1$. For example,

$$(2i)(2i) = 4i^2 = -4 \qquad (2 + i)(1 - i) = 2 - 2i + i - i^2 = 3 - i$$

Algebraic properties for addition and multiplication, such as the commutative laws, the associative laws, the distributive law, and so on, hold for complex numbers. Of these laws, the one that says every nonzero complex number has a multiplicative inverse or a reciprocal requires a closer look.

Conjugates

CONJUGATE   If $z = a + bi$ is a complex number, then its **conjugate** is denoted by

$$\bar{z} = \overline{a + bi} = a - bi$$

For example, $\overline{2 + 3i} = 2 - 3i$ and $\overline{-6 - 2i} = -6 + 2i$.

To obtain the reciprocal of a nonzero complex number z, multiply the numerator and denominator by its conjugate $\bar{z}$. Thus, if $z = a + bi$ is a nonzero complex number, then

$$\frac{1}{z} = \frac{1}{z} \cdot \frac{\bar{z}}{\bar{z}} = \frac{\bar{z}}{z\bar{z}} = \frac{a - bi}{(a + bi)(a - bi)}$$

$$= \frac{a - bi}{a^2 - b^2 i^2} = \frac{a - bi}{a^2 + b^2}$$

$$= \frac{a}{a^2 + b^2} - \frac{b}{a^2 + b^2} i$$

EXAMPLE 3   Write $1/(3 + 4i)$ in the form $a + bi$.

Solution   The idea is to multiply the numerator and denominator by the conjugate of $3 + 4i$, namely the complex number $3 - 4i$. The result is

$$\frac{1}{3 + 4i} = \frac{1}{3 + 4i} \cdot \frac{3 - 4i}{3 - 4i}$$

$$= \frac{3 - 4i}{9 - 12i + 12i - 16i^2}$$

$$= \frac{3 - 4i}{9 - 16(-1)} = \frac{3 - 4i}{25} = \frac{3}{25} - \frac{4}{25} i \qquad\qquad \blacksquare$$

To divide one complex number by another, we multiply the numerator and denominator of the ratio by the conjugate of the denominator.

EXAMPLE 4   Write each of the following in the form $a + bi$:

(a) $\dfrac{1 + 4i}{5 - 12i}$ (b) $\dfrac{2 - 3i}{4 - 3i}$

Solution   (a) $\dfrac{1 + 4i}{5 - 12i} = \dfrac{1 + 4i}{5 - 12i} \cdot \dfrac{5 + 12i}{5 + 12i}$

$$= \frac{5 + 20i + 12i + 48i^2}{25 - 144i^2}$$

$$= \frac{5 + 32i - 48}{25 + 144} = \frac{-43 + 32i}{169} = \frac{-43}{169} + \frac{32}{169} i$$

(b) $\dfrac{2 - 3i}{4 - 3i} = \dfrac{2 - 3i}{4 - 3i} \cdot \dfrac{4 + 3i}{4 + 3i}$

$$= \frac{8 - 12i + 6i - 9i^2}{16 - 9i^2}$$

$$= \frac{8 - 6i - 9(-1)}{16 - 9(-1)} = \frac{17 - 6i}{25} = \frac{17}{25} - \frac{6}{25} i \qquad\qquad \blacksquare$$

EXAMPLE 5   If $z = 2 - 3i$ and $w = 5 + 2i$, write each of the following expressions in the form $a + bi$:

(a) $z\bar{z}$ (b) $\dfrac{z}{w}$ (c) $\overline{z + w}$ (d) $z + \bar{z}$

Solution   (a) $z\bar{z} = (2 - 3i)(2 + 3i) = 4 - 6i + 6i - 9i^2 = 4 + 9 = 13$

(b) $\dfrac{z}{w} = \dfrac{z \cdot \bar{w}}{w \cdot \bar{w}} = \dfrac{(2 - 3i)(5 - 2i)}{(5 + 2i)(5 - 2i)}$

$$= \dfrac{10 - 15i - 4i + 6i^2}{25 - 4i^2}$$

$$= \dfrac{4 - 19i}{29} = \dfrac{4}{29} - \dfrac{19}{29}i$$

(c) $\overline{z + w} = \overline{(2 - 3i) + (5 + 2i)} = \overline{7 - i} = 7 + i$

(d) $z + \bar{z} = (2 - 3i) + (2 + 3i) = 4$ ≡

The conjugate of a complex number has certain general properties that we shall find useful later on.

THEOREM   The product of a complex number and its conjugate is a nonnegative real number. Thus, if $z = a + bi$,

$$z\bar{z} = a^2 + b^2 \tag{5}$$

Proof   If $z = a + bi$, then

$$z\bar{z} = (a + bi)(a - bi) = a^2 - (bi)^2 = a^2 - b^2i^2 = a^2 + b^2 \quad ≡$$

For a real number $a = a + 0i$, the conjugate is $\bar{a} = \overline{a + 0i} = a - 0i = a$. That is:

THEOREM   The conjugate of a real number is itself.   ≡

Other properties of the conjugate that are direct consequences of the definition are listed below. In each statement in which they appear, z and w represent complex numbers.

THEOREM   The conjugate of the conjugate of a complex number is the number itself:

$$(\bar{\bar{z}}) = z \tag{6}$$

The conjugate of the sum of two complex numbers equals the sum of their conjugates:

$$\overline{z + w} = \bar{z} + \bar{w} \tag{7}$$

The conjugate of the product of two complex numbers equals the product of their conjugates:

$$\overline{z \cdot w} = \bar{z} \cdot \bar{w} \tag{8}$$

≡

We leave the proofs of Equations (6), (7), and (8) as exercises.

Powers of i

POWERS OF i The **powers of** i follow a pattern that is useful to know:

$i^1 = i$	$i^2 = -1$
$i^3 = i^2 \cdot i = -i$	$i^4 = i^2 \cdot i^2 = (-1)(-1) = 1$
$i^5 = i^4 \cdot i = 1 \cdot i = i$	$i^6 = i^4 \cdot i^2 = -1$
$i^7 = i^4 \cdot i^3 = -i$	$i^8 = i^4 \cdot i^4 = 1$

And so on. Thus, the powers of i repeat every fourth power.

EXAMPLE 6   (a) $i^{27} = i^{24} \cdot i^3 = (i^4)^6 \cdot i^3 = 1^6 \cdot i^3 = -i$
(b) $i^{101} = i^{100} \cdot i^1 = (i^4)^{25} \cdot i = 1^{25} \cdot i = i$

≡

EXAMPLE 7   Write $(2 + i)^3$ in the form $a + bi$.

Solution   We use the familiar formula for $(x + a)^3$, namely,

$$(x + a)^3 = x^3 + 3ax^2 + 3a^2x + a^3$$

Thus,

$$(2 + i)^3 = 2^3 + 3 \cdot i \cdot 2^2 + 3 \cdot i^2 \cdot 2 + i^3$$
$$= 8 + 12i + 6(-1) + (-i)$$
$$= 2 + 11i$$

≡

EXERCISE 1.9   *In Problems 1–44 write each expression in the form $a + bi$.*

1. $(2 - 3i) + (6 + 8i)$ 2. $(4 + 5i) + (-8 + 2i)$
3. $(-3 + 2i) + (4 - 4i)$ 4. $(3 - 4i) + (-3 - 4i)$

5. $(2 - 5i) - (8 + 6i)$　　**6.** $(-8 + 4i) - (2 - 2i)$

7. $3(2 - 6i)$　　**8.** $-4(2 + 8i)$

9. $2i(2 - 3i)$　　**10.** $3i(-3 + 4i)$

11. $(3 - 4i)(2 + i)$　　**12.** $(5 + 3i)(2 - i)$

13. $(-6 + i)(-6 - i)$　　**14.** $(-3 + i)(3 + i)$

15. $\dfrac{10}{3 - 4i}$　　**16.** $\dfrac{13}{5 - 12i}$

17. $\dfrac{i}{2 + i}$　　**18.** $\dfrac{-2i}{2 - i}$

19. $\dfrac{6 - i}{1 + i}$　　**20.** $\dfrac{2 + 3i}{1 - i}$

21. $\left(\dfrac{1}{2} + \dfrac{\sqrt{3}}{2}i\right)^2$　　**22.** $\left(\dfrac{\sqrt{3}}{2} - \dfrac{1}{2}i\right)^2$

23. $(1 + i)^2$　　**24.** $(1 - i)^2$

25. $\dfrac{(2 - i)^2}{2 + i}$　　**26.** $\dfrac{1 + 2i}{(1 - 2i)^2}$

27. $\dfrac{1}{(3 - i)(2 + i)}$　　**28.** $\dfrac{3}{(2 + 3i)(2 - 3i)}$

29. i^{23}　　**30.** i^{14}

31. i^{-15}　　**32.** i^{-23}

33. $i^6 - 5$　　**34.** $4 + i^3$

35. $6i^3 - 4i^5$　　**36.** $4i^3 - 2i^2 + 1$

37. $(1 + i)^3$　　**38.** $(3i)^4 + 1$

39. $i^7(1 + i^2)$　　**40.** $2i^4(1 + i^2)$

41. $\dfrac{i^4 + i^3 + i^2 + i + 1}{i + 1}$　　**42.** $\dfrac{1 - i + i^2 - i^3 + i^4}{1 - i}$

43. $i^6 + i^4 + i^2 + 1$　　**44.** $i^7 + i^5 + i^3 + i$

In Problems 45–52, $z = 3 - 4i$ and $w = 8 + 3i$. Write each expression in the form $a + bi$.

45. $z + \bar{z}$　　**46.** $w - \bar{w}$

47. $z\bar{z}$　　**48.** $\overline{z - w}$

49. $\overline{z \cdot w}$　　**50.** $\bar{z} \cdot \bar{w}$

51. $\overline{\bar{z} + w}$　　**52.** $\overline{\overline{zw}}$

53. Use $z = a + bi$ and show that $z + \bar{z} = 2a$ and that $z - \bar{z} = 2bi$.

54. Use $z = a + bi$ and show that $(\bar{\bar{z}}) = z$.

55. Use $z = a + bi$ and $w = c + di$ to show that $\overline{z + w} = \bar{z} + \bar{w}$.

56. Use $z = a + bi$ and $w = c + di$ to show that $\overline{z \cdot w} = \bar{z} \cdot \bar{w}$.

≡ 1.10 SCIENTIFIC NOTATION; APPROXIMATIONS; CALCULATORS

Scientific Notation

Measurements of physical quantities can range from very small to very large. For example, the mass of a proton is approximately 0.00000000000000000000000000167 kilogram and the mass of the earth is about 5,980,000,000,000,000,000,000,000 kilograms. These numbers obviously are tedious to write down, so we use exponents to rewrite each one. When a number has been written as the product of a number x, $1 \le x < 10$, times a power of 10, it is said to be written

SCIENTIFIC NOTATION in **scientific notation**. In scientific notation:

$$\text{Mass of a proton} = 1.67 \times 10^{-27} \text{ kilogram}$$
$$\text{Mass of the earth} = 5.98 \times 10^{24} \text{ kilograms}$$

To change a positive number into scientific notation, do the following:

1. Count the number N of places the decimal point must be moved in order to arrive at a number x, $1 \le x < 10$.
2. If the original number is greater than or equal to 1, the scientific notation is $x \times 10^N$. If the original number is between 0 and 1, the scientific notation is $x \times 10^{-N}$.

EXAMPLE 1   Write each number in scientific notation.

(a) 9582   (b) 1.245   (c) 0.285   (d) 0.000561

Solution   (a) The decimal point in 9582 follows the 2. Thus, we count

$$9 \; 5 \; 8 \; 2$$
$$\quad 3 \quad 2 \quad 1$$

stopping after three moves because 9.582 is a number between 1 and 10. Since 9582 is greater than 1, we write

$$9582 = 9.582 \times 10^3$$

(b) The decimal point in 1.245 is between the 1 and the 2. Since the number is already between 1 and 10, the scientific notation for it is $1.245 \times 10^0 = 1.245$.

(c) The decimal point is between the 0 and the 2. Thus, we count

$$0 \overset{\underset{\displaystyle 1}{\longrightarrow}}{.} 2\ 8\ 5$$

stopping after one move because 2.85 is a number between 1 and 10. Since 0.285 is between 0 and 1, we write

$$0.285 = 2.85 \times 10^{-1}$$

(d) The decimal point is moved as follows:

$$0 \overset{\underset{\displaystyle 1\quad 2\quad 3\quad 4}{\longrightarrow}}{.} 0\ 0\ 0\ 5\ 6\ 1$$

Thus,
$$0.000561 = 5.61 \times 10^{-4}$$

≡

EXAMPLE 2　Write each number as a decimal.

(a) 2.1×10^4　　(b) 3.26×10^{-5}　　(c) 1×10^{-2}

Solution　(a) $2.1 \times 10^4 = 2 \underset{1\ \ 2\ \ 3\ \ 4}{.\ 1\ 0\ 0\ 0} \times 10^4 = 21{,}000$

(b) $3.26 \times 10^{-5} = 0 \underset{5\ \ 4\ \ 3\ \ 2\ \ 1}{\ 0\ 0\ 0\ 0\ 3} . 2\ \ 6 \times 10^{-5} = 0.0000326$

(c) $1 \times 10^{-2} = 0 \underset{2\ \ 1}{\ 0\ 1} . \times 10^{-2} = 0.01$

≡

On a calculator, a number such as 3.615×10^{12} would be displayed either as $\boxed{3.615 \quad 12}$ or as $\boxed{3.615 \quad E12}$, depending on the calculator.

Approximations

SIGNIFICANT DIGITS　When a number is expressed in scientific notation, the number of digits that appear are said to be **significant**. For example,

3.256×10^8	4 significant digits
8.12×10^{-6}	3 significant digits
1.5×10^4	2 significant digits
1.50×10^4	3 significant digits
3.14159	6 significant digits

In scientific measurements, the number of significant digits that appear is an indication of the degree of accuracy of the measurement.

Suppose we have measured the distance from the earth to the moon and arrived at the answer 3.7815×10^5 kilometers. But further suppose that the device we used to obtain this measurement is valid only

APPROXIMATION

to two significant digits. As a result, we want to write an **approximation** to the answer we found that contains only two significant digits. This can be accomplished in two ways:

TRUNCATION

ROUNDING

> 1. **Truncation:** Merely ignore all the digits that are not significant and write the approximation $3.7815 \times 10^5 \approx 3.7 \times 10^5$.
> 2. **Rounding:** If the digit following the final significant digit is 5 or more, add 1 to the final significant digit; otherwise, truncate. In our example, the final significant digit, 7, is followed by an 8. Thus, we added 1 to the 7, obtaining 3.8×10^5.

Some more examples follow:

ORIGINAL NUMBER (IN SCIENTIFIC NOTATION)	NUMBER OF SIGNIFICANT DIGITS TO BE USED	APPROXIMATIONS	
		BY ROUNDING	BY TRUNCATION
3.61821×10^5	4	3.618×10^5	3.618×10^5
5.0136×10^{-4}	4	5.014×10^{-4}	5.013×10^{-4}
1.0010×10^4	3	1.00×10^4	1.00×10^4
3.14159	5	3.1416	3.1415

EXAMPLE 3   Write an approximation to each number correct to three significant digits. Use (i) rounding; (ii) truncation.

(a) 52,681   (b) 0.03214

Solution   (a) $52,681 = 5.2681 \times 10^4$
   (i) By rounding, we get $5.27 \times 10^4 = 52,700$.
   (ii) By truncation, we get $5.26 \times 10^4 = 52,600$.
(b) $0.03214 = 3.214 \times 10^{-2}$
   (i) By rounding, we get $3.21 \times 10^{-2} = 0.0321$.
   (ii) By truncation, we get $3.21 \times 10^{-2} = 0.0321$.   ≡

Sometimes in approximating numbers we speak of rounding off to a number of decimal places. For example, the number 20.98752 rounded off to three decimal places is 20.988; when rounded off to four decimal places it is 20.9875. Other examples follow:

NUMBER	ROUNDED OFF TO TWO DECIMAL PLACES	ROUNDED OFF TO FOUR DECIMAL PLACES
3.14159	3.14	3.1416
0.056128	0.06	0.0561
893.46125	893.46	893.4613

Usually we use the symbol ≈ (read "approximately equal to") when we have approximated a number either by rounding or by truncation.

Calculators

Calculators are finite machines. As a result, they are incapable of handling decimals that contain too many significant digits. For example, a Texas Instruments TI57 calculator is capable of displaying only eight digits. When a number requires more than eight digits, the TI57 will write the number in scientific notation, approximating it correctly to eight significant digits by using the rounding process. To see how your calculator handles decimals, divide 2 by 3. How many digits do you see? Is the last digit a 6 or a 7? If it is a 6, your calculator truncates; if it is a 7, your calculator rounds.

When the arithmetic involved in a given problem is messy, we shall mark the problem with a [C], indicating that you should use a calculator. Calculators are not required and should not be used except on problems labeled with a [C]. The answers provided to calculator problems have been found using a TI57. Due to differences among calculators in the way they do arithmetic, your answer may vary slightly from the one given in the text.

SCIENTIFIC ARITHMETIC

In deciding on what calculator you should purchase, be sure to choose a **scientific**, as opposed to an **arithmetic**, calculator. An arithmetic calculator can only add, subtract, multiply, and divide numbers. Scientific calculators contain keys labeled ln x, log x, sin x, cos x, tan x, y^x, INV, and others, called **function keys**. As you proceed through this text, you will learn how many of the function keys are used.

FUNCTION KEYS

ALGEBRAIC SYSTEM REVERSE POLISH NOTATION

Another difference among calculators has to do with the order in which various operations are performed. Some employ an **algebraic system**, whereas others use **reverse Polish notation** (RPN). Either system is acceptable, and the choice of which system to get is a matter of individual preference. Of course, no matter what scientific calculator you purchase, be sure to study the instruction manual so that you use the calculator efficiently and correctly.

EXERCISE 1.10   *In Problems 1–8 write each number in scientific notation.*

1. 454.2	**2.** 32.14	**3.** 0.013	**4.** 0.00421
5. 32,155	**6.** 21,210	**7.** 0.000423	**8.** 0.0514

In Problems 9–16 write each number as a decimal.

9. 2.15×10^4	**10.** 6.7×10^3	**11.** 1.214×10^{-3}
12. 9.88×10^{-4}	**13.** 1.1×10^8	**14.** 4.112×10^2
15. 8.1×10^{-2}	**16.** 6.453×10^{-1}	

In Problems 17–24 tell how many significant digits are present.

17. 3.15×10^4	**18.** 2.01×10^{-3}	**19.** 1.00×10^{-4}	**20.** 3×10^4
21. 8.001×10^2	**22.** 9.15×10	**23.** 2.1	**24.** 3.42

In Problems 25–32 write an approximation to each number correct to three significant digits using (a) rounding, (b) truncation.

25. 6.789   **26.** 5.123   **27.** 189

28. 0.00264   **29.** 0.0001255   **30.** 999.2

31. 999,800   **32.** 0.00001001

In Problems 33–44 write each number as a decimal, rounded off to three decimal places.

33. 18.9526   **34.** 25.86134

35. 28.65349   **36.** 99.05229

37. 0.06291   **38.** 0.05388

C **39.** $\frac{3}{7}$ C **40.** $\frac{5}{9}$

C **41.** $\sqrt{2}$ C **42.** $\sqrt{3}$

C **43.** $\sqrt{23}$ C **44.** $\sqrt{14}$

≡ 1.11 GEOMETRY TOPICS

In this section we review some topics studied in geometry that we shall need for our study of algebra.

Pythagorean Theorem

RIGHT TRIANGLE   The Pythagorean Theorem is a statement about right triangles. A **right triangle** is one that contains a right angle—that is, an angle of 90°. The length of the side of the triangle opposite the 90° angle is called HYPOTENUSE   the **hypotenuse**; the lengths of the remaining two sides are called **legs**. LEGS   In Figure 17 we have used c to represent the hypotenuse and a and b to represent the legs. Notice the use of the symbol ∟ to show the 90° angle. We now state the Pythagorean Theorem.

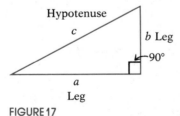

Hypotenuse
c
b Leg
90°
a
Leg

FIGURE 17

THEOREM   In a right triangle, the square of the hypotenuse is equal to the sum PYTHAGOREAN THEOREM   of the squares of the legs. That is, in a right triangle,

$$c^2 = a^2 + b^2 \qquad (1)$$

≡

We shall prove this result at the end of this section.

EXAMPLE 1   In a right triangle, one leg is of length 4 and the other of length 3. What is the length of the hypotenuse?

Solution   Since the triangle is a right triangle, we use the Pythagorean Theorem with $a = 4$ and $b = 3$ to find the length c of the hypotenuse. Thus, from Equation (1), we have

$$c^2 = a^2 + b^2$$
$$c^2 = 4^2 + 3^2 = 16 + 9 = 25$$
$$c = 5$$ ≡

The converse of the Pythagorean Theorem is also true.

THEOREM
CONVERSE OF THE
PYTHAGOREAN THEOREM
In a triangle, if the square of the length of one side equals the sum of the squares of the lengths of the other two sides, then the triangle is a right triangle. The 90° angle is opposite the longest side.   ≡

We shall prove this result at the end of this section.

EXAMPLE 2   Show that a triangle whose sides are of lengths 5, 12, and 13 is a right triangle. Identify the hypotenuse.

Solution   We square the lengths of the sides:

25,   144,   169

Notice that the sum of the first two squares (25 and 144) equals the third square (169). Hence, the triangle is a right triangle. The longest side, 13, is the hypotenuse. See Figure 18.

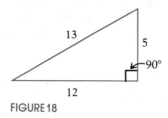

FIGURE 18   ≡

Geometry Formulas

For a rectangle of length l and width w:

$$\text{Area} = lw$$
$$\text{Perimeter} = 2l + 2w$$

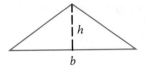

For a triangle with base b and altitude h:

$$\text{Area} = \tfrac{1}{2}bh$$

For a circle of radius R:

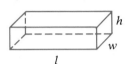

$$\text{Area} = \pi R^2$$
$$\text{Circumference} = 2\pi R$$

For a rectangular box of length l, width w, and height h:

$$\text{Volume} = lwh$$

Proof of the Pythagorean Theorem

CONGRUENT

We begin with a square, each side of length $a + b$. In this square, we can form four right triangles, each having legs equal in length to a and b. See Figure 19. Each of these triangles is **congruent** (two sides and their included angle are equal). As a result, the hypotenuse of each is the same, say c. The area of the square with side $a + b$ equals the sum of the areas of the four triangles (each of area $\tfrac{1}{2}ab$) plus the area of the square with side c. Thus,

$$(a + b)^2 = \tfrac{1}{2}ab + \tfrac{1}{2}ab + \tfrac{1}{2}ab + \tfrac{1}{2}ab + c^2$$
$$a^2 + 2ab + b^2 = 2ab + c^2$$
$$a^2 + b^2 = c^2$$

The proof is complete.

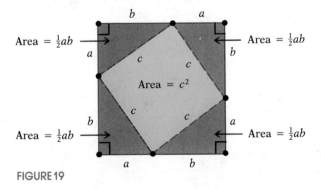

FIGURE 19

Proof of the Converse

We begin with two triangles: one a right triangle with legs a and b; the other a triangle with sides a, b, and c for which $c^2 = a^2 + b^2$ (see

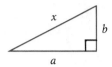

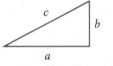

$$c^2 = a^2 + b^2$$

FIGURE 20

Figure 20). By the Pythagorean Theorem, the length x of the third side of the first triangle is

$$x^2 = a^2 + b^2$$

But $c^2 = a^2 + b^2$. Hence,

$$x^2 = c^2$$

$$x = c$$

The two triangles have the same sides and are therefore congruent; hence, corresponding angles are equal. Thus, the angle opposite side c of the second triangle equals 90°.

The proof is complete.   ≡

EXERCISE 1.11

In Problems 1–6 the lengths of the legs of a right triangle are given. Find the hypotenuse.

1. $a = 5$, $b = 12$ **2.** $a = 6$, $b = 8$
3. $a = 10$, $b = 24$ **4.** $a = 4$, $b = 3$
5. $a = 7$, $b = 24$ **6.** $a = 14$, $b = 48$

In Problems 7–14 the lengths of the sides of a triangle are given. Determine which are right triangles. For those that are, identify the hypotenuse.

7. 3, 4, 5   **8.** 6, 8, 10   **9.** 4, 5, 6
10. 2, 2, 3   **11.** 7, 24, 25   **12.** 10, 24, 26
13. 6, 4, 3   **14.** 5, 4, 7

 15. The tallest inhabited building is the Sears Tower in Chicago.* If the observation tower is 1454 feet above ground level, use the figure to determine how far a person standing on the observation tower can see (with the aid of a telescope). Use 3960 miles for the radius of the earth. [*Note:* 1 mile = 5280 feet.]

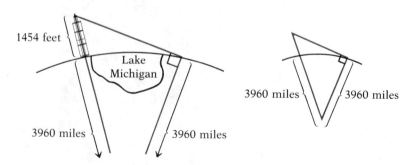

 16. The conning tower of the USS Silversides, a World War II submarine now permanently stationed in Chicago, is approximately 20 feet above sea level. How far can one see from the conning tower?

*Source: *Guinness Book of World Records.*

C **17.** A 6-foot-tall person on the beach in Fort Lauderdale, Florida, looks out onto the Atlantic Ocean. Suddenly a ship appears on the horizon. How far is the ship from shore?

C **18.** The deck of a destroyer is 100 feet above sea level. How far can a person see from the deck? How far can a person see from the bridge, which is 150 feet above sea level?

C H A P T E R   R E V I E W

VOCABULARY

variable
constant
equation
equal sign
reflexive property
symmetric property
principle of
   substitution
set
subset
counting numbers
natural numbers
ellipsis
integers
rational number
numerator
denominator
irrational numbers
real numbers
decimals
terminating decimals
repeating decimals
commutative laws
associative laws
distributive law
identity laws
inverse laws
additive inverse
multiplicative inverse
reciprocal
difference
ratio
rules for division
rules of signs
cancellation laws
product law
arithmetic of ratios
lowest terms

origin
scale
coordinate
real number line
negative real numbers
zero
positive real numbers
greater than ($>$)
less than ($<$)
greater than or equal
   to ($\geq$)
less than or equal to
   ($\leq$)
strict inequality
nonstrict inequality
inequality signs
nonnegative
   property
transitive property
addition property
multiplication
   properties
reciprocal property
trichotomy property
chain
closed interval
open interval
half-open interval
left endpoint
right endpoint
plus infinity ($+\infty$)
minus infinity ($-\infty$)
absolute value
base
exponent
power
laws of exponents
monomial

coefficient of a
   monomial
degree of a monomial
like terms
binomial
trinomial
polynomial
coefficients
terms
leading coefficient
degree of a
   polynomial
zero polynomial
standard form
subscript
special products
difference of two squares
square of a binomial
perfect square
miscellaneous
   trinomial
cube of a binomial
perfect cube
difference of two cubes
sum of two cubes
polynomial in two
   variables
polynomial in three
   variables
factors
factor over the integers
factor over the rational
   numbers
factor over the real
   numbers
prime polynomial
factored completely
factoring by grouping

rational expression	$a^{1/q}$, $a^{p/q}$	truncation
reduced to lowest terms	imaginary unit	rounding
simplified	complex number system	scientific calculator
least common multiple (LCM)	complex numbers	arithmetic calculator
complex ratio	real part	function keys
qth roots of a	imaginary part	algebraic system
principal qth root of a	pure imaginary number	reverse Polish notation (RPN)
radical	product of complex numbers	right triangle
index	conjugate	hypotenuse
radicand	powers of i	legs
square root	scientific notation	Pythagorean Theorem
cube root	significant digits	converse of Pythagorean Theorem
perfect roots	approximation	congruent triangles
rationalizing the denominator		

FILL-IN-THE-BLANK QUESTIONS

1. The _____ property of equality states that, if $a = b$, then $b = a$.

2. The _____ laws state that $a + b = b + a$ and $ab = ba$.

3. The _____ law states that $a \cdot (b + c) = a \cdot b + a \cdot c$.

4. If each side of an inequality is multiplied by a(n) _____ number, then the direction of the inequality is reversed.

5. In the expression 2^4, the number 2 is called the _____, and 4 is called the _____.

6. The polynomial $5x^3 - 3x^2 + 2x - 4$ is of degree _____; the _____ coefficient is 5.

7. In the expression $\sqrt[3]{25}$, the number 3 is called the _____; the number 25 is called the _____.

8. In the complex number $5 + 2i$, the number 5 is called the _____ part; the number 2 is called the _____ part; and the number i is called the _____ _____.

REVIEW EXERCISES

In Problems 1–16 perform the indicated operations.

1. $3 - 4 \cdot 5 + 1$

2. $8 + 4 \cdot 2 - 1$

3. $\dfrac{3}{4} - \dfrac{7}{12}$

4. $\dfrac{9}{8} - \dfrac{3}{4}$

5. $\dfrac{\frac{15}{2} - \frac{1}{4}}{\frac{2}{3}}$

6. $\dfrac{\frac{9}{2} - \frac{3}{4}}{\frac{4}{3}}$

7. $5^2 - 3^3$

8. $2^3 + 3^2 - 1$

9. $\dfrac{2^{-3} \cdot 5^0}{4^2}$

10. $\dfrac{3^{-2} \cdot 4^1}{2^{-2}}$

11. $(2\sqrt{5} - 2)(2\sqrt{5} + 2)$

12. $(2\sqrt{3} + \sqrt{2})(2\sqrt{3} - \sqrt{2})$

13. $\left(\dfrac{8}{9}\right)^{-2/3}$

14. $\left(\dfrac{4}{27}\right)^{-3/2}$

15. $(\sqrt[3]{2})^{-3}$

16. $(4\sqrt{32})^{-1/2}$

In Problems 17–20 graph the given compound inequality.

17. $x > -4$ and $x < 4$

18. $x \le -4$ and $x \ge -8$

19. $x < -4$ or $x > 0$

20. $x < -3$ or $x < 0$

In Problems 21–28 write each expression so that all exponents are positive. Assume $x > 0$ and $y > 0$.

21. $\dfrac{x^{-2}}{y^{-2}}$

22. $\left(\dfrac{x^{-1}}{y^{-3}}\right)^2$

23. $\dfrac{(x^2 y)^{-4}}{(xy)^{-3}}$

24. $\dfrac{\left(\dfrac{x}{y}\right)^2}{\left(\dfrac{y}{x}\right)^{-1}}$

25. $(25x^{-4/3}y^{-2/3})^{3/2}$

26. $(16x^{-2/3}y^{4/3})^{-3/2}$

27. $\left(\dfrac{2x^{-1/2}}{y^{-3/4}}\right)^{-4}$

28. $\left(\dfrac{8x^{-3/2}}{y^{-3}}\right)^{-2/3}$

In Problems 29–38 perform the indicated operations. Express your answer as a polynomial.

29. $(2x - 3)(x^2 - 5x + 2)$

30. $(3x + 4)(2x^2 - 8x - 1)$

31. $4(3x^3 - 2x^2 + 1) - 3(x^3 + 4x^2 - 2x - 3)$

32. $8(1 - x^2 + x^3) - 4(1 + 2x^2 - 4x^4)$

33. $(2x - 5)(3x + 2)$

34. $(1 - 2x)(1 - 4x)$

35. $(x + 1)(x + 2)(x + 3)$

36. $(x^2 - 1)(x + 1)$

37. $(x - 2y)(x + 4y)$

38. $(x - y)^2(x + y)^2$

In Problems 39–52 factor each polynomial completely (over the integers). If the polynomial cannot be factored, say it is prime.

39. $x^2 + 5x - 14$

40. $x^2 + 15x + 14$

41. $6x^2 - 5x - 6$

42. $6x^2 + x - 2$

43. $3x^2 - 15x - 42$

44. $2x^3 + 18x^2 + 28x$

45. $8x^3 + 1$

46. $27x^3 - 8$

47. $2x^3 + 3x^2 - 2x - 3$

48. $2x^3 + 3x^2 + 2x + 3$

49. $25x^2 - 4$

50. $16x^2 - 1$

51. $9x^2 + 1$

52. $x^2 - x + 1$

In Problems 53–62 perform the indicated operation and simplify.

53. $\dfrac{2x^2 + 11x + 14}{x^2 - 4}$

54. $\dfrac{x^2 - 5x - 14}{4 - x^2}$

55. $\dfrac{9x^2 - 1}{x^2 - 9} \cdot \dfrac{3x - 9}{9x^2 + 6x + 1}$

56. $\dfrac{x^2 - 25}{x^3 - 4x^2 - 5x} \cdot \dfrac{x^2 + x}{1 - x^2}$

57. $\dfrac{x+1}{x-1} - \dfrac{x-1}{x+1}$

58. $\dfrac{x}{x+1} - \dfrac{2x}{x+2}$

59. $\dfrac{3x+4}{x^2-4} - \dfrac{2x-3}{x^2+4x+4}$

60. $\dfrac{x^2}{2x^2+5x-3} + \dfrac{x^2}{2x^2-5x+2}$

61. $1 + \dfrac{1}{1+\dfrac{1}{x}}$

62. $1 - \dfrac{1}{1+\dfrac{1}{x}}$

In Problems 63–68 rationalize the denominator of each expression.

63. $\dfrac{2}{\sqrt{3}}$

64. $\dfrac{-1}{\sqrt{5}}$

65. $\dfrac{2}{1-\sqrt{2}}$

66. $\dfrac{-4}{1+\sqrt{3}}$

67. $\dfrac{1+\sqrt{5}}{1-\sqrt{5}}$

68. $\dfrac{4\sqrt{3}+2}{2\sqrt{3}+1}$

In Problems 69–72 write each expression as a single ratio in which only positive exponents and/or radicals appear.

69. $(2+x^2)^{1/2} + x \cdot \frac{1}{2}(2+x^2)^{-1/2} \cdot 2x$

70. $(x^2+4)^{2/3} + x \cdot \frac{2}{3}(x^2+4)^{-1/3} \cdot 2x$

71. $\dfrac{(x+4)^{1/2} \cdot 2x - x^2 \cdot \frac{1}{2}(x+4)^{-1/2}}{x+4}$

72. $\dfrac{(x^2+4)^{1/2} \cdot 2x - x^2 \cdot \frac{1}{2}(x^2+4)^{-1/2} \cdot 2x}{x^2+4}$

In Problems 73–82 use the complex number system and write each expression in the form $a + bi$.

73. $(6-3i) - (2+4i)$

74. $(8+3i) + (-6-2i)$

75. $4(3-i) + 3(-5+2i)$

76. $2(1+i) - 3(2-3i)$

77. $\dfrac{3}{3+i}$

78. $\dfrac{4}{2-i}$

79. i^{68}

80. i^{21}

81. $(2+3i)^3$

82. $(3-2i)^3$

83. If $u = \frac{1}{2}\left(x^3 - \dfrac{1}{x^3}\right)$, show that $\sqrt{1+u^2} = \frac{1}{2}\left(x^3 + \dfrac{1}{x^3}\right)$.

84. If $u = 2\sqrt{3}x\sqrt{3x^2+1}$, show that $1 + u^2 = (6x^2+1)^2$.

85. If $u = \frac{1}{2}\left(x^2 - \dfrac{1}{x^2}\right)$, show that $1 + u^2 = \frac{1}{4}\left(x^2 + \dfrac{1}{x^2}\right)^2$.

EQUATIONS AND INEQUALITIES

2.1 Linear Equations
2.2 Applications
2.3 Quadratic Equations
2.4 Quadratic Equations with a
     Negative Discriminant
2.5 Other Types of Equations
2.6 Linear Inequalities
2.7 Other Inequalities
2.8 Equations and Inequalities Involving
     Absolute Value
     Chapter Review

The investigation of equations and their solutions has played a central role in algebra for many hundreds of years. In fact, the Babylonians in 200 B.C. had a well-developed algebra that included a solution for quadratic equations (see the Historical Comment in Section 2.3). Of late, the study of inequalities (Sections 2.6 and 2.7) has been equally important. For example, the problem of optimizing airline scheduling leads to linear inequalities. As you go through this chapter, you will see evidence of the importance of the topics presented here for solving problems that occur in business, engineering, and the social and natural sciences.

We shall limit our discussion in this chapter to equations and inequalities containing a single variable. Later, we will study equations and inequalities containing more than one variable.

▤ 2.1 LINEAR EQUATIONS

We begin with some needed vocabulary.

EQUATION
SOLVE

An **equation** is a statement in which two expressions, called its *sides*, are equal. To "**solve** an equation" means to find all values of a specified variable in the equation for which the statement is true. Those values of the variable that result in a true statement are called

SOLUTION
ROOT

solutions or **roots** of the equation. Unless indicated otherwise, we will limit ourselves to real-number solutions.

For example, if x is a variable, then

$$x + 5 = 9, \quad x^2 + 5x = 2x - 2, \quad \frac{x^2 - 4}{x + 1} = 0, \quad \text{and} \quad x^2 + 9 = 5$$

are all equations in x. The first of these statements, $x + 5 = 9$, is true when $x = 4$ and is false for any other real number x. Thus, 4 is a solution of the equation $x + 5 = 9$. We also say that 4 **satisfies** the

SATISFIES

equation $x + 5 = 9$ because, when x is replaced by 4, a true statement results.

Sometimes an equation will have more than one solution. For example, the equation

$$\frac{x^2 - 4}{x + 1} = 0$$

has either $x = -2$ or $x = 2$ as a solution.

Sometimes we will write the solutions of an equation in set nota-

SOLUTION SET

tion. This set is called the **solution set** of the equation. For example, the solution set of the equation $x^2 - 9 = 0$ is $\{-3, 3\}$.

Some equations have no real solution. For example, the equation $x^2 + 9 = 5$ has no real solution because there is no real number whose square when added to 9 equals 5.

An equation that is satisfied for every choice of the variable for

IDENTITY

which both sides are meaningful is called an **identity**. For example, the equation

$$3x + 5 = x + 3 + 2x + 2$$

is an identity because this statement is true for any real number x.

Two or more equations that have precisely the same solutions are

EQUIVALENT EQUATIONS

called **equivalent equations**. For example, all of the following equations are equivalent because each has the solution $x = 5$:

$$2x + 3 = 13$$
$$2x = 10$$
$$x = 5$$

These three equations illustrate one method for solving many types of equations: Replace the original equation by an equivalent one, continuing until an equation with an obvious solution, such as $x = 5$, is reached. The question, though, is: "How do I obtain an equivalent equation?" In general, there are four ways to do so.

1. Interchange the two sides of the equation:

   Replace $3 = x$ by $x = 3$

2. Simplify the sides of the equation by combining like terms, eliminating parentheses, and so on:

   Replace $(x + 2) + 6 = 2x + (x + 1)$

   by $x + 8 = 3x + 1$

3. Add or subtract the same expression on both sides of the equation:

   Replace $3x - 5 = 4$

   by $(3x - 5) + 5 = 4 + 5$

4. Multiply or divide both sides of the equation by the same nonzero expression:

   Replace $\dfrac{3x}{x - 1} = \dfrac{6}{x - 1}$

   by $\dfrac{3x}{x - 1} \cdot (x - 1) = \dfrac{6}{x - 1} \cdot (x - 1)$ if $x - 1 \neq 0$

Let's look at an elementary example to illustrate these ideas.

EXAMPLE 1 Solve the equation $3x - 5 = 4$.

Solution We replace the equation by a collection of equivalent ones:

$$3x - 5 = 4$$
$$(3x - 5) + 5 = 4 + 5 \quad \text{Add 5 to both sides.}$$
$$3x = 9 \quad \text{Simplify.}$$
$$\frac{3x}{3} = \frac{9}{3} \quad \text{Divide both sides by 3.}$$
$$x = 3 \quad \text{Simplify.}$$

The last equation, $x = 3$, has the single solution 3. All the equations are equivalent, so 3 is the only solution of the original equation $3x - 5 = 4$. ≡

It is a good practice to check the solution by replacing x by 3 in the original equation:

$$3x - 5 = 4$$
$$3(3) - 5 \stackrel{?}{=} 4$$
$$9 - 5 \stackrel{?}{=} 4$$
$$4 = 4$$

The solution checks.

Sometimes, to solve an equation it is necessary to multiply or divide both sides by an expression containing a variable. Care must be exercised, because it may not be known whether the expression can equal zero. The next example illustrates what might happen if we need to multiply both sides of an equation by an expression containing a variable.

EXAMPLE 2   Solve the equation $\dfrac{3x}{x - 1} + 2 = \dfrac{3}{x - 1}$.

Solution   The two ratios that appear have the same denominator, $x - 1$. In order to simplify these ratios, we will multiply both sides by $x - 1$:

$$\frac{3x}{x - 1} + 2 = \frac{3}{x - 1}$$

$$\left(\frac{3x}{x - 1} + 2\right) \cdot (x - 1) = \frac{3}{x - 1} \cdot (x - 1)$$

$$\left[\frac{3x}{x - 1} \cdot (x - 1)\right] + [2 \cdot (x - 1)] = 3$$

$$3x + (2x - 2) = 3$$

$$5x - 2 = 3$$

$$5x = 5$$

$$x = 1$$

The solution appears to be 1. However, if we replace x by 1 in the original equation, we get

$$\frac{3x}{x - 1} + 2 = \frac{3}{x - 1}$$

$$\frac{3(1)}{1 - 1} + 2 \stackrel{?}{=} \frac{3}{1 - 1}$$

$$\frac{3}{0} + 2 \stackrel{?}{=} \frac{3}{0}$$

We may never divide by 0, so 1 cannot be a solution. It follows that the original equation has no solution; if it did, the solution would necessarily be 1 and, as we have shown, 1 is not a solution.   ▬

Let's examine what happened in Example 2. When we multiplied by $x - 1$, we were, without knowing it, multiplying by $1 - 1 = 0$. This multiplication violated Rule 4, so we did not get a succession of equivalent equations. This is precisely why, when multiplying or dividing by an expression containing a variable, you must always check any apparent solutions that are found. Apparent solutions that are not, in fact, solutions of the original equation are called **extraneous solutions**.

EXTRANEOUS SOLUTION

The next example illustrates a different outcome when both sides are divided by an expression containing a variable.

EXAMPLE 3 Solve the equation $(x + 1)(2x) = (x + 1)(2)$.

Solution We choose to divide both sides by $x + 1$ and simplify:

$$(x + 1)(2x) = (x + 1)(2)$$
$$\frac{(x + 1)(2x)}{x + 1} = \frac{(x + 1)(2)}{x + 1}$$
$$2x = 2$$
$$x = 1$$

On checking $x = 1$, we find that it is, in fact, a solution of the original equation. However, it is not the only solution; $x = -1$ is also a solution. This solution was "lost" when we divided by $x + 1$ without first checking to see whether the solutions of the equation $x + 1 = 0$ are, in fact, also solutions of the original equation. ∎

In the solution to Example 3, we could have chosen to multiply out both sides instead:

$$(x + 1)(2x) = (x + 1)(2)$$
$$2x^2 + 2x = 2x + 2$$
$$2x^2 = 2$$
$$x^2 = 1$$
$$x = 1 \quad \text{or} \quad x = -1$$

Using this method we find the two solutions.

There will be times when the only practical way to solve an equation is to divide both sides by an expression containing a variable. In such cases, remember that Rule 4 may have been violated and, as a result, you may not obtain a succession of equivalent equations. Thus, when dividing both sides of an equation by an expression containing a variable, set the expression equal to zero, solve, and check whether any solution found is also a solution of the original equation.

The experiences of Examples 2 and 3 require a special comment.

> **Warning:**
> 1. When multiplying both sides of an equation by an expression containing a variable, extraneous solutions may be introduced. Always check whether each solution does, in fact, satisfy the original equation.
> 2. When dividing both sides of an equation by an expression containing a variable, solutions of the original equation may be lost. Always set the expression you divided by equal to zero and check the resulting solutions in the original equation.

The flowchart in Figure 1 summarizes this warning.

Linear Equations

LINEAR EQUATION : **A linear equation** is an equation equivalent to one of the form

$$ax + b = 0$$

where a and b are real numbers and $a \neq 0$.

FIRST-DEGREE EQUATION   Sometimes a linear equation is called a **first-degree equation** because the left side is a polynomial in x of degree one.

It is relatively easy to solve a linear equation:

$$ax + b = 0$$
$$(ax + b) - b = 0 - b \quad \text{Subtract } b \text{ from both sides.}$$
$$ax = -b \quad \text{Simplify.}$$
$$\frac{ax}{a} = \frac{-b}{a} \quad \text{Divide both sides by } a, a \neq 0.$$
$$x = \frac{-b}{a} \quad \text{Simplify.}$$

The linear equation $ax + b = 0$ has the single solution given by the formula $x = -b/a$.

Whenever it is possible to solve a linear equation in your head, do so. For example:

The solution of $2x = 8$ is $x = 4$.

The solution of $3x - 15 = 0$ is $x = 5$.

Often, though, some rearrangement is required.

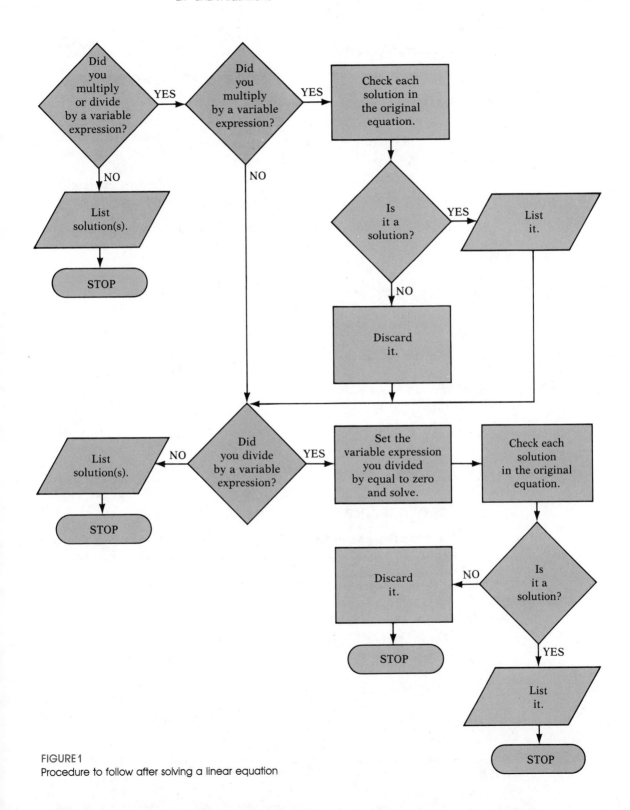

FIGURE 1
Procedure to follow after solving a linear equation

EXAMPLE 4   Solve each equation.

(a) $3(2x - 5) + 24 = 3x$ (b) $\frac{1}{2}(p + 5) - 3 = \frac{1}{3}(2p - 1)$

Solution   (a) $3(2x - 5) + 24 = 3x$

$$6x - 15 + 24 = 3x \qquad \text{Use the distributive law.}$$
$$6x + 9 = 3x \qquad \text{Combine like terms.}$$
$$6x + 9 - 3x = 3x - 3x \qquad \text{Subtract } 3x \text{ from each side.}$$
$$3x + 9 = 0 \qquad \text{Simplify.}$$
$$x = -3 \qquad \text{Mental solution}$$

(b) To clear the equation of fractions, we multiply both sides by 6, which is the least common multiple of the denominators of the fractions $\frac{1}{2}$ and $\frac{1}{3}$:

$$\frac{1}{2}(p + 5) - 3 = \frac{1}{3}(2p - 1)$$

$$6\left[\frac{1}{2}(p + 5) - 3\right] = 6\left[\frac{1}{3}(2p - 1)\right] \qquad \text{Multiply both sides by 6, the LCM of 2 and 3.}$$

$$3(p + 5) - 18 = 2(2p - 1) \qquad \begin{array}{l}\text{Use the distributive law on the left} \\ \text{and the associative law on the right.}\end{array}$$

$$3p + 15 - 18 = 4p - 2 \qquad \text{Use the distributive law.}$$
$$3p - 3 = 4p - 2 \qquad \text{Combine like terms.}$$
$$3p = 4p + 1 \qquad \text{Add 3 to each side.}$$
$$-p = 1 \qquad \text{Subtract } 4p \text{ from each side.}$$
$$p = -1 \qquad \text{Multiply both sides by } -1.$$

Equations That Lead to Linear Equations

The next two examples illustrate the solution of equations that are not linear but that, on simplification, lead to a linear equation.

EXAMPLE 5   Solve the equation $(2y + 1)(y - 1) = (y + 5)(2y - 5)$.

Solution $(2y + 1)(y - 1) = (y + 5)(2y - 5)$

$$2y^2 - y - 1 = 2y^2 + 5y - 25 \qquad \text{Multiply, and combine like terms.}$$
$$-y - 1 = 5y - 25 \qquad \text{Subtract } 2y^2 \text{ from each side.}$$
$$-y = 5y - 24 \qquad \text{Add 1 to each side.}$$
$$-6y = -24 \qquad \text{Subtract } 5y \text{ from each side.}$$
$$y = 4 \qquad \text{Divide both sides by } -6.$$

EXAMPLE 6   Solve the equation $\dfrac{3}{x - 2} = \dfrac{1}{x - 1} + \dfrac{7}{(x - 1)(x - 2)}$.

Solution   As we did in Example 4(b), we clear the equation of fractions by multiplying both sides by $(x - 1)(x - 2)$, the least common multiple of the denominators of the three fractions:

$$\frac{3}{x - 2} = \frac{1}{x - 1} + \frac{7}{(x - 1)(x - 2)}$$

$$(x - 1)(x - 2)\left(\frac{3}{x - 2}\right) = (x - 1)(x - 2)\left[\frac{1}{x - 1} + \frac{7}{(x - 1)(x - 2)}\right]$$

$$(x - 1)(3) = (x - 1)(x - 2)\frac{1}{x - 1}$$

$$+ (x - 1)(x - 2)\frac{7}{(x - 1)(x - 2)}$$

$$3x - 3 = (x - 2) + 7$$

$$3x - 3 = x + 5$$

$$2x = 8$$

$$x = 4$$

Check: Remember the previous warning. Because we multiplied both sides of the equation by an expression containing a variable, we must check the apparent solution, $x = 4$:

$$\frac{3}{x - 2} = \frac{1}{x - 1} + \frac{7}{(x - 1)(x - 2)}$$

$$\frac{3}{4 - 2} \stackrel{?}{=} \frac{1}{4 - 1} + \frac{7}{(4 - 1)(4 - 2)}$$

$$\frac{3}{2} \stackrel{?}{=} \frac{1}{3} + \frac{7}{6}$$

$$\frac{3}{2} \stackrel{?}{=} \frac{2}{6} + \frac{7}{6}$$

$$\frac{3}{2} \stackrel{?}{=} \frac{9}{6}$$

$$\frac{3}{2} = \frac{3}{2}$$

Thus, 4 is a solution—the only solution—of the original equation.   ▤

EXAMPLE 7   In the United States we measure temperature both in degrees Fahrenheit (°F) and in degrees Celsius (°C), which are related by the formula $C = (\frac{5}{9})(F - 32)$. What are the Fahrenheit temperatures corresponding to Celsius temperatures of 0°, 10°, 20°, and 30°?

Solution   We could solve four equations for F by replacing C each time by 0, 10, 20, and 30. Instead, it's much easier and faster first to solve the equation $C = \frac{5}{9}(F - 32)$ for F and then substitute in the values of C:

$$C = \frac{5}{9}(F - 32)$$

$$9C = 5(F - 32) \quad \text{Multiply by 9.}$$

$$9C = 5F - 160 \quad \text{Use the distributive law.}$$

$$5F - 160 = 9C \quad\quad\quad \text{Interchange sides.}$$

$$5F = 9C + 160 \quad \text{Add 160 to each side.}$$

$$F = \frac{9}{5}C + 32 \quad \text{Divide both sides by 5.}$$

We can now do the required arithmetic:

$$0° \text{ C:} \quad F = \frac{9}{5}(0) \ + 32 = 32° \text{ F}$$

$$10° \text{ C:} \quad F = \frac{9}{5}(10) + 32 = 50° \text{ F}$$

$$20° \text{ C:} \quad F = \frac{9}{5}(20) + 32 = 68° \text{ F}$$

$$30° \text{ C:} \quad F = \frac{9}{5}(30) + 32 = 86° \text{ F}$$

Summary

> To solve a linear equation, follow these steps:
>
> 1. If necessary, clear the equation of fractions by multiplying both sides by the least common multiple (LCM) of the denominators of all the fractions.
> 2. Remove all parentheses and simplify.
> 3. Collect all terms containing the variable on one side and all other terms on the other side.
> 4. Remember to check all apparent solutions and to look for any lost solutions if any of the rules for obtaining equivalent equations have been violated.

≡ HISTORICAL COMMENT   The solution of equations is among the oldest of mathematical activities, and efforts to systematize this activity determined much of the shape of modern mathematics. The most elementary arithmetic evolves by imperceptible steps into the solution of linear equations.

When a person solves the problem of how many apples Jim has, given that

Bob's five apples and Jim's apples together make twelve apples

by thinking

Jim's apples are the whole twelve apples less Bob's five apples

and then

Jim has seven apples,

the steps really are:

$$5 + x = 12$$
$$x = 12 - 5$$
$$x = 7$$

The word solution of this problem is the earliest form of algebra. Such problems were solved exactly this way in Babylonia in 1800 B.C. We know almost nothing of mathematical work before this date, although most authorities believe the sophistication of the earliest known texts indicates that a long period of previous development must have occurred. The method of writing out equations in words persisted for thousands of years, and although it seems extremely cumbersome to us, it was used very effectively by many generations of mathematicians. The Arabs developed a good deal of the theory of cubic equations while writing out all the equations in words. About A.D. 1500 the tendency to abbreviate words in the written equations began to lead in the direction of modern notation; for example, the Latin word *et* (meaning *and*) developed into the plus sign, +. Although the occasional use of letters to represent unknowns dates back to A.D. 1200, the practice did not become common until about A.D. 1600. Development thereafter was rapid, and by 1635 algebraic notation did not differ essentially from what we use now. ≡

EXERCISE 2.1   *In Problems 1–8 mentally solve each equation.*

1. $3x = 18$ 2. $4x = -24$

3. $5x + 10 = 0$ 4. $3x + 12 = 0$

5. $2x - 3 = 5$ 6. $3x + 4 = -8$

7. $\frac{1}{3}x = \frac{5}{12}$ 8. $\frac{2}{3}x = \frac{9}{2}$

In Problems 9–50 solve each equation.

9. $3x + 2 = x - 6$ 10. $2x - 7 = 3x + 5$

11. $2t - 6 = 3 - t$ 12. $5y + 6 = -18 - y$

13. $6 - x = 2x + 9$

14. $3 - 2x = 2 - x$

15. $3 + 2n = 5n + 7$

16. $3 - 2m = 3m + 1$

17. $2(3 + 2x) = 3(x - 4)$

18. $3(2 - x) = 2x - 1$

19. $8x - (2x + 1) = 3x - 10$

20. $5 - (2x - 1) = 10$

21. $\frac{3}{2}x + 2 = \frac{1}{2} - \frac{1}{2}x$

22. $\frac{1}{3}x = 2 - \frac{2}{3}x$

23. $\frac{1}{2}x - 4 = \frac{3}{4}x$

24. $1 - \frac{1}{2}x = 5$

25. $\frac{2}{3}p = \frac{1}{2}p + \frac{1}{3}$

26. $\frac{1}{2} - \frac{1}{3}p = \frac{4}{3}$

27. $0.9t = 0.4 + 0.1t$

28. $0.9t = 1 + t$

29. $\dfrac{x + 1}{3} + \dfrac{x + 2}{7} = 5$

30. $\dfrac{2x + 1}{3} + 16 = 3x$

31. $\dfrac{2}{y} + \dfrac{4}{y} = 3$

32. $\dfrac{4}{y} - 5 = \dfrac{5}{2y}$

33. $(x + 7)(x - 1) = (x + 1)^2$

34. $(x + 2)(x - 3) = (x - 3)^2$

35. $x(2x - 3) = (2x + 1)(x - 4)$

36. $x(1 + 2x) = (2x - 1)(x - 2)$

37. $z(z^2 + 1) = 3 + z^3$

38. $w(4 - w^2) = 8 - w^3$

39. $\dfrac{x}{x + 2} = \dfrac{1}{2}$

40. $\dfrac{3x}{x - 1} = 2$

41. $\dfrac{3}{2x - 3} = \dfrac{2}{x + 5}$

42. $\dfrac{-2}{x + 4} = \dfrac{-3}{x + 1}$

43. $(x + 2)(3x) = (x + 2)(6)$

44. $(x - 5)(2x) = (x - 5)(4)$

45. $\dfrac{6t + 7}{4t - 1} = \dfrac{3t + 8}{2t - 4}$

46. $\dfrac{8w + 5}{10w - 7} = \dfrac{4w - 3}{5w + 7}$

47. $\dfrac{2}{x - 2} = \dfrac{3}{x + 5} + \dfrac{10}{(x + 5)(x - 2)}$

48. $\dfrac{1}{2x + 3} + \dfrac{1}{x - 1} = \dfrac{1}{(2x + 3)(x - 1)}$

49. $\dfrac{2}{y + 3} + \dfrac{3}{y - 4} = \dfrac{5}{y + 6}$

50. $\dfrac{5}{5z - 11} + \dfrac{4}{2z - 3} = \dfrac{3}{5 - z}$

In Problems 51–56 solve each equation. The letters a, b, and c are constants.

51. $ax - b = c, \quad a \neq 0$

52. $1 - ax = b, \quad a \neq 0$

53. $\dfrac{x}{a} + \dfrac{x}{b} = c, \quad a \neq 0, b \neq 0$

54. $\dfrac{a}{x} + \dfrac{b}{x} = c$

55. $\dfrac{1}{x - a} + \dfrac{1}{x + a} = \dfrac{2}{x - 1}$

56. $\dfrac{b + c}{x + a} = \dfrac{b - c}{x - a}$

57. Find the number a for which $x = 4$ is a solution of the equation

$$x + 2a = 16 + ax - 6a$$

58. Find the number b for which $x = 2$ is a solution of the equation

$$x + 2b = x - 4 + 2bx$$

Problems 59–64 list some formulas that occur in applications. Solve each formula for the indicated unknown.

59. $\dfrac{1}{R} = \dfrac{1}{R_1} + \dfrac{1}{R_2}$ for R (electricity)

60. $A = P(1 + rt)$ for r (finance)

61. $F = \dfrac{mv^2}{R}$ for R (mechanics)

62. $PV = nRT$ for T (chemistry)

63. $S = \dfrac{a}{1 - r}$ for r (mathematics)

64. $v = -gt + v_0$ for t (mechanics)

65. Explain what is wrong in the following steps:

$$x = 2 \tag{1}$$
$$3x - 2x = 2 \tag{2}$$
$$3x = 2x + 2 \tag{3}$$
$$x^2 + 3x = x^2 + 2x + 2 \tag{4}$$
$$x^2 + 3x - 10 = x^2 + 2x - 8 \tag{5}$$
$$(x - 2)(x + 5) = (x - 2)(x + 4) \tag{6}$$
$$x + 5 = x + 4 \tag{7}$$
$$1 = 0 \tag{8}$$

66. Which of the following pairs of equations are equivalent?
(a) $x^2 = 9$, $x = 3$ (b) $x = \sqrt{9}$, $x = 3$
(c) $(x - 1)(x - 2) = (x - 1)^2$, $x - 2 = x - 1$

▆ 2.2 APPLICATIONS

The information in the previous section provides a tool for handling certain kinds of applied problems—that is, problems that involve an application of mathematics to some other field. Unfortunately, applied problems do not come in the form "Solve the equation . . . ." Instead, they are narratives that supply information—hopefully enough to answer the question that inevitably arises. Thus, to solve applied problems we must be able to translate the verbal description into the language of mathematics. We do this by using symbols (usually letters of the alphabet) to represent unknown quantities and then finding relationships (such as equations) that involve these symbols.

Let's look at a few examples that will help you translate words into mathematical symbols.

EXAMPLE 1　(a)　The area of a rectangle is the product of its length times its width.

Translation:　If A is used to represent the area, l the length, and w the width, then $A = lw$.

(b)　For uniform motion, the velocity of an object equals the distance traveled per unit time.

Translation:　If v is the velocity, s the distance, and t the time, then $v = s/t$.

(c)　The sum of two numbers is 50. If one number is known, how is the other one found?

Translation:　If N and M are the two numbers, then $N + M = 50$. Suppose M is the number we know. Then a formula for obtaining N, the one we don't know, is

$$N = 50 - M$$

(d)　Let x denote a number.

The number five times as large as x is $5x$.

The number three less than x is $x - 3$.

The number that exceeds x by four is $x + 4$.

The number that, when added to x, gives five, is $5 - x$.　　≡

Mathematical equations that represent real situations should be consistent in terms of the units used. In Example 1(a), if l is measured in feet, then w must also be expressed in feet and A will be expressed in square feet. In Example 1(b), if v is measured in miles per hour, then the distance s must be expressed in miles and the time t must be expressed in hours. It is a good practice to check units to be sure the units are consistent.

Although each applied problem has its unique features, we can provide a rough outline of the steps to follow in solving applied problems.

Step 1:　Read the problem carefully, perhaps two or three times. Think about the situation described; put yourself into the situation, if possible. Pay particular attention to the question being asked in order to identify what is sought.

Step 2:　Assign some symbol (variable) to represent what is being sought and, if necessary, express any remaining unknown quantities in terms of this variable.

Step 3: Make a list of all the known facts, and write down any relationships among them, especially any that involve the unknown quantity. If possible, draw an appropriately labeled picture or diagram to assist you. Sometimes a table or chart may be helpful.

Step 4: Express the ideas from Step 3 as an equation involving the unknown.

Step 5: Solve the equation for the unknown.

Step 6: Check the answer with the facts in the problem. If it agrees, congratulations. If it doesn't, don't get discouraged; try again.

Let's look at an example.

EXAMPLE 2   The sum of two numbers is 180. If one number is 30 less than the other one, find the two numbers.

Solution   Step 1:   We are being asked to find two numbers.

Step 2:   We don't know either number, but we'll let x represent the larger number. Then $x - 30$ is the other number.

Step 3:   We can make a table:

One number	Other number	Reason
x	$x - 30$	One number is 30 less than the other.

Step 4:   It follows from the table and the fact that the sum of the two numbers is 180 that

$$x + (x - 30) = 180$$

Step 5:
$$2x - 30 = 180$$
$$2x = 210$$
$$x = 105$$

From the table, one number is 105 and the other is 75.

Step 6:   The sum of 105 and 75 is 180, and 75 is 30 less than 105.

EXAMPLE 3   A friend of yours grossed $435 one week by working 52 hours. Her employer pays time and a half for all hours worked in excess of 40 hours. What is your friend's hourly wage?

Solution   Step 1:   We are looking for an hourly wage. Our answer will be in dollars per hour.

Step 2:   Let x represent the hourly wage; x is measured in dollars per hour.

Step 3:   We set up a table:

	Hourly wage	Salary
Regular hours, 40	x	$40x$
Overtime hours, 12	$1.5x$	$12(1.5x) = 18x$

Step 4:   The sum of regular salary plus overtime salary will equal $435. Thus,

$$40x + 18x = 435$$

Step 5:
$$58x = 435$$
$$x = 7.50$$

The hourly wage is $7.50 per hour.

Step 6:   Forty hours yields a salary of $40(7.50) = \$300$, and 12 hours of overtime yields a salary of $12(1.5)(7.50) = \$135$, for a total of $435.   ▤

Interest

INTEREST   The next example involves **interest**. Interest is money paid for the use of money. The total amount borrowed (whether by an individual from a bank in the form of a loan or by a bank from an individual PRINCIPAL   in the form of a savings account) is called the **principal**. The **rate of** RATE OF INTEREST   **interest**, expressed as a percent, is the amount charged for the use of the principal for a given period of time, usually on a yearly or *per annum* basis.

If a principal of P dollars is borrowed for a period of t years at a per annum interest rate r, expressed as a decimal, the interest I charged is

SIMPLE INTEREST FORMULA

$$I = Prt \qquad (1)$$

Interest charged according to Formula (1) is called **simple interest**.

EXAMPLE 4   If $500 is borrowed for 6 months at the simple interest rate of 9% per annum, what is the interest charged on this loan?

Solution   The rate of interest is given per annum, so the actual time the money is borrowed must be expressed in years. Thus, the interest charged would be the principal ($500) times the rate of interest (0.09) times the time in years ($\frac{1}{2}$):

$$\text{Interest charged} = (500)(0.09)\left(\frac{1}{2}\right) = \$22.50 \quad ≡$$

EXAMPLE 5   An investor with $70,000 decides to place part of her money in corporate bonds paying 12% per year and the rest in a certificate of deposit paying 8% per year. If she wishes to obtain a total return of 9% per year, how much should she place in each investment?

Solution   Step 1:   The question is asking for two dollar amounts: the principal to invest in the corporate bonds and the principal to invest in the certificate of deposit.

Step 2:   We'll let x represent the amount (in dollars) to be invested in the bonds. Then $70,000 - x$ is the amount that will be invested in the certificate. (Do you see why?)

Step 3:   We form the table:

	Principal ($)	Rate	Time (yr)	Interest ($)
Bonds	x	0.12	1	0.12x
Certificate	70,000 − x	0.08	1	0.08(70,000 − x)
Total	70,000	0.09	1	0.09(70,000) = 6300

Step 4:   Since the total interest from the investments is to equal 0.09(70,000) = 6300, we must have the equation

$$0.12x + 0.08(70,000 - x) = 6300$$

(Note that the units are consistent—the unit is dollars on each side.)

Step 5:
$$0.12x + 5600 - 0.08x = 6300$$
$$0.04x = 700$$
$$x = 17,500$$

Thus, the investor should place $17,500 in the bonds and $52,500 in the certificate.

Step 6:   The interest on the bond after 1 year is 0.12(17,500) = $2100; the interest on the certificate after 1 year is 0.08(52,500) = $4200. The total annual interest is $6300, the required amount.   ≡

Mixture Problems

MIXTURE PROBLEM   The next example is a type usually referred to as a **mixture problem**.

EXAMPLE 6   In a chemistry laboratory one solution contains 10% hydrochloric acid (HCl) and a second solution contains 60% HCl. How many milliliters of each should be mixed to obtain 50 milliliters of a solution containing 30% HCl?

Solution   Let x represent the number of milliliters of the 10% solution. Then $50 - x$ equals the number of milliliters of the 60% solution. See Figure 2. Based on the figure, we form the table:

	Amount (mL)	Percent HCl	Amount of pure acid (mL)
10% HCl	x	0.10	$0.10x$
60% HCl	$50 - x$	0.60	$0.60(50 - x)$
30% HCl	50	0.30	$0.30(50) = 15$

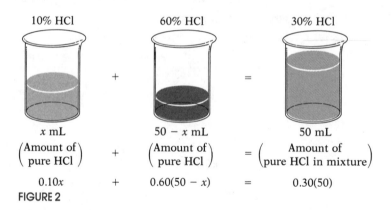

FIGURE 2

The amount of HCl in the 30% solution (15 milliliters) must equal the sum of the amounts of HCl found in the 10% solution and the 60% solution. Thus, we have the equation

$$0.10x + 0.60(50 - x) = 15$$
$$0.10x + 30 - 0.60x = 15$$
$$-0.50x = -15$$
$$x = 30 \text{ milliliters}$$

Thus, 30 milliliters of the 10% acid solution, when mixed with 20 milliliters of the 60% acid solution, yields 50 milliliters of a 30% acid solution.

To check this answer, we note that there are 0.10(30) = 3 milliliters of acid in the 10% solution and 0.60(20) = 12 milliliters of acid in the 60% solution. The 50-milliliter mixture therefore contains 15 milliliters of acid for an acid concentration of $\frac{15}{50}$ = 0.30, namely a 30% acid solution. ▬

Uniform Motion

The next example deals with moving objects.

If an object moves at an average velocity v, the distance s covered in time t is given by the formula

$$s = vt \qquad (2)$$

UNIFORM MOTION

That is, distance = velocity · time. Objects that are moving in accordance with Formula (2) are in **uniform motion**.

EXAMPLE 7 A friend of yours, a long distance runner, runs at an average velocity of 8 miles per hour. Two hours after your friend leaves your house, you leave in your car and follow the same route as your friend. If your average velocity is 40 miles per hour, how long will it be before you reach your friend? How far will each of you be from your house?

Solution We use t to represent the time (in hours) it takes the car to catch up with the runner. When this occurs the total time elapsed for the runner is $(t + 2)$ hours. Form the table:

	Velocity	Time (hr)	Distance (mi)
Runner	8	$t + 2$	$8(t + 2)$
Car	40	t	$40t$

Since the distance traveled is the same, we are led to the equation:

$$8(t + 2) = 40t$$
$$8t + 16 = 40t$$
$$32t = 16$$
$$t = \tfrac{1}{2} \text{ hour}$$

It will take you $\frac{1}{2}$ hour to reach your friend. Each of you will have gone 20 miles.

Check: The runner in 2.5 hours travels a distance of (2.5)(8) = 20 miles. The car in $\frac{1}{2}$ hour travels a distance of $(\frac{1}{2})(40)$ = 20 miles. ▬

Constant Rate Jobs

CONSTANT RATE JOBS

Our final example in this section has to do with jobs that are performed at a **constant rate**. Our assumption is that, if a job can be done in t units of time, then $1/t$ of the job is done in one unit of time. Let's look at an example.

EXAMPLE 8　　It takes Danny 4 hours to do a certain job, whereas it takes Mike 6 hours to do the same job. Assuming no gain or loss of efficiency, how long will it take them to do this job if they work together?

Solution　　In 1 hour, Danny does $\frac{1}{4}$ of the job. In 1 hour Mike does $\frac{1}{6}$ of the job. Let t be the time (in hours) it takes them to do the job together. In 1 hour, then, $1/t$ of the job is completed. We reason as follows:

$$\begin{pmatrix} \text{Part done by Danny} \\ \text{in 1 hour} \end{pmatrix} + \begin{pmatrix} \text{Part done by Mike} \\ \text{in 1 hour} \end{pmatrix} = \begin{pmatrix} \text{Part done together} \\ \text{in 1 hour} \end{pmatrix}$$

Now we form the table:

	Hours to do job	Part of job done in 1 hour
Danny	4	$\frac{1}{4}$
Mike	6	$\frac{1}{6}$
Together	t	$\frac{1}{t}$

Thus,

$$\frac{1}{4} + \frac{1}{6} = \frac{1}{t}$$

$$\frac{3 + 2}{12} = \frac{1}{t}$$

$$t = \frac{12}{5}$$

Working together, the job can be done in $\frac{12}{5}$ hours, or 2 hours, 24 minutes. ≡

EXERCISE 2.2　　*In Problems 1–8 translate each sentence into a mathematical equation. Be sure to identify the meaning of all symbols.*

1.　The area of a circle is the product of the number π times the square of the radius.

2.　The circumference of a circle is the product of the number π times twice the radius.

3. The area of a square is the square of the length of a side.

4. The perimeter of a square is four times the length of a side.

5. Force equals the product of mass times acceleration.

6. Pressure is force per unit area.

7. Work equals force times distance.

8. Kinetic energy is one-half the product of the mass times the square of the velocity.

9. Find two consecutive integers whose sum is 83.

10. Find two consecutive even integers whose sum is 66.

11. Find two consecutive integers for which the difference of their squares is 27.

12. Find two consecutive odd integers for which the difference of their squares is 48.

13. The sum of two numbers is 90. Twice one of them equals the other less 15. Find the numbers.

14. The sum of two numbers is 35. Twice one of them less 10 is the other. Find the numbers.

15. A worker who is paid time and a half for hours worked in excess of 40 had gross weekly wages of $442.00 for 48 hours worked. What is the hourly rate?

16. Rework Problem 15 if the gross weekly wages are $495 for 50 hours.

17. In an NFL football game, one team scored a total of 41 points, consisting of one safety (2 points) and two field goals (3 points each). They missed 2 extra points (1 point each) after scoring touchdowns (6 points each). How many touchdowns did they get?

18. In a basketball game, one team scored a total of 70 points and made three times as many field goals (2 points each) as free throws (1 point each). How many field goals did they have?

19. A recent retiree requires $6000 per year in extra income. She has $50,000 to invest and can invest in B-rated bonds paying 15% per year or in a certificate paying 7% per year. How much money should be invested in each to realize exactly $6000 in interest per year?

20. After 2 years, the retiree referred to in Problem 19 finds she now will require $7000 per year. Assuming the remaining information is the same, how should the money be reinvested?

21. A bank loaned out $10,000, part of it at the rate of 8% per year, and the rest at the rate of 18% per year. If the interest received from these loans totaled $1000, how much was loaned at 8%?

22. A loan officer at a bank has $100,000 to lend and is required to obtain an average return of 18% per year. If she can lend at the rate of 19% or the rate of 16%, how much can she lend at the 16% rate and still meet her requirement?

23. A nut store normally sells cashews for $4.00 per pound and peanuts for $1.50 per pound. At the end of the month, the peanuts had not sold well.

In order to sell 60 pounds of peanuts, the manager decides to mix the 60 pounds of peanuts with some cashews and sell the mixture for $2.50 per pound. How many pounds of cashews should be mixed with the peanuts to ensure no change in the profit?

24. A candy store sells boxes of candy containing caramels and cremes. Each box sells for $9.50 and holds 30 pieces of candy (all pieces are the same size). If the caramels cost $0.25 to produce and the cremes cost $0.45 to produce, how many of each should be in a box for no profit or loss to occur?

25. For a certain experiment, a student requires 100 cubic centimeters of a solution that is 8% HCl. The storeroom only has solutions that are 15% HCl and 5% HCl. How many cubic centimeters of each available solution should be mixed to get 100 cubic centimeters of 8% HCl?

26. A coffee manufacturer wants to market a new blend of coffee that will cost $2.90 per pound by mixing two coffees that sell for $2.75 and $3 per pound, respectively. What amounts of each coffee should be blended to obtain the desired mixture? [*Hint:* Assume the total weight of the desired blend is 100 pounds.]

27. The manager of the Coral Theater wants to know whether the majority of its patrons are adults or children. During a week in July, 5200 tickets were sold and the receipts totaled $11,875. The adult admission is $2.75, and the children's admission is $1.50. How many adult patrons were there?

28. A wool suit, discounted by 30% for a clearance sale, has a price tag of $199. What was the suit's original price?

29. A builder of tract homes reduced the price of a model by 15%. If the new price is $93,500, what was its original price? How much can be saved by purchasing the model?

30. A car dealer, at a year-end clearance, reduces the list price of last year's models by 15%. If a certain four-door model has a discounted price of $8000, what was its list price? How much can be saved by purchasing last year's model?

31. Mike can deliver his newspapers in 30 minutes. It takes Danny 20 minutes to do the same route. How long would it take them to deliver the newspapers working together?

32. A painter by himself can paint four rooms in 10 hours. If he hires a helper, they can do the same job together in 6 hours. If he lets the helper work alone, how long will it take for the helper to paint four rooms?

33. Going into the final exam, which will count as two tests, a student has test scores of 80, 83, 71, 61, and 95. What score does the student need on the final in order to have an average score of 80?

34. Rework Problem 33 if the final exam counts as half the final grade.

35. A tight end can run the 100-yard dash in 12 seconds. A defensive back can do it in 10 seconds. The tight end catches a pass at his own 20 yard line with the defensive back at the 15 yard line. If no other players are nearby, at what yard line will the defensive back catch up to the tight end?

36. An outside saleswoman, who uses her car for both business and pleasure, traveled 30,000 miles last year, using 900 gallons of gasoline. Her car gets 40 miles per gallon on the highway and 25 in the city. She can deduct all highway travel, but no city travel, on her taxes. How many miles should she be allowed as a business expense?

37. How much water should be added to 1 gallon of pure antifreeze to obtain a solution that is 60% antifreeze?

38. The cooling system of a certain foreign-made car has a capacity of 15 liters. If the system is filled with a mixture that is 40% antifreeze, how much of this mixture should be drained and replaced by pure antifreeze so that the system is filled with a solution that is 60% antifreeze?

39. A motorboat can maintain a constant speed of 15 miles per hour. The boat makes a trip upstream to a certain point in 20 minutes; the return trip takes 15 minutes. What is the speed of the current?

40. A car 15 feet in length overtakes a large semitrailer truck 40 feet in length. If the truck is traveling at a constant speed of 45 miles per hour, what constant speed should the car maintain to pass the truck within 550 feet? (5280 feet = 1 mile)

41. An oil tanker can be emptied by the main pump in 4 hours. An auxiliary pump can empty the tanker in 9 hours. If the main pump is started at 9 A.M., when should the auxiliary pump be started so that the tanker is emptied by noon?

42. A 20-pound bag of Economy cement mix contains 25% cement and 75% sand. How much pure cement must be added to produce a cement mix that is 40% cement?

43. A bathroom tub will fill in 15 minutes with both faucets open and the stopper in place. With both faucets closed and the stopper removed the tub will empty in 20 minutes. How long will it take for the tub to fill if both faucets are open and the stopper removed?

▤ 2.3 QUADRATIC EQUATIONS

QUADRATIC EQUATION A **quadratic equation** is an equation equivalent to one of the form

$$ax^2 + bx + c = 0 \tag{1}$$

where a, b, and c are real numbers and $a \neq 0$.

A quadratic equation written in the form $ax^2 + bx + c = 0$, is said

STANDARD FORM to be in **standard form**.

SECOND-DEGREE EQUATION Sometimes a quadratic equation is called a **second-degree equation** because the left side is a polynomial of degree two. We shall discuss three ways of solving quadratic equations: by factoring, by completing the square, and by use of the quadratic formula.

Factoring

When a quadratic equation is written in standard form, $ax^2 + bx + c = 0$, it may be possible to factor the expression on the left side as the product of two first-degree polynomials. Then, by setting each factor equal to zero and solving the resulting linear equations, we obtain the solutions of the quadratic equation. The technique of setting each factor equal to zero is a consequence of the product law, which states that if $AB = 0$, then either $A = 0$ or $B = 0$ or both equal zero.

Let's look at an example.

EXAMPLE 1   Solve the equation $x^2 - 5x + 6 = 0$.

Solution   The equation is in the standard form specified in Equation (1). The left side may be factored as

$$x^2 - 5x + 6 = 0$$
$$(x - 2)(x - 3) = 0$$

Set each factor equal to zero, and solve the resulting first-degree equations.

$$x - 2 = 0 \quad \text{or} \quad x - 3 = 0$$
$$x = 2 \qquad\qquad x = 3$$

The solution set is $\{2, 3\}$.   ≡

EXAMPLE 2   Solve each equation.

(a) $x^2 = 12 - x$ (b) $x^2 = 4x$ (c) $x^2 - 6x + 9 = 0$

Solution   (a) We put the equation in standard form by subtracting $12 - x$ from each side:

$$x^2 = 12 - x$$
$$x^2 + x - 12 = 0$$

The left side may now be factored as

$$(x + 4)(x - 3) = 0$$

so that

$$x + 4 = 0 \quad \text{or} \quad x - 3 = 0$$
$$x = -4 \qquad\qquad x = 3$$

The solution set is $\{-4, 3\}$.

(b) We put the equation in standard form and factor:

$$x^2 = 4x$$
$$x^2 - 4x = 0$$
$$x(x - 4) = 0$$
$$x = 0 \quad \text{or} \quad x - 4 = 0$$
$$x = 4$$

Thus, the solution set is $\{0, 4\}$.

(c) This equation is already in standard form, and the left side can be factored:

$$x^2 - 6x + 9 = 0$$
$$(x - 3)(x - 3) = 0$$

so that

$$x = 3 \quad \text{or} \quad x = 3$$

This equation has only the repeated solution 3.   ▤

REPEATED SOLUTION

ROOT OF MULTIPLICITY TWO

DOUBLE ROOT

When the left side factors into two linear equations with the same solution, the quadratic equation is said to have a **repeated solution**, as in Example 2(c). We also call this solution a **root of multiplicity two** or a **double root**.

Completing the Square

We begin with a preliminary result. Suppose we wish to solve the quadratic equation

$$x^2 = p \tag{2}$$

where $p \geq 0$ is a nonnegative number. We proceed as in the earlier examples:

$$x^2 - p = 0 \qquad \text{Place in standard form.}$$
$$(x - \sqrt{p})(x + \sqrt{p}) = 0 \qquad \text{Factor (over the real numbers).}$$
$$x = \sqrt{p} \quad \text{or} \quad x = -\sqrt{p} \qquad \text{Solve.}$$

Thus, we have the result:

$$\text{If } x^2 = p \text{ and } p \geq 0, \text{ then } x = \sqrt{p} \text{ or } x = -\sqrt{p}. \tag{3}$$

We usually abbreviate the solutions, $x = \sqrt{p}$ or $x = -\sqrt{p}$, of the equation $x^2 = p$ as $x = \pm\sqrt{p}$, read as "x equals plus or minus the square root of p." For example, the solutions of the equation

$$x^2 = 4$$

are

$$x = \pm\sqrt{4}$$

Since $\sqrt{4} = 2$, we find

$$x = \pm 2$$

The solution set is $\{-2, 2\}$. Do not confuse the two solutions of the equation $x^2 = 4$ with the value of the principal square root of 4, namely $\sqrt{4} = 2$.

EXAMPLE 3   Solve each equation.

(a) $x^2 = 5$ (b) $(x - 2)^2 = 16$

Solution   (a) We use the result in Equation (3) to get

$$x^2 = 5$$
$$x = \pm\sqrt{5}$$
$$x = \sqrt{5} \quad \text{or} \quad x = -\sqrt{5}$$

The solution set is $\{-\sqrt{5}, \sqrt{5}\}$.

(b) We use the result in Equation (3) to get

$$(x - 2)^2 = 16$$
$$x - 2 = \pm\sqrt{16}$$
$$x - 2 = \sqrt{16} \quad \text{or} \quad x - 2 = -\sqrt{16}$$
$$x - 2 = 4 \qquad\qquad x - 2 = -4$$
$$x = 6 \qquad\qquad\quad x = -2$$

The solution set is $\{-2, 6\}$.   ≡

COMPLETING THE SQUARE   We are now ready to **complete the square**. The idea behind this method is to "adjust" the left side of a quadratic equation, $ax^2 + bx + c = 0$, so that it becomes a perfect square—the square of a first-degree polynomial. For example, $x^2 + 6x + 9$ and $x^2 - 4x + 4$ are perfect squares because

$$x^2 + 6x + 9 = (x + 3)^2 \quad \text{and} \quad x^2 - 4x + 4 = (x - 2)^2$$

How do we "adjust" the left side? We do it by adding the appropriate number to the left side that causes a perfect square to result. For example, to make $x^2 + 6x$ a perfect square, we would add 9.

Let's look at several examples of completing the square when the coefficient of x^2 is 1:

START	ADD	RESULT
$x^2 + 4x$	4	$x^2 + 4x + 4 = (x + 2)^2$
$x^2 + 12x$	36	$x^2 + 12x + 36 = (x + 6)^2$
$x^2 - 6x$	9	$x^2 - 6x + 9 = (x - 3)^2$
$x^2 + x$	$\frac{1}{4}$	$x^2 + x + \frac{1}{4} = (x + \frac{1}{2})^2$

Do you see the pattern? Provided the coefficient of x^2 is 1, we complete the square by adding the square of one-half the coefficient of x:

START	ADD	RESULT
$x^2 + mx$	$\left(\dfrac{m}{2}\right)^2$	$x^2 + mx + \left(\dfrac{m}{2}\right)^2 = \left(x + \dfrac{m}{2}\right)^2$

The next example illustrates how the procedure of completing the square can be used to solve a quadratic equation.

EXAMPLE 4 Solve each equation by completing the square.

(a) $x^2 + 5x + 4 = 0$ (b) $2x^2 - 8x - 5 = 0$

Solution (a) We always begin by rearranging the equation so that the constant is on the right side:

$$x^2 + 5x + 4 = 0$$
$$x^2 + 5x = -4$$

Now complete the square on the left side by adding $(\frac{1}{2} \cdot 5)^2 = \frac{25}{4}$. Of course, since we want an equivalent equation, we add $\frac{25}{4}$ to *both* sides:

$$x^2 + 5x + \tfrac{25}{4} = -4 + \tfrac{25}{4}$$
$$(x + \tfrac{5}{2})^2 = \tfrac{9}{4}$$
$$x + \tfrac{5}{2} = \pm\sqrt{\tfrac{9}{4}}$$
$$x + \tfrac{5}{2} = \pm\tfrac{3}{2}$$
$$x = -\tfrac{5}{2} \pm \tfrac{3}{2}$$
$$x = -\tfrac{5}{2} + \tfrac{3}{2} = -1 \quad \text{or} \quad x = -\tfrac{5}{2} - \tfrac{3}{2} = -4$$

The solution set is $\{-4, -1\}$.

(b) First we rewrite the equation:

$$2x^2 - 8x - 5 = 0$$
$$2x^2 - 8x = 5$$

We divide by 2 so that the coefficient of x^2 is 1. (This enables us to complete the square at the next step.)

$$x^2 - 4x = \tfrac{5}{2}$$

Complete the square by adding 4 to each side:

$$x^2 - 4x + 4 = \tfrac{5}{2} + 4$$
$$(x - 2)^2 = \tfrac{13}{2}$$
$$x - 2 = \pm\sqrt{\tfrac{13}{2}}$$
$$x = 2 \pm \sqrt{\tfrac{13}{2}}$$

We choose to leave our answer in this compact form. Thus, the solution set is $\{2 - \sqrt{\tfrac{13}{2}}, 2 + \sqrt{\tfrac{13}{2}}\}$.

Note: If we wanted a decimal approximation to the solution set $\{2 - \sqrt{\tfrac{13}{2}}, 2 + \sqrt{\tfrac{13}{2}}\}$, we would use a calculator to get $\{-0.55, 4.55\}$.

≡

The Quadratic Formula

We use the method of completing the square to obtain a general formula for solving the quadratic equation

$$ax^2 + bx + c = 0 \qquad a \neq 0$$

As in Example 4, we rearrange the terms as

$$ax^2 + bx = -c$$

Since $a \neq 0$, we can divide both sides by a to get

$$x^2 + \frac{b}{a}x = -\frac{c}{a}$$

Now the coefficient of x^2 is 1. To complete the square on the left side, add the square of one-half of the coefficient of x—that is, add

$$\left(\frac{1}{2} \cdot \frac{b}{a}\right)^2 = \frac{b^2}{4a^2}$$

on each side. Then,

$$x^2 + \frac{b}{a}x + \frac{b^2}{4a^2} = \frac{b^2}{4a^2} - \frac{c}{a}$$
$$\left(x + \frac{b}{2a}\right)^2 = \frac{b^2 - 4ac}{4a^2} \qquad\qquad (4)$$

Provided $b^2 - 4ac \geq 0$, we can apply the result in Equation (3) to get

$$x + \frac{b}{2a} = \pm \sqrt{\frac{b^2 - 4ac}{4a^2}}$$

$$x = -\frac{b}{2a} \pm \frac{\sqrt{b^2 - 4ac}}{2a} = \frac{-b \pm \sqrt{b^2 - 4ac}}{2a}$$

What if $b^2 - 4ac$ is negative? Then Equation (4) states that the left expression (a real number squared) equals the right expression (a negative number). Since this occurrence is impossible for real numbers, we conclude that, if $b^2 - 4ac < 0$, the quadratic equation has no *real* solution. (We discuss quadratic equations for which the quantity $b^2 - 4ac < 0$ in detail in the next section.)

We now state the quadratic formula.

THEOREM   Consider the quadratic equation

QUADRATIC FORMULA

$$ax^2 + bx + c = 0$$

If $b^2 - 4ac < 0$, this equation has no real solution.

If $b^2 - 4ac \geq 0$, the real solution(s) of this equation is (are) given by the **quadratic formula**:

$$x = \frac{-b \pm \sqrt{b^2 - 4ac}}{2a} \tag{5}$$

≡

DISCRIMINANT   The quantity $b^2 - 4ac$ is called the **discriminant** of the quadratic equation because its value tells us whether or not the equation has real solutions. In fact, it also tells us how many solutions to expect.

For a quadratic equation $ax^2 + bx + c = 0$:

1. If $b^2 - 4ac > 0$, then there are two unequal real solutions.
2. If $b^2 - 4ac = 0$, then there is a repeated real solution—a root of multiplicity two.
3. If $b^2 - 4ac < 0$, then there is no real solution.

Thus, when asked to find the real solutions, if any, of a quadratic equation, always evaluate the discriminant to see how many real solutions there are (there may not be any).

EXAMPLE 5   Find the real solutions, if any, of each equation. Use the quadratic formula.

(a) $3x^2 - 5x + 1 = 0$ (b) $\frac{25}{2}x^2 - 30x + 18 = 0$

(c) $3x^2 - 5x = -2$ (d) $3x^2 + 2 = 4x$

Solution   (a) The equation is given in standard form, so we compare it to $ax^2 + bx + c = 0$ to find a, b, and c:

$$3x^2 - 5x + 1 = 0$$
$$ax^2 + bx + c = 0$$

With $a = 3$, $b = -5$, and $c = 1$, we evaluate the discriminant $b^2 - 4ac$:

$$b^2 - 4ac = (-5)^2 - 4(3)(1) = 25 - 12 = 13$$

Since $b^2 - 4ac > 0$, there are two real solutions, which can be found using the quadratic formula:

$$x = \frac{-b \pm \sqrt{b^2 - 4ac}}{2a} = \frac{5 \pm \sqrt{13}}{6}$$

The solution set is $\{(5 - \sqrt{13})/6, (5 + \sqrt{13})/6\}$.

(b) The equation is given in standard form. However, to simplify the arithmetic, we clear the fractions:

$$\frac{25}{2}x^2 - 30x + 18 = 0$$

$$25x^2 - 60x + 36 = 0 \quad \text{Clear fractions.}$$

$$ax^2 + bx + c = 0 \quad \text{Compare to standard form.}$$

With $a = 25$, $b = -60$, and $c = 36$, we evaluate the discriminant:

$$b^2 - 4ac = (-60)^2 - 4(25)(36) = 3600 - 3600 = 0$$

The equation has a repeated solution, which we will find by using the quadratic formula:

$$x = \frac{-b \pm \sqrt{b^2 - 4ac}}{2a} = \frac{60 \pm \sqrt{0}}{50} = \frac{60}{50} = \frac{6}{5}$$

The repeated solution is $\frac{6}{5}$.

(c) The equation, as given, is not in standard form.

$$3x^2 - 5x = -2$$

$$3x^2 - 5x + 2 = 0 \quad \text{Put in standard form.}$$

$$ax^2 + bx + c = 0 \quad \text{Compare to standard form.}$$

With $a = 3$, $b = -5$, and $c = 2$, we find

$$b^2 - 4ac = (-5)^2 - 4(3)(2) = 25 - 24 = 1$$

The two real solutions are

$$x = \frac{-b \pm \sqrt{b^2 - 4ac}}{2a} = \frac{5 \pm \sqrt{1}}{6} = \frac{5 \pm 1}{6}$$

The solution set is $\{\frac{2}{3}, 1\}$.

(d) The equation, as given, is not in standard form.

$$3x^2 + 2 = 4x$$
$$3x^2 - 4x + 2 = 0 \quad \text{Put in standard form.}$$
$$ax^2 + bx + c = 0 \quad \text{Compare to standard form.}$$

With $a = 3$, $b = -4$, and $c = 2$, we find

$$b^2 - 4ac = 16 - 24 = -8$$

Since $b^2 - 4ac < 0$, the equation has no real solution.   ≡

Sometimes the given equation can be transformed into a quadratic equation.

EXAMPLE 6 Find the real solutions, if any, of the equation

$$9 + \frac{3}{x} - \frac{2}{x^2} = 0$$

Solution In its present form the equation

$$9 + \frac{3}{x} - \frac{2}{x^2} = 0$$

is not a quadratic equation. However, it can be transformed into one by multiplying each side by x^2. The result is

$$9x^2 + 3x - 2 = 0$$

This quadratic equation in standard form is equivalent to the original one. The reason is that, even though we multiplied each side by x^2, we know that $x^2 \neq 0$. Do you see why?

Using $a = 9$, $b = 3$, and $c = -2$, the discriminant is

$$b^2 - 4ac = 9 + 72 = 81$$

Since $b^2 - 4ac > 0$, the new equation has two real solutions:

$$x = \frac{-b \pm \sqrt{b^2 - 4ac}}{2a} = \frac{-3 \pm \sqrt{81}}{18} = \frac{-3 \pm 9}{18}$$

$$x = \frac{-3 + 9}{18} = \frac{6}{18} = \frac{1}{3} \quad \text{or} \quad x = \frac{-3 - 9}{18} = \frac{-12}{18} = \frac{-2}{3}$$

The solution set is $\{-\frac{2}{3}, \frac{1}{3}\}$.   ≡

Summary

In general, when it is necessary to solve a quadratic equation, first put it in standard form:

$$ax^2 + bx + c = 0$$

Then:

1. Identify a, b, and c.
2. Evaluate the discriminant, $b^2 - 4ac$.
3. (a) If it is negative, the equation has no real solution.
   (b) If it is nonnegative, then look to see whether the left side can be factored. If you can easily spot factors, use the factoring method to solve the equation. Otherwise, use the quadratic formula or the method of completing the square.

Figure 3 provides a flowchart outlining the steps to follow for solving a quadratic equation.

Applications

Many applied problems require the solution of a quadratic equation. Let's look at one that you will probably see again in a slightly different form if you take calculus.

EXAMPLE 7   From each corner of a square piece of sheet metal, remove a square of side 9 centimeters. Turn up the edges to form an open box. If the box is to hold 144 cubic centimeters, what should be the dimensions of the sheet metal?

Solution   We use Figure 4 as a guide. We have labeled by x the length of a side of the square piece of sheet metal. The box will be of height 9 centimeters and its square base will have $x - 18$ as the length of a side. The volume (length × width × height) of the box is, therefore,

$$9(x - 18)(x - 18) = 9(x - 18)^2$$

We could place the quadratic equation in standard form and follow the steps listed in the summary. However, in this instance, it will be

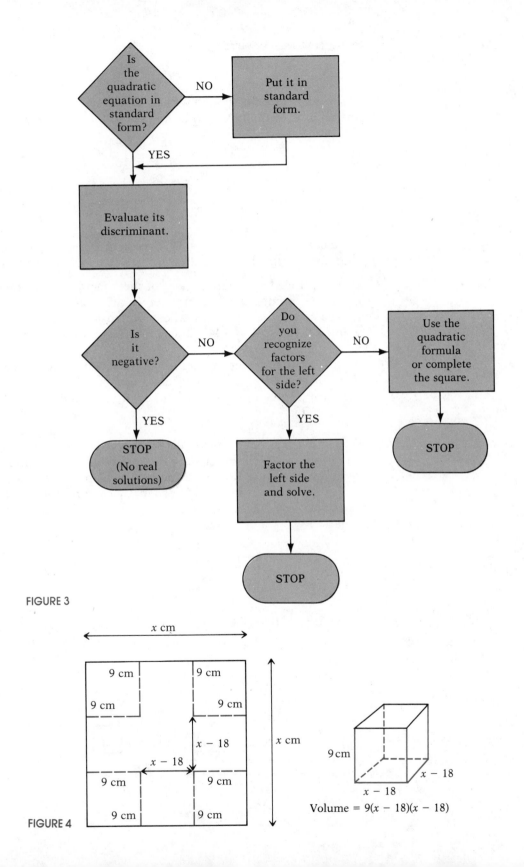

FIGURE 3

FIGURE 4

easier to proceed as though we had completed the square. Thus, since the volume of the box is to be 144 cubic centimeters, we have

$$9(x - 18)^2 = 144$$
$$(x - 18)^2 = 16$$
$$x - 18 = \pm 4$$
$$x = 18 \pm 4$$
$$x = 22 \quad \text{or} \quad x = 14$$

We discard the solution $x = 14$ (do you see why?) and conclude that the sheet metal should be 22 centimeters by 22 centimeters.

Check: If we begin with a piece of sheet metal 22 centimeters by 22 centimeters, cut out a 9-centimeter square from each corner, and fold up the edges, we get a box whose dimensions are 9 by 4 by 4 with volume $9 \times 4 \times 4 = 144$ cubic centimeters, as we wanted. ▆

EXAMPLE 8 A piece of wire 8 feet in length is to be cut into two pieces. Each of these pieces will then be bent into a square. Where should the cut in the wire be made, if the sum of the areas of these squares is to be 2 square feet?

Solution We use Figure 5 as a guide. We have labeled by x the length of one of the pieces of wire after it has been cut. The remaining piece will be of length $8 - x$. If each of the lengths is bent into a square, then one of the squares has a side of length $x/4$, and the other a side of length $(8 - x)/4$. Since the sum of the areas of these is 2, we have the equation

$$\left(\frac{x}{4}\right)^2 + \left(\frac{8 - x}{4}\right)^2 = 2$$
$$\frac{x^2}{16} + \frac{64 - 16x + x^2}{16} = 2$$
$$2x^2 - 16x + 64 = 32$$
$$2x^2 - 16x + 32 = 0 \quad \text{Put in standard form.}$$
$$x^2 - 8x + 16 = 0 \quad b^2 - 4ac = 64 - 64 = 0$$
$$(x - 4)^2 = 0$$
$$x = 4$$

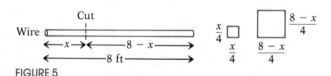

FIGURE 5

The original piece of wire should therefore be cut into two pieces each of length 4 feet.

Check: If the length of each piece of wire is 4 feet, then each piece can be formed into a square whose side is 1 foot. The area of each square is 1 square foot, so the sum of the areas is 2 square feet, as required. ≡

≡ HISTORICAL COMMENT

Problems using quadratic equations are found in the oldest known mathematical literature (not lost or destroyed). Babylonians and Egyptians were solving such problems before 1800 B.C. Euclid solved quadratic equations geometrically in his *Data* (300 B.C.), and the Hindus and Arabs gave rules for solving any quadratic equation with real roots. Because negative numbers were not freely used before A.D. 1500, there were several different types of quadratic equations, each with its own rule. Harriot (1560–1621) introduced the method of factoring to obtain solutions, and Vièta (1540–1603) introduced a method that is essentially completing the square.

Until modern times it was usual to neglect the negative roots (if there were any), and equations involving square roots of negative quantities were regarded as unsolvable until the 1500s.

≡ HISTORICAL PROBLEMS

1. *One of al-Khowârizmî's solutions* We solve $x^2 + 12x = 85$ by drawing the square below:

The area of the unshaded part is $x^2 + 12x$. Set this expression equal to 85 to get the equation $x^2 + 12x = 85$. If we add the four shaded squares, we will have a larger square of known area. Complete the solution.

2. *Vièta's method* We solve $x^2 + 12x - 85 = 0$ by letting $x = u + z$. Then

$$(u + z)^2 + 12(u + z) - 85 = 0$$
$$u^2 + (2z + 12)u + (z^2 + 12z - 85) = 0$$

Now select z so that $2z + 12 = 0$ and finish the solution.

3. *Another method to get the quadratic formula* Look at Equation (4) of Section 2.3. Rewrite the right side as $(\sqrt{b^2 - 4ac}/2a)^2$ and then subtract it from each side. The right side is now 0 and the left side is a difference of two squares. If you factor this difference of two squares, you will easily be able to get the quadratic formula, and, moreover, the quadratic expression is factored, which is sometimes useful.   ▬

EXERCISE 2.3   *In Problems 1–20 solve each equation by factoring.*

1. $x^2 = 4x$ **2.** $x^2 = -8x$

3. $x^2 - 9 = 0$ **4.** $x^2 - 1 = 0$

5. $z^2 + z - 12 = 0$ **6.** $v^2 + 7v + 12 = 0$

7. $2x^2 - 5x - 3 = 0$ **8.** $3x^2 + 5x + 2 = 0$

9. $6t^2 = 1 - t$ **10.** $6y^2 = 6 + 5y$

11. $x(x - 7) + 12 = 0$ **12.** $x(x + 1) = 12$

13. $4x^2 + 9 = 12x$ **14.** $25x^2 + 16 = 40x$

15. $6(p^2 - 1) = 5p$ **16.** $2(2u^2 - 4u) + 3 = 0$

17. $6x - 5 = \dfrac{6}{x}$ **18.** $x + \dfrac{12}{x} = 7$

19. $\dfrac{4(x - 2)}{x - 3} + \dfrac{3}{x} = \dfrac{-3}{x(x - 3)}$ **20.** $\dfrac{5}{x + 4} = 4 + \dfrac{3}{x - 2}$

In Problems 21–40 find the real solutions, if any, of each equation. Use the quadratic formula.

21. $x^2 - 4x + 2 = 0$ **22.** $x^2 + 4x + 2 = 0$

23. $x^2 - 5x - 1 = 0$ **24.** $x^2 + 5x + 3 = 0$

25. $2x^2 - 5x + 3 = 0$ **26.** $2x^2 + 5x + 3 = 0$

27. $4y^2 - y + 2 = 0$ **28.** $4t^2 + t + 1 = 0$

29. $4x^2 = 1 - 2x$ **30.** $2x^2 = 1 - 2x$

31. $4x^2 = 9x$ **32.** $5x = 4x^2$

33. $9t^2 - 6t + 1 = 0$ **34.** $4u^2 - 6u + 9 = 0$

35. $x^2 - 2x - 1 = 0$ **36.** $x^2 - 3x - 1 = 0$

37. $4 - \dfrac{1}{x} - \dfrac{2}{x^2} = 0$ **38.** $4 + \dfrac{1}{x} - \dfrac{1}{x^2} = 0$

39. $3x = 1 - \dfrac{1}{x}$ **40.** $x = 1 - \dfrac{4}{x}$

In Problems 41–48 find the real solutions, if any, of each equation. Use the quadratic formula and express any solutions rounded off to two decimal places.

41. $x^2 - 4x + 2 = 0$ **42.** $x^2 + 4x + 2 = 0$

43. $x^2 + \sqrt{3}x - 3 = 0$ **44.** $x^2 + \sqrt{2}x - 2 = 0$

45. $\pi x^2 - x - \pi = 0$ **46.** $\pi x^2 + \pi x - 2 = 0$

47. $x^2 + \pi x + \sqrt{29} = 0$ **48.** $4x^2 - 15\sqrt{2}x + 20 = 0$

In Problems 49–60 find the real solutions, if any, of each quadratic equation. Use any method.

49. $x^2 - 6 = 0$ **50.** $x^2 - 5 = 0$

51. $16x^2 - 8x + 1 = 0$ **52.** $9x^2 - 6x + 1 = 0$

53. $10x^2 - 19x - 15 = 0$ **54.** $6x^2 + 7x - 20 = 0$

55. $2 + z = 6z^2$ **56.** $2 = y + 6y^2$

57. $x^2 + \sqrt{2}x = \frac{1}{2}$ **58.** $\frac{1}{2}x^2 = \sqrt{2}x + 1$

59. $x^2 + x = 4$ **60.** $x^2 + x = 1$

In Problems 61–66 use the discriminant to tell whether each quadratic equation has two unequal real solutions, a repeated real solution, or no real solution, without solving the equation.

61. $x^2 - 5x + 7 = 0$ **62.** $x^2 + 5x + 7 = 0$

63. $9x^2 - 30x + 25 = 0$ **64.** $25x^2 - 20x + 4 = 0$

65. $3x^2 + 5x - 2 = 0$ **66.** $2x^2 - 3x - 4 = 0$

In Problems 67–72 tell what number should be added to complete the square of each expression.

67. $x^2 + 4x$ **68.** $x^2 - 2x$ **69.** $x^2 + \frac{1}{2}x$

70. $x^2 - \frac{1}{3}x$ **71.** $x^2 - \frac{2}{3}x$ **72.** $x^2 - \frac{2}{5}x$

In Problems 73–78 solve each equation by completing the square.

73. $x^2 + 4x - 21 = 0$ **74.** $x^2 - 6x = 13$

75. $x^2 - \frac{1}{2}x = \frac{3}{16}$ **76.** $x^2 + \frac{2}{3}x = \frac{1}{3}$

77. $3x^2 + x - \frac{1}{2} = 0$ **78.** $2x^2 - 3x = 1$

79. Find two consecutive odd integers whose product is 143.

80. Find two consecutive integers whose product is 306.

81. An open box is to be constructed from a square piece of sheet metal by removing a square of side 1 foot from each corner and turning up the edges. If the box is to hold 4 cubic feet, what should be the dimensions of the sheet metal?

82. Rework Problem 81 if the piece of sheet metal is a rectangle whose length is twice its width.

83. A ball is thrown vertically upward from the top of a building 96 feet above the ground with an initial velocity of 80 feet per second. The distance s (in feet) of the ball from the ground after t seconds is $s = 96 + 80t - 16t^2$.
(a) After how many seconds does the ball strike the ground?
(b) After how many seconds will the ball pass the top of the building on its way down?

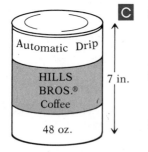

C **84.** A 3-pound Hills Bros.® coffee can requires 188.5 square inches of aluminum. If its height is 7 inches, what is its radius? (The surface area A of a right circular cylinder is $A = 2\pi R^2 + 2\pi Rh$, where R is the radius and h is the height.)

85. Mike and Danny, working together, can paint the exterior of a house in 6 days. Mike, by himself, can complete this job in 5 days less than Dan. How long will it take Mike to complete the job by himself?

86. Two different-sized pumps working together can empty a fuel tank in 5 hours. The larger pump can empty this tank in 4 hours less than the smaller one. If the larger one is out of order, how long will it take the smaller one to do the job alone?

87. A company charges $200 for each box of tools on orders of 150 or fewer boxes. If a customer orders x boxes in excess of 150, the cost for each box ordered is reduced by x dollars. If a customer's bill came to $30,625, how many boxes were ordered?

88. A contractor orders 8 cubic yards of premixed cement, all of which is to be used to pour a patio that will be 4 inches thick. If the length of the patio is specified to be twice the width, what will be the patio dimensions? (1 cubic yard = 27 cubic feet)

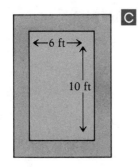

C **89.** A landscaper, who just completed a rectangular flower garden measuring 6 feet by 10 feet, orders 1 cubic yard of premixed cement, all of which is to be used to create a border of uniform width around the garden. If the border is to have a depth of 3 inches, how wide will the border be?

90. An object is propelled vertically upward with an initial velocity of 20 meters per second. The distance s (in meters) of the object from the ground after t seconds is $s = -4.9t^2 + 20t$.
(a) When will the object be 15 meters above the ground?
(b) When will it strike the ground?
(c) Will the object reach a height of 100 meters?
(d) What is the maximum height?

C **91.** A jumbo chocolate bar with a rectangular shape measures 12 centimeters in length, 7 centimeters in width, and 3 centimeters in thickness. Due to escalating costs of cocoa, management decides to reduce the volume of the bar by 10%. To accomplish this reduction, management decides the new bar should have the same 3-centimeter thickness, but the new length and width each should be reduced by the same amount. What should be the dimensions of the new candy bar?

C **92.** Rework Problem 91 if the reduction is to be 20%.

| C | **93.** | A pool in the shape of a circle measures 10 feet across. One cubic yard of concrete is to be used to create a circular border of uniform width around the pool. If the border is to have a depth of 3 inches, how wide will the border be? |

| C | **94.** | Rework Problem 93 if the depth of the border is 4 inches. |

95. A motorboat maintained a constant speed of 15 miles per hour in going 10 miles upstream and then returning. The total time for the trip was 1.5 hours. Use this information to find the speed of the current.

96. The hypotenuse of a right triangle measures 13 centimeters. Find the lengths of the legs if their sum is 17 centimeters.

97. Show that the sum of the roots of a quadratic equation is $-b/a$.

98. Show that the product of the roots of a quadratic equation is c/a.

99. *Programming exercise* Develop code that will solve a quadratic equation:

```
{Enter the coefficient of x squared}    READ (a);
{Enter the coefficient of x}            READ (b);
{Enter the constant term}               READ (c);
If b² − 4ac < 0
THEN {write no real solution}
ELSE IF b² − 4ac = 0
        THEN {write −b/2a is a double root}
        ELSE {write (−b + SQRT(b² − 4ac))/2a
              or (−b − SQRT(b² − 4ac))/2a
              is a solution};
```

≡ 2.4 QUADRATIC EQUATIONS WITH A NEGATIVE DISCRIMINANT

Quadratic equations with a negative discriminant have no real-number solution. However, if we extend our number system to allow complex numbers (refer to Section 1.9), quadratic equations will always have a solution. Since the solution to a quadratic equation involves the square root of the discriminant, we begin with a discussion of square roots of negative numbers.

Square Roots of Negative Numbers

When we first introduced square roots, we made the observation that the equation
$$x^2 = N \qquad N > 0$$

has two solutions—one positive, the other negative. To avoid confusion, we agreed that the positive solution of the equation $x^2 = N$ would be called the **principal square root of N**, and we invented the notation $\sqrt{N}$ to represent it. Thus, $\sqrt{4} = 2$, $\sqrt{9} = 3$, and so on.

PRINCIPAL SQUARE ROOT OF *N*

With these ideas as background, we give the following definition.

PRINCIPAL SQUARE ROOT OF −N   If N is a positive real number, we define the **principal square root of** $-N$, denoted by $\sqrt{-N}$, as

$$\sqrt{-N} = \sqrt{N}\, i$$

where i is the imaginary unit and $i^2 = -1$.

EXAMPLE 1   (a) $\sqrt{-1} = \sqrt{1}\, i = i$ (b) $\sqrt{-4} = \sqrt{4}\, i = 2i$
(c) $\sqrt{-8} = \sqrt{8}\, i = 2\sqrt{2}\, i$ ≡

EXAMPLE 2   Solve each equation in the complex number system.

(a) $x^2 = 4$ (b) $x^2 = -9$

Solution   (a) This equation has two solutions:

$$\sqrt{4} = 2 \quad \text{or} \quad -\sqrt{4} = -2$$

(b) This equation has two solutions:

$$\sqrt{-9} = \sqrt{9}\, i = 3i \quad \text{or} \quad -\sqrt{-9} = -\sqrt{9}\, i = -3i \qquad ≡$$

Warning: When working with square roots of negative numbers, do not set the square root of a product equal to the product of the square roots. To see why, look at this calculation: We know $\sqrt{100} = 10$. However, it is also true that $100 = (-25)(-4)$ so that

$$10 = \sqrt{100} = \sqrt{(-25)(-4)} \underset{\uparrow}{=} \sqrt{-25}\sqrt{-4}$$
$$\text{Here's the error.}$$
$$= (\sqrt{25}\, i)(\sqrt{4}\, i) = (5i)(2i) = 10i^2 = -10$$

The Quadratic Formula

Because we have defined the square root of a negative number, we can restate the quadratic formula without restriction.

THEOREM
QUADRATIC FORMULA   In the complex number system, the solutions of the quadratic equation $ax^2 + bx + c = 0$, where a, b, and c are real numbers, are given by the formula

$$x = \frac{-b \pm \sqrt{b^2 - 4ac}}{2a} \tag{1}$$

≡

EXAMPLE 3   Solve the equation $x^2 - 4x + 8 = 0$ in the complex number system.

Solution   Here $a = 1$, $b = -4$, $c = 8$, and $b^2 - 4ac = 16 - 4(8) = -16$. Using Equation (1), we find

$$x = \frac{4 \pm \sqrt{-16}}{2} = \frac{4 \pm \sqrt{16}\,i}{2} = \frac{4 \pm 4i}{2} = 2 \pm 2i$$

The equation has the solution set $\{2 - 2i, 2 + 2i\}$.

Check:

$2 + 2i$: $(2 + 2i)^2 - 4(2 + 2i) + 8 = 4 + 8i + 4i^2 - 8 - 8i + 8$
$$= 4 - 4 = 0$$

$2 - 2i$: $(2 - 2i)^2 - 4(2 - 2i) + 8 = 4 - 8i + 4i^2 - 8 + 8i + 8$
$$= 4 - 4 = 0 \qquad \blacksquare$$

The discriminant, $b^2 - 4ac$, of a quadratic equation still serves as a way to tell the character of the solutions.

> In the complex number system, consider a quadratic equation $ax^2 + bx + c = 0$ with real coefficients.
>
> 1. If $b^2 - 4ac > 0$, the equation has two unequal real solutions.
> 2. If $b^2 - 4ac = 0$, the equation has a repeated real solution— a double root.
> 3. If $b^2 - 4ac < 0$, the equation has two complex solutions that are conjugates of each other.

The third conclusion is a consequence of the fact that, if $b^2 - 4ac = -N < 0$, then by the quadratic formula, Equation (1), the solutions are

$$x = \frac{-b + \sqrt{b^2 - 4ac}}{2a} = \frac{-b + \sqrt{-N}}{2a} = \frac{-b + \sqrt{N}\,i}{2a} = \frac{-b}{2a} + \frac{\sqrt{N}}{2a}i$$

and

$$x = \frac{-b - \sqrt{b^2 - 4ac}}{2a} = \frac{-b - \sqrt{-N}}{2a} = \frac{-b - \sqrt{N}\,i}{2a} = \frac{-b}{2a} - \frac{\sqrt{N}}{2a}i$$

which are conjugates of each other.

EXAMPLE 4   Without solving, determine the character of the solution of each equation in the complex number system.

(a) $3x^2 + 4x + 5 = 0$ (b) $2x^2 + 4x + 1 = 0$

(c) $9x^2 - 6x + 1 = 0$

Solution　(a) Here $a = 3$, $b = 4$, and $c = 5$, so $b^2 - 4ac = 16 - 4(3)(5) = -44$. The solutions are complex numbers that are conjugates of each other.

(b) Here $a = 2$, $b = 4$, and $c = 1$, so $b^2 - 4ac = 16 - 8 = 8$. The solutions are two unequal real numbers.

(c) Here $a = 9$, $b = -6$, and $c = 1$, so $b^2 - 4ac = 36 - 4(9)(1) = 0$. The solution is a repeated real number—that is, a double root.

≡

EXERCISE 2.4　*In Problems 1–10 perform the indicated operations and express your answer in the form $a + bi$.*

1. $\sqrt{-4}$
2. $\sqrt{-9}$
3. $\sqrt{-25}$
4. $\sqrt{-64}$
5. $\sqrt{(3 + 4i)(4i - 3)}$
6. $\sqrt{(4 + 3i)(3i - 4)}$
7. $(\sqrt{-16} + 1)^2$
8. $(\sqrt{-4} + 2)^2$
9. $(3 - \sqrt{-4})(2 + \sqrt{-9})$
10. $(8 + \sqrt{-4})(2 - \sqrt{-9})$

In Problems 11–24 solve each equation in the complex number system.

11. $x^2 + 4 = 0$
12. $x^2 - 4 = 0$
13. $x^2 - 16 = 0$
14. $x^2 + 25 = 0$
15. $x^2 - 6x + 13 = 0$
16. $x^2 + 4x + 8 = 0$
17. $x^2 - 6x + 10 = 0$
18. $x^2 - 2x + 5 = 0$
19. $8x^2 - 4x + 1 = 0$
20. $10x^2 + 6x + 1 = 0$
21. $5x^2 + 2x + 1 = 0$
22. $13x^2 + 6x + 1 = 0$
23. $x^2 + x + 1 = 0$
24. $x^2 - x + 1 = 0$

In Problems 25–30, without solving, determine the character of the solutions of each equation in the complex number system.

25. $3x^2 - 3x + 4 = 0$
26. $2x^2 - 4x + 1 = 0$
27. $2x^2 + 3x - 4 = 0$
28. $x^2 + 2x + 6 = 0$
29. $9x^2 - 12x + 4 = 0$
30. $4x^2 + 12x + 9 = 0$

31. $2 + 3i$ is a solution of a quadratic equation with real coefficients. Find the other solution.

32. $4 - i$ is a solution of a quadratic equation with real coefficients. Find the other solution.

≡ 2.5 OTHER TYPES OF EQUATIONS

In this section we look at other types of equations, most of which can be solved using variations of techniques discussed earlier. For the rest of this chapter, we shall be working in the real number system.

Equations of the Type $x^r = a$, r a Rational Number

The solution of equations of the type $x^r = a$ was discussed in Section 1.7 on radicals. However, let's spend a little time reviewing the process once more.

EXAMPLE 1   Solve each equation.

(a) $x^3 = 6$ (b) $x^4 = 5$ (c) $x^4 = -2$

Solution   (a) If $x^3 = 6$, then $x = \sqrt[3]{6} \approx 1.82$.

(b) If $x^4 = 5$, then $(x^2)^2 = 5$, so $x^2 = \pm\sqrt{5}$. Because the square of a real number is never negative, we discard $-\sqrt{5}$, leaving $x^2 = \sqrt{5}$, from which

$$x = \pm\sqrt{\sqrt{5}} = \pm\sqrt[4]{5} \approx \pm 1.50$$

(c) If $x^4 = -2$, then $(x^2)^2 = -2$, which is impossible in the real number system, because the square of any real number is nonnegative. Thus, the equation has no real solution. ■

Table 1 summarizes the solutions of the equation $x^r = a$, when r is a positive integer.

TABLE 1

$r > 0$	a	$x^r = a$
Even integer	Negative	No real solution
Even integer	Positive	Two real solutions, $\pm\sqrt[r]{a}$
Odd integer	Positive or negative	One real solution, $\sqrt[r]{a}$

If r is a positive fraction, say m/n, raise both sides to the power n and proceed as in Example 1. One caution, though. When you do this, you are, in fact, multiplying by an expression containing a variable. Thus, the result might be a succession of equations that are not equivalent. As a result, extraneous solutions may be introduced, so a check must be made of any apparent solutions that are found.

EXAMPLE 2   Solve each equation.

(a) $x^{1/3} = 4$ (b) $x^{3/2} = 8$ (c) $x^{3/4} = -1$

Solution   (a) If $x^{1/3} = 4$, we cube both sides to get $x = 4^3 = 64$. Since $64^{1/3} = \sqrt[3]{64} = 4$, our solution checks.

(b) Since $x^{3/2} = 8$, we square both sides to get

$$(x^{3/2})^2 = 8^2$$
$$x^3 = 64$$
$$x = \sqrt[3]{64} = 4$$

Since $4^{3/2} = (\sqrt{4})^3 = 2^3 = 8$, our answer checks.

(c) Since $x^{3/4} = -1$, we raise both sides to the fourth power to get

$$(x^{3/4})^4 = (-1)^4$$
$$x^3 = 1$$
$$x = \sqrt[3]{1} = 1$$

Now $1^{3/4} = 1$, not -1, so the apparent solution 1 is extraneous and is therefore discarded. The equation has no solution. ▤

The absence of a solution in Example 2(c) should not surprise us. If we look more carefully at the equation in part (c), we observe that a real number raised to the power $\frac{3}{4}$ can never be negative. Sometimes we should look before we leap!

Examples 1 and 2 deal with $x^r = a$, where r is positive. If r happens to be negative, raise both sides to the -1 power and then proceed as in the examples.

EXAMPLE 3   Solve the equation $x^{-2/3} = \frac{1}{4}$.

Solution

$$x^{-2/3} = \frac{1}{4}$$
$$(x^{-2/3})^{-1} = \left(\frac{1}{4}\right)^{-1}$$
$$x^{2/3} = 4$$
$$(x^{2/3})^3 = 4^3$$
$$x^2 = 64$$
$$x = \pm\sqrt{64} = \pm 8$$

Check: $8^{-2/3} = (\sqrt[3]{8})^{-2} = 2^{-2} = \dfrac{1}{2^2} = \dfrac{1}{4}$

$(-8)^{-2/3} = (\sqrt[3]{-8})^{-2} = (-2)^{-2} = \dfrac{1}{(-2)^2} = \dfrac{1}{4}$

The solution set is $\{-8, 8\}$. ▤

Equations Containing Radicals

When the variable in an equation occurs in a square root, cube root, and so on—that is, when it occurs in a radical—the equation is called RADICAL EQUATION a **radical equation**. Sometimes a suitable operation will change a radical equation to one that is linear or quadratic. The most commonly used procedure is to isolate the most complicated radical on one side of the equation and then eliminate it by raising each side to a power equal to the index of the radical. Care must be taken, because, as we have seen, extraneous solutions may result. Let's look at some examples.

EXAMPLE 4 Solve the equation $\sqrt[3]{2x - 4} - 2 = 0$.

Solution The equation contains a radical whose index is 3. We isolate it on the left side:

$$\sqrt[3]{2x - 4} - 2 = 0$$
$$\sqrt[3]{2x - 4} = 2$$

Now raise each side to the third power (the index of the radical is 3) and solve:

$$(\sqrt[3]{2x - 4})^3 = 2^3$$
$$2x - 4 = 8$$
$$2x = 12$$
$$x = 6$$

Check: $\sqrt[3]{2(6) - 4} - 2 = \sqrt[3]{12 - 4} - 2 = \sqrt[3]{8} - 2 = 2 - 2 = 0$

The solution is $x = 6$. ≡

Sometimes we need to raise each side to a power more than once to solve a radical equation.

EXAMPLE 5 Solve the equation $\sqrt{2x + 3} - \sqrt{x + 2} = 2$.

Solution First, we choose to isolate the more complicated radical expression (in this case, $\sqrt{2x + 3}$) on the left side:

$$\sqrt{2x + 3} = \sqrt{x + 2} + 2$$

Now square both sides (the index of the radical is 2):

$$(\sqrt{2x + 3})^2 = (\sqrt{x + 2} + 2)^2$$
$$2x + 3 = (\sqrt{x + 2})^2 + 4\sqrt{x + 2} + 4$$
$$2x + 3 = x + 2 + 4\sqrt{x + 2} + 4$$

Because the equation still contains a radical, we combine like terms, isolate the remaining radical on the right side, and again square both sides:

$$x - 3 = 4\sqrt{x + 2}$$
$$(x - 3)^2 = 16(x + 2)$$
$$x^2 - 6x + 9 = 16x + 32$$
$$x^2 - 22x - 23 = 0$$
$$(x - 23)(x + 1) = 0$$
$$x = 23 \quad \text{or} \quad x = -1$$

The original equation appears to have the solution set $\{-1, 23\}$. However, we have not yet checked.

Check: $\quad \sqrt{2(23) + 3} - \sqrt{23 + 2} = \sqrt{49} - \sqrt{25} = 7 - 5 = 2$
$$\sqrt{2(-1) + 3} - \sqrt{-1 + 2} = \sqrt{1} - \sqrt{1} = 1 - 1 = 0$$

Thus, the equation has only one real solution, 23; the solution -1 is extraneous. ▄

Disguised Quadratic Equations

The equation $x^4 + x^2 - 12 = 0$ is not quadratic in x, but it is quadratic in x^2. That is, if we let $u = x^2$, we get $u^2 + u - 12 = 0$, a quadratic equation. This equation can be solved for u and, in turn, by using $u = x^2$, we can find the solutions x of the original equation. The original equation, although not quadratic in x, is a *disguised quadratic equation*.

If an appropriate substitution u transforms an equation into one of the form

$$au^2 + bu + c = 0 \qquad a \neq 0$$

DISGUISED QUADRATIC
EQUATION

then the original equation is called a **disguised quadratic equation** or an **equation of the quadratic type** or an **equation quadratic in form**.

The difficulty of solving a disguised quadratic equation lies in the determination that the equation is, in fact, a disguised quadratic. After you are told an equation you've been struggling with is a disguised quadratic, you'll say "Of course! Why didn't I see that?" Once you've had a lot of practice, this problem shouldn't happen too often.

EXAMPLE 6   Solve each equation.

(a) $x^4 + x^2 - 12 = 0$ (b) $x + 2\sqrt{x} - 3 = 0$

(c) $\dfrac{1}{x^4} + \dfrac{1}{x^2} - 12 = 0$

Solution (a) For the equation $x^4 + x^2 - 12 = 0$, we let $u = x^2$. Then $u^2 = x^4$, and the original equation,

$$x^4 + x^2 - 12 = 0$$

becomes $u^2 + u - 12 = 0 \quad u = x^2, u^2 = x^4$

$\qquad\qquad (u + 4)(u - 3) = 0 \quad$ Factor.

$\qquad\qquad u = -4 \quad \text{or} \quad u = 3$

But the original equation has x as the variable. Because $u = x^2$, we have

$$x^2 = -4 \quad \text{or} \quad x^2 = 3$$

The first of these has no real solution; the second has the solution set $\{-\sqrt{3}, \sqrt{3}\}$.

Check: $\qquad (\sqrt{3})^4 + (\sqrt{3})^2 - 12 = 9 + 3 - 12 = 0$

$\qquad\qquad (-\sqrt{3})^4 + (-\sqrt{3})^2 - 12 = 9 + 3 - 12 = 0$

Thus, $\{-\sqrt{3}, \sqrt{3}\}$ is the solution set of the original equation.

(b) For the equation $x + 2\sqrt{x} - 3 = 0$, let $u = \sqrt{x}$. Then $u^2 = x$, and the original equation,

$$x + 2\sqrt{x} - 3 = 0$$

becomes $u^2 + 2u - 3 = 0 \quad u = \sqrt{x}, u^2 = x$

$\qquad\qquad (u + 3)(u - 1) = 0 \quad$ Factor.

$\qquad\qquad u = -3 \quad \text{or} \quad u = 1 \quad$ Solve.

Remember that the equation we want to solve has x as a variable. Since $u = \sqrt{x}$, we have $\sqrt{x} = -3$ or $\sqrt{x} = 1$. The first of these, $\sqrt{x} = -3$, has no real solution, since the square root of a real number is never negative. The second one, $\sqrt{x} = 1$, has the solution $x = 1$.

Check: $\ 1 + 2\sqrt{1} - 3 = 1 + 2 - 3 = 0$

Thus, $x = 1$ is the only solution of the original equation.

(c) For the equation $(1/x^4) + (1/x^2) - 12 = 0$, we let $u = 1/x^2$. Then $u^2 = 1/x^4$, and the original equation,

$$\frac{1}{x^4} + \frac{1}{x^2} - 12 = 0$$

becomes $u^2 + u - 12 = 0 \quad u = \dfrac{1}{x^2}, u^2 = \dfrac{1}{x^4}$

$\qquad\qquad (u + 4)(u - 3) = 0$

$\qquad\qquad u = -4 \quad \text{or} \quad u = 3$

$\qquad\qquad \dfrac{1}{x^2} = -4 \quad \text{or} \quad \dfrac{1}{x^2} = 3$

The equation on the left has no real solution; we take reciprocals of each side of the right equation to get

$$x^2 = \tfrac{1}{3}$$

$$x = \pm\sqrt{\tfrac{1}{3}}$$

Check: $\dfrac{1}{(\pm\sqrt{\frac{1}{3}})^4} + \dfrac{1}{(\pm\sqrt{\frac{1}{3}})^2} - 12 = \dfrac{1}{\frac{1}{9}} + \dfrac{1}{\frac{1}{3}} - 12$

$$= 9 + 3 - 12 = 0$$

The original equation has the solution set $\{-\sqrt{\tfrac{1}{3}}, \sqrt{\tfrac{1}{3}}\}$.   ▬

The idea should now be clear. If an equation contains some expression and that same expression squared, make a substitution for the expression. You may get a quadratic equation.

Factorable Equations

We have already used factoring as a means of solving certain quadratic equations. This method can also be used to solve any equation that can be factored with zero on one side of the equation. The solutions are then found by setting each factor equal to zero.

EXAMPLE 7   Solve each equation.

(a) $x^{3/2} = 5x^{1/2}$ (b) $x^3 - x^2 - 4x + 4 = 0$

Solution   (a) For the equation $x^{3/2} = 5x^{1/2}$, we must first rearrange the equation to get zero on the right side:

$$x^{3/2} = 5x^{1/2}$$
$$x^{3/2} - 5x^{1/2} = 0$$

Since $x^{3/2} = x \cdot x^{1/2}$, we notice that $x^{1/2}$ is a factor of each term on the left:

$$x^{1/2}(x - 5) = 0$$
$$x^{1/2} = 0 \quad \text{or} \quad x - 5 = 0 \quad \text{Set each factor equal to zero.}$$
$$x = 0 \quad \text{or} \qquad x = 5 \quad \text{Solve.}$$

The solution set is $\{0, 5\}$.

(b) Do you recall "factoring by grouping" from Chapter 1? We group the terms of $x^3 - x^2 - 4x + 4 = 0$ as follows:

$$(x^3 - x^2) - (4x - 4) = 0$$

Factor out x^2 from the first grouping and 4 from the second:

$$x^2(x - 1) - 4(x - 1) = 0$$

This reveals the common factor $(x - 1)$, so we have

$$(x^2 - 4)(x - 1) = 0$$

$$(x - 2)(x + 2)(x - 1) = 0 \qquad \text{Factor again.}$$

$$x - 2 = 0 \quad \text{or} \quad x + 2 = 0 \quad \text{or} \quad x - 1 = 0 \quad \begin{array}{l}\text{Set each factor}\\ \text{equal to zero.}\end{array}$$

$$x = 2 \quad \text{or} \qquad x = -2 \quad \text{or} \qquad x = 1 \quad \text{Solve.}$$

The solution set is $\{-2, 1, 2\}$.

$\blacksquare$

EXERCISE 2.5

In Problems 1–26 solve each equation.

1. $x^3 = 27$ **2.** $x^3 = -8$ **3.** $x^2 = 4$ **4.** $x^2 = 9$

5. $x^2 = -4$ **6.** $x^2 = -9$ **7.** $x^{1/2} = 3$ **8.** $x^{1/2} = 5$

9. $x^{1/3} = -2$ **10.** $x^{1/5} = -1$ **11.** $y^{3/2} = 8$ **12.** $z^{2/3} = 1$

13. $x^{2/3} = -8$ **14.** $x^{3/4} = -1$ **15.** $x^{2/3} = \frac{1}{4}$ **16.** $x^{2/3} = \frac{1}{9}$

17. $t^{-1} = 2$ **18.** $s^{-2} = 25$ **19.** $x^{-2} = 4$ **20.** $x^{-3} = 8$

21. $x^{-1/5} = 2$ **22.** $x^{-1/3} = 3$ **23.** $x^{-3/2} = -\frac{1}{8}$ **24.** $x^{-3/4} = -1$

25. $x^{-2/3} = -\frac{1}{4}$ **26.** $x^{-5/2} = -9$

In Problems 27–44 solve each equation.

27. $\sqrt{2t - 1} = 1$ **28.** $\sqrt{3t + 4} = 2$

29. $\sqrt{2t - 1} = -1$ **30.** $\sqrt{3t + 4} = -2$

31. $\sqrt[3]{1 - 2x} - 3 = 0$ **32.** $\sqrt[3]{1 - 2x} - 1 = 0$

33. $\sqrt{15 - 2x} = x$ **34.** $\sqrt{12 - x} = x$

35. $x = 2\sqrt{x - 1}$ **36.** $x = 2\sqrt{-x - 1}$

37. $\sqrt{x^2 - x - 4} = x + 2$ **38.** $\sqrt{3 - x + x^2} = x - 2$

39. $\sqrt[3]{y^3 - 2y + 4} = y$ **40.** $\sqrt[3]{z^3 + z + 5} = z$

41. $\sqrt{2x + 3} - \sqrt{x + 1} = 1$ **42.** $\sqrt{3x + 7} + \sqrt{x + 2} = 1$

43. $\sqrt{3x + 1} - \sqrt{x - 1} = 2$ **44.** $\sqrt{3x - 5} - \sqrt{x + 7} = 2$

In Problems 45–64 solve each equation.

45. $x - 4\sqrt{x} = 0$ **46.** $x + 8\sqrt{x} = 0$

47. $x^{2/3} + x^{1/3} - 12 = 0$ **48.** $x^{2/3} + 7x^{1/3} + 12 = 0$

49. $2(s + 1)^2 - 5(s + 1) = 3$ **50.** $3(1 - y)^2 + 5(1 - y) + 2 = 0$

51. $x^4 - 5x^2 + 4 = 0$ **52.** $x^4 - 10x^2 + 4 = 0$

53. $t^{1/2} - 2t^{1/4} + 1 = 0$ **54.** $z^{1/2} + 2z^{1/4} + 1 = 0$

55. $4x^{1/2} - 9x^{1/4} + 4 = 0$ **56.** $x^{1/2} - 3x^{1/4} + 2 = 0$

57. $\sqrt[4]{5x^2 - 6} = x$ **58.** $\sqrt[4]{4 - 5x^2} = x$

59. $x^2 + 3x + \sqrt{x^2 + 3x} = 6$ **60.** $x^2 - 3x - \sqrt{x^2 - 3x} = 2$

61. $\dfrac{1}{(x + 1)^2} = \dfrac{1}{x + 1} + 2$ **62.** $\dfrac{1}{(x - 1)^2} + \dfrac{1}{x - 1} = 12$

63. $\left(\dfrac{v}{v+1}\right)^2 + \dfrac{2v}{v+1} = 8$ **64.** $\left(\dfrac{y}{y-1}\right)^2 = 6\left(\dfrac{y}{y-1}\right) + 7$

In Problems 65–76 solve each equation by factoring.

65. $x = 8\sqrt{x}$ **66.** $x = 6\sqrt{x}$

67. $x^{3/2} - 2x^{1/2} = 0$ **68.** $x^{3/4} - 4x^{1/4} = 0$

69. $x^3 + x^2 - 12x = 0$ **70.** $x^3 - 6x^2 - 7x = 0$

71. $x^3 + x^2 + x + 1 = 0$ **72.** $x^3 + x^2 - x - 1 = 0$

73. $x^3 - 3x^2 - 4x + 12 = 0$ **74.** $x^3 - 3x^2 - x + 3 = 0$

75. $t^6 - t^4 - t^2 + 1 = 0$ **76.** $y^6 - 4y^4 - y^2 + 4 = 0$

 *In Problems 77–82 solve each equation. Use a calculator to express solutions correct to two decimal places.*

77. $x - 4x^{1/2} + 2 = 0$ **78.** $x^{2/3} + 4x^{1/3} + 2 = 0$

79. $x^4 + \sqrt{3}x^2 - 3 = 0$ **80.** $x^4 + \sqrt{2}x^2 - 2 = 0$

81. $\pi(1 + t)^2 = \pi + 1 + t$ **82.** $\pi(1 + R)^2 = 2 + \pi(1 + R)$

83. The depth of a well can sometimes be found by dropping an object into the well and measuring the time elapsed until a sound is heard. If t_1 is the time (measured in seconds) it takes for the object to strike the bottom of the well, then t_1 will obey the equation $s = 16t_1^2$, where s is the distance (measured in feet). It follows that $t_1 = \sqrt{s}/4$. Suppose t_2 is the time it takes for the sound of impact to reach your ears. Because sound waves are known to travel at a speed of approximately 1100 feet per second, the time t_2 to travel the distance s will be $t_2 = s/1100$. Now $t_1 + t_2$ is the total time that elapses from the moment the object is dropped to the moment a sound is heard. Thus, we have the equation

$$\text{Total time elapsed} = \frac{\sqrt{s}}{4} + \frac{s}{1100}$$

Find the depth of a well if the total time elapsed from dropping a rock to hearing it hit bottom is 4 seconds. Express the answer correct to two decimal places.

▤ 2.6 LINEAR INEQUALITIES

INEQUALITY
SOLVE

SOLUTIONS

An **inequality** is a statement involving two expressions separated by one of the inequality symbols $<$, $\leq$, $>$, or $\geq$. To "**solve** an inequality" means to find all values of the variable for which the statement is true. These values are called **solutions** of the inequality.

For example, if x is a variable, then

$$x + 5 < 8, \quad 2x - 3 \geq 4, \quad x^2 - 1 \leq 3 \quad \text{and} \quad \frac{x+1}{x-2} > 0$$

are all inequalities in x.

EQUIVALENT INEQUALITIES Two inequalities having exactly the same solution set are called **equivalent**.

As with equations, one method for solving an inequality is to replace it by a series of equivalent inequalities, until an inequality with an obvious solution, such as $x < 3$, is obtained. We obtain equivalent inequalities by applying some of the same operations as those used to find equivalent equations. The addition property and the multiplication properties form the basis for the procedures listed below.

The following procedures leave the inequality symbol unchanged:

1. Simplify both sides of the inequality by combining like terms and eliminating parentheses:

$$\text{Replace} \quad (x + 2) + 6 > 2x + (x + 1)$$
$$\text{by} \qquad x + 8 > 3x + 1$$

2. Add or subtract the same expression on both sides of the inequality:

$$\text{Replace} \qquad 3x - 5 < 4$$
$$\text{by} \quad (3x - 5) + 5 < 4 + 5$$

3. Multiply or divide both sides of the inequality by the same *positive* expression:

$$\text{Replace} \quad 4x > 16 \quad \text{by} \quad \frac{4x}{4} > \frac{16}{4}$$

The following procedures reverse the sense of the inequality symbol:

1. Interchange the two sides of the inequality:

$$\text{Replace} \quad 3 < x \quad \text{by} \quad x > 3$$

2. Multiply or divide both sides of the inequality by the same *negative* expression:

$$\text{Replace} \quad -2x > 6 \quad \text{by} \quad \frac{-2x}{-2} < \frac{6}{-2}$$

LINEAR INEQUALITY A **linear inequality** is an inequality equivalent to one of the forms

$$ax + b < 0 \qquad ax + b > 0$$
$$ax + b \leq 0 \qquad ax + b \geq 0$$

where a and b are real numbers and $a \neq 0$. The remainder of this section deals with solving linear inequalities. In the next section we discuss the solution of other types of inequalities. As the examples that follow illustrate, we solve linear inequalities using the same steps as we would to solve a linear equation.

EXAMPLE 1 Solve each inequality.

(a) $3 - 2x < 5$ (b) $4x + 7 \geq 2x - 3$

Draw a graph to illustrate each solution.

Solution (a)

$$3 - 2x < 5$$

$3 - 2x - 3 < 5 - 3$	Subtract 3 from both sides.
$-2x < 2$	Simplify.
$\dfrac{-2x}{-2} > \dfrac{2}{-2}$	Divide both sides by -2. (The sense of the inequality symbol is reversed.)
$x > -1$	Simplify.

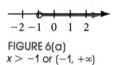

FIGURE 6(a)
$x > -1$ or $(-1, +\infty)$

The solution set consists of all numbers greater than -1. See Figure 6(a) for a graph of this solution. (Recall that an open circle denotes that -1 is not part of the graph.)

(b)

$$4x + 7 \geq 2x - 3$$

$4x + 7 - 7 \geq 2x - 3 - 7$	Subtract 7 from both sides.
$4x \geq 2x - 10$	Simplify.
$4x - 2x \geq 2x - 10 - 2x$	Subtract $2x$ from both sides.
$2x \geq -10$	Simplify.
$\dfrac{2x}{2} \geq \dfrac{-10}{2}$	Divide both sides by 2. (The sense of the inequality is unchanged.)
$x \geq -5$	Simplify.

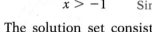

FIGURE 6(b)
$x \geq -5$ or $[-5, +\infty)$

The solution set consists of all numbers greater than or equal to -5. See Figure 6(b) for the graph of this solution. (Recall that a filled circle denotes that -5 is part of the graph.) ▤

Alternatively, we can use interval notation to write the solution set of an inequality. In this notation, the solution set of the inequality in Example 1(a) consists of all numbers in the interval $(-1, +\infty)$; the solution set of Example 1(b) consists of all numbers in the interval $[-5, +\infty)$.

EXAMPLE 2 Solve the inequality $-5 < 3x - 2 < 1$, and graph the solution set.

Solution Recall that the inequality

$$-5 < 3x - 2 < 1$$

is equivalent to the two inequalities

$$-5 < 3x - 2 \quad \text{and} \quad 3x - 2 < 1$$

We will solve each of these inequalities separately. For the first inequality,

$$\begin{aligned} -5 &< 3x - 2 && \\ -5 + 2 &< 3x - 2 + 2 && \text{Add 2 to both sides.} \\ -3 &< 3x && \text{Simplify.} \\ \frac{-3}{3} &< \frac{3x}{3} && \text{Divide both sides by 3.} \\ -1 &< x && \text{Simplify.} \end{aligned}$$

The second inequality is solved as follows:

$$\begin{aligned} 3x - 2 &< 1 && \\ 3x - 2 + 2 &< 1 + 2 && \text{Add 2 to both sides.} \\ 3x &< 3 && \text{Simplify.} \\ \frac{3x}{3} &< \frac{3}{3} && \text{Divide both sides by 3.} \\ x &< 1 && \text{Simplify.} \end{aligned}$$

The solution set of the original pair of inequalities consists of all x for which

$$-1 < x \quad \text{and} \quad x < 1$$

which may be written more compactly as the chain

$$-1 < x < 1$$

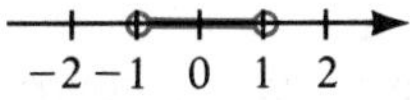

FIGURE 7
$-1 < x < 1$ or $(-1, 1)$

A graph of this solution set is given in Figure 7. ≡

We observe in the process above that the two inequalities we solved required exactly the same steps. A shortcut to solving the original inequality is to deal with the two inequalities at the same time as follows:

$$\begin{aligned} -5 < {}& 3x - 2 < 1 && \\ -5 + 2 < {}& 3x - 2 + 2 < 1 + 2 && \text{Add 2 to each part.} \\ -3 < {}& 3x < 3 && \text{Simplify.} \\ \frac{-3}{3} < {}& \frac{3x}{3} < \frac{3}{3} && \text{Divide each part by 3.} \\ -1 < {}& x < 1 && \text{Simplify.} \end{aligned}$$

We use this shortcut in the next example.

EXAMPLE 3   Solve the inequality

$$-1 \le \frac{3 - 5x}{2} \le 9$$

Graph the solution set.

Solution

$$-1 \le \frac{3 - 5x}{2} \le 9$$

$$2(-1) \le 2\left(\frac{3 - 5x}{2}\right) \le 2(9) \qquad \text{Multiply each part by 2.}$$

$$-2 \le \quad 3 - 5x \quad \le 18 \qquad \text{Simplify.}$$

$$-2 - 3 \le 3 - 5x - 3 \le 18 - 3 \qquad \text{Subtract 3 from each part.}$$

$$-5 \le \quad -5x \quad \le 15 \qquad \text{Simplify.}$$

$$\frac{-5}{-5} \ge \quad \frac{-5x}{-5} \quad \ge \frac{15}{-5} \qquad \begin{array}{l}\text{Divide each part by } -5 \text{ (change the}\\ \text{sense of each inequality symbol).}\end{array}$$

$$1 \ge \quad x \quad \ge -3 \qquad \text{Simplify.}$$

$$-3 \le \quad x \quad \le 1 \qquad \begin{array}{l}\text{Reverse the order so the numbers get}\\ \text{larger as you read from left to right.}\end{array}$$

The solution set consists of all numbers x for which $-3 \le x \le 1$. Figure 8 illustrates the graph.

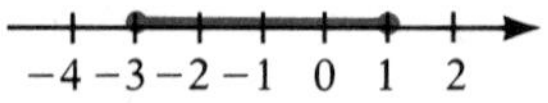

FIGURE 8
$-3 \le x \le 1$ or $[-3, 1]$

Let's look at an applied problem.

EXAMPLE 4   In electricity, Ohm's law states that $E = IR$, where E is the voltage (in volts), I is the current (in amperes), and R is the resistance (in ohms). An air-conditioning unit is rated at a resistance of 10 ohms. If the voltage varies from 110 to 120 volts, inclusive, what corresponding range of current will the air conditioner draw?

Solution   The voltage lies between 110 and 120, inclusive, so

$$110 \le \quad E \quad \le 120$$

$$110 \le \quad IR \quad \le 120 \qquad \text{Ohm's law, } E = IR$$

$$110 \le I(10) \le 120 \qquad R = 10$$

$$\frac{110}{10} \le \frac{I(10)}{10} \le \frac{120}{10} \qquad \text{Divide each part by 10.}$$

$$11 \le \quad I \quad \le 12 \qquad \text{Simplify.}$$

The air conditioner will draw between 11 and 12 amperes of current, inclusive.

≡ **HISTORICAL COMMENT** Inequalities are a relatively new component of the algebra curriculum. They have been introduced in the last 20 years for two important reasons.

First, if approximations are made in a problem, inequalities allow calculation of how serious the error is likely to be. This use is highly important for practical applications of mathematics and also is critical for the understanding of calculus [in its modern version due principally to Cauchy (1789–1857) and Weierstrass (1815–1897)].

Second, linear inequalities are the basis for solving a kind of problem that involves finding a maximum or minimum of a quantity depending on variables that are subjected to certain constraints. For example, one might wish to ship several products by truck from several different factories, each factory having a limited supply of each product, and to get enough of the products to a central point within 3 days, using the minimum amount of gas possible. Such problems, called **linear programming problems**, are discussed later. ≡

EXERCISE 2.6 *In Problems 1–4 an inequality is given. Write the equivalent inequality obtained by:*
(a) *Adding −3 to each side of the given inequality.*
(b) *Subtracting 5 from each side of the given inequality.*
(c) *Multiplying each side of the given inequality by 3.*
(d) *Multiplying each side of the given inequality by −2.*

1. $3 < 5$ **2.** $2 > 1$

3. $2x + 1 < 2$ **4.** $1 - 2x > 5$

In Problems 5–34 solve each inequality. Graph the solution set.

5. $x + 2 < 6$ **6.** $x - 5 < 2$

7. $1 - 2x \leq 3$ **8.** $2 - 3x \leq 5$

9. $3x - 7 > 2$ **10.** $2x + 5 > 1$

11. $3x - 1 \geq 3 + x$ **12.** $2x - 2 \geq 3 + x$

13. $-2(x + 3) < 6$ **14.** $-3(1 - x) < 9$

15. $4 - 3(1 - x) \leq 3$ **16.** $8 - 4(2 - x) \leq -2x$

17. $\frac{1}{2}(x - 4) > x + 8$ **18.** $3x + 4 > \frac{1}{3}(x - 2)$

19. $\frac{x}{2} \geq 1 - \frac{x}{4}$ **20.** $\frac{x}{3} \geq 2 + \frac{x}{6}$

21. $0 \leq 2x - 6 \leq 4$ **22.** $4 \leq 2x + 2 \leq 10$

23. $-6 \leq 1 - 3x \leq 2$ **24.** $-3 \leq 2 - 2x \leq 4$

25. $-3 < \frac{2x - 1}{4} < 0$ **26.** $0 < \frac{3x + 2}{2} < 4$

27. $1 < 1 - \frac{1}{2}x < 4$ **28.** $0 < 1 - \frac{1}{3}x < 1$

29. $(x + 2)(x - 3) > (x - 1)(x + 1)$ **30.** $(x - 1)(x + 1) > (x - 3)(x + 4)$

31. $x(4x + 3) \le (2x + 1)^2$ **32.** $x(9x - 5) \le (3x - 1)^2$

33. $\dfrac{1}{2} \le \dfrac{x + 1}{3} < \dfrac{3}{4}$ **34.** $\dfrac{1}{3} < \dfrac{x + 1}{2} \le \dfrac{2}{3}$

35. For a certain ideal gas, the volume V (in cubic centimeters) equals 20 times the temperature T (in degrees Celsius). If the temperature varies from 80° C to 120° C, inclusive, what is the corresponding range of the volume of the gas?

36. An investor has $1000 to invest for a period of 1 year. What range of per annum simple interest rates are needed to obtain interest that varies from $90 to $110, inclusive?

37. A realtor agrees to sell a large apartment complex according to the following commission schedule: $45,000 plus 25% of the selling price in excess of $900,000. Assuming the complex will sell at some price between $900,000 and $1,100,000, inclusive, over what range does the realtor's commission vary? How does the realtor's commission vary as a percent of selling price?

38. A used car salesperson is paid a commission of $25 plus 40% of the selling price in excess of owner's cost plus $50. The owner claims that used cars typically sell for at least owner's cost plus $70 and at most owner's cost plus $300. For each sale made, over what range can the salesperson expect the commission to vary?

39. The percentage method of withholding for federal income tax (1985)* states that a married person claiming zero dependents whose weekly wages are over $454, but not over $556, shall have $66.66 plus 25% of the excess over $454 withheld. Over what range does the amount withheld vary if the weekly wages vary from $500 to $550, inclusive?

40. Rework Problem 39 if the weekly wages vary from $475 to $525, inclusive.

C **41.** Commonwealth Edison Company's summer charge for electricity is 9.780¢ per kilowatt hour.† In addition, each monthly bill contains a customer charge of $1.94. If last summer's bills ranged from a low of $68.00 to a high of $246.52, over what range did usage vary (in kilowatt hours)?

C **42.** The Village of Oak Lawn charges homeowners $16.80 per quarter year plus $1.26 per 1000 gallons for water usage in excess of 12,000 gallons.‡ In 1985, one homeowner's quarterly bill ranged from a high of $47.04 to a low of $26.88. Over what range did water usage vary?

43. The markup over dealer's cost of a new car ranges from 12% to 18%. If the sticker price is $8800, over what range will the dealer's cost vary?

44. A standard intelligence test has an average score of 100. According to statistical theory, of the people who take a test, the 2.5% with the highest

*Source: *Employer's Tax Guide*, Department of the Treasury, Internal Revenue Service, 1985.

†Source: Commonwealth Edison Co., Chicago, Illinois, 1985.

‡Source: Village of Oak Lawn, Illinois, 1985.

STANDARD DEVIATION

scores will have scores of more than 1.95σ above the average, where σ (sigma), a number called the **standard deviation**, depends on the nature of the test. If $\sigma = 12$ for this test and there is in principle no upper limit to the score possible on the test, write the interval of possible test scores of the people in the top 2.5%.

45. In your Economics 101 class, you have scores of 70, 82, 85, and 89 on the first four of five tests. To get a grade of B, the average of the five test scores must be greater than or equal to 80 and less than 90. Solve an inequality to find the range of the score you need on the last test to get a B.

46. Repeat Problem 45 if the fifth test counts double.

47. A car that averages 25 miles per gallon has a tank that holds 20 gallons of gasoline. After a trip that covered at least 300 miles, the car ran out of gasoline. What is the range of the amount of gasoline (in gallons) that was in the tank at the start of the trip?

48. Repeat Problem 47 if the same car runs out of gasoline after a trip of no more than 250 miles.

▤ 2.7 OTHER INEQUALITIES

In this section, we solve inequalities that contain polynomials of degree two and higher as well as some that contain rational expressions. Before we can solve such inequalities, we rearrange them so that the polynomial or rational expression is on the left side and zero is on the right side. An example will show you why.

EXAMPLE 1   Solve the inequality $x^2 + x - 12 > 0$.

Solution   We factor the left side, obtaining

$$x^2 + x - 12 > 0$$
$$(x - 3)(x + 4) > 0$$

The product of two real numbers is positive either when both factors are positive or when both factors are negative.

BOTH POSITIVE	OR	BOTH NEGATIVE
$x - 3 > 0$ and $x + 4 > 0$		$x - 3 < 0$ and $x + 4 < 0$
$x > 3$ and $x > -4$		$x < 3$ and $x < -4$
The numbers x that are greater than 3 and at the same time greater than -4 are simply		The numbers x that are less than 3 and at the same time less than -4 are simply
$x > 3$	or	$x < -4$

The solution set consists of all x for which either

$$x > 3 \quad \text{or} \quad x < -4$$

See Figure 9.

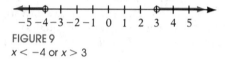

FIGURE 9
$x < -4$ or $x > 3$

We also can obtain the solution to the inequality of Example 1 by another method. The left-hand side of the inequality is factored so that it becomes $(x - 3)(x + 4) > 0$, as before. We then construct a graph that uses the solutions to the equation

$$x^2 + x - 12 = (x - 3)(x + 4) = 0$$

BOUNDARY POINTS   namely $x = 3$ and $x = -4$. These numbers, called **boundary points**, separate the real number line into three parts: $x < -4$, $-4 < x < 3$, and $x > 3$. See Figure 10(a).

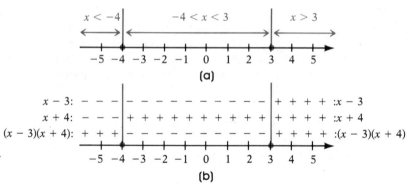

FIGURE 10

In the part of the line where $x < -4$, we deduce that both the quantities $(x - 3)$ and $(x + 4)$ are always negative, so their product must always be positive. See Figure 10(b). Therefore, $x < -4$ is a solution of the inequality. In the part of the line where $-4 < x < 3$, we deduce that $(x - 3)$ is always negative and $(x + 4)$ is always positive, so their product is always negative. We conclude that the numbers between -4 and 3 are not solutions of the inequality. In the part of the line where $x > 3$, we deduce that both the quantities $(x - 3)$ and $(x + 4)$ are always positive, so their product is always positive. Hence, numbers greater than 3 are solutions of the inequality. Table 2 summarizes these results.

TABLE 2

	Sign of $x - 3$	Sign of $x + 4$	Sign of the product $(x - 3)(x + 4)$	Conclusion
$x < -4$	$-$	$-$	$+$	$x < -4$ is a solution
$-4 < x < 3$	$-$	$+$	$-$	$-4 < x < 3$ is not a solution
$x > 3$	$+$	$+$	$+$	$x > 3$ is a solution

TEST NUMBER The signs listed in Table 2 may be more easily obtained by using what we shall call a **test number**. For example, to determine the sign of $x - 3$ when $x < -4$, select any number that is less than -4, say -5, and substitute it into the expression to be tested, in this case $x - 3$, to get $-5 - 3 = -8$. The negative result tells us that $x - 3$ is negative for all x for which $x < -4$.

As another example, in order to determine the sign of $x + 4$ when $-4 < x < 3$, we might select 0 as a test number (any number between -4 and 3 will do). The expression $x + 4$ evaluates to 4 when $x = 0$. We conclude that $x + 4$ is positive for all x for which $-4 < x < 3$.

EXAMPLE 2 Solve the inequality $x^2 \leq 4x + 12$.

Solution First we rearrange the inequality so that zero is on the right side:

$$x^2 \leq 4x + 12$$
$$x^2 - 4x - 12 \leq 0$$
$$(x - 6)(x + 2) \leq 0 \quad \text{Factor.}$$

Next we set the left side equal to zero in order to locate the boundary points:

$$(x - 6)(x + 2) = 0$$

The boundary points (solutions of the equation) are -2 and 6, and they separate the real number line into three parts:

$$x < -2, \quad -2 < x < 6, \quad \text{and} \quad x > 6$$

See Figure 11 on page 146 and Table 3.

TABLE 3

	Test number	$x - 6$	$x + 2$	$(x - 6)(x + 2)$
$x < -2$	-3	$-$	$-$	$+$
$-2 < x < 6$	0	$-$	$+$	$-$
$x > 6$	7	$+$	$+$	$+$

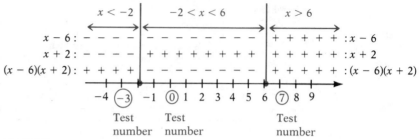

FIGURE 11

Since we want to know where the product $(x - 6)(x + 2)$ is negative, we conclude that the solutions are numbers x for which $-2 < x < 6$. However, because the original inequality is nonstrict, numbers x that satisfy the equation

$$x^2 = 4x + 12$$

are also solutions of the inequality $x^2 \leq 4x + 12$. The solutions of

$$x^2 = 4x + 12 \quad \text{or} \quad x^2 - 4x - 12 = (x - 6)(x + 2) = 0$$

are $x = 6$ or $x = -2$. Thus, the solution set of the given inequality consists of all numbers x for which

$$-2 \leq x \leq 6 \qquad\qquad \blacksquare$$

The method used in Example 2 requires that we find boundary points (solutions of an equation). When a polynomial equation has no real solutions, the polynomial is either always positive or always negative. For example, the equation

$$x^2 + 5x + 8 = 0$$

has no real solutions. (Do you see why? Its discriminant, $b^2 - 4ac = 25 - 32$, is negative.) The value of $x^2 + 5x + 8$ is therefore always positive or always negative. To see which is true, we test its value at some number (zero is the easiest). Because $0^2 + 5(0) + 8 = 8$ is positive, we conclude that $x^2 + 5x + 8 > 0$ for all x.

The boundary-point method works well for inequalities containing polynomials of any degree, provided they can be factored.

EXAMPLE 3   Solve the inequality $x^3 - 5x^2 + 6x > 0$.

Solution   Because zero is on the right-hand side, we proceed to factor. Because x is an obvious factor, we begin there:

$$x^3 - 5x^2 + 6x > 0$$
$$x(x^2 - 5x + 6) > 0$$
$$x(x - 2)(x - 3) > 0$$

The boundary points [solutions of the equation $x^3 - 5x^2 + 6x = x(x - 2)(x - 3)$] are 0, 2, and 3, and they separate the real number line into four parts, namely,

$$x < 0, \quad 0 < x < 2, \quad 2 < x < 3, \quad \text{and} \quad x > 3$$

See Figure 12 and Table 4. We want to know where the product $x(x - 2)(x - 3)$ is positive; thus the solution set consists of numbers x for which

$$0 < x < 2 \quad \text{or} \quad x > 3$$

TABLE 4

	Test number	x	$x - 2$	$x - 3$	$x(x - 2)(x - 3)$
$x < 0$	-1	$-$	$-$	$-$	$-$
$0 < x < 2$	1	$+$	$-$	$-$	$+$
$2 < x < 3$	2.5	$+$	$+$	$-$	$-$
$x > 3$	4	$+$	$+$	$+$	$+$

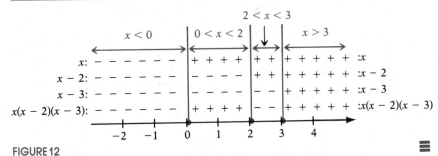

FIGURE 12

Let's use the boundary-point method to solve a rational inequality.

EXAMPLE 4   Solve the inequality

$$\frac{4x + 5}{x + 2} \geq 3$$

Solution   We rearrange terms so that zero is on the right side:

$$\frac{4x + 5}{x + 2} \geq 3$$

$$\frac{4x + 5}{x + 2} - 3 \geq 0$$

$$\frac{4x + 5 - 3(x + 2)}{x + 2} \geq 0 \quad \text{Rewrite using } x + 2 \text{ as the denominator.}$$

$$\frac{x - 1}{x + 2} \geq 0 \quad \text{Simplify.}$$

The sign of a rational expression depends on the sign of its numerator and the sign of its denominator. Thus, for a rational expression, we use as boundary points the numbers obtained by setting the numerator and the denominator equal to zero. For this example, the boundary points are 1 and -2. See Figure 13 and Table 5. The conclusions found in Figure 13 and Table 5 reveal the numbers x for which $(x - 1)/(x + 2)$ is positive. However, we want to know where the expression $(x - 1)/(x + 2)$ is positive or zero. Since $(x - 1)/(x + 2) = 0$ only if $x = 1$, we conclude that the solution set consists of all numbers x for which

$$x < -2 \quad \text{or} \quad x \geq 1$$

TABLE 5

	Test number	$x - 1$	$x + 2$	$\dfrac{x - 1}{x + 2}$
$x < -2$	-3	$-$	$-$	$+$
$-2 < x < 1$	0	$-$	$+$	$-$
$x > 1$	2	$+$	$+$	$+$

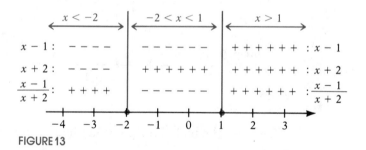

FIGURE 13

You may wonder why we didn't first multiply both sides of the inequality in Example 4 by $x + 2$ to clear the denominator. The reason is that we don't know whether $x + 2$ is positive or negative and, as a result, we don't know whether or not to reverse the sense of the inequality symbol after multiplying by $x + 2$. However, there is nothing to prevent us from multiplying both sides by $(x + 2)^2$, which is always positive, since $x \neq -2$. (Do you see why?) Then

$$\frac{4x + 5}{x + 2} \geq 3$$

$$\frac{4x + 5}{x + 2}(x + 2)^2 \geq 3(x + 2)^2$$

$$(4x + 5)(x + 2) \geq 3(x^2 + 4x + 4)$$
$$4x^2 + 13x + 10 \geq 3x^2 + 12x + 12$$
$$x^2 + x - 2 \geq 0$$
$$(x + 2)(x - 1) \geq 0$$

This last expression leads to the same solution set.

EXAMPLE 5   An object is dropped from the roof of a building 150 feet above the ground. After t seconds the object will be $150 - 16t^2$ feet above the ground. During what interval of time will the object be between 54 and 118 feet above the ground?

Solution   The inequality to be solved has the form

$$54 \leq 150 - 16t^2 \leq 118$$
$$-96 \leq \quad -16t^2 \quad \leq -32 \quad \text{Subtract 150 from each part.}$$
$$6 \geq \quad t^2 \quad \geq 2 \quad \text{Divide each part by } -16.$$
$$2 \leq \quad t^2 \quad \leq 6$$

For the situation described, the time t is assumed to be a nonnegative real number. With this restriction, the compound inequality

$$2 \leq t^2 \leq 6 \qquad t \geq 0$$

is equivalent to

$$\sqrt{2} \leq t \leq \sqrt{6}$$

which may be approximated by

$$1.41 \leq t \leq 2.45$$

For times ranging from approximately 1.41 seconds to 2.45 seconds, the object will be between 54 and 118 feet above the ground.   ▬

EXERCISE 2.7   *In Problems 1–42 solve each inequality.*

1. $(x - 3)(x + 1) < 0$

2. $(x - 1)(x + 2) < 0$

3. $x^2 - 4x > 0$

4. $x^2 + 8x > 0$

5. $x^2 - 9 < 0$

6. $x^2 - 1 < 0$

7. $x^2 + x > 12$

8. $x^2 + 7x < -12$

9. $2x^2 < 5x + 3$

10. $6x^2 < 6 + 5x$

11. $x(x - 7) > -12$

12. $x(x + 1) > 12$

13. $4x^2 + 9 < 6x$

14. $25x^2 + 16 < 40x$

15. $6(x^2 - 1) > 5x$ **16.** $2(2x^2 - 3x) > -9$

17. $(x - 1)(x^2 + x + 1) > 0$ **18.** $(x + 2)(x^2 - x + 1) > 0$

19. $(x - 1)(x - 2)(x - 3) < 0$ **20.** $(x + 1)(x + 2)(x + 3) < 0$

21. $x^3 - 2x^2 - 8x > 0$ **22.** $x^3 + 2x^2 - 8x > 0$

23. $x^3 > x$ **24.** $x^3 < 4x$

25. $x^3 > x^2$ **26.** $x^3 < 3x^2$

27. $x^4 > 1$ **28.** $x^3 > 1$

29. $\dfrac{x + 1}{x - 1} > 0$ **30.** $\dfrac{x - 3}{x + 1} > 0$

31. $\dfrac{(x - 1)(x + 1)}{x} < 0$ **32.** $\dfrac{(x - 3)(x + 2)}{x - 1} < 0$

33. $\dfrac{x - 2}{x^2 - 1} \geq 0$ **34.** $\dfrac{x + 5}{x^2 - 4} \geq 0$

35. $6x - 5 < \dfrac{6}{x}$ **36.** $x + \dfrac{12}{x} < 7$

37. $\dfrac{x + 4}{x - 2} \leq 1$ **38.** $\dfrac{x + 2}{x - 4} \geq 1$

39. $\dfrac{2x + 5}{x + 1} > \dfrac{x + 1}{x - 1}$ **40.** $\dfrac{1}{x + 2} > \dfrac{3}{x + 1}$

41. $\dfrac{x^2(3 + x)(x + 4)}{(x + 5)(x - 1)} > 0$ **42.** $\dfrac{x(x^2 + 1)(x - 2)}{(x - 1)(x + 1)} > 0$

43. For what positive numbers will the cube of a number exceed 4 times its square?

44. For what positive numbers will the square of a number exceed twice the number?

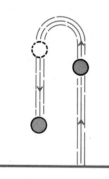

45. A ball is thrown vertically upward with an initial velocity of 80 feet per second. The distance s (in feet) of the ball from the ground after t seconds is $s = 80t - 16t^2$. For what time interval is the ball more than 96 feet above the ground?

46. Rework Problem 45 to find when the ball is less than 64 feet above the ground.

47. The monthly revenue achieved by selling x wristwatches is figured to be $x(40 - 0.2x)$ dollars. The wholesale cost of each watch is $28. How many watches must be sold each month to achieve a profit (revenue − cost) of at least $100?

48. The monthly revenue achieved by selling x boxes of candy is figured to be $x(5 - 0.05x)$ dollars. The wholesale cost of each box of candy is $1.50. How many boxes must be sold each month to achieve a profit of at least $60?

49. If $a > 0$, show that the solution set of the inequality

$$x^2 < a$$

are numbers x for which

$$-\sqrt{a} < x < \sqrt{a}$$

50. If $a > 0$, show that the solution set of the inequality

$$x^2 > a$$

are numbers x for which

$$x > \sqrt{a} \quad \text{or} \quad x < -\sqrt{a}$$

In Problems 51–58 use the results found in Problems 49–50 to solve each inequality.

51. $x^2 < 1$ **52.** $x^2 < 4$ **53.** $x^2 \geq 9$ **54.** $x^2 \geq 1$

55. $x^2 \leq 16$ **56.** $x^2 \leq 9$ **57.** $x^2 > 4$ **58.** $x^2 > 16$

▤ 2.8 EQUATIONS AND INEQUALITIES INVOLVING ABSOLUTE VALUE

The absolute value of a real number x has been defined as

$$|x| = x \quad \text{if } x \geq 0 \quad \text{and} \quad |x| = -x \quad \text{if } x < 0$$

Also, recall that geometrically the absolute value of x equals the distance from the origin to the point whose coordinate is x.

In this section we shall discuss equations and inequalities involving absolute value. In solving such equations and inequalities, the properties listed below will prove useful.

THEOREM 1. The absolute value of any real number is always nonnegative; that is,

$$|x| \geq 0 \tag{1}$$

2. The absolute value of any real number equals the principal square root of the number squared; that is,

$$|x| = \sqrt{x^2} \tag{2}$$

3. The absolute value of the product of two real numbers equals the product of their absolute values; that is,

$$|xy| = |x| \cdot |y| \tag{3}$$

TRIANGLE INEQUALITY 4. The absolute value of the sum of two real numbers never exceeds the sum of their absolute values; that is,

$$|x + y| \leq |x| + |y| \tag{4}$$

Proof   Property (1) follows directly from the definition of absolute value.

Property (2) is a consequence of the fact that, for $N \geq 0$, $\sqrt{N}$ equals the nonnegative number whose square is N. Thus, if $x \geq 0$, then $x^2 = x \cdot x$ and $\sqrt{x^2} = x$; if $x < 0$, then $-x > 0$ and $x^2 = (-x)(-x)$ so that $\sqrt{x^2} = -x$. Therefore, $\sqrt{x^2} = |x|$.

Property (3) is proved by using Property (2):

$$|xy| = \sqrt{(xy)^2} = \sqrt{x^2 y^2} = \sqrt{x^2} \cdot \sqrt{y^2} = |x| \cdot |y|$$

Property (4) is called the **triangle inequality** because it expresses algebraically the fact that the length of any side of a triangle does not exceed the sum of the lengths of the other two sides. You are asked to prove this result in Problem 45.   ■

We shall now take up the problem of solving equations and inequalities that contain absolute values.

Because there are two points whose distance from the origin is 5 units, namely -5 and 5, the equation $|x| = 5$ will have the solution set $\{-5, 5\}$. We are thus led to the following result:

THEOREM   If the absolute value of an expression equals some positive number a, then the expression itself equals either a or $-a$. Thus,

$$|u| = a \quad \text{is equivalent to} \quad u = a \quad \text{or} \quad u = -a \qquad (5)$$

■

EXAMPLE 1   Solve the equation $|x + 4| = 13$.

Solution   This follows the form of Equation (5), where $u = x + 4$. Thus, there are two possibilities:

$$x + 4 = 13 \quad \text{or} \quad x + 4 = -13$$
$$x = 9 \quad \text{or} \quad x = -17$$

Thus, the solution set is $\{-17, 9\}$.   ■

Let's look at an inequality involving absolute value.

EXAMPLE 2   Solve the inequality $|x| < 4$.

Solution   We are looking for all points whose coordinate x is a distance less than 4 units from the origin. See Figure 14 for an illustration. Because any x between -4 and 4 satisfies the condition $|x| < 4$, the solution set consists of all numbers x for which $-4 < x < 4$.

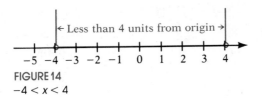

FIGURE 14
$-4 < x < 4$

We are led to the following results:

THEOREM   If a is any positive number,

$$|u| < a \quad \text{is equivalent to} \quad -a < u < a \qquad (6)$$
$$|u| \le a \quad \text{is equivalent to} \quad -a \le u \le a \qquad (7)$$

EXAMPLE 3   Solve the following inequalities, and graph their solution sets:

(a) $|2x + 4| \le 3$ (b) $|1 - 4x| < 5$

Solution   (a)

$$|2x + 4| \le 3$$ This follows the form of Equation (7); the expression $u = 2x + 4$ is inside the absolute value bars.

$$-3 \le 2x + 4 \le 3$$ Apply Equation (7).

$$-3 - 4 \le 2x + 4 - 4 \le 3 - 4$$ Subtract 4 from each part.

$$-7 \le 2x \le -1$$ Simplify.

$$\frac{-7}{2} \le \frac{2x}{2} \le \frac{-1}{2}$$ Divide each part by 2.

$$-\frac{7}{2} \le x \le -\frac{1}{2}$$ Simplify.

The solution set consists of all numbers x for which $-\frac{7}{2} \le x \le -\frac{1}{2}$. See Figure 15.

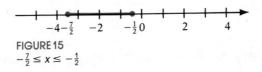

FIGURE 15
$-\frac{7}{2} \le x \le -\frac{1}{2}$

(b) $|1 - 4x| < 5$ This follows the form of Equation (6); the expression $u = 1 - 4x$ is inside the absolute value bars.

$-5 < \quad 1 - 4x \quad < 5$ Apply Equation (6).

$-5 - 1 < 1 - 4x - 1 < 5 - 1$ Subtract 1 from each part.

$-6 < \quad -4x \quad < 4$ Simplify.

$\dfrac{-6}{-4} > \quad \dfrac{-4x}{-4} \quad > \dfrac{4}{-4}$ Divide each part by -4.

$\dfrac{3}{2} > \quad x \quad > -1$ Simplify.

$-1 < \quad x \quad < \dfrac{3}{2}$ Rearrange the ordering.

The solution set consists of all numbers x for which $-1 < x < \frac{3}{2}$. See Figure 16.

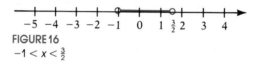

FIGURE 16
$-1 < x < \frac{3}{2}$

≡

EXAMPLE 4   Solve the inequality $|x| > 3$.

Solution   We are looking for all points whose coordinate x is a distance greater than 3 units from the origin. Figure 17 illustrates the situation. We conclude that any x less than -3 or greater than 3 satisfies the condition $|x| > 3$. Consequently, the solution set consists of all numbers x for which $x < -3$ or $x > 3$.

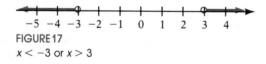

FIGURE 17
$x < -3$ or $x > 3$

≡

We may state the following results:

THEOREM   If a is any positive number,

$|u| > a$ is equivalent to $u < -a$ or $u > a$	(8)
$|u| \geq a$ is equivalent to $u \leq -a$ or $u \geq a$	(9)

≡

EXAMPLE 5   Solve the inequality $|2x - 5| > 3$, and graph the solution set.

Solution $\quad\quad |2x - 5| > 3$ This follows the form of Equation (8); the expression $u = 2x - 5$ is inside the absolute value bars.

$$2x - 5 < -3 \quad\quad \text{or} \quad\quad 2x - 5 > 3 \quad\quad \text{Apply Equation (8).}$$
$$2x - 5 + 5 < -3 + 5 \quad \text{or} \quad 2x - 5 + 5 > 3 + 5 \quad \text{Add 5 to each side.}$$
$$2x < 2 \quad\quad \text{or} \quad\quad 2x > 8 \quad\quad \text{Simplify.}$$
$$\frac{2x}{2} < \frac{2}{2} \quad\quad \text{or} \quad\quad \frac{2x}{2} > \frac{8}{2} \quad\quad \begin{array}{l}\text{Divide each side}\\ \text{by 2.}\end{array}$$
$$x < 1 \quad\quad \text{or} \quad\quad x > 4 \quad\quad \text{Simplify.}$$

The solution set consists of all numbers x for which $x < 1$ or $x > 4$. See Figure 18.

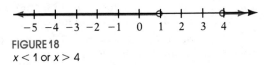

FIGURE 18
$x < 1$ or $x > 4$

Warning: A common error to be avoided is to attempt to write the solution $x < 1$ or $x > 4$ in a chain such as $1 > x > 4$, which is incorrect, since there are no numbers x for which $x < 1$ *and* $x > 4$. Another common error is to "mix" the symbols and write $1 < x > 4$, which, of course, makes no sense.

EXERCISE 2.8   *In Problems 1–22 solve each equation.*

1. $|2x| = 6$
2. $|3x| = 12$
3. $|2x + 3| = 5$
4. $|3x - 1| = 2$
5. $|1 - 4t| = 5$
6. $|1 - 2z| = 3$
7. $|-2x| = 8$
8. $|-x| = 1$
9. $|-2|x = 4$
10. $|3|x = 9$
11. $\frac{2}{3}|x| = 8$
12. $\frac{3}{4}|x| = 9$
13. $\left|\frac{x}{3} + \frac{2}{5}\right| = 2$
14. $\left|\frac{x}{2} - \frac{1}{3}\right| = 1$
15. $|u - 2| = -\frac{1}{2}$
16. $|2 - v| = -1$
17. $|x^2 - 4| = 0$
18. $|x^2 - 9| = 0$
19. $|x^2 - 2x| = 3$
20. $|x^2 + x| = 12$
21. $|x^2 + x - 1| = 1$
22. $|x^2 + 3x - 2| = 2$

In Problems 23–42 solve each inequality.

23. $|2x| < 6$
24. $|3x| < 12$
25. $|3x| > 12$
26. $|2x| > 6$
27. $|x - 2| < 1$
28. $|x + 4| < 2$
29. $|3t - 2| \le 4$
30. $|2u + 5| \le 7$
31. $|x - 1| \ge 2$
32. $|x + 3| \ge 2$
33. $|1 - 4x| < 5$
34. $|1 - 2x| < 3$
35. $|1 - 2x| > 3$
36. $|2 - 3x| > 1$
37. $|x + 2| > 0$

38. $|2 - x| > 0$ **39.** $|x + 2| > -3$ **40.** $|2 - x| > -2$

41. $|2x - 1| < 0.02$ **42.** $|3x - 2| < 0.02$

43. If $y \neq 0$, prove that $|x/y| = |x|/|y|$.

44. Show that $x \leq |x|$.

45. Prove the triangle inequality $|x + y| \leq |x| + |y|$.
[*Hint:* Expand $|x + y|^2 = (x + y)^2$, and use the result of Problem 44.]

46. Prove that $|x - y| \geq |x| - |y|$. [*Hint:* Apply the triangle inequality to $|x| = |(x - y) + y|$.]

47. Express the fact that x differs from 2 by less than $\frac{1}{2}$ as an inequality involving an absolute value. Solve for x.

48. Express the fact that x differs from -1 by less than 1 as an inequality involving an absolute value. Solve for x.

49. Express the fact that x differs from -3 by more than 2 as an inequality involving an absolute value. Solve for x.

50. Express the fact that x differs from 2 by more than 3 as an inequality involving an absolute value. Solve for x.

51. "Normal" human body temperature is 98.6° F. If a temperature x that differs from normal by at least 1.5° is considered unhealthy, write the condition for an unhealthy temperature x as an inequality involving an absolute value and solve for x.

52. In the United States, normal household voltage is 115 volts. However, it is not uncommon for actual voltage to differ from normal voltage by at most 5 volts. Express this situation as an inequality involving an absolute value. Use x as the actual voltage and solve for x.

C H A P T E R   R E V I E W

VOCABULARY

equation	mixture problem	principal square root
solve	uniform motion	of $-N$
solution	constant rate jobs	radical equation
root	quadratic equation	disguised quadratic
satisfies	standard form	equation
solution set	second-degree	inequality
identity	equation	solutions
equivalent equations	repeated solution	equivalent
extraneous solution	root of multiplicity	inequalities
linear equation	two	linear inequality
first-degree equation	double root	boundary point
interest	completing the square	test number
principal	quadratic formula	triangle inequality
rate of interest	discriminant	
simple interest	principal square root	
formula	of N	

FILL-IN-THE-BLANK QUESTIONS

1. Two equations (or inequalities) that have precisely the same solution set are called _____ .

2. An equation that is satisfied for every choice of the variable for which both sides are meaningful is called a(n) _____ .

3. To complete the square of the expression $x^2 + 5x$, you would _____ the number _____ .

4. The quantity $b^2 - 4ac$ is called the _____ of a quadratic equation. If it is _____ , the equation has no real solution.

5. If $a < 0$, then $|a| = $ _____ .

6. When a quadratic equation has a repeated solution, it is called a(n) _____ root or a root of _____ _____ .

7. When an apparent solution does not satisfy the original equation, it is called a(n) _____ solution.

REVIEW EXERCISES

In Problems 1–38 find all the real solutions, if any, of each equation. (Where they appear, a, b, m, and n are constants.)

1. $2 - \dfrac{x}{3} = 5$

2. $\dfrac{x}{4} - 2 = 4$

3. $-2(5 - 3x) + 8 = 4 + 5x$

4. $(6 - 3x) - 2(1 + x) = 6x$

5. $\dfrac{3x}{4} - \dfrac{x}{3} = \dfrac{1}{12}$

6. $\dfrac{4 - 2x}{3} + \dfrac{1}{6} = 2x$

7. $\dfrac{x}{x - 1} = \dfrac{5}{4}$

8. $\dfrac{4x - 5}{3 - 7x} = 4$

9. $x(1 - x) = 6$

10. $x(1 + x) = 2$

11. $\dfrac{1}{2}\left(x - \dfrac{1}{3}\right) = \dfrac{3}{4} - \dfrac{x}{6}$

12. $\dfrac{1 - 3x}{4} = \dfrac{x + 6}{3} + \dfrac{1}{2}$

13. $(x - 1)(2x + 3) = 3$

14. $x(2 - x) = 3(x - 4)$

15. $2x + 3 = 4x^2$

16. $1 + 6x = 4x^2$

17. $\sqrt[3]{x^2 - 1} = 2$

18. $\sqrt{1 + x^3} = 3$

19. $x(x + 1) + 2 = 0$

20. $3x^2 - x + 1 = 0$

21. $x^4 - 5x^2 + 4 = 0$

22. $3x^4 + 4x^2 + 1 = 0$

23. $\sqrt{2x - 3} + x = 3$

24. $\sqrt{2x - 1} = x - 2$

25. $x^{3/2} + 2x^{1/2} = 0$

26. $x^{2/3} - x = 0$

27. $\sqrt{x + 1} + \sqrt{x - 1} = \sqrt{2x + 1}$

28. $\sqrt{2x - 1} - \sqrt{x - 5} = 3$

29. $\sqrt{3 - 2\sqrt{x}} = \sqrt{x}$

30. $\sqrt{10 + 3\sqrt{x}} - 2 = \sqrt{x}$

31. $x^{-6} - 7x^{-3} - 8 = 0$

32. $6x^{-1} - 5x^{-1/2} + 1 = 0$

33. $x^2 + m^2 = 2mx + (nx)^2$

34. $b^2x^2 + 2ax = x^2 + a^2$

35. $10a^2x^2 - 2abx - 36b^2 = 0$

36. $\dfrac{1}{x - m} + \dfrac{1}{x - n} = \dfrac{2}{x}$

37. $\sqrt{x^2 + 3x + 7} - \sqrt{x^2 - 3x + 9} + 2 = 0$

38. $\sqrt{x^2 + 3x + 7} - \sqrt{x^2 + 3x + 9} = 2$

In Problems 39–58 solve each inequality.

39. $\dfrac{2x - 3}{5} + 1 \leq \dfrac{x}{2}$

40. $\dfrac{5 - x}{3} \leq 6x - 1$

41. $-9 \leq \dfrac{2x + 3}{-4} \leq 7$

42. $-4 < \dfrac{2x - 2}{3} < 6$

43. $6 > \dfrac{3 - 3x}{12} > 2$

44. $6 > \dfrac{5 - 3x}{2} \geq -3$

45. $2x^2 + 5x - 12 < 0$

46. $3x^2 - 2x - 1 \geq 0$

47. $\dfrac{6}{x + 2} \geq 1$

48. $\dfrac{-2}{1 - 3x} < -1$

49. $\dfrac{2x - 3}{1 - x} < 2$

50. $\dfrac{3 - 2x}{2x + 5} \geq 2$

51. $\dfrac{(x - 2)(x - 1)}{x - 3} > 0$

52. $\dfrac{x + 1}{x(x - 5)} \leq 0$

53. $\dfrac{x^2 - 8x + 12}{x^2 - 16} > 0$

54. $\dfrac{x(x^2 + x - 2)}{x^2 + 9x + 20} \leq 0$

55. $|3x + 4| < \dfrac{1}{2}$

56. $|1 - 2x| < \dfrac{1}{3}$

57. $|2x - 5| \geq 7$

58. $|3x + 1| \geq 2$

In Problems 59–68 solve each equation in the complex number system.

59. $x^2 + x + 1 = 0$

60. $x^2 - x + 1 = 0$

61. $2x^2 + x - 2 = 0$

62. $3x^2 - 2x - 1 = 0$

63. $x^2 + 3 = x$

64. $2x^2 + 1 = 2x$

65. $x(1 - x) = 6$

66. $x(1 + x) = 2$

67. $x^4 + 2x^2 - 8 = 0$

68. $x^4 - 8x^2 - 9 = 0$

69. Find k such that the equation $kx^2 + x + k = 0$ has a repeated real solution.

70. Find k such that the equation $x^2 - kx + 4 = 0$ has a repeated real solution.

71. Find k such that the equation $x^2 + kx + 1 = 0$ has no real solution.

72. Find k such that the equation $kx^2 + x + 1 = 0$ has no real solution.

73. If both sides of the equation $2x - 3 = 1$ are multiplied by 2, state whether the resulting equation is equivalent to the original equation. If not, tell whether the resulting equation contains more or fewer solutions than the original. Give reasons for your answer.

74. Repeat Problem 73 for the equation $3x^2 = 6x$ if it is divided by 3.

75. If both sides of the equation $2x - 3 = 1$ are multiplied by x, state whether the resulting equation is equivalent to the original equation. If not, tell whether the resulting equation contains more or fewer solutions than the original. Give reasons for your answer.

76. Repeat Problem 75 for the equation $3x^2 = 6x$, if it is divided by x.

77. Show that the real solutions of the equation $ax^2 + bx + c = 0$ are the negatives of the real solutions of the equation $ax^2 - bx + c = 0$. Assume $b^2 - 4ac \geq 0$.

78. Show that the real solutions of the equation $ax^2 + bx + c = 0$ are the reciprocals of the real solutions of the equation $cx^2 + bx + a = 0$. Assume $b^2 - 4ac \geq 0$.

79. The sum of the consecutive integers $1, 2, 3, \ldots, n$ is given by the formula $\frac{1}{2}n(n + 1)$. How many consecutive integers, starting with 1, must be added to get a sum of 666?

80. If a polygon of n sides has $\frac{1}{2}n(n - 3)$ diagonals, how many sides will a polygon with 65 diagonals have?

81. A bullet is fired at a target, and the sound of its impact is heard less than 3 seconds later. If the speed of sound averages 1100 feet per second, at most how far away is the target?

82. The intensity I, in candlepower, of a certain light source obeys the equation $I = 900/x^2$, where x is the distance in meters from the light. Over what range of distances can an object be placed from this light source so that the range of intensity of light is from 1600 to 3600 candlepower, inclusive?

83. A search plane has a cruising speed of 250 miles per hour and carries enough fuel for at most 5 hours of flying. If there is a wind that averages 30 miles per hour and the direction of search is with the wind one way and against it the other, how far can the search plane travel?

84. If the search plane described in Problem 83 is able to add a supplementary fuel tank that allows for an additional 2 hours of flying, how much farther can the plane extend its search?

C **85.** A life raft, set adrift from a sinking ship 150 miles offshore, travels directly toward a Coast Guard station at the rate of 5 miles per hour. At the time the raft is set adrift, a rescue helicopter is dispatched from the Coast Guard station. If the helicopter's average speed is 90 miles per hour, how long will it take the helicopter to reach the life raft?

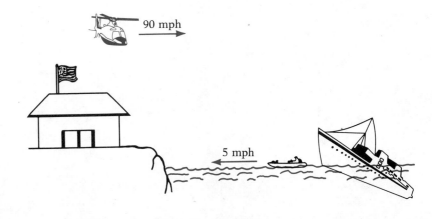

86. Two bees leave two locations 150 meters apart and fly, without stopping, back and forth between these two locations at average speeds of 3 meters per second and 5 meters per second, respectively. How long is it until the bees meet for the first time? How long is it until they meet for the second time?

C **87.** A man is walking at an average speed of 4 miles per hour alongside a railroad track. A freight train, going in the same direction at an average speed of 30 miles per hour, requires 5 seconds to pass the man. How long is the freight train? Give your answer in feet.

C **88.** One formula stating the relationship between the length l and width w of rectangles of "pleasing proportion" is $l^2 = w(l + w)$. How should a 4-by-8-foot piece of plasterboard be cut, so that the result is a rectangle of pleasing proportion with a width of 4 feet?

C **89.** A group of 20 senior citizens can charter a bus for a 1-day excursion trip for $15 per person. The charter company agrees to reduce the price of each ticket by 10 cents for each additional passenger, in excess of 20, who goes on the trip, up to a maximum of 44 passengers (the capacity of the bus). If the final bill from the charter company was $482.40, how many seniors went on the trip and how much did each pay?

GRAPHS

3.1 Rectangular Coordinates
3.2 Graphs of Equations
3.3 The Straight Line
3.4 Variation
Chapter Review

The idea of using a system of rectangular coordinates dates back to ancient times, when such a system was used for surveying and city planning. Apollonius of Perga in 200 B.C. used a form of rectangular coordinates in his work on conics, although this use does not stand out as clearly as it does in modern treatments. Sporadic use of rectangular coordinates continued until the 1600s. By that time algebra had developed sufficiently so that Descartes (1596–1650) and Fermat (1601–1665) could take the crucial step, which was the use of rectangular coordinates to translate geometry problems into algebra problems and vice versa. This step was supremely important for two reasons. First, it allowed both geometers and algebraists to gain critical new insights into their subjects, which previously had been regarded as separate but now were seen to be connected in many important ways. Second, the insights gained made possible the development of calculus, which greatly enlarged the number of areas in which mathematics could be applied and made possible a much deeper understanding of these areas.

▬ 3.1 RECTANGULAR COORDINATES

We locate a point on a real number line by using a single real number, called the *coordinate of the point*. For work in a two-dimensional plane, we locate points by using two numbers.

x-AXIS
y-AXIS
ORIGIN

We begin with two real number lines located in the same plane: one horizontal and the other vertical. We call the horizontal line the **x-axis**, the vertical line the **y-axis**, and the point of intersection the **origin** O. We assign coordinates to each point on these number lines, as described earlier (Section 1.2) and shown in Figure 1, using a convenient scale on each. (The scales are usually, but not necessarily, the same.) Thus, the origin O has a value of zero on both the x-axis and the y-axis. We follow the usual convention that points on the x-axis to the right of O are associated with positive real numbers, and those to the left of O with negative real numbers; those on the y-axis above O are associated with positive real numbers, and those below O with negative real numbers. See Figure 1 where the x-axis and y-axis are labeled. As is our custom, we use an arrow to denote the positive direction.

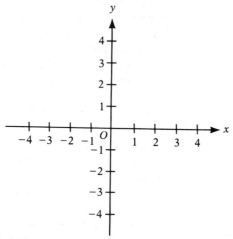

FIGURE 1

RECTANGULAR (CARTESIAN)
COORDINATE SYSTEM

xy-PLANE
COORDINATE AXES

The coordinate system described here is a **rectangular** or **Cartesian*** **coordinate system**. The plane formed by the x-axis and y-axis is sometimes called the **xy-plane**, and the x-axis and y-axis are referred to as the **coordinate axes**.

*Named after René Descartes (1596–1650), a French mathematician, philosopher, and theologian.

ORDERED PAIR

Any point P in the plane formed by the x-axis and y-axis can then be located by using an **ordered pair** (x, y) of real numbers. Let x denote the signed distance of P from the y-axis (*signed* in the sense that, if P is to the right of the y-axis, then $x > 0$, and, if P is to the left of the y-axis, then $x < 0$); and let y denote the signed distance of P from the x-axis. The ordered pair (x, y), the **coordinates** of P, then gives us enough information to locate the point P in the plane.

COORDINATES

For example, to locate the point whose coordinates are $(-3, 1)$, go out 3 units along the x-axis to the left of O and then go straight up 1 unit. We **plot** this point by placing a dot at this location. See Figure 2, in which the points with coordinates $(-3, 1), (-2, -3), (3, -2)$ and $(3, 2)$ are plotted.

PLOT

The origin has coordinates $(0, 0)$. Any point on the x-axis has coordinates of the form $(x, 0)$; any point on the y-axis has coordinates of the form $(0, y)$.

x-COORDINATE
y-COORDINATE

If (x, y) are the coordinates of the point P, then x is called the **x-coordinate** (or **abscissa**) of P and y is the **y-coordinate** (or **ordinate**) of P. We identify the point P by its coordinates (x, y) by writing $P = (x, y)$. Usually we will say "the point (x, y)" rather than "the point whose coordinates are (x, y)."

QUADRANTS

The coordinate axes divide the xy-plane into four sections, called **quadrants**. See Figure 3. In quadrant I, both the x-coordinate and the y-coordinate of all points are positive; in quadrant II, x is negative and y is positive; in quadrant III, both x and y are negative; and in quadrant IV, x is positive and y is negative. Points on the coordinate axes belong to no quadrant.

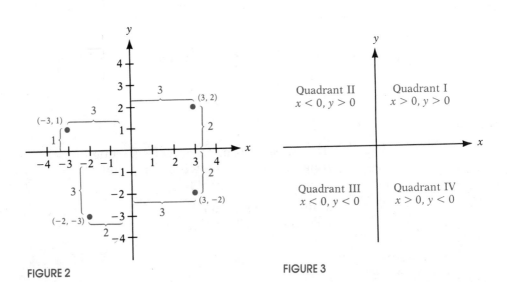

FIGURE 2

FIGURE 3

Distance between Points

If the same units of measurement—such as feet, miles, and so on—are used for both the x-axis and the y-axis, then all distances in the xy-plane can be measured using this unit of measurement.

EXAMPLE 1   Find the distance d between the points (1, 3) and (5, 6).

Solution   First we plot the points (1, 3) and (5, 6). See Figure 4(a). Then we draw a horizontal line from (1, 3) to (5, 3) and a vertical line from (5, 3) to (5, 6), forming a right triangle. See Figure 4(b). One leg of the triangle is of length 4 and the other is of length 3. By the Pythagorean Theorem, the square of the distance d we seek is

$$d^2 = 4^2 + 3^2 = 16 + 9 = 25$$

so that

$$d = 5$$

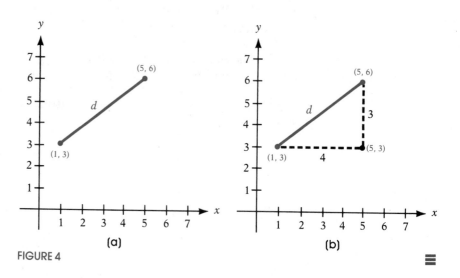

FIGURE 4

The distance formula provides a straightforward method for computing the distance between two points.

THEOREM   The distance between two points $P_1 = (x_1, y_1)$ and $P_2 = (x_2, y_2)$, which we denote by $d(P_1, P_2)$, is

DISTANCE FORMULA

$$d(P_1, P_2) = \sqrt{(x_2 - x_1)^2 + (y_2 - y_1)^2} \tag{1}$$

Thus, to compute the distance between two points, find the difference of the x-coordinates, square it, and add this to the square of the difference of the y-coordinates. The square root of this sum is the distance. See Figure 5.

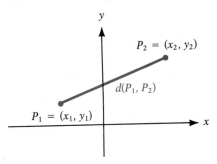

FIGURE 5
$d(P_1, P_2) = \sqrt{(x_2 - x_1)^2 + (y_2 - y_1)^2}$

Proof of the Distance Formula

Let (x_1, y_1) denote the coordinates of point P_1, and let (x_2, y_2) denote the coordinates of point P_2. Assume the line joining P_1 and P_2 is neither horizontal nor vertical. Refer to Figure 6(a). The coordinates of P_3 are (x_2, y_1). The horizontal distance from P_1 to P_3 is the absolute value of the difference of the x-coordinates, namely $|x_2 - x_1|$; the vertical distance from P_3 to P_2 is the absolute value of the difference of the y-coordinates, namely $|y_2 - y_1|$; and $d(P_1, P_2)$ is the length of the hypotenuse of a right triangle. See Figure 6(b). By the Pythagorean Theorem, it follows that

$$[d(P_1, P_2)]^2 = |x_2 - x_1|^2 + |y_2 - y_1|^2$$
$$= (x_2 - x_1)^2 + (y_2 - y_1)^2$$
$$d(P_1, P_2) = \sqrt{(x_2 - x_1)^2 + (y_2 - y_1)^2}$$

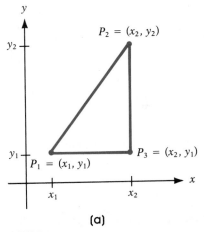

(a)

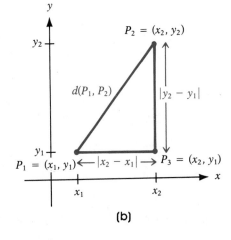

(b)

FIGURE 6

If the line joining P_1 and P_2 is horizontal, then the y-coordinate of P_1 equals the y-coordinate of P_2; that is, $y_1 = y_2$. Refer to Figure 7(a). In this case, the distance formula, Equation (1), still works because, for $y_1 = y_2$, it reduces to

$$d(P_1, P_2) = \sqrt{(x_2 - x_1)^2 + 0^2} = \sqrt{(x_2 - x_1)^2} = |x_2 - x_1|$$

A similar argument holds if the line joining P_1 and P_2 is vertical. See Figure 7(b). Thus, the distance formula is valid in all cases.

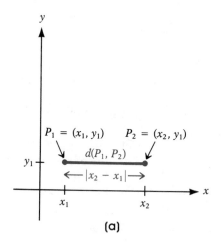

(a)

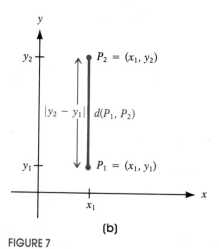

(b)

FIGURE 7

EXAMPLE 2   Find the distance d between the points $(-4, 5)$ and $(3, 2)$.

Solution Using the distance formula, Equation (1), the solution is obtained as follows:

$$d = \sqrt{[3 - (-4)]^2 + (2 - 5)^2} = \sqrt{7^2 + (-3)^2}$$
$$= \sqrt{49 + 9} = \sqrt{58} \approx 7.62 \quad \blacksquare$$

The distance between two points $P_1 = (x_1, y_1)$ and $P_2 = (x_2, y_2)$ is never a negative number. Furthermore, the distance between two points is zero only when the points are identical—that is, when $x_1 = x_2$ and $y_1 = y_2$. Also, because $(x_2 - x_1)^2 = (x_1 - x_2)^2$ and $(y_2 - y_1)^2 = (y_1 - y_2)^2$, it makes no difference whether the distance is computed from P_1 to P_2 or from P_2 to P_1—that is, $d(P_1, P_2) = d(P_2, P_1)$.

The introduction to this chapter mentioned that, by using rectangular coordinates, geometry problems can be translated into algebra problems, and vice versa. The next example shows how algebra (via the distance formula) can be used to solve some geometry problems.

EXAMPLE 3 Consider the three points $A = (-2, 1)$, $B = (2, 3)$, and $C = (3, 1)$.

(a) Plot each point and form the triangle ABC.

(b) Find the length of each side of the triangle.

(c) Verify that the triangle is a right triangle.

(d) Find the area of the triangle.

Solution (a) Points A, B, and C and the triangle ABC are plotted in Figure 8.

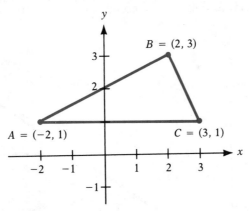

FIGURE 8

(b) $d(A, B) = \sqrt{[2 - (-2)]^2 + (3 - 1)^2} = \sqrt{16 + 4} = \sqrt{20} = 2\sqrt{5}$
$d(B, C) = \sqrt{(3 - 2)^2 + (1 - 3)^2} = \sqrt{1 + 4} = \sqrt{5}$
$d(A, C) = \sqrt{[3 - (-2)]^2 + (1 - 1)^2} = \sqrt{5^2 + 0^2} = 5$

(c) To show that the triangle is a right triangle we need to show that the sum of the squares of the lengths of two of the sides equals the square of the length of the third side. (Why is this sufficient?) Looking at Figure 8, it seems reasonable to conjecture that the right angle is at vertex B. Thus, we shall check to see whether

$$[d(A, B)]^2 + [d(B, C)]^2 = [d(A, C)]^2$$

Because, in fact,

$$[d(A, B)]^2 + [d(B, C)]^2 = (2\sqrt{5})^2 + (\sqrt{5})^2 = 20 + 5 = 25$$
$$= [d(A, C)]^2$$

it follows from the converse of the Pythagorean Theorem that triangle ABC is a right triangle.

(d) Because the right angle is at B, the sides AB and BC form the base and altitude of the triangle. Its area is therefore

$$\text{Area} = \frac{1}{2}(\text{Base})(\text{Altitude}) = \frac{1}{2}(2\sqrt{5})(\sqrt{5}) = 5 \text{ square units} \equiv$$

Midpoint Formula

MIDPOINT OF A LINE SEGMENT   We now derive a formula for the coordinates of the **midpoint of a line segment**. Let $P_1 = (x_1, y_1)$ and $P_2 = (x_2, y_2)$ be the endpoints of a line segment, and let $M = (x, y)$ be the point on the line segment that is the same distance from P_1 as it is from P_2. See Figure 9. The triangles P_1AM and MBP_2 are congruent. [Do you see why? Because two angles

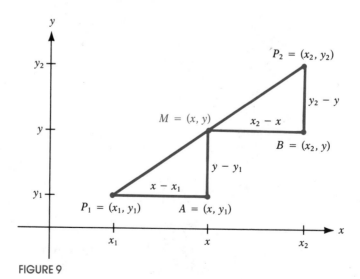

FIGURE 9

are equal and $d(P_1, M) = d(M, P_2)$, we have Angle–Side–Angle.] Hence, corresponding sides are equal in length. That is,

$$x - x_1 = x_2 - x \quad \text{and} \quad y - y_1 = y_2 - y$$
$$2x = x_1 + x_2 \qquad\qquad\quad 2y = y_1 + y_2$$
$$x = \frac{x_1 + x_2}{2} \qquad\qquad\quad y = \frac{y_1 + y_2}{2}$$

THEOREM   The midpoint (x, y) of the line segment from $P_1 = (x_1, y_1)$ to $P_2 = (x_2, y_2)$ is given by the formulas

MIDPOINT FORMULA

$$x = \frac{x_1 + x_2}{2} \quad \text{and} \quad y = \frac{y_1 + y_2}{2} \tag{2}$$

≡

Thus, to find the midpoint of a line segment, we average the x-coordinates and the y-coordinates of the endpoints.

EXAMPLE 4   Find the midpoint of the line segment from $P_1 = (-5, 3)$ to $P_2 = (3, 1)$. Plot the points P_1 and P_2, and their midpoint. Check your answer.

Solution   We apply the midpoint formula, Equation (2), using $x_1 = -5$, $x_2 = 3$, $y_1 = 3$, and $y_2 = 1$. Then the coordinates (x, y) of the midpoint are

$$x = \frac{x_1 + x_2}{2} = \frac{-5 + 3}{2} = -1 \quad \text{and} \quad y = \frac{y_1 + y_2}{2} = \frac{3 + 1}{2} = 2$$
$$M = (-1, 2)$$

See Figure 10. Because M is the midpoint, we check the answer by verifying that $d(P_1, M) = d(M, P_2)$:

$$d(P_1, M) = \sqrt{[-1 - (-5)]^2 + (2 - 3)^2} = \sqrt{16 + 1} = \sqrt{17}$$
$$d(M, P_2) = \sqrt{[3 - (-1)]^2 + (1 - 2)^2} = \sqrt{16 + 1} = \sqrt{17}$$

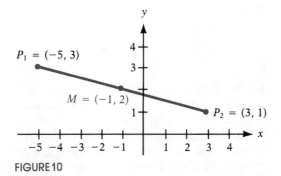

FIGURE 10

≡

EXERCISE 3.1   *In Problems 1 and 2 plot each point in the xy-plane. Tell in what quadrant or on what coordinate axis each point lies.*

1. (a) $A = (-3, 2)$ (b) $B = (6, 0)$ (c) $C = (-2, -2)$
   (d) $D = (6, 5)$ (e) $E = (0, -3)$ (f) $F = (6, -3)$

2. (a) $A = (1, 4)$ (b) $B = (-3, -4)$ (c) $C = (-3, 4)$
   (d) $D = (4, 1)$ (e) $E = (0, 1)$ (f) $F = (-3, 0)$

3. Plot the points $(2, 0)$, $(2, -3)$, $(2, 4)$, $(2, 1)$, and $(2, -1)$. Describe the set of all points of the form $(2, y)$, where y is a real number.

4. Plot the points $(0, 3)$, $(1, 3)$, $(-2, 3)$, $(5, 3)$, and $(-4, 3)$. Describe the set of all points of the form $(x, 3)$, where x is a real number.

In Problems 5–14 find the distance $d(P_1, P_2)$ between the points P_1 and P_2.

5. $P_1 = (3, -4)$, $P_2 = (3, 1)$ 6. $P_1 = (-1, 0)$, $P_2 = (2, 1)$

7. $P_1 = (-3, 2)$, $P_2 = (6, 0)$ 8. $P_1 = (2, -3)$, $P_2 = (4, 2)$

9. $P_1 = (4, -3)$, $P_2 = (6, 1)$ 10. $P_1 = (-4, -3)$, $P_2 = (2, 2)$

C 11. $P_1 = (-0.2, 0.3)$, $P_2 = (2.3, 1.1)$

C 12. $P_1 = (1.2, 2.3)$, $P_2 = (-0.3, 1.1)$

13. $P_1 = (a, b)$, $P_2 = (0, 0)$ 14. $P_1 = (a, a)$, $P_2 = (0, 0)$

In Problems 15–18 plot each point and form the triangle ABC. Verify that the triangle is a right triangle. Find its area.

15. $A = (-2, 5)$, $B = (1, 3)$, $C = (-1, 0)$

16. $A = (-2, 5)$, $B = (12, 3)$, $C = (10, -11)$

17. $A = (-5, 3)$, $B = (6, 0)$, $C = (5, 5)$

18. $A = (-6, 3)$, $B = (3, -5)$, $C = (-1, 5)$

19. Find all points on the x-axis that are 5 units from the point $(2, -3)$.

20. Find all points on the y-axis that are 5 units from the point $(-4, 4)$.

In Problems 21–30 find the midpoint of the line segment joining the points P_1 and P_2.

21. $P_1 = (3, -4)$, $P_2 = (3, 1)$ 22. $P_1 = (-1, 0)$, $P_2 = (2, 1)$

23. $P_1 = (-3, 2)$, $P_2 = (6, 0)$ 24. $P_1 = (2, -3)$, $P_2 = (4, 2)$

25. $P_1 = (4, -3)$, $P_2 = (6, 1)$ 26. $P_1 = (-4, -3)$, $P_2 = (2, 2)$

C 27. $P_1 = (-0.2, 0.3)$, $P_2 = (2.3, 1.1)$

C 28. $P_1 = (1.2, 2.3)$, $P_2 = (-0.3, 1.1)$

29. $P_1 = (a, b)$, $P_2 = (0, 0)$ 30. $P_1 = (a, a)$, $P_2 = (0, 0)$

MEDIANS   31. The **medians** of a triangle are the line segments from each vertex to the midpoint of the opposite side. Find the lengths of the medians of the triangle with vertices at $(0, 0)$, $(0, 6)$, and $(8, 0)$.

EQUILATERAL TRIANGLE   32. An **equilateral triangle** is one in which all three sides are of equal length. If two vertices of an equilateral triangle are $(4, -3)$ and $(0, 0)$, find the third vertex. How many of these triangles are possible?

In Problems 33–36 find the length of each side of the triangle determined by the three points P_1, P_2, and P_3; and state whether the triangle is an isosceles triangle, ISOSCELES TRIANGLE *a right triangle, neither of these, or both. An **isosceles triangle** is one in which at least two of the sides are of equal length.*

33. $P_1 = (2, 1)$, $P_2 = (-4, 1)$, $P_3 = (-4, -3)$
34. $P_1 = (-1, 4)$, $P_2 = (6, 2)$, $P_3 = (4, -5)$
35. $P_1 = (-2, -1)$, $P_2 = (0, 7)$, $P_3 = (3, 2)$
36. $P_1 = (7, 2)$, $P_2 = (-4, 0)$, $P_3 = (4, 6)$

Problems 38–42 use the result of Problem 37.

37. If r is a real number, prove that the coordinates of the point $P = (x, y)$ that divides the line segment from $P_1 = (x_1, y_1)$ to $P_2 = (x_2, y_2)$ in the ratio r—that is,

$$\frac{d(P_1, P)}{d(P_1, P_2)} = r$$

    are

$$x = (1 - r)x_1 + rx_2 \quad \text{and} \quad y = (1 - r)y_1 + ry_2$$

    [*Hint:* Use similar triangles.]

38. Verify that the midpoint of a line segment divides the line segment from $P_1 = (x_1, y_1)$ to $P_2 = (x_2, y_2)$ in the ratio $r = \frac{1}{2}$.

39. What point P divides the line segment from P_1 to P_2 in the ratio $r = 1$?

40. What point P divides the line segment from P_1 to P_2 in the ratio $r = 0$?

41. Find the point P on the line joining $P_1 = (1, 4)$ and $P_2 = (5, 6)$ that is twice as far from P_1 as P_2 is from P_1 and that lies on the same side of P_1 as P_2 does.

42. Find the point P on the line joining $P_1 = (0, 4)$ and $P_2 = (-1, 1)$ that is three times as far from P_1 as P_2 is from P_1.

C 43. A major league baseball "diamond" is actually a square, 90 feet on a side. What is the distance directly from home plate to second base (the diagonal of the square)?

C 44. The layout of a Little League playing field is a square, 60 feet on a side.* How far is it directly from home plate to second base (the diagonal of the square)?

45. Find the midpoint of each diagonal of a square with side of length s. Draw the conclusion that the diagonals of a square intersect at their midpoints. [*Hint:* Use $(0, 0)$, $(0, s)$, $(s, 0)$, and (s, s) as the vertices of the square.]

46. Verify that the points $(0, 0)$, $(a, 0)$, and $(a/2, \sqrt{3}a/2)$ are the vertices of an equilateral triangle. Then show that the midpoints of each side are the vertices of a second equilateral triangle.

*Source: *Little League Baseball, Official Regulations and Playing Rules,* 1983.*

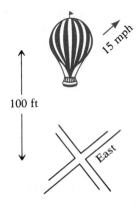

100 ft

47. An automobile and a truck leave an intersection at the same time. The automobile heads east at an average speed of 40 miles per hour while the truck heads south at an average speed of 30 miles per hour. Find an expression for their distance apart d (in miles) at the end of t hours.

48. A hot air balloon, headed due east at an average speed of 15 miles per hour and at a constant altitude of 100 feet, passes over an intersection, as shown in the figure. Find an expression for its distance d, measured in feet, from the intersection t seconds later.

▆ 3.2 GRAPHS OF EQUATIONS

In Chapter 2, we concerned ourselves mainly with equations involving a single real variable. In this section we look at equations that contain two real variables.

Illustrations play an important role in helping us to visualize the relationships that exist between two variable quantities. For example, Figure 11 shows the variation in the rate of rotation of the earth. Such illustrations are usually referred to as *graphs*. The **graph of an equation** that involves two variables x and y consists of the set of points in the xy-plane whose coordinates (x, y) satisfy the equation.

GRAPH OF AN EQUATION

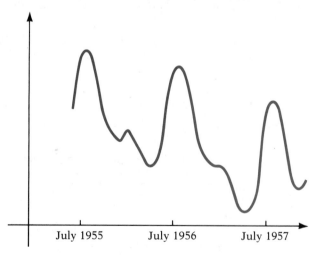

FIGURE 11
Variation in the rate of rotation of the earth

Let's look at some examples.

EXAMPLE 1   Graph the equation $y = 2x + 5$.

Solution   We want to find all points (x, y) that satisfy the equation. To locate some of these points (and thus get an idea of the pattern of the graph), we assign some numbers to x and find corresponding values for y:

IF	THEN	POINT ON GRAPH
$x = 0$	$y = 2(0) + 5 = 5$	$(0, 5)$
$x = 1$	$y = 2(1) + 5 = 7$	$(1, 7)$
$x = -5$	$y = 2(-5) + 5 = -5$	$(-5, -5)$
$x = 10$	$y = 2(10) + 5 = 25$	$(10, 25)$

By plotting these points and then connecting them, we obtain the graph of the equation (a straight line), as shown in Figure 12.

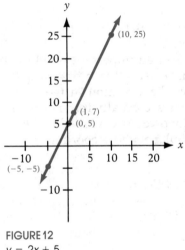

FIGURE 12
$y = 2x + 5$

EXAMPLE 2   Graph the equation $y = x^2$.

Solution   Table 1 provides several points on the graph:

TABLE 1

x	-4	-3	-2	-1	0	1	2	3	4
$y = x^2$	16	9	4	1	0	1	4	9	16
(x, y)	$(-4, 16)$	$(-3, 9)$	$(-2, 4)$	$(-1, 1)$	$(0, 0)$	$(1, 1)$	$(2, 4)$	$(3, 9)$	$(4, 16)$

PARABOLA   In Figure 13 we plot these points and connect them with a smooth curve to obtain the graph (a **parabola**).

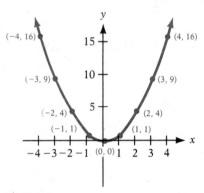

FIGURE 13
$y = x^2$

The graphs of the equations as illustrated in Figure 12 and 13 are necessarily incomplete. For example, in Figure 12 the point $(20, 45)$ is a part of the graph of $y = 2x + 5$, but it is not shown. Since the graph of $y = 2x + 5$ could be extended out as far as we please, we use arrows to indicate that the pattern shown continues. Thus, it is important when illustrating a graph always to present enough of the graph so that any viewer of the illustration will "see" the rest of it as an obvious continuation of what is actually there. How can we be sure this has been done? Part of the answer is given in this book; the rest of the answer is taken up in calculus. For the most part, we shall graph equations by locating a sufficient number of points of the graph until a pattern becomes evident; then we connect these points with a smooth curve. Shortly, we shall give various techniques that will help to graph an equation without plotting so many points.

EXAMPLE 3   Graph the equations.

    (a) $y = x^3$ (b) $x = y^2$

Solution   (a) We set up Table 2, listing several points of the graph:

TABLE 2

x	-3	-2	-1	0	1	2	3
$y = x^3$	-27	-8	-1	0	1	8	27
(x, y)	$(-3, -27)$	$(-2, -8)$	$(-1, -1)$	$(0, 0)$	$(1, 1)$	$(2, 8)$	$(3, 27)$

Figure 14(a) illustrates some of these points and the graph of $y = x^3$.

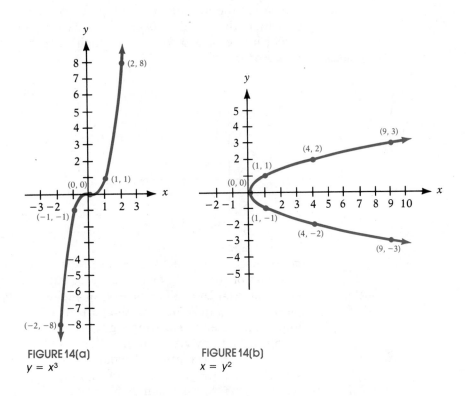

FIGURE 14(a)
$y = x^3$

FIGURE 14(b)
$x = y^2$

(b) We set up Table 3, listing several points of the graph:

TABLE 3

y	-3	-2	-1	0	1	2	3	4
$x = y^2$	9	4	1	0	1	4	9	16
(x, y)	$(9, -3)$	$(4, -2)$	$(1, -1)$	$(0, 0)$	$(1, 1)$	$(4, 2)$	$(9, 3)$	$(16, 4)$

Figure 14(b) illustrates some of these points and the graph of $x = y^2$. ≡

We said earlier that we would discuss techniques that reduce the number of points required to graph an equation. Two such techniques involve *intercepts* and *symmetry*.

Intercepts

The points, if any, at which a graph intersects the coordinate axes are called the **intercepts**. See Figure 15. The x-coordinate of a point at which the graph crosses the x-axis is an **x-intercept**, and the y-coordinate of a point at which the graph crosses the y-axis is a **y-intercept**.

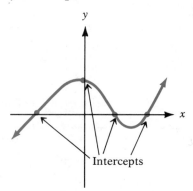

FIGURE 15

EXAMPLE 4   Find the intercepts of the graph in Figure 16. What are its x-intercepts? What are its y-intercepts?

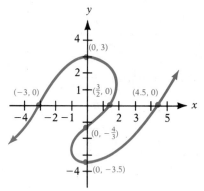

FIGURE 16

Solution   The intercepts of the graph are the points

$$(-3, 0), \quad (0, 3), \quad (\tfrac{3}{2}, 0), \quad (0, -\tfrac{4}{3}), \quad (0, -3.5), \quad (4.5, 0)$$

The x-intercepts are $-3, \tfrac{3}{2}, 4.5$; the y-intercepts are $-3.5, -\tfrac{4}{3}, 3$.   ▣

  The intercepts of the graph of an equation can be found by using the fact that points on the x-axis have y-coordinate equal to zero and points on the y-axis have x-coordinate equal to zero.

FINDING INTERCEPTS

> 1. To find the x-intercept(s) of the graph, let $y = 0$ in the equation and solve the equation for x.
> 2. To find the y-intercept(s) of the graph, let $x = 0$ in the equation and solve the equation for y.

EXAMPLE 5   Find the intercepts of the graph $y = x^2 - 4$.

Solution   To find the x-intercept(s), we let $y = 0$ to get the equation

$$x^2 - 4 = 0$$

The equation has two solutions, -2 and 2. Thus the x-intercepts are -2 and 2.

To find the y-intercept(s), we let $x = 0$ to get the equation

$$y = -4$$

Thus the y-intercept is -4.

The intercepts are, therefore, the points $(-2, 0)$, $(2, 0)$, and $(0, -4)$. These points, plus the points from Table 4, provide enough information to graph $y = x^2 - 4$. See Figure 17.

TABLE 4

x	-3	-1	1	3
$y = x^2 - 4$	5	-3	-3	5
(x, y)	$(-3, 5)$	$(-1, -3)$	$(1, -3)$	$(3, 5)$

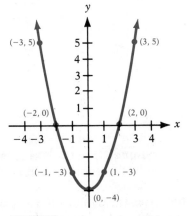

FIGURE 17
$y = x^2 - 4$

Symmetry

Another useful tool for graphing equations involves **symmetry**, particularly symmetry with respect to the x-axis, the y-axis, and the origin.

SYMMETRY WITH RESPECT TO
x-AXIS A graph is said to be **symmetric with respect to the x-axis** if, for every point (x, y) on the graph, the point $(x, -y)$ is also on the graph.

Figure 18(a) illustrates the definition. Notice that, when a graph is symmetric with respect to the x-axis, the part of the graph above the x-axis is a reflection of the part below it and vice versa.

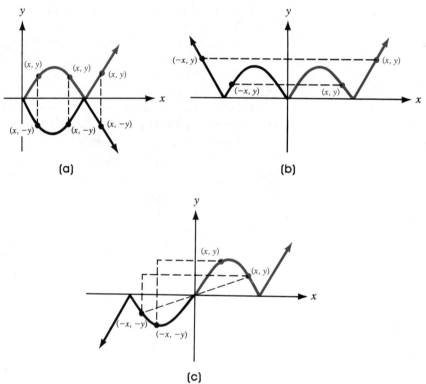

(a)

(b)

(c)

FIGURE 18

SYMMETRY WITH RESPECT TO
y-AXIS A graph is said to be **symmetric with respect to the y-axis** if, for every point (x, y) on the graph, the point $(-x, y)$ is also on the graph.

Figure 18(b) illustrates the definition. Notice that, when a graph is symmetric with respect to the y-axis, the part of the graph to the right of the y-axis is a reflection of the part to the left of it and vice versa.

SYMMETRY WITH RESPECT TO
ORIGIN
A graph is said to be **symmetric with respect to the origin** if, for every point (x, y) on the graph, the point $(-x, -y)$ is also on the graph.

Figure 18(c) illustrates the definition. Notice that symmetry with respect to the origin may be thought of as a reflection about the y-axis. followed by a reflection about the x-axis.

When the graph of an equation is symmetric, the number of points that you need to plot in order to see the pattern is reduced. For example, if the graph of an equation is symmetric with respect to the y-axis, then, once points to the right of the y-axis are plotted, an equal number of points on the graph can be obtained by reflecting them about the y-axis. Thus, before we graph an equation, we first want to determine whether it has any symmetry. The following tests are used for that purpose.

TESTS FOR SYMMETRY

To test the graph of an equation for symmetry with respect to the:

x-axis Replace y by $-y$. If an equivalent equation results, the graph of the equation is symmetric with respect to the x-axis.

y-axis Replace x by $-x$. If an equivalent equation results, the graph of the equation is symmetric with respect to the y-axis.

Origin Replace x by $-x$ and y by $-y$. If an equivalent equation results, the graph of the equation is symmetric with respect to the origin.

Let's look at an equation we have already graphed to see how these tests are used.

EXAMPLE 6 (a) To test the graph of the equation $x = y^2$ for symmetry with respect to the x-axis, we replace y by $-y$ in the equation, as follows:

$$x = y^2 \qquad \text{Original equation}$$
$$x = (-y)^2 \quad \text{Replace } y \text{ by } -y.$$
$$x = y^2 \qquad \text{Simplify.}$$

When we replace y by $-y$, the result is the same equation. Thus, the graph is symmetric with respect to the x-axis.

(b) To test the graph of the equation $x = y^2$ for symmetry with respect to the y-axis, we replace x by $-x$ in the equation:

$$x = y^2 \quad \text{Original equation}$$
$$-x = y^2 \quad \text{Replace } x \text{ by } -x.$$

Because we arrive at the equation $-x = y^2$, which is not equivalent to the original equation, we conclude that the graph is not symmetric with respect to the y-axis.

(c) To test for symmetry with respect to the origin, we replace x by $-x$ and y by $-y$:

$$x = y^2 \qquad \text{Original equation}$$
$$-x = (-y)^2 \quad \text{Replace } x \text{ by } -x, y \text{ by } -y.$$
$$-x = y^2 \qquad \text{Simplify.}$$

The resulting equation, $-x = y^2$, is not equivalent to the original equation. We conclude that the graph is not symmetric with respect to the origin. ■

Figure 19(a) illustrates the graph of $x = y^2$. In forming a table of points on the graph of $x = y^2$, we might restrict ourselves to points whose y-coordinates are positive. Once these are plotted and connected, a reflection about the x-axis (because of the symmetry) provides the rest of the graph.

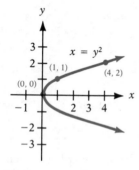

(a) Symmetry about x-axis

FIGURE 19

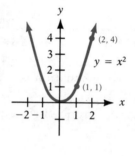

(b) Symmetry about y-axis

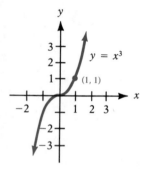

(c) Symmetry about origin

Figures 19(b) and 19(c) illustrate two equations we graphed earlier. Notice how the existence of symmetry reduces the number of points we need to plot.

EXAMPLE 7   Graph the equation $y = 1/x$.

Solution   We check for intercepts first. If we let $x = 0$, we get a zero in the denominator. Hence there is no y-intercept. If we let $y = 0$, we get the equation $1/x = 0$, which has no solution. Hence there is no x-intercept. Thus the graph of $y = 1/x$ will not cross the coordinate axes.

Next we check for symmetry:

x-axis: Replacing y by $-y$ yields $-y = 1/x$, which is not equivalent to $y = 1/x$.

y-axis: Replacing x by $-x$ yields $y = -1/x$, which is not equivalent to $y = 1/x$.

Origin: Replacing x by $-x$ and y by $-y$ yields $-y = -1/x$, which is equivalent to $y = 1/x$.

The graph is symmetric with respect to the origin.

Finally, we set up Table 5, listing several points of the graph. Because of the symmetry with respect to the origin, we use only positive values of x:

TABLE 5

x	$\frac{1}{10}$	$\frac{1}{3}$	$\frac{1}{2}$	1	2	3	10
$y = 1/x$	10	3	2	1	$\frac{1}{2}$	$\frac{1}{3}$	$\frac{1}{10}$
(x, y)	$(\frac{1}{10}, 10)$	$(\frac{1}{3}, 3)$	$(\frac{1}{2}, 2)$	$(1, 1)$	$(2, \frac{1}{2})$	$(3, \frac{1}{3})$	$(10, \frac{1}{10})$

From Table 5, we infer that, if x is a large and positive number, then $y = 1/x$ is a positive number close to zero. We also infer that, if x is a positive number close to zero, then $y = 1/x$ is a large and positive number. Armed with this information, we can graph the equation. Figure 20 illustrates some of these points and the graph of $y = 1/x$. Observe how the absence of intercepts and the existence of symmetry with respect to the origin were utilized.

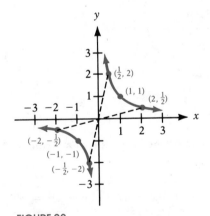

FIGURE 20

$y = \dfrac{1}{x}$

Circles

One of the advantages of a coordinate system is that it enables us to translate a geometric statement into an algebraic statement, and vice versa. Consider, for example, the following geometric statement that defines a circle.

CIRCLE   A **circle** is a set of points in the xy-plane that are a fixed distance R
RADIUS   from a fixed point (h, k). The fixed distance R is called the **radius**,
CENTER   and the fixed point (h, k) is called the **center** of the circle.

Figure 21 shows the graph of a circle. A circle, because it is a set of points, is a graph. Is there an equation having this graph? If so, what is the equation? To find the equation, we let (x, y) represent the coordinates of any point on a circle with radius R and center at (h, k). Then the distance between the points (x, y) and (h, k) must equal R. That is, by the distance formula,

$$\sqrt{(x - h)^2 + (y - k)^2} = R$$

or, equivalently,

$$(x - h)^2 + (y - k)^2 = R^2$$

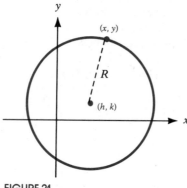

FIGURE 21

STANDARD FORM OF THE   The **standard form of the equation of a circle** with radius R and center
EQUATION OF A CIRCLE   at (h, k) is

$$(x - h)^2 + (y - k)^2 = R^2 \tag{1}$$

Conversely, by reversing the steps, we conclude: The graph of any equation of the form of Equation (1) is that of a circle with radius R and center at (h, k).

EXAMPLE 8   Graph the equation $(x + 3)^2 + (y - 2)^2 = 16$.

Solution   By comparing the given equation to the standard form of the equation of a circle, we conclude that the graph of the given equation is a circle. Moreover, the comparison yields information about the circle:

$$(x - h)^2 + (y - k)^2 = R^2$$
$$(x + 3)^2 + (y - 2)^2 = 16$$

We see that $h = -3, k = 2$, and $R = 4$. Hence, the circle has its center at $(-3, 2)$ and has a radius of 4 units. To graph this circle, we first plot the center, $(-3, 2)$. Since the radius is 4, we can locate four points on the circle by going out 4 units to the left and to the right of the center and·going 4 units up and down from the center. These four points can then be used as guides to obtain the graph. See Figure 22.

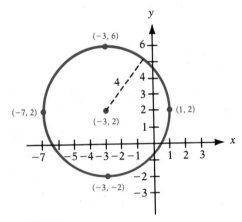

FIGURE 22

EXAMPLE 9   Write the standard form of the equation of the circle with radius 3 and center at $(1, -2)$.

Solution   Using the form of Equation (1) and substituting the values $R = 3$, $h = 1$, and $k = -2$, we have

$$(x - h)^2 + (y - k)^2 = R^2$$
$$(x - 1) + (y + 2)^2 = 9$$

The standard form of the equation of a circle of radius R with center at the origin $(0, 0)$ is

$$x^2 + y^2 = R^2$$

UNIT CIRCLE If the radius $R = 1$, the circle is the **unit circle**. See Figure 23.

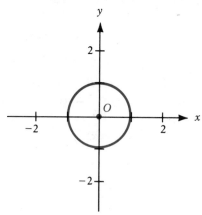

FIGURE 23
Unit circle $x^2 + y^2 = 1$

If we eliminate the parentheses from the standard form of the equation of the circle obtained in Example 9, we get

$$(x - 1)^2 + (y + 2)^2 = 9$$
$$x^2 - 2x + 1 + y^2 + 4y + 4 = 9$$

which, upon simplifying, is equivalent to

$$x^2 + y^2 - 2x + 4y - 4 = 0$$

By completing the squares on both the x-terms and the y-terms, it can be shown that any equation of the form

$$x^2 + y^2 + ax + by + c = 0$$

has a graph that is a circle, a point, or no graph at all. For example, the graph of the equation $x^2 + y^2 = 0$ is the single point $(0, 0)$. The equation $x^2 + y^2 + 5 = 0$, or $x^2 + y^2 = -5$, has no graph, because sums of squares are never negative. When its graph is a circle, the equation

$$x^2 + y^2 + ax + by + c = 0$$

GENERAL FORM OF THE EQUATION OF A CIRCLE is referred to as the **general form of the equation of a circle**.

The next example shows how to transform an equation in the general form to an equivalent equation in standard form. As we said earlier, the idea is to use the method of completing the square on both the x-terms and the y-terms.

EXAMPLE 10   Graph the equation $x^2 + y^2 + 4x - 6y + 12 = 0$.

Solution   We rearrange the equation as follows:

$$(x^2 + 4x) + (y^2 - 6y) = -12$$

Next we complete the square of each expression in parentheses. Remember that any number added on the left must also be added on the right:

$$(x^2 + 4x + 4) + (y^2 - 6y + 9) = -12 + 4 + 9$$
$$(x + 2)^2 + (y - 3)^2 = 1$$

We recognize this equation as the standard form of the equation of a circle with radius 1 and center at $(-2, 3)$. Figure 24 illustrates the graph.

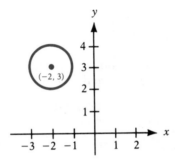

FIGURE 24
$x^2 + y^2 + 4x - 6y + 12 = 0$

EXAMPLE 11   Find the general equation of the circle whose center is at the point $(1, -2)$ and whose graph contains the point $(4, 2)$.

Solution   To find the equation of a circle, we need to know its center and its radius. Here we know the center is $(1, -2)$. Since the point $(4, 2)$ is on the graph, the radius R will equal the distance from $(4, 2)$ to the center $(1, -2)$. See Figure 25 (page 186). Thus,

$$R = \sqrt{(4 - 1)^2 + [2 - (-2)]^2} = \sqrt{9 + 16} = 5$$

The standard form of the equation of the circle is

$$(x - 1)^2 + (y + 2)^2 = 25$$

Eliminating the parentheses and rearranging terms, we get the general equation

$$x^2 + y^2 - 2x + 4y - 20 = 0$$

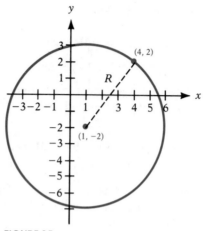

FIGURE 25
$x^2 + y^2 - 2x + 4y - 20 = 0$

Overview

The preceding discussion about circles dealt with two main types of problems that can be generalized as follows:

1. Given an equation, classify it and graph it.
2. Given a graph, or information about a graph, find its equation.

Example 10 illustrated the first type; Example 11 illustrated the second type.

This text deals mainly with the first type of problem. We shall study various equations, classify them, and graph them. The second type of problem is usually more difficult to solve than the first. Although we shall tackle such problems when it is practical to do so, we leave problems such as "fitting" equations to graphs to other courses.

EXERCISE 3.2   *In Problems 1–10 plot each point. Then plot the point that is symmetric to it with respect to:*
(a) The x-axis   (b) The y-axis   (c) The origin

1. (3, 4)	**2.** (5, 3)	**3.** (−2, 1)	**4.** (4, −2)
5. (1, 1)	**6.** (−1, −1)	**7.** (−3, −4)	**8.** (4, 0)
9. (0, −3)	**10.** (−3, 0)		

In Problems 11–30 graph each equation by plotting a sufficient number of points and connecting them with a smooth curve.

11. $y = 3x + 2$ **12.** $y = 2x - 3$
13. $y = -2x + 1$ **14.** $y = -3x + 1$

15. $3x - 2y + 6 = 0$

16. $2x - 3y + 6 = 0$

17. $y = 2x^2$

18. $y = 3x^2$

19. $y = -x^2 = (-1)x^2$

20. $y = -x^2 + 3$

21. $y = x^2 + 3$

22. $y = x^2 - 3$

23. $y = x^3 - 1$

24. $y = x^3 - 8$

25. $x^2 = 4y$

26. $x^2 = y + 2$

27. $x^2 = y + 1$

28. $x^2 = -2y$

29. $y = \sqrt{x}$

30. $x = \sqrt{y}$

If $y^2 = x$ is given, then $y = \pm\sqrt{x}$. When root in given, as written.

In Problems 31–38 write the standard form of the equation and the general form of the equation of each circle of radius R and center (h, k).

31. $R = 1$; $(h, k) = (1, -1)$

32. $R = 2$; $(h, k) = (-2, 1)$

33. $R = 2$; $(h, k) = (0, 2)$

34. $R = 3$; $(h, k) = (1, 0)$

35. $R = 5$; $(h, k) = (4, -3)$

36. $R = 4$; $(h, k) = (2, -3)$

37. $R = 2$; $(h, k) = (0, 0)$

38. $R = 3$; $(h, k) = (0, 0)$

In Problems 39–48 find the center (h, k) and radius R of each circle.

39. $x^2 + y^2 = 4$

40. $x^2 + (y - 1)^2 = 1$

41. $(x - 3)^2 + y^2 = 4$

42. $(x + 1)^2 + (y - 1)^2 = 2$

43. $x^2 + y^2 + 4x - 4y - 1 = 0$

44. $x^2 + y^2 - 6x + 2y + 9 = 0$

45. $x^2 + y^2 - x + 2y + 1 = 0$

46. $x^2 + y^2 + x + y - \frac{1}{2} = 0$

47. $2x^2 + 2y^2 - 12x + 8y - 24 = 0$

48. $2x^2 + 2y^2 + 8x + 7 = 0$

In Problems 49–52 use the graph below.

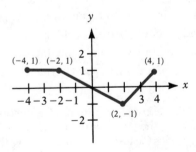

49. Add to the graph to make it symmetric with respect to the x-axis.

50. Add to the graph to make it symmetric with respect to the y-axis.

51. Add to the graph to make it symmetric with respect to the origin.

52. Add to the graph to make it symmetric with respect to the x-axis, y-axis, and origin.

In Problems 53–56 use the graph below.

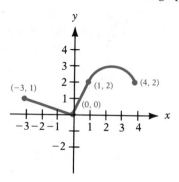

53. Add to the graph to make it symmetric with respect to the *x*-axis.

54. Add to the graph to make it symmetric with respect to the *y*-axis.

55. Add to the graph to make it symmetric with respect to the origin.

56. Add to the graph to make it symmetric with respect to the *x*-axis, *y*-axis, and origin.

In Problems 57–66 list the intercepts and test for symmetry.

57. $x^2 = y$ **58.** $y^2 = x$

59. $y = 3x$ **60.** $y = -5x$

61. $x^2 + y - 9 = 0$ **62.** $y^2 - x - 4 = 0$

63. $4x^2 + 9y^2 = 36$ **64.** $x^2 + 4y^2 = 4$

65. $y = x^3 - 27$ **66.** $y = x^4 - 1$

In Problems 67–72 find the general form of the equation of each circle.

67. Center at the origin and containing the point (2, −3)

68. Center at the point (1, 0) and containing the point (−3, 1)

69. Center at the point (2, 3) and touching the *x*-axis

70. Center at the point (−3, 1) and touching the *y*-axis

71. With endpoints of a diameter at (1, 4) and (−3, 2)

72. With endpoints of a diameter at (4, 3) and (0, 1)

C **73.** The earth is represented on a map of a portion of the solar system so that its surface is the circle with equation $x^2 + y^2 + 2x + 4y - 4091 = 0$. A satellite circles 0.6 unit above the earth with the center of its circular orbit at the center of the earth. Find the equation for the orbit of the satellite on this map.

▤ 3.3 THE STRAIGHT LINE

In this section we study a certain type of equation that contains two variables, the *linear equation*, and its graph, the *straight line*.

Slope of a Line

RUN
RISE

Consider the staircase illustrated in Figure 26. Each step contains exactly the same horizontal **run** and the same vertical **rise**. The ratio of the rise to the run, called the *slope*, is a numerical measure of the steepness of the staircase. For example, if the run is increased and the rise remains the same, the staircase becomes less steep. If the run is kept the same, but the rise is increased, the staircase becomes more steep.

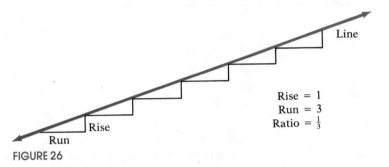

Rise = 1
Run = 3
Ratio = $\frac{1}{3}$

FIGURE 26

This important characteristic of a line, its slope, is best defined using coordinates.

SLOPE OF A LINE

Let $P = (x_1, y_1)$ and $Q = (x_2, y_2)$ be two distinct points with $x_1 \neq x_2$. The **slope** m of the nonvertical line L containing P and Q is defined by the formula

$$m = \frac{y_2 - y_1}{x_2 - x_1} \qquad (1)$$

VERTICAL LINE
UNDEFINED SLOPE

If $x_1 = x_2$, L is a **vertical line** and the slope m of L is **undefined** (since this results in division by zero).

Figure 27(a) on page 190 provides an illustration of the slope of a nonvertical line; Figure 27(b) illustrates a vertical line.

As Figure 27(a) illustrates, the slope m of a nonvertical line may be viewed as

$$m = \frac{y_2 - y_1}{x_2 - x_1} = \frac{\text{Rise}}{\text{Run}}$$

We can also express the slope m of a nonvertical line as

$$m = \frac{y_2 - y_1}{x_2 - x_1} = \frac{\text{Change in } y}{\text{Change in } x}$$

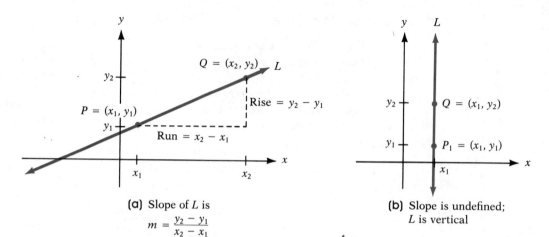

(a) Slope of L is
$$m = \frac{y_2 - y_1}{x_2 - x_1}$$

(b) Slope is undefined;
L is vertical

FIGURE 27

That is, the slope m of a nonvertical line L is the ratio of the change in the y-coordinates from P to Q to the change in the x-coordinates from P to Q.

Two comments about computing the slope of a nonvertical line:

1. Any two distinct points on the line can be used to compute the slope of the line. (See Figure 28 for a justification.)

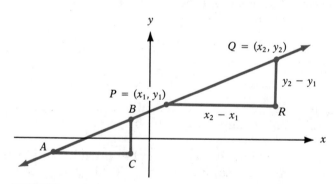

FIGURE 28

Triangles ABC and PQR are similar (equal angles). Hence, ratios of corresponding sides are proportional. Thus, slope using P and $Q = \dfrac{y_2 - y_1}{x_2 - x_1}$ = slope using A and $B = \dfrac{d(B, C)}{d(A, C)}$.

2. The slope of a line may be computed from $P = (x_1, y_1)$ to $Q = (x_2, y_2)$ or from Q to P because

$$\frac{y_2 - y_1}{x_2 - x_1} = \frac{y_1 - y_2}{x_1 - x_2}$$

EXAMPLE 1 The slope m of the line joining the points $(1, 2)$ and $(5, -3)$ may be computed as

$$m = \frac{-3 - 2}{5 - 1} = \frac{-5}{4} \quad \text{or as} \quad m = \frac{2 - (-3)}{1 - 5} = \frac{5}{-4} = \frac{-5}{4} \quad \blacksquare$$

To get a better idea of the meaning of the slope m of a line L, consider the following example.

EXAMPLE 2 Compute the slopes of the lines L_1, L_2, L_3, and L_4 containing the following pairs of points. Graph each line using the same coordinate axes.

$$
\begin{array}{lll}
L_1: & P = (2, 3) & Q_1 = (-1, -2) \\
L_2: & P = (2, 3) & Q_2 = (3, -1) \\
L_3: & P = (2, 3) & Q_3 = (5, 3) \\
L_4: & P = (2, 3) & Q_4 = (2, 5)
\end{array}
$$

Solution Let m_1, m_2, m_3, and m_4 denote the slopes of the lines L_1, L_2, L_3, and L_4, respectively. Then

$$m_1 = \frac{-2 - 3}{-1 - 2} = \frac{-5}{-3} = \frac{5}{3} \quad \text{A rise of 5 divided by a run of 3}$$

$$m_2 = \frac{-1 - 3}{3 - 2} = \frac{-4}{1} = -4$$

$$m_3 = \frac{3 - 3}{5 - 2} = \frac{0}{3} = 0$$

m_4 is undefined

The graphs of these lines are given in Figure 29.

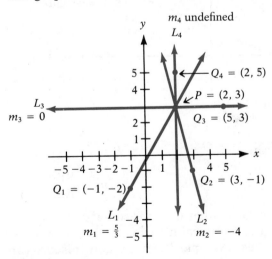

FIGURE 29

As Figure 29 indicates, when the slope m of a line is positive, the line slants upward from left to right (L_1); when the slope m is negative, the line slants downward from left to right (L_2); when the slope m is HORIZONTAL LINE   0, the line is **horizontal** (L_3); and when the slope m is undefined, the line is vertical (L_4). Figure 30 illustrates the slopes of several more lines. Note the pattern.

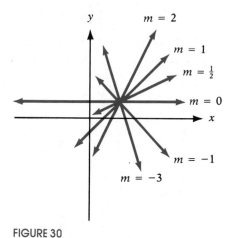

FIGURE 30

Consider the nonvertical line with slope m shown in Figure 31.

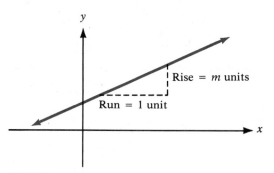

FIGURE 31
Line with slope m

Because

$$m = \frac{\text{Rise}}{\text{Run}}$$

a run of 1—that is, a horizontal movement 1 unit to the right—will require a vertical movement (rise) of m units to get back to the line. The next example illustrates how this idea can be used to graph lines.

EXAMPLE 3   Draw a graph of the line that passes through the point (3, 2) and has a slope of:

(a) $\frac{3}{4}$ (b) $-\frac{4}{5}$

Solution   (a) Slope equals rise/run. The fact that the slope is $\frac{3}{4}$ means that for every horizontal movement (run) of 4 units, there will be a vertical movement (rise) of 3 units. If we start at the given point, (3, 2), and move 4 units to the right and 3 units up, we reach the point (7, 5). By drawing the line through this point and the point (3, 2), we have the graph. See Figure 32(a).

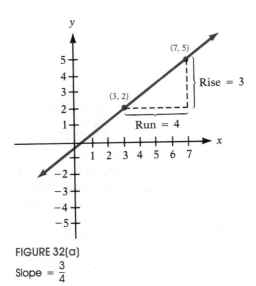

FIGURE 32(a)

Slope $= \dfrac{3}{4}$

(b) The fact that the slope is

$$-\frac{4}{5} = \frac{-4}{5} = \frac{\text{Rise}}{\text{Run}}$$

means that for every horizontal movement of 5 units, there will be a corresponding vertical movement of -4 units (a downward movement). If we start at the given point, (3, 2), and move 5 units to the right and then 4 units down, we arrive at the point (8, -2). By drawing the line through these points, we have the graph. See Figure 32(b) on page 194.

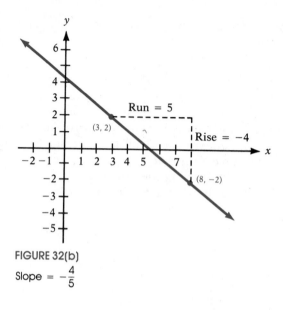

FIGURE 32(b)

Slope $= -\dfrac{4}{5}$

≡

An alternative solution to Example 3(b) would have been to set

$$-\frac{4}{5} = \frac{4}{-5} = \frac{\text{Rise}}{\text{Run}}$$

so that for every horizontal movement of -5 units (a movement to the left), there would be a corresponding vertical movement of 4 units (upward). This approach brings us to the point $(-2, 6)$ when we start at the point $(3, 2)$. The line through these two points is, of course, identical to the one in Figure 32(b).

Equations of Lines

Now that we have discussed the slope of a line, we are ready to derive equations of lines. As we shall see, there are several forms of the equation of a line. Let's start with an example.

EXAMPLE 4   Graph the equation $x = 3$.

Solution   We are looking for all points (x, y) in the plane for which $x = 3$. Thus, no matter what y-coordinate is used, the corresponding x-coordinate always equals 3. Consequently, the graph of the equation $x = 3$ is a vertical line with x-intercept 3 and undefined slope. See Figure 33.

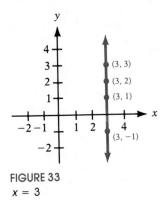

FIGURE 33
$x = 3$

Based on Example 4, we have the following result:

THEOREM   A vertical line is given by an equation of the form

EQUATION OF A VERTICAL LINE

$$x = a$$

where a is a given real number.

Now let L be a nonvertical line with slope m and containing the point (x_1, y_1). For any other point (x, y) on L, we have

$$m = \frac{y - y_1}{x - x_1} \quad \text{or} \quad y - y_1 = m(x - x_1)$$

THEOREM   An equation of a nonvertical line of slope m that passes through the point (x_1, y_1) is

POINT–SLOPE FORM OF
EQUATION OF A LINE

$$y - y_1 = m(x - x_1) \tag{2}$$

EXAMPLE 5   An equation of the line with slope 4 and passing through the point (1, 2) can be found by using the point–slope form with $m = 4$, $x_1 = 1$, and $y_1 = 2$:

$$y - y_1 = m(x - x_1)$$
$$y - 2 = 4(x - 1)$$
$$y = 4x - 2$$

See Figure 34 on page 196.

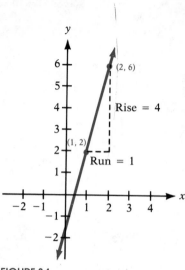

FIGURE 34
$y = 4x - 2$

≡

EXAMPLE 6   Find an equation of the horizontal line passing through the point (3, 2).

Solution   The slope of a horizontal line is zero. To get an equation, we use the point–slope form with $m = 0$, $x_1 = 3$, and $y_1 = 2$:

$$y - y_1 = m(x - x_1)$$
$$y - 2 = 0 \cdot (x - 3)$$
$$y - 2 = 0$$
$$y = 2$$

See Figure 35 for an illustration.

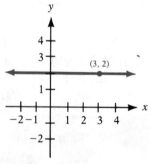

FIGURE 35
$y = 2$

≡

Based on Example 6, we have the following result:

THEOREM   A horizontal line is given by an equation of the form

EQUATION OF A HORIZONTAL
LINE

$$y = b$$

where b is a given real number.   ≡

EXAMPLE 7   Find an equation of the line L passing through the points (2, 3) and (−4, 5). Graph the line L.

Solution   Since two points are given, we first compute the slope of the line:

$$m = \frac{5 - 3}{-4 - 2} = \frac{2}{-6} = \frac{-1}{3}$$

We use the point (2, 3) and the fact that the slope $m = -\frac{1}{3}$, to get the point–slope form of the equation of the line:

$$y - 3 = -\frac{1}{3}(x - 2)$$

See Figure 36 for the graph.

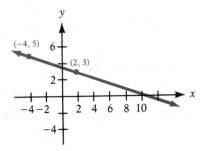

FIGURE 36
$y - 3 = -\frac{1}{3}(x - 2)$ ≡

In the solution in Example 7, we could have used the other point, (−4, 5), instead of the point (2, 3). The equation that results, although it looks different, is equivalent to the answer we obtained in the example. (Try it for yourself.)

Another form of the equation of the line in Example 7 can be obtained by multiplying both sides of its point–slope equation by 3 and collecting terms:

$$y - 3 = -\frac{1}{3}(x - 2)$$

$$3(y - 3) = 3\left(-\frac{1}{3}\right)(x - 2) \quad \text{Multiply by 3.}$$

$$3y - 9 = -1(x - 2)$$

$$3y - 9 = -x + 2$$

$$x + 3y - 11 = 0$$

GENERAL FORM OF EQUATION OF A LINE

The equation of a line L is in **general form** when it is written as

$$Ax + By + C = 0 \qquad (3)$$

where A, B, and C are three real numbers and A and B cannot both be zero.

Every line has an equation that is equivalent to its equation written in general form. For example, a vertical line whose equation is

$$x = a$$

can be written in the general form

$$1 \cdot x + 0 \cdot y - a = 0 \quad A = 1, B = 0$$

A horizontal line whose equation is

$$y = b$$

can be written in the general form

$$0 \cdot x + 1 \cdot y - b = 0 \quad A = 0, B = 1$$

Lines that are neither vertical nor horizontal have general equations of the form

$$Ax + By + C = 0 \quad A \neq 0 \text{ and } B \neq 0$$

LINEAR EQUATION

Because the equation of every line can be written in general form, we also refer to any equation in the general form, given in Equation (3), as a **linear equation**.

The next example illustrates one way of graphing a linear equation.

EXAMPLE 8   Find the intercepts of the line $2x + 3y - 6 = 0$. Graph this line.

Solution To find the point at which the graph crosses the x-axis—that is, to find the x-intercept—we need to find the number x for which $y = 0$. Thus, we let $y = 0$ to get

$$2x + 3(0) - 6 = 0$$
$$2x - 6 = 0$$
$$x = 3$$

The x-intercept is 3. To find the y-intercept, we let $x = 0$ and solve for y:

$$2(0) + 3y - 6 = 0$$
$$3y - 6 = 0$$
$$y = 2$$

The y-intercept is 2.

We now know two points on the line: $(3, 0)$ and $(0, 2)$. Because two points determine a unique line, we do not need any more information to graph the line. See Figure 37.

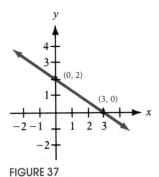

FIGURE 37

Another useful equation of a line is obtained when the slope m and y-intercept b are known. In this event we know both the slope m of the line and a point $(0, b)$ on the line; thus we may use the point–slope form from Equation (2) to obtain the following equation:

$$y - b = m(x - 0) \quad \text{or} \quad y = mx + b$$

THEOREM An equation of a line L with slope m and y-intercept b is

SLOPE–INTERCEPT FORM

$$y = mx + b \qquad (4)$$

When the equation of a line is written in slope–intercept form, it is easy to find the slope m and y-intercept b of the line. For example, suppose the equation of a line is

$$y = -2x + 3$$

Compare it to $y = mx + b$:

$$y = -2x + 3$$
$$\uparrow \qquad \uparrow$$
$$y = mx + b$$

The slope of this line is -2 and its y-intercept is 3.

Let's look at another example.

EXAMPLE 9   Find the slope m and y-intercept b of the line $2x + 4y - 8 = 0$. Graph the line.

Solution   To obtain the slope and y-intercept, we transform the equation into its slope–intercept form. Thus, we need to solve for y:

$$2x + 4y - 8 = 0$$
$$4y = -2x + 8$$
$$y = -\frac{1}{2}x + 2$$

The coefficient of x, $-\frac{1}{2}$, is the slope, and the y-intercept is 2. We can graph the line in two ways:

1. Use the fact that the y-intercept is 2 and the slope is $-\frac{1}{2}$. Then, starting at the point $(0, 2)$, go to the right 2 units and then down 1 unit to the point $(2, 1)$.

Or:

2. Locate the intercepts. Because the y-intercept is 2, we know one intercept is $(0, 2)$. To obtain the x-intercept, let $y = 0$ and solve for x. When $y = 0$, we have

$$2x + 4 \cdot 0 - 8 = 0$$
$$2x - 8 = 0$$
$$x = 4$$

Thus, the intercepts are $(4, 0)$ and $(0, 2)$.

See Figure 38.

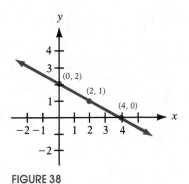

FIGURE 38

The next example illustrates a typical situation that requires the use of linear equations.

EXAMPLE 10   The National Car Rental Company has determined that the average cost of operating a vehicle is $0.41 per mile. Write an equation that relates the average cost C, in dollars, of operating a car and the number x of miles it has been driven.

Solution   If x is the number of miles the car has been driven, then the average cost C, in dollars, is $0.41x$. Thus, an equation relating C and x is

$$C = 0.41x \qquad x \geq 0$$

The average cost per mile, 0.41, is the slope of the line $C = 0.41x$. In other words, the cost increases by $0.41 for each additional mile driven. See Figure 39.

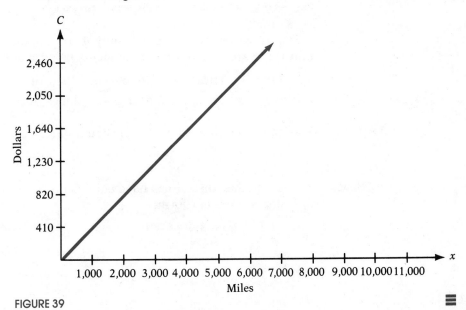

FIGURE 39

Parallel and Perpendicular Lines

PARALLEL LINES   When two lines (in a plane) have no points in common, they are said to be **parallel**. Look at Figure 40. There we have drawn two lines and have constructed two right triangles by drawing sides parallel to the coordinate axes. These lines are parallel if and only if the right triangles are similar. (Do you see why? Two angles are equal.) But the triangles are similar if and only if the sides are in proportion. This suggests the following result:

THEOREM   Two nonvertical lines are parallel if and only if their slopes are equal.

≡

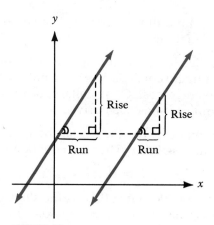

FIGURE 40
The lines are parallel if and only if the slopes are equal

The use of the words "if and only if" in the above theorem means that there are actually two statements being made:

If two nonvertical lines are parallel, then their slopes are equal.

If two nonvertical lines have equal slopes, then they are parallel.

EXAMPLE 11   Show that the lines given by the equations below are parallel:

$$L: \quad 2x + 3y - 6 = 0 \qquad M: \quad 4x + 6y = 0$$

Solution   To see whether these lines have equal slopes, we put each equation into slope–intercept form:

$$
\begin{array}{ll}
L: \quad 2x + 3y - 6 = 0 & \qquad M: \quad 4x + 6y = 0 \\
\qquad\quad 3y = -2x + 6 & \qquad\qquad 6y = -4x \\
\qquad\quad y = -\dfrac{2}{3}x + 2 & \qquad\qquad y = -\dfrac{2}{3}x \\
\qquad\quad \text{Slope} = -\dfrac{2}{3} & \qquad\qquad \text{Slope} = -\dfrac{2}{3}
\end{array}
$$

Because each line has slope $-\frac{2}{3}$, the lines are parallel. See Figure 41.

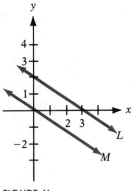

FIGURE 41
Parallel lines

When two lines intersect at a right angle (90°), they are said to be
PERPENDICULAR LINES **perpendicular**. See Figure 42.

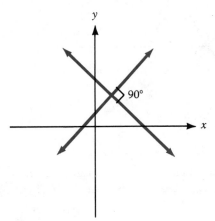

FIGURE 42
Perpendicular lines

The following result gives a condition in terms of their slope for
two lines to be perpendicular.

THEOREM Two nonvertical lines are perpendicular if and only if the product of
their slopes is -1.

Proof We shall prove here only the statement:

If two nonvertical lines are perpendicular, then the product of their
slopes is -1.

In Problem 77 you are asked to prove the second part of the theorem; that is:

If two nonvertical lines have slopes whose product is -1, then the lines are perpendicular.

Let m_1 and m_2 denote the slopes of the two lines. There is no loss in generality (that is, neither the angle nor the slopes are affected) if we situate the lines so that they meet at the origin. See Figure 43. The point $A = (1, m_2)$ is on the line having slope m_2, and the point $B = (1, m_1)$ is on the line having slope m_1. (Do you see why this must be?)

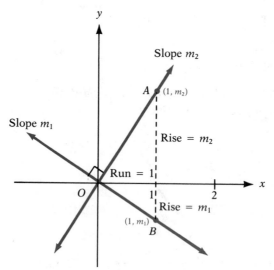

FIGURE 43

Suppose the lines are perpendicular. Then triangle OAB is a right triangle. As a result of the Pythagorean Theorem, it follows that

$$[d(O, A)]^2 + [d(O, B)]^2 = [d(A, B)]^2 \qquad (5)$$

By the distance formula, we can write each of these distances as

$$[d(O, A)]^2 = (1 - 0)^2 + (m_2 - 0)^2 = 1 + m_2^2$$
$$[d(O, B)]^2 = (1 - 0)^2 + (m_1 - 0)^2 = 1 + m_1^2$$
$$[d(A, B)]^2 = (1 - 1)^2 + (m_2 - m_1)^2 = m_2^2 - 2m_1m_2 + m_1^2$$

Using these facts in Equation (5), we get

$$(1 + m_2^2) + (1 + m_1^2) = m_2^2 - 2m_1m_2 + m_1^2$$

which, upon simplification, can be written as

$$m_1m_2 = -1$$

Thus, if the lines are perpendicular, the product of their slopes is -1.
≡

You may find it easier to remember the condition for two non-vertical lines to be perpendicular by observing that the equality $m_1 m_2 = -1$ means that m_1 and m_2 are negative reciprocals of each other; that is, either $m_1 = -1/m_2$ or $m_2 = -1/m_1$.

EXAMPLE 12   If a line has slope $\frac{3}{2}$, any line having slope $-\frac{2}{3}$ is perpendicular to it.
≡

EXAMPLE 13   Find the general form of the equation of a line passing through the point $(1, -2)$ and perpendicular to the line $x + 3y - 6 = 0$.

Solution   We first put the equation of the given line into slope–intercept form to find its slope:

$$x + 3y - 6 = 0$$
$$3y = -x + 6$$
$$y = -\frac{1}{3}x + 2$$

The given line has slope $-\frac{1}{3}$. Any line perpendicular to this line will have slope 3. Because we require the point $(1, -2)$ to be on this line with slope 3, we use the point–slope form of the equation to get

$$y + 2 = 3(x - 1)$$

This equation is equivalent to the general form

$$3x - y - 5 = 0$$
≡

EXERCISE 3.3   *In Problems 1–10 plot the points and compute the slope of the line containing them. Graph the line.*

1. $(2, 3)$, $(1, 0)$
2. $(1, 2)$, $(3, 4)$
3. $(-2, 3)$, $(2, 1)$
4. $(-1, 1)$, $(2, 3)$
5. $(-3, -1)$, $(2, -1)$
6. $(4, 2)$, $(-5, 2)$
7. $(-1, 2)$, $(-1, -2)$
8. $(2, 0)$, $(2, 2)$
C 9. $(\sqrt{2}, 3)$, $(1, \sqrt{3})$
C 10. $(-2\sqrt{2}, 0)$, $(4, \sqrt{5})$

In Problems 11–18 graph the line passing through the point P and having slope m.

11. $P = (1, 2)$; $m = 2$
12. $P = (2, 1)$; $m = 3$
13. $P = (2, 4)$; $m = -3$
14. $P = (1, 3)$; $m = -2$
15. $P = (-1, 3)$; $m = 0$
16. $P = (2, -4)$; $m = 0$
17. $P = (0, 3)$; slope undefined
18. $P = (-2, 0)$; slope undefined

In Problems 19–42 find a general equation for the line with the given properties.

19. Slope = 2;   passing through $(-2, 3)$
20. Slope = 3;   passing through $(4, -3)$
21. Slope = $-\frac{2}{3}$;   passing through $(1, -1)$
22. Slope = $\frac{1}{2}$;   passing through $(3, 1)$
23. Passing through $(1, 3)$ and $(-1, 2)$
24. Passing through $(-3, 4)$ and $(2, 5)$
25. Slope = -3; y-intercept = 3
26. Slope = -2; y-intercept = -2
27. x-intercept = 2; y-intercept = -1
28. x-intercept = -4; y-intercept = 4
29. Slope undefined;   passing through $(1, 4)$
30. Slope undefined;   passing through $(2, 1)$
31. Parallel to the line $y = 3x$;   passing through $(-1, 2)$
32. Parallel to the line $y = -2x$;   passing through $(-1, 2)$
33. Parallel to the line $2x - y + 2 = 0$;   passing through $(0, 0)$
34. Parallel to the line $x - 2y + 5 = 0$;   passing through $(0, 0)$
35. Parallel to the line $x = 5$;   passing through $(4, 2)$
36. Parallel to the line $y = 5$;   passing through $(4, 2)$
37. Perpendicular to the line $y = \frac{1}{2}x + 4$;   passing through $(1, -2)$
38. Perpendicular to the line $y = 2x - 3$;   passing through $(1, -2)$
39. Perpendicular to the line $2x + y - 2 = 0$;   passing through $(-3, 0)$
40. Perpendicular to the line $x - 2y + 5 = 0$;   passing through $(0, 4)$
41. Perpendicular to the line $x = 5$;   passing through $(3, 4)$
42. Perpendicular to the line $y = 5$;   passing through $(3, 4)$

In Problems 43–54 find the slope and y-intercept of each line. Graph the line.

43. $y = 2x + 3$ **44.** $y = -3x + 4$ **45.** $\frac{1}{2}y = x - 1$
46. $\frac{1}{3}x + y = 2$ **47.** $2x - 3y = 6$ **48.** $3x + 2y = 6$
49. $x + y = 1$ **50.** $x - y = 2$ **51.** $x = -4$
52. $y = -1$ **53.** $2y - 3x = 0$ **54.** $x + y = 0$

In Problems 55–58 find the general equation of each line.

55.

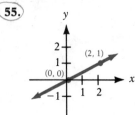

56.

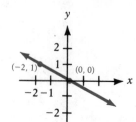

57.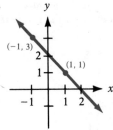

58.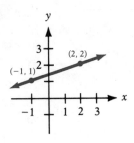

59. Find the general equation of the x-axis.

60. Find the general equation of the y-axis.

61. Use slopes to show that the triangle whose vertices are $(-2, 5)$, $(1, 3)$, and $(-1, 0)$ is a right triangle.

62. Use slopes to show that the quadrilateral whose vertices are $(1, -1)$, $(4, 1)$, $(2, 2)$, and $(5, 4)$ is a parallelogram.

63. Use slopes to show that the quadrilateral whose vertices are $(-1, 0)$, $(2, 3)$, $(1, -2)$, and $(4, 1)$ is a rectangle.

64. Show that an equation for a line with nonzero x- and y-intercepts can be written as

$$\frac{x}{a} + \frac{y}{b} = 1$$

INTERCEPT FORM

where a is the x-intercept and b is the y-intercept. This is called the **intercept form** of the equation of a line.

65. The relationship between Celsius (°C) and Fahrenheit (°F) for measuring temperature is linear. Find an equation relating °C and °F if 0° C corresponds to 32° F and 100° C corresponds to 212° F. Use the equation to find the Celsius measure of 70° F.

66. The Kelvin (K) scale for measuring temperature is obtained by adding 273 to the Celsius temperature.
(a) Write an equation relating K and °C.
(b) Write an equation relating K and °F (see Problem 65).

FAMILY OF LINES

67. The equation $2x - y + C = 0$ defines a **family of lines**, one line for each value of C. On one set of coordinate axes, graph the members of the family when $C = -4$, $C = 0$, and $C = 2$. Can you draw a conclusion from the graph about each member of the family?

68. Rework Problem 67 for the family of lines $Cx + y + 4 = 0$.

69. Each Sunday a newspaper agency sells x copies of a certain newspaper for $1.00 per copy. The cost to the agency of each newspaper is $0.50. The agency pays a fixed cost for storage, delivery, and so on, of $100 per Sunday. Write an equation that relates the profit P, in dollars, to the number x of copies sold. Graph this equation.

70. Repeat Problem 69 if the cost to the agency is $0.45 per copy and the fixed cost is $125 per Sunday.

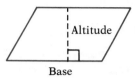

Altitude

Base

In Problems 71–76 find the area of the parallelogram with the given vertices. [*Hint: The area of a parallelogram is the product of the base times the altitude to that base. See the figure.*]

71. (0, 0), (4, 0), (1, 3), (5, 3)
72. (0, 0), (0, 2), (3, 1), (3, −1)
73. (2, 1), (4, 2), (2, 3), (4, 4)
74. (1, 2), (2, 0), (3, 0), (2, 2)
75. (1, 1), (3, 2), (2, −1), (4, 0)
76. (1, 3), (0, 1), (3, 2), (2, 0)

77. Prove that, if two nonvertical lines have slopes whose product is −1, then the lines are perpendicular. [*Hint: Use the converse of the Pythagorean Theorem.*]

▰ 3.4 VARIATION

Relationships that exist between quantities are often expressed using language such as the following:

> Force is proportional to acceleration.
>
> For an ideal gas held at a constant temperature, pressure and volume are inversely proportional.
>
> The force of attraction between two heavenly bodies is inversely proportional to the square of the distance between them.
>
> Revenue is directly proportional to sales.

VARIATION Each of these statements illustrates the idea of **variation**, or how one quantity varies in relation to another quantity. Quantities may vary directly, inversely, or jointly.

Direct Variation

VARIES DIRECTLY Let x and y denote two quantities. Then y **varies directly** as x if there is a nonzero number k such that

$$y = kx$$

CONSTANT OF PROPORTIONALITY
DIRECTLY PROPORTIONAL

The number k is called the **constant of proportionality**. When y varies directly as x (or with x), we say that y is **directly proportional** to x.

The graph in Figure 44 illustrates the relationship between y and x if y varies directly as x and $k > 0$, $x \geq 0$. Note that the constant of proportionality is, in fact, the slope of the line.

If we know that two quantities vary directly, then knowing the value of each quantity in one instance enables us to write a formula that is true in all cases.

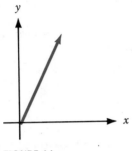

FIGURE 44
$y = kx, k > 0; x \geq 0$

EXAMPLE 1 For a certain gas enclosed in a container of fixed volume, the pressure P (in pounds per square inch) varies directly as the temperature T (in degrees Celsius). If the pressure is found to be 10 pounds per square inch at a temperature of 30° C, find a formula that relates pressure P to temperature T. Then find the pressure P when $T = 120°$ C.

Solution Because P varies directly as T, we know that

$$P = kT$$

for some number k. Because $P = 10$ when $T = 30$,

$$10 = k(30)$$

$$k = \frac{1}{3}$$

Thus, in all cases,

$$P = \frac{1}{3}T$$

In particular, when $T = 120°$ C, we find

$$P = \frac{1}{3}(120) = 40 \text{ pounds per square inch}$$

Figure 45 illustrates the relationship between the pressure P and the temperature T.

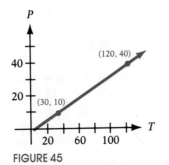

FIGURE 45

Inverse Variation

VARIES INVERSELY   Let x and y denote two quantities. Then y **varies inversely** as x if there is a nonzero constant k such that

$$y = \frac{k}{x}$$

INVERSELY PROPORTIONAL   In the case of inverse variation, we say that y is **inversely proportional** to x.

The graph in Figure 46 illustrates the relationship between y and x if y varies inversely as x and $k > 0$, $x > 0$.

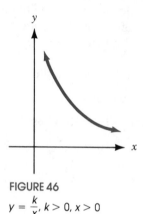

FIGURE 46
$y = \dfrac{k}{x}, k > 0, x > 0$

EXAMPLE 2   The weight W that can be safely supported by a 2-by-4-inch piece of lumber varies inversely as its length l. Experiments indicate that the maximum weight a 10-foot piece of 2-by-4-inch pine can support is 500 pounds. Write a general formula relating the safe weight W (in pounds) to length l (in feet). Find the maximum weight W that can be safely supported by a length of 25 feet.

Solution   Because W varies inversely as l, we know that

$$W = \frac{k}{l}$$

for some number k. Because $W = 500$ when $l = 10$, we have

$$500 = \frac{k}{10}$$

$$k = 5000$$

Thus, in all cases,

$$W = \frac{5000}{l}$$

In particular, the maximum weight W that can be safely supported by a piece of lumber 25 feet in length is

$$W = \frac{5000}{25} = 200 \text{ pounds}$$

Figure 47 illustrates the relationship between the weight W and the length l.

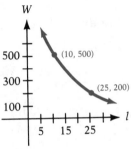

FIGURE 47

In direct or inverse variation, the quantities that vary can be raised to powers. For example, the square of the period T of a planet varies directly as the cube of its mean distance a from the sun. That is, $T^2 = ka^3$, where k is the constant of proportionality.

Joint Variation and Combined Variation

VARIES JOINTLY

COMBINED VARIATION

When a variable quantity Q is proportional to the product of two or more other variables, then we say that Q **varies jointly** as (or with) these quantities. Finally, combinations of direct and/or inverse variation may occur. This is usually referred to as **combined variation**.
  Let's look at an example.

EXAMPLE 3   The loss of heat through a wall varies jointly as the area of the wall and the temperature difference between the inside and outside, and inversely with the thickness of the wall. Write an equation that relates these quantities.

Solution   We begin by assigning symbols to represent the quantities:

L = Heat loss T = Temperature difference
A = Area of wall d = Thickness of wall

Then,

$$L = k\frac{AT}{d}$$

where k is the constant of proportionality.

EXAMPLE 4   The force F of the wind on a plane surface positioned at a right angle to the direction of the wind varies jointly as the area A of the surface and the square of the speed v of the wind. A wind of 30 miles per hour blowing on a window measuring 4 feet by 5 feet has a force of 150 pounds. What is the force on a window measuring 3 feet by 4 feet caused by a wind of 50 miles per hour?

Solution   Since F varies jointly as A and v^2, we have

$$F = kAv^2$$

where k is the constant of proportionality. We are told that $F = 150$ when $v = 30$ and $A = 4 \cdot 5 = 20$. Thus,

$$150 = k(20)(900)$$

$$k = \frac{1}{120}$$

The general formula is therefore

$$F = \frac{1}{120}Av^2$$

For a wind of 50 miles per hour blowing on a window whose area is $A = 3 \cdot 4 = 12$ square feet, the force F is

$$F = \frac{1}{120}(12)(2500) = 250 \text{ pounds}$$

EXERCISE 3.4   *In Problems 1–12 write a general formula to describe each variation.*

1. y varies directly as x; $y = 5$ when $x = 10$

2. v varies directly as t; $v = 16$ when $t = 1$

3. A varies directly as x^2; $A = 4\pi$ when $x = 2$

4. V varies directly as x^3; $V = 36\pi$ when $x = 3$

5. F varies inversely as d^2; $F = 10$ when $d = 5$

6. y varies inversely as $\sqrt{x}$; $y = 4$ when $x = 9$

7. z varies directly as the sum of the squares of x and y; $z = 5$ when $x = 3$ and $y = 4$

8. T varies jointly as the cube root of x and the square of d; $T = 18$ when $x = 8$ and $d = 3$

9. M varies directly as the square of d and inversely as the square root of x; $M = 24$ when $x = 9$ and $d = 4$

10. z varies directly as the sum of the cube of x and the square of y; $z = 1$ when $x = 2$ and $y = 3$

11. The square of T varies directly as the cube of a and inversely as the square of d; $T = 2$ when $a = 1$ and $d = 4$

12. The cube of z varies directly as the sum of the squares of x and y; $z = 2$ when $x = 4$ and $y = 9$

In Problems 13–20 write an equation that relates the quantities.

13. The volume V of a sphere varies directly as the cube of its radius R. The constant of proportionality is $4\pi/3$.

14. The square of the hypotenuse c of a right triangle varies directly as the sum of the squares of the legs a and b. The constant of proportionality is 1.

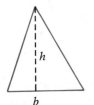

15. The area A of a triangle varies jointly as the lengths of the base b and the height h. The constant of proportionality is $\frac{1}{2}$.

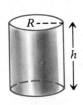

16. The perimeter p of a rectangle varies directly as the sum of the lengths of its sides l and w. The constant of proportionality is 2.

17. The volume V of a right circular cylinder varies jointly as the square of its radius R and its height h. The constant of proportionality is π.

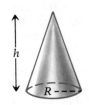

18. The volume V of a right circular cone varies jointly as the square of its radius R and its height h. The constant of proportionality is $\pi/3$.

19. The force F (in newtons) of attraction between two bodies varies jointly as their masses m and M (in kilograms) and inversely as the square of the distance d (in meters) between them. The constant of proportionality is $G = 6.67 \times 10^{-11}$.

PERIOD
SIMPLE

20. The **period** of a pendulum is the time required for one oscillation; the pendulum is usually referred to as **simple** when the angle made to the vertical is less than 5 degrees. The period T of a simple pendulum (in seconds) varies directly as the square root of its length l (in feet). The constant of proportionality is $2\pi/\sqrt{32}$.

21. The distance s an object falls is directly proportional to the square of the time t of the fall. If an object falls 16 feet in 1 second, how far will it fall in 3 seconds? How long will it take an object to fall 64 feet?

22. The velocity v of a falling object is directly proportional to the time t of the fall. If, after 2 seconds, the velocity of the object is 64 feet per second, what will its velocity be after 3 seconds?

23. The elongation E of a spring balance varies directly as the applied weight W. If $E = 3$ when $W = 20$, find E when $W = 15$.

24. The rate of vibration of a string under constant tension varies inversely as the length of the string. If a string is 48 inches long and vibrates 256 times per second, what is the length of a string that vibrates 576 times per second?

C 25. The weight of a body above the surface of the earth varies inversely as the square of the distance from the center of the earth. If a certain body weighs 55 pounds when it is 4×10^3 miles from the center of the earth, how much will it weigh when it is 4.4×10^3 miles from the center?

C 26. The force exerted by the wind on a plane surface varies jointly as the area of the surface and the square of the velocity of the wind. If the force on an area of 20 square feet is 11 pounds when the wind velocity is 22 miles per hour, find the force on a surface area of 47.125 square feet when the wind velocity is 36.5 miles per hour.

C 27. The horsepower that a shaft can safely transmit varies jointly as its speed (in revolutions per minute, rpm) and the cube of its diameter. If a shaft of a certain material 3 inches in diameter can transmit 36 horsepower at 75 rpm, what diameter must the shaft have in order to transmit 45 horsepower at 125 rpm?

28. The weight of a body varies inversely as the square of its distance from the center of the earth. Assuming that the radius of the earth is 4000 miles, how much would a man weigh at an altitude of 1 mile above the earth's surface if he weighs 180 pounds on the earth's surface?

29. The kinetic energy K of a moving object varies jointly as its mass m and the square of its velocity v. If an object weighing 25 pounds and moving with a velocity of 100 feet per second has a kinetic energy of 400 foot-pounds, find its kinetic energy when the velocity is 150 feet per second.

C 30. The electrical resistance of a wire varies directly as the length of the wire and inversely as the square of the diameter of the wire. If a wire 432 feet long and 4 millimeters in diameter has a resistance of 1.24 ohms, find the length of a wire of the same material whose resistance is 1.44 ohms and whose diameter is 3 millimeters.

31. The stress in the material of a pipe subject to internal pressure varies jointly as the internal pressure and the internal diameter of the pipe and inversely as the thickness of the pipe. The stress is 100 pounds per square inch when the diameter is 5 inches, the thickness is 0.75 inch, and the internal pressure is 25 pounds per square inch. Find the stress when the internal pressure is 40 pounds per square inch, if the diameter is 8 inches and the thickness is 0.50 inch.

C 32. The maximum safe load for a horizontal rectangular beam varies jointly as the width of the beam and the square of the thickness of the beam and inversely as its length. If an 8-foot beam will support up to 750 pounds when the beam is 2 inches wide and 4 inches thick, what is the maximum safe load in a similar beam 10 feet long, 2 inches wide, and 6 inches thick?

33. The resistance in ohms of a circular conductor varies directly as the length of the conductor and inversely as the square of the radius of the conductor. If 50 feet of a wire with a radius of 6×10^{-3} inch has a resistance of 10 ohms, what would be the resistance of 100 feet of the same wire if the radius is increased to 7×10^{-3} inch?

34. The volume V of an ideal gas varies directly with the temperature T and inversely with the pressure P. Write an equation relating V, T, and P, using k as the constant of proportionality. If a cylinder contains oxygen at a temperature of 20° C and a pressure of 15 atmospheres in a volume of 100 liters, what is the constant of proportionality k? If a piston is lowered into the cylinder decreasing the volume occupied by the gas to 80 liters and raising the temperature to 25° C, what is the gas pressure?

Problems 36–40 use the result obtained in Problem 35.

35. The speed v required of a satellite to maintain a near-earth circular orbit is directly proportional to the square root of the distance R of the satellite from the center of the earth.* The constant of proportionality is $\sqrt{g}$, where g is the acceleration of gravity for earth. Write an equation that shows the relationship between v and R. (The radius of the earth is approximately 4000 miles.)

C **36.** What speed is required to maintain a communications satellite in a circular orbit 500 miles above the earth's surface?

C **37.** Find the speed of a satellite that moves in a circular orbit 100 miles above the earth's surface.

C **38.** Find the distance of a satellite from the surface of the earth as it moves around the earth in a circular orbit at a constant speed of 18,630 miles per hour.

C **39.** A weather satellite orbits the earth in a circle every 1.5 hours. How high is it above the earth?

C **40.** Some communications satellites remain stationary above a fixed point on the equator of the earth's surface. How high are such satellites and what is their common speed? Assume the earth turns once every 24 hours.

Problems 42–46 use the result obtained in Problem 41.

41. The force F (in newtons) required to maintain an object in a circular path varies jointly with the mass m (in kilograms) of the object and the square of the speed v (in meters per second) of the object and inversely with the radius R (in meters) of the circular path. The constant of proportionality is 1. Write an equation relating F, m, v, and R.

*Near-earth orbits are at least 100 miles above the earth's surface (out of the earth's atmosphere) and up to an altitude of approximately 15,000 miles. The effect of the gravitational attraction of other bodies is ignored. Although the acceleration of gravity at such altitudes is somewhat less than $g \approx 32$ feet per second per second $\approx 79,036$ miles per hour per hour, we shall ignore this discrepancy in our calculations.

C 42. A motorcycle that weighs 150 kilograms is driven at a constant speed of 120 kilometers per hour on a circular track with a radius of 100 meters. To keep the motorcycle from skidding, what frictional force must be exerted by the tires on the track?

43. If the speed of the motorcycle described in Problem 42 is increased by 10%, by how much is the frictional force of the tires increased?

44. If the radius of the track described in Problem 42 is cut in half, how much slower should the motorcycle be driven to maintain the same frictional force?

45. A girl is spinning a bucket of water in a horizontal plane at the end of a rope of length L. If she triples the speed of the bucket, how many times as hard must she pull on the rope?

46. If the girl in Problem 45 doubles the length of the rope and maintains the same speed for the bucket, will she have to pull on the rope more or less? How much?

C H A P T E R R E V I E W

VOCABULARY

x-axis
y-axis
origin
rectangular coordinate system
Cartesian coordinate system
xy-plane
coordinate axes
ordered pair
coordinates
plot
x-coordinate
y-coordinate
quadrants
distance formula
midpoint of a line segment
midpoint formula
graph of an equation
parabola
intercepts

x-intercept
y-intercept
symmetry with respect to the x-axis
symmetry with respect to the y-axis
symmetry with respect to the origin
tests for symmetry
circle
radius
center
standard form of the equation of a circle
unit circle
general form of the equation of a circle
run
rise
slope of a line
vertical line
undefined slope

horizontal line
equation of a vertical line
point–slope form
equation of a horizontal line
general form of the equation of a line
linear equation
slope–intercept form
parallel lines
perpendicular lines
variation
varies directly
constant of proportionality
directly proportional
varies inversely
inversely proportional
varies jointly
combined variation

FILL-IN-THE-BLANK QUESTIONS

1. If (x, y) are the coordinates of a point P in the xy-plane, then x is called the _____ of P and y is the _____ of P.

2. If three points P, Q, and R all lie on a line and if $d(P, Q) = d(Q, R)$, then Q is called the _____ of the line segment from P to R.

3. If, for every point (x, y) on a graph, the point $(-x, y)$ is also on the graph, then the graph is symmetric with respect to the _____.

4. The set of points in the xy-plane that are a fixed distance from a fixed point is called a(n) _____. The fixed distance is called the _____; the fixed point is called the _____.

5. The slope of a vertical line is _____; the slope of a horizontal line is _____.

6. Two nonvertical lines have slopes m_1 and m_2, respectively. The lines are parallel if _____; the lines are perpendicular if _____ _____.

7. If z varies jointly as x^2 and y^3 and inversely as $\sqrt{t}$, then $z =$ _____ _____, where k is the constant of proportionality.

REVIEW EXERCISES

In Problems 1–10 find a general equation of the line having the given characteristics.

1. Slope $= -2$; passing through $(2, -1)$
2. Slope $= 0$; passing through $(-3, 4)$
3. Slope undefined; passing through $(-3, 4)$
4. x-intercept $= 2$; passing through $(4, -5)$
5. y-intercept $= -2$; passing through $(5, -3)$
6. Passing through $(3, -4)$ and $(2, 1)$
7. Parallel to the line $2x - 3y + 4 = 0$; passing through $(-5, 3)$
8. Parallel to the line $x + y - 2 = 0$; passing through $(1, -3)$
9. Perpendicular to the line $x + y - 2 = 0$; passing through $(1, -3)$
10. Perpendicular to the line $3x - y + 4 = 0$; passing through $(-2, 2)$

In Problems 11–16 graph each line, labeling the x-intercept and y-intercept.

11. $4x - 5y + 20 = 0$
12. $3x + 4y - 12 = 0$
13. $\frac{1}{2}x - \frac{1}{3}y + \frac{1}{6} = 0$
14. $-\frac{3}{4}x + \frac{1}{2}y = 0$
15. $\sqrt{2}x + \sqrt{3}y = \sqrt{6}$
16. $\frac{x}{3} + \frac{y}{4} = 1$

In Problems 17–20 find the center and radius of each circle.

17. $x^2 + y^2 - 2x + 4y - 4 = 0$
18. $x^2 + y^2 + 4x - 4y - 1 = 0$
19. $3x^2 + 3y^2 - 6x + 12y = 0$
20. $2x^2 + 2y^2 - 4x = 0$

In Problems 21–28 list the intercepts and test for symmetry.

21. $2x = 3y^2$
22. $y = 5x$
23. $4x^2 + y^2 = 1$
24. $x^2 - 9y^2 = 9$
25. $y = x^4 + 2x^2 + 1$
26. $y = x^3 - x$
27. $x^2 + x + y^2 + 2y = 0$
28. $x^2 + 4x + y^2 - 2y = 0$

C **29.** The area of an equilateral triangle varies directly as the square of the length of a side. If the area of the equilateral triangle whose sides are of length 1 centimeter is $\sqrt{3}/4$, find the length s of each side of an equilateral triangle whose area A is 16 square centimeters.

30. In a vibrating string, the pitch varies directly as the square root of the tension of the string. If a certain string vibrates 300 times per second under a tension of 9 pounds, find the tension required to cause the string to vibrate 400 times per second.

C **31.** Kepler's third law of planetary motion states that the square of the period T of revolution of a planet is proportional to the cube of its mean distance from the sun. If the mean distance of the earth from the sun is 93 million miles, what is the mean distance a of the planet Mercury from the sun, given that Mercury has a "year" of 88 days?

C **32.** Use Problem 31 to find the mean distance of the planet Jupiter from the sun, given that Jupiter circles the sun every $5\sqrt{5}$ years.

33. Show that the midpoint of the hypotenuse of a right triangle is the same distance from each of the three vertices. [*Hint:* Use the figure in the margin.]

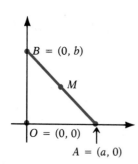

34. Show that the line joining the midpoints of two sides of a triangle is parallel to the third side.

35. Show that the points $A = (3, 4)$, $B = (1, 1)$ and $C = (-2, 3)$ are the vertices of an isosceles triangle.

36. Show that the points $A = (-2, 0)$, $B = (-4, 4)$, and $C = (8, 5)$ are the vertices of a right triangle in two ways:
(a) By using the converse of the Pythagorean Theorem.
(b) By using the slopes of the lines joining the vertices.

37. Show that the points $A = (2, 5)$, $B = (6, 1)$, and $C = (8, -1)$ lie on a straight line by using slopes.

38. Show that the points $A = (1, 5)$, $B = (2, 4)$, and $C = (-3, 5)$ lie on a circle with center at $(-1, 2)$. What is the radius of this circle?

39. The endpoints of the diameter of a circle are $(-3, 2)$ and $(5, -6)$. Find the center and radius of the circle. Write the general equation of this circle.

40. Find two numbers y such that the distance from $(-3, 2)$ to $(5, y)$ is 10.

FUNCTIONS AND THEIR GRAPHS

4.1 Functions
4.2 The Graph of a Function
4.3 Graphing Techniques
4.4 Operations on Functions; Composite
     Functions
4.5 Inverse Functions
     Chapter Review

Perhaps the most central idea in mathematics is the notion of a *function*. This important chapter deals with what a function is, how to graph functions, and how to perform operations on functions.

The word *function* apparently was introduced by Descartes in 1637. For him, a function simply meant any positive integral power of a variable *x*. Leibniz, who always emphasized the geometric, used the word to denote any quantity associated with a curve, such as the coordinates of a point on the curve. Euler employed the word to mean any equation or formula involving variables and constants. His idea of a function is the one most often used today in courses that precede calculus. However, the use of functions in investigating heat flow equations led to a very broad definition, due to Lejeune Dirichlet (1805–1859), which described a function as a rule or correspondence between two sets.

■ 4.1 FUNCTIONS

In many applications, a correspondence often exists between two sets of numbers. For example, the revenue R resulting from the sale of x items selling for $10 each is $R = 10x$ dollars. If we know how many items have been sold, then we can calculate the revenue by using the rule $R = 10x$. This rule is an example of a *function*.

As another example, if an object is dropped from a height of 64 feet above the ground, the distance s in feet of the object from the ground after t seconds is given by the formula $s = 64 - 16t^2$. When $t = 0$ seconds, the object is $s = 64$ feet above the ground. After 1 second, the object is $s = 64 - 16(1)^2 = 48$ feet above the ground. After 2 seconds, the object strikes the ground. The formula $s = 64 - 16t^2$ provides a way of finding the distance s when the time t, $0 \leq t \leq 2$, is prescribed. There is a correspondence between each time t in the interval $0 \leq t \leq 2$ and the distance s. We say that the distance s is a *function* of the time t because:

1. There is a correspondence between the set of times and the set of distances.
2. There is exactly one distance s obtained for a prescribed time t in the interval $0 \leq t \leq 2$.

Let's now look at the definition of a function.

Definition of a Function

FUNCTION   Let X and Y be two sets of real numbers.* A **function** from X into Y is a rule or a correspondence that associates with each element of X

DOMAIN   a unique element of Y. The set X is called the **domain** of the function. For each element x in X, the corresponding element y in Y is called

VALUE   the **value** of the function at x, or the **image** of x. The set of all images

IMAGE   of the elements of the domain is called the **range** of the function.

RANGE

See Figure 1.

Since there may be some elements in Y that are not the image of some x in X, it follows that the range of a function may be a subset of Y. Look again at Figure 1.

*The two sets X and Y also could be sets of complex numbers; in this case, we have defined a complex function. In the broad definition (due to Lejeune Dirichlet), X and Y can be any two sets.

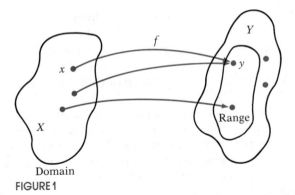

Domain

FIGURE 1

EXAMPLE 1   Consider the function: $y = 2x - 5$ $1 \le x \le 6$

The domain $1 \le x \le 6$ specifies that the number x is restricted to the real numbers from 1 to 6, inclusive. The rule $y = 2x - 5$ specifies that the number x is to be multiplied by 2 and then 5 is to be subtracted from the result to get y. For example, the value of the function at $x = \frac{3}{2}$—that is, the image of $x = \frac{3}{2}$—is $y = 2 \cdot \frac{3}{2} - 5 = -2$.   ≡

Functions are often denoted by letters such as f, F, g, G, and so on. If f is a function, then, for each number x in its domain, the corresponding image in the range is designated by the symbol $f(x)$, read "f of x." We refer to $f(x)$ as the **value of f at the number x**. Thus, $f(x)$ is the number that results when x is given and the rule for f is applied; $f(x)$ does *not* mean "f times x."

VALUE OF A FUNCTION

We list below a summary of some important items to remember about a function f.

1. $f(x)$ is the image of x or the value of f at x when the rule f is applied to an x in the domain.
2. To each x in the domain of f, there is one and only one image $f(x)$ in the range.
3. f is the symbol we use to denote the function. It is symbolic of a domain and a rule we use to get from an x in the domain to $f(x)$ in the range.

Domain of a Function

Usually the domain of a function f is not specified; instead, only a rule or equation defining the function is given. In such cases we automatically assume that the domain of f is the largest set of real numbers for which the rule makes sense or, more precisely, for which the value $f(x)$ can be computed as a real number. The range will then consist of all the images of the numbers in the domain.

EXAMPLE 2   The functions f, g, and F are defined by:

(a) $f(x) = x^2$ (b) $g(x) = \dfrac{1}{x}$ (c) $F(x) = \sqrt{x}$

Find the domain and range of each function.

Solution   (a) The operation of squaring a number can be performed on any real number x. Therefore, the domain of f is the set of all real numbers. What is the result when real numbers are squared? We get nonnegative real numbers. The range of f is therefore the set of nonnegative real numbers.

(b) We can divide 1 by any nonzero real number. Hence, the domain of g is all real numbers except zero. The range of g is also the set of nonzero real numbers.

(c) Because the square root of a negative number is not defined, the domain of F consists of all nonnegative real numbers. Because the square root of a nonnegative number is itself nonnegative, the range of F is the set of nonnegative real numbers.   ▬

Figure 2 illustrates the functions in Example 2 for selected choices of x. Because each of these functions is defined by an equation, we can obtain the image corresponding to a particular number by substituting the number for x in the equation.

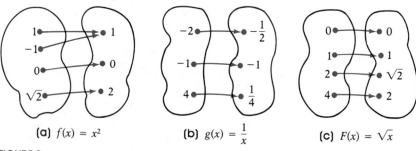

(a) $f(x) = x^2$ (b) $g(x) = \dfrac{1}{x}$ (c) $F(x) = \sqrt{x}$

FIGURE 2

It is often difficult to find the range of a function. Therefore, we shall usually be content to find just the domain of a function when only the rule for the function is given.

EXAMPLE 3   Find the domain of each function defined by:

(a) $f(x) = \dfrac{3}{x^2 - 4}$ (b) $g(x) = \sqrt{4 - 3x}$

Solution   (a) The rule f tells us to divide 3 by $x^2 - 4$. Division by zero is not allowed; therefore, the denominator $x^2 - 4$ can never be zero, so x can never equal 2 or -2. The domain of the function $f(x) = 3/(x^2 - 4)$ is the set of all real numbers except 2 and -2.

(b) The rule g tells us to take the square root of $4 - 3x$. But only nonnegative numbers have real square roots. Hence, we require that

$$4 - 3x \geq 0$$
$$-3x \geq -4$$
$$x \leq \tfrac{4}{3}$$

The domain of g consists of all real numbers less than or equal to $\tfrac{4}{3}$.

f(x) EXISTS
f(x) DOES NOT EXIST

If x is in the domain of a function f, we shall say that **f is defined at x** or that **$f(x)$ exists**. If x is not in the domain of f, we say that **f is not defined at x** or that **$f(x)$ does not exist**. For example, if $f(x) = x/(x^2 - 1)$, then $f(0)$ exists, but $f(1)$ and $f(-1)$ do not exist. Do you see why?

When we use functions in a practical setting, the domain may be restricted by physical or geometric considerations. For example, the domain of the function f defined by $f(x) = x^2$ is the set of all real numbers. However, if f is used as the rule for obtaining the area of a square when the length x of a side is known, then we would restrict the domain of f to the positive real numbers, because the length of a side is never zero or negative.

One other comment: Usually we write "the function $y = f(x)$" when we mean "the function f defined by the equation $y = f(x)$." Although this usage is not entirely correct, it is rather common and should not cause any confusion.

Independent Variable; Dependent Variable

Consider a function $y = f(x)$. The number x that appears here is called
INDEPENDENT VARIABLE   the **independent variable** because it can be assigned any of the permissible numbers from the domain. The number y is called the
DEPENDENT VARIABLE   **dependent variable** because its value depends on the number x.

Any choice of a symbol can be used to represent the independent and dependent variables. For example, if f is the "cubing function," then f can be defined by $f(x) = x^3$ or by $f(t) = t^3$ or by $f(z) = z^3$. All three rules are identical—each tells us to cube the independent variable. In practice the choice of symbols for the independent and dependent variables is based on common usage.

EXAMPLE 4
(a) The cost per square foot to build a house is $70. Express the cost C as a function of the square footage x.

(b) Express the area of a circle as a function of its radius.

Solution
(a) The cost C of building a house containing x square feet is $70x$ dollars. A function expressing this relationship is

$$C(x) = 70x$$

where x is the independent variable and C is the dependent variable. In this setting, the domain consists of positive real numbers x.

(b) We know that the formula for the area A of a circle of radius R is $A = \pi R^2$. If we use R to represent the independent variable and A to represent the dependent variable, the function expressing this relationship is

$$A(R) = \pi R^2$$

In this setting, the domain consists of positive real numbers R.

Function Notation

ARGUMENT
The independent variable x of a function $y = f(x)$ is sometimes called the **argument** of the function f. Thinking of the independent variable x as an argument can sometimes make it easier to apply the rule of the function. For example, suppose f is the function defined by $f(x) = x^3$. Here x is the argument of the function f. Thus, the function $f(x) = x^3$ is the rule that tells us to cube the argument. Then $f(2)$ means to cube the argument 2; $f(-1)$ means to cube -1; $f(a)$ means to cube the number a; $f(x + h)$ means to cube the quantity $(x + h)$. In actually evaluating $f(2)$, for example, we would replace x by 2 and compute the result: $f(2) = 2^3 = 8$.

EXAMPLE 5
For the function G, defined by $G(x) = 2x^2 - 3x$, evaluate:

(a) $G(3)$ (b) $G(-1)$
(c) $G(-x)$ (d) $-G(x)$
(e) $G(h)$ (f) $G(x + h)$
(g). $G(x) + G(h)$

Solution
(a) We replace x by 3 in the rule for G to get

$$G(3) = 2(3)^2 - 3(3) = 18 - 9 = 9$$

(b) $G(-1) = 2(-1)^2 - 3(-1) = 2 + 3 = 5$
(c) $G(-x) = 2(-x)^2 - 3(-x) = 2x^2 + 3x$

(d) $-G(x) = -(2x^2 - 3x) = -2x^2 + 3x$

(e) $G(h) = 2(h)^2 - 3(h) = 2h^2 - 3h$

(f) $G(x + h) = 2(x + h)^2 - 3(x + h)$
$$= 2(x^2 + 2xh + h^2) - 3x - 3h$$
$$= 2x^2 + 4xh + 2h^2 - 3x - 3h$$

(g) $G(x) + G(h) = 2x^2 - 3x + 2h^2 - 3h$

≡

FUNCTION NOTATION

Example 5 illustrates **function notation**. An important use of function notation is to find

$$\frac{f(x + h) - f(x)}{h} \qquad h \neq 0 \qquad (1)$$

DIFFERENCE QUOTIENT

This expression, called the **difference quotient** of f, is used often in calculus.

EXAMPLE 6   Find the difference quotient of the function f defined by

$$f(x) = 2x^2 - x + 1$$

Solution   It is helpful to proceed in steps.

Step 1:   First we calculate $f(x + h)$:

$$f(x + h) = 2(x + h)^2 - (x + h) + 1$$
$$= 2(x^2 + 2xh + h^2) - x - h + 1$$
$$= 2x^2 + 4xh + 2h^2 - x - h + 1$$

Step 2:   Now subtract $f(x)$ from this result:

$$f(x + h) - f(x) = 2x^2 + 4xh + 2h^2 - x - h + 1 - \overbrace{(2x^2 - x + 1)}$$

Be careful to subtract the quantity $f(x)$ ⟶

$$= 2x^2 + 4xh + 2h^2 - x - h + 1 - 2x^2 + x - 1$$
$$= 4xh + 2h^2 - h$$

Step 3:   Now divide by h:

$$\frac{f(x + h) - f(x)}{h} = \frac{4xh + 2h^2 - h}{h}$$

$$\underset{\underset{\text{Factor numerator}}{\uparrow}}{=} \frac{\cancel{h}(4x + 2h - 1)}{\cancel{h}}$$

$$\underset{\underset{\text{Cancellation law}}{\uparrow}}{=} 4x + 2h - 1$$

≡

Other Notations

Sometimes, **arrow notation** is used to signify a function. Suppose f is a function defined by the equation $y = f(x)$. In arrow notation this would be written as

$$f : x \rightarrow y \quad \text{or} \quad f : x \rightarrow f(x)$$

We read this notation as: "f maps x onto y" or "f maps x onto $f(x)$."

Another notation for functions, **ordered-pair notation**, is discussed in the next section.

Calculators

Most calculators have special keys that enable you to find the value of many functions. On your calculator you should be able to find the squaring function, $f(x) = x^2$; the square-root function, $f(x) = \sqrt{x}$; the reciprocal function, $f(x) = 1/x$; and many others that will be discussed later in this book, such as $\ln x$, $\log x$, and so on. When you enter x and then press one of these function keys, you get the value of that function at x. Try it yourself with the following functions.

C EXAMPLE 7   (a) $f(x) = x^2$; $f(1.234) \approx 1.522756$

(b) $g(x) = \dfrac{1}{x}$; $g(1.234) \approx 0.8103728$

(c) $F(x) = \sqrt{x}$; $F(1.234) \approx 1.1108555$ ≡

Summary

For your convenience, we list here some of the important vocabulary introduced in this section, with a brief description of each term.

Function	A rule or correspondence between two sets of real numbers so that each number in the first set, the **domain**, has corresponding to it exactly one number in the second set. The **range** is the set of values of the function for x's in the domain.
Function notation	$y = f(x)$ f: a symbol for the rule that defines the function x: the **argument** or **independent variable** y: the **dependent variable** $f(x)$: the **value of the function at x** or the **image of x**

Unspecified domain   If a function f is defined by an equation and no domain is specified, then the domain will be taken to be the largest collection of real numbers for which the rule can be applied.

EXERCISE 4.1   *In Problems 1–8 find the following for each function:*

(a) $f(0)$ (b) $f(1)$ (c) $f(-1)$ (d) $f(2)$

1. $f(x) = -3x^2 + 2x - 4$

2. $f(x) = 2x^2 + x - 1$

3. $f(x) = \dfrac{x}{x^2 + 1}$

4. $f(x) = \dfrac{x^2 - 1}{x + 4}$

5. $f(x) = |x| + 4$

6. $f(x) = \sqrt{x^2 + x}$

7. $f(x) = \dfrac{2x + 1}{3x - 5}$

8. $f(x) = 1 - \dfrac{1}{(x + 2)^2}$

In Problems 9–18 find the following for each function:

(a) $f(-x)$ (b) $-f(x)$ (c) $f(2x)$ (d) $f(x - 3)$ (e) $f(1/x)$

9. $f(x) = 2x + 3$

10. $f(x) = 4 - x$

11. $f(x) = 2x^2 - 4$

12. $f(x) = x^3 + 1$

13. $f(x) = x^3 - 3x$

14. $f(x) = x^2 + x$

15. $f(x) = \dfrac{x}{x^2 + 1}$

16. $f(x) = \dfrac{x^2}{x^2 + 1}$

17. $f(x) = |x|$

18. $f(x) = \dfrac{1}{x}$

In Problems 19–30 find the domain of each function.

19. $f(x) = 2x + 1$

20. $f(x) = 3x^2 - 2$

21. $f(x) = \dfrac{x}{x^2 + 1}$

22. $f(x) = \dfrac{x^2}{x^2 + 1}$

23. $g(x) = \dfrac{x}{x^2 - 1}$

24. $h(x) = \dfrac{x}{x - 1}$

25. $F(x) = \dfrac{x - 2}{x^3 + x}$

26. $G(x) = \sqrt{1 - x}$

27. $f(x) = \sqrt{x^2 - 9}$

28. $f(x) = \dfrac{1}{\sqrt{x^2 - 4}}$

29. $p(x) = \sqrt{\dfrac{x - 2}{x - 1}}$

30. $q(x) = \sqrt{x^2 - x - 2}$

In Problems 31–40 find the difference quotient,

$$\frac{f(x + h) - f(x)}{h} \qquad h \neq 0$$

for each function. Be sure to simplify.

31. $f(x) = 3$

32. $f(x) = 2x$

33. $f(x) = 1 - 3x$

34. $f(x) = x^2 + 1$

35. $f(x) = 3x^2 - 2x$

36. $f(x) = 4x - 2x^2$

37. $f(x) = x^3 - x$

38. $f(x) = x^3 + x$

39. $f(x) = \dfrac{1}{x}$

40. $f(x) = \dfrac{1}{x^2}$

41. For the function $f(x) = \sqrt{x}$ show that

$$\frac{f(x + h) - f(x)}{h} = \frac{1}{\sqrt{x + h} + \sqrt{x}} \qquad h \neq 0$$

[*Hint:* Multiply numerator and denominator by $\sqrt{x + h} + \sqrt{x}$.]

42. For the function $f(x) = \sqrt{x + 3}$ show that

$$\frac{f(x + h) - f(x)}{h} = \frac{1}{\sqrt{x + 3 + h} + \sqrt{x + 3}} \qquad h \neq 0$$

43. If $f(x) = 2x^3 + Ax^2 + 4x - 5$ and if $f(2) = 3$, what is the value of A?

44. If $f(x) = 3x^2 - Bx + 4$ and if $f(-1) = 10$, what is the value of B?

45. If $f(x) = (3x + 8)/(2x - A)$ and if $f(0) = 2$, what is the value of A?

46. If $f(x) = (2x - B)/(3x + 4)$ and if $f(2) = \frac{1}{2}$, what is the value of B?

47. If $f(x) = (2x - A)/(x - 3)$ and if $f(4) = 0$, what is the value of A? Where is f not defined?

48. If $f(x) = (x - B)/(x - A)$ and if $f(2) = 0$ and $f(1)$ is undefined, what are the numbers A and B?

49. If $f(x) = (3x + 2)/(2x - 3)$, find $f(1/x)$ if $x \neq 0$.

50. If $f(x) = (2 - x)/(1 + 2x)$, find $f(3x)$.

C **51.** If a rock falls from a height of 20 meters on earth, the height H (in meters) after x seconds is approximately

$$H(x) = 20 - 4.9x^2$$

(a) What is the height of the rock when $x = 1$ second? $x = 1.1$ seconds? $x = 1.2$ seconds? $x = 1.3$ seconds?

(b) When does the rock strike the ground?

C **52.** If a rock falls from a height of 20 meters on the planet Jupiter, its height H (in meters) after x seconds is approximately

$$H(x) = 20 - 13x^2$$

(a) What is the height of the rock when $x = 1$ second? $x = 1.1$ seconds? $x = 1.2$ seconds?

(b) When does the rock strike the ground?

53. Express the area A of a rectangle as a function of the length x, if the length is twice the width of the rectangle.

54. Express the area A of an isosceles right triangle as a function of the length x of one of the two equal sides.

55. Express the gross salary G of a person who earns \$5 per hour as a function of the number x of hours worked.

56. A commissioned salesperson earns \$100 base pay plus \$10 per item sold. Express the gross salary G as a function of the number x of items sold.

57. A page with dimensions of 11 inches by 7 inches has a border of uniform width x surrounding the printed matter of the page. Write a formula for the area A of the printed part as a function of the width x of the border. Give the domain and range of A.

C **58.** An airplane crosses the Atlantic Ocean (3000 miles) with an airspeed of 500 miles per hour. The cost C (in dollars) per passenger is

$$C(x) = 100 + \frac{x}{10} + \frac{36,000}{x}$$

where x is the ground speed (airspeed $\pm$ wind).
(a) What is the cost per passenger for quiescent (no wind) conditions?
(b) What is the cost per passenger with a head wind of 50 miles per hour?
(c) What is the cost per passenger with a tail wind of 100 miles per hour?
(d) What is the cost per passenger with a head wind of 100 miles per hour?

C **59.** The period T (in seconds) of a simple pendulum is a function of its length l (in feet) defined by the equation

$$T(l) = 2\pi\sqrt{\frac{l}{g}}$$

where $g \approx 32.2$ feet per second per second is the acceleration due to gravity. Use a calculator to determine the period of a pendulum whose length is 1 foot. By how much does the period increase if the length is increased to 2 feet?

▥ 4.2 THE GRAPH OF A FUNCTION

In applications, a graph often demonstrates more clearly the relationship between two variables than, say, a table would. For example, Figure 3 (page 230) shows OPEC crude-oil production in millions of barrels per day (vertical axis) from January 1981 through August 1983 (horizontal axis). It is easy to see from the graph that production was falling during 1981 and that production was rising in 1983. It is also easy to see that production was at the lowest level some time at the beginning of 1983. Tables, on the other hand, do not have such advantages. This is but one of the reasons why graphs are preferred.

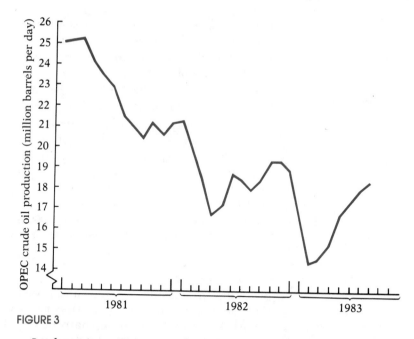

FIGURE 3

Look again at Figure 3. The graph shows that, for each time on the horizontal axis, there is exactly one production amount on the vertical axis. Thus we have the graph of a function, although the exact rule for getting from time to production amount is not known.

GRAPH OF A FUNCTION   The **graph of a function** f is defined as the graph of the equation $y = f(x)$. A function will always have a graph. However, not every collection of points in the xy-plane is the graph of a function. Remember, for a function $y = f(x)$, each number x in the domain of f has one and only one image $y = f(x)$. Thus, the graph of a function f cannot contain two points with the same x-coordinate and different y-coordinates. Therefore, the graph of a function must satisfy the **vertical-line test**.

THEOREM   A set of points in the xy-plane is the graph of a function if and only
VERTICAL-LINE TEST   if no vertical line contains more than one point of the set.   ☰

In other words, if any vertical line intersects a graph at more than one point, the graph is not the graph of a function.

EXAMPLE 1   Which of the graphs in Figure 4 are graphs of functions?

Solution   The graphs in Figures 4(a) and 4(b) are graphs of functions, because no vertical line contains more than one point of each of these graphs. The graphs in Figures 4(c) and 4(d) are not graphs of functions, because some vertical line intersects each graph in more than one point.

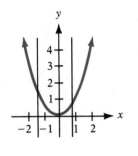

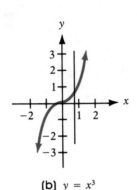

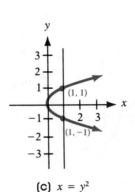

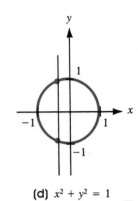

(a) $y = x^2$

(b) $y = x^3$

(c) $x = y^2$

(d) $x^2 + y^2 = 1$

FIGURE 4

Ordered Pairs

ORDERED PAIR

The preceding discussion provides an alternative way to think of a function. We may consider a function f as a set of **ordered pairs** (x, y) or $(x, f(x))$, in which no two pairs have the same first element. The set of all first elements is the domain and the set of all second elements is the range of the function. Thus, there is associated with each element x in the domain a unique element y in the range. An example is the set of all ordered pairs (x, y) such that $y = x^2$. Some of the pairs in this set are

$$(2, 2^2) = (2, 4) \qquad (0, 0^2) = (0, 0)$$
$$(-2, (-2)^2) = (-2, 4) \qquad (\tfrac{1}{2}, (\tfrac{1}{2})^2) = (\tfrac{1}{2}, \tfrac{1}{4})$$

In this set no two pairs have the same *first* element (even though there are pairs that have the same *second* element). This set is the squaring function, which associates with each real number x the value x^2. Look again at Figure 4(a).

The ordered pairs (x, y) for which $y^2 = x$ do not represent a function because there are ordered pairs with the same first element but different second elements. For example, $(1, 1)$ and $(1, -1)$ are ordered pairs obeying the relationship $y^2 = x$ with the same first element but different second elements. Look again at Figure 4(c).

The next example illustrates how to determine the domain and range of a function if its graph is given.

EXAMPLE 2

Let f be a function whose graph is given in Figure 5 (page 232). Some points on the graph are labeled.

(a) What is the value of the function when $x = -6$, $x = -4$, $x = 0$, and $x = 6$?

(b) What is the domain of f? (c) What is the range of f?

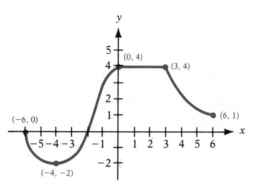

FIGURE 5

Solution   (a) Since $(-6, 0)$ is on the graph of f, the y-coordinate 0 must be the value of f at the x-coordinate -6; that is, $f(-6) = 0$. In a similar way we find that, when $x = -4$, then $y = -2$, or $f(-4) = -2$; when $x = 0$, then $y = 4$, or $f(0) = 4$; and, when $x = 6$, then $y = 1$, or $f(6) = 1$.

(b) To determine the domain of f, we notice that the points on the graph of f all have x-coordinates between -6 and 6, inclusive; and, for each number x between -6 and 6, there is a point $(x, f(x))$ on the graph. Thus, the domain of f is $-6 \le x \le 6$.

(c) The points on the graph all have y-coordinates between -2 and 4, inclusive; and, for each such number y, there is at least one number x in the domain. Hence, the range of f is $-2 \le y \le 4$. ∎

Increasing and Decreasing Functions

Consider again the graph given in Figure 5. If you look from left to right along the graph of this function, you will notice that parts of the graph are rising, parts are falling, and parts are horizontal. In such cases, the function is described as *increasing*, *decreasing*, and *stationary*, respectively. More precise definitions follow.

INCREASING FUNCTION   A function f is (strictly) **increasing** on an interval I if, for any choice of x_1 and x_2 in I, with $x_1 < x_2$, we have $f(x_1) < f(x_2)$.

DECREASING FUNCTION   A function f is (strictly) **decreasing** on an interval I if, for any choice of x_1 and x_2 in I, with $x_1 < x_2$, we have $f(x_1) > f(x_2)$.

STATIONARY FUNCTION   A function f is **stationary** on an interval I if, for all choices of x in I the values $f(x)$ are equal.

Figure 6 illustrates the definitions given above.

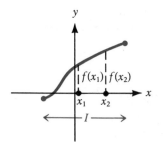

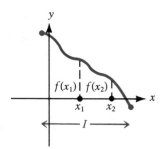

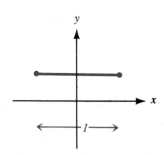

(a) For $x_1 < x_2$ in I,
 $f(x_1) < f(x_2)$;
 f is increasing

(b) For $x_1 < x_2$ in I,
 $f(x_1) > f(x_2)$;
 f is decreasing

(c) Values of f are equal;
 f is stationary

FIGURE 6

To answer the question of where the function is increasing, where it is decreasing, and where it is stationary, we use an inequality involving the independent variable x or an interval of x-coordinates.

EXAMPLE 3   Where is the function in Figure 5 increasing? Where is it decreasing? Where is it stationary?

Solution   The graph is rising for $-4 \leq x \leq 0$; that is, the function is increasing on the interval $[-4, 0]$. It is decreasing for $-6 \leq x \leq -4$ and for $3 \leq x \leq 6$; that is, on the intervals $[-6, -4]$ and $[3, 6]$. It is stationary for $0 \leq x \leq 3$; that is, on the interval $[0, 3]$.   ▤

Even and Odd Functions

EVEN FUNCTION   A function f is **even** if, for every number x in its domain, the number $-x$ is also in the domain and

$$f(-x) = f(x)$$

ODD FUNCTION   A function f is **odd** if, for every number x in its domain, the number $-x$ is also in the domain and

$$f(-x) = -f(x)$$

Refer to Section 3.2 where the tests for symmetry are listed. The following results are then evident:

THEOREM   A function is even if and only if its graph is symmetric with respect to the y-axis.

A function is odd if and only if its graph is symmetric with respect to the origin.   ▤

EXAMPLE 4   Tell whether each graph given in Figure 7 is the graph of an even function, an odd function, or a function that is neither even nor odd.

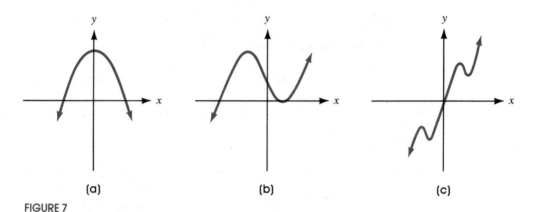

(a)   (b)   (c)

FIGURE 7

Solution   The graph in Figure 7(a) is that of an even function, because the graph is symmetric with respect to the y-axis. The function whose graph is given in Figure 7(b) is neither even nor odd, because the graph is neither symmetric with respect to the y-axis nor symmetric with respect to the origin. The function whose graph is given in Figure 7(c) is odd, because its graph is symmetric with respect to the origin.

≡

EXAMPLE 5   Tell whether each of the following functions is even, odd, or neither:

(a) $f(x) = x^2 - 5$ (b) $g(x) = x^3 - 1$
(c) $h(x) = 5x^3 - x$ (d) $F(x) = |x|$

Solution   (a) We replace x by $-x$. Then,

$$f(-x) = (-x)^2 - 5 = x^2 - 5$$

Because $f(-x) = f(x)$, we conclude that f is an even function.

(b) We replace x by $-x$. Then,

$$g(-x) = (-x)^3 - 1 = -x^3 - 1$$

Because $g(-x) \neq g(x)$ and $g(-x) \neq -g(x)$, we conclude that g is neither even nor odd.

(c) We replace x by $-x$. Then,

$$h(-x) = 5(-x)^3 - (-x) = -5x^3 + x$$

Because $h(-x) = -h(x)$, h is an odd function.

(d) We replace x by $-x$. Then,

$$F(-x) = |-x| = |x|$$

Because $F(-x) = F(x)$, F is an even function. ▪

Important Functions

We now give names to some of the functions we've encountered.

LINEAR FUNCTION

$$f(x) = mx + b \qquad m \text{ and } b \text{ are real numbers}$$

The domain of the **linear function** f consists of all real numbers. The graph of this function is a nonvertical straight line with slope m and y-intercept b. A linear function is increasing if $m > 0$, decreasing if $m < 0$, and stationary if $m = 0$.

CONSTANT FUNCTION

$$f(x) = b \qquad b \text{ is a real number}$$

The **constant function** is a special linear function ($m = 0$). Its domain is the set of all real numbers; its range is the set consisting of the single number b. Its graph is a horizontal line whose y-intercept is b. The constant function is an even function whose graph is stationary over its domain. See Figure 8.

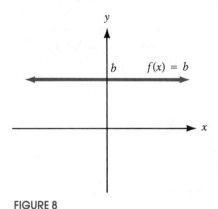

FIGURE 8
Constant function

IDENTITY FUNCTION

$$f(x) = x$$

The **identity function** is also a special linear function. Its domain and its range are the set of all real numbers. Its graph is a line whose slope is $m = 1$ and whose y-intercept is zero. The line consists of all points for which the *x*-coordinate equals the *y*-coordinate. The identity function is an odd function that is increasing over its domain. See Figure 9.

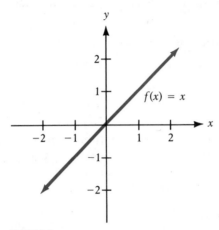

FIGURE 9
Identity function

SQUARING FUNCTION

$$f(x) = x^2$$

The domain of the **squaring function** f is the set of all real numbers; its range is the set of nonnegative real numbers. The graph of this function is a parabola, whose intercept is at (0, 0). The squaring function is an even function that is decreasing on the interval $(-\infty, 0]$ and increasing on the interval $[0, +\infty)$. See Figure 10.

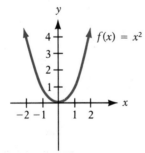

FIGURE 10
Squaring function

CUBING FUNCTION

$$f(x) = x^3$$

The domain and the range of the **cubing function** are the set of all real numbers. The intercept of the graph is at (0, 0). The cubing function is odd and is increasing on the interval $(-\infty, +\infty)$. See Figure 11.

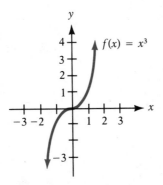

FIGURE 11
Cubing function

SQUARE-ROOT FUNCTION

$$f(x) = \sqrt{x}$$

The domain and the range of the **square-root function** are the set of nonnegative real numbers. The intercept of the graph is at $(0, 0)$. The square-root function is neither even nor odd and is increasing on the interval $[0, +\infty)$. See Figure 12.

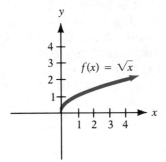

FIGURE 12
Square-root function

ABSOLUTE-VALUE FUNCTION

$$f(x) = |x|$$

The domain of the **absolute-value function** is the set of all real numbers; its range is the set of nonnegative real numbers. The intercept of the graph is at (0, 0). If $x \geq 0$, then $f(x) = x$ and the graph of f is part of the line $y = x$; if $x < 0$, then $f(x) = -x$ and the graph of f is part of the line $y = -x$. The absolute-value function is an even function; it is decreasing on the interval $(-\infty, 0]$ and increasing on the interval $[0, +\infty)$. See Figure 13.

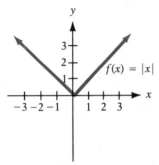

FIGURE 13
Absolute-value function

BRACKET x The symbol $[\![x]\!]$, read "**bracket x**," stands for the largest integer less than or equal to x. For example, $[\![1]\!] = 1$, $[\![2.5]\!] = 2$, $[\![\frac{1}{2}]\!] = 0$, $[\![-\frac{3}{4}]\!] = -1$. This type of correspondence occurs frequently enough in mathematics that we give it a name.

GREATEST-INTEGER FUNCTION

$$f(x) = [\![x]\!] = \text{Greatest integer less than or equal to } x$$

The domain of the **greatest-integer function** is the set of all real numbers; its range is the set of integers. The y-intercept of the graph is at zero. The greatest-integer function is neither even nor odd. It is stationary on every interval of the form $[k, k + 1)$, for k an integer. In Figure 14 we use a solid dot to indicate, for example, that, at $x = 1$, the value of f is $f(1) = 1$; we use an open circle to illustrate that the function does not assume the value zero at $x = 1$.

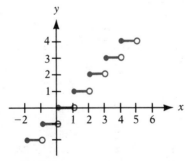

FIGURE 14
Greatest-integer function, or bracket function

STEP FUNCTION

From the graph of the greatest-integer function, we can see why it is sometimes called a **step function**. This function exhibits, at $x = 0$, $x = \pm 1$, $x = \pm 2$, and so on, what is called a *discontinuity*; that is, at integer values the graph suddenly "steps" from one value to another without taking on any of the intermediate values. For example, to the left of $x = 3$, the y-coordinates are 2 and, to the right of $x = 3$, the y-coordinates are 3.

The functions we have discussed so far are basic. Whenever you encounter one of them, you should see a mental picture of its graph. For example, if you encounter the function $f(x) = x^2$, you should see in your mind's eye a picture like Figure 10.

Piecewise-Defined Functions

Sometimes a function is defined by a rule consisting of two or more equations. The choice of which equation to use depends on the value of the independent variable x. For example, the absolute-value function $f(x) = |x|$ is actually defined by two equations: $f(x) = x$ if $x \geq 0$ and $f(x) = -x$ if $x < 0$. For convenience, we generally combine these equations into one expression as

$$f(x) = |x| = \begin{cases} x & \text{if } x \geq 0 \\ -x & \text{if } x < 0 \end{cases}$$

PIECEWISE-DEFINED FUNCTIONS

When functions are defined by more than one equation, they are called **piecewise defined**.

Let's look at another example of a piecewise-defined function.

EXAMPLE 6

For the function f given below, find $f(0)$, $f(1)$, and $f(2)$; and graph the function f.

$$f(x) = \begin{cases} -x + 1 & \text{if } -1 \leq x < 1 \\ 2 & \text{if } x = 1 \\ x & \text{if } x > 1 \end{cases}$$

Solution

To find $f(0)$, we observe that, when $x = 0$, the equation for f is given by $f(x) = -x + 1$. So,

$$f(0) = -0 + 1 = 1$$

When $x = 1$, the equation for f is $f(x) = 2$. Thus,

$$f(1) = 2$$

When $x = 2$, the equation for f is $f(x) = x$. So

$$f(2) = 2$$

The domain of f consists of all real numbers greater than or equal to -1. To graph f, we graph "each piece." Thus, we first graph the line $y = -x + 1$, and then erase all but the part for which $-1 \leq x < 1$. Then we plot the point $(1, 2)$, because, when $x = 1$, $f(x) = 2$. Finally, we graph the line $y = x$, and then erase all but the part for which $x > 1$. See Figure 15. The range, as we see from the graph, consists of all real numbers y for which $y > 0$.

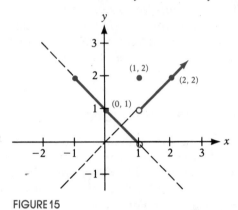

FIGURE 15

EXAMPLE 7   Holders of credit cards issued by banks, department stores, oil companies, and so on, receive bills each month that state minimum amounts that must be paid by a certain due date. The minimum due depends on the total amount owed. One such credit card company uses the following rules: For a bill of less than \$10, the entire amount is due. For a bill of at least \$10 but less than \$500, the minimum due is \$10. There is a minimum due of \$30 on a bill of at least \$500 but less than \$1000, a minimum of \$50 on a bill of at least \$1000 but less than \$1500, and a minimum of \$70 on bills of \$1500 or more. The function f that describes the minimum payment due on a bill of x dollars is

$$f(x) = \begin{cases} x & \text{if } 0 \leq x < 10 \\ 10 & \text{if } 10 \leq x < 500 \\ 30 & \text{if } 500 \leq x < 1000 \\ 50 & \text{if } 1000 \leq x < 1500 \\ 70 & \text{if } 1500 \leq x \end{cases}$$

To graph this function f, we proceed as follows: For $0 \leq x < 10$, draw the graph of $y = x$; for $10 \leq x < 500$, draw the graph of the constant function $y = 10$; for $500 \leq x < 1000$, draw the graph of the constant function $y = 30$; and so on. The graph of f is given in Figure 16.

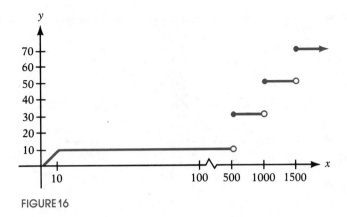

FIGURE 16

The card holder may pay any amount between the minimum due and the total owed. The organization issuing the card charges the card holder interest of 1.5% per month for the first $1000 owed and 1% per month on any unpaid balance over $1000. Thus, if $g(x)$ is the amount of interest charged per month on a balance of x, then $g(x) = 0.015x$ for $0 \leq x \leq 1000$. The amount of the unpaid balance above $1000 is $x - 1000$. If the balance due is $x > 1000$, then the interest is $0.015(1000) + 0.01(x - 1000) = 15 + 0.01x - 10 = 5 + 0.01x$, so

$$g(x) = \begin{cases} 0.015x & \text{if } 0 \leq x \leq 1000 \\ 5 + 0.01x & \text{if } x > 1000 \end{cases}$$

See Figure 17.

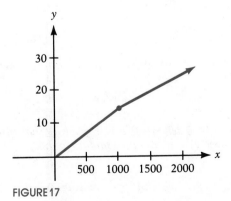

FIGURE 17

EXERCISE 4.2   *In Problems 1–12 use the graph of the function f given below.*

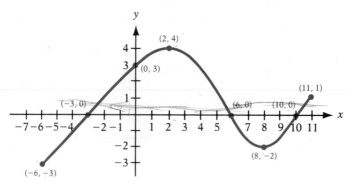

1. Find $f(0)$ and $f(2)$.   2.   Find $f(8)$ and $f(-3)$.
3. Is $f(2)$ positive or negative?   4.   Is $f(8)$ positive or negative?
5. For what numbers x is $f(x) = 0$?
6. For what numbers x is $f(x) > 0$?
7. What is the domain of f?   8.   What is the range of f?
9. Where is f increasing?   10.   Where is f decreasing?
11. How often does the line $y = \frac{1}{2}$ intersect the graph?
12. How often does the line $y = 3$ intersect the graph?
13. Is the point $(3, 14)$ on the graph of the function whose equation is given below?

$$f(x) = \frac{x + 2}{x - 6}$$

14. Is the point $(1, \frac{3}{5})$ on the graph of the function whose equation is given below?

$$f(x) = \frac{x^2 + 2}{x + 4}$$

15. Is the point $(-1, \frac{3}{2})$ on the graph of the function whose equation is given below?

$$f(x) = \frac{3x^2}{x^4 + 1}$$

16. Is the point $(\frac{1}{2}, -\frac{2}{3})$ on the graph of the function whose equation is given below?

$$f(x) = \frac{2x}{x - 2}$$

In Problems 17–30 tell whether the graph is that of a function by using the vertical-line test. If it is, use the graph to find:

(a) *Its domain and range*
(b) *The intervals on which it is increasing, decreasing, or stationary*
(c) *Whether it is even, odd, or neither*
(d) *The intercepts, if any*

17.

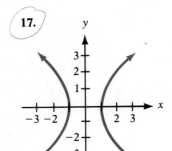

18.

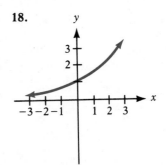

19.

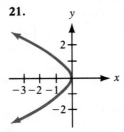

20.

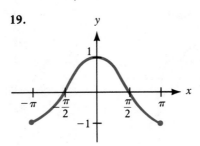

21.

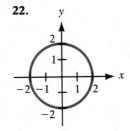

22.

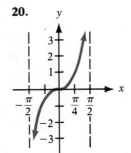

23.

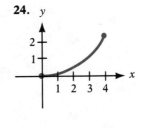

24.

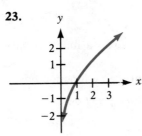

25.

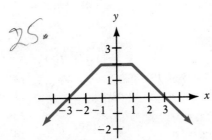

26.

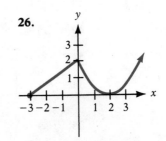

27.

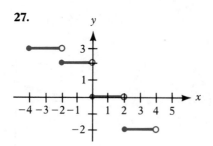

28.

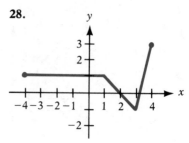

29.

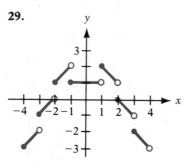

30.

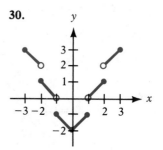

In Problems 31–42 tell whether each function is even, odd, or neither without drawing a graph.

31. $f(x) = 2x^3$

32. $f(x) = x^4 - x^2$

33. $g(x) = 2x^2 - 5$

34. $h(x) = 3x^3 + 2$

35. $F(x) = \sqrt[3]{x}$

36. $G(x) = \sqrt{x}$

37. $f(x) = x + |x|$

38. $f(x) = \sqrt[3]{2x^2 + 1}$

39. $g(x) = \dfrac{1}{x^2}$

40. $h(x) = \dfrac{x}{x^2 - 1}$

41. $h(x) = \dfrac{x^3}{3x^2 - 9}$

42. $F(x) = \dfrac{x}{|x|}$

43. How many x-intercepts can a function have if it is increasing on its domain? Explain.

44. How many y-intercepts can a function have? Explain.

In Problems 45–64 graph each function.

45. $f(x) = 3x - 3$

46. $f(x) = 4 - 2x$

47. $g(x) = x^2 - 4$

48. $h(x) = x^2 + 4$

49. $F(x) = 2x^3$

50. $G(x) = \dfrac{2}{x}$

51. $f(x) = |x - 2|$

52. $f(x) = \sqrt{x} + 2$

53. $f(x) = \begin{cases} 1 & \text{if } x \text{ is an integer} \\ -1 & \text{if } x \text{ is not an integer} \end{cases}$

54. $f(x) = \begin{cases} x & \text{if } x \geq 1 \\ 1 & \text{if } x < 1 \end{cases}$

55. $f(x) = \begin{cases} 2x & \text{if } x \neq 0 \\ 0 & \text{if } x = 0 \end{cases}$

56. $f(x) = \begin{cases} 3x & \text{if } x \neq 0 \\ 4 & \text{if } x = 0 \end{cases}$

57. $f(x) = \begin{cases} 3x + 2 & \text{if } x \neq -1 \\ 1 & \text{if } x = -1 \end{cases}$

58. $f(x) = \begin{cases} 2x - 1 & \text{if } x \neq 2 \\ 3 & \text{if } x = 2 \end{cases}$

59. $f(x) = \begin{cases} -2x - 3 & \text{if } x < 0 \\ x - 3 & \text{if } 0 \leq x < 5 \end{cases}$

60. $f(x) = \begin{cases} x & \text{if } x \leq 2 \\ -1/x & \text{if } 2 < x \end{cases}$

61. $f(x) = \begin{cases} x + 3 & \text{if } -2 \leq x < 0 \\ 1 & \text{if } x = 0 \\ \frac{1}{2}x^2 & \text{if } x > 0 \end{cases}$

62. $f(x) = \begin{cases} 4 - x & \text{if } x \leq 0 \\ x^2 - 2 & \text{if } 0 < x \end{cases}$

63. $f(x) = [\![x - \frac{1}{2}]\!]$

64. $f(x) = [\![2x]\!]$

In Problems 65–72 tell whether or not the set of ordered pairs (x, y) defined by each equation is a function.

65. $y = x^2 + 2x$

66. $y = x^3 - 3x$

67. $y = \dfrac{2}{x}$

68. $y = \dfrac{3}{x} - 3$

69. $y^2 = 1 - x^2$

70. $y = \pm\sqrt{1 - 2x}$

71. $x^2 + y = 1$

72. $x + 2y^2 = 1$

73. Let f denote any function with the property that, whenever x is in its domain, then so is $-x$. Define the functions $E(x)$ and $O(x)$ to be

$$E(x) = \frac{1}{2}[f(x) + f(-x)] \qquad O(x) = \frac{1}{2}[f(x) - f(-x)]$$

(a) Show that $E(x)$ is an even function.
(b) Show that $O(x)$ is an odd function.
(c) Show that $f(x) = E(x) + O(x)$.
(d) Draw the conclusion that any such function f can be written as the sum of an even function and an odd function.

74. A trucking company transports goods between Chicago and New York, a distance of 960 miles. The company's policy is to charge, for each pound, $0.50 per mile for the first 100 miles, $0.40 per mile for the next 300 miles, $0.25 per mile for the next 400 miles, and no charge for the remaining 160 miles. Graph the relationship between the cost of transportation in dollars and mileage over the entire 960-mile route. Find the cost as a function of mileage for hauls between 100 and 400 miles from Chicago. Find the cost as a function of mileage for hauls between 400 and 800 miles from Chicago.

75. An economy car rented from National Car Rental® on a weekly basis in California costs $95 per week.* Extra days cost $24 per day until the daily rate exceeds the weekly rate, in which case the weekly rate applies. Find the cost C of renting an economy car as a piecewise-defined function of the number x of days used, where $7 \leq x \leq 14$. Graph this function. (*Note:* Any part of a day counts as a full day.)

76. Rework Problem 75 for a luxury car, which costs $219 on a weekly basis with extra days at $45 per day.

*Source: National Car Rental, February 1985.

77. The 1984 Tax Rate Schedule X for single taxpayers is given below. Write a piecewise-defined function T where $T(x)$ is the tax due on taxable income of x dollars.

SCHEDULE X   Single Taxpayers

Taxable Income, x		Tax Due, T(x)	of the amount
Over—	But not over—		over—
$0	$2,300	—0—	
2,300	3,400	........11%	$2,300
3,400	4,400	$121 + 12%	3,400
4,400	6,500	241 + 14%	4,400
6,500	8,500	535 + 15%	6,500
8,500	10,800	835 + 16%	8,500
10,800	12,900	1,203 + 18%	10,800
12,900	15,000	1,581 + 20%	12,900
15,000	18,200	2,001 + 23%	15,000
18,200	23,500	2,737 + 26%	18,200
23,500	28,800	4,115 + 30%	23,500
28,800	34,100	5,705 + 34%	28,800
34,100	41,500	7,507 + 38%	34,100
41,500	55,300	10,319 + 42%	41,500
55,300	81,800	16,115 + 48%	55,300
81,800	....	28,835 + 50%	81,800

78. Rework Problem 77 for the 1984 Tax Rate Schedule Y for married taxpayers filing joint returns and qualifying widows and widowers.

SCHEDULE Y   Married Filing Joint Returns and Qualifying Widows and Widowers

Taxable Income, x		Tax Due, T(x)	of the amount
Over—	But not over—		over—
$0	$3,400	—0—	
3,400	5,500	........11%	$3,400
5,500	7,600	$231 + 12%	5,500
7,600	11,900	483 + 14%	7,600
11,900	16,000	1,085 + 16%	11,900
16,000	20,200	1,741 + 18%	16,000
20,200	24,600	2,497 + 22%	20,200
24,600	29,900	3,465 + 25%	24,600
29,900	35,200	4,790 + 28%	29,900
35,200	45,800	6,274 + 33%	35,200
45,800	60,000	9,772 + 38%	45,800
60,000	85,600	15,168 + 42%	60,000
85,600	109,400	25,920 + 45%	85,600
109,400	162,400	36,630 + 49%	109,400
162,400	....	62,600 + 50%	162,400

▤ 4.3 GRAPHING TECHNIQUES

At this stage if you were asked to graph any of the functions defined by $y = x^2$, $y = x^3$, $y = x$, $y = \sqrt{x}$, or $y = |x|$, your response should be "Yes, I recognize these functions and know the general shapes of their graphs." (If this is not your answer, you should review the previous section, "Important Functions," and Figures 9–13.)

Sometimes we are asked to graph a function that is "almost" one we already know how to graph. In this section we look at some of these functions and develop techniques for graphing them.

Vertical Shifts

If a number c is added to or subtracted from the right side of the equation $y = f(x)$ of a function f, the graph of the new function $y = f(x) + c$ is the graph of f **shifted vertically** upward (if $c > 0$) or downward (if $c < 0$). Let's look at an example.

VERTICAL SHIFT

EXAMPLE 1   Use the graph of $f(x) = x^2$ to obtain the graph of:

(a) $g(x) = x^2 + 3$
(b) $h(x) = x^2 - 4$

Solution   The graph of $f(x) = x^2$ is familiar to us—we've sketched it several times.

(a) To see why a vertical shift occurs, we begin by obtaining some points on the graphs of f and g. For example, when $x = 0$, then $y = f(0) = 0$ on f and $y = g(0) = 3$ on g. When $x = 1$, then $y = f(1) = 1$ on f and $y = g(1) = 4$ on g. Table 1 lists these and a few other points on each graph:

TABLE 1

x	-2	-1	0	1	2
$y = f(x) = x^2$	4	1	0	1	4
$y = g(x) = x^2 + 3$	7	4	3	4	7

We conclude that the graph of g is identical to that of f except it is shifted up 3 units. See Figure 18(a).

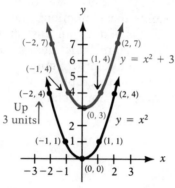

FIGURE 18(a)

(b) Table 2 lists some points on the graphs of f and h:

TABLE 2

x		-2	-1	0	1	2
$y = f(x) = x^2$		4	1	0	1	4
$y = h(x) = x^2 - 4$		0	-3	-4	-3	0

The graph of h is identical to that of f except it is shifted down 4 units. See Figure 18(b).

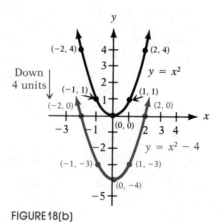

FIGURE 18(b)

Horizontal Shifts

HORIZONTAL SHIFT

If a number c is added to (or subtracted from) the argument x of a function f, the graph of the new function $g(x) = f(x + c)$ is the same as the graph of f except it is **shifted horizontally** left (if $c > 0$) or right (if $c < 0$). Let's look at an example.

EXAMPLE 2 Use the graph of $f(x) = |x|$ to obtain the graph of:

(a) $g(x) = |x - 2|$ (b) $h(x) = |x + 4|$

Solution The graph of $f(x) = |x|$ is familiar to us—it is in the shape of a V. The bottom of the V is at the origin $(0, 0)$. See Figure 19(a).

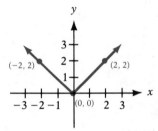

FIGURE 19(a)
$y = |x|$

(a) The function $g(x) = |x - 2|$ is basically an absolute-value function. Its graph is still in the shape of a V, but the bottom of the V is now at the point $(2, 0)$. Table 3 lists some points on each graph:

TABLE 3

x	-2	0	2	4
$y = f(x) = |x|$	2	0	2	4
$y = g(x) = |x - 2|$	4	2	0	2

We conclude that the graph of g is identical to that of f, except that it is shifted 2 units to the right. See Figure 19(b).

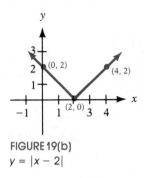

FIGURE 19(b)
$y = |x - 2|$

(b) The function $h(x) = |x + 4|$ is basically an absolute-value function. Its graph is V-shaped, with the bottom of the V at $(-4, 0)$. Thus, its graph is the same as that of f except it is shifted 4 units to the left. Do you see why? See Figure 19(c).

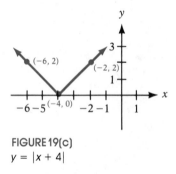

FIGURE 19(c)
$y = |x + 4|$

Vertical and horizontal shifts are sometimes combined.

EXAMPLE 3   Graph the function $f(x) = (x - 1)^3 + 3$.

Solution   We graph f in steps. First we note that the rule for f *cubes* something. Thus, we begin with the graph of $y = x^3$. See Figure 20(a). Next, to get the graph of $y = (x - 1)^3$, we shift the graph of $y = x^3$ horizontally 1 unit to the right. See Figure 20(b). Finally, to get the graph of $y = (x - 1)^3 + 3$, we shift the graph of $y = (x - 1)^3$ vertically up 3 units. See Figure 20(c). Note the three points that have been plotted on each graph. Using "key" points such as these can be helpful in keeping track of just what is taking place.

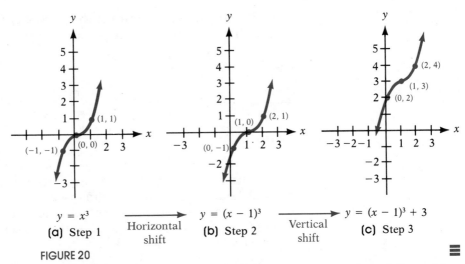

$y = x^3$ ——Horizontal shift—→ $y = (x - 1)^3$ ——Vertical shift—→ $y = (x - 1)^3 + 3$

(a) Step 1   **(b)** Step 2   **(c)** Step 3

FIGURE 20

In Example 3, had the vertical shift been done first, followed by the horizontal shift, the result would have been the same. Try it for yourself.

Change in Vertical Scale

When the right side of the equation $y = f(x)$ of a function f is multiplied by a positive number k, the graph of the new function $y = kf(x)$ is the same as the graph of the function $y = f(x)$ except that the scale used on the y-axis (vertical axis) is changed by a multiple of k. The next example will clarify this idea.

EXAMPLE 4   Use the graph of $y = f(x) = x^2$ to obtain the graph of:

(a) $y = g(x) = 3x^2$ (b) $y = h(x) = \frac{1}{2}x^2$

Solution   The graph of $y = f(x) = x^2$ is given in Figure 21(a).

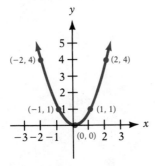

FIGURE 21(a)
$y = f(x) = x^2$

(a) To see the relationship between the graphs of f and g, we form Table 4, listing points on each graph:

TABLE 4

x	-2	-1	0	1	2
$y = f(x) = x^2$	4	1	0	1	4
$y = g(x) = 3x^2$	12	3	0	3	12

In each case, the y-coordinate of a point on the graph of g is 3 times as large as the corresponding y-coordinate on the graph of f. Figure 21(b) shows the graph of g. Notice that it is identical to that of f except that the scale used on the y-axis has been changed by a factor of 3.

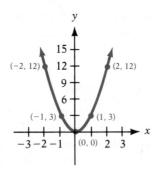

FIGURE 21(b)
$y = g(x) = 3x^2$

(b) Based on the discussion in part (b), the graph of $h(x) = \frac{1}{2}x^2$ is the same as the graph of $f(x) = x^2$, except that the vertical scale is changed by a factor of $\frac{1}{2}$. See Table 5 and Figure 21(c).

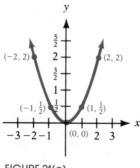

FIGURE 21(c)
$y = h(x) = \frac{1}{2}x^2$

TABLE 5

x		-2	-1	0	1	2
$y = f(x) = x^2$		4	1	0	1	4
$y = h(x) = \frac{1}{2}x^2$		2	$\frac{1}{2}$	0	$\frac{1}{2}$	2

Figure 22 illustrates the graphs of the functions $f(x) = x^2$, $g(x) = 3x^2$, and $h(x) = \frac{1}{2}x^2$ using the *same* scale. Note how the graph of $y = x^2$ has been "compressed" in Figure 22(b) and "stretched" in Figure 22(c), resulting in a sort of "accordion" effect.

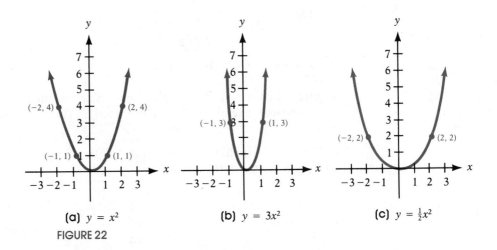

(a) $y = x^2$

(b) $y = 3x^2$

(c) $y = \frac{1}{2}x^2$

FIGURE 22

Change in Horizontal Scale

If the argument x of a function $y = f(x)$ is multiplied by a positive number k, the graph of the new function $y = f(kx)$ is the same as the graph of f, except that the scale used on the x-axis (horizontal axis) is changed by a multiple of $1/k$. To see why this happens, we look at the following example.

EXAMPLE 5   Use the graph of $f(x) = \sqrt{x}$ to obtain the graph of $g(x) = \sqrt{2x}$.

Solution   The graph of f is familiar to us. See Figure 23(a), page 254. Now we form Table 6, which lists some points on the graph of f and the graph of g.

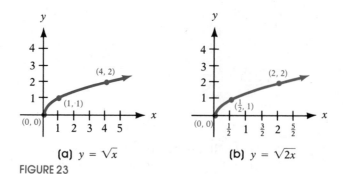

(a) $y = \sqrt{x}$ **(b)** $y = \sqrt{2x}$

FIGURE 23

TABLE 6

x	0	1	4
$y = f(x) = \sqrt{x}$	0	1	2
x	0	$\frac{1}{2}$	2
$y = g(x) = \sqrt{2x}$	0	1	2

Figure 23(b) shows the graph of g. Notice that the graphs of f and g are identical, except that the scale used on the x-axis has changed by a factor of $\frac{1}{2}$.   ▤

Figure 24 illustrates the graphs of the functions $f(x) = \sqrt{x}$ and $g(x) = \sqrt{2x}$ using the *same* scale.

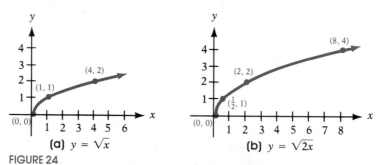

(a) $y = \sqrt{x}$ **(b)** $y = \sqrt{2x}$

FIGURE 24

To summarize, we have two ways to graph $y = kf(x)$ [and $y = f(kx)$], $k > 0$, when the graph of $y = f(x)$ is known:

1. *Change the scale.* Then the graphs of $y = f(x)$ and $y = kf(x)$ [and $y = f(kx)$] look exactly alike, but the scale used on each graph is different.
2. *Keep the scale.* Then the graphs of $y = f(x)$ and $y = kf(x)$ [and $y = f(kx)$] will be stretched or compressed versions of each other, but each graph will use the same scale.

Reflection about the x-Axis

REFLECTION When the right side of the equation $y = f(x)$ of a function f is multiplied by -1, the graph of the new function $y = -f(x)$ is the **reflection** about the x-axis of the graph of the function f.

EXAMPLE 6 Graph the function $f(x) = -x^2$.

Solution We begin with the graph of $y = x^2$. See Figure 25(a). For each point (x, y) on the graph of $y = x^2$, the point $(x, -y)$ is on the graph of $y = -x^2$, as indicated in the table:

x	-2	-1	0	1	2
$y = x^2$	4	1	0	1	4
$y = -x^2$	-4	-1	0	-1	-4

Thus, we can draw the graph of $y = -x^2$ by reflecting the graph of $y = x^2$ about the x-axis. See Figure 25(b).

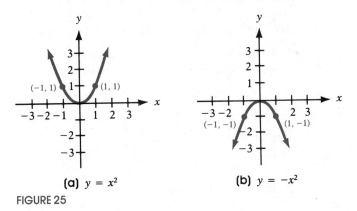

(a) $y = x^2$　　　　**(b)** $y = -x^2$

FIGURE 25

Reflection about the y-Axis

When the graph of the function $y = f(x)$ is known, the graph of the new function $y = f(-x)$ is the reflection about the y-axis of the graph of the function f.

EXAMPLE 7 Graph the function $f(x) = \sqrt{-x}$.

Solution First, notice that the domain of f consists of all real numbers x for which $-x \geq 0$, or for which $x \leq 0$. To get the graph of $f(x) = \sqrt{-x}$, we begin with the graph of $y = \sqrt{x}$. See Figure 26(a). Based on the discussion in Example 6, for each point (x, y) on the graph of $y = \sqrt{x}$, the point $(-x, y)$ is on the graph of $y = \sqrt{-x}$. Thus, we get the graph of $y = \sqrt{-x}$ by reflecting the graph of $y = \sqrt{x}$ about the y-axis. See Figure 26(b).

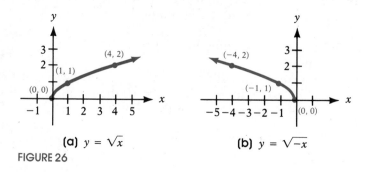

(a) $y = \sqrt{x}$ **(b)** $y = \sqrt{-x}$

FIGURE 26

The example that follows combines some of the procedures outlined in this section to get the required graph.

EXAMPLE 8   Graph the function $f(x) = \sqrt{1 - x} + 2$.

Solution   We shall use the following steps to get the graph of $y = \sqrt{1 - x} + 2$:

1. $y = \sqrt{x}$ Square-root function
2. $y = \sqrt{x + 1}$ Replace x by $x + 1$; shift left 1 unit.
3. $y = \sqrt{-x + 1} = \sqrt{1 - x}$ Replace x by $-x$; reflect about y-axis.
4. $y = \sqrt{1 - x} + 2$ Shift up 2 units.

See Figure 27.

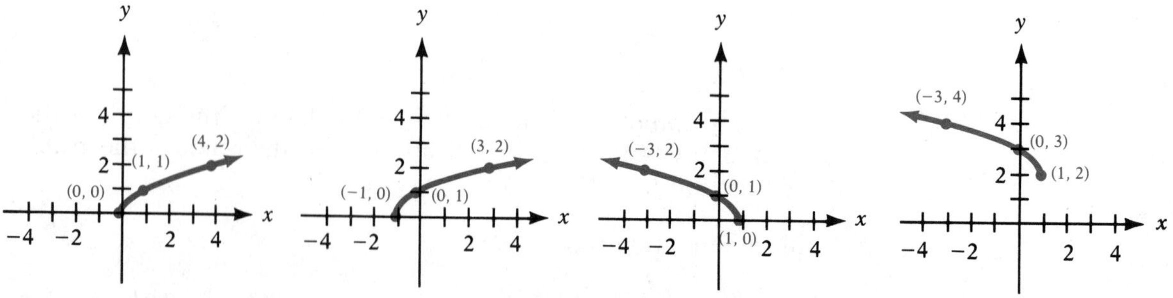

(a) $y = \sqrt{x}$ $\xrightarrow{\text{Horizontal shift}}$ **(b)** $y = \sqrt{x + 1}$ $\xrightarrow{\text{Reflection about } y\text{-axis}}$ **(c)** $y = \sqrt{1 - x}$ $\xrightarrow{\text{Vertical shift}}$ **(d)** $y = \sqrt{1 - x} + 2$

FIGURE 27

A different ordering of the steps used to solve Example 8 is given below:

1. $y = \sqrt{x}$ Square-root function
2. $y = \sqrt{-x}$ Reflect about y-axis.
3. $y = \sqrt{1 - x}$ Replace x by $x - 1$ or $-x$ by $-(x - 1) = 1 - x$; shift right 1 unit.
4. $y = \sqrt{1 - x} + 2$ Shift up 2 units.

EXAMPLE 9   Show the steps required to graph $y = \sqrt{2x - 1}$.

Solution   1. $y = \sqrt{x}$ Square-root function
2. $y = \sqrt{2x}$ Replace x by $2x$; change horizontal scale by $\frac{1}{2}$.
3. $y = \sqrt{2(x - \frac{1}{2})} = \sqrt{2x - 1}$ Replace x by $x - \frac{1}{2}$; shift right $\frac{1}{2}$ unit. ▤

A different solution to Example 9 is:

1. $y = \sqrt{x}$ Square-root function
2. $y = \sqrt{x - 1}$ Replace x by $x - 1$; shift right 1 unit.
3. $y = \sqrt{2x - 1}$ Replace x by $2x$; change horizontal scale by $\frac{1}{2}$.

Summary of Graphing Techniques

The table below summarizes the graphing procedures we've discussed.

TO GRAPH:	DRAW THE GRAPH OF f AND:
Vertical shifts	
$y = f(x) + c, \quad c > 0$	Raise it c units.
$y = f(x) - c, \quad c > 0$	Lower it c units.
Horizontal shifts	
$y = f(x + c), \quad c > 0$	Shift it to the left c units.
$y = f(x - c), \quad c > 0$	Shift it to the right c units.
Change in vertical scale	
$y = kf(x), \quad k > 0$	Change the vertical scale by a factor of k.
Change in horizontal scale	
$y = f(kx), \quad k > 0$	Change the horizontal scale by a factor of $1/k$.
Reflection about the x-axis	
$y = -f(x)$	Reflect it about the x-axis.
Reflection about the y-axis	
$y = f(-x)$	Reflect it about the y-axis.

EXERCISE 4.3   *In Problems 1–20 graph each function using the techniques of shifting, change of scale, and/or reflecting.*

1. $f(x) = x + 5$
2. $f(x) = x - 3$
3. $g(x) = x^2 - 1$
4. $h(x) = \sqrt{x} + 2$
5. $F(x) = |x + 1|$
6. $G(x) = (x - 2)^2$
7. $f(x) = \frac{1}{2}(x + 3)^2$
8. $f(x) = 2x^2 + 4$
9. $g(x) = \sqrt{x - 1} - 2$
10. $h(x) = |x + 3| - 4$
11. $F(x) = -x^2 + 2$
12. $G(x) = -|x| - 2$
13. $f(x) = -\frac{1}{2}x^2 + 2$
14. $f(x) = -3\sqrt{x} + 1$
15. $g(x) = 2\sqrt{x - 3}$
16. $h(x) = -x^3 + 2$
17. $F(x) = [\![x - 1]\!]$
18. $G(x) = 2[\![x]\!]$
19. $f(x) = -\frac{1}{2}(x + 1)^3 + 1$
20. $f(x) = 2|x - 3| + 4$

In Problems 21–26 the graph of a function f is illustrated. Use the graph of f as the first step toward graphing each of the following functions:

(a) $F(x) = f(x) + 3$
(b) $G(x) = f(x + 2)$
(c) $P(x) = -f(x)$
(d) $Q(x) = \frac{1}{2}f(x)$
(e) $g(x) = f(-x)$
(f) $h(x) = 3f(x)$
(g) $r(x) = f(\frac{1}{2}x)$
(h) $H(x) = 3f(2x)$

21.

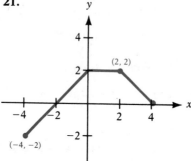

22.

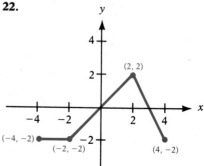

23.

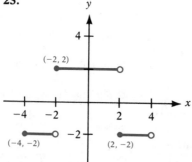

24.

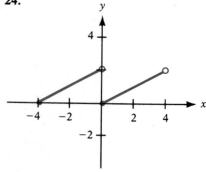

25.

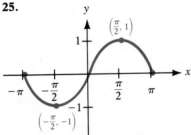

26.

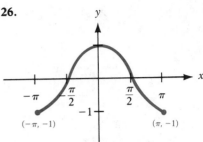

In Problems 27–32 complete the square of each quadratic expression. Then graph each function using the technique of shifting.

27. $f(x) = x^2 + 2x$ $(x+2)$ **28.** $f(x) = x^2 - 6x$

29. $f(x) = x^2 - 8x + 1$ **30.** $f(x) = x^2 + 4x + 2$

31. $f(x) = x^2 + x + 1$ **32.** $f(x) = x^2 - x + 1$

FAMILY OF PARABOLAS   **33.** The equation $y = (x - c)^2$ defines a **family of parabolas**, one parabola for each value of c. On one set of coordinate axes, graph the members of the family for $c = 0$, $c = 3$, and $c = -2$.

34. Repeat Problem 33 for the family of parabolas $y = x^2 + c$.

35. The relationship between Celsius (°C) and Fahrenheit (°F) for measuring temperature is given by the equation

$$F = \frac{9}{5}C + 32$$

The relationship between Celsius (°C) and Kelvin (K) is $K = C + 273$. Graph the equation $F = \frac{9}{5}C + 32$, using degrees Fahrenheit on the y-axis and degrees Celsius on the x-axis. Use the techniques introduced in this section to obtain the graph showing the relationship between Kelvin and Fahrenheit temperatures.

C   **36.** The period T, in seconds, of a simple pendulum is a function of its length l, in feet, defined by the equation

$$T = 2\pi\sqrt{\frac{l}{g}}$$

where $g \approx 32.2$ feet per second per second is the acceleration due to gravity. Graph this function using T on the y-axis and l on the x-axis. On the same coordinate axes, graph the functions:

(a) $T = 2\pi\sqrt{\dfrac{l + 2}{g}}$ (b) $T = 2\pi\sqrt{\dfrac{4l}{g}}$

Discuss how changes in the length l affect the period T of the pendulum.

▤ 4.4 OPERATIONS ON FUNCTIONS; COMPOSITE FUNCTIONS

In this section we introduce some operations on functions. We shall see that functions, like numbers, can be added, subtracted, multiplied, and divided.

If f and g are functions:

SUM FUNCTION   their **sum** $f + g$ is defined by

$$(f + g)(x) = f(x) + g(x)$$

DIFFERENCE FUNCTION   their **difference** $f - g$ is defined by

$$(f - g)(x) = f(x) - g(x)$$

PRODUCT FUNCTION   their **product** $f \cdot g$ is defined by

$$(f \cdot g)(x) = f(x) \cdot g(x)$$

QUOTIENT FUNCTION   and their **quotient** is defined by

$$\left(\frac{f}{g}\right)(x) = \frac{f(x)}{g(x)}$$

In each case, the domain of the resulting function consists of the numbers x that are common to the domains of f and g, but the numbers x for which $g(x) = 0$ must be excluded from the domain of the quotient f/g.

Thus, the sum function, $f + g$, is defined as the sum of the values of the functions f and g, and so on.

EXAMPLE 1   Let f and g be two functions defined as

$$f(x) = \sqrt{x + 2} \quad \text{and} \quad g(x) = \sqrt{x - 3}$$

Find the following and, in each case, determine the domain:

(a) $(f + g)(x)$ (b) $(f - g)(x)$ (c) $(f \cdot g)(x)$ (d) $\left(\dfrac{f}{g}\right)(x)$

Solution   (a) $(f + g)(x) = f(x) + g(x) = \sqrt{x + 2} + \sqrt{x - 3}$

(b) $(f - g)(x) = f(x) - g(x) = \sqrt{x + 2} - \sqrt{x - 3}$

(c) $(f \cdot g)(x) = f(x) \cdot g(x) = (\sqrt{x + 2})(\sqrt{x - 3}) = \sqrt{(x + 2)(x - 3)}$

(d) $\left(\dfrac{f}{g}\right)(x) = \dfrac{f(x)}{g(x)} = \dfrac{\sqrt{x + 2}}{\sqrt{x - 3}} = \sqrt{\dfrac{x + 2}{x - 3}}$

The domain of f consists of all numbers x for which $x \geq -2$; the domain of g consists of all numbers x for which $x \geq 3$. The numbers

x common to both these domains are those for which $x \geq 3$. As a result, the numbers x for which $x \geq 3$ comprise the domain of the sum function $f + g$, the difference function $f - g$, and the product function $f \cdot g$. For the quotient function f/g, we must exclude from this set the number 3, because the denominator, g, has the value zero when $x = 3$. Thus, the domain of f/g consists of all x for which $x > 3$.

It is sometimes helpful to view a complicated function as the sum, difference, product, or quotient of simpler functions. For example:

$F(x) = x^2 + \sqrt{x}$ is the sum of $f(x) = x^2$ and $g(x) = \sqrt{x}$.
$H(x) = (x^2 - 1)/(x^2 + 1)$ is the quotient of $f(x) = x^2 - 1$ and $g(x) = x^2 + 1$.

One use of this view of functions is to obtain a graph, although the method is not used too frequently. The next example illustrates this graphing technique when the function to be graphed is the sum of two simpler functions. In this instance, the method used is called **adding y-coordinates**.

ADDING y-COORDINATES

EXAMPLE 2   Graph the function $F(x) = x + \sqrt{x}$.

Solution   First, we notice that the domain of F is $x \geq 0$. Next, we graph the two functions $f(x) = x$ and $g(x) = \sqrt{x}$ for $x \geq 0$. See Figures 28(a) and 28(b). To plot a point $(x, F(x))$ on the graph of F, we select a nonnegative number x and add the y-coordinates $f(x)$ and $g(x)$ to get the y-coordinate $F(x) = f(x) + g(x)$. For example, when $x = 1$, then $f(1) = 1$, $g(1) = 1$, and $F(1) = f(1) + g(1) = 1 + 1 = 2$. When $x = 4$, then $f(4) = 4$, $g(4) = 2$, and $F(4) = f(4) + g(4) = 4 + 2 = 6$; and so on. Figure 28(c) illustrates the graph of F.

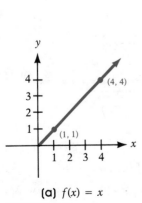

(a) $f(x) = x$

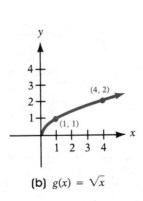

(b) $g(x) = \sqrt{x}$

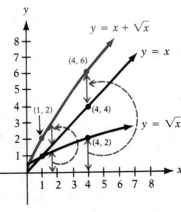

(c) $F(x) = x + \sqrt{x}$

FIGURE 28

Composite Functions

Consider the function $y = (2x + 3)^2$. If we write $y = f(u) = u^2$ and $u = g(x) = 2x + 3$, then, by a substitution process, we can obtain the original function, namely $y = f(u) = f(g(x)) = (2x + 3)^2$. This process COMPOSITION   is called **composition**. In general, suppose that f and g are two functions, and suppose that x is a number in the domain of g. By evaluating g at x, we get $g(x)$. If $g(x)$ is in the domain of f, then we may evaluate f at $g(x)$ and thereby obtain the value $f(g(x))$. If we do this for all x such that x is in the domain of g and $g(x)$ is in the domain of f, the resulting correspondence from x to $f(g(x))$ is called a *composite function*. See Figure 29.

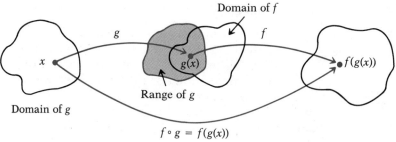

$$f \circ g = f(g(x))$$

FIGURE 29

COMPOSITE FUNCTION   Given the two functions f and g, the **composite function**, denoted by $f \circ g$ (read "f composed with g") is defined by

$$(f \circ g)(x) = f(g(x))$$

where the domain of $f \circ g$ is the set of all numbers x in the domain of g such that $g(x)$ is in the domain of f.

Figure 30 provides a second illustration of the definition. Some examples will give you the idea.

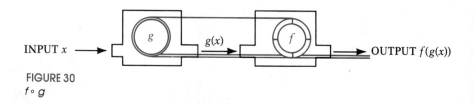

FIGURE 30
$f \circ g$

EXAMPLE 3   Suppose $f(x) = 2x^2 - 3$ and $g(x) = 4x$. Find:

(a) $(f \circ g)(1)$ (b) $(g \circ f)(1)$ (c) $(f \circ f)(-2)$ (d) $(g \circ g)(-1)$

Solution   (a) $(f \circ g)(1) = f(g(1)) = f(4) = 2 \cdot 16 - 3 = 29$
$$\uparrow$$
$$g(1) = 4$$

(b) $(g \circ f)(1) = g(f(1)) = g(-1) = 4 \cdot (-1) = -4$
$$\uparrow$$
$$f(1) = -1$$

(c) $(f \circ f)(-2) = f(f(-2)) = f(5) = 2 \cdot 25 - 3 = 47$
$$\uparrow$$
$$f(-2) = 5$$

(d) $(g \circ g)(-1) = g(g(-1)) = g(-4) = 4 \cdot (-4) = -16$
$$\uparrow$$
$$g(-1) = -4$$

EXAMPLE 4   Suppose $f(x) = \sqrt{x}$ and $g(x) = x^3 - 1$. Find the following composite functions, and then find the domain of each composite function:

(a) $f \circ g$ (b) $g \circ f$ (c) $f \circ f$ (d) $g \circ g$

Solution   (a) The function f is the square-root function, so the composite function $f \circ g = f(g(x))$ means to take the square root of $g(x)$; thus,
$$(f \circ g)(x) = f(g(x)) = \sqrt{g(x)} = \sqrt{x^3 - 1}$$
The domain of $f \circ g$ is the interval $[1, +\infty)$ and is found by determining those x in the domain of g for which $x^3 - 1 \geq 0$.

(b) The function g tells us to cube x and then subtract 1. The composite function $(g \circ f)(x) = g(f(x))$ tells us to cube $f(x)$ and then subtract 1. Thus,
$$(g \circ f)(x) = g(f(x)) = [f(x)]^3 - 1 = (\sqrt{x})^3 - 1 = x^{3/2} - 1$$
The domain of $g \circ f$ is $[0, +\infty)$.

(c) $(f \circ f)(x) = f(f(x)) = \sqrt{f(x)} = \sqrt{\sqrt{x}} = \sqrt[4]{x}$
The domain of $f \circ f$ is $[0, +\infty)$.

(d) $(g \circ g)(x) = g(g(x)) = [g(x)]^3 - 1 = (x^3 - 1)^3 - 1$
The domain of $g \circ g$ is the set of all real numbers.

Examples 4(a) and 4(b) illustrate that, in general, $f \circ g \neq g \circ f$. However, sometimes $f \circ g$ does equal $g \circ f$, as shown in the next example.

EXAMPLE 5   If $f(x) = 3x - 4$, find a function g such that $(f \circ g)(x) = x$ for every x in the domain of f. Then show that $f \circ g = g \circ f$.

Solution   We seek a function g for which

$$(f \circ g)(x) = f(g(x)) = x$$

Because

$$f(x) = 3x - 4$$

the function g will obey the equation

$$3g(x) - 4 = x$$

We can solve for $g(x)$ to get

$$g(x) = \frac{x + 4}{3}$$

Now we compute $f \circ g$ and $g \circ f$:

$$(f \circ g)(x) = f(g(x)) = 3g(x) - 4 = 3\left(\frac{x + 4}{3}\right) - 4$$

$$= x + 4 - 4 = x$$

$$(g \circ f)(x) = g(f(x)) = \frac{f(x) + 4}{3} = \frac{3x - 4 + 4}{3}$$

$$= \frac{3x}{3} = x$$

That is, in this case $f \circ g = g \circ f$.   ≡

In the next section, we shall see that there is an important relationship between functions f and g for which $f \circ g = g \circ f$.

Calculus Application

Some techniques in calculus require that we be able to determine the components of a composite function. For example, the function $H(x) = \sqrt{x + 1}$ is the composition of the functions f and g, where $f(x) = \sqrt{x}$ and $g(x) = x + 1$, because $H(x) = (f \circ g)(x) = f(g(x)) = \sqrt{g(x)} = \sqrt{x + 1}$.

EXAMPLE 6   Find functions f and g such that $f \circ g = H$ if $H(x) = (x^2 + 1)^{50}$.

Solution   The function H takes $x^2 + 1$ and raises it to the power 50. A natural choice (there are others) for f is, therefore, to raise x to the power 50. The choice of g is to square x and add 1. Let's try these choices. Let $f(x) = x^{50}$ and $g(x) = x^2 + 1$. Then

$$(f \circ g)(x) = f(g(x)) = [g(x)]^{50} = (x^2 + 1)^{50} = H(x)$$

See Figure 31.

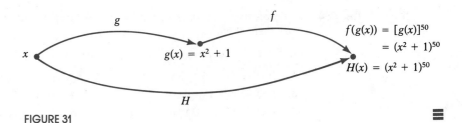

$$f(g(x)) = [g(x)]^{50}$$
$$= (x^2 + 1)^{50}$$
$$H(x) = (x^2 + 1)^{50}$$

FIGURE 31

Other functions f and g may be found for which $f \circ g = H$. For example, if $f(x) = x^2$ and $g(x) = (x^2 + 1)^{25}$, then

$$(f \circ g)(x) = f(g(x)) = [g(x)]^2 = [(x^2 + 1)^{25}]^2 = (x^2 + 1)^{50}$$

Thus, although the functions f and g found as a solution to Example 6 are not unique, there is usually a "natural" selection for f and g—one that comes to mind first. In the meantime, it is sufficient to be able to write *some* functions f and g whose composite is a given function H. You will most likely find that your selection is the natural one.

The natural selection referred to above will enable you to use your calculator most efficiently. Let's look again at Example 6. To calculate the value of H at, say 3, we would use a calculator in the following way:

$$H(x) = (x^2 + 1)^{50}$$

2		x^2		$+$	1	$=$		y^x	5	0		$=$

Display: | 2 | | 4 | | 5 | | 8.8818 E 34 |

INPUT 2	SQUARE 2	ADD ONE	RAISE TO 50TH POWER
$x = 2$	$g(x) = x^2$	$+ 1$	$f(x) = x^{50}$
	$g(2) = 2^2$	$+ 1$	$f(g(2)) = [g(2)]^{50} = 5^{50} \approx 8.8818 \text{ E } 34$

$$H(x) = (f \circ g)(x) = f(g(x))$$
$$H(2) = (f \circ g)(2) = f(g(2)) = 5^{50} \approx 8.8818 \text{ E } 34$$

EXAMPLE 7   Find functions f and g such that $f \circ g = H$ if $H(x) = 1/(x + 1)$.

Solution   If we let $f(x) = 1/x$ and $g(x) = x + 1$, we find that

$$(f \circ g)(x) = f(g(x)) = \frac{1}{g(x)} = \frac{1}{x + 1} = H(x)$$

EXERCISE 4.4   *In Problems 1–10, for the given functions f and g, find the functions below, and state their domains:*

(a) $f + g$ (b) $f - g$ (c) $f \cdot g$ (d) f/g

1. $f(x) = 3x - 4$, $g(x) = 2x + 3$ 2. $f(x) = 2x - 1$, $g(x) = 3x + 2$

3. $f(x) = x - 1$, $g(x) = 2x^2$ 4. $f(x) = 2x^2 + 3$, $g(x) = 4x^3 + 1$

5. $f(x) = \sqrt{x}$, $g(x) = 3x - 5$ 6. $f(x) = |x|$, $g(x) = x$

7. $f(x) = 1 + \dfrac{1}{x}$, $g(x) = \dfrac{1}{x}$ 8. $f(x) = 2x^2 - x$, $g(x) = 2x^2 + x$

9. $f(x) = \dfrac{2x + 3}{3x - 2}$, $g(x) = \dfrac{x}{3x - 2}$ 10. $f(x) = \sqrt{x - 1}$, $g(x) = \dfrac{1}{x}$

11. Given $f(x) = 3x + 1$ and $(f + g)(x) = 6 - \frac{1}{2}x$, find the function g.

12. Given $f(x) = 1/x$ and $(f/g)(x) = (x + 1)/(x^2 - x)$, find the function g.

In Problems 13–16 use the method of adding y-coordinates to graph each function on the interval [0, 2].

13. $f(x) = |x| + x^2$ 14. $f(x) = |x| + \sqrt{x}$

15. $f(x) = x^3 + x$ 16. $f(x) = x^3 + x^2$

In Problems 17–26, for the given functions f and g, find:

(a) $(f \circ g)(4)$ (b) $(g \circ f)(2)$ (c) $(f \circ f)(1)$ (d) $(g \circ g)(0)$

17. $f(x) = 2x$, $g(x) = 3x^2 + 1$ 18. $f(x) = 3x + 2$, $g(x) = 2x^2 - 1$

19. $f(x) = 4x^2 - 3$, $g(x) = 3 - \frac{1}{2}x^2$ 20. $f(x) = 2x^2$, $g(x) = 1 - 3x^2$

21. $f(x) = \sqrt{x}$, $g(x) = 2x$ 22. $f(x) = \sqrt{x + 1}$, $g(x) = 3x$

23. $f(x) = |x|$, $g(x) = \dfrac{1}{x^2 + 1}$ 24. $f(x) = |x - 2|$, $g(x) = \dfrac{3}{x^2 + 2}$

25. $f(x) = \dfrac{3}{x^2 + 1}$, $g(x) = \sqrt{x}$ 26. $f(x) = x^3$, $g(x) = \dfrac{2}{x^2 + 1}$

In Problems 27–40 functions f and g are given. In each problem find:

(a) $f \circ g$ (b) $g \circ f$ (c) $f \circ f$ (d) $g \circ g$

27. $f(x) = 2x + 1$, $g(x) = 3x$ 28. $f(x) = -x$, $g(x) = 2x - 3$

29. $f(x) = 3x + 1$, $g(x) = x^2$ 30. $f(x) = \sqrt{x + 1}$, $g(x) = x + 4$

31. $f(x) = \sqrt{x}$, $g(x) = x^2 - 1$ 32. $f(x) = \sqrt{x + 1}$, $g(x) = \dfrac{1}{x^2}$

33. $f(x) = \dfrac{x - 1}{x + 1}$, $g(x) = \dfrac{1}{x}$ 34. $f(x) = x + \dfrac{1}{x}$, $g(x) = x^2$

35. $f(x) = x^2$, $g(x) = \sqrt{x}$ 36. $f(x) = 2x + 4$, $g(x) = \dfrac{1}{2}x - 2$

37. $f(x) = \dfrac{1}{2x + 3}$, $g(x) = 2x + 3$ 38. $f(x) = \dfrac{x + 1}{x - 1}$, $g(x) = \dfrac{x - 1}{x + 1}$

39. $f(x) = ax + b$, $g(x) = cx + d$ 40. $f(x) = \dfrac{ax + b}{cx + d}$, $g(x) = mx$

In Problems 41–48, for each function f, find a function g such that $(f \circ g)(x) =$
x for every x in the domain of f.

could be anything
not just x.

41. $f(x) = 3x$ **42.** $f(x) = 2x$

43. $f(x) = x^3$ **44.** $f(x) = x + 5$

45. $f(x) = 2x - 6$ **46.** $f(x) = 4 - 3x$

47. $f(x) = ax + b, \quad a \neq 0$ **48.** $f(x) = \dfrac{1}{x}$

49. If $f(x) = 2x^3 - 3x^2 + 4x - 1$ and $g(x) = 2$, find $(f \circ g)(x)$ and $(g \circ f)(x)$.

50. If $f(x) = x/(x - 1)$, find $(f \circ f)(x)$.

In Problems 51–54 use $f(x) = x^2$, $g(x) = \sqrt{x} + 2$ and $h(x) = 1 - 3x$ to find the
indicated composite function.

51. $f \circ (g \circ h)$ **52.** $(f \circ g) \circ h$

53. $(f + g) \circ h$ **54.** $f \circ h + g \circ h$

In Problems 55–62 let $f(x) = x^2$, $g(x) = 3x$, and $h(x) = \sqrt{x} + 1$. Express each
function as a composite of f, g, and/or h.

55. $F(x) = 9x^2$ **56.** $G(x) = 3x^2$

57. $H(x) = |x| + 1$ **58.** $p(x) = 3\sqrt{x} + 3$

59. $q(x) = x + 2\sqrt{x} + 1$ **60.** $R(x) = 9x$

61. $P(x) = x^4$ **62.** $Q(x) = \sqrt{\sqrt{x} + 1} + 1$

In Problems 63–70 find functions f and g so that $f \circ g = H$.

63. $H(x) = (2x + 5)^3$ **64.** $H(x) = (1 - x^2)^{3/2}$

65. $H(x) = \sqrt{x^2 + x + 1}$ **66.** $H(x) = \dfrac{1}{1 + x^2}$

67. $H(x) = \left(1 - \dfrac{1}{x^2}\right)^2$ **68.** $H(x) = |2x^2 + 3|$

69. $H(x) = [\![x^2 + 1]\!]$ **70.** $H(x) = (4 - x^2)^{-4}$

71. The surface area S, in square meters, of a hot air balloon is given by

$$S(R) = 4\pi R^2$$

where R is the radius of the balloon, in meters. If the radius R is increasing with time t, in seconds, according to the formula $R(t) = \frac{2}{3}t^3$, $t \geq 0$, find the surface area S of the balloon as a function of the time t.

72. The volume V, in cubic meters, of the hot air balloon described in Problem 71 is given by $V(R) = \frac{4}{3}\pi R^3$. If the radius R is the same function of t as in Problem 71, find the volume V as a function of the time t.

73. The number N of cars produced at a certain factory in 1 day after t hours of operation is given by $N(t) = 100t - 5t^2$, $0 \leq t \leq 10$. If the cost C, in dollars, of producing x cars is $C(x) = 5000 + 6000x$, find the cost C as a function of the time t of operation of the factory.

☰ 4.5   INVERSE FUNCTIONS

We begin the discussion of inverse functions with the idea of a *one-to-one function*.

One-to-One Functions

ONE-TO-ONE FUNCTION   A function f is said to be **one-to-one** if, for any choice of numbers x_1 and x_2, $x_1 \neq x_2$, in the domain of f, then $f(x_1) \neq f(x_2)$.

In other words, for a one-to-one function $y = f(x)$, no two ordered pairs (x, y) can have the same second element. Do you see why? For a function $y = f(x)$, the ordered pairs (x, y) always have different first elements. Thus, if (x_1, y_1) and (x_2, y_2) are two ordered pairs of f, we must have $x_1 \neq x_2$. If f is also one-to-one, then we must also have $y_1 \neq y_2$.

EXAMPLE 1   (a) The function $f(x) = |x|$ is not one-to-one because, for example, the distinct elements 2 and -2 of the domain have the same value 2 in the range. That is, because $f(2) = f(-2)$, the ordered pairs $(2, 2)$ and $(-2, 2)$ have the same second element.

(b) The function $g(x) = 3x$ is one-to-one because all the ordered pairs (x, y), where $y = 3x$, have different second elements.   ☰

If the graph of a function f is known, there is a simple test, called the **horizontal-line test**, to determine whether or not f is one-to-one.

THEOREM   If any horizontal line intersects the graph of a function f at more than
HORIZONTAL-LINE TEST   one point, then f is not one-to-one.   ☰

The reason this test works is clear based on the graph given in Figure 32, where the horizontal line $y = h$ intersects the graph at two points. The result is that the two ordered pairs (x_1, h) and (x_2, h) have the same second element.

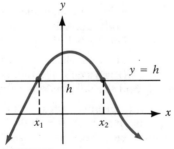

FIGURE 32
$x_1 \neq x_2$, but $f(x_1) = f(x_2) = h$; not a one-to-one function

EXAMPLE 2   For each function, use the graph to determine whether the function is one-to-one.

(a) $f(x) = x^2$ (b) $g(x) = x^3$

Solution   (a) Figure 33(a) illustrates the horizontal-line test for $f(x) = x^2$. The horizontal line $y = 1$ meets the graph of f twice, at $(1, 1)$ and at $(-1, 1)$, so f is not one-to-one.

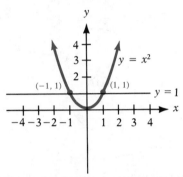

FIGURE 33(a)
A horizontal line intersects the graph twice; thus, f is not one-to-one

(b) Figure 33(b) illustrates the horizontal-line test for $g(x) = x^3$. Because each horizontal line will intersect the graph of g exactly once, it follows that g is one-to-one.

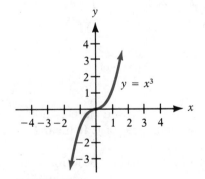

FIGURE 33(b)
No horizontal line intersects the graph more than once; thus, g is one-to-one   ≡

Let's look more closely at the one-to-one function $g(x) = x^3$. This function is an increasing function. Because an increasing (or decreasing) function will always have unequal y values for unequal x values, it follows that a function that is increasing (or decreasing) on its domain is also a one-to-one function.

THEOREM   An increasing (decreasing) function is a one-to-one function.   ≡

Inverse of a Function

Let f be a one-to-one function defined by $y = f(x)$. Then for each x in its domain, there is exactly one y in its range; furthermore, to each y in the range there corresponds exactly one x in the domain. The correspondence from the range of f onto the domain of f is, therefore, also a function, which we call the **inverse of f** and symbolize by the notation f^{-1}. Figures 34 and 35 illustrate f and f^{-1}.

INVERSE OF f

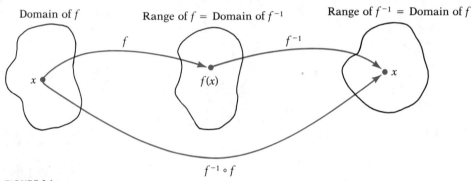

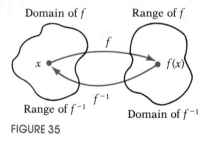

Domain of f Range of f = Domain of f^{-1} Range of f^{-1} = Domain of f

FIGURE 34
$(f^{-1} \circ f)(x) = f^{-1}(f(x)) = x$

Domain of f Range of f

Range of f^{-1} Domain of f^{-1}

FIGURE 35

Be careful! The -1 used here is not an exponent. Thus, f^{-1} does *not* mean the reciprocal of f; it means the inverse of f.

THEOREM   The inverse of a one-to-one function is unique.

Proof   If y is in the range of a one-to-one function, then, at some number x, the value of f will equal y—that is, $y = f(x)$. Because f is one-to-one, this number x is unique. Because $x = f^{-1}(y)$, the function f^{-1} is unique. ∎

This discussion leads us to the following conclusion:

THEOREM   Let $y = f(x)$ be a one-to-one function. The inverse function f^{-1} is the unique function defined on the range of f for which $x = f^{-1}(y)$.

Furthermore,

$$f^{-1}(f(x)) = x \quad \text{and} \quad f(f^{-1}(y)) = y$$

≣

Based on this theorem, we conclude that

Domain of f = Range of f^{-1} Range of f = Domain of f^{-1}

Look again at Figure 35 to visualize the relationship. If we start with x, apply f, and then apply f^{-1}, we get x back. If we start with y, apply f^{-1}, and then apply f, we get the number y back. To put it simply, whatever f does, f^{-1} undoes, and vice versa:

$$\boxed{\text{Input } x} \xrightarrow{\text{Apply } f} \boxed{f(x)} \xrightarrow{\text{Apply } f^{-1}} \boxed{f^{-1}(f(x)) = x}$$

$$\boxed{\text{Input } y} \xrightarrow{\text{Apply } f^{-1}} \boxed{f^{-1}(y)} \xrightarrow{\text{Apply } f} \boxed{f(f^{-1}(y)) = y}$$

The relationship between a function f and its inverse f^{-1} can be used as verification that f^{-1} is, in fact, the inverse of f.

Let's look at some examples.

EXAMPLE 3   (a) We verify that the inverse of $g(x) = x^3$ is $g^{-1}(y) = \sqrt[3]{y}$ by showing that $g^{-1}(g(x)) = x$:

$$g^{-1}(g(x)) = \sqrt[3]{g(x)} = \sqrt[3]{x^3} = x$$

(b) The inverse of $h(x) = 3x$ is $h^{-1}(y) = \frac{1}{3}y$, because

$$h^{-1}(h(x)) = \frac{1}{3}h(x) = \frac{1}{3}(3x) = x$$

(c) The inverse of $f(x) = 2x + 3$ is $f^{-1}(y) = \frac{1}{2}(y - 3)$ because

$$f^{-1}(f(x)) = \frac{1}{2}[f(x) - 3] = \frac{1}{2}(2x + 3 - 3) = x$$ ≣

Alternatively, we could have verified the inverses of Example 3 by showing instead that

$$g(g^{-1}(y)) = y, \quad h(h^{-1}(y)) = y, \quad \text{and} \quad f(f^{-1}(y)) = y$$

We shall use this fact in Example 4 to show how to find the inverse of the one-to-one function of Example 3(c).

EXAMPLE 4   Find the inverse of $f(x) = 2x + 3$.

Solution   First we note that the graph of the function $y = f(x) = 2x + 3$ is a line with slope 2, so it is an increasing function. We conclude that f is one-to-one and, therefore, that f has an inverse. Suppose f^{-1} is the inverse of f. Then

$$f(f^{-1}(y)) = y$$

But $f(x) = 2x + 3$. Thus,

$$2f^{-1}(y) + 3 = y \tag{1}$$

$$2f^{-1}(y) = y - 3 \quad \text{Solve for } f^{-1}(y). \tag{2}$$

$$f^{-1}(y) = \frac{1}{2}(y - 3) \tag{3}$$

Because the symbol traditionally used to represent the independent variable of a function is x, we shall follow the usual practice and replace y by x in $f^{-1}(y) = \frac{1}{2}(y - 3)$. Thus, the inverse of $f(x) = 2x + 3$ is $f^{-1}(x) = \frac{1}{2}(x - 3)$. ▰

Let's review the procedure we used to find f^{-1} in Example 4. First, we note that f is one-to-one, so we know it has a unique inverse. Now let's look at the steps in Equations (1)–(3). Because $y = f(x) = 2x + 3$, we can write these steps as

$$2x + 3 = y$$

$$2x = y - 3$$

$$x = \frac{1}{2}(y - 3)$$

Now, if we replace x by y and y by x, we get

$$y = \frac{1}{2}(x - 3)$$

The inverse of f is $f^{-1}(x) = \frac{1}{2}(x - 3)$. The solution to Example 3(c) verifies this answer.

We outline the steps to follow for finding the inverse of a function below:

Step 1: Determine whether the function is one-to-one.

Step 2: If it is, then, in the equation $y = f(x)$ that defines f, solve for x. You will now actually have $x = f^{-1}(y)$.

Step 3: Replace x by y and y by x in $x = f^{-1}(y)$ to get $y = f^{-1}(x)$.

Step 4: The equation $y = f^{-1}(x)$ defines the inverse function f^{-1} of f.

Step 5: Check your answer by verifying that

$$f^{-1}(f(x)) = x$$

These steps require some comment. If, in Step 1, the function f is incorrectly identified as one-to-one (when, in fact, it isn't), then the solution for x in Step 2 will not be unique. When this happens, f has no inverse. For example, the function $f(x) = x^2$ is not one-to-one. If we let $y = x^2$ and solve for x, we get

$$y = x^2$$
$$x = \pm\sqrt{y}$$

If we interchange x and y, we get

$$y = \pm\sqrt{x}$$

which is not a function. Thus, no inverse exists.

The requirement in Step 2 that we solve for x may not be easy to do; in fact, it may be impossible to do. For example, if the function f defined by $y = x^{13} - 2x^3 + x - 1$ is one-to-one, it cannot be easily solved for x in terms of y. In such cases, the best we can do is to acknowledge the existence of an inverse; however, we will not be able to write an equation of the inverse.

Geometric Interpretation

If we sketched the graphs of f and f^{-1} of Example 4 (illustrated in Figure 36), we notice an interesting fact:

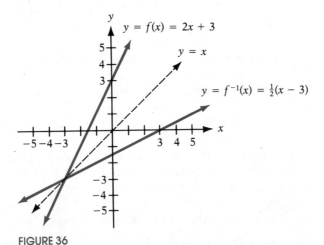

FIGURE 36

THEOREM The graphs of f and f^{-1} are symmetric with respect to the line $y = x$. ▬

The graph of f^{-1} is the reflection of the graph of f about the line $y = x$. Each graph is a mirror image of the other, the mirror being the line $y = x$.

Proof   If (a, b) is a point on the graph of the function f defined by $y = f(x)$, then $b = f(a)$. But this means $a = f^{-1}(b)$, and so (b, a) is a point on the graph of f^{-1}. Now look at Figure 37. The shaded triangles are congruent (Side–Side–Side). Hence, triangle OPM is congruent to triangle OQM. (Do you see why? Side–Angle–Side, where one side is the common side, the other side has length $d(O, P) = d(O, Q) = \sqrt{a^2 + b^2}$, and angles POM and QOM are their respective included angles.) As a result, (a, b) and (b, a) are the same distance from M, which lies on the line $y = x$. Thus, the graphs of f and f^{-1} are symmetric with respect to the line $y = x$.

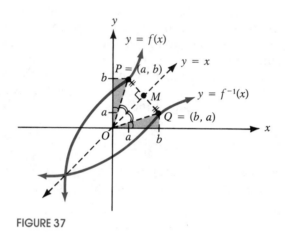

FIGURE 37

Not every function f is one-to-one. Sometimes, though, an appropriate restriction on the domain of a function will yield a new function that is one-to-one. Let's look at an example of this common practice.

EXAMPLE 5   Find the inverse of $y = f(x) = x^2$ if $x \geq 0$.

Solution   The function $f(x) = x^2$ is not one-to-one. [Refer to Example 2(a).] However, if we restrict f to only that part of its domain for which $x \geq 0$, we have a new function that is increasing and therefore is one-to-one.

Let's follow the steps outlined above to find f^{-1}:

Step 1:   The function defined by $y = x^2$, $x \geq 0$, is one-to-one.

Step 2:   In the equation $y = x^2$, $x \geq 0$, solve for x to get

$$x = \sqrt{y} \qquad \text{The minus sign is excluded because } x \geq 0.$$

Step 3:   Replace x by y and y by x to get

$$y = \sqrt{x} \qquad x \geq 0$$

Step 4: The inverse function f^{-1} is defined by the equation

$$y = f^{-1}(x) = \sqrt{x} \qquad x \geq 0$$

Step 5: *Check:* $f^{-1}(f(x)) = \sqrt{x^2} = |x| = x$

$$\uparrow$$
$$x \geq 0$$

Figure 38 illustrates the graphs of $f(x) = x^2$, $x \geq 0$, and $f^{-1}(x) = \sqrt{x}$, $x \geq 0$.

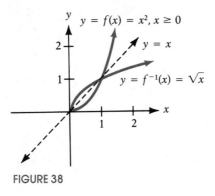

FIGURE 38

Calculators

We noted earlier that many calculators contain keys that allow you to find the value of a function. These same calculators will usually also have a key labeled $\boxed{\text{INV}}$ or $\boxed{\text{INVERSE}}$ that enables you to calculate the value of the inverse function. (If the actual inverse is present as a function key, such as $\sqrt{x}$ and x^2, the inverse key is usually disengaged for such functions.)

Try the following experiment if you have a calculator with an inverse key:

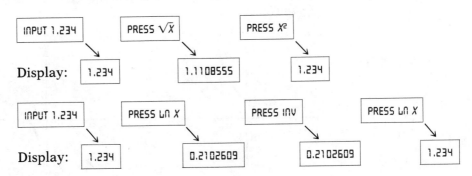

EXERCISE 4.5   *In Problems 1–6 the graph of a function f is given. Use the horizontal-line test to determine whether f is one-to-one.*

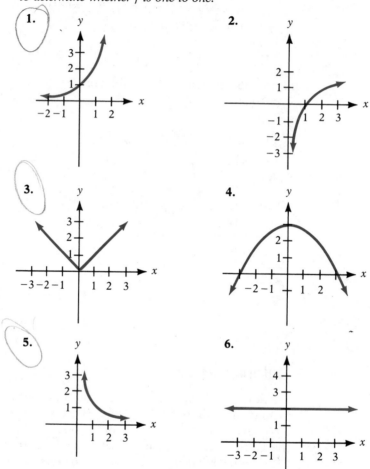

In Problems 7–12 the graph of a one-to-one function f is given. Find the graph of the inverse function f^{-1}. For convenience (and as a hint) the graph of $y = x$ is also given.

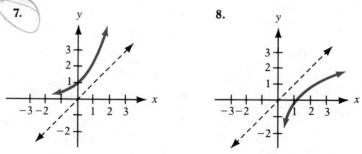

9.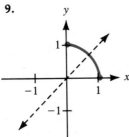

10.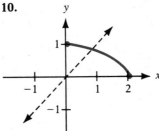

11.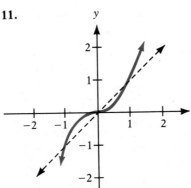

12.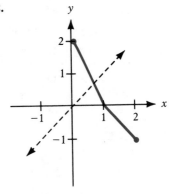

In Problems 13–22 show that the functions f and g are inverses of each other by showing that f(g(x)) = x.

13. $f(x) = 3x - 4$; $g(x) = \frac{1}{3}(x + 4)$

14. $f(x) = 1 - 2x$; $g(x) = -\frac{1}{2}(x - 1)$

15. $f(x) = 4x - 8$; $g(x) = \frac{x}{4} + 2$

16. $f(x) = 2x + 6$; $g(x) = \frac{1}{2}x - 3$

17. $f(x) = x^3 - 8$; $g(x) = \sqrt[3]{x + 8}$

18. $f(x) = (x - 2)^2$, $x \geq 2$; $g(x) = \sqrt{x} + 2$, $x \geq 0$

19. $f(x) = \frac{1}{x}$; $g(x) = \frac{1}{x}$

20. $f(x) = x$; $g(x) = x$

21. $f(x) = \frac{2x + 3}{x + 4}$; $g(x) = \frac{4x - 3}{2 - x}$

22. $f(x) = \frac{x - 5}{2x + 3}$; $g(x) = \frac{3x + 5}{1 - 2x}$

In Problems 23–36 the function f is one-to-one. Find its inverse, and verify your answer.

23. $f(x) = 4x + 2$

24. $f(x) = 1 - 3x$

25. $f(x) = x^3 - 1$

26. $f(x) = x^3 + 1$

27. $f(x) = \dfrac{4}{x}$ **28.** $f(x) = -\dfrac{3}{x}$

29. $f(x) = \dfrac{x}{x - 1},\quad x > 1$ **30.** $f(x) = \dfrac{1}{1 - x},\quad x > 1$

31. $f(x) = x^2 + 4,\quad x \geq 0$ **32.** $f(x) = (x - 1)^2,\quad x \geq 1$

33. $f(x) = x^2 + 9,\quad x \leq 0$ **34.** $f(x) = (x + 2)^2,\quad x \leq -2$

35. $f(x) = x^4,\quad x \geq 0$ **36.** $f(x) = x^5 - 1$

37. Find the inverse of the linear function $f(x) = mx + b,\ m \neq 0$.

38. Find the inverse of the function $f(x) = \sqrt{R^2 - x^2},\ 0 \leq x \leq R$.

39. Can an even function be one-to-one? Explain.

40. Is every odd function one-to-one? Explain.

41. A function f has an inverse. If the graph of f lies in the first quadrant, in what quadrant does the graph of f^{-1} lie?

42. A function f has an inverse. If the graph of f lies in the second quadrant, in what quadrant does the graph of f^{-1} lie?

43. To convert from x degrees Celsius to y degrees Fahrenheit, we use the formula $y = f(x) = \frac{9}{5}x + 32$. To convert from u degrees Fahrenheit to v degrees Celsius, we use the formula $v = g(u) = \frac{5}{9}(u - 32)$. Show that f and g are inverse functions.

44. The demand for corn obeys the equation $p(x) = 300 - 50x$, where p is the price per bushel, in dollars, and x is the number produced, in millions of bushels. Express the production amount x as a function of the price p.

45. The period T, in seconds, of a simple pendulum is a function of its length l, in feet, given by $T(l) = 2\pi\sqrt{l/g}$, where $g \approx 32.2$ feet per second per second is the acceleration due to gravity. Express the length l as a function of the period T.

C H A P T E R   R E V I E W

VOCABULARY

function	arrow notation	squaring function
domain	ordered-pair notation	cubing function
value	graph of a function	square-root function
image	vertical-line test	absolute-value
range	ordered pair	function
value of a function	increasing function	bracket x
$f(x)$ exists	decreasing function	greatest-integer
$f(x)$ does not exist	stationary function	function
independent variable	even function	step function
dependent variable	odd function	piecewise-defined
argument	linear function	function
function notation	constant function	vertical shift
difference quotient	identity function	horizontal shift

change in vertical
   scale
change in horizontal
   scale
reflection about x-axis
reflection about y-axis

sum function
difference function
product function
quotient function
adding y-coordinates

composition
composite function
one-to-one function
horizontal-line test
inverse of f

FILL-IN-THE-BLANK QUESTIONS

1. If f is a function defined by the equation $y = f(x)$, then x is called the _____ variable and y is the _____ variable.

2. A set of points in the xy-plane is the graph of a function if and only if no _____ line contains more than one point of the set.

3. A(n) _____ function f is one for which $f(-x) = f(x)$ for every x in the domain of f; a(n) _____ function f is one for which $f(-x) = -f(x)$ for every x in the domain of f.

4. Suppose the graph of a function f is known. Then the graph of $y = f(x - 2)$ may be obtained by a(n) _____ shift of the graph of f to the _____ a distance of 2 units.

5. If $f(x) = x + 1$ and $g(x) = x^3$, then _____ $= (x + 1)^3$.

6. If every horizontal line intersects the graph of a function f at no more than one point, then f is a(n) _____ function.

7. If f^{-1} denotes the inverse of a function f, then the graphs of f and f^{-1} are symmetric with respect to the line _____.

REVIEW EXERCISES

1. Given that f is a linear function, $f(4) = -2$, and $f(1) = 4$, write the equation that defines f.

2. Given that g is a linear function with slope $= -2$ and $g(-2) = 2$, write the equation that defines g.

3. A function f is defined by the equation

$$f(x) = \frac{Ax + 5}{6x - 2}$$

If $f(1) = 4$, find A.

4. A function g is defined by the equation

$$g(x) = \frac{A}{x} + \frac{8}{x^2}$$

If $g(-1) = 0$, find A.

In Problems 5–10 find the domain of each function.

5. $f(x) = \dfrac{x}{x - 2}$

6. $g(x) = \sqrt{x + 2}$

7. $h(x) = \dfrac{\sqrt{x}}{|x|}$

8. $F(x) = \dfrac{1}{x^2 - 3x - 4}$

9. $G(x) = \begin{cases} |x| & \text{if } -1 \le x \le 1 \\ 1/x & \text{if } x > 1 \end{cases}$

10. $H(x) = \begin{cases} 1/x & \text{if } 0 < x < 4 \\ \sqrt{x - 4} & \text{if } 4 \le x \le 8 \end{cases}$

In Problems 11–16 graph each function.

11. $f(x) = \sqrt{x + 2}$

12. $f(x) = \sqrt{2 - x}$

13. $f(x) = 1 - x^2$

14. $f(x) = 4 - x^3$

15. $F(x) = \begin{cases} x^2 + 4 & \text{if } x < 0 \\ 4 - x^2 & \text{if } x \ge 0 \end{cases}$

16. $H(x) = \begin{cases} |1 - x| & \text{if } 0 \le x \le 2 \\ [\![x - 1]\!] & \text{if } x > 2 \end{cases}$

17. (a) Tell which of the graphs below are graphs of functions.
    (b) Tell which of the graphs below are graphs of one-to-one functions.

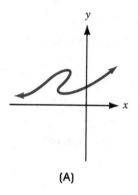

(A)

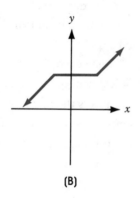

(B)

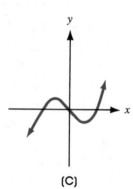

(C)

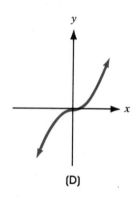

(D)

18. Use the graph of the function f shown below to find:
    (a) The domain and range of f
    (b) The intervals on which f is increasing
    (c) The intervals on which f is stationary
    (d) The intercepts of f

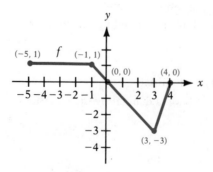

In Problems 19–24 determine whether the given function is even, odd, or neither.

19. $f(x) = x^3 - x$

20. $g(x) = \dfrac{1 + x^2}{1 + x^4}$

21. $h(x) = \dfrac{1}{x^4} + \dfrac{1}{x^2} + 1$

22. $F(x) = \sqrt{1 - x^3}$

23. $G(x) = 1 - x + x^3$

24. $H(x) = 1 + x + x^2$

In Problems 25–30 the function f is one-to-one. Find the inverse of each function, and verify your answer.

25. $f(x) = \dfrac{2x + 3}{x - 2}$

26. $f(x) = \dfrac{2 - x}{1 + x}$

27. $f(x) = \dfrac{1}{x - 1}$

28. $f(x) = \sqrt{x - 2}$

29. $f(x) = \dfrac{3}{x^{1/3}}$

30. $f(x) = x^{1/3} + 1$

In Problems 31–36, for the given functions f and g, find:

(a) $(f \circ g)(2)$ (b) $(g \circ f)(-2)$ (c) $(f \circ f)(4)$ (d) $(g \circ g)(-1)$

31. $f(x) = 3x - 5,\ g(x) = 1 - 2x^2$ 32. $f(x) = 4 - x,\ g(x) = 1 + x^2$

33. $f(x) = \sqrt{x + 2},\ g(x) = 2x^2 + 1$ 34. $f(x) = 1 - 3x^2,\ g(x) = \sqrt{1 - x}$

35. $f(x) = \dfrac{1}{x^2 + 4},\ g(x) = 3x - 2$ 36. $f(x) = \dfrac{2}{1 + 2x^2},\ g(x) = 3x$

In Problems 37–42 find $f \circ g$, $g \circ f$, $f \circ f$, and $g \circ g$ for each pair of functions.

37. $f(x) = \dfrac{2 - x}{x},\ g(x) = 3x + 2$ 38. $f(x) = \dfrac{x}{x + 1},\ g(x) = \dfrac{x}{x - 1}$

39. $f(x) = 3x^2 + x + 1,\ g(x) = |3x|$ 40. $f(x) = \sqrt{3x},\ g(x) = 1 + x + x^2$

41. $f(x) = \dfrac{x + 1}{x - 1},\ g(x) = \dfrac{1}{x}$ 42. $f(x) = \sqrt{x^2 - 3},\ g(x) = \sqrt{3 - x^2}$

43. For the graph of the function f shown below:
    (a) Draw the graph of $y = f(-x)$.
    (b) Draw the graph of $y = -f(x)$.
    (c) Draw the graph of $y = f(x + 2)$.
    (d) Draw the graph of $y = f(x) + 2$.
    (e) Draw the graph of $y = f(2 - x)$.
    (f) Draw the graph of f^{-1}.

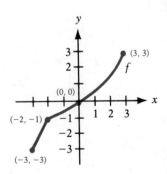

44. Repeat Problem 43 for the graph of the function g shown below.

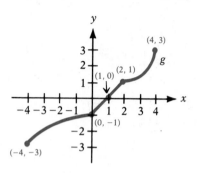

45. The temperature T of the air is approximately a linear function of the altitude h for altitudes within 6 miles of the surface of the earth. If the surface temperature is 30° C and the temperature at 10,000 meters is 5° C, find the equation of the function $T = T(h)$.

46. The speed v, in feet per second, of a car is a linear function of the time t, in seconds, for $10 \leq t \leq 30$. If, after each second, the speed of the car has increased by 5 feet per second and if, after 20 seconds, the speed is 80 feet per second, how fast is the car going after 30 seconds? Find the equation of the function $v = v(t)$.

POLYNOMIAL AND RATIONAL FUNCTIONS

5.1   Quadratic Functions
5.2   Polynomial Functions
5.3   Dividing Polynomials; Synthetic
        Division
5.4   The Zeros of a Polynomial Function
5.5   Approximating the Zeros of a
        Polynomial Function
5.6   Rational Functions
5.7   Complex Polynomials; Fundamental
        Theorem of Algebra
        Chapter Review

In the previous chapter, we graphed linear functions $f(x) = ax + b$, $a \neq 0$; the squaring function $f(x) = x^2$; and the cubing function $f(x) = x^3$. Each of these functions belongs to the class of functions called **polynomial functions.** In this chapter we study polynomial functions, placing special emphasis on their graphs. This emphasis will require that we learn techniques for evaluating polynomials (Section 5.3) and for solving polynomial equations (Sections 5.4 and 5.5). Then we will discuss **rational functions**, which are ratios of polynomial functions. The final section deals with polynomials having coefficients that are complex numbers.

■ 5.1 QUADRATIC FUNCTIONS

QUADRATIC FUNCTION   A **quadratic function** is a function of the form

$$f(x) = ax^2 + bx + c \qquad (1)$$

in which a, b, and c are real numbers and $a \neq 0$. The domain of a quadratic function consists of all real numbers.

Many applications require a knowledge of quadratic functions. For example, suppose the equation that relates the number x of units sold and the price p per unit is given by

$$x = 15{,}000 - 750p$$

Then the revenue R derived from selling x units at the price p per unit is

$$R = xp = (15{,}000 - 750p)p = -750p^2 + 15{,}000p$$

Figure 1 illustrates the graph of this revenue function, whose domain is $0 \leq p \leq 20$, since both x and p must be nonnegative.

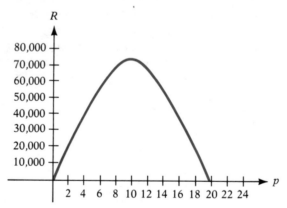

FIGURE 1
Graph of a revenue function: $R = -750p^2 + 15{,}000p$

A second situation in which a quadratic function appears has to do with the motion of a projectile. Based on Newton's second law of motion (force equals mass times acceleration, $F = ma$), it can be shown that the path of a projectile propelled upward at an inclination to the horizontal is the graph of some quadratic function. See Figure 2 for an illustration.

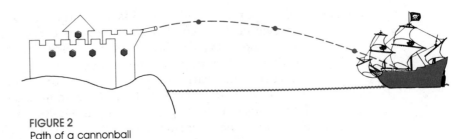

FIGURE 2
Path of a cannonball

Graphing Quadratic Functions

We already know how to graph quadratic functions. For example, based on the discussion of change of scale and reflection in Section 4.3, we know how to graph quadratic functions of the form $f(x) = ax^2$, $a \neq 0$. Figure 3 illustrates the graphs of $f(x) = ax^2$ for $a = 1$, $a = 3$, and $a = \frac{1}{2}$, drawn on the same set of coordinate axes. Observe that the choice of a larger value of a in $f(x) = ax^2$ results in a "thinner" or "narrower" graph—a compression.

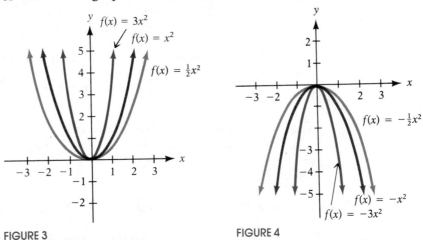

FIGURE 3

FIGURE 4

Of course, the graphs of $f(x) = ax^2$ for $a < 0$ are merely the reflection about the x-axis of the corresponding graphs of $f(x) = |a|x^2$. See Figure 4.

The graphs in Figures 3 and 4 are typical of the graphs of all quadratic functions, which we call **parabolas**.* Refer to Figure 5 (page 286) where two parabolas are pictured. The one on the left **opens upward** and has a lowest point; the one on the right **opens downward** and has a highest point. The lowest or highest point of a parabola is called the **vertex**. The dashed vertical line passing through the vertex

PARABOLAS

OPENS UPWARD

OPENS DOWNWARD

VERTEX

*We shall study parabolas using a geometric definition in the chapter on conics.

AXIS OF SYMMETRY   in each parabola in Figure 5 is called the **axis of symmetry** (usually abbreviated to **axis**) of the parabola. Because the parabola is symmetric about its axis, the axis of symmetry of a parabola can be used to advantage in graphing the parabola.

Figure 5 illustrates the graphs of a quadratic function $f(x) = ax^2 + bx + c$, $a \neq 0$. Notice that the coordinate axes are not included. Depending on the values of a, b, and c, the axes could be placed anywhere. The important fact is that, except possibly for a change of scale, the shape of the graph of a quadratic function will look like one of the parabolas in Figure 5.

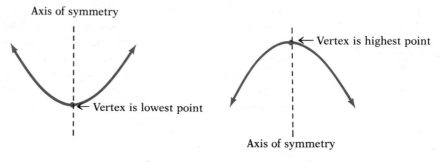

(a) Opens upward   (b) Opens downward
FIGURE 5
Graphs of a quadratic function, $f(x) = ax^2 + bx + c$, $a \neq 0$

Let's graph a quadratic function $f(x) = ax^2 + bx + c$, $a \neq 0$, using techniques from Section 4.3. In so doing, we shall write the function f in the form $f(x) = a(x - h)^2 + k$.

EXAMPLE 1   Graph the function $f(x) = 2x^2 + 8x + 5$.

Solution   We begin by completing the square on the right side:

$f(x) = 2x^2 + 8x + 5$

$f(x) = 2(x^2 + 4x) + 5$ Factor out the 2 from $2x^2 + 8x$.

$f(x) = 2(x^2 + 4x + 4) + 5 - 8$ Complete the square of $2(x^2 + 4x)$ by adding and subtracting 8. Look closely at this step.

$f(x) = 2(x + 2)^2 - 3$

The graph of f can be obtained in three stages. Look at Figure 6. Now compare this graph to the graph in Figure 5(a). The graph of $f(x) = 2x^2 + 8x + 5$ is a parabola that opens upward and has its vertex (lowest point) at $(-2, -3)$. Its axis of symmetry is the line $x = -2$.

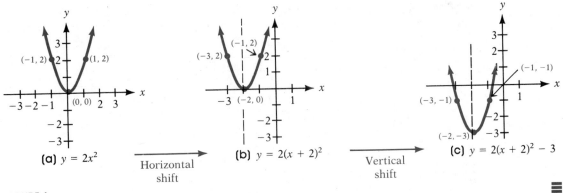

FIGURE 6

The method used in Example 1 can be used to graph any quadratic function

$$f(x) = ax^2 + bx + c$$

$$f(x) = ax^2 + bx + c$$

$$f(x) = a\left(x^2 + \frac{b}{a}x\right) + c \qquad \text{Factor out } a \text{ from } ax^2 + bx.$$

$$f(x) = a\left(x^2 + \frac{b}{a}x + \frac{b^2}{4a^2}\right) + c - a\left(\frac{b^2}{4a^2}\right) \qquad \text{Complete the square by adding and subtracting } a(b^2/4a^2). \text{ Look closely at this step!}$$

$$f(x) = a\left(x + \frac{b}{2a}\right)^2 + c - \frac{b^2}{4a}$$

$$f(x) = a\left(x + \frac{b}{2a}\right)^2 + \frac{4ac - b^2}{4a}$$

If we let $h = -b/2a$ and let $k = (4ac - b^2)/4a$, this last equation can be rewritten as

$$f(x) = a(x - h)^2 + k$$

The graph of f is the parabola $y = ax^2$ shifted horizontally h units and vertically k units. As a result, the vertex is at (h, k), and the graph opens upward if $a > 0$ and downward if $a < 0$. The axis is the vertical line $x = h$. The result at the top of the next page summarizes these conclusions.

THEOREM   The quadratic function

$$f(x) = ax^2 + bx + c$$

can be written in the equivalent form

$$f(x) = a(x - h)^2 + k \qquad (2)$$

where

$$h = \frac{-b}{2a} \quad \text{and} \quad k = \frac{4ac - b^2}{4a}$$

Its graph is a parabola with vertex at the point (h, k); the parabola opens upward if $a > 0$ and opens downward if $a < 0$; the axis is the vertical line $x = h$. ▬

   In almost every case, it is easier to obtain the vertex of a quadratic function f by remembering that its x-coordinate is $h = -b/2a$. The y-coordinate can then be found by evaluating f at $-b/2a$.
   These results are summarized below:

$$f(x) = ax^2 + bx + c$$

$$\text{Vertex} = \left(\frac{-b}{2a}, f\!\left(\frac{-b}{2a}\right) \right)$$

$$\text{Axis: The line } x = \frac{-b}{2a} \qquad (3)$$

Graph opens upward if $a > 0$

Graph opens downward if $a < 0$

EXAMPLE 2   Without graphing, locate the vertex and axis of the parabola defined by $f(x) = -3x^2 + 6x + 1$. Does it open upward or downward?

Solution   For this quadratic function, $a = -3$, $b = 6$, and $c = 1$. The x-coordinate of the vertex is

$$\frac{-b}{2a} = \frac{-6}{-6} = 1$$

The y-coordinate of the vertex is therefore

$$f\!\left(\frac{-b}{2a}\right) = f(1) = -3 + 6 + 1 = 4$$

The vertex is located at the point $(1, 4)$. The axis of symmetry is the line $x = 1$. Finally, because $a = -3 < 0$, the parabola opens downward. ▬

The information we gathered in Example 2, together with the location of the intercepts, usually provides enough information to sketch the graph.

EXAMPLE 3 Use the information from Example 2 and the location of intercepts to graph $f(x) = -3x^2 + 6x + 1$.

Solution The y-intercept is found by letting $x = 0$. Thus the y-intercept is $f(0) = 1$. The x-intercepts are found by letting $y = f(x) = 0$. This results in the equation

$$-3x^2 + 6x + 1 = 0$$

The discriminant $b^2 - 4ac = (6)^2 - 4(-3)(1) = 36 + 12 = 48 > 0$, so the equation has two real solutions. Using the quadratic formula, we find the solutions:

$$x = \frac{-b + \sqrt{b^2 - 4ac}}{2a} = \frac{-6 + \sqrt{48}}{-6} = \frac{-6 + 4\sqrt{3}}{-6} \approx -0.15$$

and

$$x = \frac{-b - \sqrt{b^2 - 4ac}}{2a} = \frac{-6 - \sqrt{48}}{-6} = \frac{-6 - 4\sqrt{3}}{-6} \approx 2.15$$

The x-intercepts are approximately -0.15 and 2.15.

The graph is illustrated in Figure 7. Notice how we used the y-intercept and the axis of symmetry, $x = 1$, to obtain the additional point $(2, 1)$ on the graph.

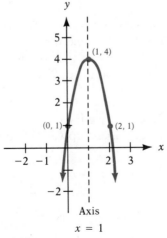

FIGURE 7
$f(x) = -3x^2 + 6x + 1$

We now know two ways to graph a quadratic function:

1. Complete the square and apply shifting techniques (Example 1).
2. Use the results given in Display (3) to find the vertex and axis of symmetry and to determine whether the graph opens upward or downward. Then locate the y-intercept and the x-intercepts, if there are any (Examples 2 and 3).

EXAMPLE 4   Graph each quadratic function by determining whether its graph opens upward or downward and by finding its vertex, axis of symmetry, y-intercept, and x-intercepts, if any.

(a) $f(x) = x^2 - 6x + 9$ (b) $g(x) = 2x^2 + x + 1$

Solution   (a) For $f(x) = x^2 - 6x + 9$, we have $a = 1$, $b = -6$, and $c = 9$. Since $a = 1 > 0$, the parabola opens upward. The x-coordinate of the vertex is

$$\frac{-b}{2a} = \frac{-(-6)}{2 \cdot 1} = 3$$

The y-coordinate of the vertex is

$$f(3) = 9 - 6 \cdot 3 + 9 = 0$$

So the vertex is at $(3, 0)$. The axis of symmetry is the line $x = 3$. The y-intercept is $f(0) = 9$. The x-intercept(s), if any, obey the equation

$$x^2 - 6x + 9 = 0$$
$$(x - 3)^2 = 0$$
$$x = 3$$

Thus, there is one x-intercept, at 3. Since the vertex $(3, 0)$ lies on the x-axis, the graph will touch the x-axis at the x-intercept. Also, the y-intercept is at $(0, 9)$, so that by using the axis of symmetry, we can locate another point $(6, 9)$ on the graph. See Figure 8(a) for the graph.

(b) For $g(x) = 2x^2 + x + 1$, we have $a = 2$, $b = 1$, and $c = 1$. Since $a = 2 > 0$, the graph opens upward. The x-coordinate of the vertex is

$$\frac{-b}{2a} = -\frac{1}{4}$$

The y-coordinate of the vertex is

$$f\left(-\frac{1}{4}\right) = 2\left(\frac{1}{16}\right) + \left(-\frac{1}{4}\right) + 1 = \frac{7}{8}$$

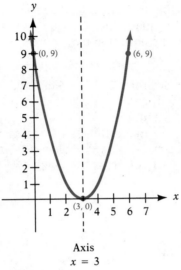

FIGURE 8(a)

$f(x) = x^2 - 6x + 9$

So the vertex is at $(-\frac{1}{4}, \frac{7}{8})$. The axis of symmetry is the line $x = -\frac{1}{4}$. The y-intercept is $f(0) = 1$. The x-intercept(s), if any, obey the equation

$$2x^2 + x + 1 = 0$$

Since the discriminant $b^2 - 4ac = 1 - 8 = -7 < 0$, this equation has no real solution, and so the graph has no x-intercepts. We use the point $(0, 1)$ and the axis of symmetry $x = -\frac{1}{4}$ to locate the point $(-\frac{1}{2}, 1)$ on the graph. See Figure 8(b).

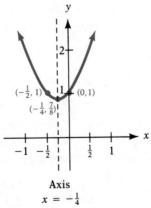

FIGURE 8(b)

$g(x) = 2x^2 + x + 1$

The preceding examples lead to the following connection between the graph of a quadratic function $f(x) = ax^2 + bx + c$ and the discriminant of the equation $ax^2 + bx + c = 0$:

1. If the discriminant $b^2 - 4ac > 0$, the graph of $f(x) = ax^2 + bx + c$ will cross the x-axis in two places.
2. If the discriminant $b^2 - 4ac = 0$, the graph of $f(x) = ax^2 + bx + c$ will touch the x-axis at its vertex.
3. If the discriminant $b^2 - 4ac < 0$, the graph of $f(x) = ax^2 + bx + c$ will not cross the x-axis.

Applications

We have already seen that the graph of a quadratic function $f(x) = ax^2 + bx + c$ is a parabola with vertex at $(-b/2a, f(-b/2a))$. This vertex is the highest point on the graph if $a < 0$ and is the lowest point on the graph if $a > 0$. If the vertex is the highest point ($a < 0$), then MAXIMUM VALUE $f(-b/2a)$ is the **maximum value** of f. If the vertex is the lowest point MINIMUM VALUE ($a > 0$), then $f(-b/2a)$ is the **minimum value** of f. These notions give rise to many applications.

EXAMPLE 5   The manufacturer of a digital watch has found that, when they are sold at a price of p dollars per unit, the revenue R (in dollars) as a function of the price p is

$$R(p) = -750p^2 + 15,000p$$

What price should be established for each watch in order to maximize revenue? If this price is charged, what is the maximum revenue?

Solution   The revenue R is

$$R(p) = -750p^2 + 15,000p$$
$$= ap^2 + bp + c$$

The function R is a quadratic function with $a = -750$, $b = 15,000$, and $c = 0$. Because $a < 0$, the vertex is the highest point of the parabola. The revenue R is therefore a maximum when

$$p = \frac{-b}{2a} = \frac{-15,000}{2(-750)} = \frac{-15,000}{-1500} = \$10$$

The maximum revenue R is

$$R(10) = -750(10)^2 + 15,000(10) = \$75,000 \qquad \equiv$$

EXAMPLE 6   A company charges $200 for each box of tools on orders of 150 or fewer boxes. The cost to the buyer on every box is reduced by $1 for each box ordered in excess of 150. For what size order is revenue maximum? What is the maximum revenue?

Solution   For an order of exactly 150 boxes, the company's revenue is

$$\$200(150) = \$30{,}000$$

For an order of 160 boxes (which is 10 in excess of 150), the per-box charge is $200 - 10(1) = 190$ and the revenue is

$$\$190(160) = \$30{,}400$$

To solve the problem, let x denote the number of boxes sold. The revenue R is

$$R = (\text{Number of boxes})(\text{Cost per box})$$
$$= x(\text{Cost per box})$$

If $x \geq 150$, the charge per box is

$$200 - 1\left(\begin{array}{c}\text{Number of boxes}\\ \text{in excess of 150}\end{array}\right) = 200 - 1(x - 150) = 350 - x$$

Hence, the revenue R is

$$R = x(350 - x) = -x^2 + 350x \qquad x \geq 150$$

which is a quadratic function with $a = -1$, $b = 350$, and $c = 0$. Because $a < 0$, the vertex is the highest point. The revenue R is therefore a maximum when

$$x = \frac{-b}{2a} = \frac{-350}{-2} = 175$$

The maximum revenue is

$$R(175) = 175(350 - 175) = \$30{,}625$$

The company would want to set 175 as the maximum number of boxes a person could purchase on this plan, because revenue to the company starts to decrease for orders in excess of 175.   ▤

EXAMPLE 7   A cruise ship leaves the Port of Miami heading due east at a constant speed of 5 knots (1 knot = 1 nautical mile per hour). At 5:00 P.M., the cruise ship is 5 nautical miles due south of a cabin cruiser that is moving south at a constant speed of 10 knots. At what time are the two ships closest?

Solution   We begin with an illustration depicting the relative position of each ship at 5:00 P.M. See Figure 9(a). After a time t (in hours) has passed, the cruise ship has moved east $5t$ nautical miles, and the cabin cruiser

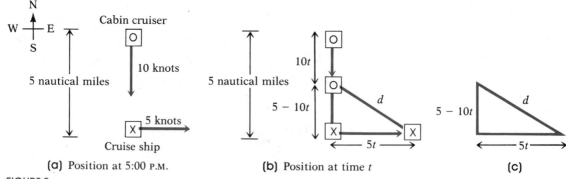

(a) Position at 5:00 P.M.

(b) Position at time t

(c)

FIGURE 9

has moved south $10t$ nautical miles. Figure 9(b) illustrates the relative position of each ship after t hours. Figure 9(c) shows a right triangle extracted from Figure 9(b). By the Pythagorean Theorem, the square of the distance d between the ships after time t is

$$d^2 = (5 - 10t)^2 + (5t)^2$$
$$= 125t^2 - 100t + 25$$

Now, the distance d is a minimum when d^2 is a minimum. Because d^2 is a quadratic function of t, it follows that d^2, and hence d, is a minimum when

$$t = \frac{-b}{2a} = \frac{100}{2(125)} = \frac{2}{5} \text{ hour}$$

Thus, the ships are closest after $\frac{2}{5}(60) = 24$ minutes—that is, at 5:24 P.M. ≡

In a suspension bridge the main cables are of parabolic shape because, if the total weight of a bridge is uniformly distributed along its length, the only cable shape that will bear the load evenly is that of a parabola.

EXAMPLE 8   A suspension bridge with weight uniformly distributed along its length has twin towers that extend 100 meters above the road surface and are 400 meters apart. The cables are parabolic in shape and touch the road surface at the center of the bridge. Find the height of the cables at a point 100 meters from the center. (The road is assumed to be horizontal.)

Solution   We begin by choosing the placement of the coordinate axes so that the x-axis coincides with the road surface and the origin coincides with the center of the bridge. As a result, the twin towers will be vertical (height 100 meters) and located 200 meters from the center. Also, the cable, which has the shape of a parabola, will extend from the towers, open upward, and have its vertex at $(0, 0)$. As illustrated in Figure 10, the choice of placement of the axes enables us to identify the equation of the parabola as $y = ax^2$, $a > 0$. We also can see that the points (200, 100) and (−200, 100) are on the graph. Based on these facts, we can find the value of a in $y = ax^2$:

$$y = ax^2$$
$$100 = a(200)^2$$
$$a = \frac{1}{400}$$

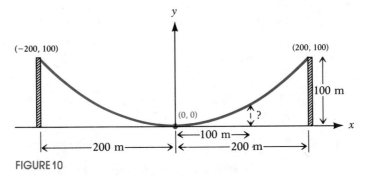

FIGURE 10

The equation of the parabola is therefore

$$y = \frac{1}{400}x^2$$

The height of the cable when $x = 100$ is

$$y = \frac{1}{400}(100)^2 = 25$$

Thus, the cable is 25 meters high at a distance of 100 meters from the center of the bridge.   ≡

EXERCISE 5.1   *In Problems 1–20 graph the function f by starting with the graph of $y = x^2$ and using shifting, change of scale, and/or reflections.*

1. $f(x) = \frac{1}{4}x^2$
2. $f(x) = 2x^2$
3. $f(x) = \frac{1}{4}x^2 - 2$
4. $f(x) = 2x^2 - 3$
5. $f(x) = \frac{1}{4}x^2 + 2$
6. $f(x) = 2x^2 + 4$
7. $f(x) = x^2 + 1$
8. $f(x) = x^2 - 2$
9. $f(x) = -x^2 + 1$
10. $f(x) = -x^2 - 2$
11. $f(x) = \frac{1}{3}x^2 + 1$
12. $f(x) = 2x^2 - 2$
13. $f(x) = x^2 + 4x + 2$
14. $f(x) = x^2 - 6x - 1$
15. $f(x) = 2x^2 - 4x + 1$
16. $f(x) = 3x^2 + 6x$
17. $f(x) = -x^2 - 2x$
18. $f(x) = -2x^2 + 6x + 2$
19. $f(x) = \frac{1}{2}x^2 + x - 1$
20. $f(x) = \frac{2}{3}x^2 + \frac{4}{3}x - 1$

In Problems 21–34 graph each quadratic function by determining whether its graph opens upward or downward and by finding its vertex, axis of symmetry, y-intercept, and x-intercepts, if any.

21. $f(x) = x^2 + 2x - 3$

22. $f(x) = x^2 - 2x - 8$

23. $f(x) = -x^2 - 3x + 4$

24. $f(x) = -x^2 + x + 2$

25. $f(x) = x^2 + 2x + 1$

26. $f(x) = -x^2 + 4x - 4$

27. $f(x) = 2x^2 - x + 2$

28. $f(x) = 4x^2 - 2x + 1$

29. $f(x) = -2x^2 + 2x - 3$

30. $f(x) = -3x^2 + 3x - 2$

C **31.** $f(x) = 3x^2 - 6x + 2$

C **32.** $f(x) = 2x^2 - 5x + 3$

C **33.** $f(x) = -4x^2 - 6x + 2$

C **34.** $f(x) = 3x^2 - 8x + 2$

In Problems 35–40 determine whether each quadratic function has a maximum value or a minimum value and then find the value.

35. $f(x) = 6x^2 + 12x - 3$

36. $f(x) = 4x^2 - 8x + 2$

37. $f(x) = -x^2 + 10x - 4$

38. $f(x) = -2x^2 + 8x + 3$

39. $f(x) = -3x^2 + 12x + 1$

40. $f(x) = 4x^2 - 4x$

41. On one set of coordinate axes, graph the family of parabolas $f(x) = x^2 + 2x + c$, for $c = -3$, $c = 0$, and $c = 1$. Can you describe the characteristics of a member of this family?

42. Repeat the directions of Problem 41 for the family $f(x) = x^2 + cx$.

43. Find two positive numbers whose sum is 30 and whose product is a maximum.

44. Find two numbers whose difference is 50 and whose product is a minimum.

45. Suppose the manufacturer of a gas clothes dryer has found that, when they are sold at a price of p dollars per unit, the revenue R (in dollars) is

$$R = -4p^2 + 4000p$$

What price should be established for each dryer to maximize revenue? What is the maximum revenue?

46. A car rental agency has 24 identical cars. The owner of the agency finds that, at a price of $10 per day, all the cars can be rented. However, for each $1 increase in rental, one of the cars is not rented. What should be charged to maximize income?

47. An aircraft carrier maintains a constant speed of 10 knots, heading due north. At 4:00 P.M., the ship's radar detects a destroyer 100 nautical miles due east of the carrier. If the destroyer is heading due west at 20 knots, when will the two ships be closest?

C **48.** An air traffic controller sees two aircraft flying at the same altitude on his screen. One, a Piper Cub, is headed due west at 150 miles per hour. The other, a Lear jet, is 5 miles due north of the Piper and is headed due south at 400 miles per hour. How close do the two aircraft come to each other?

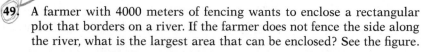

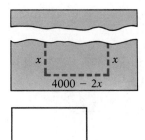

49. A farmer with 4000 meters of fencing wants to enclose a rectangular plot that borders on a river. If the farmer does not fence the side along the river, what is the largest area that can be enclosed? See the figure.

50. A farmer with 3000 meters of fencing wants to enclose a rectangular plot that borders on a straight highway. If the farmer does not fence the side along the highway, what is the largest area that can be enclosed?

51. A farmer with 30,000 meters of fencing wants to enclose a rectangular field and then divide it into two plots with a fence parallel to one of the sides. (See the figure.) What is the largest area that can be enclosed?

52. A farmer wants to enclose 6000 square meters of land in a rectangular plot and then divide it into two plots with a fence parallel to one of the sides. What dimensions of the rectangular plot will require the least amount of fence?

53. A charter flight club charges its members $200 per year. But, for each new member in excess of 60, the charge for every member is reduced by $2. What number of members leads to a maximum revenue?

54. A projectile is fired at an inclination of 45° to the horizontal with an initial velocity of v_0 feet per second. If the starting point is the origin, the x-axis is horizontal, and the y-axis is vertical, then the height y (in feet) after a horizontal distance x has been traversed is

$$y = -\frac{32}{v_0^2}x^2 + x$$

Find the maximum height in terms of the initial velocity v_0. If the initial velocity is doubled, what happens to the maximum height? Assuming the ground is flat, how far from the starting point will the projectile land if the initial velocity is 64 feet per second?

55. A suspension bridge with weight uniformly distributed along its length has twin towers that extend 75 meters above the road surface and are 300 meters apart. The cables are parabolic in shape and are suspended from the tops of the towers. The cables touch the road surface at the center of the bridge. Find the height of the cables at a point 80 meters from the center. Assume the road is horizontal.

56. A parabolic arch has a span of 120 feet and a maximum height of 25 feet. Choose suitable rectangular coordinate axes and find the equation of the parabola. Then calculate the height of the arch at points 10 feet, 20 feet, and 40 feet from the center.

▉ 5.2 POLYNOMIAL FUNCTIONS

POLYNOMIAL FUNCTION OF
DEGREE n

A polynomial function of degree n is a function of the form

$$f(x) = a_n x^n + a_{n-1} x^{n-1} + \cdots + a_1 x + a_0 \qquad (1)$$

where $a_n, a_{n-1}, \ldots, a_1, a_0$ are real numbers, $a_n \neq 0$, and n is a nonnegative integer. The domain consists of all real numbers.

Thus, a polynomial function is a function whose rule is given by a polynomial in one variable (refer to Section 1.4).

EXAMPLE 1 Tell which of the following are polynomial functions. For those that are, state the degree; for those that are not, tell why not.

(a) $f(x) = 2 - 3x^4$　　(b) $g(x) = \sqrt{x}$　　(c) $h(x) = \dfrac{x^2 - 2}{x^3 - 1}$

(d) $F(x) = 0$　　　　　　(e) $G(x) = 8$

Solution
(a) f is a polynomial function of degree four.

(b) g is not a polynomial function. The variable x is raised to the $\frac{1}{2}$ power, which is not a nonnegative integer.

(c) h is not a polynomial function. It is the ratio of two polynomials, and the polynomial in the denominator is of positive degree.

(d) F is the zero polynomial function; it is not assigned a degree.

(e) G is a nonzero constant function, a polynomial function of degree zero. ▄

We have already discussed in detail polynomial functions of degrees zero, one, and two. See Table 1 for a summary of the characteristics of the graphs of these polynomial functions.

TABLE 1

DEGREE	FORM	NAME	GRAPH
No degree	$f(x) = 0$	Zero function	The x-axis
0	$f(x) = a_0,\ \ a_0 \neq 0$	Constant function	Horizontal line with y-intercept a_0
1	$f(x) = a_1x + a_0,\ \ a_1 \neq 0$	Linear function	Nonvertical line with slope a_1 and y-intercept a_0
2	$f(x) = a_2x^2 + a_1x + a_0,\ \ a_2 \neq 0$	Quadratic function	Parabola: graph opens upward if $a_2 > 0$; graph opens downward if $a_2 < 0$

Power Functions

We begin our study of polynomial functions by considering a special kind of polynomial function.

POWER FUNCTION OF
DEGREE *n*

A **power function of degree *n*** is a function of the form

$$f(x) = ax^n \qquad (2)$$

where a is a real number, $a \neq 0$, and $n > 0$ is an integer.

The graph of a power function of degree one, $f(x) = ax$, is a straight line, with slope a, that passes through the origin. The graph of a power function of degree two, $f(x) = ax^2$, is a parabola with vertex at the origin that opens upward if $a > 0$ and downward if $a < 0$.

If we know how to graph a power function of the form $f(x) = x^n$, then a change of scale and perhaps a reflection about the x-axis will enable us to obtain the graph of $g(x) = ax^n$. Consequently, we shall concentrate on graphing power functions of the form $f(x) = x^n$.

We begin with power functions of even degree of the form $f(x) = x^n$, $n \geq 2$ and n even. The domain of f is the set of all real numbers, and the range is the set of nonnegative real numbers. Such a power function is an even function (do you see why?) and hence its graph is symmetric with respect to the y-axis. Its graph always contains the origin and the points $(-1, 1)$ and $(1, 1)$.

If $n = 2$, the graph is the familiar parabola that opens upward with vertex at the origin. If $n \geq 4$, the graph of $f(x) = x^n$, n even, will be closer to the x-axis than the parabola $y = x^2$ if $-1 < x < 1$ and will be farther from the x-axis than the parabola $y = x^2$ if $x < -1$ or if $x > 1$. Figure 11 illustrates this conclusion. Figure 12 shows the graph of $y = x^4$ and the graph of $y = x^8$ on the same set of coordinate axes.

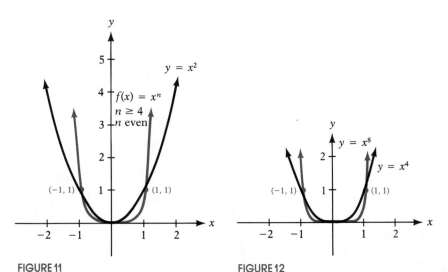

FIGURE 11   FIGURE 12

It is correct to conclude that, as n increases, the graph of $f(x) = x^n$, $n \geq 2$ and n even, tends to flatten out near the origin and to increase very rapidly when x is far from zero. For large n it may appear that near the origin the graph coincides with the x-axis; however, it does not, because the graph touches the x-axis only at the origin (see Table 2). Also, for large n, it may appear that, for $x < -1$ or for $x > 1$, the graph is vertical. It is not; rather it is increasing very rapidly in those intervals.

TABLE 2

	$x = 0.1$	$x = 0.3$	$x = 0.5$
$f(x) = x^8$	10^{-8}	0.0000656	0.0039063
$f(x) = x^{20}$	10^{-20}	$3.487 \cdot 10^{-11}$	0.000001
$f(x) = x^{40}$	10^{-40}	$1.216 \cdot 10^{-21}$	$9.095 \cdot 10^{-13}$

Let's consider power functions of odd degree of the form $f(x) = x^n$, $n \geq 3$ and n odd. The domain and range of f are the set of real numbers. Such a power function is an odd function (do you see why?) and hence its graph is symmetric with respect to the origin. Its graph always contains the origin and the points $(-1, -1)$ and $(1, 1)$.

The graph of $f(x) = x^n$ when $n = 3$ has been illustrated several times. It is repeated in Figure 13. If $n \geq 5$, the graph of $f(x) = x^n$, n odd, will be closer to the x-axis than that of $y = x^3$ if $-1 < x < 1$ and will be farther from the x-axis than that of $y = x^3$ if $x < -1$ or if $x > 1$. Figure 13 also illustrates this conclusion. Figure 14 shows

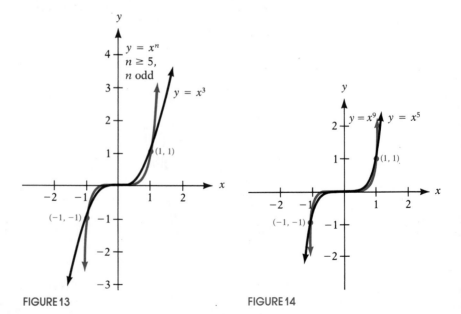

FIGURE 13   FIGURE 14

the graph of $y = x^5$ and the graph of $y = x^9$ on the same set of coordinate axes.

It is correct to conclude that, as n increases, the graph of $f(x) = x^n$, $n \geq 3$ and n odd, tends to flatten out near the origin and to become nearly vertical when x is far from zero.

The methods of shifting, change of scale, and reflection studied in Section 4.3, when used in conjunction with the facts just presented, enable us to graph a varied selection of polynomials.

EXAMPLE 2   Graph each function:

   (a) $f(x) = 1 - x^5$ (b) $f(x) = \frac{1}{2}(x - 1)^4$ (c) $f(x) = (x + 1)^4 - 3$

Solution   (a) Figure 15 shows the required stages.

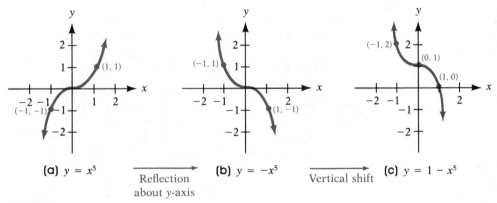

(a) $y = x^5$ Reflection about y-axis   **(b)** $y = -x^5$ Vertical shift   **(c)** $y = 1 - x^5$

FIGURE 15

   (b) Figure 16 shows the required stages.

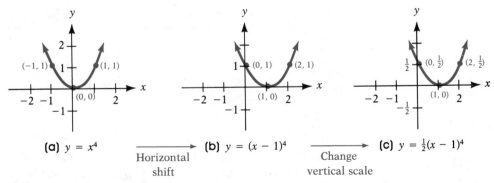

(a) $y = x^4$ Horizontal shift   **(b)** $y = (x - 1)^4$ Change vertical scale   **(c)** $y = \frac{1}{2}(x - 1)^4$

FIGURE 16

(c) Figure 17 shows the required stages.

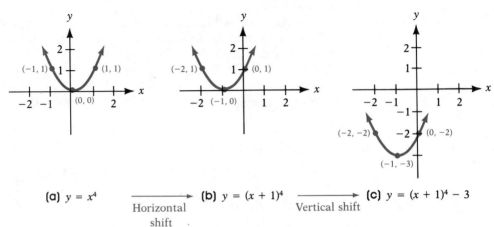

(a) $y = x^4$ —Horizontal shift→ (b) $y = (x + 1)^4$ —Vertical shift→ (c) $y = (x + 1)^4 - 3$

FIGURE 17

Graphing Other Polynomials

To graph most polynomial functions of degree three or higher requires techniques beyond the scope of this text. In calculus classes one learns that the graph of every polynomial function is both smooth and continuous, like the graph illustrated in Figure 18(a). Such graphs will never contain sharp corners or gaps like the graph illustrated in Figure 18(b).

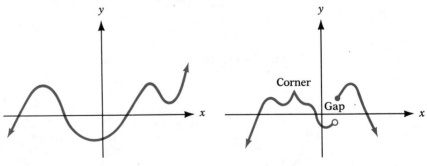

(a) Graph of a polynomial function—smooth, continuous

(b) Cannot be the graph of a polynomial function

FIGURE 18

Suppose it is possible to find all the x-intercepts of a polynomial. These x-intercepts then divide the x-axis into open intervals, and, on each such interval, the graph of the polynomial will be either above or below the x-axis. Let's look at an example.

EXAMPLE 3 (a) Find the x-intercepts of the polynomial

$$f(x) = x^4 - 3x^3 - 4x^2$$

(b) Use the x-intercepts to find the numbers x for which the graph of f is above the x-axis and the numbers x for which the graph of f is below the x-axis.

Solution (a) The x-intercepts obey the equation

$$f(x) = x^4 - 3x^3 - 4x^2 = 0$$

We note that x^2 is a factor of each term in this equation, so we solve the equation by factoring:

$$x^2(x^2 - 3x - 4) = 0$$
$$x^2(x - 4)(x + 1) = 0$$
$$x^2 = 0 \quad \text{or} \quad x - 4 = 0 \quad \text{or} \quad x + 1 = 0$$
$$x = 0 \qquad\qquad x = 4 \qquad\qquad x = -1$$

The x-intercepts are -1, 0, and 4.

(b) The three x-intercepts divide the x-axis into four parts:

$$x < -1, \quad -1 < x < 0, \quad 0 < x < 4, \quad x > 4$$

See Figure 19. Since the graph of f crosses (or touches) the x-axis only at $x = -1$, $x = 0$, and $x = 4$, it follows that the graph of f is either above or below the x-axis in each of these intervals. To determine which is the case, we use the factored form of f, namely $f(x) = x^2(x - 4)(x + 1)$, and the idea of a test number. (Test numbers are discussed in Section 2.7.) Table 3 illustrates the result.

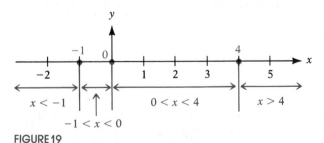

FIGURE 19

TABLE 3

TEST NUMBER	x	SIGN OF x^2	SIGN OF $x + 1$	SIGN OF $x - 4$	SIGN OF $f(x)$	POINTS ON GRAPH
-2	$x < -1$	$+$	$-$	$-$	$+$	Above x-axis
$-\frac{1}{2}$	$-1 < x < 0$	$+$	$+$	$-$	$-$	Below x-axis
1	$0 < x < 4$	$+$	$+$	$-$	$-$	Below x-axis
5	$x > 4$	$+$	$+$	$+$	$+$	Above x-axis

Thus, the graph of f is above the x-axis for $x < -1$ and for $x > 4$, and the graph is below the x-axis for $-1 < x < 0$ and $0 < x < 4$. $\blacksquare$

What more can we determine about the graph of the function $f(x)$ in Example 3? The y-intercept is $f(0) = 0$. Tests for symmetry with respect to the x-axis, y-axis, and origin fail. From Table 3, we know the graph will cross the x-axis at $x = -1$ and at $x = 4$; it touches the x-axis at $x = 0$. Figure 20(a) shows what we know about the graph so far. To find more points on the graph, we could evaluate f at, say, -2, $-\frac{1}{2}$, 1, 3, and 5. [In fact, we used four of these numbers as test numbers in Table 3 to decide whether $f(x)$ was positive or negative on the interval.] We find $f(-2) = 24$, $f(-\frac{1}{2}) = -\frac{9}{16}$, $f(1) = -6$, $f(3) = -36$, and $f(5) = 150$. Figure 20(b) shows the graph with these additional points. Notice how we scaled the y-axis to reflect this new information about the graph of f.

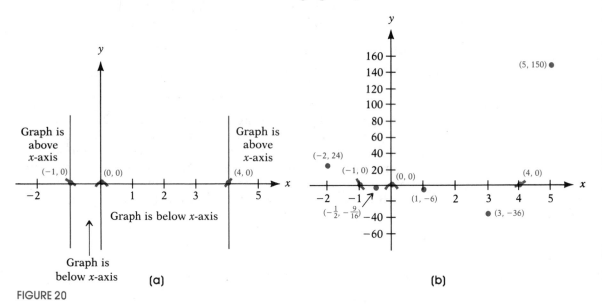

FIGURE 20

We are still missing information about just how low or high the graph actually goes on each interval. Also, we cannot be sure of the general shape of the graph. In fact, based only on what we know now, any of the graphs given in Figure 21 could be the graph of $f(x) = x^4 - 3x^3 - 4x^2$. Calculus provides the tools needed to determine that the low points are $(-0.7, -0.7)$ and $(2.9, -36.1)$, and the graph in Figure 21(b) is the graph of f. Needless to say, you will not be required to find the low points (or high points). For now, we shall connect the points on the graph with a smooth curve. Of course, plotting a few additional points on the graph can be very helpful.

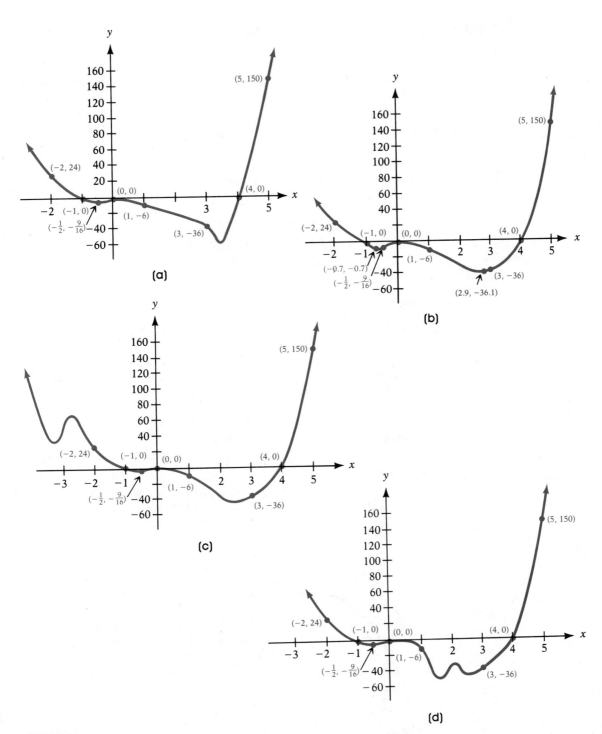

(a)

(b)

(c)

(d)

FIGURE 21

EXAMPLE 4　(a) Find the x-intercepts of the polynomial

$$f(x) = x^3 + x^2 - 12x$$

(b) Use the x-intercepts to find the numbers x for which the graph of f is above the x-axis and the numbers x for which the graph of f is below the x-axis.

(c) Obtain several other points on the graph and connect them with a smooth curve.

Solution　(a) The x-intercepts obey the equation

$$f(x) = x^3 + x^2 - 12x = 0$$
$$x(x^2 + x - 12) = 0$$
$$x(x + 4)(x - 3) = 0$$
$$x = 0 \quad \text{or} \quad x + 4 = 0 \quad \text{or} \quad x - 3 = 0$$
$$x = -4 \qquad x = 3$$

The x-intercepts are -4, 0, and 3.

(b) The three x-intercepts divide the x-axis into four parts:

$$x < -4, \quad -4 < x < 0, \quad 0 < x < 3, \quad x > 3$$

To determine the sign of $f(x)$ in each part, we find the value of f at the test numbers -5, -2, 1, and 4:

$$f(-5) = -40, \quad f(-2) = 20, \quad f(1) = -10, \quad f(4) = 32$$

Based on this, the graph of f lies below the x-axis if $x < -4$ or if $0 < x < 3$; the graph lies above the x-axis if $-4 < x < 0$ or if $x > 3$.

(c) Using these points and the other information we've gathered, we obtain the graph of $f(x) = x^3 + x^2 - 12x$. See Figure 22. [Again,

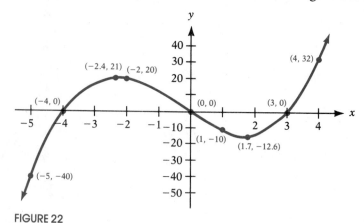

FIGURE 22

the low point $(1.7, -12.6)$ and the high point $(-2.4, 21)$ were obtained using calculus.] ≡

Examples 3 and 4 point out the importance of being able to solve polynomial equations in order to obtain information about the graph. The ability to evaluate polynomials at a given number is equally important. The next two sections deal with various techniques for solving polynomial equations and evaluating polynomials.

Summary

To sketch the graph of a polynomial function $y = f(x)$, follow these steps:

> 1. (a) Find the x-intercepts, if any, by solving the equation $f(x) = 0$.
>   (b) Find the y-intercept by letting $x = 0$ and finding the value of $f(0)$.
> 2. Use the x-intercepts to find the numbers x for which the graph of f is above and for which it is below the x-axis. During this step, several additional points on the graph are located.
> 3. Connect the points with a smooth curve.

EXERCISE 5.2   *In Problems 1–10 tell which functions are polynomial functions. For those that are, state the degree. For those that are not, tell why not.*

1. $f(x) = 2x - x^3$ 2. $f(x) = 3x^2 - 4x^4$

3. $g(x) = \dfrac{1 - x^2}{2}$ 4. $h(x) = 3 - \dfrac{1}{2}x$

5. $f(x) = 1 - \dfrac{1}{x}$ 6. $f(x) = x(x - 1)$

7. $g(x) = x^{3/2} - x^2 + 2$ 8. $h(x) = \sqrt{x}(\sqrt{x} - 1)$

9. $F(x) = 5x^4 - \pi x^3 + \dfrac{1}{2}$ 10. $F(x) = \dfrac{x^2 - 5}{x^3}$

In Problems 11–18 use the graph of $y = x^4$ to graph each function.

11. $f(x) = (x - 1)^4$ 12. $f(x) = x^4 + 3$

13. $f(x) = \dfrac{1}{2}x^4$ 14. $f(x) = -x^4$

15. $f(x) = 2(x + 1)^4 + 1$ 16. $f(x) = 3 - (x + 2)^4$

17. $f(x) = -\dfrac{1}{2}(x - 2)^4 - 1$ 18. $f(x) = 1 - 2(x + 1)^4$

In Problems 19–30 find the intercepts of each polynomial f(x). Find the numbers x for which the graph of f is above the x-axis and those for which it is below the x-axis. Obtain several other points on the graph and connect them with a smooth curve.

19. $f(x) = x^2 - 4x$
20. $f(x) = x^3 - x$
21. $f(x) = x^3 - 2x^2 - 3x$
22. $f(x) = x^3 + 2x^2 - 3x$
23. $f(x) = x^3 - 4x$
24. $f(x) = x^4 - x^2$
25. $f(x) = x^4 - 2x^3 - 3x^2$
26. $f(x) = x^4 + 2x^3 - 3x^2$
27. $f(x) = x^2(x - 2)(x^2 + 3)$
28. $f(x) = x(x - 4)(x + 2)$
29. $f(x) = (x + 1)^2(x - 3)$
30. $f(x) = (x - 1)^3(x + 1)$

5.3 DIVIDING POLYNOMIALS; SYNTHETIC DIVISION

Long Division

The process of long division for dividing one polynomial by another should be familiar to you. Just to be sure, let's look at an example.

EXAMPLE 1

$$
\begin{array}{r}
x^2 - 2x - 3 \quad \leftarrow \text{Quotient} \\
x^2 - x + 1 \overline{)\, x^4 - 3x^3 \qquad\quad + 2x - 5} \quad \leftarrow \text{Dividend} \\
\end{array}
$$

Divisor

$$
\begin{array}{r}
x^4 - x^3 + x^2 \\
\hline
\text{Subtract} \quad -2x^3 - x^2 + 2x - 5 \\
-2x^3 + 2x^2 - 2x \\
\hline
\text{Subtract} \quad -3x^2 + 4x - 5 \\
-3x^2 + 3x - 3 \\
\hline
\text{Subtract} \quad x - 2 \quad \leftarrow \text{Remainder}
\end{array}
$$

Thus,

$$\frac{x^4 - 3x^3 + 2x - 5}{x^2 - x + 1} = x^2 - 2x - 3 + \frac{x - 2}{x^2 - x + 1}$$

The long-division process carried out in Example 1 yielded the quotient $x^2 - 2x - 3$ and the remainder $x - 2$ when the dividend $x^4 - 3x^3 + 2x - 5$ was divided by the divisor $x^2 - x + 1$. The process ends when, after subtracting, we obtain a polynomial, the remainder, that is either the zero polynomial or a polynomial whose degree is less than the degree of the divisor.

The work we did in Example 1 may be checked by verifying that

$$(\text{Divisor})(\text{Quotient}) + \text{Remainder} = \text{Dividend}$$

Thus,

$$\text{(Divisor)(Quotient)} + \text{Remainder} = (x^2 - x + 1)(x^2 - 2x - 3) + x - 2$$
$$= x^4 - 2x^3 - 3x^2 - x^3 + 2x^2 + 3x$$
$$+ x^2 - 2x - 3 + x - 2$$
$$= x^4 - 3x^3 + 2x - 5 = \text{Dividend}$$

This checking routine is the basis for a famous theorem called the **Division Algorithm* for Polynomials**, which we now state without proof.

THEOREM

DIVISION ALGORITHM FOR POLYNOMIALS

If $f(x)$ and $g(x)$ denote polynomial functions and if $g(x)$ is not the zero polynomial, then there are unique polynomial functions $q(x)$ and $r(x)$ such that

$$\frac{f(x)}{g(x)} = q(x) + \frac{r(x)}{g(x)} \qquad \text{or} \qquad f(x) = g(x)q(x) + r(x) \qquad (1)$$

where $r(x)$ is either the zero polynomial or a polynomial of degree less than that of $g(x)$.

DIVIDEND

DIVISOR

QUOTIENT

REMAINDER

In Equation (1), $f(x)$ is the **dividend**, $g(x)$ is the **divisor**, $q(x)$ is the **quotient**, and $r(x)$ is the **remainder**.

If the divisor $g(x)$ is a first-degree polynomial of the form

$$g(x) = x - c \qquad c \text{ a real number}$$

then the remainder $r(x)$ is either the zero polynomial or a polynomial of degree less than one—that is, $r(x)$ is of degree zero. Thus, for such divisors, the remainder is some number, say, R, and we may write

$$f(x) = (x - c)q(x) + R \qquad (2)$$

This equation is an identity in x and so is true for all real numbers x. In particular, it is true when $x = c$. Thus, if $x = c$, then Equation (2) becomes

$$f(c) = (c - c)q(c) + R$$
$$f(c) = R$$

Thus Equation (2) takes the form

$$f(x) = (x - c)q(x) + f(c) \qquad (3)$$

---

*A systematic process in which certain steps are repeated a finite number of times is called an **algorithm**. Thus, long division is an algorithm.

We have proved the following result, called the **Remainder Theorem**:

THEOREM

REMAINDER THEOREM

Let f be a polynomial function. If $f(x)$ is divided by $x - c$, then the remainder is $f(c)$. ≡

EXAMPLE 2   Find the remainder if $f(x) = x^3 - 4x^2 + 2x - 5$ is divided by:

(a) $x - 3$ (b) $x + 2$

Solution   (a) We could use long division. However, it is much easier to use the Remainder Theorem, which says the remainder is

$$f(3) = (3)^3 - 4(3)^2 + 2(3) - 5 = 27 - 36 + 6 - 5 = -8$$

(b) To find the remainder when $f(x)$ is divided by $x + 2 = x - (-2)$, we evaluate

$$f(-2) = (-2)^3 - 4(-2)^2 + 2(-2) - 5 = -8 - 16 - 4 - 5 = -33$$

Thus, the remainder is -33. ≡

An important and useful consequence of the Remainder Theorem is the **Factor Theorem**.

THEOREM

FACTOR THEOREM

Let f be a polynomial function. Then $x - c$ is a factor of $f(x)$ if and only if $f(c) = 0$. ≡

The Factor Theorem actually consists of two separate statements:

1. If $f(c) = 0$, then $x - c$ is a factor of $f(x)$.
2. If $x - c$ is a factor of $f(x)$, then $f(c) = 0$.

Thus, the proof requires two parts.

Proof   1. Suppose $f(c) = 0$. Then, by Equation (3), we have

$$f(x) = (x - c)q(x)$$

for some polynomial $q(x)$. That is, $x - c$ is a factor of $f(x)$.
2. Suppose $x - c$ is a factor of $f(x)$. Then there is a polynomial function q such that

$$f(x) = (x - c)q(x)$$

Replacing x by c, we find

$$f(c) = (c - c)q(c) = 0 \cdot q(c) = 0$$

This completes the proof. ≡

One use of the Factor Theorem is to determine whether a polynomial has a particular factor.

EXAMPLE 3 Use the Factor Theorem to determine whether

$$f(x) = 2x^3 - x^2 + 2x - 3$$

has the factor:

(a) $x - 1$ (b) $x + 3$

Solution (a) Because $x - 1$ is of the form $x - c$, with $c = 1$, we find the value of $f(1)$:

$$f(1) = 2(1)^3 - (1)^2 + 2(1) - 3 = 2 - 1 + 2 - 3 = 0$$

By the Factor Theorem, $x - 1$ is a factor of $f(x)$.

(b) To test the factor $x + 3$, we first need to write it in the form $x - c$. Since $x + 3 = x - (-3)$ we find the value of $f(-3)$:

$$f(-3) = 2(-3)^3 - (-3)^2 + 2(-3) - 3 = -54 - 9 - 6 - 3 = -72$$

Because $f(-3) \neq 0$, we conclude from the Factor Theorem that $x - (-3) = x + 3$ is not a factor of $f(x)$. ■

Synthetic Division

SYNTHETIC DIVISION To find the quotient as well as the remainder when a polynomial function f of degree one or higher is divided by $g(x) = x - c$, a shortened version of the long-division process, called **synthetic division**, makes the task simpler.

To see how synthetic division works, we'll divide the polynomial $f(x) = 2x^3 - x^2 + 3$ by $g(x) = x - 3$. First, in long division, we have

$$
\require{enclose}
\begin{array}{r}
2x^2 + 5x\ + 15 \\
x - 3\enclose{longdiv}{2x^3 -\ \ x^2\qquad\ \ + 3} \\
\underline{2x^3 - 6x^2} \\
5x^2 \\
\underline{5x^2 - 15x} \\
15x\ +\ \ 3 \\
\underline{15x\ -\ 45} \\
48
\end{array}
$$

The process of synthetic division arises from rewriting the long division in a more compact form, using simpler notation. For example, in the long division above, the circled terms are not really necessary because they are identical to the terms directly above them.

With these terms removed, we have

$$
\begin{array}{r}
2x^2 + 5x \;\; + 15 \qquad\qquad \\
x - 3\overline{\smash{)}2x^3 - \;\; x^2 \qquad\qquad + \;\; 3} \\
-\;6x^2 \qquad\qquad\qquad \\
\overline{5x^2} \qquad\qquad\qquad \\
-\;15x \qquad\quad \\
\overline{15x + \;\; 3} \\
-\;45 \\
\overline{48}
\end{array}
$$

Most of the x's that appear in this process can also be removed, provided we are careful about positioning each coefficient. In this regard, we will need to use 0 as the coefficient of x in the dividend, because that power of x is missing. Now we have

$$
\begin{array}{r}
2x^2 + 5x + 15 \qquad\qquad\quad \\
x - 3\overline{\smash{)}2 \quad\;\; -1 \qquad\;\; 0 \qquad 3} \\
-6 \qquad\qquad\qquad\qquad \\
\overline{\boxed{5}} \qquad\qquad\qquad\quad \\
-15 \qquad\quad \\
\overline{\boxed{15} \quad 3} \\
-45 \\
\overline{\boxed{48}}
\end{array}
$$

We can make this display more compact by moving the lines up until the circled numbers align horizontally:

$$
\begin{array}{r r r r l}
2x^2 + 5x + 15 \qquad\quad & & & & \text{Row 1} \\
x - 3\overline{\smash{)}2 \quad\; -1} & 0 & 3 & & \text{Row 2} \\
-6 & -15 & -45 & & \text{Row 3} \\
\bigcirc \quad\;\; 5 & 15 & 48 & & \text{Row 4}
\end{array}
$$

Now, if we place the leading coefficient of the quotient, 2, in the circled position, the first three numbers in Row 4 are precisely the coefficients of the quotient, and the last number in Row 4 is the remainder. Thus, Row 1 is not really needed, so we can compress the process to three rows, where the bottom row contains the coefficients of both the quotient and the remainder. We will now also drop the x from the divisor $x - 3$:

$$-3\overline{)2 \quad -1 \quad \;\;0 \quad \;\;3}$$
$$\quad\quad -6 \quad -15 \quad -45$$
$$\overline{\quad 2 \quad \;\;5 \quad \;\;15 \quad \;\;48} \;\; \text{Bottom row}$$
$$\quad\quad \underbrace{\text{Quotient}} \quad \underbrace{\text{Remainder}}$$

$$\overbrace{2x^2 + 5x + 15}, \overbrace{R = 48}$$

The entries in the bottom row give the quotient and the remainder. Let's go through another example step by step.

EXAMPLE 4   Use synthetic division to find the quotient and remainder when

$$f(x) = 3x^4 + 8x^2 - 7x + 4 \quad \text{is divided by} \quad g(x) = x - 1$$

Solution   Step 1:   Write the dividend in descending powers of x. Then copy the coefficients, remembering to insert a zero for any missing powers of x:

$$3 \quad 0 \quad 8 \quad -7 \quad 4 \quad \text{Row 1}$$

Step 2:   To the left of this display, enter the divisor $x - 1$, and insert the usual division symbol:

$$x + 1\overline{)3 \quad 0 \quad 8 \quad -7 \quad 4} \quad \text{Row 1}$$

Step 3:   Drop the x from $x - 1$. Bring the 3 down two rows, and enter it in Row 3:

$$+1\overline{)3 \quad 0 \quad 8 \quad -7 \quad 4} \quad \text{Row 1}$$
$$\downarrow \quad\quad\quad\quad\quad\quad\quad \text{Row 2}$$
$$3 \quad\quad\quad\quad\quad\quad\quad\quad \text{Row 3}$$

Step 4:   Multiply the latest entry in Row 3 by -1 and place the result in Row 2, but over one column to the right:

$$+1\overline{)3 \quad \;\;0 \quad 8 \quad -7 \quad 4} \quad \text{Row 1}$$
$$\quad -3 \quad\quad\quad\quad\quad \text{Row 2}$$
$$3 \quad\quad\quad\quad\quad\quad\quad \text{Row 3}$$

Stupid Bugger!

Step 5:   Subtract the entry in Row 2 from the entry above it in Row 1, and enter the answer in Row 3:

$$+1\overline{)3 \quad \;\;0 \quad 8 \quad -7 \quad 4} \quad \text{Row 1}$$
$$\quad +3 \quad\quad\quad\quad\quad \text{Row 2}$$
$$3 \quad 3 \quad\quad\quad\quad\quad\quad \text{Row 3}$$

Step 6: Repeat Steps 4 and 5 until no more entries are available in Row 1:

$$-1 \overline{)\begin{array}{ccccc} 3 & 0 & 8 & -7 & 4 \end{array}} \qquad \text{Row 1}$$

$$\underline{\qquad -3 \quad -3 \quad -11 \quad -4 \quad} \text{(Subtract)} \quad \text{Row 2}$$

$$\begin{array}{ccccc} 3 & 3 & 11 & 4 & 8 \end{array} \qquad \text{Row 3}$$

Step 7: The final entry in Row 3, an 8, is the remainder; the other entries in Row 3—3, 3, 11, and 4—are the coefficients (in descending order) of a polynomial whose degree is one less than that of the dividend; this is the quotient. Thus,

$$\text{Quotient} = 3x^3 + 3x^2 + 11x + 4 \qquad \text{Remainder} = 8$$

Check:

$$(\text{Divisor})(\text{Quotient}) + \text{Remainder} = (x - 1)(3x^3 + 3x^2 + 11x + 4) + 8$$
$$= 3x^4 + 3x^3 + 11x^2 + 4x - 3x^3 - 3x^2 - 11x - 4 + 8$$
$$= 3x^4 + 8x^2 - 7x + 4 = \text{Dividend} \qquad \equiv$$

Let's do an example in which all seven steps are combined.

EXAMPLE 5   Use synthetic division to show that $g(x) = x + 3$ is a factor of

$$f(x) = 2x^5 + 5x^4 - 2x^3 + 2x^2 - 2x + 3$$

Solution   The divisor is $x + 3$, so the Row 3 entries will be multiplied by 3, entered in Row 2, and subtracted from Row 1:

$$3 \overline{)\begin{array}{cccccc} 2 & 5 & -2 & 2 & -2 & 3 \end{array}}$$
$$\underline{\qquad 6 \quad -3 \quad 3 \quad -3 \quad 3 \quad}$$
$$\begin{array}{cccccc} 2 & -1 & 1 & -1 & 1 & 0 \end{array}$$

Because the remainder is 0, it follows that $f(-3) = 0$. Hence, by the Factor Theorem, $x - (-3) = x + 3$ is a factor of $f(x)$. $\qquad \equiv$

One important use of synthetic division is to find the value of a polynomial.

EXAMPLE 6   Use the Remainder Theorem and synthetic division to find the value of $f(x) = -3x^4 + 2x^3 - x + 1$ at $x = -2$; that is, find $f(-2)$.

Solution   The Remainder Theorem tells us that the value of a polynomial function at c equals the remainder when the polynomial is divided by $x - c$. This remainder is the final entry of the third row in the process of synthetic division. We want $f(-2)$, so we divide by $x - (-2) = x + 2$:

$$2\overline{)\begin{array}{ccccc} -3 & 2 & 0 & -1 & 1 \\ & -6 & 16 & -32 & 62 \\ \hline -3 & 8 & -16 & 31 & -61 \end{array}}$$

The quotient is $q(x) = -3x^3 + 8x^2 - 16x + 31$; the remainder is $R = -61$. Because the remainder was found to be -61, it follows from the Remainder Theorem that $f(-2) = -61$.   ▤

As Example 6 illustrates, we can use the process of synthetic division to find the value of a polynomial function at a number c as an alternative to merely substituting c for x. Compare the work required in Example 6 with the arithmetic involved in substituting:

$$f(-2) = -3(-2)^4 + 2(-2)^3 - (-2) + 1$$
$$= -3(16) + 2(-8) + 2 + 1$$
$$= -48 - 16 + 2 + 1 = -61$$

As you can see, finding $f(-2)$ is somewhat easier using synthetic division.

Sometimes neither substitution nor synthetic division avoids the need for messy calculations. Consider the problem of evaluating $f(x) = 3x^5 - 5x^4 + 0.2x^3 - 1.5x^2 + 2x - 6$ at $x = 1.2$. Here, a third method—using the nested form of a polynomial—is more helpful.

Nested Form of a Polynomial

Consider the polynomial

$$f(x) = 3x^3 - 5x^2 + 2x - 7$$

We can factor $f(x)$ as follows:

$$\begin{aligned} f(x) &= 3x^3 - 5x^2 + 2x - 7 \\ &= (3x^3 - 5x^2 + 2x) - 7 \\ &= (3x^2 - 5x + 2)x - 7 \quad \text{Factor } x \text{ from parentheses.} \\ &= [(3x^2 - 5x) + 2]x - 7 \quad \text{Regroup.} \\ &= [(3x - 5)x + 2]x - 7 \quad \text{Factor } x \text{ from parentheses.} \end{aligned}$$

NESTED FORM   Notice that this form of the polynomial contains only linear expressions. A polynomial function written in this way is said to be in **nested form**.

Let's look at some other examples.

EXAMPLE 7   Write each polynomial in nested form:

(a) $f(x) = 2x^2 - 3x + 5$ (b) $f(x) = 5x^3 - 6x^2 + 2$

(c) $f(x) = -5x^4 + 3x^3 - 2x^2 + 10x - 8$

Solution　(a) We proceed in steps as follows:

$$f(x) = (2x^2 - 3x) + 5 = (2x - 3)x + 5$$

The expression $(2x - 3)x + 5$ is the nested form of $2x^2 - 3x + 5$.

(b)
$$f(x) = (5x^3 - 6x^2) + 2 = (5x^2 - 6x)x + 2$$
$$= [(5x - 6)x]x + 2$$

The expression $[(5x - 6)x]x + 2$ is the nested form of $5x^3 - 6x^2 + 2$.

(c)
$$f(x) = (-5x^4 + 3x^3 - 2x^2 + 10x) - 8$$
$$= (-5x^3 + 3x^2 - 2x + 10)x - 8$$
$$= [(-5x^2 + 3x - 2)x + 10]x - 8$$
$$= \{[(-5x + 3)x - 2]x + 10\}x - 8 \qquad \blacksquare$$

The advantage of evaluating a polynomial in nested form is that this method avoids the need to raise a number to a power, which on a calculator or computer can cause serious round-off errors. Further, computers can perform the operation of addition much faster than the operation of multiplication, and the nested form requires fewer multiplications than the ordinary form of a polynomial. In Example 7(b), to evaluate $f(x) = 5x^3 - 6x^2 + 2$ in its ordinary form requires five multiplications and two additions:

Multiplication
$$\{[(5 \cdot x) \cdot x] \cdot x\} - 6 \cdot x \cdot x + 2$$
Addition

In nested form, it requires three multiplications and two additions:

Multiplication
$$[(5 \cdot x - 6) \cdot x] \cdot x + 2$$
Addition

Thus, to avoid errors and to speed up calculations, many computers evaluate polynomials by using the nested form.

C　EXAMPLE 8　Use the nested form and a calculator to evaluate

$$f(x) = 0.5x^3 - 1.2x^2 + 5.1x - 6.2$$

at $x = 1.3$.

Solution　We write f in nested form as

$$f(x) = [(0.5x - 1.2)x + 5.1]x - 6.2$$

We start inside the parentheses by multiplying 0.5 by $x = 1.3$. Then subtract 1.2. Multiply the result by $x = 1.3$ and add 5.1. Multiply this result by $x = 1.3$ and subtract 6.2. The value is

$$f(1.3) = -0.4995$$

On a calculator, you would proceed as follows:

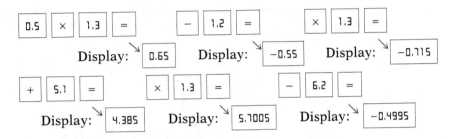

Notice that no memory key was used in this process.

Summary

We now have three ways to find the value of a polynomial function $f(x)$ at a number c.

1. Replace x by the number c to find $f(c)$.
2. Use synthetic division and the Remainder Theorem to find $f(c)$.
3. Write $f(x)$ in nested form and use a calculator to find $f(c)$.

EXERCISE 5.3

In Problems 1–12 find the quotient $q(x)$ and the remainder $r(x)$ when $f(x)$ is divided by $g(x)$. Check your work by verifying that (Divisor)(Quotient) + Remainder = Dividend.

1. $f(x) = 4x^3 - 2x^2 - x + 1$; $g(x) = x^2 + x + 1$
2. $f(x) = -5x^3 + 2x^2 - x + 2$; $g(x) = x^2 - x - 1$
3. $f(x) = -2x^4 + x^3 - 3x^2 + 2$; $g(x) = x + 3$
4. $f(x) = 4x^3 - 3x^2 + 2$; $g(x) = x - 3$
5. $f(x) = 3x^5 - 5x^4 + 2x - 4$; $g(x) = x^3 + 2$
6. $f(x) = 4x^4 - 2x^3 + x^2 - 1$; $g(x) = x^2 - 2$
7. $f(x) = 1 - x^5$; $g(x) = x^2 - 1$

8. $f(x) = 1 - x^4$; $g(x) = x^2 + 1$

9. $f(x) = x^4 - c^4$; $g(x) = x - c$

10. $f(x) = x^4 + c^4$; $g(x) = x + c$

11. $f(x) = 5x^3 + 3x + 3$; $g(x) = 2x^2 + 1$

12. $f(x) = 4x^4 + 5x - 6$; $g(x) = 4x - 3$

In Problems 13–22 use synthetic division to find the quotient q(x) and remainder R when f(x) is divided by g(x).

13. $f(x) = 3x^3 + 2x^2 - x + 3$; $g(x) = x - 3$

14. $f(x) = -4x^3 + 2x^2 - x + 1$; $g(x) = x + 2$

15. $f(x) = x^5 - 4x^3 + x$; $g(x) = x + 3$

16. $f(x) = x^4 + x^2 + 2$; $g(x) = x - 2$

17. $f(x) = 4x^6 - 3x^4 + x^2 + 5$; $g(x) = x - 1$

18. $f(x) = x^5 + 5x^3 - 10$; $g(x) = x + 1$

C 19. $f(x) = 0.1x^3 + 0.2x$; $g(x) = x + 1.1$

C 20. $f(x) = 0.1x^2 - 0.2$; $g(x) = x + 2.1$

21. $f(x) = x^5 - 1$; $g(x) = x - 1$

22. $f(x) = x^5 + 1$; $g(x) = x + 1$

In Problems 23–32 use the Factor Theorem and synthetic division to determine whether x − c is a factor of f(x).

23. $f(x) = 3x^3 + 2x^2 - 3x + 3$; $c = 3$

24. $f(x) = -4x^3 + 3x^2 + 2$; $c = -2$

25. $f(x) = 3x^4 - 6x^3 - 5x + 10$; $c = 2$

26. $f(x) = 4x^4 - 15x^2 - 4$; $c = 2$

27. $f(x) = 3x^6 - 82x^3 + 27$; $c = -3$

28. $f(x) = 2x^6 - 18x^4 + x^2 - 9$; $c = -3$

29. $f(x) = 4x^6 - 64x^4 + x^2 - 15$; $c = -4$

30. $f(x) = x^6 - 16x^4 + x^2 - 16$; $c = -4$

31. $f(x) = 2x^4 - x^3 + 2x - 1$; $c = \frac{1}{2}$

32. $f(x) = 3x^4 + x^3 - 3x + 1$; $c = -\frac{1}{3}$

In Problems 33–38 use the Remainder Theorem and synthetic division to find f(c).

33. $f(x) = 3x^4 - 2x^2 + 1$; $c = 2$

34. $f(x) = -2x^3 + 2x^2 + 3$; $c = -2$

35. $f(x) = 4x^5 - 3x^3 + 2x - 1$; $c = -1$

36. $f(x) = -3x^4 + 3x^3 - 2x^2 + 5$; $c = -1$

37. $f(x) = 9x^{17} - 8x^{10} + 9x^8 + 5$; $c = 1$

38. $f(x) = 10x^{15} + 4x^{12} - 2x^5 + x^2$; $c = -1$

In Problems 39–48 write each polynomial in nested form.

39. $f(x) = 3x^3 + 2x^2 - 3x + 3$ **40.** $f(x) = -4x^3 + 3x^2 + 2$

41. $f(x) = 3x^4 - 6x^3 - 5x + 10$ **42.** $f(x) = 4x^4 - 15x^2 - 4$

43. $f(x) = 3x^6 - 82x^3 + 27$ **44.** $f(x) = 2x^6 - 18x^4 + x^2 - 9$

45. $f(x) = 4x^6 - 64x^4 + x^2 - 15$ **46.** $f(x) = x^6 - 16x^4 + x^2 - 16$

47. $f(x) = 2x^4 - x^3 + 2x - 1$ **48.** $f(x) = 3x^4 + x^3 - 3x + 1$

C   *In Problems 49–58 use the nested form and a calculator to evaluate each polynomial at $x = 1.2$. Avoid using any memory key.*

49. $f(x) = 3x^3 + 2x^2 - 3x + 3$ **50.** $f(x) = -4x^3 + 3x^2 + 2$

51. $f(x) = 3x^4 - 6x^3 - 5x + 10$ **52.** $f(x) = 4x^4 - 15x^2 - 4$

53. $f(x) = 3x^6 - 82x^3 + 27$ **54.** $f(x) = 2x^6 - 18x^4 + x^2 - 9$

55. $f(x) = 4x^6 - 64x^4 + x^2 - 15$ **56.** $f(x) = x^6 - 16x^4 + x^2 - 16$

57. $f(x) = 2x^4 - x^3 + 2x - 1$ **58.** $f(x) = 3x^4 + x^3 - 3x + 1$

59. Find k such that $f(x) = x^3 - kx^2 + kx + 2$ has the factor $x - 2$.

60. Find k such that $f(x) = x^4 - kx^3 + kx^2 + 1$ has the factor $x + 2$.

61. What is the remainder when $f(x) = 2x^{20} - 8x^{10} + x - 2$ is divided by $x - 1$?

62. What is the remainder when $f(x) = -3x^{17} + x^9 - x^5 + 2x$ is divided by $x + 1$?

63. Use the Factor Theorem to prove that $x - c$ is a factor of $x^n - c^n$, for any positive integer n.

64. Use the Factor Theorem to prove that $x + c$ is a factor of $x^n + c^n$, if $n \geq 1$ is an odd integer.

65. An IBM AT microcomputer finds powers by multiplication. Suppose each multiplication of two numbers requires 33,333 nanoseconds and each addition or subtraction requires 500 nanoseconds. If all other times are disregarded, how long will it take the computer to find the value of $f(x) = 2x^3 - 6x^2 + 4x - 10$ at $x = 2.013$ by:
(a) Replacing x by 2.013 in the expression for $f(x)$?
(b) Replacing x by 2.013 in the nested form for $f(x)$?

66. Using the microcomputer described in Problem 65, how long would it take by method (a) and by method (b) to find the value of $f(x) = ax^3 + bx^2 + cx + d$ for 5000 values of x?

67. *Programming exercise* Write a program that simulates synthetic division and divides a polynomial by $x - c$. Your input should consist of the coefficients of the polynomial, in order from highest to lowest, followed by the number c. Your output should consist of the numbers that would appear in the third row of the process of synthetic division.

68. *Programming exercise* Write a program that will evaluate a polynomial by using the nested form. Your input should consist of the coefficients of the polynomial, in order from highest to lowest, followed by the number at which the polynomial is to be evaluated. Test your program on the polynomials given in Problems 49–58.

≣ 5.4 THE ZEROS OF A POLYNOMIAL FUNCTION

If f denotes a function and if r is a real number for which

$$f(r) = 0$$

ZERO OF A FUNCTION
then r is called a **zero of** f. Thus, the zeros of a function are the real solutions or roots of the equation $f(x) = 0$. The Factor Theorem tells us that if r is a zero of f, then $x - r$ is a factor of f. If the same factor $x - r$ occurs more than once, then r is called a **repeated** or **multiple zero** of f. More precisely, we have the following definition.

ZERO OF MULTIPLICITY m
If $(x - r)^m$ is a factor of a polynomial f and $(x - r)^{m+1}$ is not a factor of f, then r is called a **zero of multiplicity** m of f.

EXAMPLE 1
For the polynomial

$$f(x) = 5(x - 2)(x + 3)^2\left(x - \frac{1}{2}\right)^4$$

2 is a zero of multiplicity one

−3 is a zero of multiplicity two

$\frac{1}{2}$ is a zero of multiplicity four   ≣

The zeros of a function are the x-intercepts of its graph. For polynomial functions, we have seen the importance of locating the zeros for graphing. In most cases, however, the zeros of a polynomial function are difficult to find. There are no nice formulas like the quadratic formula available to us for polynomials of degree higher than two. Although formulas do exist for solving any third- or fourth-degree polynomial equation, they are somewhat complicated. (If you are interested in learning about them, consult a book on the theory of equations.) It has been proved that no general formulas exist for polynomial equations of degree five or higher. In this section, we shall learn some ways of detecting information about the character of the zeros, which, in turn, may help us find them or at least isolate them.

Our first result concerns the number of zeros a polynomial function can have. In counting the zeros of a polynomial, we count each zero as many times as its multiplicity.

THEOREM
NUMBER OF ZEROS
A polynomial function cannot have more zeros than its degree.

Proof
The proof is based on the Factor Theorem. If r is a zero of a polynomial function f, then $f(r) = 0$ and, hence, $x - r$ is a factor of $f(x)$. Thus,

each zero corresponds to a factor of degree one. Because f cannot have more first-degree factors than its degree, the result follows. ≡

The next result, called *Descartes' Rule of Signs*, provides information about the number *and location* of the zeros of a polynomial function. This, in turn, helps us to know where to look for the zeros. Descartes' Rule of Signs, which assumes that the polynomial is written in descending powers of x, requires that we count the number of variations in sign of the coefficients of $f(x)$ and $f(-x)$.

For example, the polynomial function

$$f(x) = -3x^7 + 4x^4 + 3x^2 - 2x - 1$$
$$= -3x^7 + 0x^6 + 0x^5 + 4x^4 + 0x^3 + 3x^2 - 2x - 1$$

$$\underbrace{\qquad}_{-\text{ to }+} \qquad \underbrace{\qquad}_{+\text{ to }-}$$

has two variations in the sign of the coefficient. Notice that we ignored the zero coefficients in $0x^6$, $0x^5$, and $0x^3$ in counting the number of variations in sign of $f(x)$. Replacing x by $-x$, we get

$$f(-x) = 3x^7 + 4x^4 + 3x^2 + 2x - 1$$

$$\underbrace{\qquad}_{+\text{ to }-}$$

which has one variation in sign.

THEOREM

DESCARTES' RULE OF SIGNS

Let f denote a polynomial function.

The number of positive zeros of f either equals the number of variations in sign of the coefficients of $f(x)$ or else equals that number less some even integer.

The number of negative zeros of f either equals the number of variations in sign of the coefficients of $f(-x)$ or else equals that number less some even integer. ≡

We shall not prove Descartes' Rule of Signs. Let's see how it is used.

EXAMPLE 2   Discuss the zeros of

$$f(x) = 3x^6 - 4x^4 + 3x^3 + 2x^2 - x - 3$$

Solution   There are at most six zeros because the equation is of degree six. Since there are three variations in sign of the coefficients of $f(x)$, by Descartes' Rule of Signs we expect either three or one positive zero(s). To continue, we look at $f(-x)$.

$$f(-x) = 3x^6 - 4x^4 - 3x^3 + 2x^2 + x - 3$$

There are three variations in sign, so we expect either three or one negative zero(s). ≡

Although we haven't actually found the zeros, we know something about the number of zeros and how many might be positive or negative. The next result, which you are asked to prove in Exercise 5.4 (Problem 70), is called the **Rational Zeros Theorem**. It provides information about the rational zeros of a polynomial with integer coefficients.

THEOREM
RATIONAL ZEROS THEOREM

Let f be a polynomial function of degree one or higher of the form

$$f(x) = a_n x^n + a_{n-1} x^{n-1} + \cdots + a_1 x + a_0$$

where each coefficient is an integer. If p/q, in lowest terms, is a rational zero of f, then p must be a factor of a_0 and q must be a factor of a_n.

EXAMPLE 3   List the potential rational zeros of

$$f(x) = 3x^5 - 2x^4 - 15x^3 + 10x^2 + 12x - 8$$

Solution   Because f has integer coefficients, we may proceed to use the Rational Zeros Theorem. First, we list all the integers p that are factors of $a_0 = -8$ and all the integers q that are factors of $a_5 = 3$:

$$p: \quad \pm 1, \pm 2, \pm 4, \pm 8$$
$$q: \quad \pm 1, \pm 3$$

Now we form all possible ratios p/q:

$$\frac{p}{q}: \quad \pm 1, \pm 2, \pm 4, \pm 8, \pm \frac{1}{3}, \pm \frac{2}{3}, \pm \frac{4}{3}, \pm \frac{8}{3}$$

If f has a rational zero, it will be found in this list, which contains 16 possibilities.

Be sure you understand what the Rational Zeros Theorem says: For a polynomial with integer coefficients, *if* there is a rational zero, it is one of those listed. There may not be any rational zeros. Synthetic division may be used to test each potential rational zero to determine whether it is indeed a zero. To make the work easier, the integers are usually tested first. Let's continue this example.

EXAMPLE 4   Continue working with Example 3 to find the zeros of

$$f(x) = 3x^5 - 2x^4 - 15x^3 + 10x^2 + 12x - 8$$

Solution   We gather all the information we can about the zeros:

1. There are at most five zeros.

2. By Descartes' Rule of Signs, there are three or one positive zero(s). Also, because

$$f(-x) = -3x^5 - 2x^4 + 15x^3 + 10x^2 - 12x - 8$$

there are two or zero negative zeros.

3. Now we use our list of potential rational zeros from Example 3: $\pm 1, \pm 2, \pm 4, \pm 8, \pm \frac{1}{3}, \pm \frac{2}{3}, \pm \frac{4}{3}, \pm \frac{8}{3}$. We can test the potential rational zero 1 using synthetic division. This means we want to determine whether $x - 1$ is a factor:

$$
\begin{array}{r|rrrrrr}
-1)3 & -2 & -15 & 10 & 12 & -8 \\
 & -3 & -1 & 14 & 4 & -8 \\
\hline
3 & 1 & -14 & -4 & 8 & 0
\end{array}
$$

The remainder is zero. Thus, $x - 1$ is a factor and 1 is a zero. The entries in the bottom row of this synthetic division can be used to factor f:

$$
\begin{aligned}
f(x) &= 3x^5 - 2x^4 - 15x^3 + 10x^2 + 12x - 8 \\
&= (x - 1)(3x^4 + x^3 - 14x^2 - 4x + 8)
\end{aligned}
$$

Because every zero of the quotient

$$q_1(x) = 3x^4 + x^3 - 14x^2 - 4x + 8$$

DEPRESSED EQUATION

is also a zero of f (do you see why?), this equation is referred to as a **depressed equation** of $f(x)$. The depressed equation is of lower degree than the original equation. Thus, it is usually easier to find the zeros of the depressed equation than to continue working with the original equation.

4. Quotient q_1 has two or zero positive zeros, and because

$$q_1(-x) = 3x^4 - x^3 - 14x^2 + 4x + 8$$

q_1 has two or zero negative zeros.

5. The potential rational zeros of q_1 are the same as those listed earlier for f.

6. We choose to test 1 again because it could be a repeated root:

$$
\begin{array}{r|rrrrr}
-1)3 & 1 & -14 & -4 & 8 \\
 & -3 & -4 & 10 & 14 \\
\hline
3 & 4 & -10 & -14 & -6
\end{array}
$$

The remainder tells us that 1 is not a zero of q_1. Now we'll test -1 by determining whether $x + 1$ is a factor:

$$
\begin{array}{r|rrrrr}
1)3 & 1 & -14 & -4 & 8 \\
 & 3 & -2 & -12 & 8 \\
\hline
3 & -2 & -12 & 8 & 0
\end{array}
$$

We find that $x + 1$ is a factor; thus, -1 is a zero of q_1, and we have

$$f(x) = (x - 1)(x + 1)(3x^3 - 2x^2 - 12x + 8)$$

7. We work now with the depressed equation

$$q_2(x) = 3x^3 - 2x^2 - 12x + 8$$

Quotient q_2 has two or zero positive zeros, and because

$$q_2(-x) = -3x^3 - 2x^2 + 12x + 8$$

q_2 has one negative zero. The list of potential rational zeros is still unchanged. However, because 1 was not a zero of q_1, it cannot be a zero of q_2. Also, the fact that -1 is a zero of q_1 does not mean it cannot also be a zero of q_2 (that is, it could be a repeated root of q_1). We know there is one negative zero (which, of course, may not be rational), so we test -1 once more to determine whether it is a root of q_2:

$$\begin{array}{r|rrrr}
1) & 3 & -2 & -12 & 8 \\
   &   & 3 & -5 & -7 \\
\hline
   & 3 & -5 & -7 & 15
\end{array}$$

It is not. Next, we choose to test -2:

$$\begin{array}{r|rrrr}
2) & 3 & -2 & -12 & 8 \\
   &   & 6 & -16 & 8 \\
\hline
   & 3 & -8 & 4 & 0
\end{array}$$

We find that $x + 2$ is a factor, thus, -2 is a zero, and we have

$$f(x) = (x - 1)(x + 1)(x + 2)(3x^2 - 8x + 4) \tag{1}$$

8. The new depressed equation of f, $q_3(x) = 3x^2 - 8x + 4 = 0$, is a quadratic equation with a discriminant of $b^2 - 4ac = (-8)^2 - 4(3)(4) = 16$. Therefore, $q_3(x)$ has two real solutions, which can be found by factoring:

$$3x^2 - 8x + 4 = 0$$
$$(3x - 2)(x - 2) = 0$$
$$3x - 2 = 0 \quad \text{or} \quad x - 2 = 0$$
$$x = \frac{2}{3} \quad \text{or} \quad x = 2$$

The zeros of f are -2, -1, $\frac{2}{3}$, 1, and 2.   ≡

In Example 4, we found the zeros of the polynomial function $f(x)$. In so doing, we also factored f over the real numbers. Starting at

Equation (1), we found the factored form of f over the real numbers to be

$$
\begin{aligned}
f(x) &= 3x^5 - 2x^4 - 15x^3 + 10x^2 + 12x - 8 \\
&= (x - 1)(x + 1)(x + 2)(3x^2 - 8x + 4) \\
&= (x - 1)(x + 1)(x + 2)(3x - 2)(x - 2) \\
&= 3(x - 1)(x + 1)(x + 2)\left(x - \frac{2}{3}\right)(x - 2)
\end{aligned}
$$

EXAMPLE 5 Use the factored form of

$$
f(x) = 3x^5 - 2x^4 - 15x^3 + 10x^2 + 12x - 8
$$

to graph f.

Solution Using the zeros of f and the factored form of f, we can construct Table 4. A calculator and the nested form of f were used to evaluate f at the test numbers (except $x = 0$).

TABLE 4

x	TEST NUMBER	VALUE OF f	GRAPH OF f
$x < -2$	-3	$f(-3) = -440$	Below x-axis
$-2 < x < -1$	$-\frac{3}{2}$	$f(-\frac{3}{2}) \approx 14.2$	Above x-axis
$-1 < x < \frac{2}{3}$	0	$f(0) = -8$	Below x-axis
$\frac{2}{3} < x < 1$	0.8	$f(0.8) \approx 0.5$	Above x-axis
$1 < x < 2$	$\frac{3}{2}$	$f(\frac{3}{2}) \approx -5.5$	Below x-axis
$x > 2$	3	$f(3) = 280$	Above x-axis

Figure 23 gives an idea of the graph of f.

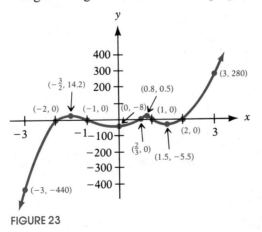

FIGURE 23

EXAMPLE 6   Use Descartes' Rule of Signs and the Rational Zeros Theorem to find the real zeros of the polynomial function

$$g(x) = x^5 - x^4 - x^3 + x^2 - 2x + 2$$

Use the zeros to factor g over the real numbers. Then graph g.

Solution   1. There are at most five zeros.
2. There are four or two or zero positive zeros. Also, because

$$g(-x) = -x^5 - x^4 + x^3 + x^2 + 2x + 2$$

there is one negative zero.
3. The potential rational zeros of g are ± 1, ± 2. We'll test 1 by evaluating $g(1)$:

$$g(1) = 1 - 1 - 1 + 1 - 2 + 2 = 0$$

Thus, 1 is a zero, so $x - 1$ is a factor and

$$g(x) = (x - 1)(x^4 - x^2 - 2)$$

4. The depressed equation $q_1(x) = x^4 - x^2 - 2$ is a disguised quadratic equation that can be factored:

$$x^4 - x^2 - 2 = 0$$
$$(x^2 - 2)(x^2 + 1) = 0$$
$$x^2 - 2 = 0 \quad \text{or} \quad x^2 + 1 = 0$$
$$x = \pm\sqrt{2}$$

Because $x^2 + 1 = 0$ has no real solution, the depressed equation has only the two zeros $\sqrt{2}$ and $-\sqrt{2}$.

Thus, the zeros of g are $-\sqrt{2}$, 1, and $\sqrt{2}$. The factored form of g over the real numbers is

$$g(x) = x^5 - x^4 - x^3 + x^2 - 2x + 2$$
$$= (x - 1)(x^4 - x^2 + 2)$$
$$= (x - 1)(x^2 - 2)(x^2 + 1)$$
$$= (x - 1)(x - \sqrt{2})(x + \sqrt{2})(x^2 + 1)$$

Now we construct Table 5:

TABLE 5

x	TEST NUMBER	VALUE OF g	GRAPH OF g
$x < -\sqrt{2}$	-2	$g(-2) = -30$	Below x-axis
$-\sqrt{2} < x < 1$	-1	$g(-1) = 4$	Above x-axis
$1 < x < \sqrt{2}$	1.1	$g(1.1) \approx -0.17$	Below x-axis
$x > \sqrt{2}$	2	$g(2) = 10$	Above x-axis

Figure 24 gives an idea of the graph of g.

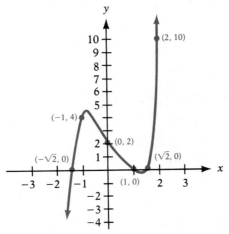

FIGURE 24

The quadratic factor $x^2 + 1$ that appears in the factored form of

$$g(x) = x^5 - x^4 - x^3 + x^2 - 2x + 2$$

is called *irreducible*, because the polynomial $x^2 + 1$ cannot be factored over the real numbers.

IRREDUCIBLE QUADRATIC   In general, a quadratic factor $ax^2 + bx + c$ is said to be **irreducible** if it cannot be factored over the real numbers—that is, if it is prime over the reals.

Refer back to Example 4, where the factored form of $f(x)$ was given as

$$f(x) = 3(x - 1)(x + 1)(x + 2)\left(x - \frac{2}{3}\right)(x - 2)$$

The factored form of the polynomial g from Example 6 is

$$g(x) = (x - 1)(x - \sqrt{2})(x + \sqrt{2})(x^2 + 1)$$

The polynomial function f has five real zeros; its factored form contains five linear factors. The polynomial function g has three real zeros; its factored form contains three linear factors and one irreducible quadratic factor. The next result tells us what to expect when we factor a polynomial.

THEOREM   Every polynomial function can be uniquely factored into a product of linear factors and irreducible quadratic factors.

We shall prove this result in Section 5.7, and, in fact, we shall draw several additional conclusions about the zeros of a polynomial function. For now, though, we summarize the results obtained thus far.

Summary

To obtain information about the (real) zeros of a polynomial function, follow these steps:

> 1. Use the degree of the polynomial to determine the maximum number of zeros.
> 2. Use Descartes' Rule of Signs to determine the possible number of positive zeros and negative zeros.
> 3. (a) If the polynomial has integer coefficients, use the Rational Zeros Theorem to identify those rational numbers that potentially can be zeros.
>   (b) Use synthetic division to test each potential rational zero.
>   (c) Each time a zero (and thus a factor) is found, repeat Steps 2 and 3 on the depressed equation.
> 4. In attempting to locate the zeros, remember to use (if possible) the factoring techniques you already know (special products, disguised quadratics, factoring by grouping, and so on).

If these procedures fail to locate all the zeros, you may have to be satisfied with "estimating" or "approximating" the zeros—the subject of the next section.

≡ **HISTORICAL COMMENT**

Formulas for the solution of third- and fourth-degree polynomial equations exist, and, while not very practical, they do have an interesting history.

In the 1500s in Italy, mathematical contests were a popular pastime, and persons possessing methods for solving problems kept them secret. (Solutions that were published were already common knowledge.) Nicolo of Brescia (1499–1557), commonly referred to as Tartaglia (the stammerer), had the secret for solving cubic (third-degree) equations, which gave him a decided advantage in the contests. Girolamo Cardano (1501–1576) found out that Tartaglia had the secret, and, being interested in cubics, he requested it from Tartaglia. The reluctant Tartaglia hesitated for some time but finally, swearing Cardano to secrecy with midnight oaths by candlelight, told him the secret. Cardano then published the solution in his book *Ars Magna* (1545), giving Tartaglia the credit but rather compromising the secrecy. Tartaglia exploded into bitter recriminations, and each wrote pamphlets that reflected on the other's mathematics, moral character, and ancestry.

The quartic (fourth-degree) equation was solved by Cardano's student Lodovico Ferrari, and this solution also was included, with credit and this time with permission, in the *Ars Magna*.

Attempts were made to solve the fifth-degree equation in similar ways, all of which failed. About 1800, Ruffini, Abel, and Galois found ways to show that it is not possible to solve fifth-degree equations by formula, but the proofs required the introduction of new methods. Galois' methods eventually developed into a large part of modern algebra.

The following problems develop the Tartaglia–Cardano solution of the cubic equation and show why it is not altogether practical.

≡ **HISTORICAL PROBLEMS**

Problems 1–4 provide the general idea behind the Tartaglia–Cardano formula for solving cubic equations.

1. In the equation

$$x^3 + px + q = 0$$

   replace x by $H + K$. Let $3HK = -p$, and show that

$$H^3 + K^3 = -q$$

   [*Hint:* $3H^2K + 3HK^2 = 3HKx$]

2. Based on Problem 1, we have the two equations

$$3HK = -p \quad \text{and} \quad H^3 + K^3 = -q$$

   Solve for K in $3HK = -p$ and substitute into $H^3 + K^3 = -q$. Then show that

$$H = \sqrt[3]{\frac{-q}{2} + \sqrt{\frac{q^2}{4} + \frac{p^3}{27}}}$$

   [*Hint:* Look for a disguised quadratic equation.]

3. Use the solution for H from Problem 2 and the equation $H^3 + K^3 = -q$ to show that

$$K = \sqrt[3]{\frac{-q}{2} - \sqrt{\frac{q^2}{4} + \frac{p^3}{27}}}$$

4. Use the results from Problems 1–3 to show that the solution of the equation

$$x^3 + px + q = 0$$

   is

$$x = \sqrt[3]{\frac{-q}{2} + \sqrt{\frac{q^2}{4} + \frac{p^3}{27}}} + \sqrt[3]{\frac{-q}{2} - \sqrt{\frac{q^2}{4} + \frac{p^3}{27}}}$$

5. Use the result of Problem 4 to solve the equation

$$x^3 - 6x - 9 = 0$$

6. Use the result of Problem 4 to solve the equation

$$x^3 + 3x - 14 = 0$$

7. Use the methods of this chapter to solve the equation

$$x^3 + 3x - 14 = 0$$

8. *Requires complex numbers* Show that the formula derived in Problem 4 leads to the cube roots of a complex number when applied to the equation

$$x^3 - 6x + 4 = 0$$

Use the methods of this chapter to solve the equation.   ≣

EXERCISE 5.4 *In Problems 1–10, for each polynomial function, list each zero and its multiplicity.*

1. $f(x) = 3(x - 4)(x + 5)^2$

2. $f(x) = 4(x + 1)(x - 3)^3$

3. $f(x) = 4(x^2 + 1)(x - 2)^3$

4. $f(x) = 2(x - 3)(x + 4)^3$

5. $f(x) = -2(x + \frac{1}{2})^2(x^2 + 4)^2$

6. $f(x) = (x - \frac{1}{3})^2(x - 1)^3$

7. $f(x) = (x - 5)^3(x + 4)^2$

8. $f(x) = (x + \sqrt{3})^2(x - 2)^4$

9. $f(x) = 3(x^2 + 4)(x^2 + 9)^2$

10. $f(x) = -2(x^2 + 1)^3$

In Problems 11–14 tell the maximum number of zeros for each polynomial function.

11. $f(x) = -4x^7 + x^3 - 1$

12. $f(x) = 5x^4 - 2x^2 + x - 2$

13. $f(x) = 2x^6 - 3x^2 - x + 1$

14. $f(x) = -3x^5 + 4x^4 + 2$

In Problems 15–22 use Descartes' Rule of Signs to determine how many positive and how many negative zeros each polynomial function might have. Do not attempt to find the actual zeros.

15. $f(x) = 3x^3 - 2x^2 + x + 2$

16. $f(x) = -x^3 - x^2 + x + 1$

17. $f(x) = -x^4 + x^2 - 1$

18. $f(x) = x^4 + 5x^3 - 2$

19. $f(x) = x^5 + x^4 + x^2 + x + 1$

20. $f(x) = x^5 - x^4 + x^3 - x^2 + x - 1$

21. $f(x) = x^6 - 1$

22. $f(x) = x^6 + 1$

In Problems 23–34 list the potential rational zeros of each polynomial function. Do not attempt to find the actual zeros.

23. $f(x) = x^4 - 3x^3 + x^2 - x + 1$

24. $f(x) = x^5 - x^4 + 2x^2 + 2$

25. $f(x) = x^5 - 6x^2 + 9x - 3$

26. $f(x) = 2x^5 - x^4 - x^2 + 1$

27. $f(x) = -2x^3 - x^2 + x + 1$

28. $f(x) = 3x^4 - x^2 + 2$

29. $f(x) = 3x^4 - x^2 + 2$

30. $f(x) = -4x^3 + x^2 + x + 2$

31. $f(x) = 2x^5 - x^3 + 2x^2 + 4$

32. $f(x) = 3x^5 - x^2 + 2x + 3$

33. $f(x) = 6x^4 + 2x^3 - x^2 + 2$

34. $f(x) = -6x^3 - x^2 + x + 3$

In Problems 35–56 use Descartes' Rule of Signs and the Rational Zeros Theorem to find all the real zeros of each polynomial function. Use the zeros to factor f over the real numbers.

35. $f(x) = x^3 + 2x^2 - 5x - 6$

36. $f(x) = x^3 + 8x^2 + 11x - 20$

37. $f(x) = 2x^3 - x^2 + 2x - 1$

38. $f(x) = 2x^3 + x^2 + 2x + 1$

39. $f(x) = x^4 + x^2 - 2$

40. $f(x) = x^4 - 3x^2 - 4$

41. $f(x) = 4x^4 + 7x^2 - 2$

42. $f(x) = 4x^4 + 15x^2 - 4$

43. $f(x) = x^4 + x^3 - 3x^2 - x + 2$

44. $f(x) = x^4 - x^3 - 6x^2 + 4x + 8$

45. $f(x) = 4x^5 - 8x^4 - x + 2$

46. $f(x) = 4x^5 + 12x^4 - x - 3$

47. $f(x) = x^4 - x^3 + 2x^2 - 4x - 8$

48. $f(x) = 2x^3 + 3x^2 + 2x + 3$

49. $f(x) = 3x^3 + 4x^2 - 7x + 2$

50. $f(x) = 2x^3 - 3x^2 - 3x - 5$

51. $f(x) = 3x^3 - x^2 - 15x + 5$

52. $f(x) = 2x^3 - 11x^2 + 10x + 8$

53. $f(x) = x^4 + 4x^3 + 2x^2 - x + 6$

54. $f(x) = x^4 - 2x^3 + 10x^2 - 18x + 9$

55. $f(x) = x^3 - \frac{2}{3}x^2 + \frac{8}{3}x + 1$

56. $f(x) = x^3 + \frac{3}{2}x^2 + 3x - 2$

In Problems 57–66 find the intercepts of each polynomial function f(x). Find the numbers x for which the graph of f is above the x-axis and those for which it is below the x-axis. Obtain several other points on the graph, and connect them with a smooth curve. [Hint: Use the factored form of f (see Problems 37–46).]

57. $f(x) = 2x^3 - x^2 + 2x - 1$

58. $f(x) = 2x^3 + x^2 + 2x + 1$

59. $f(x) = x^4 + x^2 - 2$

60. $f(x) = x^4 - 3x^2 - 4$

61. $f(x) = 4x^4 + 7x^2 - 2$

62. $f(x) = 4x^4 + 15x^2 - 4$

63. $f(x) = x^4 + x^3 - 3x^2 - x + 2$

64. $f(x) = x^4 - x^3 - 6x^2 + 4x + 8$

65. $f(x) = 4x^5 - 8x^4 - x + 2$

66. $f(x) = 4x^5 + 12x^4 - x - 3$

67. What is the length of the edge of a cube if, after a slice 1 inch thick is cut from one side, the volume remaining is 294 cubic inches?

68. What is the length of the edge of a cube if its volume could be doubled by an increase of 6 centimeters in one edge, an increase of 12 centimeters in a second edge, and a decrease of 4 centimeters in the third edge?

69. Let $f(x)$ be a polynomial function whose coefficients are integers. Suppose that r is a (real) zero of f and that the leading coefficient of f is 1. Use the Rational Zeros Theorem to show that r is either an integer or an irrational number.

70. Prove the Rational Zeros Theorem. [*Hint:* Let p/q, where p and q have no common factors except 1 and -1, be a solution of the polynomial $f(x) = a_nx^n + a_{n-1}x^{n-1} + \cdots + a_1x + a_0$, whose coefficients are all integers. Show that $a_np^n + a_{n-1}p^{n-1}q + \cdots + a_1pq^{n-1} + a_0q^n = 0$. Now, because p is a factor of the first n terms of this equation, p also must be a factor of the term a_0q^n. Since p is not a factor of q (why?), p must be a factor of a_0. Similarly, q must be a factor of a_n.]

▤ 5.5 APPROXIMATING THE ZEROS OF A POLYNOMIAL FUNCTION

Sometimes the procedures we discussed in Section 5.4 yield limited information about the zeros of a polynomial. Let's look at an example.

EXAMPLE 1 Discuss the zeros of $f(x) = x^5 - x^3 - 1$.

Solution 1. f has at most five zeros.
2. f has one positive zero. Because

$$f(-x) = -x^5 + x^3 - 1$$

f has two or zero negative zeros.
3. The potential rational zeros are ± 1, neither of which is an actual zero. We conclude that f has one positive irrational zero and perhaps two negative irrational zeros. ▤

To obtain more information about the zeros of the polynomial of Example 1, we need some additional results.

Upper and Lower Bounds

The search for the zeros of a polynomial function can be reduced somewhat if upper and lower bounds to the zeros can be found. A **UPPER BOUND** number M is an **upper bound** to the zeros of a polynomial f if no zero **LOWER BOUND** of f exceeds M. The number m is a **lower bound** if no zero of f is less than m.

Thus, if m is a lower bound and M is an upper bound to the zeros of a polynomial f, then

$$m \leq \text{Any zero of } f \leq M$$

One immediate advantage of knowing the values of a lower bound m and an upper bound M is that, for polynomials with integer coefficients, it may allow you to eliminate some potential rational zeros, namely any that lie outside the interval $[m, M]$. The next result tells us how to locate lower and upper bounds.

THEOREM Let f denote a polynomial function whose leading coefficient is **BOUNDS ON ZEROS** positive.

If $M \geq 0$ is a real number and if the third row in the process of synthetic division of f by $x - M$ contains only numbers that are positive or zero, then M is an upper bound to the zeros of f.

If $m \leq 0$ is a real number and if the third row in the process of synthetic division of f by $x - m$ contains numbers that are alternately positive (or zero) and negative (or zero), then m is a lower bound to the zeros of f.   ≡

We shall give only an outline of the proof of the first part of the theorem.

Proof
(Outline)

Suppose M is a nonnegative real number and the third row in the process of synthetic division of the polynomial f by $x - M$ contains only numbers that are positive or zero. Then there is a quotient q and a remainder R so that

$$f(x) = (x - M)q(x) + R$$

where the coefficients of $q(x)$ are positive or zero and the remainder $R \geq 0$. Then, for any $x > M$, we must have

$$x - M > 0, \quad q(x) > 0, \quad \text{and} \quad R \geq 0$$

so that $f(x) > 0$. That is, there is no zero of f larger than M.   ≡

EXAMPLE 2   Find upper and lower bounds to the zeros of $f(x) = x^5 - x^3 - 1$.

Solution   To get an upper bound to the zeros, the usual practice is to start with 1, and continue with 2, 3, . . . , until the third row of the process of synthetic division yields only numbers that are positive or zero. Thus, we begin by dividing $f(x)$ by $x - 1$:

$$
\begin{array}{r|rrrrrr}
-1) & 1 & 0 & -1 & 0 & 0 & -1 \\
& & -1 & -1 & 0 & 0 & 0 \\
\hline
& 1 & 1 & 0 & 0 & 0 & -1 \\
\end{array}
$$

$$
\begin{array}{r|rrrrrr}
-2) & 1 & 0 & -1 & 0 & 0 & -1 \\
& & -2 & -4 & -6 & -12 & -24 \\
\hline
& 1 & 2 & 3 & 6 & 12 & 23 \\
\end{array}
$$

The third row has only positive numbers; thus, 2 is an upper bound. To get a lower bound to the zeros, we start with -1, and continue with $-2, -3, \ldots$, until the third row of the process of synthetic division yields numbers that alternate in sign:

$$
\begin{array}{r|rrrrrr}
1) & 1 & 0 & -1 & 0 & 0 & -1 \\
& & 1 & 1 & -1 & 0 & 0 \\
\hline
& 1 & 1 & 0 & 0 & 0 & -1 \\
\end{array}
$$

Because the entries do alternate in sign (remember 0 can be counted as positive or negative as needed), -1 is a lower bound. Thus, the zeros of f lie between -1 and 2. ≡

In Example 2 we found that the zeros of $f(x) = x^5 - x^3 - 1$ lie in the interval $[-1, 2]$. However, remember that in finding the lower bound -1 and the upper bound 2, we tested only integers. Were we to test other positive numbers less than 2 and other negative numbers greater than -1, we might be able to "fine-tune" the bounds and find a smaller interval containing the zeros of f. However, the effort required to do this is usually not worth it, since integer-valued upper and lower bounds provide enough practical information about the zeros.

Intermediate Value Theorem

The next result, called the **Intermediate Value Theorem**, is based on the fact that the graph of a polynomial function is continuous—that is, contains no "jumps" or "gaps."

THEOREM
INTERMEDIATE VALUE THEOREM

Let f denote a polynomial function. If $a < b$ and if $f(a)$ and $f(b)$ are of opposite sign, then there is at least one zero of f between a and b. ≡

Although the proof of this result requires advanced methods in calculus, it is easy to "see" why the result is true. Look at Figure 25.

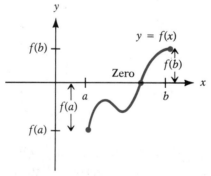

FIGURE 25
If $f(a) < 0$ and $f(b) > 0$, there is a zero between a and b

EXAMPLE 3   Show that

$$f(x) = x^5 - x^3 - 1$$

has a zero between 1 and 2.

Solution   We know from Example 1 that f has exactly one positive zero. Look back at the solution to Example 2, where we used synthetic division to divide f by $x - 1$ and then by $x - 2$. There we see that

$$f(1) = -1 \quad \text{and} \quad f(2) = 23$$

Because $f(1) < 0$ and $f(2) > 0$, it follows from the Intermediate Value Theorem that f has a zero between 1 and 2.   ■

Note that the zero we now know to lie between 1 and 2 is irrational, because we found in Example 1 that the only possible rational zeros are -1 and 1.

We can use the Intermediate Value Theorem to get a better approximation of the zero of a function f as follows:

ESTIMATING THE ZEROS OF A FUNCTION

1. Find two consecutive integers a and $a + 1$ such that f has a zero between them.
2. Divide the interval $[a, a + 1]$ into ten equal subintervals.
3. Evaluate f at each endpoint until the Intermediate Value Theorem applies; that interval then contains a zero.
4. Repeat the process starting at Step 2 until the desired accuracy is achieved.

[C] EXAMPLE 4   Find the positive zero of $f(x) = x^5 - x^3 - 1$ to within 0.01.

Solution   From Example 3 we know the positive zero is between 1 and 2. We divide the interval $[1, 2]$ into ten equal subintervals:

$$[1, 1.1], \quad [1.1, 1.2], \quad [1.2, 1.3], \quad [1.3, 1.4], \quad [1.4, 1.5],$$
$$[1.5, 1.6], \quad [1.6, 1.7], \quad [1.7, 1.8], \quad [1.8, 1.9], \quad [1.9, 2]$$

Now we find the value of f at each endpoint until the Intermediate Value Theorem applies. The easiest method is to write $f(x)$ in nested form and use a calculator. Thus, we write

$$f(x) = x^5 - x^3 - 1 = (x^2 - 1) \cdot x \cdot x \cdot x - 1 = (x \cdot x - 1) \cdot x \cdot x \cdot x - 1$$
$$f(1.0) = -1$$
$$f(1.1) = -0.72049$$
$$f(1.2) = -0.23968$$
$$f(1.3) = 0.51593$$

We can stop here and conclude that the zero is between 1.2 and 1.3. Now we divide the interval [1.2, 1.3] into ten equal subintervals and proceed to evaluate f at each endpoint:

$$f(1.20) = -0.23968$$
$$f(1.21) = -0.1778185$$
$$f(1.22) = -0.1131398$$
$$f(1.23) = -0.0455613$$
$$f(1.24) = 0.025001$$

We conclude that the zero lies between 1.23 and 1.24, and so we have approximated it to within 0.01.   ≡

There are many other numerical techniques for approximating the zeros of a polynomial. The one outlined in Example 4, a variation of the bisection method, has the advantages that it will always work, that it can be programmed rather easily on a computer, and that each time it is used another decimal place of accuracy is achieved.

EXERCISE 5.5   *In Problems 1–6 use the Intermediate Value Theorem to show that each polynomial function has a zero in the given interval.*

1. $f(x) = 8x^4 - 2x^2 + 5x - 1$; $[0, 1]$
2. $f(x) = x^4 + 8x^3 - x^2 + 2$; $[-1, 0]$
3. $f(x) = 2x^3 + 6x^2 - 8x + 2$; $[-5, -4]$
4. $f(x) = 3x^3 - 10x + 9$; $[-3, -2]$

C 5. $f(x) = x^5 - x^4 + 7x^3 - 7x^2 - 18x + 18$; $[1.4, 1.5]$
C 6. $f(x) = x^5 - 3x^4 - 2x^3 + 6x^2 + x + 2$; $[1.7, 1.8]$

In Problems 7–12 find upper and lower bounds to the zeros of each polynomial function.

7. $f(x) = 2x^3 + x^2 - 1$
8. $f(x) = 3x^3 - 2x^2 + x + 4$
9. $f(x) = x^3 + 5x^2 + 11x + 10$
10. $f(x) = 2x^3 + x^2 + 11x - 6$
11. $f(x) = x^4 + 3x^3 - 5x^2 + 9$
12. $f(x) = 2x^4 - 12x^3 + 27x^2 - 54x + 81$

C   *In Problems 13–16 each polynomial function has exactly one positive root. Use the method of Example 4 to approximate the zero to within 0.01.*

13. $f(x) = x^3 + x^2 + x - 4$
14. $f(x) = 2x^4 + x^2 - 1$
15. $f(x) = 2x^4 - 3x^3 - 4x^2 - 8$
16. $f(x) = 3x^3 - 2x^2 - 20$

17. *Programming exercise* Write a computer program that will estimate the positive zero of a polynomial function to any desired degree of accuracy. Input should consist of the coefficients of the polynomial, in order from

highest to lowest, followed by the degree N of accuracy wanted (that is, the zero is to be estimated to within 10^{-N}), followed by two consecutive integers between which the zero lies. Output will consist of two decimal numbers between which the zero lies. The program should contain a subroutine that writes the polynomial in nested form.

18. *Programming exercise* Modify the program in Problem 17 to include a subroutine that will locate the two consecutive integers between which the zero lies.

≡ 5.6 RATIONAL FUNCTIONS

Ratios of integers are called *rational numbers*. Ratios of polynomial functions are called *rational functions*.

RATIONAL FUNCTION   A **rational function** is a function of the form

$$R(x) = \frac{p(x)}{q(x)}$$

where p and q are polynomial functions and q is not the zero polynomial. The domain consists of all real numbers except those for which the denominator q is zero.

We shall assume throughout this section that rational functions are written in lowest terms so that p and q contain no common factors.
For a rational function $R(x) = p(x)/q(x)$, the zeros, if any, of the numerator are the x-intercepts of the graph of R and so will play a major role in graphing R. The zeros of the denominator of R—that is, the numbers x, if any, for which $q(x) = 0$—although not in the domain of R, also play a major role in the graph of R.

EXAMPLE 1   (a) The domain of $R(x) = \dfrac{2x^2 - 4}{x + 5}$ consists of all real numbers except -5.

(b) The domain of $R(x) = \dfrac{1}{x^2 - 4}$ consists of all real numbers except -2 and 2.

(c) The domain of $R(x) = \dfrac{x^3}{x^2 + 1}$ consists of all real numbers.

(d) The domain of $R(x) = \dfrac{-x^2 + 2}{3}$ consists of all real numbers.   ≡

EXAMPLE 2   Graph each function:

(a) $R(x) = \dfrac{1}{x}$ (b) $H(x) = \dfrac{1}{x^2}$

Solution   (a) The domain of $R(x) = p(x)/q(x) = 1/x$ consists of all real numbers except zero. Thus, the graph has no y-intercept because x can never equal zero. The graph has no x-intercept because the equation $p(x) = 0$ has no solution. Therefore, the graph of R will not cross the coordinate axes. Because

$$R(-x) = \frac{1}{-x} = -\frac{1}{x} = -R(x)$$

R is an odd function and, hence, its graph is symmetric with respect to the origin. Table 6 shows the behavior of $R(x) = 1/x$ for selected positive numbers x (we will use symmetry to obtain the graph of R when $x < 0$):

TABLE 6

x	$\frac{1}{1000}$	$\frac{1}{100}$	$\frac{1}{10}$	$\frac{1}{2}$	1	2	3	10	100
$R(x) = 1/x$	1000	100	10	2	1	$\frac{1}{2}$	$\frac{1}{3}$	$\frac{1}{10}$	$\frac{1}{100}$

Look at the entries in Table 6. As the values of x get closer to zero, the values $R(x)$ become larger and larger positive numbers. When this happens, we say that R is **unbounded in the positive direction**. We symbolize this by writing $R \to +\infty$ (read "R **approaches plus infinity**"). Look again at Table 6. As $x \to +\infty$, the values of $R(x)$ get closer to zero. Figure 26(a) illustrates the graph.

UNBOUNDED IN THE POSITIVE
DIRECTION
APPROACHES PLUS INFINITY
$(\to +\infty)$

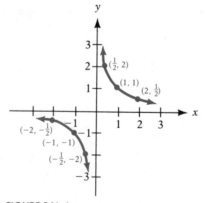

FIGURE 26(a)

$R(x) = \dfrac{1}{x}$

(b) The domain of $H(x) = p(x)/q(x) = 1/x^2$ consists of all real numbers x except zero. Thus, the graph has no y-intercept because x can never equal zero. The graph has no x-intercept because the equation $p(x) = 0$ has no solution. Therefore, the graph of H will not cross the coordinate axes. Because

$$H(-x) = \frac{1}{(-x)^2} = \frac{1}{x^2} = H(x)$$

H is an even function and, hence, its graph is symmetric with respect to the y-axis. Table 7 shows the behavior of $H(x) = 1/x^2$ for selected positive numbers x (we will use symmetry to obtain the graph of H when $x < 0$):

TABLE 7

x	$\frac{1}{100}$	$\frac{1}{10}$	$\frac{1}{2}$	1	2	10	100
$H(x) = 1/x^2$	10,000	100	4	1	$\frac{1}{4}$	$\frac{1}{100}$	$\frac{1}{10,000}$

As x gets closer and closer to zero, we see that $H \to +\infty$. Also, as $x \to +\infty$, the values $H(x)$ get closer and closer to zero. Figure 26(b) illustrates the graph.

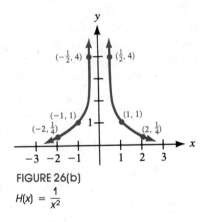

FIGURE 26(b)

$H(x) = \dfrac{1}{x^2}$

Sometimes the techniques of shifting, change of scale, and reflection can be used to graph a rational function.

EXAMPLE 3   Graph the rational function

$$R(x) = \frac{1}{x-2} + 1$$

Solution   First, we take note of the fact that the domain of R consists of all real numbers except $x = 2$. To graph R, we start with the graph of $y = 1/x$. See Figure 27 for the steps.

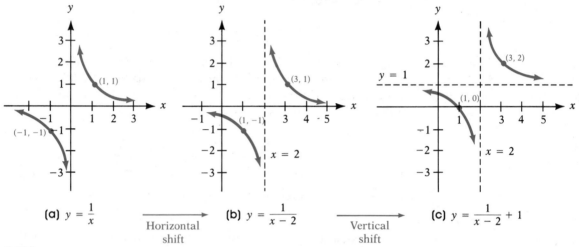

(a) $y = \dfrac{1}{x}$ → Horizontal shift →   (b) $y = \dfrac{1}{x - 2}$ → Vertical shift →   (c) $y = \dfrac{1}{x - 2} + 1$

FIGURE 27

Asymptotes

In Figure 27(c), notice that, as the values of x become more negative— that is, as x becomes **unbounded in the negative direction** ($x \to -\infty$, read "x **approaches minus infinity**")—the values $R(x)$ get closer and closer to 1. In fact, we can conclude the following from Figure 27(c):

UNBOUNDED IN THE NEGATIVE DIRECTION

APPROACHES MINUS INIFINITY ($\to -\infty$)

1. As $x \to -\infty$, the values $R(x)$ get closer to 1.
2. As x gets closer to 2, the values $R(x) \to -\infty$ if $x < 2$, and the values $R(x) \to +\infty$ if $x > 2$.
3. As $x \to +\infty$, the values $R(x)$ get closer to 1.

This behavior of the graph is depicted by the dashed vertical line $x = 2$ and the dashed horizontal line $y = 1$. These lines are called *asymptotes* of the graph. The definition follows.

Let F denote a function.

HORIZONTAL ASYMPTOTE

If, as $x \to -\infty$ or as $x \to +\infty$, the values of $F(x)$ get closer to some fixed number L, then the line $y = L$ is a **horizontal asymptote** of the graph of F.

VERTICAL ASYMPTOTE

If, as x gets closer to some number c, the values $|F(x)| \to +\infty$, then the line $x = c$ is a **vertical asymptote** of the graph of F.

Even though asymptotes of a function are not part of the graph of a function, they provide information about the way the graph looks far away from the origin.

Figure 28 illustrates some of the possibilities.

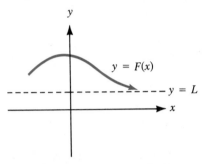

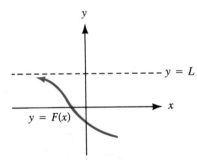

(a) As $x \to +\infty$, the values of $F(x)$ get closer to L; $y = L$ is a horizontal asymptote

(b) As $x \to -\infty$, the values of $F(x)$ get closer to L; $y = L$ is a horizontal asymptote

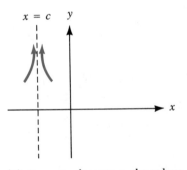

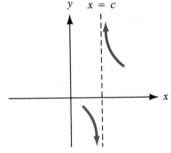

(c) As x gets closer to c, the values $|F(x)| \to +\infty$; $x = c$ is a vertical asymptote

(d) As x gets closer to c, the values $|F(x)| \to +\infty$; $x = c$ is a vertical asymptote

FIGURE 28

OBLIQUE ASYMPTOTE   Thus, an asymptote is a line that the points P on the graph of a function keep getting closer and closer to as P moves further and further away from the origin. If an asymptote is neither horizontal nor vertical, it is called **oblique**. Figure 29 (page 342) shows an oblique asymptote.

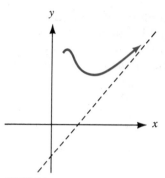

FIGURE 29
Oblique asymptote

Now, let's look at some ways to find the asymptotes of rational functions.

The vertical asymptotes, if any, of a rational function $R(x) = p(x)/q(x)$ are as easy to find as the zeros of the denominator $q(x)$. Suppose c is a zero of the denominator of the rational function R. Then $x - c$ is a factor of the denominator. Now, as x gets closer to c, the values of $x - c$ will get closer to zero, causing the ratio to become unbounded—that is, causing $|R(x)| \to +\infty$. Based on the definition, we conclude that the line $x = c$ is a vertical asymptote.

THEOREM
LOCATING VERTICAL ASYMPTOTES

A rational function $R(x) = p(x)/q(x)$ in lowest terms will have a vertical asymptote $x = c$ for each zero c of the denominator q.   ▤

EXAMPLE 4   Find the vertical asymptotes, if any, of the graph of each rational function.

(a) $R(x) = \dfrac{x}{x^2 - 4}$ (b) $F(x) = \dfrac{x + 3}{x - 1}$ (c) $H(x) = \dfrac{x^2}{x^2 + 1}$

Solution   (a) The zeros of the denominator satisfy the equation $x^2 - 4 = 0$, which has the solution set $\{-2, 2\}$. Hence, the lines $x = -2$ and $x = 2$ are the vertical asymptotes of the graph of R.

(b) The lone zero of the denominator is 1. Hence, the line $x = 1$ is the only vertical asymptote of the graph of F.

(c) The denominator has no zeros. Hence, the graph of H has no vertical asymptotes.   ▤

As Example 4 points out, rational functions can have no vertical asymptotes, one vertical asymptote, or more than one vertical asymptote. However, the graph of a rational function will never intersect any of its vertical asymptotes. Do you see why?

The procedure for finding horizontal and oblique asymptotes is somewhat more involved. To find such asymptotes, we need to know how the values of a function behave as $x \to -\infty$ or as $x \to +\infty$.

PROPER RATIONAL FUNCTION

If the rational function is **proper**—that is, if the degree of the numerator is less than the degree of the denominator—then as $x \to -\infty$ or as $x \to +\infty$, the values of $R(x)$ get closer to zero. Consequently, the line $y = 0$ (the x-axis) is a horizontal asymptote of the graph.

IMPROPER RATIONAL FUNCTION

If the rational function is **improper**—that is, if the degree of the numerator is greater than or equal to the degree of the denominator—we must use long division to write the rational function as the sum of a polynomial plus a proper rational function. (The Division Algorithm guarantees that this can be done.) Now, if the polynomial that results is of the form $ax + b$, then the line $y = ax + b$ is an asymptote of the graph. It is horizontal if $a = 0$, and it is oblique if $a \neq 0$. If the polynomial that results has degree greater than one, then the rational function has no horizontal and no oblique asymptote. The next example will help make all this clear.

EXAMPLE 5

Find the horizontal and oblique asymptotes, if any, of the graph of each rational function.

(a) $R(x) = \dfrac{x - 12}{4x^2 + x + 1}$

(b) $H(x) = \dfrac{3x^4 - x^2}{x^3 - x^2 + 1}$

(c) $R(x) = \dfrac{8x^2 - x + 2}{4x^2 - 1}$

(d) $G(x) = \dfrac{2x^5 - x^3 + 2}{x^3 - 1}$

Solution

(a) The rational function

$$R(x) = \frac{x - 12}{4x^2 + x + 1}$$

is proper because the degree of its numerator is one and the degree of its denominator is two. Consequently, the graph of $R(x)$ will have the line $y = 0$ (the x-axis) for a horizontal asymptote.

(b) The rational function

$$H(x) = \frac{3x^4 - x^2}{x^3 - x^2 + 1}$$

is improper, so we use long division:

$$
\require{enclose}
\begin{array}{r}
3x + 3 \\
x^3 - x^2 + 1 \enclose{longdiv}{3x^4 - x^2 } \\
\underline{3x^4 - 3x^3 + 3x} \\
3x^3 - x^2 - 3x \\
\underline{3x^3 - 3x^2 + 3} \\
2x^2 - 3x - 3
\end{array}
$$

Thus,

$$H(x) = \frac{3x^4 - x^2}{x^3 - x^2 + 1} = 3x + 3 + \frac{2x^2 - 3x - 3}{x^3 - x^2 + 1}$$

Now we have written the improper rational function $H(x)$ as the sum of the polynomial $3x + 3$ and the proper rational function $(2x^2 - 3x - 3)/(x^3 - x^2 + 1)$. We conclude that the rational function has the oblique asymptote $y = 3x + 3$.

(c) The rational function

$$R(x) = \frac{8x^2 - x + 2}{4x^2 - 1}$$

is improper, so we use long division:

$$
\begin{array}{r}
2 \\
4x^2 - 1 \overline{)8x^2 - x + 2} \\
\underline{8x^2 - 2} \\
-x + 4
\end{array}
$$

Thus,

$$R(x) = \frac{8x^2 - x + 2}{4x^2 - 1} = 2 + \frac{-x + 4}{4x^2 - 1}$$

Now $R(x)$ is written as the sum of the polynomial 2 and the proper rational function $(-x + 4)/(4x^2 - 1)$. We conclude that $y = 2$ is a horizontal asymptote of the graph.

(d) The rational function

$$G(x) = \frac{2x^5 - x^3 + 2}{x^3 - 1}$$

is improper, so we use long division:

$$
\begin{array}{r}
2x^2 - 1 \\
x^3 - 1 \overline{)2x^5 - x^3 + 2} \\
\underline{2x^5 - 2x^2} \\
-x^3 + 2x^2 + 2 \\
\underline{-x^3 + 1} \\
2x^2 + 1
\end{array}
$$

Thus,

$$G(x) = \frac{2x^5 - x^3 + 2}{x^3 - 1} = 2x^2 - 1 + \frac{2x^2 + 1}{x^3 - 1}$$

Now $G(x)$ is written as the sum of the polynomial $2x^2 - 1$ and the proper rational function $(2x^2 + 1)/(x^3 - 1)$. Since the polynomial is of degree two, we conclude that the graph of $G(x)$ has no horizontal and no oblique asymptote. ▤

Based on Example 5, a rational function can have a horizontal asymptote, an oblique asymptote, or no horizontal and no oblique asymptote. We shall see in a subsequent example that it is possible for the graph of a rational function to intersect its horizontal or oblique asymptotes.

Example 5 also serves to clarify conditions under which a rational function will have horizontal or oblique asymptotes.

Consider the rational function

$$R(x) = \frac{p(x)}{q(x)} = \frac{a_n x^n + a_{n-1} x^{n-1} + \cdots + a_1 x + a_0}{b_m x^m + b_{m-1} x^{m-1} + \cdots + b_1 x + b_0}$$

in which the degree of the numerator is n and the degree of the denominator is m.

1. If $n < m$, the line $y = 0$ (the x-axis) is a horizontal asymptote of the graph of R.
2. If $n = m$, then the line $y = a_n/b_m$ is a horizontal asymptote of the graph of R.
3. If $n = m + 1$ (that is, if the degree of the numerator is one more than the degree of the denominator), the graph of $R(x)$ will have an oblique asymptote.
4. If $n > m + 1$, the graph of $R(x)$ has no horizontal and no oblique asymptote.

Now we're ready to begin graphing rational functions.

Graphing Rational Functions

We commented earlier that calculus provides the tools required to graph a polynomial function accurately. The same holds true for rational functions. However, we can gather together quite a bit of information about such graphs and can put together an idea of the graph.

In the examples that follow, we will discuss the graph of a rational function $R(x) = p(x)/q(x)$ by using the following steps:

Step 1: Locate the intercepts, if any, of the graph. The x-intercepts, if any, of $R(x) = p(x)/q(x)$ obey the equation $p(x) = 0$; the y-intercept, if there is one, is $R(0)$.

Step 2: Test for symmetry. Replace x by $-x$ in $R(x)$. If $R(-x) = R(x)$, there is symmetry with respect to the y-axis; if $R(-x) = -R(-x)$, there is symmetry with respect to the origin.

Step 3: Locate the vertical asymptotes, if any. Find the real zeros of the denominator polynomial $q(x)$.

Step 4: Locate the horizontal asymptotes, if any.

Step 5: Determine where the graph is above the x-axis and where the graph is below the x-axis.

Step 6: Plot some additional points and sketch the graph.

EXAMPLE 6 Discuss the graph of the rational function $R(x) = \dfrac{x-1}{x^2-4}$.

Solution First we factor both the numerator and the denominator of R:

$$R(x) = \frac{x-1}{(x+2)(x-2)}$$

Step 1: We locate the x-intercepts by finding the zeros of the numerator. By inspection, 1 is the only x-intercept. The y-intercept is $R(0) = \frac{1}{4}$.

Step 2: Because

$$R(-x) = \frac{-x-1}{x^2-4}$$

we conclude that R is neither even nor odd. Thus, no symmetry is present.

Step 3: We locate the vertical asymptotes by finding the zeros of the denominator, $x^2 - 4$. The graph of R thus has two vertical asymptotes: the lines $x = -2$ and $x = 2$.

Step 4: The degree of the numerator is less than the degree of the denominator, so the line $y = 0$ (the x-axis) is a horizontal asymptote of the graph.

Step 5: The zero of the numerator, 1, and the zeros of the denominator, -2 and 2, divide the x-axis into four parts:

$$x < -2, \quad -2 < x < 1, \quad 1 < x < 2, \quad x > 2$$

Now we can construct Table 8.

TABLE 8

x	TEST NUMBER	VALUE OF R	GRAPH OF R
$x < -2$	-3	$R(-3) = -0.8$	Below x-axis
$-2 < x < 1$	0	$R(0) = \frac{1}{4}$	Above x-axis
$1 < x < 2$	$\frac{3}{2}$	$R(\frac{3}{2}) = -\frac{2}{7}$	Below x-axis
$x > 2$	3	$R(3) = 0.4$	Above x-axis

Step 6:   Now we are ready to put all the information together to sketch the graph. In Figure 30(a) we have plotted the points found in Table 8. Since the x-axis is a horizontal asymptote and the graph lies below the x-axis for $x < -2$, we can sketch a portion of the graph by placing a small arrow to the far left and under the x-axis. Since the line $x = -2$ is a vertical asymptote and the graph lies below the x-axis for $x < -2$, we continue the sketch with an arrow placed well below the x-axis and approaching the line $x = -2$ on the left. Similar explanations account for the positions of the other portions of the graph. In particular, note how we use the facts that the graph lies above the x-axis for $-2 < x < 1$ and below for $1 < x < 2$, and that $(1, 0)$ is an intercept, to draw the conclusion that the graph crosses the x-axis at $(1, 0)$. Figure 30(b) shows the complete sketch.

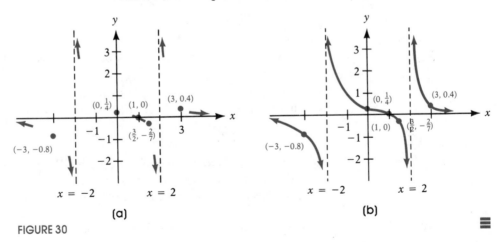

(a)   (b)

FIGURE 30

EXAMPLE 7   Discuss the graph of the rational function $R(x) = \dfrac{x^2 - 1}{x}$.

Solution   Step 1:   The graph has two x-intercepts: -1 and 1. There is no y-intercept.

Step 2: Since $R(-x) = -R(x)$, the function is odd and the graph is symmetric with respect to the origin.

Step 3: The graph of $R(x)$ has the line $x = 0$ (the y-axis) as a vertical asymptote.

Step 4: The degree of the numerator is one more than the degree of the denominator, so the graph of $R(x)$ will have an oblique asymptote. To find it, we use long division:

$$x \overline{)x^2 - 1}$$
$$\underline{x^2 }$$
$$- 1$$

Thus,

$$R(x) = \frac{x^2 - 1}{x} = x + \frac{-1}{x}$$

The line $y = x$ is an oblique asymptote of the graph.

Step 5: The zeros of the numerator are -1 and 1; the denominator has the zero 0. Thus, we divide the x-axis into four parts:

$$x < -1, \quad -1 < x < 0, \quad 0 < x < 1, \quad x > 1$$

and construct Table 9.

TABLE 9

x	TEST NUMBER	VALUE OF R	GRAPH OF R
$x < -1$	-2	$R(-2) = -\frac{3}{2}$	Below x-axis
$-1 < x < 0$	$-\frac{1}{2}$	$R(-\frac{1}{2}) = \frac{3}{2}$	Above x-axis
$0 < x < 1$	$\frac{1}{2}$	$R(\frac{1}{2}) = -\frac{3}{2}$	Below x-axis
$x > 1$	2	$R(2) = \frac{3}{2}$	Above x-axis

Step 6: Figure 31(a) shows a partial graph. We need to be careful about the oblique asymptote $y = x$, since the graph might intersect it. To see if it does, we see if $R(x)$ can equal x:

$$\frac{x^2 - 1}{x} = x$$
$$x^2 - 1 = x^2$$
$$-1 = 0 \quad \text{Impossible}$$

We conclude that the equation $(x^2 - 1)/x = x$ has no solution, so the graph of $R(x)$ does not intersect the line $y = x$. Figure 31(b) shows the complete graph.

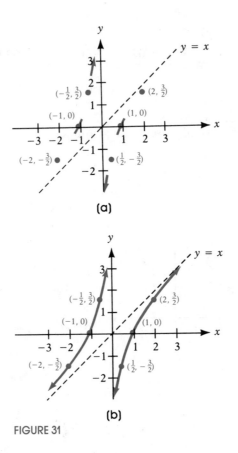

FIGURE 31

EXAMPLE 8   Discuss the graph of the rational function $R(x) = \dfrac{x^4 + 1}{x^2}$.

Solution   Step 1:   The graph has no x-intercepts and no y-intercepts.

Step 2:   Since $R(-x) = R(x)$, the function is even and the graph is symmetric with respect to the y-axis.

Step 3:   The graph of $R(x)$ has the line $x = 0$ (the y-axis) as a vertical asymptote.

Step 4:   The degree of the numerator exceeds the degree of the denominator by two, so the graph has no horizontal and no oblique asymptote.

Step 5:   The numerator has no zeros and the denominator has one zero, 0. Thus, we divide the x-axis into two parts: $x < 0$ and $x > 0$. See Table 10.

TABLE 10

x	TEST NUMBER	VALUE OF R	GRAPH OF R
$x < 0$	-1	$R(-1) = 2$	Above x-axis
$x > 0$	1	$R(1) = 2$	Above x-axis

Step 6: Figure 32 shows the graph.

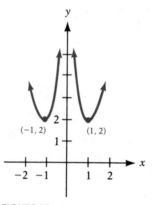

FIGURE 32

EXAMPLE 9   Discuss the graph of the rational function $R(x) = \dfrac{3x^2 - 3x}{x^2 + x - 12}$.

Solution   We factor R to get

$$R(x) = \frac{3x(x - 1)}{(x + 4)(x - 3)}$$

Step 1: The graph has two x-intercepts: 0 and 1. The y-intercept is $R(0) = 0$.

Step 2: No symmetry is present.

Step 3: The graph of R has two vertical asymptotes: $x = -4$ and $x = 3$.

Step 4: Since the degree of the numerator equals the degree of the denominator, the graph has a horizontal asymptote. To find it, we use long division:

$$
\begin{array}{r}
3 \\
x^2 + x - 12 \overline{\smash{)}3x^2 - 3x } \\
\underline{3x^2 + 3x - 36} \\
-6x + 36
\end{array}
$$

Thus,

$$R(x) = \frac{3x^2 - 3x}{x^2 + x - 12} = 3 + \frac{-6x + 36}{x^2 + x - 12}$$

Thus, the graph of R has the horizontal asymptote $y = 3$.

Step 5: The zeros of the numerator, 0 and 1, and the zeros of the denominator, -4 and 3, divide the x-axis into five parts:

$$x < -4, \quad -4 < x < 0, \quad 0 < x < 1, \quad 1 < x < 3, \quad x > 3$$

Now we can construct Table 11.

TABLE 11

x	TEST NUMBER	VALUE OF R	GRAPH OF R
$x < -4$	-5	$R(-5) = 11.25$	Above x-axis
$-4 < x < 0$	-1	$R(-1) = -\frac{1}{2}$	Below x-axis
$0 < x < 1$	$\frac{1}{2}$	$R(\frac{1}{2}) = \frac{1}{15}$	Above x-axis
$1 < x < 3$	2	$R(2) = -1$	Below x-axis
$x > 3$	4	$R(4) = 4.5$	Above x-axis

Step 6: Figure 33(a) on page 352 shows a partial graph. Notice that we have not yet used the fact that the line $y = 3$ is a horizontal asymptote, because the graph might cross (or touch) it. To find out whether it does, we solve the equation

$$R(x) = 3$$
$$\frac{3x^2 - 3x}{x^2 + x - 12} = 3$$
$$3x^2 - 3x = 3x^2 + 3x - 36$$
$$-6x = -36$$
$$x = 6$$

Thus, the graph crosses or touches the line $y = 3$ only at $x = 6$. To see whether the graph, in fact, crosses or touches the line $y = 3$, we plot an additional point to the right of $(6, 3)$. We'll use $x = 7$ to find $R(7) = \frac{63}{22} < 3$. Thus, the graph crosses $y = 3$ at $x = 6$. Because $y = 3$ is an asymptote of the graph, the graph will approach the line $y = 3$ from above as $x \to -\infty$ and the graph will approach the line $y = 3$ from below as $x \to +\infty$. See Figure 33(b). The actual graph is provided in Figure 33(c).

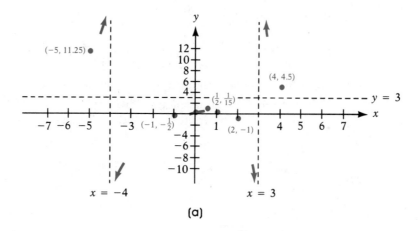

(a)

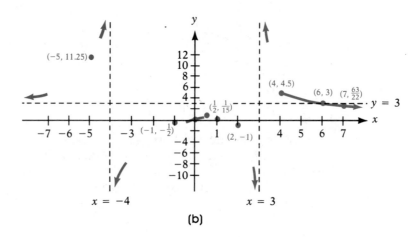

(b)

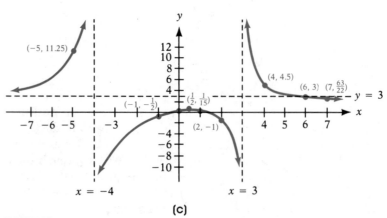

(c)

FIGURE 33

EXERCISE 5.6   *In Problems 1–10 find the domain of each rational function.*

1. $R(x) = \dfrac{x}{x - 2}$

2. $R(x) = \dfrac{5x^2}{3 - x}$

3. $H(x) = \dfrac{-4x^2}{(x - 2)(x + 1)}$

4. $G(x) = \dfrac{1}{(x + 2)(4 - x)}$

5. $F(x) = \dfrac{3x(x - 1)}{2x^2 - 5x - 3}$

6. $Q(x) = \dfrac{-x(1 - x)}{3x^2 + 5x - 2}$

7. $R(x) = \dfrac{x}{x^3 - 8}$

8. $R(x) = \dfrac{x}{x^4 - 1}$

9. $H(x) = \dfrac{3x^2 + x}{x^2 + 4}$

10. $G(x) = \dfrac{x - 3}{x^4 + 1}$

In Problems 11–20 graph each rational function using the methods of shifting, change of scale, and reflection.

11. $R(x) = \dfrac{1}{(x - 1)^2}$

12. $R(x) = \dfrac{3}{x}$

13. $H(x) = \dfrac{-2}{x + 1}$

14. $G(x) = \dfrac{2}{(x + 2)^2}$

15. $R(x) = \dfrac{1}{x^2 + 4x + 4}$

16. $R(x) = \dfrac{1}{x - 1} + 1$

17. $F(x) = 1 - \dfrac{1}{x}$

18. $Q(x) = 1 + \dfrac{1}{x}$

19. $R(x) = \dfrac{x^2 - 4}{x^2}$

20. $R(x) = \dfrac{x - 4}{x}$

In Problems 21–30 find the vertical, horizontal, and oblique asymptotes, if any, of each rational function. Do not graph.

21. $R(x) = \dfrac{x}{x + 1}$

22. $R(x) = \dfrac{2x + 3}{x - 4}$

23. $H(x) = \dfrac{x^4 + 2x^2 + 1}{3x^2 - x + 1}$

24. $G(x) = \dfrac{-x^2 + 1}{x + 5}$

25. $T(x) = \dfrac{x^3}{x^4 - 1}$

26. $P(x) = \dfrac{4x^5}{x^3 - 1}$

27. $Q(x) = \dfrac{5 - x^2}{3x^4}$

28. $F(x) = \dfrac{-2x^2 + 1}{2x^3 + 4x^2}$

29. $R(x) = \dfrac{3x^4 + 1}{x^3 + 5x}$

30. $R(x) = \dfrac{6x^2 + x - 12}{3x^2 + 5x - 2}$

In Problems 31–50 discuss the graph of each rational function.

31. $R(x) = \dfrac{x + 1}{x(x + 4)}$

32. $R(x) = \dfrac{x}{(x - 1)(x + 2)}$

33. $R(x) = \dfrac{3x + 3}{2x + 4}$

34. $R(x) = \dfrac{2x + 4}{x - 1}$

35. $R(x) = \dfrac{3}{x^2 - 4}$

36. $R(x) = \dfrac{6}{x^2 - x - 6}$

37. $P(x) = \dfrac{x^4 + x^2 + 1}{x^2 - 1}$

38. $Q(x) = \dfrac{x^4 - 1}{x^2 - 4}$

39. $H(x) = \dfrac{x^3 - 1}{x^2 - 9}$

40. $G(x) = \dfrac{x^3 + 1}{x^2 + 2x}$

41. $R(x) = \dfrac{x^2}{x^2 + x - 6}$

42. $R(x) = \dfrac{x^2 + x - 12}{x^2 - 4}$

43. $G(x) = \dfrac{x}{x^2 - 4}$

44. $G(x) = \dfrac{3x}{x^2 - 1}$

45. $R(x) = \dfrac{3}{(x - 1)(x^2 - 4)}$

46. $R(x) = \dfrac{-4}{(x + 1)(x^2 - 9)}$

47. $H(x) = 4\dfrac{x^2 - 1}{x^4 - 16}$

48. $H(x) = \dfrac{x^2 + 1}{x^4 - 1}$

49. $F(x) = \dfrac{x^2 - 3x - 4}{x + 2}$

50. $F(x) = \dfrac{x^2 + 3x + 2}{x - 1}$

51. If the graph of a rational function R has the vertical asymptote $x = 4$, then the factor $x - 4$ must be present in the denominator of R. Explain why.

52. If the graph of a rational function R has the horizontal asymptote $y = 2$, then the degree of the numerator of R equals the degree of the denominator of R. Explain why.

▤ 5.7 COMPLEX POLYNOMIALS; FUNDAMENTAL THEOREM OF ALGEBRA

COMPLEX VARIABLE
COMPLEX FUNCTION

DOMAIN
INDEPENDENT VARIABLE
DEPENDENT VARIABLE
COMPLEX POLYNOMIAL
FUNCTION

A variable z in the complex number system is referred to as a **complex variable**. A **complex function** f is a rule that assigns to each complex number z in a set D a unique complex number $w = f(z)$. Here, D is called the **domain** of f, z is called the **independent (complex) variable**, and w is the **dependent (complex) variable**. A **complex polynomial function** $f(z)$ of degree n is a complex function of the form

$$f(z) = a_n z^n + a_{n-1} z^{n-1} + \cdots + a_1 z + a_0 \tag{1}$$

LEADING COEFFICIENT

ZERO

where $a_n, a_{n-1}, \ldots, a_1, a_0$ are complex numbers, $a_n \neq 0$, and n is a nonnegative integer. Here, a_n is called the **leading coefficient** of f. A complex number r is called a (complex) **zero** of a complex function f if $f(r) = 0$.

The Division Algorithm for Polynomials (see Section 5.3) is true for complex polynomials. As a result, the Remainder Theorem and Factor Theorem are also true for complex numbers. In fact, the process of synthetic division works for complex polynomials.

EXAMPLE 1   Use synthetic division and the Factor Theorem to show that $1 + 2i$ is a zero of

$$f(z) = (1 + i)z^2 + (2 - i)z + (3 - 4i)$$

Solution   We use synthetic division and divide $f(z)$ by $z - (1 + 2i)$:

$$
\begin{array}{r}
-(1 + 2i)\overline{)1 + i \quad 2 - i \quad 3 - 4i} \\
1 - 3i \quad 3 - 4i \\
\hline
1 + i \quad 1 + 2i \quad 0
\end{array}
$$

Thus, $z - (1 + 2i)$ is a factor, and $1 + 2i$ is a zero.   ≡

Of course, we could have shown that $1 + 2i$ is a zero of the polynomial $f(z)$ in Example 1 by using substitution as follows:

$$f(1 + 2i) = (1 + i)(1 + 2i)^2 + (2 - i)(1 + 2i) + (3 - 4i)$$
$$= -7 + i + 4 + 3i + 3 - 4i = 0$$

In Chapter 2 we discovered that some quadratic equations have no real solutions, but that in the complex number system, every quadratic equation has a solution—either real or complex. The next result, proved by Karl Friedrich Gauss (1777–1855) when he was 22 years of age,* gives an extension to complex polynomials. In fact, this result is so important and useful, it has become known as the **Fundamental Theorem of Algebra**.

THEOREM

FUNDAMENTAL THEOREM OF
ALGEBRA

Every complex polynomial function of degree $n \geq 1$ has at least one complex zero.   ≡

We shall not prove this result, as the proof is beyond the scope of this book. However, using the Fundamental Theorem of Algebra and the Factor Theorem, we can prove the following result:

THEOREM   Every complex polynomial function $f(z)$ of degree $n \geq 1$ can be factored into n linear factors of the form

$$f(z) = a_n(z - r_1)(z - r_2) \cdots \cdots (z - r_n) \tag{2}$$

where $a_n, r_1, r_2, \ldots, r_n$ are complex numbers.

Proof   Let

$$f(z) = a_n z^n + a_{n-1} z^{n-1} + \cdots + a_1 z + a_0$$

*In all, Gauss gave four different proofs of this theorem, the first one in 1799 being the subject of his doctoral dissertation.

By the Fundamental Theorem of Algebra, f has at least one zero, call it r_1. Then, by the Factor Theorem, $z - r_1$ is a factor, and

$$f(z) = (z - r_1)q_1(z)$$

where $q_1(z)$ is a complex polynomial of degree $n - 1$ whose leading coefficient is a_n. Again, by the Fundamental Theorem of Algebra, the complex polynomial $q_1(z)$ has at least one zero, say r_2. By the Factor Theorem, $q_1(z)$ has the factor $z - r_2$, so that

$$q_1(z) = (z - r_2)q_2(z)$$

where $q_2(z)$ is a complex polynomial of degree $n - 2$ whose leading coefficient is a_n. Consequently,

$$f(z) = (z - r_1)(z - r_2)q_2(z)$$

Repeating this argument n times, we finally arrive at

$$f(z) = (z - r_1)(z - r_2) \cdot \cdots \cdot (z - r_n)q_n(z)$$

where $q_n(z)$ is a complex polynomial of degree $n - n = 0$ whose leading coefficient is a_n. Thus, $q_n(z) = a_n z^0 = a_n$, and so,

$$f(z) = a_n(z - r_1)(z - r_2) \cdot \cdots \cdot (z - r_n) \qquad \blacksquare$$

Based on Equation (2), a complex polynomial $f(z)$ of degree n has n linear factors. These n linear factors need not be distinct; some may be repeated more than once. When a linear factor $z - r$ appears exactly m times in the factored form $f(z)$, then r is called a **zero of multiplicity** ZERO OF MULTIPLICITY m **m** of f. Thus, if a zero of multiplicity m is counted m times, it follows that a complex polynomial $f(z)$ of degree $n \geq 1$ has exactly n zeros.

EXAMPLE 2   The complex polynomial function

$$f(z) = (2 + i)(z - 5)^3(z + i)^2[z - (3 + i)]^4(z - i)$$

of degree ten has $2 + i$ as leading coefficient. Its zeros are listed below.

$$
\begin{array}{rl}
5: & \text{Multiplicity 3} \\
-i: & \text{Multiplicity 2} \\
3 + i: & \text{Multiplicity 4} \\
i: & \underline{\text{Multiplicity 1}} \\
\text{Degree:} & \qquad 10
\end{array}
$$

$\blacksquare$

EXAMPLE 3   Form a complex polynomial $f(z)$ of degree three that has the zeros

$$
\begin{array}{rl}
1 + i: & \text{Multiplicity 1} \\
-i: & \text{Multiplicity 2}
\end{array}
$$

Solution   Since $1 + i$ is a zero of multiplicity 1 and $-i$ is a zero of multiplicity 2, then $z - (1 + i)$ and $(z + i)^2$ are factors of f. Thus, $f(z)$ is of the form

$$\begin{aligned} f(z) &= [z - (1 + i)](z + i)^2 \\ &= [z - (1 + i)](z^2 + 2iz - 1) \\ &= z^3 + (-1 + i)z^2 + (1 - 2i)z + 1 + i \end{aligned}$$

Although there are other complex polynomials having the three required zeros, the only ones of degree three will be $f(z)$ or $kf(z)$, where $k \neq 0$ is any complex number.   ≡

Complex Polynomials with Real Coefficients

We can use the Fundamental Theorem of Algebra to obtain valuable information about the zeros of complex polynomials whose coefficients are real numbers.

THEOREM   Let $f(z)$ be a complex polynomial whose coefficients are real numbers.
CONJUGATE PAIRS   If $r = a + bi$ is a zero of f, then the complex conjugate $\bar{r} = a - bi$ also is a zero of f.

In other words, for complex polynomials whose coefficients are real numbers, the zeros occur in conjugate pairs.

Proof   Let

$$f(z) = a_n z^n + a_{n-1} z^{n-1} + \cdots + a_1 z + a_0$$

where $a_n, a_{n-1}, \ldots, a_1, a_0$ are real numbers and $a_n \neq 0$. If r is a zero of f, then $f(r) = 0$, so that

$$a_n r^n + a_{n-1} r^{n-1} + \cdots + a_1 r + a_0 = 0$$

We take the conjugate of both sides to get

$$\overline{a_n r^n + a_{n-1} r^{n-1} + \cdots + a_1 r + a_0} = \overline{0}$$

$$\overline{a_n r^n} + \overline{a_{n-1} r^{n-1}} + \cdots + \overline{a_1 r} + \overline{a_0} = \overline{0} \qquad \text{The conjugate of a sum equals the sum of the conjugates (see Section 1.9).}$$

$$\overline{a_n}(\bar{r})^n + \overline{a_{n-1}}(\bar{r})^{n-1} + \cdots + \overline{a_1}\bar{r} + \overline{a_0} = \overline{0} \qquad \text{The conjugate of a product equals the product of the conjugates.}$$

$$a_n(\bar{r})^n + a_{n-1}(\bar{r})^{n-1} + \cdots + a_1\bar{r} + a_0 = 0 \qquad \text{The conjugate of a real number equals the real number.}$$

This last equation states that $f(\bar{r}) = 0$; that is, $\bar{r}$ is a zero of f.   ≡

The value of this result should be clear. Once we know that, say $3 + 4i$ is a zero, we know that $3 - 4i$ also is a zero. This result has an important corollary.

COROLLARY   A complex polynomial $f(z)$ of odd degree with real coefficients has at least one zero that is a real number.

Proof   Because complex zeros occur as conjugate pairs in a complex polynomial with real coefficients, there will always be an even number of zeros that are not real numbers. Consequently, since f is of odd degree, one of its zeros has to be a real number. ≡

For example, the polynomial $f(z) = z^5 - 3z^4 + 4z^3 - 5$ has at least one zero that is a real number since f is of degree five (odd) and has real coefficients.

Now we prove the theorem we conjectured earlier in Section 5.4.

THEOREM   Every polynomial function with real coefficients can be uniquely factored over the real numbers into a product of linear factors and irreducible quadratic factors.

Proof   Every complex polynomial $f(z)$ of degree n has exactly n zeros and can be factored into a product of n linear factors. If its coefficients are real, then those zeros that are complex numbers will always occur as conjugate pairs. As a result, if $r = a + bi$ is a complex zero, then so is $\bar{r} = a - bi$. Consequently, the linear factors, $z - r$ and $z - \bar{r}$ of $f(z)$, when multiplied, give rise to

$$(z - r)(z - \bar{r}) = z^2 - (r + \bar{r})z + r\bar{r} = z^2 - 2az + a^2 + b^2$$

This second-degree polynomial has real coefficients and is irreducible (over the real numbers). Thus, the factors of f are either linear or irreducible quadratic factors. ≡

EXAMPLE 4   A polynomial $f(z)$ of degree five whose coefficients are real numbers has the zeros $1, 5i$, and $1 + i$. Find the remaining two zeros and find $f(z)$.

Solution   Since complex zeros appear as conjugate pairs, it follows that $-5i$, the conjugate of $5i$, and $1 - i$, the conjugate of $1 + i$, are the two missing zeros. The Factor Theorem says that if r is a zero, then $z - r$ is a factor. Thus, $z - 1, z - 5i, z + 5i, z - (1 + i)$, and $z - (1 - i)$ are factors of $f(z)$, so that

$$\begin{aligned}
f(z) &= (z - 1)(z - 5i)(z + 5i)[z - (1 + i)][z - (1 - i)] \\
&= (z - 1)(z^2 + 25)[(z - 1)^2 + 1] \\
&= (z - 1)(z^2 + 25)(z^2 - 2z + 2) \\
&= z^5 - 3z^4 + 29z^3 - 77z^2 + 100z - 50
\end{aligned}$$

Although there are other polynomials having the three required zeros, the only ones of degree five whose coefficients are real numbers are $f(z)$ or $kf(z)$, where $k \neq 0$ is some real number. ∎

EXERCISE 5.7   *In Problems 1–6 evaluate each complex polynomial function $f(z)$ at $z = 1 + i$.*

1. $f(z) = iz - 2$
2. $f(z) = 2z - i$
3. $f(z) = 3z^2 - z$
4. $f(z) = (4 + i)z^2 + 5 - 2i$
5. $f(z) = z^3 + iz - 1 + i$
6. $f(z) = iz^3 - 2z^2 + 1$

In Problems 7–12 use synthetic division to find the value of $f(r)$.

7. $f(z) = 5z^5 - iz^4 + 2; \quad r = 1 + i$
8. $f(z) = iz^4 + (2 + i)z^2 - z; \quad r = 1 - i$
9. $f(z) = (1 + i)z^4 - z^3 + iz; \quad r = 2 - i$
10. $f(z) = 2iz^3 + 8z^2 - 4iz + 1; \quad r = 2 + i$
11. $f(z) = iz^5 + iz^3 + iz; \quad r = 1 + 2i$
12. $f(z) = z^4 + z^2 + 1; \quad r = 1 - 2i$

In Problems 13–18 form a complex polynomial $f(z)$ having the given degree and the given zeros.

13. Degree three;   zeros: $1 + 2i$, multiplicity one; 3, multiplicity two
14. Degree three;   zeros: $-i$, multiplicity one; $1 + 2i$, multiplicity two
15. Degree three;   zeros: 2, multiplicity one; $-i$, multiplicity one; $1 + i$, multiplicity one
16. Degree three;   zeros: i, multiplicity one; $4 - i$, multiplicity one; $2 + i$, multiplicity one
17. Degree four;   zeros: 3, multiplicity two; $-i$, multiplicity two
18. Degree four;   zeros: 1, multiplicity three; $1 + i$, multiplicity one

In Problems 19–24 information is given about a complex polynomial $f(z)$ whose coefficients are real numbers. Find the missing zeros of f and find $f(z)$.

19. Degree three;   zeros: 2, $1 - i$
20. Degree three;   zeros: 1, $2 + i$
21. Degree four;   zeros: i, $1 + i$
22. Degree four;   zeros: 1, 2, $2 + i$
23. Degree five;   zeros: 1, i, $2i$
24. Degree five;   zeros: 0, 1, 2, i

In Problems 25 and 26 tell why the facts given are contradictory.

25. $f(z)$ is a complex polynomial of degree three whose coefficients are real numbers; its zeros are $1 + i$, $1 - i$, and $2 + i$.
26. $f(z)$ is a complex polynomial of degree three whose coefficients are real numbers; its zeros are 2, i, and $1 + i$.

27. $f(z)$ is a complex polynomial of degree four whose coefficients are real numbers; three of its zeros are 2, $1 + 2i$, and $1 - 2i$. Explain why the remaining zero must be a real number.

28. $f(z)$ is a complex polynomial of degree four whose coefficients are real numbers; two of its zeros are -3 and $4 - i$. Explain why one of the remaining zeros must be a real number. Write down one of the missing zeros.

29. Find all the zeros of $f(z) = z^3 - 1$.

30. Find all the zeros of $f(z) = z^4 - 1$.

C H A P T E R   R E V I E W

VOCABULARY

quadratic function
parabolas
graph opens upward
graph opens
   downward
vertex
axis of symmetry
maximum value
minimum value
polynomial function of
   degree n
power function of
   degree n
Division Algorithm for
   Polynomials
dividend
divisor
quotient
remainder
Remainder Theorem
Factor Theorem
synthetic division
nested form of a
   polynomial function

zero of a function
repeated or multiple
   zero
number of zeros
Descartes' Rule of
   Signs
Rational Zeros
   Theorem
depressed equation
irreducible quadratic
upper bound
lower bound
bounds on zeros
Intermediate Value
   Theorem
estimating the zeros of
   a function
rational function
unbounded in the
   positive direction
approaches plus
   infinity ($\rightarrow +\infty$)
unbounded in the
   negative direction

approaches minus
   infinity ($\rightarrow -\infty$)
horizontal asymptote
vertical asymptote
oblique asymptote
proper rational
   function
improper rational
   function
complex variable
complex function
domain
independent variable
dependent variable
complex polynomial
   function
leading coefficient
zero
Fundamental
   Theorem of Algebra
zero of multiplicity m
conjugate pairs

FILL-IN-THE BLANK QUESTIONS

1. The graph of a quadratic function is called a(n) _____. Its lowest or highest point is called the _____.

2. In the process of long division,

   (Divisor)(Quotient) + _____ = _____

3. When a polynomial function f is divided by $x - c$, the remainder is

   _____.

4. A polynomial function f has the factor $x - c$ if and only if

    _____.

5. A number r for which $f(r) = 0$ is called a(n) _____ of the function f.

6. The polynomial function $f(x) = x^5 - 2x^3 + x^2 + x - 1$ has either

    _____ or _____ positive zeros; it has _____

    or _____ negative zeros.

7. The possible rational zeros of $f(x) = 2x^5 - x^3 + x^2 - x + 1$ are

    _____.

8. The line _____ is a horizontal asymptote of $R(x) = \dfrac{x^3 - 1}{x^3 + 1}$.

9. The line _____ is a vertical asymptote of $R(x) = \dfrac{x^3 - 1}{x^3 + 1}$.

10. If $3 + 4i$ is a zero of a polynomial of degree five with real coefficients, then so is _____.

REVIEW EXERCISES

In Problems 1–10 graph each quadratic function by determining whether its graph opens upward or downward, and by finding its vertex, axis of symmetry, y-intercept, and x-intercepts, if any.

1. $f(x) = (x - 2)^2 + 2$
2. $f(x) = (x + 1)^2 - 4$
3. $f(x) = \frac{1}{4}x^2 - 16$
4. $f(x) = -\frac{1}{2}x^2 + 2$
5. $f(x) = -4x^2 + 4x$
6. $f(x) = 9x^2 - 6x + 3$
7. $f(x) = \frac{9}{2}x^2 + 3x + 1$
8. $f(x) = -x^2 + x + \frac{1}{2}$
9. $f(x) = 3x^2 + 4x - 1$
10. $f(x) = -2x^2 - x + 4$

In Problems 11–16 graph each function using the techniques of shifting, change of scale, and reflection.

11. $f(x) = (x + 2)^3$
12. $f(x) = -x^3 + 3$
13. $f(x) = -(x - 1)^4$
14. $f(x) = (x - 1)^4 - 2$
15. $f(x) = (x - 1)^4 + 2$
16. $f(x) = (1 - x)^3$

In Problems 17–20 use synthetic division to find the quotient q(x) and remainder R when f(x) is divided by g(x).

17. $f(x) = 8x^3 - 2x^2 + x - 4$; $g(x) = x - 1$
18. $f(x) = 2x^3 + 4x^2 - 6x + 5$; $g(x) = x - 2$
19. $f(x) = x^4 - 2x^3 + x - 1$; $g(x) = x + 2$
20. $f(x) = x^4 - x^2 + 3x$; $g(x) = x + 1$

In Problems 21–24 write each polynomial in nested form. Use a calculator to evaluate each polynomial at $x = 1.5$. Avoid using any memory key.

21. $f(x) = 8x^3 - 2x^2 + x - 4$
22. $f(x) = 2x^3 + 4x^2 - 6x + 5$
23. $f(x) = x^4 - 2x^3 + x - 1$
24. $f(x) = x^4 - x^2 + 3x$

In Problems 25–34 use Descartes' Rule of Signs and the Rational Zeros Theorem to find all the real zeros of each polynomial function. Use the zeros to factor f over the real numbers.

25. $f(x) = x^3 - 3x^2 - 6x + 8$

26. $f(x) = x^3 - x^2 - 10x - 8$

27. $f(x) = 4x^3 + 4x^2 - 7x + 2$

28. $f(x) = 4x^3 - 4x^2 - 7x - 2$

29. $f(x) = x^4 - 4x^3 + 9x^2 - 20x + 20$

30. $f(x) = x^4 + 6x^3 + 11x^2 + 12x + 18$

31. $f(x) = 2x^4 + 2x^3 - 11x^2 + x - 6$

32. $f(x) = 3x^4 + 3x^3 - 17x^2 + x - 6$

33. $f(x) = 2x^4 + 7x^3 + x^2 - 7x - 3$

34. $f(x) = 2x^4 + 7x^3 - 5x^2 - 28x - 12$

In Problems 35–44 find the intercepts of each polynomial f(x). Find the numbers x for which the graph of f is above the x-axis and those for which it is below the x-axis. Obtain several other points on the graph and connect them with a smooth curve. [Hint: Use the answers found in Problems 25–34.]

35. $f(x) = x^3 - 3x^2 - 6x + 8$

36. $f(x) = x^3 - x^2 - 10x - 8$

37. $f(x) = 4x^3 + 4x^2 - 7x + 2$

38. $f(x) = 4x^3 - 4x^2 - 7x - 2$

39. $f(x) = x^4 - 4x^3 + 9x^2 - 20x + 20$

40. $f(x) = x^4 + 6x^3 + 11x^2 + 12x + 18$

41. $f(x) = 2x^4 + 2x^3 - 11x^2 + x - 6$

42. $f(x) = 3x^4 + 3x^3 - 17x^2 + x - 6$

43. $f(x) = 2x^4 + 7x^3 + x^2 - 7x - 3$

44. $f(x) = 2x^4 + 7x^3 - 5x^2 - 28x - 12$

In Problems 45–48 use the Intermediate Value Theorem to show that each polynomial has a zero in the given interval.

45. $f(x) = 3x^3 - x - 1;$ $[0, 1]$

46. $f(x) = 2x^3 - x^2 - 3;$ $[1, 2]$

47. $f(x) = 8x^4 - 4x^3 - 2x - 1;$ $[0, 1]$

48. $f(x) = 3x^4 + 4x^3 - 8x - 2;$ $[1, 2]$

In Problems 49–52 find upper and lower bounds to the zeros of each polynomial function.

49. $f(x) = 2x^3 - x^2 - 4x + 2$

50. $f(x) = 2x^3 + x^2 - 10x - 5$

51. $f(x) = 2x^3 - 7x^2 - 10x + 35$

52. $f(x) = 3x^3 - 7x^2 - 6x + 14$

C   *In Problems 53–56 each polynomial has exactly one positive zero. Approximate the zero to within 0.01.*

53. $f(x) = x^3 - x - 2$

54. $f(x) = 2x^3 - x^2 - 3$

55. $f(x) = 8x^4 - 4x^3 - 2x - 1$

56. $f(x) = 3x^4 + 4x^3 - 8x - 2$

In Problems 57–66 discuss each rational function following the six steps outlined in Section 5.6.

57. $R(x) = \dfrac{2x - 6}{x}$

58. $R(x) = \dfrac{4 - x}{x}$

59. $H(x) = \dfrac{x + 2}{x(x - 2)}$

60. $H(x) = \dfrac{x}{x^2 - 1}$

61. $R(x) = \dfrac{x^2}{(x - 1)^2}$

62. $R(x) = \dfrac{(x - 3)^2}{x^2}$

63. $F(x) = \dfrac{x^3}{x^2 - 4}$

64. $F(x) = \dfrac{3x^3}{(x - 1)^2}$

65. $R(x) = \dfrac{2x^4}{(x - 1)^2}$

66. $R(x) = \dfrac{x^4}{x^2 - 9}$

In Problems 67–70 form a complex polynomial $f(z)$ having the given degree and the given zeros.

67. Degree four; zeros: 1, multiplicity two; i, multiplicity one; 2, multiplicity one

68. Degree four; zeros: i, multiplicity two; 3, multiplicity two

69. Degree three; zeros: $1 + i$, 2, 3, each of multiplicity one

70. Degree three; zeros: 1, $1 + i$, $1 + 2i$, each of multiplicity one

In Problems 71–74 information is given about a complex polynomial $f(z)$ whose coefficients are real numbers. Find the missing zeros of f and find $f(z)$.

71. Degree three; zeros: $1 + i$, 1

72. Degree three; zeros: $3 - 4i$, 2

73. Degree four; zeros: i, $1 + i$

74. Degree four; zeros: 1, 2, $1 + i$

75. Find the quotient and remainder if $x^4 + 2x^3 - 7x^2 - 8x + 12$ is divided by $(x - 2)(x - 1)$.

76. Find the quotient and remainder if $x^4 + 2x^3 - 4x^2 - 5x - 6$ is divided by $(x - 2)(x + 3)$.

In Problems 77–80 solve each equation in the complex number system.

77. $x^3 - x^2 - 8x + 12 = 0$

78. $x^3 - 3x^2 - 4x + 12 = 0$

79. $3x^4 - 4x^3 + 4x^2 - 4x + 1 = 0$

80. $x^4 + 4x^3 + 2x^2 - 8x - 8 = 0$

81. Find the value of $f(x) = 12x^6 - 8x^4 + 1$ at $x = 4$.

82. Find the value of $f(x) = -16x^3 + 18x^2 - x + 2$ at $x = -2$.

83. Is $\frac{1}{3}$ a zero of $f(x) = 2x^3 + 3x^2 - 6x + 7$?

84. Is $\frac{1}{3}$ a zero of $f(x) = 4x^3 - 5x^2 - 3x + 1$?

85. Is $\frac{3}{5}$ a zero of $f(x) = 2x^6 - 5x^4 + x^3 - x + 1$?

86. Is $\frac{2}{3}$ a zero of $f(x) = x^7 + 6x^5 - x^4 + x + 2$?

In Problems 87 and 88 use Descartes' Rule of Signs to determine how many positive and how many negative zeros each polynomial function might have. Do not attempt to find the actual zeros.

87. $f(x) = 12x^8 - x^7 + 6x^4 - x^3 + x - 3$

88. $f(x) = -6x^5 + x^4 + 2x^3 - x + 1$

89. List all the potential rational zeros of

$$f(x) = 12x^8 - x^7 + 6x^4 - x^3 + x - 3$$

90. List all the potential rational zeros of

$$f(x) = -6x^5 + x^4 + 2x^3 - x + 1$$

91. Find upper and lower bounds to the zeros of

$$f(x) = 4x^5 - 3x^4 + 8x^2 + x + 2$$

92. Find upper and lower bounds to the zeros of

$$f(x) = 8x^6 - x^4 + 6x^2 + 24x + 15$$

93. Find the point on the line $y = x$ that is closest to the point $(3, 1)$. [*Hint:* Find the minimum value of the function $f(x) = d^2$, where d is the distance from $(3, 1)$ to a point on the line.]

94. Find the point on the line $y = x + 1$ that is closest to the point $(4, 1)$.

95. A horizontal bridge is in the shape of a parabolic arch. Given the information shown in the illustration, what is the height h of the arch 2 feet from shore?

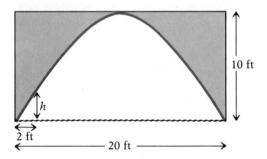

EXPONENTIAL AND LOGARITHMIC FUNCTIONS

6.1 Exponential Functions
6.2 Compound Interest
6.3 Logarithms
6.4 Logarithmic Functions
6.5 Growth and Decay
   Chapter Review

Until now, our study of functions has concentrated primarily on polynomial and rational functions. These functions belong to the class of **algebraic functions**—that is, functions that can be expressed in terms of sums, differences, products, quotients, powers, or roots of polynomials. Functions that are not algebraic are termed **transcendental** (they transcend, or go beyond, algebraic functions).

In this chapter we continue our study of functions with the exponential and logarithmic functions. These functions, which occur frequently in a wide variety of applications, belong to the class of transcendental functions.

▤ 6.1 EXPONENTIAL FUNCTIONS

In Chapter 1 we gave a definition for raising a real number a to a rational power. Based on that discussion, we gave meaning to expressions of the form

$$a^r$$

where the base a is a positive real number and the exponent r is a rational number.

But what is the meaning of a^q, where the base a is a positive real number and the exponent q is an irrational number? Although a rigorous definition requires methods discussed in calculus, the basis for the definition is easy to follow: Select a rational number r that is formed by truncating (removing) all but a finite number of digits from the irrational number q. Then

$$a^q \approx a^r$$

For example, take the irrational number $\pi = 3.14159\ldots$. Then, an approximation to a^π is

$$a^\pi \approx a^{3.14} = a^{314/100} = a^{157/50} = (\sqrt[50]{a})^{157}$$

in which the digits after the hundredths position have been removed. A better approximation would be

$$a^\pi \approx a^{3.1415} = a^{31,415/10,000} = (\sqrt[2000]{a})^{6283}$$

in which the digits after the ten-thousandths position have been removed. Continuing in this way, we can obtain approximations to a^π to any desired degree of accuracy.

✐ Most calculators that do more than just arithmetic operations have a $\boxed{y^x}$ key. To use this key, first enter the base y, then press the $\boxed{y^x}$ key, then enter x, and then press the $\boxed{=}$ key.

C EXAMPLE 1 Using a calculator with a $\boxed{y^x}$ key, evaluate:

(a) $2^{1.4}$ (b) $2^{1.41}$ (c) $2^{1.414}$ (d) $2^{1.4142}$ (e) $2^{\sqrt{2}}$

Solution (a) $2^{1.4} \approx 2.6390158$ (b) $2^{1.41} \approx 2.6573716$
(c) $2^{1.414} \approx 2.6647497$ (d) $2^{1.4142} \approx 2.6651191$
(e) $2^{\sqrt{2}} \approx 2.6651441$ ▤

Based on the ideas just presented, we now have an intuitive meaning of a^x, where x is a real number. The familiar laws of exponents hold for real exponents.

THEOREM If s, t, $a > 0$, and $b > 0$ are real numbers, then

LAWS OF EXPONENTS

$$a^s \cdot a^t = a^{s+t} \qquad (a^s)^t = a^{st} \qquad (ab)^s = a^s \cdot b^s$$

$$1^s = 1, \qquad a^{-s} = \frac{1}{a^s} = \left(\frac{1}{a}\right)^s \qquad a^0 = 1 \tag{1}$$

We are now ready for the following definition.

EXPONENTIAL FUNCTION

An **exponential function** is a function of the form

$$f(x) = a^x$$

where a is a positive real number and $a \neq 1$. The domain of f is the set of all real numbers.

We exclude the base $a = 1$ because this function is, in fact, the constant function $f(x) = 1^x = 1$. We also need to exclude bases that are negative because, otherwise, we would have to exclude many values of x from the domain, such as $x = \frac{1}{2}$, $x = \frac{3}{4}$, and so on. [Recall that $(-2)^{1/2}$, $(-3)^{3/4}$, and so on, are not defined.]

Because $f(x) = a^x$ is a function, it follows that:

$$\text{If } u = v, \text{ then } a^u = a^v \tag{2}$$

We shall have occasion to use this fact later.

The Graphs of Exponential Functions

First, let's obtain the graph of a particular exponential function.

EXAMPLE 2   Graph the exponential function $f(x) = 2^x$.

Solution   The domain of $f(x) = 2^x$ consists of all real numbers. We begin by locating some points on the graph of $f(x) = 2^x$. See Table 1, where a calculator with a $\boxed{y^x}$ key was used to find some of the entries.

TABLE 1

x	-10	-5	-4	-3	-2
$f(x) = 2^x$	$2^{-10} \approx 0.00098$	$2^{-5} \approx 0.031$	$2^{-4} = 0.0625$	$2^{-3} = 0.125$	$2^{-2} = 0.25$
x	-1	-0.5	0	0.5	1
$f(x) = 2^x$	$2^{-1} = 0.5$	$2^{-0.5} \approx 0.707$	$2^0 = 1.0$	$2^{0.5} \approx 1.414$	$2^1 = 2.0$
x	2	$\sqrt{7} \approx 2.6458$	3	4	10
$f(x) = 2^x$	$2^2 = 4.0$	$2^{\sqrt{7}} \approx 6.2582$	$2^3 = 8$	$2^4 = 16$	$2^{10} = 1024$

Notice that $2^x > 0$ for all x. From this we conclude that the graph has no x-intercepts; in fact, it follows that the graph will lie above the x-axis. As Table 1 indicates, the y-intercept is 1. Table 1 also indicates that, as $x \to -\infty$, the value of $f(x) = 2^x$ gets closer and closer to zero. Thus, the x-axis is a horizontal asymptote to the graph as $x \to -\infty$. Look again at Table 1. As $x \to +\infty$, $f(x) = 2^x$ grows very quickly. In fact, the graph of $f(x) = 2^x$ is rising very rapidly. Thus, f is an increasing function and hence is one-to-one. Using the information just gathered, we plot some of the points from Table 1 and connect them with a smooth curve. See Figure 1.

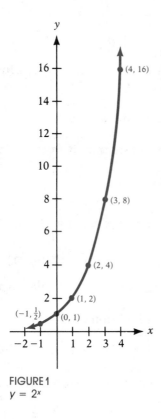

FIGURE 1
$y = 2^x$

As we shall see, graphs that look like the one in Figure 1 occur very frequently in a diverse variety of situations. For example, look at the graph in Figure 2, which illustrates the amount of money paid to doctors in the Medicare program. Researchers might conclude from this graph that Medicare payments "behave exponentially"; that is, the graph exhibits "rapid growth." We shall have more to say about situations that lead to exponential graphs later in the chapter. For now, we continue to seek properties of the exponential functions.

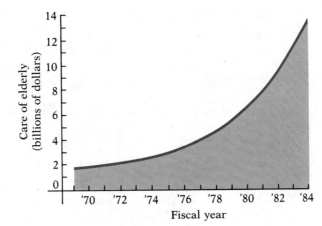

FIGURE 2
Medicare payments to doctors
Source: Office of Management and Budget

The graph of $f(x) = 2^x$ in Figure 1 is typical of all exponential functions that have a base larger than 1. Such functions are increasing functions and hence are one-to-one. Their graphs lie above the x-axis, pass through the point $(0, 1)$, and thereafter rise rapidly as $x \to +\infty$. As $x \to -\infty$, the x-axis is a horizontal asymptote. There are no vertical asymptotes. Finally, the graphs are smooth, with no corners or gaps. Figure 3 illustrates the graphs of two more exponential functions whose bases are larger than 1. Notice that the choice of a larger base results in a steeper graph.

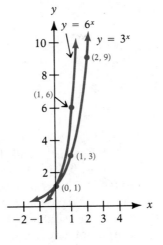

FIGURE 3

To graph an exponential function whose base is less than 1, we use the fact that

$$\left(\frac{1}{a}\right)^x = \frac{1}{a^x} = a^{-x}$$

and the idea of a reflection about the y-axis.

EXAMPLE 3   Graph the exponential function $f(x) = (\frac{1}{2})^x$.

Solution   First, we observe that

$$f(x) = \left(\frac{1}{2}\right)^x = 2^{-x}$$

Thus, the graph of $f(x) = (\frac{1}{2})^x$ is the reflection about the y-axis of the graph of $y = 2^x$. See Figure 4.

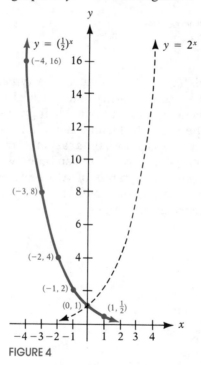

FIGURE 4

The graph of $f(x) = (\frac{1}{2})^x$ in Figure 4 is typical of all exponential functions that have a base between 0 and 1. Such functions are decreasing, one-to-one functions. Their graphs lie above the x-axis and pass through the point (0, 1). The graphs rise rapidly as $x \to -\infty$. As $x \to +\infty$, the x-axis is a horizontal asymptote. There are no vertical asymptotes. Finally, the graphs are smooth, with no corners or gaps. Figure 5 illustrates the graphs of two more exponential functions whose bases are between 0 and 1. Notice that the choice of a base closer to zero results in a steeper graph.

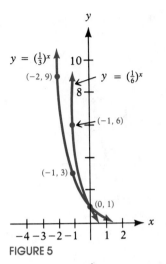

FIGURE 5

The techniques of shifting, change of scale, and reflection may be used to graph many exponential functions.

EXAMPLE 4 Graph each function:

(a) $f(x) = 2^{-x} - 3$ (b) $g(x) = -(2^{x-3})$

Solution (a) Figure 6 shows the various steps.

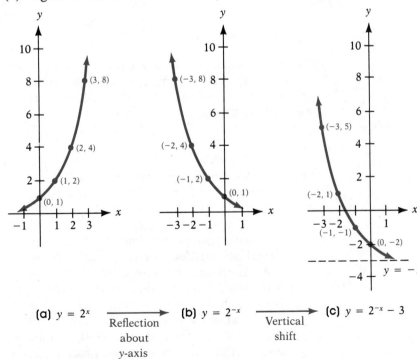

(a) $y = 2^x$ $\xrightarrow{\text{Reflection about } y\text{-axis}}$ **(b)** $y = 2^{-x}$ $\xrightarrow{\text{Vertical shift}}$ **(c)** $y = 2^{-x} - 3$

FIGURE 6

(b) Figure 7 shows the various steps.

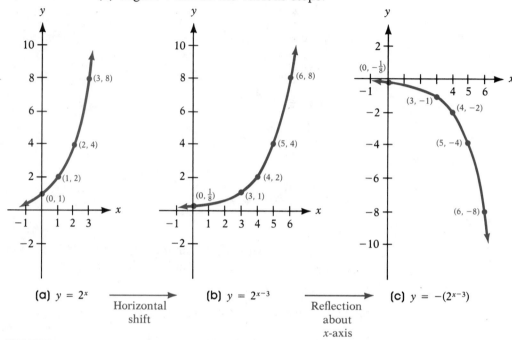

(a) $y = 2^x$ → Horizontal shift → **(b)** $y = 2^{x-3}$ → Reflection about x-axis → **(c)** $y = -(2^{x-3})$

FIGURE 7

The Base e

NUMBER e As we shall see shortly, many problems that occur in nature require the use of an exponential function whose base is a certain irrational number, symbolized by the letter e. Correct to five decimal places, the **number e** is approximated as

$$e \approx 2.71828$$

Let's look now at one way of arriving at this important number e. (We'll see another way in the next section.) We begin by carefully graphing $y = 2^x$ and $y = 3^x$ on the same set of coordinate axes. See Figure 8(a). The dashed lines shown are the **tangent lines** (from the Latin *tangere*, meaning *to touch*) to each graph at the point (0, 1). (Intuitively, these are lines that just "touch" the graph at a point; calculus is needed to give a precise definition of a tangent.)

TANGENT LINES   As indicated, the slope of the tangent line to the graph of $y = 2^x$ is less than 1, and the slope of the tangent line to the graph of $y = 3^x$ is more than 1. It is reasonable to conclude that there is an exponential function, whose graph lies between that of $y = 2^x$ and $y = 3^x$, for which the tangent line at (0, 1) has slope equal to 1. Look at Figure 8(b). This exponential function is the one whose base is the number e.

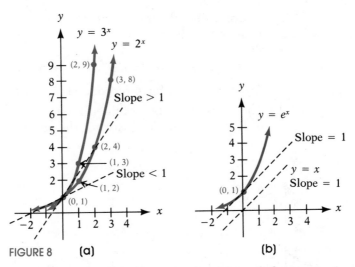

FIGURE 8   **(a)**   **(b)**

The function $f(x) = e^x$ occurs with such frequency in applications that it is usually referred to as *the* exponential function. Indeed, many calculators have the key $\boxed{E^x}$ or $\boxed{EXP(X)}$, which may be used to evaluate the exponential function for a given value of x.

If your calculator does not have this key but does have an $\boxed{INV}$ key and an $\boxed{LN\ X}$ key, you can display the number e as follows:

The reason this works will become clear in Section 6.3. Now use the $\boxed{Y^x}$ key to find e^x (or refer to Table I in the back of this book).

There are many applications involving the exponential function. Let's look at one.

C   **EXAMPLE 5**   The proportion R of people who respond to a newspaper advertisement for a new product and purchase the item advertised after t days is
$$R = 0.5 - e^{-0.3t}$$

(a) What proportion has responded after 5 days? After 10 days?

(b) Graph this function, measuring R along the y-axis and t along the x-axis.

(c) What is the highest proportion of people expected to respond?

Solution   (a) When $t = 5$, we have
$$R = 0.5 - e^{(-0.3)(5)} = 0.5 - e^{-1.5} \approx 0.27687$$

About 28% will have responded after 5 days. When $t = 10$, we have

$$R = 0.5 - e^{(-0.3)(10)} = 0.5 - e^{-3} \approx 0.45021$$

About 45% will have responded after 10 days.

(b) The graph may be obtained in steps, beginning with the graph of $R = e^t$. Figure 9 illustrates the various steps.

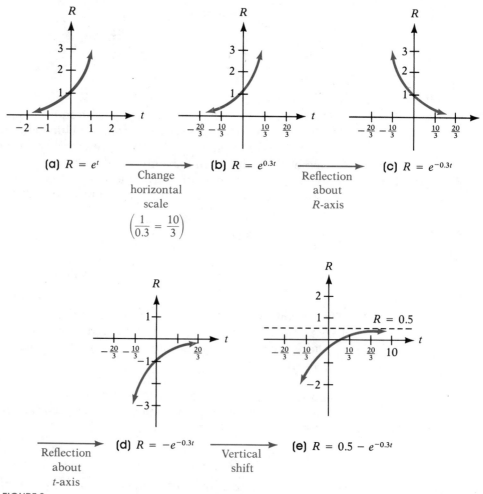

FIGURE 9

(c) As Figure 9(e) indicates, as time goes on, the graph approaches (asymptotically) a response level of $0.5 = 50\%$.

Exponential Equations

Equations that involve terms of the form a^x, $a > 0$, $a \neq 1$, are often referred to as **exponential equations**. Such equations sometimes can be solved by appropriately applying the laws of exponents and the following fact:

EXPONENTIAL EQUATIONS

$$\text{If } a^u = a^v, \text{ then } u = v \qquad (3)$$

The result in Equation (3) is a consequence of the fact that exponential functions are one-to-one.

EXAMPLE 6 Solve the equations:

(a) $3^{x+1} = 81$ (b) $e^{-x^2} = (e^x)^2 \cdot \dfrac{1}{e^3}$

Solution   (a) In order to apply Equation (3), each side of the equality must be written with the same base. Thus, we need to express 81 as a power of 3:

$$3^{x+1} = 81$$
$$3^{x+1} = 3^4$$

Now apply Equation (3) to get

$$x + 1 = 4$$
$$x = 3$$

(b) We use some laws of exponents first to get the same base e on each side:

$$e^{-x^2} = (e^x)^2 \cdot \frac{1}{e^3}$$
$$e^{-x^2} = e^{2x} \cdot e^{-3}$$
$$e^{-x^2} = e^{2x-3}$$

Now apply Equation (3) to get

$$-x^2 = 2x - 3$$
$$x^2 + 2x - 3 = 0$$
$$(x + 3)(x - 1) = 0$$
$$x = -3 \quad \text{or} \quad x = 1$$ ≡

We hasten to point out that most exponential equations are not as easily solved as those in Example 6. Consider, for example, the equation $x + e^x = 2$. Approximate solutions to such equations are usually found by numerical methods studied in calculus. Other types of exponential equations that can be solved using algebra are discussed in Section 6.3.

EXERCISE 6.1

C In Problems 1–6 approximate each number using a calculator with a $\boxed{y^x}$ key.

1. (a) $3^{2.2}$ (b) $3^{2.23}$ (c) $3^{2.236}$ (d) $3^{\sqrt{5}}$

2. (a) $5^{1.7}$ (b) $5^{1.73}$ (c) $5^{1.732}$ (d) $5^{\sqrt{3}}$

3. (a) $2^{3.14}$ (b) $2^{3.141}$ (c) $2^{3.1415}$ (d) 2^{π}

4. (a) $2^{2.7}$ (b) $2^{2.71}$ (c) $2^{2.718}$ (d) 2^{e}

5. (a) $3.1^{2.7}$ (b) $3.14^{2.71}$ (c) $3.141^{2.718}$ (d) π^{e}

6. (a) $2.7^{3.1}$ (b) $2.71^{3.14}$ (c) $2.718^{3.141}$ (d) e^{π}

C In Problems 7–8 graph each function by carefully preparing a table listing points on the graph.

7. $f(x) = 3^x$ 8. $f(x) = 5^x$

In Problems 9–18 use the results of Problems 7 and 8 to graph each function.

9. $f(x) = (\frac{1}{3})^x$ 10. $f(x) = (\frac{1}{5})^x$ 11. $f(x) = 3^{x+1}$

12. $f(x) = 5^{x-2}$ 13. $f(x) = 3^x + 4$ 14. $f(x) = 5^x - 3$

15. $f(x) = 3^{-x} - 2$ 16. $f(x) = 5^{3-x} + 2$ 17. $f(x) = 9^x$

18. $f(x) = 5^{x/2}$

In Problems 19–30 solve each equation.

19. $2^{2x+1} = 4$ 20. $5^{1-2x} = \frac{1}{5}$ 21. $3^{x^3} = 9^x$

22. $4^{x^2} = 2^x$ 23. $8^{x^2-2x} = \frac{1}{2}$ 24. $9^{-x} = \frac{1}{3}$

25. $2^x \cdot 8^{-x} = 4^x$ 26. $(\frac{1}{2})^{1-x} = 4$ 27. $2^{2x} - 2^x - 12 = 0$

28. $3^{2x} + 3^x - 2 = 0$ 29. $3^{2x} + 3^{x+1} - 4 = 0$ 30. $4^x - 3 \cdot 2^x = 0$

C 31. The proportion R of viewers who respond to a television commercial for a new product after t days is expected to obey the equation

$$R = 0.7 - e^{-0.2t}$$

(a) What proportion is expected to respond after 10 days? After 20 days?

(b) Graph this function, measuring R along the y-axis and t along the x-axis.

(c) What is the highest proportion of viewers expected to respond?

C 32. Rework Problem 31 if the equation is changed to

$$R = 0.7 - e^{-0.4t}$$

C 33. Rework Problem 31 if the equation is changed to

$$R = 0.9 - e^{-0.2t}$$

C 34. Rework Problem 31 if the equation is changed to

$$R = 0.9 - e^{-0.1t}$$

C **35.** The equation governing the amount of current I (in amperes) after time t (in seconds) in a single RL circuit consisting of a resistance R (in ohms), an inductance L (in henrys), and an electromotive force E (in volts) is

$$I = \frac{E}{R}(1 - e^{-(R/L)t})$$

(a) If $E = 120$ volts, $R = 10$ ohms, and $L = 5$ henrys, how much current is available after 0.01 second? After 0.5 second?
(b) Graph this function measuring I along the y-axis and t along the x-axis.
(c) What is the maximum current?

C **36.** Rework Problem 35 if $E = 120$ volts, $R = 5$ ohms, and $L = 10$ henrys.

C **37.** The annual profit P of a company due to the sales of a particular item after it has been on the market x years is determined to be

$$P = \$40{,}000 + \$60{,}000\left[1 - \left(\tfrac{1}{2}\right)^x\right] = 100{,}000 - 60{,}000\left(\tfrac{1}{2}\right)^x$$

(a) How much profit is earned after 5 years? After 10 years?
(b) Graph this function, measuring P along the y-axis and x along the x-axis.
(c) What is the maximum profit the company expects from this item?

C **38.** The demand for a new product increases rapidly at first, and then levels off. The percentage P of actual purchases of this product after it has been on the market t months is

$$P = 90 - 80\left(\tfrac{1}{4}\right)^t$$

(a) What is the percentage of purchases of the product after 5 months? After 10 months?
(b) Graph this function, measuring P along the y-axis and t along the x-axis.
(c) What is the maximum percentage of purchases of the product?

C **39.** If $f(x) = e^x$, compute $f(3) - f(2)$.

C **40.** If $f(x) = e^{-x}$, compute $[f(5) - f(2)]/3$.

C **41.** Compute the value of $[1 + (1/n)]^n$ for $n = 10, 100, 1000, 10{,}000$, and $100{,}000$. Compare each result with e.

C **42.** Compute the value of

$$2 + \frac{1}{2!} + \frac{1}{3!} + \cdots + \frac{1}{n!}$$

for $n = 4, 6, 8$, and 10. Compare each result with e.

HYPERBOLIC SINE FUNCTION **43.** The **hyperbolic sine function**, designated by $\sinh x$, is defined as

$$\sinh x = \frac{1}{2}(e^x - e^{-x})$$

(a) Show that $f(x) = \sinh x$ is an odd function.
(b) Graph $y = e^x$ and $y = e^{-x}$ on the same set of coordinate axes, and use the method of subtracting y-coordinates to obtain a graph of $f(x) = \sinh x$.

44. The **hyperbolic cosine function**, designated by cosh x, is defined as

$$\cosh x = \frac{1}{2}(e^x + e^{-x})$$

(a) Show that $f(x) = \cosh x$ is an even function.
(b) Graph $y = e^x$ and $y = e^{-x}$ on the same set of coordinate axes, and use the method of adding y-coordinates to obtain a graph of $f(x) = \cosh x$.
(c) Use parts (a) and (b) and Problem 43 to show that

$$(\cosh x)^2 - (\sinh x)^2 = 1$$

for every x.

45. If $f(x) = a^x$, show that

$$\frac{f(x + h) - f(x)}{h} = a^x\left(\frac{a^h - 1}{h}\right)$$

46. If $f(x) = a^x$, show that $f(A + B) = f(A) \cdot f(B)$.
47. If $f(x) = a^x$, show that $f(-x) = 1/f(x)$.
48. If $f(x) = a^x$, show that $f(\alpha x) = [f(x)]^\alpha$.

C **49.** *Historical problem* Fermat conjectured that the function

$$f(x) = 2^{(2^x)} + 1$$

for $x = 1, 2, 3, \ldots$ would always have a value equal to a prime number. Euler showed that this formula fails for $x = 5$. Determine the prime numbers produced by f for $x = 1, 2, 3, 4$. Then show that $f(5) = 641 \times 6,700,417$, which is not prime.

▤ 6.2 COMPOUND INTEREST

When the interest due at the end of a payment period is added to the principal so that the interest computed at the end of the next payment period is based on this new principal amount (old principal + interest), the interest is said to have been **compounded**. Thus, **compound**
interest is interest paid on previously earned interest.

EXAMPLE 1   A savings and loan association pays interest of 8% per annum compounded quarterly on a certain savings plan. If $1000 is deposited in such a plan and the interest is left to accumulate, how much is in the account after 1 year?

Solution   We use the simple interest formula, $I = Prt$ (see Section 2.2). After the first quarter (of a year), the interest earned is

$$I = Prt = (\$1000)(0.08)\left(\frac{1}{4}\right) = \$20$$

The new principal is $P + I = \$1000 + \$20 = \$1020$. At the end of the second quarter, the interest on this principal is

$$I = (\$1020)(0.08)\left(\frac{1}{4}\right) = \$20.40$$

After the third quarter, the interest on the new principal of ($\$1020 + \$20.40) = \$1040.40$ is

$$I = (\$1040.40)(0.08)\left(\frac{1}{4}\right) \approx \$20.81$$

Finally, after the fourth quarter, the interest is

$$I = (\$1061.21)(0.08)\left(\frac{1}{4}\right) \approx \$21.22$$

Thus, after 1 year, the account contains $1082.43.　■

The pattern of the calculations performed in Example 1 leads to a general formula for compound interest. To fix our ideas, let P represent the principal to be invested at a per-annum interest rate r, which is compounded n times per year. (For computing purposes, r is expressed as a decimal.) The interest earned after each compounding period is the principal times r/n. Thus, the amount A after one compounding period is

$$A = P + P\left(\frac{r}{n}\right) = P\left(1 + \frac{r}{n}\right)$$

After two compounding periods, the amount A, based on the new principal $P(1 + r/n)$, is

$$A = \underbrace{P\left(1 + \frac{r}{n}\right)}_{\substack{\text{New} \\ \text{principal}}} + \underbrace{P\left(1 + \frac{r}{n}\right)\left(\frac{r}{n}\right)}_{\substack{\text{Interest on} \\ \text{new principal}}} = P\left(1 + \frac{r}{n}\right)\left(1 + \frac{r}{n}\right) = P\left(1 + \frac{r}{n}\right)^2$$

After three compounding periods,

$$A = P\left(1 + \frac{r}{n}\right)^2 + P\left(1 + \frac{r}{n}\right)^2\left(\frac{r}{n}\right) = P\left(1 + \frac{r}{n}\right)^2\left(1 + \frac{r}{n}\right) = P\left(1 + \frac{r}{n}\right)^3$$

Continuing in this way, after n compounding periods (1 year),

$$A = P\left(1 + \frac{r}{n}\right)^n$$

Because t years will contain $n \cdot t$ compounding periods, after t years we have

$$A = P\left(1 + \frac{r}{n}\right)^{nt}$$

THEOREM   The amount A after t years due to a principal P invested at an annual interest rate r compounded n times per year is

COMPOUND INTEREST
FORMULA

$$A = P\left(1 + \frac{r}{n}\right)^{nt} \tag{1}$$

≡

C **EXAMPLE 2**   Investing $1000 at an annual rate of 10% compounded annually, quarterly, monthly, and daily will yield the following amounts after 1 year:

Annual compounding: $A = P(1 + r) = (\$1000)(1 + 0.10) = \$1100.00$

Quarterly compounding: $A = P\left(1 + \frac{r}{4}\right)^{4} = (\$1000)(1 + 0.025)^4 \approx \$1103.81$

Monthly compounding: $A = P\left(1 + \frac{r}{12}\right)^{12} = (\$1000)(1 + 0.00833)^{12} \approx \$1104.71$

Daily compounding: $A = P\left(1 + \frac{r}{365}\right)^{365} = (\$1000)(1 + 0.000274)^{365} \approx \$1105.16$

≡

From Example 2, we can see that the effect of compounding more frequently is that the amount after 1 year is higher: $1000 compounded 4 times a year at 10% results in $1103.81; $1000 compounded 12 times a year at 10% results in $1104.71; and $1000 compounded 365 times a year at 10% results in $1105.16. This leads to the following question: What would happen to the amount after 1 year if the number of times the interest is compounded were increased without bound?

Let's find the answer. Suppose P is the principal, r is the per-annum interest rate, and n is the number of times the interest is compounded each year. The amount after 1 year is

$$A = P\left(1 + \frac{r}{n}\right)^{n}$$

Now suppose the number n of times the interest is compounded per year gets larger and larger; that is, suppose $n \rightarrow +\infty$. Table 2 compares $(1 + r/n)^n$, for large values of n, to e^r for $r = 0.05$, $r = 0.10$, $r = 0.15$, and $r = 1$. The larger n gets, the closer $(1 + r/n)^n$ gets to e^r, a fact proven in many calculus books.* Thus, no matter how frequent the compounding, the amount after 1 year has the definite ceiling Pe^r.

*Sometimes e is defined as the number that $(1 + 1/n)^n$ approaches as $n \rightarrow +\infty$.

TABLE 2

	$\left(1 + \dfrac{r}{n}\right)^n$			
	$n = 100$	$n = 1000$	$n = 10{,}000$	e^r
$r = 0.05$	1.0512579	1.05127	1.051271	1.0512711
$r = 0.10$	1.1051157	1.1051654	1.1051703	1.1051709
$r = 0.15$	1.1617037	1.1618212	1.1618329	1.1618342
$r = 1$	2.7048138	2.7169239	2.7181459	2.7182818

When interest is compounded so that the amount after 1 year is Pe^r, we say the interest is **compounded continuously.**

THEOREM

CONTINUOUS COMPOUNDING

The amount A after t years due to a principal P invested at an annual interest rate r compounded continuously is

$$A = Pe^{rt} \tag{2}$$

 EXAMPLE 3 The amount A that results from investing a principal P of $1000 at an annual rate r of 10% compounded continuously for a time t of 1 year is

$$A = \$1000e^{0.10} = \$1105.17$$

EFFECTIVE RATE OF INTEREST

The **effective rate of interest** is the equivalent annual simple rate of interest that would yield the same amount as compounding after 1 year. For example, based on Example 3, a principal of $1000 will result in $1105.17 at a rate of 10% compounded continuously. To get this same amount using a simple rate of interest would require that interest of $1105.17 − $1000.00 = $105.17 be earned on the principal. Since $105.17 is 10.517% of $1000, a simple rate of interest of 10.517% is needed to equal 10% compounded continuously. Thus, the effective rate of interest of 10% compounded continuously is 10.517%.

Based on the results of Examples 2 and 3, we find the following comparisons:

	ANNUAL RATE	EFFECTIVE RATE
Annual compounding	10%	10%
Quarterly compounding	10%	10.381%
Monthly compounding	10%	10.471%
Daily compounding	10%	10.516%
Continuous compounding	10%	10.517%

[C] **EXAMPLE 4** On January 2, 1987, $2000 is placed in an Individual Retirement Account (IRA) that will pay interest of 10% per annum compounded continuously. What will the IRA be worth on January 1, 2007?

Solution The amount A after 20 years is

$$A = Pe^{rt} = \$2,000e^{(0.10)(20)} = \$14,778.11$$ ≡

PRESENT VALUE When people engaged in finance speak of the "time value of money," they are usually referring to the **present value** of money. The present value of A dollars to be received at a future date is the principal you would need to invest now so that it would grow to A dollars in the specified time period. Thus, the present value of money to be received at a future date is always less than the amount to be received, since the amount to be received will equal the present value (money invested now) *plus* the interest accrued over the time period.

We use the compound interest formula, Equation (1), to get a formula for present value. If V is the present value of A dollars to be received after t years at a per-annum interest rate r compounded n times per year, then by Equation (1),

$$A = V\left(1 + \frac{r}{n}\right)^{nt}$$

To solve for V, we divide both sides by $(1 + r/n)^{nt}$, and the result is

$$\frac{A}{(1 + r/n)^{nt}} = V \quad \text{or} \quad V = A\left(1 + \frac{r}{n}\right)^{-nt}$$

THEOREM The present value V of A dollars to be received after t years assuming
PRESENT VALUE FORMULAS a per-annum interest rate r compounded n times per year is

$$V = A\left(1 + \frac{r}{n}\right)^{-nt} \tag{3}$$

If the interest is compounded continuously, then

$$V = Ae^{-rt} \tag{4}$$

≡

You are asked to derive Formula (4) in Problem 31.

EXAMPLE 5 A zero-coupon (noninterest-bearing) bond can be redeemed in 10 years for $1000. How much should you be willing to pay for it now if you want a return of:

(a) 8% compounded monthly?

(b) 7% compounded continuously?

Solution   (a) We are seeking the present value of $1000. Thus, we use Formula (3) with $A = \$1000$, $n = 12$, $r = 0.08$, and $t = 10$:

$$V = A\left(1 + \frac{r}{n}\right)^{-nt}$$

$$= \$1000\left(1 + \frac{0.08}{12}\right)^{-12(10)}$$

$$\approx \$450.52$$

(b) We use Formula (4) with $A = \$1000$, $r = 0.07$, and $t = 10$:

$$V = Ae^{-rt}$$

$$= \$1000e^{-(0.07)(10)}$$

$$\approx \$496.59$$

EXERCISE 6.2

[C] *In Problems 1–10 find the amount that results from each investment.*

1. $100 invested at 8% compounded monthly after a period of 2 years
2. $50 invested at 6% compounded daily after a period of 3 years
3. $500 invested at 10% compounded quarterly after a period of $2\frac{1}{2}$ years
4. $300 invested at 12% compounded monthly after a period of $1\frac{1}{2}$ years
5. $600 invested at 5% compounded daily after a period of 3 years
6. $700 invested at 9% compounded quarterly after a period of 2 years
7. $10 invested at 11% compounded continuously after a period of 2 years
8. $40 invested at 7% compounded continuously after a period of 3 years
9. $100 invested at 10% compounded continuously after a period of $2\frac{1}{4}$ years
10. $100 invested at 12% compounded continuously after a period of $3\frac{3}{4}$ years

[C] *In Problems 11–20 find the principal needed now to get each amount; that is, find the present value.*

11. To get $100 after 2 years at 8% compounded monthly
12. To get $75 after 3 years at 10% compounded quarterly
13. To get $1000 after $2\frac{1}{2}$ years at 6% compounded daily
14. To get $800 after $3\frac{1}{2}$ years at 7% compounded monthly
15. To get $600 after 2 years at 12% compounded quarterly
16. To get $300 after 4 years at 14% compounded daily
17. To get $80 after $3\frac{1}{4}$ years at 9% compounded continuously
18. To get $800 after $2\frac{1}{2}$ years at 8% compounded continuously
19. To get $400 after 1 year at 10% compounded continuously
20. To get $1000 after 1 year at 12% compounded continuously

C **21.** A zero-coupon bond can be redeemed in 20 years for $10,000. How much should you be willing to pay for it now if you want a return of:
(a) 10% compounded monthly?
(b) 12% compounded continuously?

C **22.** Rework Problem 21 if the bond can be redeemed in 10 years for $5000.

C **23.** You invest $2000 in a bond trust that pays 9% interest compounded semiannually. Your friend invests $2000 in a Certificate of Deposit that pays $8\frac{1}{2}$% compounded continuously. Who has more money after 20 years, you or your friend?

C **24.** Rework Problem 23 if the bond trust pays 10% interest compounded semiannually whereas the Certificate of Deposit pays only 8% compounded continuously.

C **25.** Suppose you have access to an investment vehicle that will pay 10% interest compounded continuously. Which is better: to be given $1000 now or to be given $1325 after 3 years?

C **26.** Rework Problem 25 if the 10% interest is compounded annually.

C **27.** You have just purchased a house for $150,000, with the seller holding a second mortgage of $50,000. Five years from now you promise to pay the seller $50,000 plus all accrued interest. The seller offers you three interest options on the second mortgage:
(a) Simple interest at 12% per annum
(b) $11\frac{1}{2}$% interest compounded monthly
(c) $11\frac{1}{4}$% interest compounded continuously
Which option is best; that is, which one results in the least interest on the loan?

C **28.** Refer to Problem 27. If you promise to pay the seller $50,000 plus all accrued interest 3 years from now, which option is best?

C **29.** Financial Federal Savings Bank, in the spring of 1985, advertised the following investment options:

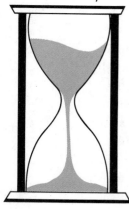

Time is Money

FIXED-TERM CERTIFICATE INVESTMENT OPTIONS

MATURITY TERM	RATE	YIELD
91 days	8.50%	8.77%
6 months	9.15%	9.47%
1 year	9.50%	9.84%
$1\frac{1}{2}$ years	9.80%	10.17%
$2\frac{1}{2}$ years	10.15%	10.54%
3 years	10.35%	10.76%
Passbook savings account, no maturity restrictions		
Regular savings	5.50%	5.65%

For the 1 year fixed-term certificate of 9.50%, determine whether the yield of 9.84% is an effective rate for 9.50% compounded continuously, compounded daily (365 days), compounded monthly, or compounded quarterly. Assume a principal deposit of $100.00.

C 30. Rework Problem 29 for the regular savings account of 5.50% showing a yield of 5.65%.

31. Derive the formula $V = Ae^{-rt}$ for the present value V of A dollars if the interest rate r is compounded continuously.

≣ 6.3 LOGARITHMS

As the following definition states, a *logarithm* is merely a name for the solution of an exponential equation.

Consider the exponential equation

$$a^x = M \tag{1}$$

LOGARITHM TO THE BASE
a OF M

where a and M are positive real numbers, and $a \neq 1$. The solution x of this equation is called the **logarithm to the base a of M** and is symbolized as

$$x = \log_a M \tag{2}$$

Thus, $\log_a M$ is an exponent—the exponent to which the base must be raised to obtain M. For example, to calculate $\log_3 9$, we must answer the question: "To what power must 3 be raised to obtain 9?" Because the answer is 2, we know that $\log_3 9 = 2$.

Equations (1) and (2) are equivalent. You may find it helpful to use the following device to remember the definition:

$$\log_a M = x$$

a to the power x
equals M

Notice, in the definition of $\log_a M$, that both a and M are positive. Logarithms of negative numbers are not defined.

To evaluate a logarithm, we use the equivalent exponential equation.

EXAMPLE 1   Evaluate:

(a) $\log_e 1$ (b) $\log_2 8$ (c) $\log_3(\tfrac{1}{3})$ (d) $\log_5 5$

Solution   (a) Because $e^0 = 1$, we have $\log_e 1 = 0$.
           (b) Because $2^3 = 8$, we have $\log_2 8 = 3$.
           (c) Because $3^{-1} = \frac{1}{3}$, we have $\log_3(\frac{1}{3}) = -1$.
           (d) Because $5^1 = 5$, we have $\log_5 5 = 1$.

EXAMPLE 2   Change each exponential expression to an equivalent expression involving a logarithm.

(a) $1.2^3 = m$ (b) $e^x = 9$ (c) $a^4 = 24$

Solution   We use the fact that Equations (1) and (2) are equivalent:

(a) If $1.2^3 = m$, then $3 = \log_{1.2} m$.
(b) If $e^x = 9$, then $x = \log_e 9$.
(c) If $a^4 = 24$, then $4 = \log_a 24$.

EXAMPLE 3   Change each logarithmic expression to an equivalent expression involving an exponent.

(a) $\log_a 4 = 5$ (b) $\log_e x = -3$ (c) $\log_3 5 = x$

Solution   (a) If $\log_a 4 = 5$, then $a^5 = 4$.
           (b) If $\log_e x = -3$, then $e^{-3} = x$.
           (c) If $\log_3 5 = x$, then $3^x = 5$.

Properties of Logarithms

Logarithms have some useful properties that can be derived directly from the definition and the laws of exponents.

THEOREM   In the properties listed below, M and a are positive real numbers,
PROPERTIES OF LOGARITHMS   with $a \neq 1$, and r is any real number.
    The logarithm to the base a of 1 equals 0. That is,

$$\log_a 1 = 0 \tag{3}$$

The logarithm to the base a of a equals 1. That is,

$$\log_a a = 1 \tag{4}$$

The number $\log_a M$ is the exponent to which a must be raised to obtain M. That is,

$$a^{\log_a M} = M \tag{5}$$

The logarithm to the base a of a raised to a power equals that power. That is,

$$\log_a a^r = r \tag{6}$$

≡

EXAMPLE 4   (a) $\log_{\sqrt{3}} 1 = 0$ (b) $\log_{1/2} \dfrac{1}{2} = 1$

(c) $\sqrt{2}^{\log_{\sqrt{2}} \pi} = \pi$ (d) $\log_{0.2} 0.2^{-\sqrt{2}} = -\sqrt{2}$ ≡

To gain experience in working with logarithms, we derive Properties (3)–(6).

Proof of Property (3) $\log_a 1 = 0$ because $a^0 = 1$

Proof of Property (4) $\log_a a = 1$ because $a^1 = a$

Proof of Property (5)   Let $x = \log_a M$. Change this logarithmic expression to the equivalent exponential expression

$$a^x = M$$

Now replace x by $\log_a M$ to get

$$a^{\log_a M} = M$$

Proof of Property (6)   Let $x = a^r$. Change this exponential expression to the equivalent logarithmic expression

$$\log_a x = r$$

Now replace x by a^r to get

$$\log_a a^r = r$$ ≡

There are some other useful properties of logarithms.

THEOREM   In the properties listed below, M, N, and a are positive real numbers, with $a \neq 1$, and r is any real number.

LOG OF A PRODUCT EQUALS THE SUM OF THE LOGS

LOG OF A QUOTIENT EQUALS THE DIFFERENCE OF THE LOGS

$$\log_a MN = \log_a M + \log_a N \tag{7}$$

$$\log_a \frac{M}{N} = \log_a M - \log_a N \tag{8}$$

$$\log_a \frac{1}{N} = -\log_a N \tag{9}$$

$$\log_a M^r = r \log_a M \tag{10}$$

≡

We shall derive Properties (7) and (10) and leave the derivations of Properties (8) and (9) as exercises (see Problems 113 and 114).

Proof of Property (7)   Let $A = \log_a M$ and let $B = \log_a N$. These expressions are equivalent to the exponential expressions

$$a^A = M \quad \text{and} \quad a^B = N$$

Now,

$$\log_a MN = \log_a a^A a^B = \log_a a^{A+B} \qquad \text{Law of exponents}$$

$$= A + B \qquad\qquad \text{Property (6) of logarithms}$$

$$= \log_a M + \log_a N$$

Proof of Property (10)   Let $A = \log_a M$. This expression is equivalent to

$$a^A = M$$

Now,

$$\log_a M^r = \log_a (a^A)^r = \log_a a^{rA} \quad \text{Law of exponents}$$

$$= rA \qquad \text{Property (6) of logarithms}$$

$$= r \log_a M$$

$\equiv$

One use of Properties (7)–(10) is to write sums and/or differences of logarithms with the same base as a single logarithm.

EXAMPLE 5   Write each of the following as a single logarithm:

(a) $\log_a 7 + 4 \log_a 3$ (b) $\frac{2}{3} \log_a 8 - \log_a (3^4 - 8)$

(c) $\log_a x + \log_a 9 + \log_a (x^2 + 1) - \log_a 5$

Solution   (a) $\log_a 7 + 4 \log_a 3 = \log_a 7 + \log_a 3^4$ Property (10)

$$= \log_a 7 + \log_a 81$$

$$= \log_a (7 \cdot 81) \qquad \text{Property (7)}$$

$$= \log_a 567$$

(b) $\frac{2}{3} \log_a 8 - \log_a (3^4 - 8) = \log_a 8^{2/3} - \log_a (81 - 8)$ Property (10)

$$= \log_a 4 - \log_a 73$$

$$= \log_a \frac{4}{73} \qquad\qquad \text{Property (8)}$$

(c) $\log_a x + \log_a 9 + \log_a (x^2 + 1) - \log_a 5$

$$= \log_a 9x + \log_a (x^2 + 1) - \log_a 5$$

$$= \log_a [9x(x^2 + 1)] - \log_a 5$$

$$= \log_a \frac{9x(x^2 + 1)}{5} \qquad \blacksquare$$

Logarithms can be used to transform products into sums, quotients into differences, and powers into factors.

EXAMPLE 6 (a) Write $\log_a (x\sqrt{x^2 + 1})$ as a sum of logarithms. Express all powers as factors.

(b) Write

$$\log_a \frac{x^2}{(x - 1)^3}$$

as a difference of logarithms. Express all powers as factors.

(c) Write

$$\log_a \frac{x^3\sqrt{x^2 + 1}}{(x + 1)^4}$$

as a sum and difference of logarithms. Express all powers as factors.

Solution (a) $\log_a (x\sqrt{x^2 + 1}) = \log_a x + \log_a \sqrt{x^2 + 1}$

$$= \log_a x + \log_a (x^2 + 1)^{1/2}$$

$$= \log_a x + \frac{1}{2} \log_a (x^2 + 1)$$

(b) $\log_a \dfrac{x^2}{(x - 1)^3} = \log_a x^2 - \log_a (x - 1)^3 = 2 \log_a x - 3 \log_a (x - 1)$

(c) $\log_a \dfrac{x^3\sqrt{x^2 + 1}}{(x + 1)^4} = \log_a (x^3\sqrt{x^2 + 1}) - \log_a (x + 1)^4$

$$= \log_a x^3 + \log_a \sqrt{x^2 + 1} - \log_a (x + 1)^4$$

$$= \log_a x^3 + \log_a (x^2 + 1)^{1/2} - \log_a (x + 1)^4$$

$$= 3 \log_a x + \frac{1}{2} \log_a (x^2 + 1) - 4 \log_a (x + 1) \qquad \blacksquare$$

There remain two other properties of logarithms we need to know.

THEOREM   In the properties below, M, N, and a are positive real numbers, with $a \neq 1$.

$$\text{If } M = N, \text{ then } \log_a M = \log_a N \qquad (11)$$
$$\text{If } \log_a M = \log_a N, \text{ then } M = N \qquad (12)$$

≡

We shall derive both these properties.

Proof of Property (11)   Let $M = N$. If

$$x = \log_a M$$

then

$$a^x = M$$

Since $M = N$, we also have

$$a^x = N \quad \text{or} \quad x = \log_a N$$

Thus,

$$\log_a M = \log_a N$$

Proof of Property (12)   Let

$$x = \log_a M = \log_a N$$

From $x = \log_a M$, we have

$$M = a^x$$

From $x = \log_a N$, we have

$$N = a^x$$

Therefore, $M = N$.   ≡

Properties (11) and (12) are useful for solving logarithmic equations.

Logarithmic Equations

LOGARITHMIC EQUATIONS   Equations that contain terms of the form $\log_a x$, where a is a positive real number, with $a \neq 1$, are often called **logarithmic equations**.

Let's see how we can use the properties of logarithms to solve logarithmic equations.

EXAMPLE 7   Solve each equation:

(a) $2 \log_5 x = \log_5 9$ (b) $\log_3(4x - 7) = 2$

(c) $\log_5 x + \log_5(2x - 3) = 1$

Solution   (a) $2 \log_5 x = \log_5 9$

$\log_5 x^2 = \log_5 9$ Property (10)

$x^2 = 9$ Property (12)

$x = 3$ or $x = -3$

Recall that logarithms of negative numbers are not defined, so, in the expression $2 \log_5 x$, x must be positive. Therefore, -3 is extraneous and we discard it. This leaves 3 as the only solution.

(b) We change the expression to exponential form to solve:

$$\log_3(4x - 7) = 2$$
$$4x - 7 = 3^2$$
$$4x - 7 = 9$$
$$4x = 16$$
$$x = 4$$

(c) We need to express the left side as a single logarithm. Then, we will change the expression to exponential form.

$$\log_5 x + \log_5(2x - 3) = 1$$
$$\log_5[x(2x - 3)] = 1 \quad \text{Property (7)}$$
$$x(2x - 3) = 5^1 = 5$$
$$2x^2 - 3x - 5 = 0$$
$$(2x - 5)(x + 1) = 0$$
$$x = \frac{5}{2} \quad \text{or} \quad x = -1$$

The solution $x = -1$ is extraneous because x must be larger than $\frac{3}{2}$. (Do you see why?) Thus, the equation has one solution, $\frac{5}{2}$. ∎

As Example 7 illustrates, some care must be taken when solving logarithmic equations. Be sure to check each apparent solution in the original equation and discard any that are extraneous. Remember that, in $\log_a M$, a and M are positive and $a \neq 1$.

Common Logarithms; Natural Logarithms

COMMON LOGARITHMS   Until recently, **common logarithms**—that is, logarithms to the base 10—were used to facilitate arithmetic computations. The next example illustrates this application.

EXAMPLE 8   Given that $\log_{10} 2 = 0.3010$ and $\log_{10} 3 = 0.4771$,* compute:

(a) $\log_{10} 4$ (b) $\log_{10} 6$ (c) $\log_{10} 200$ (d) $\log_{10} 15$

*These values are found in Table II in the back of the book.

Solution (a) $\log_{10}4 = \log_{10}2^2 = 2\log_{10}2 = 2(0.3010) = 0.6020$

(b) $\log_{10}6 = \log_{10}(2 \cdot 3) = \log_{10}2 + \log_{10}3 = 0.3010 + 0.4771$
$= 0.7781$

(c) $\log_{10}200 = \log_{10}(2 \cdot 10^2) = \log_{10}2 + \log_{10}10^2$
$= 0.3010 + 2 = 2.3010$

(d) $\log_{10}15 = \log_{10}\dfrac{30}{2} = \log_{10}30 - \log_{10}2 = \log_{10}(3 \cdot 10) - \log_{10}2$
$= \log_{10}3 + \log_{10}10 - \log_{10}2$
$= 0.4771 + 1 - 0.3010 = 1.1761$ ≣

NATURAL LOGARITHMS

The widespread availability of handheld calculators has made this particular use of common logarithms less important than it once was. However, **natural logarithms**—that is, logarithms to the base $e = 2.718\ldots$—remain very important because they arise in the study of many natural phenomena.

log

ln

Common logarithms are usually abbreviated by writing **log**, with the base understood to be 10. Natural logarithms are abbreviated by **ln** (from the Latin, *logarithmus naturalis*), with the base understood to be e.

Most calculators have both a ⌊LOG⌋ key and an ⌊ln⌋ key to calculate the common logarithm and the natural logarithm of a number. For your convenience, Tables II and III in the back of this book give values of $\log x$ and $\ln x$ for selected numbers x. To calculate logarithms having a base other than 10 or e, we employ the **change-of-base formula**.

THEOREM   If $a \neq 1$, $b \neq 1$, and M are positive real numbers, then

CHANGE-OF-BASE FORMULA

$$\log_a M = \frac{\log_b M}{\log_b a} \qquad (13)$$

Proof   We derive this formula as follows: Let $x = \log_a M$. Then $a^x = M$ so that

$$\log_b a^x = \log_b M \quad \text{Property (11)}$$
$$x \log_b a = \log_b M \quad \text{Property (10)}$$
$$\log_a M \cdot \log_b a = \log_b M \quad x = \log_a M$$
$$\log_a M = \frac{\log_b M}{\log_b a}$$

≣

In practice, the change-of-base formula uses either $b = 10$ or $b = e$. Thus,

$$\log_a M = \frac{\log M}{\log a} \qquad \log_a M = \frac{\ln M}{\ln a} \qquad (14)$$

C EXAMPLE 9 Calculate:

(a) $\log_5 89$ (b) $\log_{\sqrt{2}} \sqrt{5}$

Solution (a)

$$\log_5 89 = \frac{\log 89}{\log 5} \approx \frac{1.94939}{0.69897} = 2.7889$$

$$\log_5 89 = \frac{\ln 89}{\ln 5} \approx \frac{4.4886}{1.6094} = 2.7889$$

(b)

$$\log_{\sqrt{2}} \sqrt{5} = \frac{\log \sqrt{5}}{\log \sqrt{2}} = \frac{\frac{1}{2} \log 5}{\frac{1}{2} \log 2} \approx \frac{0.69897}{0.30103} = 2.3219$$

$$\log_{\sqrt{2}} \sqrt{5} = \frac{\ln \sqrt{5}}{\ln \sqrt{2}} = \frac{\frac{1}{2} \ln 5}{\frac{1}{2} \ln 2} \approx \frac{1.6094}{0.6931} = 2.3219 \qquad \equiv$$

Logarithms are very useful for solving certain types of exponential equations. Let's see how.

C EXAMPLE 10 Solve for x:

(a) $2^x = 5$ (b) $5^{x-2} = 3^{3x+2}$

Solution (a) We write the exponential equation as the equivalent logarithmic equation:

$$2^x = 5$$

$$x = \log_2 5 = \frac{\ln 5}{\ln 2} \approx 2.3219$$

$$\uparrow$$

Change-of-base formula (14)

Alternatively, we can solve the equation $2^x = 5$ by taking the natural logarithm (or common logarithm) of each side. Taking the natural logarithm,

$$2^x = 5$$

$$\ln 2^x = \ln 5$$

$$x \ln 2 = \ln 5$$

$$x = \frac{\ln 5}{\ln 2}$$

$$x \approx 2.3219$$

(b) Because the bases are different, we cannot use the method shown in Section 6.1. However, if we take the natural logarithm of each side and apply appropriate properties of logarithms, we are led to an equation in x:

$$5^{x-2} = 3^{3x+2}$$
$$\ln 5^{x-2} = \ln 3^{3x+2}$$
$$(x - 2)\ln 5 = (3x + 2)\ln 3$$
$$(\ln 5)x - 2 \ln 5 = (3 \ln 3)x + 2 \ln 3$$
$$(\ln 5 - 3 \ln 3)x = 2 \ln 3 + 2 \ln 5$$
$$x = \frac{2(\ln 3 + \ln 5)}{\ln 5 - 3 \ln 3} \approx \frac{2(2.7081)}{-1.6864} = -3.2116 \quad \blacksquare$$

≡ HISTORICAL COMMENT Logarithms were invented about 1590 by John Napier (1550–1617) and Jobst Bürgi (1552–1632), working independently. Napier, whose work had the greater influence, was a Scottish lord, a secretive man whose neighbors were inclined to believe him to be in league with the devil. His approach to logarithms was quite different from ours; it was based on the relationship between arithmetic and geometric series (see the chapter on induction and sequences), and not on the inverse-function relationship of logarithms to exponential functions (see Section 6.4). Napier's tables, published in 1614, listed what would now be called natural logarithms of sines and were rather difficult to use. A London professor, Henry Briggs, became interested in the tables and visited Napier. In their conversations, they developed the idea of common logarithms, and Briggs then converted Napier's tables into tables of common logarithms, which were published in 1617. Their importance for calculation was immediately recognized and, by 1650, they were being printed in China. They remained an important calculation tool until the advent of the inexpensive hand-held calculator about 1972, which has decreased their calculational, but not their theoretical, importance.

A side effect of the invention of logarithms was the popularization of the decimal system of notation for real numbers. ≡

EXERCISE 6.3 *In Problems 1–10 change each exponential expression to an equivalent expression involving a logarithm.*

1. $9 = 3^2$
2. $16 = 4^2$
3. $a^2 = 1.6$
4. $a^3 = 2.1$
5. $1.1^2 = M$
6. $2.2^3 = N$
7. $2^x = 7.2$
8. $3^x = 4.6$
9. $x^{\sqrt{2}} = \pi$
10. $x^\pi = e$

In Problems 11–20 change each logarithmic expression to an equivalent expression involving an exponent.

11. $\log_2 8 = 3$ **12.** $\log_3(\frac{1}{9}) = -2$ **13.** $\log_a 3 = 6$

14. $\log_b 4 = 2$ **15.** $\log_3 2 = x$ **16.** $\log_2 6 = x$

17. $\log_2 M = 1.3$ **18.** $\log_3 N = 2.1$ **19.** $\log_{\sqrt{2}} \pi = x$

20. $\log_\pi x = \frac{1}{2}$

In Problems 21–30 evaluate each logarithm.

21. $\log_2 1$ **22.** $\log_8 8$ **23.** $\log_5 25$ **24.** $\log_3(\frac{1}{9})$

25. $\log_{1/2} 16$ **26.** $\log_{1/3} 9$ **27.** $\log_{10} \sqrt{10}$ **28.** $\log_5 \sqrt[3]{25}$

29. $\log_{\sqrt{2}} 4$ **30.** $\log_{\sqrt{3}} 9$

In Problems 31–40 write each expression as a single logarithm.

31. $3 \log_5 u + 4 \log_5 v$ **32.** $\log_3 u^2 - \log_3 v$

33. $\log_{1/2} \sqrt{x} - \log_{1/2} x^3$ **34.** $\log_2 \frac{1}{x} + \log_2 \frac{1}{x^2}$

35. $\ln \frac{x}{x-1} + \ln \frac{x+1}{x} - \ln(x^2 - 1)$

36. $\log \frac{x^2 + 2x - 3}{x^2 - 4} - \log \frac{x^2 + 7x + 6}{x + 2}$

37. $8 \log_2 \sqrt{3x - 2} - \log_2 \frac{4}{x} + \log_2 4$

38. $21 \log_3 \sqrt[3]{x} + \log_3 9x^2 - \log_5 25$

39. $2 \log_a 5x^3 - \frac{1}{2} \log_a(2x + 3)$ **40.** $\frac{1}{3} \log(x^3 + 1) + \frac{1}{2} \log(x^2 + 1)$

In Problems 41–50 write each expression as a sum and/or difference of logarithms. Express powers as factors.

41. $\ln x^2 \sqrt{1 - x}$ **42.** $\ln x \sqrt{1 + x^2}$ **43.** $\log_2 \frac{x^3}{x - 3}$

44. $\log_5 \frac{\sqrt[3]{x^2 + 1}}{x^2 - 1}$ **45.** $\log \frac{x(x + 2)}{(x + 3)^2}$ **46.** $\log \frac{x^3 \sqrt{x + 1}}{(x - 2)^2}$

47. $\ln \left[\frac{x^2 - x - 2}{(x + 4)^2} \right]^{1/3}$ **48.** $\ln \left[\frac{(x - 4)^2}{x^2 - 1} \right]^{2/3}$

49. $\ln \frac{5x\sqrt{1 - 3x}}{3(x - 4)^3 \sqrt{3 - 4x}}$

50. $\ln \frac{5x^2 \sqrt[3]{1 - x}}{4(x + 1)^2}$

In Problems 51–70 solve each equation.

51. $\frac{1}{2} \log_3 x = 2$ **52.** $2 \log_4 x = 3$

53. $3 \log_2(x - 1) + \log_2 4 = 5$ **54.** $2 \log_3(x + 4) - \log_3 9 = 2$

55. $\log_{10}x + \log_{10}(x + 15) = 2$ **56.** $\log_4x + \log_4(x - 3) = 1$

57. $\log_x4 = 2$ **58.** $\log_x\frac{1}{8} = 3$

59. $\log_3(x - 1)^2 = 2$ **60.** $\log_2(x + 4)^3 = 6$

61. $\log_{1/2}(3x + 1)^{1/3} = -2$ **62.** $\log_{1/3}(1 - 2x)^{1/2} = -1$

63. $\log_a(x - 1) - \log_a(x + 6) = \log_a(x - 2) - \log_a(x + 3)$

64. $\log_ax + \log_a(x - 2) = \log_a(x + 4)$

65. $\log_{1/3}(x^2 + x) - \log_{1/3}(x^2 - x) = -1$

66. $\log_4(x^2 - 9) - \log_4(x + 3) = 3$

67. $\log_3|3x - 2| = 2$ **68.** $\log_2|3 - 2x| = 3$

69. $\log_28^x = -3$ **70.** $\log_33^x = -1$

In Problems 71–82 use the facts that $\ln 2 = 0.6931$ *and* $\ln 3 = 1.0986$ *to calculate the value of each expression.*

71. $\ln 6$ **72.** $\ln\frac{2}{3}$ **73.** $\ln 1.5$ **74.** $\ln 0.5$

75. $\ln 2e$ **76.** $\ln\frac{3}{e}$ **77.** $\ln 12$ **78.** $\ln 24$

79. $\ln\sqrt[5]{18}$ **80.** $\ln\sqrt[4]{48}$ **81.** $\log_23$ **82.** $\log_32$

83. Show that $\log_a(x + \sqrt{x^2 - 1}) + \log_a(x - \sqrt{x^2 - 1}) = 0$

84. Show that $\log_a(\sqrt{x} + \sqrt{x - 1}) + \log_a(\sqrt{x} - \sqrt{x - 1}) = 0$

Ⓒ *In Problems 85–92 use a calculator that has either a* $\boxed{\text{LOG}}$ *key or an* $\boxed{\text{LN}}$ *key to evaluate each logarithm.*

85. $\log_321$ **86.** $\log_518$ **87.** $\log_{1/3}71$ **88.** $\log_{1/2}15$

89. $\log_{\sqrt{2}}7$ **90.** $\log_{\sqrt{5}}8$ **91.** $\log_\pi e$ **92.** $\log_\pi\sqrt{2}$

Ⓒ *In Problems 93–104 solve for x using a calculator that has either a* $\boxed{\text{LOG}}$ *key or an* $\boxed{\text{LN}}$ *key.*

93. $2^x = 10$ **94.** $3^x = 14$ **95.** $8^{-x} = 1.2$

96. $2^{-x} = 1.5$ **97.** $3^{1-2x} = 4^x$ **98.** $2^{x+1} = 5^{1-2x}$

99. $\left(\frac{3}{5}\right)^x = 7^{1-x}$ **100.** $\left(\frac{4}{3}\right)^{1-x} = 5^x$ **101.** $1.2^x = (0.5)^{-x}$

102. $(0.3)^{1+x} = 1.7^{2x-1}$ **103.** $\pi^{1-x} = e^x$ **104.** $e^{x+3} = \pi^x$

105. Find the value of $\log_23 \cdot \log_34 \cdot \log_45 \cdot \log_56 \cdot \log_67 \cdot \log_78$.

106. Find the value of $\log_24 \cdot \log_46 \cdot \log_68$.

107. Find the value of $\log_23 \cdot \log_34 \cdot \cdots \cdot \log_n(n + 1) \cdot \log_{n+1}2$.

108. Find the value of $\log_22 \cdot \log_24 \cdot \cdots \cdot \log_22^n$.

Ⓒ **109.** How long does it take for an investment to double in value if it is invested at 8% per annum compounded monthly? Compounded continuously?

C **110.** How long does it take for an investment to double in value if it is invested at 10% per annum compounded monthly? Compounded continuously?

C **111.** If you have $100 to invest at 8% per annum compounded monthly, how long will it be before the amount is $150? If the compounding is continuous, how long will it be?

C **112.** If you have $100 to invest at 10% per annum compounded monthly, how long will it be before the amount is $175? If the compounding is continuous, how long will it be?

113. Show that $\log_a(M/N) = \log_a M - \log_a N$, where a, M, and N are positive real numbers, with $a \neq 1$.

114. Show that $\log_a(1/N) = -\log_a N$, where a and N are positive real numbers, with $a \neq 1$.

≣ 6.4 LOGARITHMIC FUNCTIONS

LOGARITHMIC FUNCTION

A **logarithmic function** with base a, where $a \neq 1$ is a positive real number, is a function of the form

$$f(x) = \log_a x$$

The domain of $f(x) = \log_a x$ consists of the positive real numbers, $x > 0$.

It should not be surprising that the logarithmic function and the exponential function are somehow related. Now let's see what that relationship is. Suppose g represents an exponential function with base a and f represents a logarithmic function with base a. Then

$$g(x) = a^x \qquad f(x) = \log_a x \qquad x > 0$$

Using properties of logarithms, we find that

$$f(g(x)) = \log_a g(x) = \log_a a^x = x \qquad \text{for all real numbers } x$$

and

$$g(f(x)) = a^{f(x)} = a^{\log_a x} = x \qquad \text{for all positive real numbers } x$$

We conclude that the functions f and g are inverses of each other.

Based on the discussion in Section 4.5 about the properties of a function and its inverse, we now know two facts about logarithmic functions:

The range of a logarithmic function consists of all real numbers because the domain of the exponential function consists of all real numbers.

The graph of a logarithmic function $f(x) = \log_a x$ is the reflection about the line $y = x$ of the graph of the exponential function $g(x) = a^x$. See Figure 10.

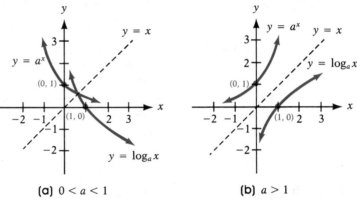

(a) $0 < a < 1$ **(b)** $a > 1$

FIGURE 10

Some other facts about a logarithmic function $f(x) = \log_a x$ are also now apparent:

The x-intercept of the graph is 1 since $f(1) = \log_a 1 = 0$. There is no y-intercept.

The y-axis is a vertical asymptote of the graph.

A logarithmic function is increasing if $a > 1$ and is decreasing if $0 < a < 1$.

The graph is smooth, with no corners or gaps.

NATURAL LOGARITHM FUNCTION

If the base of a logarithmic function is the number e, then we have the **natural logarithm function** $f(x) = \ln x$. Its graph is shown in Figure 11.

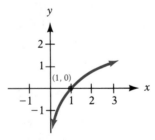

FIGURE 11

$y = \ln x$

EXAMPLE 1   Graph each logarithmic function by starting with the graph of $y = \ln x$ in Figure 11.

(a) $g(x) = \ln \dfrac{1}{x} = -\ln x$ (b) $h(x) = \ln(x + 2)$

(c) $k(x) = \ln(1 - x)$

Solution   (a) See Figure 12(a).

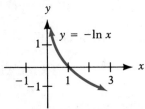

FIGURE 12(a)
Reflection about x-axis of $y = \ln x$

(b) See Figure 12(b).

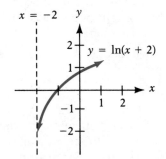

FIGURE 12(b)
Horizontal shift of $y = \ln x$

(c) We use the following steps [See Figure 12(c)]:

$$y = \ln x$$

$$y = \ln(-x) \qquad \qquad \text{Reflection about } y\text{-axis}$$

$$y = \ln[-(x - 1)] = \ln(1 - x) \quad \text{Replace } x \text{ by } x - 1;$$
$$\text{shift right 1 unit}$$

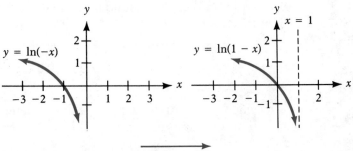

Horizontal
shift

FIGURE 12(c)
Reflection about y-axis of $y = \ln x$; followed by a horizontal shift

Remember that the domain of a logarithmic function consists of the *positive* real numbers.

EXAMPLE 2   Find the domain of each logarithmic function:

(a) $f(x) = \log_2(1 - x)$ (b) $g(x) = \ln\dfrac{1 + x}{1 - x}$ (c) $h(x) = \log_{1/2}|x|$

Solution   (a) The domain of f consists of all x for which $1 - x > 0$; that is, all $x < 1$.

(b) The domain of g is restricted to

$$\frac{1 + x}{1 - x} > 0$$

Using techniques introduced in Chapter 2 for solving inequalities, we find that the domain of g consists of all x between -1 and 1; that is, $-1 < x < 1$.

(c) Since $|x| > 0$ provided $x \neq 0$, the domain of h consists of all nonzero real numbers.   ≡

Applications

Common logarithms often appear in the measurement of quantities because they provide a way to scale down positive numbers that vary from very small to very large. For example, if a certain quantity can take on values from $0.0000000001 = 10^{-10}$ to $10,000,000,000 = 10^{10}$, the common logarithms of such numbers would be between -10 and 10.

INTENSITY OF A SOUND WAVE   Our first application utilizes a logarithmic scale to measure the loudness of a sound. Physicists define the **intensity of a sound wave** as the amount of energy the wave transmits through a given area.

For example, the least intense sound that a human ear can detect at a frequency of 100 hertz is about 10^{-12} watt per square meter. The LOUDNESS   **loudness** $L(x)$, measured in **decibels** (so-named in honor of Alexander DECIBELS   Graham Bell), of a sound of intensity x, measured in watts per square meter, is defined as

$$L(x) = 10 \log \frac{x}{I_0} \tag{1}$$

where $I_0 = 10^{-12}$ watt per square meter is the least intense sound that a human ear can detect. If we let $x = I_0$ in Equation (1), we get

$$L(I_0) = 10 \log \frac{I_0}{I_0} = 10 \log 1 = 0$$

Thus, at the threshold of human hearing, the loudness is 0 decibels. Table 3 gives the loudness of some common sounds.

TABLE 3 Loudness of common sounds, in decibels

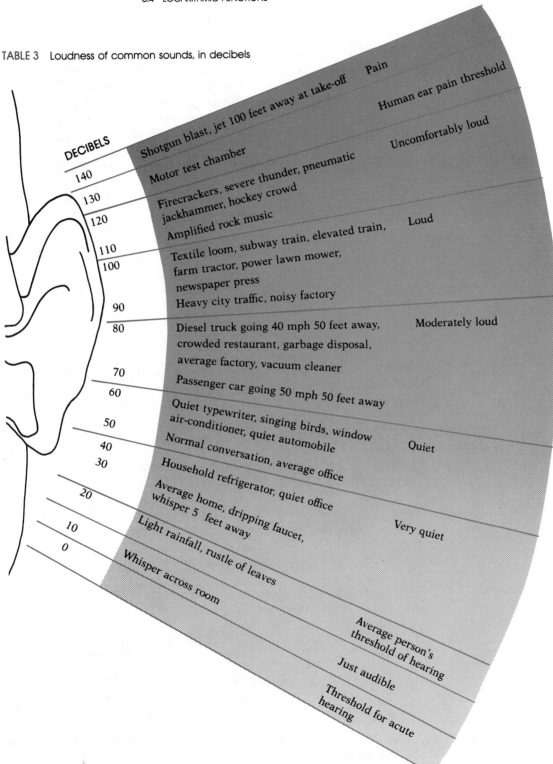

DECIBELS

Pain

	Human ear pain threshold
Shotgun blast, jet 100 feet away at take-off	
140	Uncomfortably loud
Motor test chamber	
130	
Firecrackers, severe thunder, pneumatic	
120	jackhammer, hockey crowd
Amplified rock music	
110	Loud
100	Textile loom, subway train, elevated train,
	farm tractor, power lawn mower,
	newspaper press
90	Heavy city traffic, noisy factory
80	Diesel truck going 40 mph 50 feet away, Moderately loud
	crowded restaurant, garbage disposal,
	average factory, vacuum cleaner
70	Passenger car going 50 mph 50 feet away
60	
	Quiet typewriter, singing birds, window
50	air-conditioner, quiet automobile
40	Normal conversation, average office Quiet
30	Household refrigerator, quiet office
	Average home, dripping faucet,
20	whisper 5 feet away
	Very quiet
10	Light rainfall, rustle of leaves
0	Whisper across room

Average person's
threshold of hearing

Just audible

Threshold for acute
hearing

Note that a decibel is not a linear unit like the meter. For example, a noise level of 10 decibels is 10 times as great as a noise level of 0 decibels. [If $L(x) = 10$, then $x = 10I_0$.] A noise level of 20 decibels is 100 times as great as a noise level of 0 decibels. [If $L(x) = 20$, then $x = 100I_0$.] A noise level of 30 decibels is 1000 times as great as a noise level of 0 decibels, and so on.

C **EXAMPLE 3** Use Table 3 to find the intensity of the sound of a dripping faucet.

Solution From Table 3, we see that the loudness of the sound of dripping water is 30 decibels. Thus, by Equation (1), its intensity x obeys

$$30 = 10 \log \frac{x}{I_0}$$

$$3 = \log \frac{x}{I_0}$$

$$\frac{x}{I_0} = 10^3$$

$$x = 1000I_0$$

where $I_0 = 10^{-12}$ watt per square meter. Thus, the intensity of the sound of a dripping faucet is 1000 times as great as a noise level of 0 decibels; that is, such a sound has an intensity of $1000 \cdot 10^{-12} = 10^{-9}$ watt per square meter. ■

EXAMPLE 4 Use Table 3 to determine the loudness of a jackhammer (in use), if it is known that this sound is 10 times as intense as the sound due to heavy city traffic.

Solution The sound due to heavy city traffic has a loudness of 90 decibels. Its intensity, therefore, is the value of x in the equation

$$90 = 10 \log \frac{x}{I_0}$$

A sound 10 times as intense as x has loudness $L(10x)$. Thus, the loudness of the jackhammer is

$$L(10x) = 10 \log \frac{10x}{I_0} = 10\left(\log 10 + \log \frac{x}{I_0}\right)$$

$$= 10 \log 10 + 10 \log \frac{x}{I_0}$$

$$= 10 + 90 = 100 \text{ decibels}$$ ■

Our second application uses a logarithmic scale to measure the magnitude of an earthquake.

RICHTER SCALE   The **Richter scale*** is one way of converting seismographic readings into numbers that provide an easy reference for measuring the magnitude M of an earthquake. All earthquakes are compared to a so-called **zero-level earthquake** whose seismographic reading measures

ZERO-LEVEL EARTHQUAKE   0.001 millimeter at a distance of 100 kilometers from the epicenter. An earthquake whose seismographic reading measures x millimeters

MAGNITUDE   has **magnitude** $M(x)$ given by

$$M(x) = \log \frac{x}{x_0} \qquad (2)$$

where $x_0 = 10^{-3}$ is the reading of a zero-level earthquake the same distance from its epicenter.

EXAMPLE 5   What is the magnitude of an earthquake whose seismographic reading is 0.1 millimeter at a distance of 100 kilometers from its epicenter?

Solution   If $x = 0.1$, the magnitude $M(x)$ of this earthquake is

$$M(0.1) = \log \frac{x}{x_0} = \log \frac{0.1}{0.001} = \log \frac{10^{-1}}{10^{-3}} = \log 10^2 = 2$$

This earthquake thus measures 2.0 on the Richter scale.   ■

INTENSITY OF AN EARTHQUAKE   Based on Formula (2), we define the **intensity of an earthquake** as the ratio of x to x_0. For example, the intensity of the earthquake described in Example 5 is $\frac{0.1}{0.001} = 10^2 = 100$. That is, it is 100 times as intense as a zero-level earthquake.

EXAMPLE 6   The devastating San Francisco earthquake of 1906 measured 8.9 on the Richter scale. How did the intensity of that earthquake compare to the Casiguran Bay earthquake described in Figure 13 (page 404), which measured 6.2 on the Richter scale?

Solution   Let x_1 and x_2 denote the seismographic readings, respectively, of the 1906 San Francisco earthquake and the Casiguran Bay earthquake. Then, based on Formula (2),

$$8.9 = \log \frac{x_1}{x_0} \qquad 6.2 = \log \frac{x_2}{x_0}$$

Consequently,

$$\frac{x_1}{x_0} = 10^{8.9} \qquad \frac{x_2}{x_0} = 10^{6.2}$$

*Named after the American scientist C. F. Richter, who devised it in 1935. Richter died September 30, 1985.

2 earthquakes rock Philippines

MANILA, Philippines (UPI)—Two earthquakes struck a wide area of the Philippines's main island of Luzon within hours of each other Wednesday, but there were no immediate reports of damage or casualties, officials said.

The first temblor struck at 12:16 a.m. (10:16 a.m. CST Tuesday) and was centered in the waters of Casiguran Bay 131 miles northeast of Manila, the Manila Geophysical Division said. The U.S. Geological Survey in Golden, Colo., said the quake measured 6.2 on the Richter scale.

Seismologist Ulpiano Trillanes said the quake shook buildings but added he did not think it was strong enough to cause damage.

Corporal Rodolfo Mendoza of the Baguio city police said there were no reports of casualties in the mountain resort 120 miles north of the capital.

The USGS said that about nine hours later, at 9:07 a.m. Wednesday (7:07 p.m. CST Tuesday), a second quake measuring 6.0 on the Richter scale occurred about 125 miles northeast of Manila.

Police in the northern city of Laoag reported government employees fled buildings in panic after the second temblor struck.

FIGURE 13
Source: United Press International, April 24, 1985

The 1906 San Francisco earthquake was $10^{8.9}$ times as intense as a zero-level earthquake. The Casiguran Bay earthquake was $10^{6.2}$ times as intense as a zero-level earthquake. Thus,

$$\frac{x_1}{x_2} = \frac{10^{8.9}x_0}{10^{6.2}x_0} = 10^{2.7} \approx 501$$

$$x_1 \approx 501x_2$$

Hence, the San Francisco earthquake was 501 times as intense as the Casiguran Bay earthquake. ≡

Example 6 demonstrates that the relative intensity of two earthquakes can be found by raising 10 to a power equal to the difference of their Richter-scale readings.

EXERCISE 6.4

 In Problems 1 and 2 graph each function by carefully preparing a table listing points on the graph.

1. $f(x) = \log_2 x$ **2.** $g(x) = \log_5 x$

In Problems 3–16 use the results of Problems 1 and 2 to graph each function.

3. $f(x) = \log_2(x + 4)$ **4.** $g(x) = \log_5(x - 3)$

5. $f(x) = \log_2(-x)$ **6.** $g(x) = \log_5(-x)$

7. $f(x) = \log_2(3 - x)$ **8.** $g(x) = \log_5(4 - x)$

9. $f(x) = \log_2\dfrac{1}{x}$ **10.** $g(x) = \log_5\dfrac{1}{x}$

11. $f(x) = \log_2 4x$

12. $g(x) = \log_5 25x$

13. $f(x) = \log_2 x^2$

14. $g(x) = \log_5 x^2$

15. $f(x) = \log_{1/2} x$

16. $g(x) = \log_{1/5} x$

In Problems 17–24 find the domain of each function.

17. $f(x) = \ln(3 - x)$

18. $g(x) = \ln(x^2 - 1)$

19. $F(x) = \log_2 x^2$

20. $H(x) = \log_5 x^3$

21. $h(x) = \log_{1/2}(-x)$

22. $G(x) = \log_{1/2}\dfrac{1}{x}$

23. $f(x) = \dfrac{1}{\ln x}$

24. $g(x) = \ln(x - 5)$

25. Find a such that the graph of $f(x) = \log_a x$ contains the point $(2, 2)$.

26. Find a such that the graph of $f(x) = \log_a x$ contains the point $(\frac{1}{2}, -4)$.

27. Find the domain of $f(x) = \log_a x^2$ and the domain of $g(x) = 2 \log_a x$. Since $\log_a x^2 = 2 \log_a x$, how do you reconcile the fact that the domains are unequal?

28. If $f(x) = \log_a x$, show that

$$\frac{f(x + h) - f(x)}{h} = \log_a\left(1 + \frac{h}{x}\right)^{1/h} \qquad h \neq 0$$

29. If $f(x) = \log_a x$, show that $-f(x) = \log_{1/a} x$.

30. If $f(x) = \log_a x$, show that $f(1/x) = -f(x)$.

31. If $f(x) = \log_a x$, show that $f(AB) = f(A) + f(B)$.

32. If $f(x) = \log_a x$, show that $f(x^\alpha) = \alpha f(x)$.

In Problems 33–42 express y as a function of x. The constant C is a positive number.

33. $\ln y = \ln x + \ln C$

34. $\ln y = \ln(x + C)$

35. $\ln y = \ln x + \ln(x + 1) + \ln C$

36. $\ln y = 2 \ln x - \ln(x + 1) + \ln C$

37. $\ln y = 3x + \ln C$

38. $\ln y = -2x + \ln C$

39. $\ln(y - 3) = -4x + \ln C$

40. $\ln(y + 4) = 5x + \ln C$

41. $3 \ln y = \dfrac{1}{2} \ln(2x + 1) - \dfrac{1}{3} \ln(x + 4) + \ln C$

42. $2 \ln y = -\dfrac{1}{2} \ln x + \dfrac{1}{3} \ln(x^2 + 1) + \ln C$

43. Find the loudness of a dishwasher that operates at an intensity of 10^{-5} watt per square meter. Express your answer in decibels.

44. Find the loudness of a diesel engine that operates at an intensity of 10^{-3} watt per square meter. Express your answer in decibels.

[C] **45.** With engines at full throttle, a Boeing 727 jetliner produces noise at an intensity of 0.15 watt per square meter. Find the loudness of these engines in decibels.

C **46.** A whisper produces noise at an intensity of $10^{-9.8}$ watt per square meter. What is the loudness of a whisper in decibels?

47. For humans, the threshold of pain due to sound averages 130 decibels. What is the intensity of such a sound in watts per square meter?

C **48.** If one sound is 50 times as intense as another, what is the difference in the loudness of the two sounds? Express your answer in decibels.

49. Find the magnitude of an earthquake whose seismograph reading is 10.0 millimeters at a distance of 100 kilometers from its epicenter.

C **50.** Find the magnitude of an earthquake whose seismograph reading is 1210 millimeters at a distance of 100 kilometers from its epicenter.

C **51.** The Mexico City earthquake of 1978 registered 7.85 on the Richter scale. What would a seismograph 100 kilometers from the epicenter have measured for this earthquake? How does this earthquake compare in intensity to the 1906 San Francisco earthquake, which registered 8.9 on the Richter scale?

52. Two earthquakes differ by 1.0 when measured on the Richter scale. How would the seismographic readings differ at a distance of 100 kilometers from the epicenter? How do their intensities compare?

ALTITUDE *In Problems 53 and 54 use the following result: If x is the atmospheric pressure, measured in millimeters of mercury, then the formula for the* **altitude** *h(x), measured in meters above sea level, is*

$$h(x) = (30T + 8000)\log \frac{P_0}{x}$$

where T is the temperature, measured in degrees Celsius, and P_0 is the atmospheric pressure at sea level, which is approximately 760 millimeters of mercury.

C **53.** At what height is an aircraft whose instruments record an outside temperature of 0° C and a barometric pressure of 300 millimeters of mercury?

C **54.** What is the atmospheric pressure, in millimeters of mercury, on Mt. Everest, altitude approximately 8900 meters, if the air temperature is 5° C?

▦ 6.5 GROWTH AND DECAY

Many natural phenomena have been found to follow the law that the amount A varies with time t according to the rule

$$A = A_0 e^{kt} \tag{1}$$

where A_0 is the original amount and $k \neq 0$ is a constant.

If $k > 0$, then Equation (1) states that the amount A is increasing over time; if $k < 0$, the amount A is decreasing over time. In either case, when an amount A varies over time according to Equation (1), it is said to follow the **exponential law** or the **law of uninhibited growth** ($k > 0$) **or decay** ($k < 0$).

EXPONENTIAL LAW
LAW OF UNINHIBITED GROWTH
LAW OF UNINHIBITED DECAY

For example, we saw in Section 6.2 that continuously compounded interest follows the law of uninhibited growth. In this section we shall look at three additional phenomena that follow the exponential law.

Biology

MITOSIS

Mitosis, or division of cells, is a universal process in the growth of living organisms such as amebas, plants, human skin cells, and many others. (Human nerve cells are one of the few exceptions.) Based on an ideal situation in which no cells die and no by-products are produced, the number of cells present at a given time follows the law of uninhibited growth. Actually, however, after enough time has passed, growth at an exponential rate will cease due to the influence of factors such as lack of living space, dwindling food supply, and so on. The law of uninhibited growth, though, does accurately reflect the early stages of the mitosis process.

The mitosis process begins with a culture containing N_0 cells. Each cell in the culture will grow for a certain period of time and then divide into two identical cells. We assume that the time needed for each cell to divide in two is constant and does not change as the number of cells increases. These cells will then grow, and eventually each will divide in two, and so on. A formula that will give the number N of cells in the culture after a time t has passed is

$$N(t) = N_0 e^{kt} \qquad (2)$$

where k is a positive constant.

C   EXAMPLE 1

A colony of bacteria increases according to the law of uninhibited growth. If the number of bacteria doubles in 3 hours, how long will it take for the size of the colony to triple?

Solution

Using Formula (2), the number N of cells at a time t is

$$N(t) = N_0 e^{kt}$$

where N_0 is the initial number of bacteria present and k is a positive number. We first seek the number k. The number of cells doubles in 3 hours; thus we have

$$N(3) = 2N_0$$

But $N(3) = N_0 e^{k(3)}$, so that

$$N_0 e^{k(3)} = 2N_0$$

$$e^{3k} = 2$$

$$3k = \ln 2$$

$$k = \frac{1}{3} \ln 2 \approx \frac{1}{3}(0.6931) = 0.2310$$

Formula (2) for this growth process is therefore

$$N(t) = N_0 e^{0.2310t}$$

The time t needed for the size of the colony to triple requires that $N = 3N_0$. Thus, we substitute $3N_0$ for N to get

$$3N_0 = N_0 e^{0.2310t}$$

$$3 = e^{0.2310t}$$

$$0.2310t = \ln 3$$

$$t = \frac{1}{0.2310} \ln 3 \approx \frac{1.0986}{0.2310} = 4.756 \text{ hours}$$

It will take about 4.756 hours for the size of the colony to triple. ▀

Radioactive Decay

Radioactive materials follow the law of uninhibited decay. Thus, the amount A of a radioactive material present at time t is given by the formula

$$A = A_0 e^{kt} \tag{3}$$

where A_0 is the original amount of radioactive material and k is a negative number.

HALF-LIFE   All radioactive substances have a specific **half-life**, which is the
CARBON DATING   time required for half of the particular substance to decay. In **carbon dating**, we use the fact that all living organisms contain two kinds of carbon, carbon-12 (a stable carbon) and carbon-14 (a radioactive carbon, with a half-life of 5600 years). While an organism is living, the ratio of carbon-12 to carbon-14 is constant. But, when an organism dies, the amount of carbon-12 present remains unchanged, whereas the amount of carbon-14 begins to decrease. This change in the amount of carbon-14 present relative to the amount of carbon-12 present makes it possible to calculate when an organism died.

EXAMPLE 2 The skull of an animal found in an archaeological dig was found to contain approximately 15% of the original amount of carbon-14. If the half-life of carbon-14 is 5600 years, approximately when did the animal die?

Solution Using Equation (3), the amount A of carbon-14 present at time t is

$$A = A_0 e^{kt}$$

where A_0 is the original amount of carbon-14 present and k is a negative number. We first seek the number k. To find it, we use the fact that, after 5600 years, half of the original amount of carbon-14 remains. Thus,

$$\frac{1}{2}A_0 = A_0 e^{k(5600)}$$

$$\frac{1}{2} = e^{5600k}$$

$$5600k = \ln \frac{1}{2}$$

$$k = \frac{1}{5600} \ln \frac{1}{2} \approx -0.000124$$

Formula (3) is therefore

$$A = A_0 e^{-0.000124t}$$

If the amount A of carbon-14 now present is 15% of the original amount, it follows that

$$0.15A_0 = A_0 e^{-0.000124t}$$
$$0.15 = e^{-0.000124t}$$
$$-0.000124t = \ln 0.15$$
$$t = \frac{1}{-0.000124} \ln 0.15 \approx 15{,}300 \text{ years}$$

The animal died about 15,300 years ago. ≡

Newton's Law of Cooling

NEWTON'S LAW OF COOLING **Newton's Law of Cooling*** states that the temperature of a heated object will decrease exponentially over time toward the temperature of the surrounding medium. That is, the temperature u of a heated object at a given time t obeys the law

$$u = T + (u_0 - T)e^{kt} \qquad (4)$$

*Named after Sir Isaac Newton (1642–1727), one of the cofounders of calculus.

where T is the constant temperature of the surrounding medium, u_0 is the initial temperature of the heated object, and k is a negative number.

EXAMPLE 3   An object is heated to 100° C and is then allowed to cool in a room whose air temperature is 30° C. If after 5 minutes the temperature of the object is 80° C, when will its temperature be 50° C?

Solution   Using Equation (4), with $T = 30$ and $u_0 = 100$, the temperature (in degrees Celsius) of the object at time t (in minutes) is

$$u = 30 + (100 - 30)e^{kt} = 30 + 70e^{kt} \tag{5}$$

where k is a negative number. To find k, we use the fact that $u = 80$ when $t = 5$. Then,

$$80 = 30 + 70e^{k(5)}$$
$$50 = 70e^{5k}$$
$$e^{5k} = \frac{50}{70}$$
$$5k = \ln\frac{5}{7}$$
$$k = \frac{1}{5}\ln\frac{5}{7} \approx -0.0673$$

Formula (5) is therefore

$$u = 30 + 70e^{-0.0673t}$$

Now we want to find t when $u = 50°$ C:

$$50 = 30 + 70e^{-0.0673t}$$
$$20 = 70e^{-0.0673t}$$
$$e^{-0.0673t} = \frac{20}{70}$$
$$-0.0673t = \ln\frac{2}{7}$$
$$t = \frac{1}{-0.0673}\ln\frac{2}{7} \approx 18.6 \text{ minutes}$$

Thus, the temperature of the object will be 50° C after about 18.6 minutes.   ≡

[C] EXERCISE 6.5   1. The size P of a certain insect population at time t, in days, obeys the equation $P = 500e^{0.02t}$. After how many days will the population reach 1000? 2000?

2. The number N of bacteria present in a culture at time t, in hours, obeys the equation $N = 1000e^{0.01t}$. After how many hours will the population equal 1500? 2000?

3. Strontium-90 is a radioactive material that decays according to the law $A = A_0e^{-0.0244t}$, where A_0 is the initial amount present and A is the amount present at time t, in years. What is the half-life of strontium-90?

4. Iodine-131 is a radioactive material that decays according to the law $A = A_0e^{-0.087t}$, where A_0 is the initial amount present and A is the amount present at time t, in days. What is the half-life of iodine-131?

5. Use the information in Problem 3 to determine how long it takes for 100 grams of strontium-90 to decay to 10 grams.

6. Use the information in Problem 4 to find out how long it takes for 100 grams of iodine-131 to decay to 10 grams.

7. The population of a colony of mosquitoes obeys the law of uninhibited growth. If there are 1000 mosquitoes initially, and there are 1800 after 1 day, what is the size of the colony after 3 days? How long is it until there are 10,000 mosquitoes?

8. A culture of bacteria obeys the law of uninhibited growth. If there are 500 bacteria cells present initially, and there are 800 after 1 hour, how many will be present in the culture after 5 hours? How long is it until there are 20,000 bacteria?

9. The population of a southern city follows the exponential law. If the population doubled in size over an 18 month period and the current population is 10,000, what will the population be 2 years from now?

10. The population of a midwestern city follows the exponential law. If the population decreased from 900,000 to 800,000 from 1984 to 1986, what will the population be in 1988?

11. The half-life of radium is 1690 years. If 10 grams are present now, how much will be present in 50 years?

12. The half-life of radioactive potassium is 1.3 billion years. If 10 grams are present now, how much will be present in 100 years? In 1000 years?

13. A piece of charcoal is found to contain 30% of the carbon-14 it originally had. When did the tree from which the charcoal came die? Use 5600 years as the half-life of carbon-14.

14. A fossilized leaf contains 70% of its normal amount of carbon-14. How old is the fossil?

15. A pizza baked at 450° F is removed from the oven at 5:00 P.M. into a room that is a constant 70° F. After 5 minutes, the pizza is at 300° F. At what time can you begin eating the pizza if you want its temperature to be 135° F?

16. A thermometer reading 72° F is placed in a refrigerator where the temperature is a constant 38° F. If the thermometer reads 60° F after 2 minutes, what will it read after 7 minutes? How long will it take before the thermometer reads 39° F?

17. A thermometer reading 8° C is brought into a room with a constant temperature of 35° C. If the thermometer reads 15° C after 3 minutes, what will it read after being in the room for 5 minutes? For 10 minutes? [*Hint:* You need to construct a formula similar to Equation (4).]

18. A frozen steak has a temperature of 28° F. It is placed in a room with a constant temperature of 70° F. After 10 minutes, the temperature of the steak has risen to 35° F. What will the temperature of the steak be after 30 minutes? How long will it take the steak to thaw, if *thawing* means the steak is at a temperature of 45° F? [See hint given for Problem 17.]

19. Salt (NaCl) decomposes in water into sodium (Na^+) and chloride (Cl^-) ions according to the law of uninhibited decay. If the initial amount of salt is 25 kilograms and, after 10 hours, 15 kilograms of salt are left, how much salt is left after 1 day? How long does it take until $\frac{1}{2}$ kilogram of salt is left?

20. The voltage of a certain condenser decreases over time according to the law of uninhibited decay. If the initial voltage is 40 volts, and 2 seconds later it is 10 volts, what is the voltage after 5 seconds?

21. The concentration of alcohol in a person's blood is measurable. Recent medical research suggests that the risk R (given as a percentage) of having a car accident obeys the law of uninhibited growth,

$$R = 1.5e^{ka}$$

where a is the variable concentration of alcohol in the blood and k is a constant. Suppose a concentration of alcohol in the blood of 0.11 results in a 10% risk of an accident. What is the risk, then, if the concentration is 0.17? What concentration of alcohol corresponds to a risk of 100%? If anyone whose risk of having an accident is 20% or more should not drive, what concentration of alcohol in the blood should be used to test a person's ability to continue driving?

22. Rework Problem 21 if the law changes to $R = 1.0e^{ka}$.

23. The equation governing the amount of current I (in amperes) after time t (in seconds) in a simple RL circuit consisting of a resistance R (in ohms), an inductance L (in henrys), and an electromotive force E (in volts) is

$$I = \frac{E}{R}(1 - e^{-(R/L)t})$$

If $E = 12$ volts, $R = 10$ ohms, and $L = 5$ henrys, how long does it take to obtain a current of 0.5 ampere? Of 1.0 ampere? Graph this function, measuring I along the y-axis and t along the x-axis.

24. Rework Problem 23 if $E = 12$ volts, $R = 5$ ohms, and $L = 10$ henrys.

25. Psychologists sometimes use the function

$$L(t) = A(1 - e^{-kt})$$

to measure the amount L learned at time t. The number A represents the amount to be learned and the number k measures the rate of learning. Suppose a student has an amount A of 200 vocabulary words to learn. A psychologist determines that the student learned 20 vocabulary words after 5 minutes. Determine the rate of learning k. Approximately how many words will the student have learned after 10 minutes? After 15 minutes? How long does it take for the student to learn 180 words?

26. Rework Problem 25 for a student who has to learn 200 vocabulary words but learns only 15 words after 5 minutes.

C H A P T E R   R E V I E W

VOCABULARY

algebraic functions
transcendental
  functions
laws of exponents
exponential function
number e
tangent lines
exponential equations
compounded
compound interest
compound interest
  formula
continuous
  compounding
effective rate of
  interest
present value
present value formulas

logarithm to the base
  a of M
properties of
  logarithms
log of a product
log of a quotient
logarithmic equations
common logarithms
natural logarithms
log
ln
change-of-base formula
logarithmic function
natural logarithm
  function
intensity of a sound
  wave

loudness
decibels
Richter scale
zero-level earthquake
magnitude
intensity of an
  earthquake
exponential law
law of uninhibited
  growth
law of uninhibited
  decay
mitosis
half-life
carbon dating
Newton's Law of
  Cooling

FILL-IN-THE-BLANK QUESTIONS

1. The graph of every exponential function $f(x) = a^x$, $a > 0$, $a \neq 1$, passes through the point _____.

2. If the graph of an exponential function $f(x) = a^x$, $a > 0$, $a \neq 1$, is decreasing, then its base must be less than _____.

3. If $3^x = 3^4$, then $x =$ _____.

4. The logarithm of a product equals the _____ of the logarithms.

5. For every base, the logarithm of _____ equals zero.

6. If $\log_8 M = \log_5 7 / \log_5 8$, then $M =$ _____.

7. The domain of the logarithmic function $f(x) = \log_a x$ consists of

_____.

8. The graph of every logarithmic function $f(x) = \log_a x$, $a > 0$, $a \neq 1$,

passes through the point _____.

9. If the graph of a logarithmic function $f(x) = \log_a x$, $a > 0$, $a \neq 1$, is increasing, then its base must be larger than _____.

10. If $\log_3 x = \log_3 7$, then $x =$ _____.

REVIEW EXERCISES

In Problems 1–6 evaluate each expression.

1. $\log_2 \dfrac{1}{8}$

2. $\log_3 81$

3. $\ln e^{\sqrt{2}}$

4. $e^{\ln 0.1}$

5. $2^{\log_2 0.4}$

6. $\log_2 2^{\sqrt{3}}$

In Problems 7–12 write each expression as a single logarithm.

7. $3 \log_4 x^2 + \dfrac{1}{2} \log_4 \sqrt{x}$

8. $-2 \log_3 \dfrac{1}{x} + \dfrac{1}{3} \log_3 \sqrt{x}$

9. $\ln \dfrac{x-1}{x} + \ln \dfrac{x}{x+1} - \ln(x^2 - 1)$

10. $\log(x^2 - 9) - \log(x^2 + 7x + 12)$

11. $2 \log 2 + 3 \log x - \dfrac{1}{2}[\log(x + 3) + \log(x - 2)]$

12. $\dfrac{1}{2} \ln(x^2 + 1) - 4 \ln \dfrac{1}{2} - \dfrac{1}{2}[\ln(x - 4) + \ln x]$

In Problems 13–20 find y as a function of x. The constant C is a positive number.

13. $\ln y = 2x^2 + \ln C$

14. $\ln(y - 3) = \ln 2x^2 + \ln C$

15. $\dfrac{1}{2} \ln y = 3x^2 + \ln C$

16. $\ln 2y = \ln(x + 1) + \ln(x + 2) + \ln C$

17. $\ln(y - 3) + \ln(y + 3) = x + C$

18. $\ln(y - 1) + \ln(y + 1) = -x + C$

19. $e^{y+C} = x^2 + 4$

20. $e^{3y-C} = (x + 4)^2$

In Problems 21–30 graph each function. Begin each problem either with the graph of $y = e^x$ or with the graph of $y = \ln x$.

21. $f(x) = e^{-x}$

22. $f(x) = \ln(-x)$

23. $f(x) = 1 - e^x$

24. $f(x) = 3 + \ln x$

25. $f(x) = 3e^x$

26. $f(x) = \dfrac{1}{2} \ln x$

27. $f(x) = e^{|x|}$

28. $f(x) = \ln|x|$

29. $f(x) = 3 - e^{-x}$

30. $f(x) = 4 - \ln(-x)$

In Problems 31–50 solve each equation.

31. $4^{1-2x} = 2$

32. $8^{6+3x} = 4$

33. $3^{x^2+x} = \sqrt{3}$

34. $4^{x-x^2} = \dfrac{1}{2}$

35. $\log_x 64 = -3$

36. $\log_{\sqrt{2}} x = -6$

37. $\log_{\sqrt{3}} 9\sqrt{3} = x$

38. $\log_x 3 = \dfrac{1}{5}$

C **39.** $5^x = 3^{x+2}$

C **40.** $5^{x+2} = 7^{x-2}$

41. $9^{2x} = 27^{3x-4}$

42. $25^{2x} = 5^{x^2-12}$

43. $8 = 4^{x^2} \cdot 2^{5x}$

44. $2^x \cdot 5 = 10^x$

45. $\log_6(x + 3) + \log_6(x + 4) = 1$

46. $\log_{10}(7x - 12) = 2 \log_{10} x$

C **47.** $e^{1-x} = 5$

C **48.** $e^{1-2x} = 4$

C **49.** $2^{3x} = 3^{2x+1}$

C **50.** $2^{x^3} = 3^{x^2}$

C **51.** First Colonial Bankshares Corporation advertised the following **IRA** investment plans in the spring of 1985. Assuming continuous compounding, what was the annual rate of interest they offered?

Target IRA Plans

FOR EACH $5,000 MATURITY VALUE DESIRED

Deposit:	At a term of:
$620.17	20 years
$1,045.02	15 years
$1,760.92	10 years
$2,967.26	5 years

C **52.** See Problem 51. First Colonial Bankshares claims that $4000 invested today will have a value of over $32,000 in 20 years. Use the answer found in Problem 51 to find the actual value of the $4000 in 20 years. Assume continuous compounding.

C **53.** Refer back to Example 6 and Figure 13 (Section 6.4). How did the intensities of the two Philippine earthquakes compare?

54. On September 9, 1985, the western suburbs of Chicago experienced a mild earthquake that registered 3.0 on the Richter scale. How did this earthquake compare in intensity to the great San Francisco earthquake of 1906, which registered 8.9 on the Richter scale?

55. The bones of a prehistoric man found in the desert of New Mexico contain approximately 5% of the original amount of carbon-14. If the half-life of carbon-14 is 5600 years, approximately how long ago did the man die?

56. A skillet is removed from an oven whose temperature is 450° F and placed in a room whose temperature is 70° F. After 5 minutes, the temperature of the skillet is 400° F. How long will it be until its temperature is 150° F?

SYSTEMS OF EQUATIONS AND INEQUALITIES

7.1 Solutions of Systems of Linear
     Equations: Substitution
7.2 Solutions of Systems of Linear
     Equations: Elimination
7.3 Solutions of Systems of Linear
     Equations: Matrices
7.4 Solutions of Systems of Linear
     Equations: Determinants
7.5 Solutions of Systems of Nonlinear
     Equations
7.6 Solutions of Systems of Linear
     Inequalities
     Chapter Review

In Chapter 2 we studied ways to solve equations and inequalities containing one variable. In this chapter we take up the problem of solving equations and inequalities containing two or more variables. As the chapter contents suggest, there are a variety of ways to solve such problems.

The method of substitution for solving equations in several unknowns goes back to ancient times.

The method of elimination, though it had existed for centuries, was put into systematic order by Karl Friedrich Gauss (1777–1855) and by Camille Jordan (1838–1922). This method is now used for solving large systems by computer.

The theory of matrices was begun in 1857 by Arthur Cayley (1821–1895), though only later were matrices used as we use them in this chapter. Matrices have become a very flexible instrument, useful in almost all areas of mathematics. You will meet them again in the next chapter.

The method of determinants was invented by Seki Kōwa (1642–1708) in 1683 in Japan and by Gottfried Wilhelm Leibniz (1646–1716) in 1693 in Germany. Both used them only in relation to linear equations. Cramer's Rule is named after Gabriel Cramer (1704–1752), a professor in Geneva, who popularized the use of determinants for solving linear equations.

▤ 7.1 SOLUTIONS OF SYSTEMS OF LINEAR EQUATIONS: SUBSTITUTION

SYSTEM OF EQUATIONS

A **system of equations** is a collection of two or more equations, each containing one or more variables. Example 1 gives some samples of systems of equations.

EXAMPLE 1 (a) $\begin{cases} 2x + y = 5 & (1) \\ -4x + 6y = -2 & (2) \end{cases}$ Two equations containing two variables, x and y

(b) $\begin{cases} x + y^2 = 5 & (1) \\ 2x + y = 4 & (2) \end{cases}$ Two equations containing two variables, x and y

(c) $\begin{cases} x + y + z = 6 & (1) \\ 3x - 2y + 4z = 9 & (2) \\ x - y - z = 0 & (3) \end{cases}$ Three equations containing three variables, x, y, and z

(d) $\begin{cases} x + y + z = 5 & (1) \\ x - y = 2 & (2) \end{cases}$ Two equations containing three variables, x, y, and z

(e) $\begin{cases} x + y + z = 6 & (1) \\ 2x + 2z = 4 & (2) \\ y + z = 2 & (3) \\ x = 4 & (4) \end{cases}$ Four equations containing three variables, x, y, and z

▤

We will use a brace to remind us that we are dealing with a system of equations. We also will find it convenient to number each equation in the system.

SOLUTION

A **solution** of a system of equations consists of values for the variables that make each equation of the system a true statement. To

SOLVE

solve a system of equations means to find all solutions of the system.

For example, $x = 2$, $y = 1$ is a solution of the system in Example 1(a) because

$$2(2) + 1 = 5 \quad \text{and} \quad -4(2) + 6(1) = -2$$

A solution of the system in Example 1(b) is $x = 1$, $y = 2$ because

$$1 + 2^2 = 5 \quad \text{and} \quad 2(1) + 2 = 4$$

Another solution of the system in Example 1(b) is $x = \frac{11}{4}$, $y = -\frac{3}{2}$, which you can check for yourself. A solution of the system in Example 1(c) is $x = 3$, $y = 2$, $z = 1$ because

$\begin{cases} x + y + z = 6 & (1) \\ 3x - 2y + 4z = 9 & (2) \\ x - y - z = 0 & (3) \end{cases}$ $\begin{cases} 3 + 2 + 1 = 6 & (1) \\ 3(3) - 2(2) + 4(1) = 9 & (2) \\ 3 - 2 - 1 = 0 & (3) \end{cases}$ $x = 3, y = 2, z = 1$

Note that $x = 3$, $y = 3$, $z = 0$ is not a solution of the system in Example 1(c):

$$\begin{cases} x + y + z = 6 & (1) \\ 3x - 2y + 4z = 9 & (2) \\ x - y - z = 0 & (3) \end{cases} \qquad \begin{cases} 3 + 3 + 0 = 6 & (1) \quad x = 3, y = 3, z = 0 \\ 3(3) - 2(3) + 4(0) = 3 \neq 9 & (2) \\ 3 - 3 - 0 = 0 & (3) \end{cases}$$

Although these values satisfy the first and third equations, they do not satisfy the second equation. Any solution of the system must satisfy *each* equation of the system.

When a system of equations has at least one solution, it is said to be **consistent**; otherwise, it is called **inconsistent**.

CONSISTENT SYSTEM

INCONSISTENT SYSTEM

LINEAR EQUATION

An equation is said to be **linear** if it is equivalent to an equation of the form

$$a_1 x_1 + a_2 x_2 + \cdots + a_n x_n = b$$

where $x_1, x_2, \ldots, x_n$ are n distinct variables, $a_1, a_2, \ldots, a_n$, b are constants, and at least one of the a's is not zero.

Some examples of linear equations are

$$2x + 3y = 2 \qquad 5x - 2y + 3z = 10 \qquad 8x + 8y - 2z + 5w = 0$$

SYSTEM OF LINEAR EQUATIONS

If each equation in a system of equations is linear, then we have a **system of linear equations**. Thus, the systems in Examples 1(a), (c), (d), and (e) are linear, whereas the system in Example 1(b) is nonlinear. We shall concentrate on solving linear systems in Sections 7.1–7.4 and will take up nonlinear systems in Section 7.5.

The Method of Substitution

Let's now discuss one way to solve a system of linear equations.

EXAMPLE 2   Solve

$$\begin{cases} 2x + y = 5 & (1) \\ -4x + 6y = -2 & (2) \end{cases}$$

Solution   We solve the first equation for y, obtaining

$$y = 5 - 2x \qquad\qquad (1)$$

We substitute this result for y in the second equation and solve for the one remaining variable:

$$-4x + 6y = -2$$
$$-4x + 6(5 - 2x) = -2$$
$$-4x + 30 - 12x = -2$$
$$-16x = -32$$
$$x = 2$$

Once we know $x = 2$, we can easily find the value of y by substituting 2 for x in one of the original equations (we'll use the first one):

$$2x + y = 5$$
$$2(2) + y = 5$$
$$y = 1$$

The solution of the system is $x = 2$, $y = 1$. We checked this solution earlier.   ≡

In the solution to Example 2, once we found $x = 2$, we could have found y by using Equation (1), the one we obtained at the beginning when we solved for y.

$$y = 5 - 2x = 5 - 2(2) = 1$$
$$\uparrow$$
$$x = 2$$

SUBSTITUTION METHOD
The method used to solve the system in Example 2 is called **substitution**. The steps to be used are outlined below.

Step 1: Pick one of the equations and solve for one of the variables in terms of the remaining variables.

Step 2: Substitute the result in the remaining equations.

Step 3: If one equation in one variable results, solve this equation. Otherwise, repeat Step 1 until a single equation with one variable remains.

Step 4: Find the values of the remaining variables.

Step 5: Check the solution found.

EXAMPLE 3   Solve:

(a) $\begin{cases} 3x - 2y = 5 & (1) \\ 5x - y = 6 & (2) \end{cases}$ (b) $\begin{cases} 2x - 3y = 7 & (1) \\ 4x + 5y = 3 & (2) \end{cases}$

(c) $\begin{cases} x + y + z = 6 & (1) \\ 2x + 2z = 4 & (2) \\ y + z = 2 & (3) \\ x = 4 & (4) \end{cases}$

Solution (a) Step 1: After looking at the two equations, we conclude that it is easiest to solve for the variable y in the second equation:

$$5x - y = 6$$
$$y = 5x - 6$$

Step 2: We substitute this result into the first equation:

$$3x - 2y = 5$$
$$3x - 2(5x - 6) = 5$$
$$-7x + 12 = 5$$
$$-7x = -7$$
$$x = 1$$

Step 3: Because we now have one equation with one variable, namely $x = 1$, we proceed to Step 4.

Step 4: Knowing $x = 1$, we can find y from the equation

$$y = 5x - 6 = 5(1) - 6 = -1$$

Step 5: *Check:* $3(1) - 2(-1) = 3 + 2 = 5$ and
$$5(1) - (-1) = 5 + 1 = 6$$

The solution of the system is $x = 1$, $y = -1$.

(b) Step 1: In looking over the system, we conclude that there is no way to solve for one of the variables without introducing fractions. We'll solve for the variable x in the first equation:

$$2x - 3y = 7$$
$$2x = 3y + 7$$
$$x = \tfrac{3}{2}y + \tfrac{7}{2}$$

Step 2: We substitute this result for x in the second equation:

$$4x + 5y = 3$$
$$4(\tfrac{3}{2}y + \tfrac{7}{2}) + 5y = 3$$
$$6y + 14 + 5y = 3$$
$$11y + 14 = 3$$
$$11y = -11$$
$$y = -1$$

Step 3:

Step 4: $x = \tfrac{3}{2}y + \tfrac{7}{2} = \tfrac{3}{2}(-1) + \tfrac{7}{2} = \tfrac{4}{2} = 2$

Step 5: *Check:* $2(2) - 3(-1) = 4 + 3 = 7$ and
$$4(2) + 5(-1) = 8 - 5 = 3$$

The solution is $x = 2$, $y = -1$.

(c) **Step 1:**   In looking over the system

$$\begin{cases} x + y + z = 6 & (1) \\ \quad 2x + 2z = 4 & (2) \\ \quad y + z = 2 & (3) \\ \quad x = 4 & (4) \end{cases}$$

we see that we can use the fourth equation and proceed to Step 2.

Step 2:   Substitute $x = 4$ in the remaining equations:

$$\begin{cases} 4 + y + z = 6 & (1) \\ \quad 2(4) + 2z = 4 & (2) \\ \quad y + z = 2 & (3) \end{cases}$$

$$\begin{cases} y + z = 2 & (1) \\ \quad 2z = -4 & (2) \\ y + z = 2 & (3) \end{cases}$$

Step 3:   We solve the second equation $2z = -4$ for the variable z:

$$2z = -4$$
$$z = -2$$

Step 4:   Notice that the first and third equations are identical, so we substitute $z = -2$:

$$y + (-2) = 2$$
$$y = 4$$

Step 5:   *Check:*   We now check the solution $x = 4, y = 4, z = -2$:

$$\begin{cases} 4 + 4 - 2 = 6 \\ 2(4) + 2(-2) = 8 - 4 = 4 \\ 4 + (-2) = 2 \\ 4 = 4 \end{cases}$$

The solution checks.   ≡

Two Linear Equations Containing Two Variables

We can view the problem of solving a system of two linear equations containing two variables as a geometry problem. The graph of each equation in such a system is a straight line. Thus, a system of two equations containing two variables represents a pair of lines. The lines either (*1*) intersect or (*2*) are parallel or (*3*) are **coincident** (that is, identical).

COINCIDENT LINES

1. If the lines intersect, then the system of equations has one solution, given by the point of intersection. The system is consistent.
2. If the lines are parallel, then the system of equations has no solution, because the lines never intersect. The system is inconsistent.
3. If the lines are coincident, then the system of equations has infinitely many solutions, represented by the totality of points on the line. The system is consistent.

Figure 1 illustrates these conclusions.

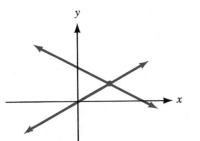
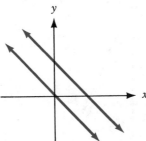

(a) Intersecting lines; system has one solution

(b) Parallel lines; system has no solution

(c) Coincident lines; system has infinitely many solutions

FIGURE 1

We have already seen several examples in which the system of equations represents two lines that intersect. The next example illustrates what happens when the method of substitution is used to solve a system of equations that has no solution.

EXAMPLE 4   Solve

$$\begin{cases} 2x + y = 5 & (1) \\ 4x + 2y = 8 & (2) \end{cases}$$

Solution   We choose to solve the first equation for y:

$$2x + y = 5$$
$$y = 5 - 2x$$

Substituting in the second equation, we get

$$4x + 2y = 8$$
$$4x + 2(5 - 2x) = 8$$
$$4x + 10 - 4x = 8$$
$$0 \cdot x = -2$$

This equation has no solution. Thus, we conclude that the system itself has no solution and is therefore inconsistent.

Figure 2 illustrates the pair of lines whose equations form the system in Example 4. Notice that the graphs of the two equations will be lines, each with slope -2; one will have a y-intercept of 5, the other a y-intercept of 4. Thus the lines are parallel and have no point of intersection. This geometric statement is equivalent to the algebraic statement that the system has no solution.

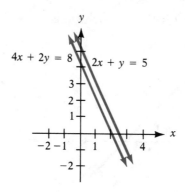

FIGURE 2

The next example is of a system with infinitely many solutions.

EXAMPLE 5   Solve

$$\begin{cases} 2x + y = 4 & (1) \\ -6x - 3y = -12 & (2) \end{cases}$$

Solution   We choose to solve the first equation for y:

$$2x + y = 4$$
$$y = 4 - 2x$$

Substituting in the second equation, we get

$$-6x - 3y = -12$$
$$-6x - 3(4 - 2x) = -12$$
$$-6x - 12 + 6x = -12$$
$$0 \cdot x = 0$$

The result is an identity, and so any value of x is a solution. Thus, the solution of the system is

$$y = 4 - 2x$$

where x can be any real number. In other words, the system has infinitely many solutions.   ▬

Figure 3 illustrates the situation presented in Example 5. Notice that the graphs of the two equations are lines, each with slope -2 and each with y-intercept 4. Thus, the lines are coincident. This geometric statement is equivalent to the algebraic statement that the system has infinitely many solutions. For the system in Example 5, we can write down some of the infinitude of solutions by assigning values to x and then finding $y = 4 - 2x$. Thus:

If $x = 4$, then $y = -4$, and $x = 4, y = -4$ is a solution.

If $x = 0$, then $y = 4$, and $x = 0, y = 4$ is a solution.

If $x = \frac{1}{2}$, then $y = 3$, and $x = \frac{1}{2}, y = 3$ is a solution.

These solutions, of course, are points on the line in Figure 3. We can write the solution of the system as $y = 4 - 2x$ or as $2x + y = 4$.

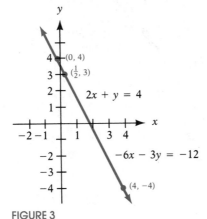

FIGURE 3

EXAMPLE 6   A-1 Car Hire charges \$10 per day plus 12¢ per mile for the rental of a standard car. The same type of car at E-Z Rent costs \$12 per day plus 10¢ per mile.

(a) What mileage results in equal rental charges for the two firms?

(b) Use a graph to decide which firm to use if the number of miles to be driven is known and cost is the deciding factor.

Solution   (a) Let C denote the cost, in dollars, of each rental, and let x be the miles driven. Then we can form the system of equations

$$\begin{cases} C = 10 + 0.12x & \text{For A-1 Car Hire} \\ C = 12 + 0.10x & \text{For E-Z Rent} \end{cases}$$

We solve the system using substitution. Then

$$10 + 0.12x = 12 + 0.10x$$
$$0.02x = 2$$
$$x = 100 \text{ miles}$$

To check, we see whether the costs are equal:

$$C = 10 + 0.12(100) = 10 + 12 = 22$$
$$C = 12 + 0.10(100) = 12 + 10 = 22$$

Thus, a daily usage of 100 miles results in equal costs of $22 for both companies.

(b) We graph each equation in the system, measuring C along the vertical axis. See Figure 4. The point where the lines intersect is the solution we found in part (a). From the graph it is clear that, for $0 < x < 100$, the cost at A-1 Car Hire is less; whereas, for mileage in excess of 100 miles, the cost at E-Z Rent is less.

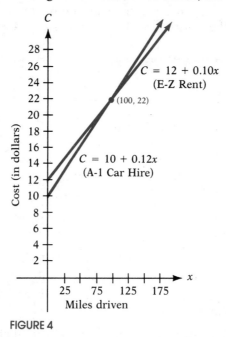

FIGURE 4

Three Linear Equations Containing Three Variables

Just as with a system of two linear equations containing two unknowns, a system of three linear equations containing three unknowns also has either exactly one solution, or no solution, or infinitely many solutions.

EXAMPLE 7   Solve

$$\begin{cases} x + y + z = 6 & (1) \\ 3x - 2y + 4z = 9 & (2) \\ 2x + y - 2z = 6 & (3) \end{cases}$$

Solution We choose to solve for z in the first equation:

$$x + y + z = 6$$
$$z = 6 - x - y$$

Substituting for z in the second and third equations, we get

$$\begin{cases} 3x - 2y + 4z = 9 & (2) \\ 2x + y - 2z = 6 & (3) \end{cases}$$

$$\begin{cases} 3x - 2y + 4(6 - x - y) = 9 & (2) \\ 2x + y - 2(6 - x - y) = 6 & (3) \end{cases}$$

$$\begin{cases} -x - 6y = -15 & (2) \\ 4x + 3y = 18 & (3) \end{cases}$$

The result is a system of two equations containing two variables, x and y. Now use the method of substitution on this system. Thus, solving for x in equation (2), we get

$$-x - 6y = -15$$
$$x = -6y + 15$$

Substituting for x in equation (3), $4x + 3y = 18$, we find

$$4(-6y + 15) + 3y = 18$$
$$-21y = -42$$
$$y = 2$$

With $y = 2$, we can use $x = -6y + 15$ to find $x = 3$. Finally, with $x = 3$ and $y = 2$, we return to the original system. We can find z using any of the equations, so we choose the first one:

$$x + y + z = 6$$
$$3 + 2 + z = 6$$
$$z = 1$$

The solution of the system is $x = 3, y = 2, z = 1$.

Check:

$$\begin{cases} 3 + 2 + 1 = 6 \\ 3(3) - 2(2) + 4(1) = 9 - 4 + 4 = 9 \\ 2(3) + 2 - 2(1) = 6 + 2 - 2 = 6 \end{cases}$$

The solution checks.

EXAMPLE 8 Solve

$$\begin{cases} x + y - z = 5 & (1) \\ -x + y = 4 & (2) \\ 2x - z = 1 & (3) \end{cases}$$

Solution We'll begin by solving for z in the third equation:

$$2x - z = 1$$
$$-z = -2x + 1$$
$$z = 2x - 1$$

Now we substitute for z in the first equation. (Note that the variable z is missing from the second equation.)

$$x + y - (2x - 1) = 5$$
$$-x + y = 4$$

This result is identical to the second equation, so we have a system of two equations containing two variables that has an infinite number of solutions:

$$\begin{cases} -x + y = 4 & (1) \\ -x + y = 4 & (2) \end{cases}$$

We can write the solution as $y = x + 4$, where x is any real number. Then $z = 2x - 1$, where x is any real number, so the solution of the original system is

$$\begin{cases} y = x + 4 \\ z = 2x - 1 \end{cases} \quad \text{where } x \text{ is any real number} \qquad \blacksquare$$

Some of the solutions of the original system in Example 8 are $x = 0, y = 4, z = -1; x = 1, y = 5, z = 1; x = 2, y = 6, z = 3$.

Another way to express the solution to Example 8 can be obtained by writing the solutions of the system

$$\begin{cases} -x + y = 4 & (1) \\ -x + y = 4 & (2) \end{cases}$$

as $x = y - 4$, where y is any real number. Then

$$z = 2x - 1 = 2(y - 4) - 1 = 2y - 9$$

and the solution of the original system is

$$\begin{cases} x = y - 4 \\ z = 2y - 9 \end{cases} \quad \text{where } y \text{ is any real number}$$

Of course, this expression for the solution to Example 8 gives the same infinitude of solutions as our first solution. Using this expression, some of the solutions are $y = 4, x = 0, z = -1; y = 5, x = 1, z = 1; y = 6, x = 2, z = 3$.

EXERCISE 7.1 *In Problems 1–10 verify that the values of the variables listed are solutions of the system of equations.*

1. $\begin{cases} 2x - y = 5 \\ 5x + 2y = 8 \end{cases}$
   $x = 2, y = -1$

2. $\begin{cases} 3x + 2y = 2 \\ x - 7y = -30 \end{cases}$
   $x = -2, y = 4$

3. $\begin{cases} 3x - 4y = 4 \\ x - 3y = \frac{1}{2} \end{cases}$
   $x = 2, y = \frac{1}{2}$

4. $\begin{cases} 2x + y = 0 \\ 5x - 4y = -\frac{13}{2} \end{cases}$
   $x = -\frac{1}{2}, y = 1$

5. $\begin{cases} x^2 - y^2 = 3 \\ xy = 2 \end{cases}$
   $x = 2, y = 1$

6. $\begin{cases} x^2 - y^2 = 3 \\ xy = 2 \end{cases}$
   $x = -2, y = -1$

7. $\begin{cases} \dfrac{x}{1 + x} + 3y = 6 \\ x + 9y^2 = 36 \end{cases}$
   $x = 0, y = 2$

8. $\begin{cases} \dfrac{x}{x - 1} + y = 5 \\ 3x - y = 3 \end{cases}$
   $x = 2, y = 3$

9. $\begin{cases} 3x + 3y + 2z = 4 \\ x - y - z = 0 \\ 2y - 3z = -8 \end{cases}$
   $x = 1, y = -1, z = 2$

10. $\begin{cases} 4x - z = 7 \\ 8x + 5y - z = 0 \\ -x - y + 5z = 6 \end{cases}$
    $x = 2, y = -3, z = 1$

In Problems 11–46 solve each system of equations using the method of substitution. If the system has no solution, say it is inconsistent.

11. $\begin{cases} x + y = 8 \\ x - y = 4 \end{cases}$

12. $\begin{cases} x + 2y = 5 \\ x + y = 3 \end{cases}$

13. $\begin{cases} 5x - y = 13 \\ 2x + 3y = 12 \end{cases}$

14. $\begin{cases} x + 3y = 5 \\ 2x - 3y = -8 \end{cases}$

15. $\begin{cases} 3x = 24 \\ x + 2y = 0 \end{cases}$

16. $\begin{cases} 4x + 5y = -3 \\ -2y = -4 \end{cases}$

17. $\begin{cases} 3x - 6y = 24 \\ 5x + 4y = 12 \end{cases}$

18. $\begin{cases} 2x + 4y = 16 \\ 3x - 5y = -9 \end{cases}$

19. $\begin{cases} 2x + y = 1 \\ 4x + 2y = 6 \end{cases}$

20. $\begin{cases} x - y = 5 \\ -3x + 3y = 2 \end{cases}$

21. $\begin{cases} 2x - 4y = -2 \\ 3x + 2y = 3 \end{cases}$

22. $\begin{cases} 3x + 3y = 3 \\ 4x + 2y = \frac{8}{3} \end{cases}$

23. $\begin{cases} x + 2y = 4 \\ 2x + 4y = 8 \end{cases}$

24. $\begin{cases} 3x - y = 7 \\ 9x - 3y = 21 \end{cases}$

25. $\begin{cases} 2x - 3y = -1 \\ 10x + 10y = 5 \end{cases}$

26. $\begin{cases} 3x - 2y = 0 \\ 5x + 10y = 4 \end{cases}$

27. $\begin{cases} 2x + 3y = 6 \\ x - y = \frac{1}{2} \end{cases}$

28. $\begin{cases} \frac{1}{2}x + y = -2 \\ x - 2y = 8 \end{cases}$

29. $\begin{cases} 2x + 3y = 5 \\ 4x + 6y = 10 \end{cases}$

30. $\begin{cases} 2x + 3y = 5 \\ 4x + 6y = 6 \end{cases}$

31. $\begin{cases} 3x - 5y = 3 \\ 15x + 5y = 21 \end{cases}$

32. $\begin{cases} 2x - y = -1 \\ x + \frac{1}{2}y = \frac{3}{2} \end{cases}$

33. $\begin{cases} x - y = 6 \\ 2x - 3z = 16 \\ 2y + z = 4 \end{cases}$

34. $\begin{cases} 2x + y = -4 \\ -2y + 4z = 0 \\ 3x - 2z = -11 \end{cases}$

35. $\begin{cases} x - 2y + 3z = 7 \\ 2x + y + z = 4 \\ -3x + 2y - 2z = -10 \end{cases}$

36. $\begin{cases} 2x + y - 3z = 0 \\ -2x + 2y + z = -7 \\ 3x - 4y - 3z = 7 \end{cases}$

37. $\begin{cases} x - y - z = 1 \\ 2x + 3y + z = 2 \\ 3x + 2y = 0 \end{cases}$

38. $\begin{cases} 2x - 3y - z = 0 \\ -x + 2y + z = 5 \\ 3x - 4y - z = 1 \end{cases}$

39. $\begin{cases} x - y - z = 1 \\ -x + 2y - 3z = -4 \\ 3x - 2y - 7z = 0 \end{cases}$

40. $\begin{cases} 2x - 3y - z = 0 \\ 3x + 2y + 2z = 2 \\ x + 5y + 3z = 2 \end{cases}$

41. $\begin{cases} 2x - 2y + 3z = 6 \\ 4x - 3y + 2z = 0 \\ -2x + 3y - 7z = 1 \end{cases}$

42. $\begin{cases} 3x - 2y + 2z = 6 \\ 7x - 3y + 2z = -1 \\ 2x - 3y + 4z = 0 \end{cases}$

43. $\begin{cases} x + y - z = 6 \\ 3x - 2y + z = -5 \\ x + 3y - 2z = 14 \end{cases}$

44. $\begin{cases} x - y + z = -4 \\ 2x - 3y + 4z = -15 \\ 5x + y - 2z = 12 \end{cases}$

45. $\begin{cases} x + 2y - z = -3 \\ 2x - 4y + z = -7 \\ -2x + 2y - 3z = 4 \end{cases}$

46. $\begin{cases} x + 4y - 3z = -8 \\ 3x - y + 3z = 12 \\ x + y + 6z = 1 \end{cases}$

47. Solve

$$\begin{cases} \dfrac{1}{x} + \dfrac{1}{y} = 8 \\[2mm] \dfrac{3}{x} - \dfrac{5}{y} = 0 \end{cases}$$

[*Hint:* Let $u = 1/x$ and $v = 1/y$, and solve for u and v. Then $x = 1/u$ and $y = 1/v$.]

48. Solve

$$\begin{cases} \dfrac{4}{x} - \dfrac{3}{y} = 0 \\[2mm] \dfrac{6}{x} + \dfrac{3}{2y} = 2 \end{cases}$$

49. The sum of two numbers is 81. The difference of twice one and three times the other is 62. Find the two numbers.

50. The difference of two numbers is 40. Six times the smaller one less the larger one is 5. Find the two numbers.

51. The perimeter of a rectangular room is 90 feet. Find the dimensions of the room if the length is twice the width.

52. The length of fence required to enclose a rectangular field is 3000 meters. What are the dimensions of the field if it is known that the difference between its length and width is 50 meters?

53. Four large cheeseburgers and two chocolate shakes cost a total of $7.90. Two shakes cost 15¢ more than one cheeseburger. What is the cost of a cheeseburger? A shake?

54. A movie theater charges $4.00 for adults and $1.50 for children under 12. On a day when 325 people paid an admission, the total receipts were $1025. How many who paid were adults? How many were under 12?

55. With a tail wind, a small Piper aircraft can fly 600 miles in 3 hours. Against this same wind, the Piper can fly the same distance in 4 hours. Find the average wind speed and the average airspeed of the Piper.

56. The average airspeed of a single-engine aircraft is 150 miles per hour. If the aircraft flies the same distance in 2 hours with the wind as it flew in 3 hours against the wind, what was the wind speed?

57. A store sells cashews for $5.00 per pound and peanuts for $1.50 per pound. The manager decides to mix 30 pounds of peanuts with some cashews and sell the mixture for $3.00 per pound. How many pounds of cashews should be mixed with the peanuts so that the mixture will produce the same revenue as would selling the nuts separately?

58. Rework Problem 57 for a mixture using 40 pounds of peanuts.

59. Find real numbers b and c such that the parabola $y = x^2 + bx + c$ passes through the points $(1, 2)$ and $(-1, 3)$.

60. Find real numbers b and c such that the parabola $y = x^2 + bx + c$ passes through the points (x_1, y_1) and (x_2, y_2).

61. Find real numbers a, b, and c such that the parabola $y = ax^2 + bx + c$ passes through the points $(-1, 4)$, $(2, 3)$, and $(0, 1)$.

62. Find real numbers a, b, and c such that the parabola $y = ax^2 + bx + c$ passes through the points (x_1, y_1), (x_2, y_2), and (x_3, y_3).

63. Solve

$$\begin{cases} y = m_1x + b_1 \\ y = m_2x + b_2 \end{cases}$$

where $m_1 \neq m_2$.

64. Solve

$$\begin{cases} y = m_1x + b_1 \\ y = m_2x + b_2 \end{cases}$$

where $m_1 = m_2 = m$ and $b_1 \neq b_2$.

65. Solve

$$\begin{cases} y = m_1x + b_1 \\ y = m_2x + b_2 \end{cases}$$

where $m_1 = m_2 = m$ and $b_1 = b_2 = b$.

▤ 7.2 SOLUTIONS OF SYSTEMS OF LINEAR EQUATIONS: ELIMINATION

A second method for solving a system of linear equations is the method of elimination. This method is usually preferred over substitution if substitution leads to fractions or if the system contains more than two variables. Elimination also provides the necessary motivation for solving systems using matrices, the subject of the next section.

The idea behind elimination is to keep replacing the original equations in the system with equivalent equations until a system of equations with an obvious solution is reached. When we proceed in this way we obtain **equivalent systems of equations**. The rules for obtaining equivalent equations are the same as those studied in Chapter 2. However, we may also interchange any two equations of the system and/or replace any equation in the system by the sum or difference of that equation and any other equation in the system.

EQUIVALENT SYSTEMS OF EQUATIONS

RULES FOR OBTAINING AN EQUIVALENT SYSTEM OF EQUATIONS

1. Interchange any two equations of the system.
2. Multiply (or divide) each side of an equation by the same nonzero constant.
3. Replace any equation in the system by the sum (or difference) of that equation and any other equation in the system.

An example will give you the idea. As you work through the example, pay particular attention to the pattern being followed.

EXAMPLE 1   Solve the system of equations

$$\begin{cases} 2x + 3y = 1 & (1) \\ -x + y = -3 & (2) \end{cases}$$

Solution   If we multiply each side of the second equation by 2, we get the equivalent system

$$\begin{cases} 2x + 3y = 1 & (1) \\ -2x + 2y = -6 & (2) \end{cases}$$

If we replace the second equation of this system by the sum of the two equations, we get the equivalent system

$$\begin{cases} 2x + 3y = 1 & (1) \\ 5y = -5 & (2) \end{cases}$$

Now we multiply the second equation by $\frac{1}{5}$, obtaining the equivalent system

$$\begin{cases} 2x + 3y = 1 & (1) \\ \qquad\quad y = -1 & (2) \end{cases}$$

From this system we see the obvious solution $y = -1$. We use this value for y in the first equation to get

$$\begin{cases} 2x + 3(-1) = 1 & (1) \\ \qquad\qquad\; y = -1 & (2) \end{cases}$$

$$\begin{cases} 2x = 4 & (1) \\ \;\; y = -1 & (2) \end{cases}$$

$$\begin{cases} x = 2 & (1) \\ y = -1 & (2) \end{cases}$$

Thus, the solution of the original system is $x = 2, y = -1$.   ≡

METHOD OF ELIMINATION

BACK-SUBSTITUTE

The procedure used in Example 1 is called the **method of elimination.*** Notice the pattern of the solution. First, we eliminated the variable x from the second equation. Then we **back-substituted**; that is, we substituted the value found for y back into the first equation to find x.

Let's do some more examples.

EXAMPLE 2   Use the method of elimination to solve each system of equations.

(a) $\begin{cases} \frac{1}{3}x + 5y = -4 \\ 2x + 3y = 3 \end{cases}$ (b) $\begin{cases} 2x + 3y = 6 \\ \frac{2}{3}x + \;\; y = 2 \end{cases}$ (c) $\begin{cases} 5x - 2y = 5 \\ \;\; x - \frac{2}{5}y = 2 \end{cases}$

Solution   (a) We begin by multiplying each side of the first equation by 3 in order to remove the fraction $\frac{1}{3}$.

$$\begin{cases} \frac{1}{3}x + 5y = -4 & (1) \\ 2x + 3y = 3 & (2) \end{cases}$$

$$\begin{cases} \;\; x + 15y = -12 & (1) \\ 2x + \;\; 3y = 3 & (2) \end{cases}$$

Thinking ahead, we decide to multiply each side of the first equation by -2 because then the sum of the two equations will result in an equation with the variable x eliminated. (Note that we also could have multiplied each side of the first equation by 2 and

*This method for solving a system of equations is also sometimes called the add/subtract method.

then replaced the first equation by the difference of the two equations. However, because subtracting is more likely to result in a calculation error, we shall follow the safer practice of adding equations.)

$$\begin{cases} -2x - 30y = 24 & (1) \\ 2x + 3y = 3 & (2) \end{cases}$$

$$\begin{cases} -27y = 27 & (1) \\ 2x + 3y = 3 & (2) \end{cases}$$ Replace the first equation by the sum of the two equations.

$$\begin{cases} y = -1 & (1) \\ 2x + 3y = 3 & (2) \end{cases}$$ Multiply each side of the first equation by $-\frac{1}{27}$.

$$\begin{cases} y = -1 & (1) \\ 2x + 3(-1) = 3 & (2) \end{cases}$$ Replace y by -1.

$$\begin{cases} y = -1 & (1) \\ 2x = 6 & (2) \end{cases}$$

$$\begin{cases} y = -1 & (1) \\ x = 3 & (2) \end{cases}$$

The solution of the original system is $x = 3, y = -1$.

(b) $$\begin{cases} 2x + 3y = 6 & (1) \\ \frac{2}{3}x + y = 2 & (2) \end{cases}$$

$$\begin{cases} 2x + 3y = 6 & (1) \\ -2x - 3y = -6 & (2) \end{cases}$$ Multiply each side of the second equation by -3.

$$\begin{cases} 2x + 3y = 6 & (1) \\ 0 = 0 & (2) \end{cases}$$ Replace the second equation with the sum of the two equations.

The original system is thus equivalent to a system containing one equation. This means any values of x and y for which $2x + 3y = 6$ are solutions. For example, $x = 3, y = 0; x = 0, y = 2; x = -3, y = 4; x = 6, y = -2$ are some of the solutions of the original system. There are, in fact, infinitely many values of x and y for which $2x + 3y = 6$, so the original system has infinitely many solutions. We will write the solutions of the original system as

$$y = -\tfrac{2}{3}x + 2$$

where x can be any real number, or as

$$x = -\tfrac{3}{2}y + 3$$

where y can be any real number.

(c) $\begin{cases} 5x - 2y = 5 & (1) \\ x - \frac{2}{5}y = 2 & (2) \end{cases}$

$\begin{cases} 5x - 2y = 5 & (1) \\ -5x + 2y = -10 & (2) \end{cases}$ Multiply each side of the second equation by -5.

$\begin{cases} 5x - 2y = 10 & (1) \\ 0x + 0y = -5 & (2) \end{cases}$ Replace the second equation by the sum of the two equations.

The second equation has no solution. Hence, because the original system is equivalent to a system having no solution, the original system is inconsistent. ≡

In working through Example 2, you may have noticed that sometimes we replaced the first equation by the sum of the two equations and sometimes we replaced the second one. Actually, it makes no difference which equation is replaced, provided the one being replaced is added to (or subtracted from) another equation in the system. However, it is important to use a systematic approach when using the method of elimination.

Let's see how elimination works on a system of three equations containing three variables.

EXAMPLE 3 Use the method of elimination to solve the system of equations

$$\begin{cases} x + y - z = -1 & (1) \\ 4x - 3y + 2z = 16 & (2) \\ 2x - 2y - 3z = 5 & (3) \end{cases}$$

Solution For a system of three equations, we attempt to eliminate one variable at a time, using a pair of equations. We begin by multiplying each side of the first equation by -2, in anticipation of eliminating the variable x by adding the first and third equations:

$$\begin{cases} -2x - 2y + 2z = 2 & (1) \\ 4x - 3y + 2z = 16 & (2) \\ 2x - 2y - 3z = 5 & (3) \end{cases}$$

$$\begin{cases} -2x - 2y + 2z = 2 & (1) \\ 4x - 3y + 2z = 16 & (2) \\ -4y - z = 7 & (3) \end{cases}$$ Replace the third equation by the sum of the first and third equations.

We now seek to eliminate the same variable x from the second equation.

$$\begin{cases} -4x - 4y + 4z = 4 & (1) \\ 4x - 3y + 2z = 16 & (2) \\ -4y - z = 7 & (3) \end{cases}$$

  (1) Multiply each side of the first equation in the previous system by 2.

$$\begin{cases} -4x - 4y + 4z = 4 & (1) \\ -7y + 6z = 20 & (2) \\ -4y - z = 7 & (3) \end{cases}$$

  (2) Replace the second equation by the sum of the first and second equations.

We now seek to eliminate y from the second and third equations:

$$\begin{cases} -4x - 4y + 4z = 4 & (1) \\ -28y + 24z = 80 & (2) \\ 28y + 7z = -49 & (3) \end{cases}$$

  (2) Multiply each side by 4.
  (3) Multiply each side by -7.

$$\begin{cases} -4x - 4y + 4z = 4 & (1) \\ -28y + 24z = 80 & (2) \\ 31z = 31 & (3) \end{cases}$$

  (3) Replace the third equation by the sum of the second and third equations.

$$\begin{cases} -4x - 4y + 4z = 4 & (1) \\ -28y + 24z = 80 & (2) \\ z = 1 & (3) \end{cases}$$

  (3) Multiply each side by $\frac{1}{31}$.

$$\begin{cases} -4x - 4y + 4 = 4 & (1) \\ -28y + 24 = 80 & (2) \\ z = 1 & (3) \end{cases}$$

  (1) Back-substitute; replace z by 1.
  (2) Back-substitute; replace z by 1.

$$\begin{cases} -4x - 4y = 0 & (1) \\ y = -2 & (2) \\ z = 1 & (3) \end{cases}$$

$$\begin{cases} -4x + 8 = 0 & (1) \\ y = -2 & (2) \\ z = 1 & (3) \end{cases}$$

  Back-substitute; replace y by -2.

$$\begin{cases} x = 2 & (1) \\ y = -2 & (2) \\ z = 1 & (3) \end{cases}$$

The solution of the original system is $x = 2$, $y = -2$, $z = 1$. (You should check this.) ▤

Look back over the solution given in Example 3, noting the pattern of seeking to make the third equation contain only the variable z, followed by making the second equation contain only the variable y and the first equation contain only the variable x. Although which variables to isolate is your choice, the methodology remains the same for all systems.

EXERCISE 7.2   *In Problems 1–36 solve each system of equations using the method of elimination. If the system has no solution, say it is inconsistent.*

1. $\begin{cases} x + y = 8 \\ x - y = 4 \end{cases}$

2. $\begin{cases} x + 2y = 5 \\ x + y = 3 \end{cases}$

3. $\begin{cases} 5x - y = 13 \\ 2x + 3y = 12 \end{cases}$

4. $\begin{cases} x + 3y = 5 \\ 2x - 3y = -8 \end{cases}$

5. $\begin{cases} 3x = 24 \\ x + 2y = 0 \end{cases}$

6. $\begin{cases} 4x + 5y = -3 \\ -2y = -4 \end{cases}$

7. $\begin{cases} 3x - 6y = 24 \\ 5x + 4y = 12 \end{cases}$

8. $\begin{cases} 2x + 4y = 16 \\ 3x - 5y = -9 \end{cases}$

9. $\begin{cases} 2x + y = 1 \\ 4x + 2y = 6 \end{cases}$

10. $\begin{cases} x - y = 5 \\ -3x + 3y = 2 \end{cases}$

11. $\begin{cases} 2x - 4y = -2 \\ 3x + 2y = 3 \end{cases}$

12. $\begin{cases} 3x + 3y = 3 \\ 4x + 2y = \frac{8}{3} \end{cases}$

13. $\begin{cases} x + 2y = 4 \\ 2x + 4y = 8 \end{cases}$

14. $\begin{cases} 3x - y = 7 \\ 9x - 3y = 21 \end{cases}$

15. $\begin{cases} 2x - 3y = -1 \\ 10x + 10y = 5 \end{cases}$

16. $\begin{cases} 3x - 2y = 0 \\ 5x + 10y = 4 \end{cases}$

17. $\begin{cases} 2x + 3y = 6 \\ x - y = \frac{1}{2} \end{cases}$

18. $\begin{cases} \frac{1}{2}x + y = -2 \\ x - 2y = 8 \end{cases}$

19. $\begin{cases} 2x + 3y = 5 \\ 4x + 6y = 10 \end{cases}$

20. $\begin{cases} 2x + 3y = 5 \\ 4x + 6y = 6 \end{cases}$

21. $\begin{cases} 3x - 5y = 3 \\ 15x + 5y = 21 \end{cases}$

22. $\begin{cases} 2x - y = -1 \\ x - \frac{1}{2}y = \frac{3}{2} \end{cases}$

23. $\begin{cases} x - y = 6 \\ 2x - 3z = 16 \\ 2y + z = 4 \end{cases}$

24. $\begin{cases} 2x + y = -4 \\ -2y + 4z = 0 \\ 3x - 2z = -11 \end{cases}$

25. $\begin{cases} x - 2y + 3z = 7 \\ 2x + y + z = 4 \\ -3x + 2y - 2z = -10 \end{cases}$

26. $\begin{cases} 2x + y - 3z = 0 \\ -2x + 2y + z = -7 \\ 3x - 4y - 3z = 7 \end{cases}$

27. $\begin{cases} x - y - z = 1 \\ 2x + 3y + z = 2 \\ 3x + 2y = 0 \end{cases}$

28. $\begin{cases} 2x - 3y - z = 0 \\ -x + 2y + z = 5 \\ 3x - 4y - z = 1 \end{cases}$

29. $\begin{cases} x - y - z = 1 \\ -x + 2y - 3z = -4 \\ 3x - 2y - 7z = 0 \end{cases}$ **30.** $\begin{cases} 2x - 3y - z = 0 \\ 3x + 2y + 2z = 2 \\ x + 5y + 3z = 2 \end{cases}$

31. $\begin{cases} 2x - 2y + 3z = 6 \\ 4x - 3y + 2z = 0 \\ -2x + 3y - 7z = 1 \end{cases}$ **32.** $\begin{cases} 3x - 2y + 2z = 6 \\ 7x - 3y + 2z = -1 \\ 2x - 3y + 4z = 0 \end{cases}$

33. $\begin{cases} x + y - z = 6 \\ 3x - 2y + z = -5 \\ x + 3y - 2z = 14 \end{cases}$ **34.** $\begin{cases} x - y + z = -4 \\ 2x - 3y + 4z = -15 \\ 5x + y - 2z = 12 \end{cases}$

35. $\begin{cases} x + 2y - z = -3 \\ 2x - 4y + z = -7 \\ -2x + 2y - 3z = 4 \end{cases}$ **36.** $\begin{cases} x + 4y - 3z = -8 \\ 3x - y + 3z = 12 \\ x + y + 6z = 1 \end{cases}$

37. A chemistry laboratory can be used by 38 students at one time. The laboratory has 16 work stations, some set up for 2 students each, the others set up for 3 students each. How many are there of each kind of work station?

38. One group of people purchased 10 hot dogs and 5 soft drinks at a cost of $10.00. A second group bought 8 hot dogs and 4 soft drinks at a cost of $8.00. What is the cost of a single hot dog? A single soft drink?

39. The grocery store we use does not mark prices on its goods. My wife went to this store, bought three 1-pound packages of bacon and two cartons of eggs, and paid a total of $7.45. Not knowing she went to the store, I also went to the same store, purchased two 1-pound packages of bacon and three cartons of eggs, and paid a total of $6.45. Now we want to return two 1-pound packages of bacon and two cartons of eggs. How much will be refunded?

40. A swimmer requires 3 hours to swim 15 miles downstream. The return trip upstream takes 5 hours. Find the average speed of the swimmer in still water. How fast is the current of the stream? (Assume the speed of the swimmer is the same in each direction.)

41. The sum of three numbers is 48. The sum of the two larger numbers is three times the smallest. The sum of the two smaller numbers is 6 more than the largest. Find the numbers.

42. A coin collection consists of 50 coins—nickels, dimes, and quarters. If the collection has a face value of $3.25 and there are 5 more dimes than there are nickels, how many of each coin are in the collection?

43. A Broadway theater has 500 seats, divided into orchestra, main, and balcony seating. Orchestra seats sell for $50, main seats for $35, and balcony seats for $25. If all the seats are sold, the gross revenue to the theater is $17,100. If all the main and balcony seats are sold, but only half the orchestra seats are sold, the gross revenue is $14,600. How many are there of each kind of seat?

7.3 SOLUTIONS OF SYSTEMS OF LINEAR EQUATIONS: MATRICES

The systematic approach of the method of elimination for solving a system of linear equations provides a third method of solution that involves a simplified notation.

Consider the system of linear equations

$$\begin{cases} x + 4y = 14 \\ 3x - 2y = 0 \end{cases}$$

If we choose not to write the symbols used for the variables, we can represent this system as

$$\begin{bmatrix} 1 & 4 & | & 14 \\ 3 & -2 & | & 0 \end{bmatrix}$$

where it is understood that the first column represents the coefficients of the variable x, the second column the coefficients of y, and the third column the constants on the right side of the equal signs. The vertical line serves as a reminder of the equal signs. The large square brackets are the traditional symbols used for a matrix in algebra.

MATRIX   A **matrix** is defined as a rectangular array of numbers,

$$\begin{array}{c} \\ \text{Row 1} \\ \text{Row 2} \\ \vdots \\ \text{Row } i \\ \vdots \\ \text{Row } m \end{array} \begin{bmatrix} \begin{array}{ccccc} \text{Column 1} & \text{Column 2} & & \text{Column } j & \text{Column } n \\ a_{11} & a_{12} & \cdots & a_{1j} & \cdots & a_{1n} \\ a_{21} & a_{22} & \cdots & a_{2j} & \cdots & a_{2n} \\ \vdots & \vdots & & \vdots & & \vdots \\ a_{i1} & a_{i2} & \cdots & a_{ij} & \cdots & a_{in} \\ \vdots & \vdots & & \vdots & & \vdots \\ a_{m1} & a_{m2} & \cdots & a_{mj} & \cdots & a_{mn} \end{array} \end{bmatrix} \quad (1)$$

ROW INDEX
COLUMN INDEX
ENTRIES

Each number a_{ij} of the matrix has two indices: the **row index** i and the **column index** j. The matrix shown in Equation (1) has m rows and n columns. The numbers a_{ij} are usually referred to as the **entries** of the matrix.

We will discuss matrices in more detail in the next chapter. For now, we shall utilize matrix notation to represent a system of linear equations.

EXAMPLE 1 Write the matrix representation of each system of equations.

(a) $\begin{cases} 3x - 4y = -6 & (1) \\ 2x - 3y = -5 & (2) \end{cases}$

(b) $\begin{cases} 2x - y + z = 0 & (1) \\ x + z - 1 = 0 & (2) \\ x + 2y - 8 = 0 & (3) \end{cases}$

Solution (a) The matrix representation is

$$\left[\begin{array}{cc|c} 3 & -4 & -6 \\ 2 & -3 & -5 \end{array}\right]$$

(b) Care must be taken that the system is written with the coefficients of all variables present (if any variable is missing, its coefficient is zero) and with all constants to the right of the equal signs. Thus, we need to rearrange the given system as follows:

$$\begin{cases} 2x - y + z = 0 & (1) \\ x + z - 1 = 0 & (2) \\ x + 2y - 8 = 0 & (3) \end{cases}$$

$$\begin{cases} 2x - y \quad + z \quad = 0 & (1) \\ x + 0 \cdot y + z \quad = 1 & (2) \\ x + 2y \quad + 0 \cdot z = 8 & (3) \end{cases}$$

The matrix representation is

$$\left[\begin{array}{ccc|c} 2 & -1 & 1 & 0 \\ 1 & 0 & 1 & 1 \\ 1 & 2 & 0 & 8 \end{array}\right]$$

≡

In working with the matrix representation of a system of equations, the matrices found in Example 1,

$$\left[\begin{array}{cc|c} 3 & -4 & -6 \\ 2 & -3 & -5 \end{array}\right] \quad \text{and} \quad \left[\begin{array}{ccc|c} 2 & -1 & 1 & 0 \\ 1 & 0 & 1 & 1 \\ 1 & 2 & 0 & 8 \end{array}\right]$$

AUGMENTED MATRICES usually are referred to as the **augmented matrices** of the system of equations. The matrices below, which do not include the constants to the right of the equal signs, are called the **coefficient matrices** of COEFFICIENT MATRICES the system:

$$\left[\begin{array}{cc} 3 & -4 \\ 2 & -3 \end{array}\right] \quad \text{and} \quad \left[\begin{array}{ccc} 2 & -1 & 1 \\ 1 & 0 & 1 \\ 1 & 2 & 0 \end{array}\right]$$

EXAMPLE 2   Write the system of linear equations corresponding to each augmented matrix.

(a) $\begin{bmatrix} 5 & 2 & | & 13 \\ -3 & 1 & | & -10 \end{bmatrix}$ (b) $\begin{bmatrix} 3 & -1 & -1 & | & 7 \\ 2 & 0 & 2 & | & 8 \\ 0 & 1 & 1 & | & 0 \end{bmatrix}$

Solution   (a) The matrix has two rows and so represents a system of two equations. The two columns to the left of the vertical bar indicate the system has two variables. If x and y are used to denote these variables, the system of equations is

$$\begin{cases} 5x + 2y = 13 & (1) \\ -3x + y = -10 & (2) \end{cases}$$

(b) This matrix represents a system of three equations containing three variables. If x, y, and z are the three variables, this system is

$$\begin{cases} 3x - y - z = 7 & (1) \\ 2x \quad\quad + 2z = 8 & (2) \\ \quad\quad y + z = 0 & (3) \end{cases}$$

■

Row Operations on a Matrix

Consider the system of equations

$$\begin{cases} 4x - 3y = 11 & (1) \\ 3x + 2y = 4 & (2) \end{cases}$$

We shall use a version of the method of elimination to solve the system. First, multiply each side of the second equation by -1 and add it to the first equation. Replace the first equation by the result.

$$\begin{cases} x - 5y = 7 & (1) \\ 3x + 2y = 4 & (2) \end{cases}$$

Multiply each side of the first equation by -3 and add it to the second equation. Replace the second equation by the result.

$$\begin{cases} x - 5y = 7 & (1) \\ 0 \cdot x + 17y = -17 & (2) \end{cases}$$

Multiply each side of the second equation by $\frac{1}{17}$.

$$\begin{cases} x - 5y = 7 & (1) \\ \quad\quad y = -1 & (2) \end{cases}$$

Now we back-substitute $y = -1$ into the first equation to get

$$x - 5y = 7$$
$$x - 5(-1) = 7$$
$$x = 2$$

The solution of the system above and of the original system is $x = 2$, $y = -1$.

The pattern of solution shown above provides a systematic way to solve any system of equations. The idea is to start with the augmented matrix of the system,

$$\begin{bmatrix} 4 & -3 & | & 11 \\ 3 & 2 & | & 4 \end{bmatrix} \qquad \begin{cases} 4x - 3y = 11 & \text{(1)} \\ 3x + 2y = 4 & \text{(2)} \end{cases}$$

and eventually arrive at the matrix

$$\begin{bmatrix} 1 & 0 & | & 2 \\ 0 & 1 & | & -1 \end{bmatrix} \qquad \begin{cases} x = 2 & \text{(1)} \\ y = -1 & \text{(2)} \end{cases}$$

Let's go through the procedure again, this time starting with the augmented matrix and keeping the final augmented matrix given above in mind:

$$\begin{bmatrix} 4 & -3 & | & 11 \\ 3 & 2 & | & 4 \end{bmatrix} \qquad \begin{cases} 4x - 3y = 11 & \text{(1)} \\ 3x + 2y = 4 & \text{(2)} \end{cases}$$

As before, we start by multiplying each side of the second equation by -1 and adding it to the first equation. This is equivalent to multiplying each entry in the second row of the matrix by -1, adding the result to the corresponding entries in row one, and replacing row one by these entries. The result of this step is that the number 1 appears in row one, column one:

$$\begin{bmatrix} 1 & -5 & | & 7 \\ 3 & 2 & | & 4 \end{bmatrix} \qquad \begin{cases} x - 5y = 7 & \text{(1)} \\ 3x + 2y = 4 & \text{(2)} \end{cases}$$

Multiply each entry in the first row by -3, add the result to the entries in the second row, and replace the second row by these entries. The result of this step is that the number 0 appears in row two, column one:

$$\begin{bmatrix} 1 & -5 & | & 7 \\ 0 & 17 & | & -17 \end{bmatrix} \qquad \begin{cases} x - 5y = 7 & \text{(1)} \\ 0 \cdot x + 17y = -17 & \text{(2)} \end{cases}$$

Multiply each entry in the second row by $\frac{1}{17}$. The result of this step is that the number 1 appears in row two, column two:

$$\begin{bmatrix} 1 & -5 & | & 7 \\ 0 & 1 & | & -1 \end{bmatrix} \qquad \begin{cases} x - 5y = 7 & \text{(1)} \\ y = -1 & \text{(2)} \end{cases}$$

Finally, multiply each entry in the second row by 5, add the result to the entries in the first row, and replace the first row by these entries. The result of this is that the number 0 appears in row one, column two:

$$\begin{bmatrix} 1 & 0 & | & 2 \\ 0 & 1 & | & -1 \end{bmatrix} \qquad \begin{cases} x = 2 & (1) \\ y = -1 & (2) \end{cases}$$

ROW OPERATIONS

The manipulations just performed on the augmented matrix are called **row operations**. There are three basic row operations:

1. The interchange of any two rows.
2. The replacement of a row by a nonzero multiple of that row.
3. The replacement of a row by the sum of that row and a multiple of some other row.

These three row operations correspond to the three rules given earlier for obtaining an equivalent system of equations. Thus, when a row operation is performed on a matrix, the resulting matrix represents a system of equations equivalent to the system represented by the original matrix.

For example, consider the augmented matrix

$$\begin{bmatrix} 1 & 2 & | & 3 \\ 4 & -1 & | & 2 \end{bmatrix}$$

Suppose we want to apply a row operation to this matrix that results in a matrix whose entry in row two, column one is a 0. The row operation to use is:

Multiply each entry in row one by -4 and add the result to the corresponding entries in row two. $\qquad$ (2)

If we use R_2 to represent the new entries in row two and use r_1 and r_2 to represent the original entries in row one and row two, respectively, then we can represent the row operation in Statement (2) by

$$R_2 = -4r_1 + r_2$$

Then,

$$\begin{bmatrix} 1 & 2 & | & 3 \\ 4 & -1 & | & 2 \end{bmatrix} \rightarrow \begin{bmatrix} 1 & 2 & | & 3 \\ -4(1) + 4 & -4(2) + (-1) & | & -4(3) + 2 \end{bmatrix} = \begin{bmatrix} 1 & 2 & | & 3 \\ 0 & -9 & | & -10 \end{bmatrix}$$
$$\uparrow$$
$$R_2 = -4r_1 + r_2$$

As desired, we now have the entry 0 in row two, column one.

EXAMPLE 3   Apply the row operation $R_2 = -3r_1 + r_2$ to the augmented matrix

$$\begin{bmatrix} 1 & -2 & | & 2 \\ 3 & -5 & | & 9 \end{bmatrix}$$

Solution   The row operation $R_2 = -3r_1 + r_2$ tells us that the entries in row two are to be replaced by the entries obtained after multiplying each entry in row one by -3 and adding the result to the corresponding entries in row two. Thus,

$$\begin{bmatrix} 1 & -2 & | & 2 \\ 3 & -5 & | & 9 \end{bmatrix} \rightarrow \begin{bmatrix} 1 & -2 & | & 2 \\ -3(1) + 3 & (-3)(-2) + (-5) & | & -3(2) + 9 \end{bmatrix} = \begin{bmatrix} 1 & -2 & | & 2 \\ 0 & 1 & | & 3 \end{bmatrix}$$

$$\uparrow$$
$$R_2 = -3r_1 + r_2$$

$\equiv$

EXAMPLE 4   Using the matrix

$$\begin{bmatrix} 1 & -2 & | & 2 \\ 0 & 1 & | & 3 \end{bmatrix}$$

find a row operation that will result in a matrix with a 0 in row one, column two.

Solution   We want a 0 in row one, column two. This result can be accomplished by multiplying row two by 2 and adding the result to row one. That is, we apply the row operation $R_1 = 2r_2 + r_1$:

$$\begin{bmatrix} 1 & -2 & | & 2 \\ 0 & 1 & | & 3 \end{bmatrix} \rightarrow \begin{bmatrix} 2(0) + 1 & 2(1) + (-2) & | & 2(3) + 2 \\ 0 & 1 & | & 3 \end{bmatrix} = \begin{bmatrix} 1 & 0 & | & 8 \\ 0 & 1 & | & 3 \end{bmatrix}$$

$$\uparrow$$
$$R_1 = 2r_2 + r_1$$

$\equiv$

A word about the notation we have introduced. A row operation, such as $R_1 = 2r_2 + r_1$, changes the entries in row one. Note also that, to change the entries in a given row, we multiply the entries in some other row by the appropriate number and *add* the results to the original entries of the row to be changed.

Now let's see how we use row operations to solve a system of linear equations.

EXAMPLE 5   Solve

$$\begin{cases} 4x + 3y = 11 & (1) \\ x - 3y = -1 & (2) \end{cases}$$

Solution First, we write the augmented matrix that represents this system:

$$\left[\begin{array}{cc|c} 4 & 3 & 11 \\ 1 & -3 & -1 \end{array}\right]$$

The first step requires getting the entry 1 in row one, column one. An interchange of row one and row two is the easiest way to do this:

$$\left[\begin{array}{cc|c} 1 & -3 & -1 \\ 4 & 3 & 11 \end{array}\right]$$

Next, we want a 0 under the entry 1 in column one. We use the row operation $R_2 = -4r_1 + r_2$:

$$\left[\begin{array}{cc|c} 1 & -3 & -1 \\ 4 & 3 & 11 \end{array}\right] \rightarrow \left[\begin{array}{cc|c} 1 & -3 & -1 \\ 0 & 15 & 15 \end{array}\right]$$
$$\uparrow$$
$$R_2 = -4r_1 + r_2$$

Now we want the entry 1 in row two, column two. We use $R_2 = \frac{1}{15}r_2$:

$$\left[\begin{array}{cc|c} 1 & -3 & -1 \\ 0 & 15 & 15 \end{array}\right] \rightarrow \left[\begin{array}{cc|c} 1 & -3 & -1 \\ 0 & 1 & 1 \end{array}\right]$$
$$\uparrow$$
$$R_2 = \frac{1}{15}r_2$$

Using $y = 1$ (from the second row of the matrix), we back-substitute to get
$$x - 3(1) = -1$$
$$x = 2$$

The solution of the system is $x = 2, y = 1$. ≡

The steps we used to solve the system of linear equations in Example 5 can be summarized as follows:

MATRIX METHOD FOR
SOLVING A SYSTEM OF LINEAR
EQUATIONS

Step 1: Write the augmented matrix that represents the system.

Step 2: Perform row operations that place the entry 1 in row one, column one.

Step 3: Perform row operations that leave the entry 1 in row one, column one unchanged, while causing 0's to appear below it in column one.

Step 4: Perform row operations that place the entry 1 in row two, column two and leave the entries in columns to the left unchanged. If it is impossible to place a 1 in row two, column two, then proceed to place a 1 in row two, column three. Once a 1 is in place, perform row operations to place 0's under it.

Step 5: Now repeat Step 4, placing a 1 in the next row, but one column to the right. Continue until the bottom row or the vertical bar is reached.

Step 6: If any rows are obtained that contain only 0's on the left side of the vertical bar, then place such rows at the bottom of the matrix.

ECHELON FORM

After Steps 1–6 have been completed, the matrix is said to be in **echelon form**. A little thought should convince you that a matrix is in echelon form when:

1. The entry in row one, column one is 1, and 0's appear below it.
2. The first nonzero entry in each row after the first row is a 1, 0's appear below it, and it appears to the right of the first nonzero entry in any row above.
3. Any rows that contain all 0's to the left of the vertical bar appear at the bottom.

Two advantages of solving a system of equations by writing the augmented matrix in echelon form are:

1. The process is algorithmic; that is, it consists of repetitive steps so that it can be programmed on a computer.
2. The process works on any system of linear equations, no matter how many equations or variables are present.

The next example shows how to write a matrix in echelon form.

EXAMPLE 6   Solve

$$\begin{cases} x - y + z = 8 & (1) \\ 2x + 3y - z = -2 & (2) \\ 3x - 2y - 9z = 9 & (3) \end{cases}$$

Solution   Step 1: The augmented matrix of the system is

$$\begin{bmatrix} 1 & -1 & 1 & | & 8 \\ 2 & 3 & -1 & | & -2 \\ 3 & -2 & -9 & | & 9 \end{bmatrix}$$

Step 2: Because the entry 1 is already present in row one, column one, we can go to Step 3.

Step 3: Perform the row operations

$$R_2 = -2r_1 + r_2 \qquad R_3 = -3r_1 + r_3$$

Each of these leaves the entry 1 in row one, column one unchanged, while causing 0's to appear under it:

$$\begin{bmatrix} 1 & -1 & 1 & | & 8 \\ 2 & 3 & -1 & | & -2 \\ 3 & -2 & -9 & | & 9 \end{bmatrix} \rightarrow \begin{bmatrix} 1 & -1 & 1 & | & 8 \\ 0 & 5 & -3 & | & -18 \\ 0 & 1 & -12 & | & -15 \end{bmatrix}$$

$$\begin{array}{c} \uparrow \\ R_2 = -2r_1 + r_2 \\ R_3 = -3r_1 + r_3 \end{array}$$

Step 4: The easiest way to obtain the entry 1 in row two, column two without altering column one is to interchange row two and row three (another way would be to multiply row two by $\frac{1}{5}$, but this introduces fractions):

$$\begin{bmatrix} 1 & -1 & 1 & | & 8 \\ 0 & 1 & -12 & | & -15 \\ 0 & 5 & -3 & | & -18 \end{bmatrix}$$

To get 0's under the 1 in row two, column two, perform the row operation $R_3 = -5r_2 + r_3$:

$$\begin{bmatrix} 1 & -1 & 1 & | & 8 \\ 0 & 1 & -12 & | & -15 \\ 0 & 5 & -3 & | & -18 \end{bmatrix} \rightarrow \begin{bmatrix} 1 & -1 & 1 & | & 8 \\ 0 & 1 & -12 & | & -15 \\ 0 & 0 & 57 & | & 57 \end{bmatrix}$$

$$\begin{array}{c} \uparrow \\ R_3 = -5r_2 + r_3 \end{array}$$

Step 5: Continuing, we place a 1 in row three, column three by using $R_3 = \frac{1}{57}r_3$:

$$\begin{bmatrix} 1 & -1 & 1 & | & 8 \\ 0 & 1 & -12 & | & -15 \\ 0 & 0 & 57 & | & 57 \end{bmatrix} \rightarrow \begin{bmatrix} 1 & -1 & 1 & | & 8 \\ 0 & 1 & -12 & | & -15 \\ 0 & 0 & 1 & | & 1 \end{bmatrix}$$

$$\begin{array}{c} \uparrow \\ R_3 = \frac{1}{57}r_3 \end{array}$$

Because we have reached the bottom row, the matrix is in echelon form and we can stop.

Using $z = 1$ (from the third row of the matrix), we back-substitute to get

$$\begin{cases} x - y + 1 = 8 & \text{From row one of the matrix} \\ y - 12(1) = -15 & \text{From row two of the matrix} \end{cases}$$

Thus, we get $y = -3$, and back-substituting into $x - y = 7$, we find $x = 4$. The solution of the system is $x = 4, y = -3, z = 1$.

REDUCED ECHELON FORM   Sometimes it is advantageous to write a matrix in **reduced echelon form**. In this form, row operations are used to obtain entries that are 0 above (as well as below) the leading 1 in a row. For example, the echelon form obtained in the solution to Example 6 is

$$\begin{bmatrix} 1 & -1 & 1 & | & 8 \\ 0 & 1 & -12 & | & -15 \\ 0 & 0 & 1 & | & 1 \end{bmatrix}$$

To write this matrix in reduced echelon form, we proceed as follows:

$$\begin{bmatrix} 1 & -1 & 1 & | & 8 \\ 0 & 1 & -12 & | & -15 \\ 0 & 0 & 1 & | & 1 \end{bmatrix} \rightarrow \begin{bmatrix} 1 & 0 & -11 & | & -7 \\ 0 & 1 & -12 & | & -15 \\ 0 & 0 & 1 & | & 1 \end{bmatrix}$$

$$\uparrow$$
$$R_1 = r_2 + r_1$$

$$\rightarrow \begin{bmatrix} 1 & 0 & 0 & | & 4 \\ 0 & 1 & 0 & | & -3 \\ 0 & 0 & 1 & | & 1 \end{bmatrix}$$

$$\uparrow$$
$$R_1 = 11r_3 + r_1$$
$$R_2 = 12r_3 + r_2$$

The matrix is now written in reduced echelon form. The advantage of writing the matrix in this form is that the solution to the system, namely $x = 4$, $y = -3$, $z = 1$, is readily found, without the need to back-substitute.

   The matrix method for solving a system of linear equations also identifies systems that have infinitely many solutions and systems that are inconsistent. Let's see how.

EXAMPLE 7   Solve

$$\begin{cases} 6x - y - z = 4 & (1) \\ -12x + 2y + 2z = -8 & (2) \\ 5x + y - z = 3 & (3) \end{cases}$$

Solution   We start with the augmented matrix of the system:

$$\begin{bmatrix} 6 & -1 & -1 & | & 4 \\ -12 & 2 & 2 & | & -8 \\ 5 & 1 & -1 & | & 3 \end{bmatrix} \rightarrow \begin{bmatrix} 1 & -2 & 0 & | & 1 \\ -12 & 2 & 2 & | & -8 \\ 5 & 1 & -1 & | & 3 \end{bmatrix} \rightarrow \begin{bmatrix} 1 & -2 & 0 & | & 1 \\ 0 & -22 & 2 & | & 4 \\ 0 & 11 & -1 & | & -2 \end{bmatrix}$$

$$\uparrow$$
$$R_1 = -1r_3 + r_1$$

$$\uparrow$$
$$R_2 = 12r_1 + r_2$$
$$R_3 = -5r_1 + r_3$$

Obtaining a 1 in row two, column two without altering column one can be accomplished only by $R_2 = -\frac{1}{22}r_2$ or by $R_3 = \frac{1}{11}r_3$. (Do you see why?) We shall use the first of these:

$$\begin{bmatrix} 1 & -2 & 0 & | & 1 \\ 0 & -22 & 2 & | & 4 \\ 0 & 11 & -1 & | & -2 \end{bmatrix} \rightarrow \begin{bmatrix} 1 & -2 & 0 & | & 1 \\ 0 & 1 & -\frac{1}{11} & | & -\frac{2}{11} \\ 0 & 11 & -1 & | & -2 \end{bmatrix}$$

$$\uparrow$$
$$R_2 = -\tfrac{1}{22}r_2$$

$$\rightarrow \begin{bmatrix} 1 & -2 & 0 & | & 1 \\ 0 & 1 & -\frac{1}{11} & | & -\frac{2}{11} \\ 0 & 0 & 0 & | & 0 \end{bmatrix}$$

$$\uparrow$$
$$R_3 = -11r_2 + r_3$$

This matrix is in echelon form. Because the bottom row consists entirely of 0's, the system actually consists of only two equations:

$$\begin{cases} x - 2y = 1 & (1) \\ y - \frac{1}{11}z = -\frac{2}{11} & (2) \end{cases}$$

We shall back-substitute the solution for y from the second equation, namely $y = \frac{1}{11}z - \frac{2}{11}$, to get

$$x = 2y + 1 = 2(\tfrac{1}{11}z - \tfrac{2}{11}) + 1 = \tfrac{2}{11}z + \tfrac{7}{11}$$

Thus, the original system is equivalent to the system

$$\begin{cases} x = \frac{2}{11}z + \frac{7}{11} & (1) \\ y = \frac{1}{11}z - \frac{2}{11} & (2) \end{cases}$$

Let's look at the situation. The original system of three equations is equivalent to a system containing two equations. This means any values of x, y, z that satisfy both

$$x = \tfrac{2}{11}z + \tfrac{7}{11} \quad \text{and} \quad y = \tfrac{1}{11}z - \tfrac{2}{11}$$

will be solutions. For example,

$$z = 0, x = \tfrac{7}{11}, y = -\tfrac{2}{11};$$
$$z = 1, x = \tfrac{9}{11}, y = -\tfrac{1}{11}; \quad z = -1, x = \tfrac{5}{11}, y = -\tfrac{3}{11}$$

are some of the solutions of the original system. There are, in fact, infinitely many values of x, y, and z, for which the two equations are satisfied. That is, the original system has infinitely many solutions.

We will write the solution of the original system as

$$\begin{cases} x = \frac{2}{11}z + \frac{7}{11} \\ y = \frac{1}{11}z - \frac{2}{11} \end{cases}$$

where z can be any real number.

We also can find the solution by writing the augmented matrix in reduced echelon form. Starting with the echelon form, we have

$$\begin{bmatrix} 1 & -2 & 0 & | & 1 \\ 0 & 1 & -\frac{1}{11} & | & -\frac{2}{11} \\ 0 & 0 & 0 & | & 0 \end{bmatrix} \rightarrow \begin{bmatrix} 1 & 0 & -\frac{2}{11} & | & \frac{7}{11} \\ 0 & 1 & -\frac{1}{11} & | & -\frac{2}{11} \\ 0 & 0 & 0 & | & 0 \end{bmatrix}$$

$$\uparrow$$
$$R_1 = 2r_2 + r_1$$

The matrix on the right is in reduced echelon form. The corresponding system of equations is

$$\begin{cases} x - \frac{2}{11}z = \frac{7}{11} & (1) \\ y - \frac{1}{11}z = -\frac{2}{11} & (2) \end{cases}$$

or equivalently,

$$\begin{cases} x = \frac{2}{11}z + \frac{7}{11} & (1) \\ y = \frac{1}{11}z - \frac{2}{11} & (2) \end{cases}$$

where z can be any real number.   ≡

EXAMPLE 8   Solve

$$\begin{cases} x + y + z = 6 \\ 2x - y - z = 3 \\ x + 2y + 2z = 0 \end{cases}$$

Solution   The augmented matrix is

$$\begin{bmatrix} 1 & 1 & 1 & | & 6 \\ 2 & -1 & -1 & | & 3 \\ 1 & 2 & 2 & | & 0 \end{bmatrix} \rightarrow \begin{bmatrix} 1 & 1 & 1 & | & 6 \\ 0 & -3 & -3 & | & -9 \\ 0 & 1 & 1 & | & -6 \end{bmatrix}$$

$$\uparrow$$
$$R_2 = -2r_1 + r_2$$
$$R_3 = -1r_1 + r_3$$

$$\rightarrow \begin{bmatrix} 1 & 1 & 1 & | & 6 \\ 0 & 1 & 1 & | & -6 \\ 0 & -3 & -3 & | & -9 \end{bmatrix}$$

$$\uparrow$$
Interchange rows two and three.

$$\rightarrow \begin{bmatrix} 1 & 1 & 1 & \bigm| & 6 \\ 0 & 1 & 1 & \bigm| & -6 \\ 0 & 0 & 0 & \bigm| & -27 \end{bmatrix}$$

$$\uparrow$$
$$R_3 = 3r_2 + r_3$$

This matrix is in echelon form. The bottom row is equivalent to the equation

$$0x + 0y + 0z = -27$$

which has no solution. Hence, the original system is inconsistent.

■

The matrix method is especially effective for systems of equations for which the number of equations and the number of variables are unequal. Here, too, such a system either is inconsistent or is consistent. If it is consistent, it will have either exactly one solution or else infinitely many solutions.

Let's look at a system of four equations containing three variables.

EXAMPLE 9 Solve

$$\begin{cases} x - 2y + z = 0 & (1) \\ 2x + 2y - 3z = -3 & (2) \\ y - z = -1 & (3) \\ -x + 4y + 2z = 13 & (4) \end{cases}$$

Solution The augmented matrix is

$$\begin{bmatrix} 1 & -2 & 1 & \bigm| & 0 \\ 2 & 2 & -3 & \bigm| & -3 \\ 0 & 1 & -1 & \bigm| & -1 \\ -1 & 4 & 2 & \bigm| & 13 \end{bmatrix} \rightarrow \begin{bmatrix} 1 & -2 & 1 & \bigm| & 0 \\ 0 & 6 & -5 & \bigm| & -3 \\ 0 & 1 & -1 & \bigm| & -1 \\ 0 & 2 & 3 & \bigm| & 13 \end{bmatrix}$$

$$\uparrow$$
$$R_2 = -2r_1 + r_2$$
$$R_4 = r_1 + r_4$$

$$\rightarrow \begin{bmatrix} 1 & -2 & 1 & \bigm| & 0 \\ 0 & 1 & -1 & \bigm| & -1 \\ 0 & 6 & -5 & \bigm| & -3 \\ 0 & 2 & 3 & \bigm| & 13 \end{bmatrix}$$

$$\uparrow$$
Interchange rows two and three.

$$\rightarrow \begin{bmatrix} 1 & -2 & 1 & | & 0 \\ 0 & 1 & -1 & | & -1 \\ 0 & 0 & 1 & | & 3 \\ 0 & 0 & 5 & | & 15 \end{bmatrix}$$

↑
$R_3 = -6r_2 + r_3$
$R_4 = -2r_2 + r_4$

$$\rightarrow \begin{bmatrix} 1 & -2 & 1 & | & 0 \\ 0 & 1 & -1 & | & -1 \\ 0 & 0 & 1 & | & 3 \\ 0 & 0 & 0 & | & 0 \end{bmatrix}$$

↑
$R_4 = -5r_3 + r_4$

$$\rightarrow \begin{bmatrix} 1 & 0 & -1 & | & -2 \\ 0 & 1 & -1 & | & -1 \\ 0 & 0 & 1 & | & 3 \\ 0 & 0 & 0 & | & 0 \end{bmatrix}$$

↑
$R_1 = 2r_2 + r_1$

$$\rightarrow \begin{bmatrix} 1 & 0 & 0 & | & 1 \\ 0 & 1 & 0 & | & 2 \\ 0 & 0 & 1 & | & 3 \\ 0 & 0 & 0 & | & 0 \end{bmatrix}$$

↑
$R_1 = r_3 + r_1$
$R_2 = r_3 + r_2$

The matrix is now in reduced echelon form, and we can see that the solution is

$$x = 1, y = 2, z = 3$$ ▤

EXAMPLE 10   A chemistry laboratory has three containers of nitric acid, HNO_3. One container holds a solution with a concentration of 10% HNO_3, the second holds 20% HNO_3, and the third holds 40% HNO_3. How many liters of each solution should be mixed to obtain 100 liters of a solution whose concentration is 25% HNO_3?

Solution   Let x, y, and z represent the number of liters of 10%, 20%, and 40% concentrations of HNO_3, respectively. We want 100 liters in all, and the concentration of HNO_3 from each solution must sum to 25% of 100 liters; thus, we find that

$$\begin{cases} x + y + z = 100 \\ 0.10x + 0.20y + 0.40z = 0.25(100) \end{cases}$$

Now, the augmented matrix is

$$\begin{bmatrix} 1 & 1 & 1 & | & 100 \\ 0.10 & 0.20 & 0.40 & | & 25 \end{bmatrix} \rightarrow \begin{bmatrix} 1 & 1 & 1 & | & 100 \\ 0 & 0.10 & 0.30 & | & 15 \end{bmatrix}$$

$$\uparrow$$
$$R_2 = -0.10r_1 + r_2$$

$$\rightarrow \begin{bmatrix} 1 & 1 & 1 & | & 100 \\ 0 & 1 & 3 & | & 150 \end{bmatrix}$$

$$\uparrow$$
$$R_2 = 10r_2$$

$$\rightarrow \begin{bmatrix} 1 & 0 & -2 & | & -50 \\ 0 & 1 & 3 & | & 150 \end{bmatrix}$$

$$\uparrow$$
$$R_1 = -1r_2 + r_1$$

The matrix is now in reduced echelon form. The final matrix represents the system

$$\begin{cases} x - 2z = -50 & (1) \\ y + 3z = 150 & (2) \end{cases}$$

which has infinitely many solutions given by

$$\begin{cases} x = 2z - 50 & (1) \\ y = -3z + 150 & (2) \end{cases}$$

where z is any real number. However, the practical considerations of this problem require us to restrict the solutions to

$$x \geq 0, y \geq 0, z \geq 0$$

Furthermore, we require

$$z \geq 25, z \leq 50$$

because otherwise $x < 0$ or $y < 0$. Some of the possible solutions are given in Table 1. The final determination of what solution the laboratory will pick very likely depends on availability, cost differences, and other considerations.

TABLE 1

LITERS OF 10% SOLUTION	LITERS OF 20% SOLUTION	LITERS OF 40% SOLUTION	LITERS OF 25% SOLUTION
0	75	25	100
10	60	30	100
12	57	31	100
16	51	33	100
26	36	38	100
38	18	44	100
46	6	48	100
50	0	50	100

EXERCISE 7.3   *In Problems 1–10 write the matrix representation of the given system of equations.*

1. $\begin{cases} x - 3y = 5 \\ 4x + y = 6 \end{cases}$

2. $\begin{cases} 3x + y = 1 \\ x - 2y = 5 \end{cases}$

3. $\begin{cases} 2x + 3y - 6 = 0 \\ 4x - 6y + 2 = 0 \end{cases}$

4. $\begin{cases} 9x - y = 0 \\ 3x - y - 4 = 0 \end{cases}$

5. $\begin{cases} 0.01x - 0.03y = 0.06 \\ 0.13x + 0.10y = 0.20 \end{cases}$

6. $\begin{cases} \frac{4}{3}x - \frac{3}{2}y = \frac{3}{4} \\ -\frac{1}{4}x + \frac{1}{3}y = \frac{2}{3} \end{cases}$

7. $\begin{cases} x - y + z = 10 \\ 3x + 2y = 5 \\ x + y + 2z = 2 \end{cases}$

8. $\begin{cases} 5x - y - z = 0 \\ x + y = 5 \\ 2x - 3z = 2 \end{cases}$

9. $\begin{cases} x + y - z = 2 \\ 3x - 2y = 2 \end{cases}$

10. $\begin{cases} 2x + 3y - 4z = 0 \\ x - 5z + 2 = 0 \end{cases}$

In Problems 11–14 state the row operation used to transform the matrix on the left into the one on the right.

11. $\begin{bmatrix} 2 & 3 & | & 5 \\ 3 & 1 & | & 4 \end{bmatrix} \rightarrow \begin{bmatrix} -1 & 2 & | & 1 \\ 3 & 1 & | & 4 \end{bmatrix}$

12. $\begin{bmatrix} 4 & 5 & | & -1 \\ 2 & 2 & | & 0 \end{bmatrix} \rightarrow \begin{bmatrix} 4 & 5 & | & -1 \\ 1 & 1 & | & 0 \end{bmatrix}$

13. $\begin{bmatrix} 1 & 4 & | & -3 \\ 0 & 1 & | & -1 \end{bmatrix} \rightarrow \begin{bmatrix} 1 & 0 & | & 1 \\ 0 & 1 & | & -1 \end{bmatrix}$

14. $\begin{bmatrix} 4 & 8 & | & -4 \\ 2 & 2 & | & 0 \end{bmatrix} \rightarrow \begin{bmatrix} 0 & 4 & | & -4 \\ 2 & 2 & | & 0 \end{bmatrix}$

In Problems 15–20 perform the indicated row operation on the matrix

$$\begin{bmatrix} 1 & 2 & 3 & | & 0 \\ 2 & 4 & 3 & | & 3 \\ -3 & 2 & 1 & | & -2 \end{bmatrix}$$

15. $R_2 = -2r_1 + r_2$

16. $R_3 = 3r_1 + r_3$

17. $R_2 = -2r_3 + r_2$

18. $R_2 = \frac{1}{2}r_2$

19. $R_3 = r_2 + r_3$

20. $R_1 = r_2 + r_1$

In Problems 21–68 solve each system of equations using matrices (row operations). If the system has no solution, say it is inconsistent.

21. $\begin{cases} x + y = 8 \\ x - y = 4 \end{cases}$

22. $\begin{cases} x + 2y = 5 \\ x + y = 3 \end{cases}$

23. $\begin{cases} 5x - y = 13 \\ 2x + 3y = 12 \end{cases}$

24. $\begin{cases} x + 3y = 5 \\ 2x - 3y = -8 \end{cases}$

25. $\begin{cases} 3x = 24 \\ x + 2y = 0 \end{cases}$

26. $\begin{cases} 4x + 5y = -3 \\ -2y = -4 \end{cases}$

27. $\begin{cases} 3x - 6y = 24 \\ 5x + 4y = 12 \end{cases}$

28. $\begin{cases} 2x + 4y = 16 \\ 3x - 5y = -9 \end{cases}$

29. $\begin{cases} 2x + y = 1 \\ 4x + 2y = 6 \end{cases}$

30. $\begin{cases} x - y = 5 \\ -3x + 3y = 2 \end{cases}$

31. $\begin{cases} 2x - 4y = -2 \\ 3x + 2y = 3 \end{cases}$

32. $\begin{cases} 3x + 3y = 3 \\ 4x + 2y = \frac{8}{3} \end{cases}$

33. $\begin{cases} x + 2y = 4 \\ 2x + 4y = 8 \end{cases}$

34. $\begin{cases} 3x - y = 7 \\ 9x - 3y = 21 \end{cases}$

35. $\begin{cases} 2x - 3y = -1 \\ 10x + 10y = 5 \end{cases}$

36. $\begin{cases} 3x - 2y = 0 \\ 5x + 10y = 4 \end{cases}$

37. $\begin{cases} 2x + 3y = 6 \\ x - y = \frac{1}{2} \end{cases}$

38. $\begin{cases} \frac{1}{2}x + y = -2 \\ x - 2y = 8 \end{cases}$

39. $\begin{cases} 2x + 3y = 5 \\ 4x + 6y = 10 \end{cases}$

40. $\begin{cases} 2x + 3y = 5 \\ 4x + 6y = 6 \end{cases}$

41. $\begin{cases} 3x - 5y = 3 \\ 15x + 5y = 21 \end{cases}$

42. $\begin{cases} 2x - y = -1 \\ x + \frac{1}{2}y = \frac{3}{2} \end{cases}$

43. $\begin{cases} x - y = 6 \\ 2x - 3z = 16 \\ 2y + z = 4 \end{cases}$

44. $\begin{cases} 2x + y = -4 \\ -2y + 4z = 0 \\ 3x - 2z = -11 \end{cases}$

45. $\begin{cases} x - 2y + 3z = 7 \\ 2x + y + z = 4 \\ -3x + 2y - 2z = -10 \end{cases}$

H 46. $\begin{cases} 2x + y - 3z = 0 \\ -2x + 2y + z = -7 \\ 3x - 4y - 3z = 7 \end{cases}$

47. $\begin{cases} x - y - z = 1 \\ 2x + 3y + z = 2 \\ 3x + 2y = 0 \end{cases}$

48. $\begin{cases} 2x - 3y - z = 0 \\ -x + 2y + z = 5 \\ 3x - 4y - z = 1 \end{cases}$

49. $\begin{cases} x - y - z = 1 \\ -x + 2y - 3z = -4 \\ 3x - 2y - 7z = 0 \end{cases}$

50. $\begin{cases} 2x - 3y - z = 0 \\ 3x + 2y + 2z = 2 \\ x + 5y + 3z = 2 \end{cases}$

51. $\begin{cases} 2x - 2y + 3z = 6 \\ 4x - 3y + 2z = 0 \\ -2x + 3y - 7z = 1 \end{cases}$

52. $\begin{cases} 3x - 2y + 2z = 6 \\ 7x - 3y + 2z = -1 \\ 2x - 3y + 4z = 0 \end{cases}$

53. $\begin{cases} x + y - z = 6 \\ 3x - 2y + z = -5 \\ x + 3y - 2z = 14 \end{cases}$

54. $\begin{cases} x - y + z = -4 \\ 2x - 3y + 4z = -15 \\ 5x + y - 2z = 12 \end{cases}$

55. $\begin{cases} x + 2y - z = -3 \\ 2x - 4y + z = -7 \\ -2x + 2y - 3z = 4 \end{cases}$

56. $\begin{cases} x + 4y - 3z = -8 \\ 3x - y + 3z = 12 \\ x + y + 6z = 1 \end{cases}$

57. $\begin{cases} 3x + y - z = \frac{2}{3} \\ 2x - y + z = 1 \\ 4x + 2y = \frac{8}{3} \end{cases}$

58. $\begin{cases} x + y = 1 \\ 2x - y + z = 1 \\ x + 2y + z = \frac{8}{3} \end{cases}$

59. $\begin{cases} x + y + z + w = 4 \\ 2x - y + z = 0 \\ 3x + 2y + z - w = 6 \\ x - 2y - 2z + 2w = -1 \end{cases}$

60. $\begin{cases} x + y + z + w = 4 \\ -x + 2y + z = 0 \\ 2x + 3y + z - w = 6 \\ -2x + y - 2z + 2w = -1 \end{cases}$

61. $\begin{cases} x + 2y + z = 1 \\ 2x - y + 2z = 2 \\ 3x + y + 3z = 3 \end{cases}$

62. $\begin{cases} x + 2y - z = 3 \\ 2x - y + 2z = 6 \\ x - 3y + 3z = 4 \end{cases}$

63. $\begin{cases} x - y + z = 5 \\ 3x + 2y - 2z = 0 \end{cases}$

64. $\begin{cases} 2x + y - z = 4 \\ -x + y + 3z = 1 \end{cases}$

65. $\begin{cases} 2x + 3y - z = 3 \\ x - y - z = 0 \\ -x + y + z = 0 \\ x + y + 3z = 5 \end{cases}$

66. $\begin{cases} x - 3y + z = 1 \\ 2x - y - 4z = 0 \\ x - 3y + 2z = 1 \\ x - 2y = 5 \end{cases}$

67. $\begin{cases} 4x + y + z - w = 4 \\ x - y + 2z + 3w = 3 \end{cases}$

68. $\begin{cases} -4x + y = 5 \\ 2x - y + z - w = 5 \\ z + w = 4 \end{cases}$

69. Find the parabola $y = ax^2 + bx + c$ that passes through the points $(1, 2), (-2, -7)$, and $(2, -3)$.

70. Find the parabola $y = ax^2 + bx + c$ that passes through the points $(1, -1), (3, -1)$, and $(-2, 14)$.

71. Find the function $f(x) = ax^3 + bx^2 + cx + d$ for which $f(-3) = -112$, $f(-1) = -2$, $f(1) = 4$, and $f(2) = 13$.

72. Find the function $f(x) = ax^3 + bx^2 + cx + d$ for which $f(-2) = -10$, $f(-1) = 3$, $f(1) = 5$, and $f(3) = 15$.

73. A chemistry laboratory has three containers of sulfuric acid, H_2SO_4. One container holds a solution with a concentration of 15% H_2SO_4, the second holds 25% H_2SO_4, and the third holds 50% H_2SO_4. How many liters of each solution should be mixed to obtain 100 liters of a solution with a concentration of 40% H_2SO_4? Construct a table similar to Table 1 illustrating some of the possible combinations.

74. Three painters, Mike, Dan, and Katy, working together can paint the exterior of a home in 10 hours. Dan and Katy together have painted a similar house in 15 hours. One day all three worked on this same kind of house for 4 hours, after which Katy left. Mike and Dan required 8 more hours to finish. Assuming no gain or loss in efficiency, how long should it take each person to complete such a job alone?

75. Consider the system of equations

$$\begin{cases} a_1x + b_1y = c_1 \\ a_2x + b_2y = c_2 \end{cases}$$

If $D = a_1b_2 - a_2b_1 \neq 0$, use matrices to show that the solution is

$$x = \frac{1}{D}(c_1b_2 - c_2b_1)$$

$$y = \frac{1}{D}(a_1c_2 - a_2c_1)$$

76. For the system in Problem 75, suppose $D = a_1b_2 - a_2b_1 = 0$. Use matrices to show that the system is inconsistent if either $a_1c_2 \neq a_2c_1$ or $b_1c_2 \neq b_2c_1$ and has infinitely many solutions if both $a_1c_2 = a_2c_1$ and $b_1c_2 = b_2c_1$.

77. *Programming exercise* Write a computer program that will solve any system of linear equations.

≡ 7.4 SOLUTIONS OF SYSTEMS OF LINEAR EQUATIONS: DETERMINANTS

In the preceding section, we described a method of using matrices to solve any system of linear equations. This section deals with yet another method for solving systems of linear equations; however, it can only be used when the number of equations equals the number of variables. Although the method will work for any system (provided the number of equations equals the number of unknowns), it is usually practical only for systems of two equations containing two variables or three equations containing three variables. This method, called
CRAMER'S RULE **Cramer's Rule**, is based on the concept of a *determinant*.

2 by 2 Determinants

If a, b, c, and d are four numbers, the symbol

$$D = \begin{vmatrix} a & b \\ c & d \end{vmatrix}$$

2 BY 2 DETERMINANT is called a **2 by 2 determinant**. Its **value** is the number $ad - bc$; that
VALUE is,

$$D = \begin{vmatrix} a & b \\ c & d \end{vmatrix} = ad - bc \qquad (1)$$

A device thay may be helpful for remembering the value of a 2 by 2 determinant is the following:

$$\begin{vmatrix} a & b \\ c & d \end{vmatrix} \begin{matrix} \nearrow bc \\ \searrow ad \end{matrix} = ad - bc$$

Minus

EXAMPLE 1
$$\begin{vmatrix} 3 & -2 \\ 6 & 1 \end{vmatrix} = (3)(1) - (6)(-2) = 3 - (-12) = 15$$

Let's now see the role that a 2 by 2 determinant plays in the solution of a system of two equations containing two variables. Consider the system

$$\begin{cases} ax + by = s & (1) \\ cx + dy = t & (2) \end{cases} \qquad (2)$$

We shall use the method of elimination to solve this system.

Provided $d \neq 0$ and $b \neq 0$, this system is equivalent to the system

$$\begin{cases} adx + bdy = sd & (1) \quad \text{Multiply by } d. \\ bcx + bdy = tb & (2) \quad \text{Multiply by } b. \end{cases}$$

On subtracting the second equation from the first equation, we get

$$\begin{cases} (ad - bc)x + 0 \cdot y = sd - tb & (1) \\ bcx \qquad + bdy = tb & (2) \end{cases}$$

The first equation can be rewritten, using the determinant notation, as

$$\begin{vmatrix} a & b \\ c & d \end{vmatrix} x = \begin{vmatrix} s & b \\ t & d \end{vmatrix}$$

If $D = \begin{vmatrix} a & b \\ c & d \end{vmatrix} = ad - bc \neq 0$, we can solve for x to get

$$x = \frac{\begin{vmatrix} s & b \\ t & d \end{vmatrix}}{\begin{vmatrix} a & b \\ c & d \end{vmatrix}} = \frac{\begin{vmatrix} s & b \\ t & d \end{vmatrix}}{D} \qquad (3)$$

Return now to the system in Equation (2). Provided $a \neq 0$ and $c \neq 0$, the system is equivalent to

$$\begin{cases} acx + bcy = cs & (1) \quad \text{Multiply by } c. \\ acx + ady = at & (2) \quad \text{Multiply by } a. \end{cases}$$

On subtracting the first equation from the second equation, we get

$$\begin{cases} acx + \quad bcy \quad = cs & (1) \\ 0 \cdot x + (ad - bc)y = at - cs & (2) \end{cases}$$

The second equation can be rewritten, using the determinant notation, as

$$\begin{vmatrix} a & b \\ c & d \end{vmatrix} y = \begin{vmatrix} a & s \\ c & t \end{vmatrix}$$

If $D = \begin{vmatrix} a & b \\ c & d \end{vmatrix} = ad - bc \neq 0$, we can solve for y to get

$$y = \frac{\begin{vmatrix} a & s \\ c & t \end{vmatrix}}{\begin{vmatrix} a & b \\ c & d \end{vmatrix}} = \frac{\begin{vmatrix} a & s \\ c & t \end{vmatrix}}{D} \tag{4}$$

Equations (3) and (4) lead us to the following result. The solution to the system of equations

CRAMER'S RULE FOR TWO
EQUATIONS CONTAINING
TWO VARIABLES

$$\begin{cases} ax + by = s & (1) \\ cx + dy = t & (2) \end{cases} \tag{5}$$

is given by

$$x = \frac{\begin{vmatrix} s & b \\ t & d \end{vmatrix}}{\begin{vmatrix} a & b \\ c & d \end{vmatrix}}, \qquad y = \frac{\begin{vmatrix} a & s \\ c & t \end{vmatrix}}{\begin{vmatrix} a & b \\ c & d \end{vmatrix}} \tag{6}$$

provided that $D = \begin{vmatrix} a & b \\ c & d \end{vmatrix} = ad - bc \neq 0.$

In the derivation given for Cramer's Rule above, we assumed that none of the numbers $a, b, c,$ and d were 0. In Problem 56 at the end of this section you will be asked to complete the proof under the less stringent conditions that $D = ad - bc \neq 0$.

Now look carefully at the pattern in Cramer's Rule. The denominator in the solution, Display (6), is the determinant of the coefficients of the variables:

$$\begin{cases} ax + by = s \\ cx + dy = t \end{cases} \qquad D = \begin{vmatrix} a & b \\ c & d \end{vmatrix}$$

In the solution for x, the numerator is the determinant, denoted by D_x, formed by replacing the first column entries (the coefficients of x) in D by the constants on the right side of the equal sign:

$$D_x = \begin{vmatrix} s & b \\ t & d \end{vmatrix}$$

In the solution for y, the numerator is the determinant, denoted by D_y, formed by replacing the second column entries (the coefficients of y) in D by the constants on the right side of the equal sign:

$$D_y = \begin{vmatrix} a & s \\ c & t \end{vmatrix}$$

Cramer's Rule then states that, if $D \neq 0$,

$$x = \frac{D_x}{D}, \qquad y = \frac{D_y}{D} \tag{7}$$

EXAMPLE 2   Use Cramer's Rule, if applicable, to solve the system

$$\begin{cases} 3x - 2y = 4 & (1) \\ 6x + y = 13 & (2) \end{cases}$$

Solution   The determinant D of the coefficients of the variables is

$$D = \begin{vmatrix} 3 & -2 \\ 6 & 1 \end{vmatrix} = (3)(1) - (6)(-2) = 15$$

Because $D \neq 0$, Cramer's Rule (7) can be used:

$$x = \frac{D_x}{D} = \frac{\begin{vmatrix} 4 & -2 \\ 13 & 1 \end{vmatrix}}{15} = \frac{30}{15} = 2, \qquad y = \frac{D_y}{D} = \frac{\begin{vmatrix} 3 & 4 \\ 6 & 13 \end{vmatrix}}{15} = \frac{15}{15} = 1$$

The solution is $x = 2, y = 1$.   ≡

If, in attempting to use Cramer's Rule, the determinant D of the coefficients of the variables equals 0, so that Cramer's Rule is not applicable, then the system either is inconsistent or has infinitely many solutions. (Refer to Problem 76 in Exercise 7.3.)

3 by 3 Determinants

In order to use Cramer's Rule to solve a system of three equations containing three variables, we need to be able to find the value of a 3 by 3 determinant.

3 BY 3 DETERMINANT   A **3 by 3 determinant** is an expression of the form

$$\begin{vmatrix} a_{11} & a_{12} & a_{13} \\ a_{21} & a_{22} & a_{23} \\ a_{31} & a_{32} & a_{33} \end{vmatrix} \tag{8}$$

in which $a_{11}, a_{12}, \ldots$ are nine real numbers.

As with matrices, we use a double subscript to identify an entry by indicating its row and column numbers. For example, the entry a_{23} is in row two, column three.

The value of a 3 by 3 determinant is defined in terms of 2 by 2 determinants by the formula

$$
\begin{vmatrix} a_{11} & a_{12} & a_{13} \\ a_{21} & a_{22} & a_{23} \\ a_{31} & a_{32} & a_{33} \end{vmatrix} = a_{11} \overset{\text{Minus}}{\begin{vmatrix} a_{22} & a_{23} \\ a_{32} & a_{33} \end{vmatrix}} - a_{12} \begin{vmatrix} a_{21} & a_{23} \\ a_{31} & a_{33} \end{vmatrix} + a_{13} \begin{vmatrix} a_{21} & a_{22} \\ a_{31} & a_{32} \end{vmatrix} \quad (9)
$$

	2 by 2 determinant left after removing row and column containing a_{11}	2 by 2 determinant left after removing row and column containing a_{12}	2 by 2 determinant left after removing row and column containing a_{13}

Be sure to take note of the minus sign that appears with the second term—it's easy to forget it! Formula (9) for finding the value of a 3 by 3 determinant is best remembered by noting that each entry in row one is multiplied by the 2 by 2 determinant that remains after the row and column containing the entry has been removed, as indicated below:

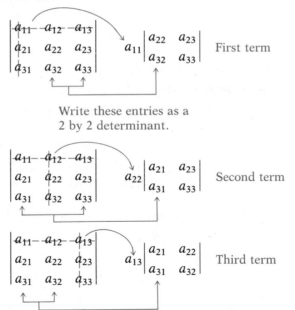

$a_{11} \begin{vmatrix} a_{22} & a_{23} \\ a_{32} & a_{33} \end{vmatrix}$ First term

Write these entries as a 2 by 2 determinant.

$a_{22} \begin{vmatrix} a_{21} & a_{23} \\ a_{31} & a_{33} \end{vmatrix}$ Second term

$a_{13} \begin{vmatrix} a_{21} & a_{22} \\ a_{31} & a_{32} \end{vmatrix}$ Third term

Now insert the minus sign before the middle expression and add:

$$
\begin{vmatrix} a_{11} & a_{12} & a_{13} \\ a_{21} & a_{22} & a_{23} \\ a_{31} & a_{32} & a_{33} \end{vmatrix} = a_{11} \overset{\text{Minus}}{\begin{vmatrix} a_{22} & a_{23} \\ a_{32} & a_{33} \end{vmatrix}} - a_{12} \begin{vmatrix} a_{21} & a_{23} \\ a_{31} & a_{33} \end{vmatrix} + a_{13} \begin{vmatrix} a_{21} & a_{22} \\ a_{31} & a_{32} \end{vmatrix}
$$

Let's do an example using numbers.

EXAMPLE 3   Find the value of the 3 by 3 determinant

$$\begin{vmatrix} 3 & 4 & -1 \\ 4 & 6 & 2 \\ 8 & -2 & 3 \end{vmatrix}$$

Solution   We follow the pattern in Formula (9). Remember to insert the minus sign preceding the middle expression.

Remember the minus sign.
↓

$$\begin{vmatrix} 3 & 4 & -1 \\ 4 & 6 & 2 \\ 8 & -2 & 3 \end{vmatrix} = 3\begin{vmatrix} 6 & 2 \\ -2 & 3 \end{vmatrix} - 4\begin{vmatrix} 4 & 2 \\ 8 & 3 \end{vmatrix} + (-1)\begin{vmatrix} 4 & 6 \\ 8 & -2 \end{vmatrix}$$

$$= 3(18 + 4) - 4(12 - 16) + (-1)(-8 - 48)$$
$$= 3(22) - 4(-4) + (-1)(-56)$$
$$= 66 + 16 + 56 = 138$$ ≡

Systems of Three Equations Containing Three Variables

Consider the system of three equations containing three variables,

$$\begin{cases} a_{11}x + a_{12}y + a_{13}z = c_1 \\ a_{21}x + a_{22}y + a_{23}z = c_2 \\ a_{31}x + a_{32}y + a_{33}z = c_3 \end{cases} \tag{10}$$

CRAMER'S RULE FOR THREE EQUATIONS CONTAINING THREE VARIABLES   If the determinant D of the coefficients of the variables is not 0— that is, if

$$D = \begin{vmatrix} a_{11} & a_{12} & a_{13} \\ a_{21} & a_{22} & a_{23} \\ a_{31} & a_{32} & a_{33} \end{vmatrix} \neq 0$$

then the solution of the system in Display (10) is given by

$$x = \frac{D_x}{D}, \qquad y = \frac{D_y}{D}, \qquad z = \frac{D_z}{D}$$

where

$$D_x = \begin{vmatrix} c_1 & a_{12} & a_{13} \\ c_2 & a_{22} & a_{23} \\ c_3 & a_{32} & a_{33} \end{vmatrix} \qquad D_y = \begin{vmatrix} a_{11} & c_1 & a_{13} \\ a_{21} & c_2 & a_{23} \\ a_{31} & c_3 & a_{33} \end{vmatrix} \qquad D_z = \begin{vmatrix} a_{11} & a_{12} & c_1 \\ a_{21} & a_{22} & c_2 \\ a_{31} & a_{32} & c_3 \end{vmatrix}$$

The similarity of this pattern and the pattern observed earlier for a system of two equations containing two variables should be apparent.

EXAMPLE 4 Use Cramer's Rule, if applicable, to solve the system

$$\begin{cases} 2x + y - z = 3 & (1) \\ -x + 2y + 4z = -3 & (2) \\ x - 2y - 3z = 4 & (3) \end{cases}$$

Solution The value of the determinant D of the coefficients of the unknowns is

$$D = \begin{vmatrix} 2 & 1 & -1 \\ -1 & 2 & 4 \\ 1 & -2 & -3 \end{vmatrix} = 2\begin{vmatrix} 2 & 4 \\ -2 & -3 \end{vmatrix} - 1\begin{vmatrix} -1 & 4 \\ 1 & -3 \end{vmatrix} + (-1)\begin{vmatrix} -1 & 2 \\ 1 & -2 \end{vmatrix}$$

$$= 2(2) - 1(-1) + (-1)(0)$$

$$= 4 + 1 = 5$$

Because $D \neq 0$, we proceed to find the value of D_x, D_y, and D_z:

$$D_x = \begin{vmatrix} 3 & 1 & -1 \\ -3 & 2 & 4 \\ 4 & -2 & -3 \end{vmatrix} = 3\begin{vmatrix} 2 & 4 \\ -2 & -3 \end{vmatrix} - 1\begin{vmatrix} -3 & 4 \\ 4 & -3 \end{vmatrix} + (-1)\begin{vmatrix} -3 & 2 \\ 4 & -2 \end{vmatrix}$$

$$= 3(2) - 1(-7) + (-1)(-2) = 15$$

$$D_y = \begin{vmatrix} 2 & 3 & -1 \\ -1 & -3 & 4 \\ 1 & 4 & -3 \end{vmatrix} = 2\begin{vmatrix} -3 & 4 \\ 4 & -3 \end{vmatrix} - 3\begin{vmatrix} -1 & 4 \\ 1 & -3 \end{vmatrix} + (-1)\begin{vmatrix} -1 & -3 \\ 1 & 4 \end{vmatrix}$$

$$= 2(-7) - 3(-1) + (-1)(-1)$$

$$= -14 + 3 + 1 = -10$$

$$D_z = \begin{vmatrix} 2 & 1 & 3 \\ -1 & 2 & -3 \\ 1 & -2 & 4 \end{vmatrix} = 2\begin{vmatrix} 2 & -3 \\ -2 & 4 \end{vmatrix} - 1\begin{vmatrix} -1 & -3 \\ 1 & 4 \end{vmatrix} + 3\begin{vmatrix} -1 & 2 \\ 1 & -2 \end{vmatrix}$$

$$= 2(2) - 1(-1) + 3(0) = 5$$

As a result,

$$x = \frac{D_x}{D} = \frac{15}{5} = 3, \qquad y = \frac{D_y}{D} = \frac{-10}{5} = -2, \qquad z = \frac{D_z}{D} = \frac{5}{5} = 1$$

The solution is $x = 3, y = -2, z = 1$. ■

If the determinant of the coefficients of the unknowns of a system of three linear equations containing three variables is 0, then Cramer's Rule is not applicable. In such a case, the system either is inconsistent or has infinitely many solutions.

More about Determinants

Determinants have several properties that are sometimes helpful for obtaining their value. We shall list some of them here.

THEOREM   The value of a determinant changes sign if any two rows (or any two columns) are interchanged.   (11)

Proof for 2 by 2 Determinants

$$\begin{vmatrix} a & b \\ c & d \end{vmatrix} = ad - bc \quad \text{and} \quad \begin{vmatrix} b & a \\ d & c \end{vmatrix} = bc - ad = -(ad - bc) \quad ■$$

EXAMPLE 5

$$\begin{vmatrix} 3 & 4 \\ 1 & 2 \end{vmatrix} = 6 - 4 = 2 \qquad \begin{vmatrix} 1 & 2 \\ 3 & 4 \end{vmatrix} = 4 - 6 = -2 \qquad ■$$

THEOREM   If all the entries in any row (or any column) equal 0, the value of the determinant is 0.   (12)

Proof for 3 by 3 Determinants   Consider a 3 by 3 determinant. If the 0's appear in row one, then each term in Formula (9) has a factor of 0, and hence the value is 0. If the 0's appear in a row other than the first, we can interchange the row containing the 0's with row one. The value of the new determinant, 0, equals -1 times the value of the original determinant. Hence, the original determinant has a value of 0. If the 0's appear in a column, say column one, we have

$$\begin{vmatrix} 0 & a & b \\ 0 & c & d \\ 0 & e & f \end{vmatrix} = 0\begin{vmatrix} c & d \\ e & f \end{vmatrix} - a\begin{vmatrix} 0 & d \\ 0 & f \end{vmatrix} + b\begin{vmatrix} 0 & c \\ 0 & e \end{vmatrix}$$

$$= 0 - a \cdot 0 + b \cdot 0 = 0$$

This completes the proof. ■

THEOREM   If any two rows (or any two columns) of a determinant have corresponding entries that are equal, the value of   (13) the determinant is 0. ≡

You are asked to prove this result for a 3 by 3 determinant in which the entries in column one equal the entries in column three in Problem 59 at the end of this section.

EXAMPLE 6

$$\begin{vmatrix} 1 & 2 & 3 \\ 1 & 2 & 3 \\ 4 & 5 & 6 \end{vmatrix} = 1 \begin{vmatrix} 2 & 3 \\ 5 & 6 \end{vmatrix} - 2 \begin{vmatrix} 1 & 3 \\ 4 & 6 \end{vmatrix} + 3 \begin{vmatrix} 1 & 2 \\ 4 & 5 \end{vmatrix}$$

$$= 1(-3) - 2(-6) + 3(-3)$$

$$= -3 + 12 - 9 = 0 \qquad ≡$$

THEOREM   If any row (or any column) of a determinant is multiplied by a nonzero number k, the value of the determinant is   (14) also changed by a factor of k. ≡

You are asked to prove this result for a 3 by 3 determinant using row two in Problem 58 at the end of this section.

EXAMPLE 7

$$\begin{vmatrix} 1 & 2 \\ 4 & 6 \end{vmatrix} = 6 - 8 = -2$$

$$\begin{vmatrix} k & 2k \\ 4 & 6 \end{vmatrix} = 6k - 8k = -2k = k(-2) = k \begin{vmatrix} 1 & 2 \\ 4 & 6 \end{vmatrix} \qquad ≡$$

THEOREM   If any row (or any column) of a determinant is multiplied by a nonzero number k and the result is added to the entries of another row (or column), the value of the deter-   (15) minant remains unchanged. ≡

In Problem 60 at the end of this section, you are asked to prove this result for a 3 by 3 determinant using row two and row one.

EXAMPLE 8

$$\begin{vmatrix} 3 & 4 \\ 5 & 2 \end{vmatrix} = \begin{vmatrix} -7 & 0 \\ 5 & 2 \end{vmatrix} = -14$$

↑
Multiply row two by -2
and add to row one. ≡

Our final example illustrates how these results can be combined to find the value of a determinant.

EXAMPLE 9   Find the value of

$$\begin{vmatrix} 3 & 6 & 4 \\ -1 & 2 & 5 \\ 4 & 4 & 0 \end{vmatrix}$$

Solution

$$\begin{vmatrix} 3 & 6 & 4 \\ -1 & 2 & 5 \\ 4 & 4 & 0 \end{vmatrix} = 4 \begin{vmatrix} 3 & 6 & 4 \\ -1 & 2 & 5 \\ 1 & 1 & 0 \end{vmatrix}$$

Row three has the factor 4; use Theorem (14).

$$= -4 \begin{vmatrix} 1 & 1 & 0 \\ -1 & 2 & 5 \\ 3 & 6 & 4 \end{vmatrix}$$

Interchange rows one and three; change sign; use Theorem (11).

$$= -4 \left[1 \begin{vmatrix} 2 & 5 \\ 6 & 4 \end{vmatrix} - 1 \begin{vmatrix} -1 & 5 \\ 3 & 4 \end{vmatrix} + 0 \begin{vmatrix} -1 & 2 \\ 3 & 6 \end{vmatrix} \right]$$

Apply Formula (9). Do you see now why we interchanged rows one and three?

$$= -4[(-22 + 19)] = 12$$ ≡

EXERCISE 7.4   *In Problems 1–10 find the value of each determinant.*

1. $\begin{vmatrix} 3 & 1 \\ 4 & 2 \end{vmatrix}$ **2.** $\begin{vmatrix} 6 & 1 \\ 5 & 2 \end{vmatrix}$ **3.** $\begin{vmatrix} 6 & 4 \\ -1 & 3 \end{vmatrix}$

4. $\begin{vmatrix} 8 & -3 \\ 4 & 2 \end{vmatrix}$ **5.** $\begin{vmatrix} -3 & -1 \\ 4 & 2 \end{vmatrix}$ **6.** $\begin{vmatrix} -4 & 2 \\ -5 & 3 \end{vmatrix}$

7. $\begin{vmatrix} 3 & 4 & 2 \\ 1 & -1 & 5 \\ 1 & 2 & -2 \end{vmatrix}$ **8.** $\begin{vmatrix} 1 & 3 & -2 \\ 6 & 1 & -5 \\ 8 & 2 & 3 \end{vmatrix}$ **9.** $\begin{vmatrix} 4 & -1 & 2 \\ 6 & -1 & 0 \\ 1 & -3 & 4 \end{vmatrix}$

10. $\begin{vmatrix} 3 & -9 & 4 \\ 1 & 4 & 0 \\ 8 & -3 & 1 \end{vmatrix}$

In Problems 11–38 solve each system of equations using Cramer's Rule, if it is applicable. If it is not, say so.

11. $\begin{cases} x + y = 8 \\ x - y = 4 \end{cases}$ **12.** $\begin{cases} x + 2y = 5 \\ x + y = 3 \end{cases}$

13. $\begin{cases} 5x - y = 13 \\ 2x + 3y = 12 \end{cases}$ **14.** $\begin{cases} x + 3y = 5 \\ 2x - 3y = -8 \end{cases}$

15. $\begin{cases} 3x = 24 \\ x + 2y = 0 \end{cases}$ **16.** $\begin{cases} 4x + 5y = -3 \\ -2y = -4 \end{cases}$

17. $\begin{cases} 3x - 6y = 24 \\ 5x + 4y = 12 \end{cases}$

18. $\begin{cases} 2x + 4y = 16 \\ 3x - 5y = -9 \end{cases}$

19. $\begin{cases} 3x - 2y = 4 \\ 6x - 4y = 0 \end{cases}$

20. $\begin{cases} -x + 2y = 5 \\ 4x - 8y = 6 \end{cases}$

21. $\begin{cases} 2x - 4y = -2 \\ 3x + 2y = 3 \end{cases}$

22. $\begin{cases} 3x + 3y = 3 \\ 4x + 2y = \frac{8}{3} \end{cases}$

23. $\begin{cases} 2x - 3y = -1 \\ 10x + 10y = 5 \end{cases}$

24. $\begin{cases} 3x - 2y = 0 \\ 5x + 10y = 4 \end{cases}$

25. $\begin{cases} 2x + 3y = 6 \\ x - y = \frac{1}{2} \end{cases}$

26. $\begin{cases} \frac{1}{2}x + y = -2 \\ x - 2y = 8 \end{cases}$

27. $\begin{cases} 3x - 5y = 3 \\ 15x + 5y = 21 \end{cases}$

28. $\begin{cases} 2x - y = -1 \\ x + \frac{1}{2}y = \frac{3}{2} \end{cases}$

29. $\begin{cases} x + y - z = 6 \\ 3x - 2y + z = -5 \\ x + 3y - 2z = 14 \end{cases}$

30. $\begin{cases} x - y + z = -4 \\ 2x - 3y + 4z = -15 \\ 5x + y - 2z = 12 \end{cases}$

31. $\begin{cases} x + 2y - z = -3 \\ 2x - 4y + z = -7 \\ -2x + 2y - 3z = 4 \end{cases}$

32. $\begin{cases} x + 4y - 3z = -8 \\ 3x - y + 3z = 12 \\ x + y + 6z = 1 \end{cases}$

33. $\begin{cases} x - 2y + 3z = 1 \\ 3x + y - 2z = 0 \\ 2x - 4y + 6z = 2 \end{cases}$

34. $\begin{cases} x - y + 2z = 5 \\ 3x + 2y = 4 \\ -2x + 2y - 4z = -10 \end{cases}$

35. $\begin{cases} x + 2y - z = 0 \\ 2x - 4y + z = 0 \\ -2x + 2y - 3z = 0 \end{cases}$

36. $\begin{cases} x + 4y - 3z = 0 \\ 3x - y + 3z = 0 \\ x + y + 6z = 0 \end{cases}$

37. $\begin{cases} x - 2y + 3z = 0 \\ 3x + y - 2z = 0 \\ 2x - 4y + 6z = 0 \end{cases}$

38. $\begin{cases} x - y + 2z = 0 \\ 3x + 2y = 0 \\ -2x + 2y - 4z = 0 \end{cases}$

39. Solve

$$\begin{cases} \dfrac{1}{x} + \dfrac{1}{y} = 8 \\ \dfrac{3}{x} - \dfrac{5}{y} = 0 \end{cases}$$

[*Hint:* Let $u = 1/x$ and $v = 1/y$ and solve for u and v.]

40. Solve

$$\begin{cases} \dfrac{4}{x} - \dfrac{3}{y} = 0 \\ \dfrac{6}{x} + \dfrac{3}{2y} = 2 \end{cases}$$

In Problems 41–46 solve for x.

41. $\begin{vmatrix} x & x \\ 4 & 3 \end{vmatrix} = 5$

42. $\begin{vmatrix} x & 1 \\ 3 & x \end{vmatrix} = -2$

43. $\begin{vmatrix} x & 1 & 1 \\ 4 & 3 & 2 \\ -1 & 2 & 5 \end{vmatrix} = 2$

44. $\begin{vmatrix} 3 & 2 & 4 \\ 1 & x & 5 \\ 0 & 1 & -2 \end{vmatrix} = 0$

45. $\begin{vmatrix} x & 2 & 3 \\ 1 & x & 0 \\ 6 & 1 & -2 \end{vmatrix} = 7$

46. $\begin{vmatrix} x & 1 & 2 \\ 1 & x & 3 \\ 0 & 1 & 2 \end{vmatrix} = -4x$

In Problems 47–54 use properties of determinants to find the value of each determinant if it is known that

$$\begin{vmatrix} x & y & z \\ u & v & w \\ 1 & 2 & 3 \end{vmatrix} = 4$$

47. $\begin{vmatrix} 1 & 2 & 3 \\ u & v & w \\ 1 & 2 & 3 \end{vmatrix}$

48. $\begin{vmatrix} x & y & z \\ u & v & w \\ 2 & 4 & 6 \end{vmatrix}$

49. $\begin{vmatrix} x & y & z \\ -3 & -6 & -9 \\ u & v & w \end{vmatrix}$

50. $\begin{vmatrix} 1 & 2 & 3 \\ x-u & y-v & z-w \\ u & v & w \end{vmatrix}$

51. $\begin{vmatrix} 1 & 2 & 3 \\ x-3 & y-6 & z-9 \\ 2u & 2v & 2w \end{vmatrix}$

52. $\begin{vmatrix} x & y & z-x \\ u & v & w-u \\ 1 & 2 & 2 \end{vmatrix}$

53. $\begin{vmatrix} 1 & 2 & 3 \\ 2x & 2y & 2z \\ u-1 & v-2 & w-3 \end{vmatrix}$

54. $\begin{vmatrix} x+3 & y+6 & z+9 \\ 3u-1 & 3v-2 & 3w-3 \\ 1 & 2 & 3 \end{vmatrix}$

55. Show that
$$\begin{vmatrix} x^2 & x & 1 \\ y^2 & y & 1 \\ z^2 & z & 1 \end{vmatrix} = (y-z)(x-y)(x-z)$$

56. Complete the proof of Cramer's Rule for two equations containing two variables. [*Hint:* In Display (5), page 459, if $a = 0$, then $b \neq 0$ and $c \neq 0$, so $D = -bc \neq 0$. Now show that Display (6) provides a solution of the system when $a = 0$. There are then three remaining cases: $b = 0$, $c = 0$, and $d = 0$.]

57. Interchange columns one and three of a 3 by 3 determinant. Show that the value of the new determinant is -1 times the value of the original determinant.

58. Multiply each entry in row two of a 3 by 3 determinant by the number k, $k \neq 0$. Show that the value of the new determinant is k times the value of the original determinant.

59. Prove that a 3 by 3 determinant in which the entries in column one equal those in column three has the value 0.

60. Prove that if the second row of a 3 by 3 determinant is multiplied by k, $k \neq 0$, and the result is added to the entries in row one, then there is no change in the value of the determinant.

▤ 7.5 SOLUTIONS OF SYSTEMS OF NONLINEAR EQUATIONS

There is no general methodology for solving a system of nonlinear equations. There are times when substitution is best; other times, elimination is best; and there are times when neither of these methods works. Experience and a certain degree of imagination are your allies here.

Before we begin, two comments are in order:

First, if the system contains two variables and if the equations in the system are easy to graph, then graph them. By graphing each equation in the system, we can get an idea of how many solutions a system has and where their approximate locations are.

Second, extraneous solutions can creep in when solving nonlinear systems, so it is imperative that all apparent solutions be checked.

EXAMPLE 1 Solve the system of equations

$$\begin{cases} 3x - y = -2 & \text{(1)} \quad \text{A line} \\ 2x^2 - y = 0 & \text{(2)} \quad \text{A parabola} \end{cases}$$

Solution First, we notice that the system contains two variables and that we know how to graph each equation. In Figure 5 we see that the system apparently has two solutions.

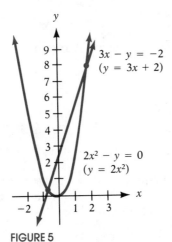

FIGURE 5

We shall use substitution to solve the system. The first equation can be easily solved for y:

$$3x - y = -2$$
$$y = 3x + 2$$

Substituting this value for y in the second equation gives

$$2x^2 - y = 0$$
$$2x^2 - (3x + 2) = 0$$
$$2x^2 - 3x - 2 = 0$$
$$(2x + 1)(x - 2) = 0$$
$$2x + 1 = 0 \quad \text{or} \quad x - 2 = 0$$
$$x = -\frac{1}{2} \quad \text{or} \quad x = 2$$

Using these values for x in $y = 3x + 2$, we find

$$y = 3\left(-\frac{1}{2}\right) + 2 = \frac{1}{2} \quad \text{or} \quad y = 3(2) + 2 = 8$$

The apparent solutions are $x = -\frac{1}{2}$, $y = \frac{1}{2}$ and $x = 2$, $y = 8$.

Check: For $x = -\frac{1}{2}$, $y = \frac{1}{2}$:

$$3\left(-\frac{1}{2}\right) - \frac{1}{2} = -\frac{3}{2} - \frac{1}{2} = -2 \qquad (1)$$
$$2\left(-\frac{1}{2}\right)^2 - \frac{1}{2} = 2\left(\frac{1}{4}\right) - \frac{1}{2} = 0 \qquad (2)$$

For $x = 2$, $y = 8$:

$$\begin{cases} 3(2) - 8 = 6 - 8 = -2 & (1) \\ 2(2)^2 - 8 = 2(4) - 8 = 0 & (2) \end{cases}$$

Each solution checks.   ≣

Our next solution illustrates how the method of elimination works for nonlinear systems.

EXAMPLE 2   Solve

$$\begin{cases} x^2 + y^2 = 13 & (1) \quad \text{A circle} \\ x^2 - y = 7 & (2) \quad \text{A parabola} \end{cases}$$

Solution   First, we graph each equation. See Figure 6. Based on the graph, we expect four solutions. By subtracting the second equation from the first, the variable x is eliminated, leaving

$$y^2 + y = 6$$

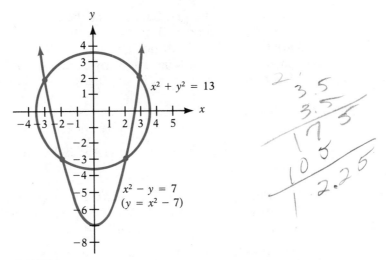

FIGURE 6

This quadratic equation in y is easily solved by factoring:

$$y^2 + y - 6 = 0$$
$$(y + 3)(y - 2) = 0$$
$$y = -3 \quad \text{or} \quad y = 2$$

We use these values for y in the second equation to find x. If $y = 2$, then $x^2 = y + 7 = 9$ and $x = 3$ or -3. If $y = -3$, then $x^2 = y + 7 = 4$ and $x = 2$ or -2. Thus, we have four solutions: $x = 3$, $y = 2$; $x = -3$, $y = 2$; $x = 2$, $y = -3$; and $x = -2$, $y = -3$. You should verify that, in fact, these four solutions also satisfy the first equation, so that all four are solutions of the system. These four solutions, $(3, 2)$, $(-3, 2)$, $(2, -3)$, and $(-2, -3)$, are the four points of intersection of the graphs. Look again at Figure 6. ▄

EXAMPLE 3 Solve

$$\begin{cases} x^2 - y^2 = 1 & (1) \\ x^3 - y^2 = x & (2) \end{cases}$$

Solution Because these equations are not so easy to graph, we omit their graphs. We use elimination, subtracting the second equation from the first, to get

$$x^2 - x^3 = 1 - x$$
$$x^2(1 - x) = 1 - x$$
$$x^2(1 - x) - (1 - x) = 0 \qquad \text{Factor } x^2 - 1 \quad (1 - x)$$
$$(x^2 - 1)(1 - x) = 0$$
$$x^2 - 1 = 0 \quad \text{or} \quad 1 - x = 0$$
$$x = \pm 1 \quad \text{or} \quad x = 1$$

We now use the first equation of the system to get y. If $x = 1$, then $1 - y^2 = 1$ and $y = 0$. If $x = -1$, then $1 - y^2 = 1$ and $y = 0$. There are two apparent solutions: $x = 1, y = 0$ and $x = -1, y = 0$. Because each of these solutions also satisfies the second equation, the system has two solutions: $x = 1, y = 0$ and $x = -1, y = 0$. ∎

EXAMPLE 4 Solve

$$\begin{cases} x^2 + x + y^2 - 3y + 2 = 0 & (1) \\ x + 1 + \dfrac{y^2 - y}{x} = 0 & (2) \end{cases}$$

Solution We multiply the second equation by x to eliminate the fraction. The result will be an equivalent system because x cannot be 0 (look at the second equation to see why).

$$\begin{cases} x^2 + x + y^2 - 3y + 2 = 0 & (1) \\ x^2 + x + y^2 - y = 0 & (2) \end{cases}$$

Now subtract the second equation from the first to eliminate x. The result is

$$-2y + 2 = 0$$
$$y = 1$$

To find x, we use the first equation. Then,

$$x^2 + x + 1 - 3 + 2 = 0$$
$$x^2 + x = 0$$
$$x(x + 1) = 0$$
$$x = 0 \quad \text{or} \quad x = -1$$

Because x cannot be 0, the solution $x = 0$ is extraneous, and we discard it.

We now check $x = -1, y = 1$:

$$\begin{cases} (-1)^2 + (-1) + 1^2 - 3(1) + 2 = 1 - 1 + 1 - 3 + 2 = 0 & (1) \\ -1 + 1 + \dfrac{1^2 - 1}{-1} = 0 + \dfrac{0}{-1} = 0 & (2) \end{cases}$$

Thus, the only solution to the system is $x = -1, y = 1$. ∎

The next examples illustrate two of the more imaginative ways to solve systems of nonlinear equations.

EXAMPLE 5 Solve

$$\begin{cases} 4x^2 - 9xy - 28y^2 = 0 & (1) \\ 16x^2 - 4xy = 16 & (2) \end{cases}$$

Solution   We take note of the fact that the first equation can be factored:

$$4x^2 - 9xy - 28y^2 = 0$$
$$(4x + 7y)(x - 4y) = 0$$

This results in the two equations

$$4x + 7y = 0 \qquad \text{or} \quad x - 4y = 0$$
$$x = -\frac{7}{4}y \quad \text{or} \qquad x = 4y$$

We substitute each of these values for x in the second equation:

$$16x^2 - 4xy = 16 \qquad\qquad 16x^2 - 4xy = 16$$
$$16\left(-\frac{7}{4}y\right)^2 - 4\left(-\frac{7}{4}y\right)y = 16 \qquad 16(4y)^2 - 4(4y)y = 16$$
$$49y^2 + 7y^2 = 16 \qquad\qquad 16 \cdot 16y^2 - 16y^2 = 16$$
$$56y^2 = 16 \qquad\qquad\qquad 15y^2 = 1$$
$$7y^2 = 2 \qquad\qquad\qquad\qquad y^2 = \frac{1}{15}$$
$$y^2 = \frac{2}{7} \qquad\qquad\qquad\qquad y = \pm\frac{\sqrt{15}}{15}$$
$$y = \pm\sqrt{\frac{2}{7}} = \pm\frac{\sqrt{14}}{7} \qquad x = 4y = \pm\frac{4\sqrt{15}}{15}$$
$$x = -\frac{7}{4}y = \mp\frac{\sqrt{14}}{4}$$

You should verify for yourself that, in fact, the four solutions $x = -\sqrt{14}/4$, $y = \sqrt{14}/7$; $x = \sqrt{14}/4$, $y = -\sqrt{14}/7$; $x = 4\sqrt{15}/15$, $y = \sqrt{15}/15$; and $x = -4\sqrt{15}/15$, $y = -\sqrt{15}/15$ are actually solutions of the system.   ≣

EXAMPLE 6   Solve

$$\begin{cases} 3xy - 2y^2 = -2 & (1) \\ 9x^2 + 4y^2 = 10 & (2) \end{cases}$$

Solution   We multiply the first equation by 2 and add the result to the second equation to eliminate the y^2 terms:

$$\begin{cases} 6xy - 4y^2 = -4 & (1) \\ 9x^2 + 4y^2 = 10 & (2) \end{cases}$$

$$9x^2 + 6xy = 6$$
$$3x^2 + 2xy = 2 \quad \text{Divide each side by 3.}$$

Solve for y in this equation to get

$$y = \frac{2 - 3x^2}{2x} \qquad x \neq 0 \tag{1}$$

Now substitute for y in the second equation of the system:

$$9x^2 + 4y^2 = 10$$

$$9x^2 + 4\left(\frac{2 - 3x^2}{2x}\right)^2 = 10$$

$$9x^2 + \frac{(4 - 12x^2 + 9x^4)}{x^2} = 10$$

$$9x^4 + 4 - 12x^2 + 9x^4 = 10x^2$$

$$18x^4 - 22x^2 + 4 = 0$$

$$9x^4 - 11x^2 + 2 = 0$$

This quadratic equation (in x^2) can be factored:

$$(9x^2 - 2)(x^2 - 1) = 0$$

$$9x^2 - 2 = 0 \qquad \text{or} \quad x^2 - 1 = 0$$

$$x^2 = \frac{2}{9} \qquad \text{or} \qquad x^2 = 1$$

$$x = \pm\frac{\sqrt{2}}{3} \quad \text{or} \qquad x = \pm 1$$

To find y, we use Equation (1):

$$\text{If } x = \frac{\sqrt{2}}{3}, \, y = \frac{2 - 3x^2}{2x} = \frac{2 - \frac{2}{3}}{2(\sqrt{2}/3)} = \frac{4}{2\sqrt{2}} = \sqrt{2}$$

$$\text{If } x = -\frac{\sqrt{2}}{3}, \, y = \frac{2 - 3x^2}{2x} = \frac{2 - \frac{2}{3}}{-2(\sqrt{2}/3)} = \frac{4}{-2\sqrt{2}} = -\sqrt{2}$$

$$\text{If } x = 1, \, y = \frac{2 - 3x^2}{2x} = \frac{2 - 3}{2} = -\frac{1}{2}$$

$$\text{If } x = -1, \, y = \frac{2 - 3x^2}{2x} = \frac{2 - 3}{-2} = \frac{1}{2}$$

The system has four solutions. Check them for yourself.   ≡

≡ HISTORICAL COMMENT   Recall that, in the beginning of this section, we said imagination and experience are important in solving simultaneous nonlinear equations. Indeed, these kinds of problems lead into some of the deepest and most difficult parts of modern mathematics. Look again at the graphs from Examples 1 and 2 of this section for a moment (Figures 5 and 6). We see that Example 1 has two solutions, and Example 2 has four solutions. One might conjecture that the number of solutions

344-3311

is equal to the product of the degrees of the equations involved. This conjecture was indeed made by Etienne Bezout (1739–1783), but working out the details took about 150 years. It turns out that to come up with the correct number of intersections, we must count not only the complex-number intersections, but also those intersections that, in a certain sense, lie at infinity. For example, a parabola and a line lying on the axis of the parabola intersect at the vertex and at infinity. These matters are part of the study of algebraic geometry.

≡ HISTORICAL PROBLEM

1. A papyrus dating back to 1950 B.C. contains the following problem: A given surface area of 100 units of area shall be represented as the sum of two squares whose sides are to each other as $1 : \frac{3}{4}$. Solve for the sides by solving the system of equations

$$\begin{cases} x^2 + y^2 = 100 \\ x = \frac{3}{4}y \end{cases}$$

≡

EXERCISE 7.5   *In Problems 1–8 graph each equation of the system and then solve the system.*

1. $\begin{cases} x + 2y + 3 = 0 \\ x^2 + y^2 = 5 \end{cases}$

2. $\begin{cases} 3x - y - 3 = 0 \\ x + 2y^2 - 2 = 0 \end{cases}$

3. $\begin{cases} x^2 + y^2 = 4 \\ y^2 - x = 4 \end{cases}$

4. $\begin{cases} x^2 + y^2 = 16 \\ x^2 - 2y = 8 \end{cases}$

5. $\begin{cases} x^2 + y^2 = 36 \\ x + y = 8 \end{cases}$

6. $\begin{cases} x^2 + y^2 = 4 \\ 2x - y + 4 = 0 \end{cases}$

7. $\begin{cases} xy = 4 \\ x^2 + y^2 = 8 \end{cases}$

8. $\begin{cases} x^2 = y \\ xy = 1 \end{cases}$

In Problems 9–36 solve each system. Use any method you wish.

9. $\begin{cases} 2x^2 + y^2 = 18 \\ xy = 4 \end{cases}$

10. $\begin{cases} x^2 - y^2 = 14 \\ x + y = 7 \end{cases}$

11. $\begin{cases} 3x - y = 1 \\ x^2 + 4y^2 = 17 \end{cases}$

12. $\begin{cases} x^2 - 4y^2 = 16 \\ 2y - x = 2 \end{cases}$

13. $\begin{cases} x + y + 1 = 0 \\ x^2 + y^2 + 6y - x = 7 \end{cases}$

14. $\begin{cases} 2x^2 - xy + y^2 = 8 \\ xy = 4 \end{cases}$

15. $\begin{cases} 4x^2 - 3xy + 9y^2 = 15 \\ 2x + 3y = 5 \end{cases}$

16. $\begin{cases} 2y^2 - 3xy + 6y + 2x + 4 = 0 \\ 2x - 3y + 4 = 0 \end{cases}$

17. $\begin{cases} x^2 - 4y^2 + 7 = 0 \\ 3x^2 + y^2 = 31 \end{cases}$

18. $\begin{cases} 3x^2 - 2y^2 + 5 = 0 \\ 2x^2 - y^2 + 2 = 0 \end{cases}$

19. $\begin{cases} 7x^2 - 3y^2 + 5 = 0 \\ 3x^2 + 5y^2 = 12 \end{cases}$

20. $\begin{cases} x^2 - 3y^2 + 1 = 0 \\ 2x^2 - 7y^2 + 5 = 0 \end{cases}$

21. $\begin{cases} x^2 + 2xy = 10 \\ 3x^2 - xy = 2 \end{cases}$

22. $\begin{cases} 5xy + 13y^2 + 36 = 0 \\ xy + 7y^2 = 6 \end{cases}$

23. $\begin{cases} x^2 + 2y^2 = 16 \\ 4x^2 - y^2 = 24 \end{cases}$ **24.** $\begin{cases} 4x^2 + 3y^2 = 4 \\ 2x^2 - 6y^2 = -3 \end{cases}$

25. $\begin{cases} \dfrac{5}{x^2} - \dfrac{2}{y^2} + 3 = 0 \\[2mm] \dfrac{3}{x^2} + \dfrac{1}{y^2} = 7 \end{cases}$ **26.** $\begin{cases} \dfrac{2}{x^2} - \dfrac{3}{y^2} + 1 = 0 \\[2mm] \dfrac{6}{x^2} - \dfrac{7}{y^2} + 2 = 0 \end{cases}$

27. $\begin{cases} \dfrac{1}{x^4} + \dfrac{6}{y^4} = 6 \\[2mm] \dfrac{2}{x^4} - \dfrac{2}{y^4} = 19 \end{cases}$ **28.** $\begin{cases} \dfrac{1}{x^4} - \dfrac{1}{y^4} = 1 \\[2mm] \dfrac{1}{x^4} + \dfrac{1}{y^4} = 4 \end{cases}$

29. $\begin{cases} x^2 - 3xy + 2y^2 = 0 \\ x^2 + xy = 6 \end{cases}$ **30.** $\begin{cases} x^2 - xy - 2y^2 = 0 \\ xy + x + 6 = 0 \end{cases}$

31. $\begin{cases} xy - x^2 + 3 = 0 \\ 3xy - 4y^2 = 2 \end{cases}$ **32.** $\begin{cases} 5x^2 + 4xy + 3y^2 = 36 \\ x^2 + xy + y^2 = 9 \end{cases}$

33. $\begin{cases} x^3 - y^3 = 26 \\ x - y = 2 \end{cases}$ **34.** $\begin{cases} x^3 + y^3 = 26 \\ x + y = 2 \end{cases}$

35. $\begin{cases} y^2 + y + x^2 - x - 2 = 0 \\ y + 1 + \dfrac{x-2}{y} = 0 \end{cases}$ **36.** $\begin{cases} x^3 - 2x^2 + y^2 + 3y - 4 = 0 \\ x - 2 + \dfrac{y^2 - y}{x^2} = 0 \end{cases}$

37. The sum of two numbers is 8 and the sum of their squares is 36. Find the numbers.

38. The sum of two numbers is 24 and the difference of their squares is 48. Find the numbers.

39. The product of two numbers is 7 and the sum of their squares is 50. Find the numbers.

40. The product of two numbers is 12 and the difference of their squares is 7. Find the numbers.

41. The difference of two numbers is the same as their product, and the sum of their reciprocals is 5. Find the numbers.

42. The sum of two numbers is the same as their product, and the difference of their reciprocals is 3. Find the numbers.

43. A rectangular piece of cardboard, whose area is 216 square centimeters, is made into a box by cutting a 2-centimeter square from each corner and turning up the sides. If the box is to have a volume of 224 cubic centimeters, what size cardboard should you start with?

44. Two circles have perimeters that add up to 12π centimeters and areas that add up to 20π square centimeters. Find the radius of each circle.

45. The altitude of an isosceles triangle drawn to its base is 3 centimeters and its perimeter is 18 centimeters. Find the length of its base.

46. In a 21-meter race between a tortoise and a hare, the tortoise leaves 9 minutes before the hare. The hare, by running at an average speed of 0.5 meter per hour faster than the tortoise, crosses the finish line 3 minutes before the tortoise. What are the average speeds of the tortoise and the hare?

47. A tangent line to a graph is a line that touches the graph in exactly one point. We can find the equation of the tangent line to the circle $x^2 + y^2 = 25$ at the point $(3, 4)$ by insisting that the system

$$\begin{cases} x^2 + y^2 = 25 \\ \quad\quad y = mx + b \end{cases}$$

has only one solution. Use the substitution method to get a quadratic equation in x. This quadratic equation has one solution if its discriminant equals 0. The result will be an equation involving m and b. Solve this equation simultaneously with the equation $4 = m(3) + b$ to find m and b. [The latter equation comes about because the tangent line must pass through the point $(3, 4)$.] The tangent line is then $y = mx + b$. State this equation.

In Problems 48–54 use the idea behind Problem 47 to find the equation of the line tangent to each graph at the given point.

48. $x^2 + y^2 = 10$ at $(1, 3)$ **49.** $y = x^2 + 2$ at $(1, 3)$

50. $x^2 + y = 5$ at $(-2, 1)$ **51.** $2x^2 + 3y^2 = 14$ at $(1, 2)$

52. $3x^2 + y^2 = 7$ at $(-1, 2)$ **53.** $x^2 - y^2 = 3$ at $(2, 1)$

54. $2y^2 - x^2 = 14$ at $(2, 3)$

55. Find formulas for the length l and width w of a rectangle in terms of its area A and perimeter p.

56. Find formulas for the base b and one of the equal sides l of an isosceles triangle in terms of its altitude h and perimeter p.

In Problems 57–62 graph each equation and find the point(s) of intersection, if any.

57. The line $x + 2y = 0$ and the circle $(x - 1)^2 + (y - 1)^2 = 5$

58. The line $x + 2y + 6 = 0$ and the circle $(x + 1)^2 + (y + 1)^2 = 5$

59. The circle $(x - 1)^2 + (y + 2)^2 = 4$ and the parabola $y^2 + 4y - x + 1 = 0$

60. The circle $(x + 2)^2 + (y - 1)^2 = 4$ and the parabola $y^2 - 2y - x - 5 = 0$

61. The graph of $y = 4/(x - 3)$ and the circle $x^2 - 6x + y^2 + 1 = 0$

62. The graph of $y = 4/(x + 2)$ and the circle $x^2 + 4x + y^2 - 4 = 0$

63. Earlier we concluded that, if r_1 and r_2 are two solutions of a quadratic equation $ax^2 + bx + c = 0$, then

$$r_1 + r_2 = -\frac{b}{a} \quad\quad r_1 r_2 = \frac{c}{a}$$

Solve this system of equations for r_1 and r_2.

▰ 7.6 SOLUTIONS OF SYSTEMS OF LINEAR INEQUALITIES

In Chapter 3, we discussed linear equations in two variables—namely, equations of the form

$$Ax + By + C = 0$$

where A and B are not both 0, and A, B, and C are real numbers. The graph of a linear equation is, of course, a line. In this section we study the graphs of linear inequalities in two variables.

LINEAR INEQUALITY   A **linear inequality** in two variables is an expression of one of the following forms:

$$\begin{array}{ll} Ax + By + C < 0 & Ax + By + C > 0 \\ Ax + By + C \leq 0 & Ax + By + C \geq 0 \end{array} \tag{1}$$

where A and B are not both 0, and A, B, and C are real numbers. The
STRICT INEQUALITY   top two inequalities in Display (1) are referred to as **strict** inequal-
NONSTRICT INEQUALITY ities; the remaining two are **nonstrict**. For example, each of the expressions

$$x + 3y - 5 < 0 \qquad 3x - y + 10 \geq 0$$

is a linear inequality. The one on the left is strict; the one on the right is nonstrict.

GRAPH OF A LINEAR   The **graph of a linear inequality** in two variables x and y is the set
INEQUALITY   of all points (x, y) whose coordinates satisfy the inequality.
Let's look at an example.

EXAMPLE 1   Graph the linear inequality

$$3x + y - 6 \leq 0$$

Solution   We begin with the associated problem of the graph of the linear equality

$$3x + y - 6 = 0$$

formed by replacing (for now) the $\leq$ symbol with an $=$ sign. The graph of the linear equation is a line. See Figure 7(a). This line is part of the graph of the inequality we seek because the inequality is non-strict. (Do you see why? We are seeking points for which $3x + y - 6$ is less than *or equal to* 0.)

Now let's test a few randomly selected points such as $(4, -1)$, $(5, 5)$, $(-1, 2)$, and $(-2, -2)$, to see whether they belong to the graph of the inequality:

	$3x + y - 6$	CONCLUSION
$(4, -1)$	$3(4) + (-1) - 6 = 5 > 0$	Does not belong to graph
$(5, 5)$	$3(5) + 5 - 6 = 14 > 0$	Does not belong to graph
$(-1, 2)$	$3(-1) + 2 - 6 = -7 < 0$	Belongs to graph
$(-2, -2)$	$3(-2) + (-2) - 6 = -14 < 0$	Belongs to graph

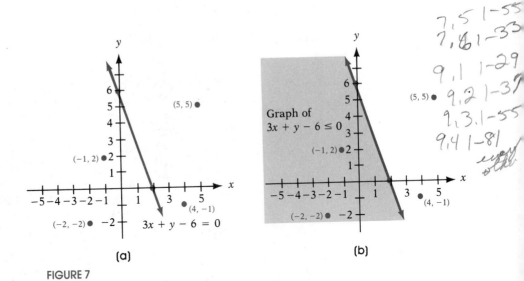

FIGURE 7

Look again at Figure 7(a). Notice that the two points that belong to the graph both lie on the same side of the line, and the two points that do not belong to the graph lie on the opposite side. As it turns out, this is always the case. Thus, the graph we seek consists of all points that lie on the same side of the line as do $(-1, 2)$ and $(-2, -2)$. The graph we seek is the shaded region in Figure 7(b). ≡

HALF-PLANE
BOUNDARY

The collection of points that comprise the graph of a linear inequality is called a **half-plane**, and the line L is called its **boundary**. See Figure 8. As shown there, if the equation of the boundary line is $Ax + By + C = 0$, then it divides the plane into two half-planes: one for which $Ax + By + C > 0$, the other for which $Ax + By + C < 0$.

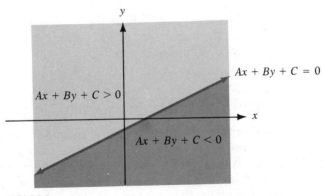

FIGURE 8

Let's outline the steps to follow to graph a linear inequality.

Step 1: Graph the line L that results from replacing the inequality symbol in the linear inequality with an $=$ sign. This line is the boundary of the graph.
(a) If the linear inequality is strict, the line L will not belong to the graph of the linear inequality. (In this case, we will show the line as a dashed line.)
(b) If the linear inequality is nonstrict, the line L will belong to the graph of the linear inequality. (In this case, we will show the line as a solid line.)

Step 2: Select a test point P that is not on the line L graphed in Step 1.

Step 3: (a) If the coordinates of this point P satisfy the linear inequality, then so do all points on the same side of the line L as the point P.
(b) If the coordinates of this point P do not satisfy the linear inequality, then all points on the opposite side of the line L from the point P satisfy the linear inequality.

Step 4: Shade the half-plane that belongs to the graph.

EXAMPLE 2   Graph

$$3x + y + 3 > 0$$

Solution   First we graph the boundary $3x + y + 3 = 0$; see Figure 9(a). Note that we have used a dashed line because the points on the line do not satisfy the inequality. The point $(0, 0)$ is not on the graph of the boundary, so we can use it as a test point:

$$(0, 0): \quad 3(0) + 0 + 3 = 3 > 0$$

The origin obeys the linear inequality. Thus, all points in the half-plane on the same side of the boundary as $(0, 0)$ are solutions. See the shaded region of Figure 9(b).   ≡

Systems of Linear Inequalities in Two Variables

GRAPH OF A SYSTEM OF
LINEAR INEQUALITIES
The **graph of a system of linear inequalities** in two variables x and y is the set of all points (x, y) that satisfy *each* of the linear inequalities in the system. Thus, the graph of a system of linear inequalities can be obtained by graphing each inequality individually and then determining where they all intersect.

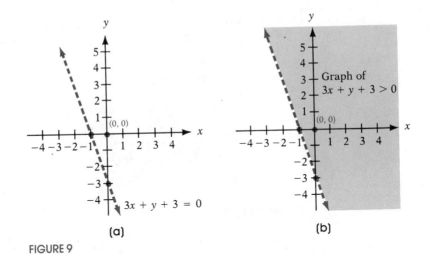

FIGURE 9

EXAMPLE 3 Graph the system

$$\begin{cases} x + y \geq 2 \\ 2x - y \leq 4 \end{cases}$$

Solution First, we graph the inequality $x + y \geq 2$. See the shaded region in Figure 10(a). Next, we graph the inequality $2x - y \leq 4$. See the shaded region in Figure 10(b). Now superimpose the two graphs. See Figure 10(c). Those points that are in both shaded regions—the overlapping darker region in Figure 10(c)—are the solutions we seek to the system because they satisfy each linear inequality.

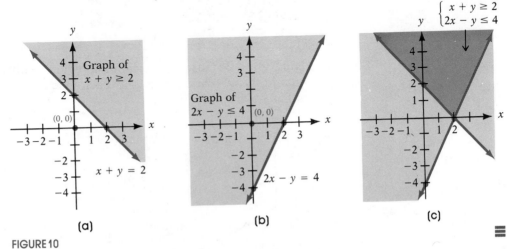

FIGURE 10

EXAMPLE 4 Graph each system:

(a) $\begin{cases} x + y \leq 2 \\ x + y \geq 0 \end{cases}$ (b) $\begin{cases} 2x - y \geq 2 \\ 2x - y \geq 0 \end{cases}$ (c) $\begin{cases} x + 2y \leq 2 \\ x + 2y \geq 6 \end{cases}$

Solution   (a) See Figure 11(a). The overlapping darker shaded region between the two boundaries is the graph of the system.

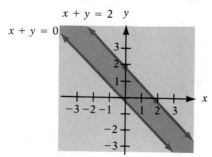

FIGURE 11(a)

(b) See Figure 11(b). The overlapping darker shaded region is the graph of the system. Note that the graph of the system is identical to the graph of the single inequality $2x - y \geq 2$.

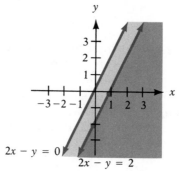

FIGURE 11(b)

(c) See Figure 11(c). Because no overlapping region results, there are no points in the xy-plane that satisfy each inequality. Hence, the system has no graph.

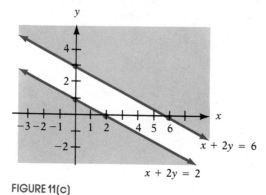

FIGURE 11(c)

EXAMPLE 5   Graph each system:

(a) $\begin{cases} x + y \geq 3 \\ 2x + y \geq 4 \\ x \geq 0 \\ y \geq 0 \end{cases}$ (b) $\begin{cases} x + y \leq 6 \\ 2x + y \leq 8 \\ x \geq 0 \\ y \geq 0 \end{cases}$

Solution   (a) The two inequalities $x \geq 0$ and $y \geq 0$ require that the graph be in quadrant I. Thus, we concentrate on the other two inequalities. See the overlapping darker shaded region in Figure 12(a).

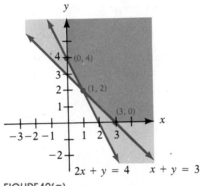

FIGURE 12(a)

(b) See the overlapping darker shaded region in Figure 12(b). Note that the inequalities $x \geq 0$ and $y \geq 0$ again require that the graph of the system be in quadrant I.

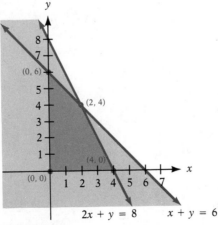

FIGURE 12(b)

UNBOUNDED GRAPH   The graph of the system of linear inequalities in Figure 12(a) is said to be **unbounded** because it extends indefinitely in a particular direction. On the other hand, the graph of the system of linear inequalities

BOUNDED GRAPH   in Figure 12(b) is said to be **bounded** because it can be contained within some circle of sufficiently large radius.

   Notice in Figures 12(a) and 12(b) that those points belonging to the graph that are also points of intersection of boundary lines have been

VERTICES   plotted. Such points are referred to as **vertices** of the graph. Thus, the system graphed in Figure 12(a) has three vertices: $(0, 4)$, $(1, 2)$, and $(3, 0)$. The system graphed in Figure 12(b) has four vertices: $(0, 0)$, $(0, 6)$, $(2, 4)$, and $(4, 0)$.

   To find the vertex $(2, 4)$ in Figure 12(b) for the system in Example 5(b), we found the point of intersection of the two lines

$$x + y = 6 \quad \text{and} \quad 2x + y = 8$$

   These ideas will be used later in developing the method of solution of linear programming problems—one of the important uses made of linear inequalities.

EXERCISE 7.6   *In Problems 1–12 graph each linear inequality.*

1. $x \geq 0$ 2. $y \geq 0$

3. $x + y \geq 4$ 4. $x + y \leq 2$

5. $2x + y \geq 6$ 6. $3x + 2y \leq 6$

7. $4x - y \geq 4$ 8. $y - 2x \leq 2$

9. $3x - 2y \geq -6$ 10. $4x - y \leq 8$

11. $3x + 4y \geq 0$ 12. $2x - 3y \leq 0$

In Problems 13–24 graph each system of inequalities.

13. $\begin{cases} x + y \leq 2 \\ 2x + y \geq 4 \end{cases}$ 14. $\begin{cases} 3x - y \geq 6 \\ x + 2y \leq 2 \end{cases}$

15. $\begin{cases} 2x - y \leq 4 \\ 3x + 2y \geq -6 \end{cases}$ 16. $\begin{cases} 4x - 5y \leq 0 \\ 2x - y \geq 2 \end{cases}$

17. $\begin{cases} 2x - 3y \leq 0 \\ 3x + 2y \leq 6 \end{cases}$ 18. $\begin{cases} 4x - y \geq 2 \\ x + 2y \geq 2 \end{cases}$

19. $\begin{cases} x - 2y \leq 6 \\ 2x - 4y \geq 0 \end{cases}$ 20. $\begin{cases} x + 4y \leq 8 \\ x + 4y \geq 4 \end{cases}$

21. $\begin{cases} 2x + y \geq -2 \\ 2x + y \geq 2 \end{cases}$ 22. $\begin{cases} x - 4y \leq 4 \\ x - 4y \geq 0 \end{cases}$

23. $\begin{cases} 2x + 3y \geq 6 \\ 2x + 3y \leq 0 \end{cases}$ 24. $\begin{cases} 2x + y \geq 0 \\ 2x + y \geq 2 \end{cases}$

In Problems 25–34 graph each system of inequalities. Tell whether the graph is bounded or unbounded and label the vertices.

25. $\begin{cases} x \geq 0 \\ y \geq 0 \\ 2x + y \leq 6 \\ x + 2y \leq 6 \end{cases}$

[handwritten: $y = -2x+6$ $2y = -x+6$ $y = \frac{1}{2}x+3$]

26. $\begin{cases} x \geq 0 \\ y \geq 0 \\ x + y \geq 4 \\ 2x + 3y \geq 6 \end{cases}$

27. $\begin{cases} x \geq 0 \\ y \geq 0 \\ x + y \geq 2 \\ 2x + y \geq 4 \end{cases}$

28. $\begin{cases} x \geq 0 \\ y \geq 0 \\ 3x + y \leq 6 \\ 2x + y \leq 2 \end{cases}$

29. $\begin{cases} x \geq 0 \\ y \geq 0 \\ x + y \geq 2 \\ 2x + 3y \leq 12 \\ 3x + y \leq 12 \end{cases}$

30. $\begin{cases} x \geq 0 \\ y \geq 0 \\ x + y \geq 2 \\ x + y \leq 10 \\ 2x + y \leq 3 \end{cases}$

31. $\begin{cases} x \geq 0 \\ y \geq 0 \\ x + y \geq 2 \\ x + y \leq 8 \\ 2x + y \leq 10 \end{cases}$

32. $\begin{cases} x \geq 0 \\ y \geq 0 \\ x + y \geq 2 \\ x + y \leq 8 \\ x + 2y \geq 1 \end{cases}$

33. $\begin{cases} x \geq 0 \\ y \geq 0 \\ x + 2y \geq 1 \\ x + 2y \leq 10 \end{cases}$

[handwritten: $2y = -\frac{x}{2} + 20$ $2y = -\frac{x}{2} + 5$]

34. $\begin{cases} x \geq 0 \\ y \geq 0 \\ x + 2y \geq 1 \\ x + 2y \leq 10 \\ x + y \geq 2 \\ x + y \leq 8 \end{cases}$

C H A P T E R   R E V I E W

VOCABULARY

system of equations
solution
solve
consistent system
inconsistent system
linear equation
system of linear
   equations
substitution method
coincident lines
equivalent systems of
   equations
method of elimination
back-substitute

matrix
row index
column index
entries
augmented matrices
coefficient matrices
row operations
matrix method
echelon form
reduced echelon form
Cramer's Rule
2 by 2 determinant
value

3 by 3 determinant
linear inequality
strict inequality
nonstrict inequality
graph of a linear
   inequality
half-plane
boundary
graph of a system of
   linear inequalities
unbounded graph
bounded graph
vertices

FILL-IN-THE-BLANK QUESTIONS

1. If a system of equations has no solution, it is said to be _____.
2. An n by n rectangular array of numbers is called a(n) _____.
3. Cramer's Rule uses _____ to solve a system of linear equations.
4. The matrix used to represent a system of linear equations is called a(n) _____ matrix.
5. The graph of a linear inequality is called a(n) _____.

REVIEW EXERCISES

In Problems 1–10 solve each system of equations using the method of substitution. If the system has no solution, say it is inconsistent.

1. $\begin{cases} 2x - y = 5 \\ 5x + 2y = 8 \end{cases}$

2. $\begin{cases} 2x + 3y = 2 \\ 7x - y = 3 \end{cases}$

3. $\begin{cases} 3x - 4y = 4 \\ x - 3y = \frac{1}{2} \end{cases}$

4. $\begin{cases} 2x + y = 0 \\ 5x - 4y = -\frac{13}{2} \end{cases}$

5. $\begin{cases} x - 2y - 4 = 0 \\ 3x + 2y - 4 = 0 \end{cases}$

6. $\begin{cases} x - 3y + 5 = 0 \\ 2x + 3y - 5 = 0 \end{cases}$

7. $\begin{cases} y = 2x - 5 \\ x = 3y + 4 \end{cases}$

8. $\begin{cases} x = 5y + 2 \\ y = 5x + 2 \end{cases}$

9. $\begin{cases} x - y + 4 = 0 \\ \frac{1}{2}x + \frac{1}{6}y + \frac{2}{3} = 0 \end{cases}$

10. $\begin{cases} x + \frac{1}{4}y = 2 \\ y + 4x + 2 = 0 \end{cases}$

In Problems 11–20 solve each system of equations using the method of elimination. If the system has no solution, say it is inconsistent.

11. $\begin{cases} x - 2y - 8 = 0 \\ 2x + 2y - 10 = 0 \end{cases}$

12. $\begin{cases} x - 3y + 5 = 0 \\ 2x + 3y - 5 = 0 \end{cases}$

13. $\begin{cases} y - 2x = 11 \\ 2y - 3x = 18 \end{cases}$

14. $\begin{cases} 3x - 4y - 12 = 0 \\ 5x + 2y + 6 = 0 \end{cases}$

15. $\begin{cases} 2x + 3y - 13 = 0 \\ 3x - 2y = 0 \end{cases}$

16. $\begin{cases} 4x + 5y = 21 \\ 5x + 6y = 42 \end{cases}$

17. $\begin{cases} 3x - 2y = 8 \\ x - \frac{2}{3}y = 12 \end{cases}$

18. $\begin{cases} 2x + 5y = 10 \\ 4x + 10y = 15 \end{cases}$

19. $\begin{cases} x + 2y - z = 6 \\ 2x - y + 3z = -13 \\ 3x - 2y + 3z = -16 \end{cases}$

20. $\begin{cases} x + 5y - z = 2 \\ 2x + y + z = 7 \\ x - y + 2z = 11 \end{cases}$

In Problems 21–30 solve each system using matrices. If the system has no solution, say it is inconsistent.

21. $\begin{cases} 3x - 2y = 1 \\ 10x + 10y = 5 \end{cases}$

22. $\begin{cases} 3x + 2y = 6 \\ x - y = -\frac{1}{2} \end{cases}$

23. $\begin{cases} 5x + 6y - 3z = 6 \\ 4x - 7y - 2z = -3 \\ 3x + y - 7z = 1 \end{cases}$

24. $\begin{cases} 2x + y + z = 5 \\ 4x - y - 3z = 1 \\ 8x + y - z = 5 \end{cases}$

25. $\begin{cases} x - 2z = 1 \\ 2x + 3y = -3 \\ 4x - 3y - 4z = 3 \end{cases}$

26. $\begin{cases} x + 2y - z = 2 \\ 2x - 2y + z = -1 \\ 6x + 4y + 3z = 5 \end{cases}$

27. $\begin{cases} x - y + z = 0 \\ x - y - 5z - 6 = 0 \\ 2x - 2y + z - 1 = 0 \end{cases}$

28. $\begin{cases} 4x - 3y + 5z = 0 \\ 2x + 4y - 3z = 0 \\ 6x + 2y + z = 0 \end{cases}$

29. $\begin{cases} x - y - z - t = 1 \\ 2x + y + z + 2t = 3 \\ x - 2y - 2z - 3t = 0 \\ 3x - 4y + z + 5t = -3 \end{cases}$

30. $\begin{cases} x - 3y + 3z - t = 4 \\ x + 2y - z = -3 \\ x + 3z + 2t = 3 \\ x + y + 5z = 6 \end{cases}$

In Problems 31–36 find the value of each determinant.

31. $\begin{vmatrix} 3 & 4 \\ 1 & 3 \end{vmatrix}$

32. $\begin{vmatrix} -4 & 0 \\ 1 & 3 \end{vmatrix}$

33. $\begin{vmatrix} 1 & 4 & 0 \\ -1 & 2 & 6 \\ 4 & 1 & 3 \end{vmatrix}$

34. $\begin{vmatrix} 2 & 3 & 10 \\ 0 & 1 & 5 \\ -1 & 2 & 3 \end{vmatrix}$

35. $\begin{vmatrix} 2 & 1 & -3 \\ 5 & 0 & 1 \\ 2 & 6 & 0 \end{vmatrix}$

36. $\begin{vmatrix} -2 & 1 & 0 \\ 1 & 2 & 3 \\ -1 & 4 & 2 \end{vmatrix}$

In Problems 37–42 use Cramer's Rule, if applicable, to solve each system.

37. $\begin{cases} x - 2y = 4 \\ 3x + 2y = 4 \end{cases}$

38. $\begin{cases} x - 3y = -5 \\ 2x + 3y = 5 \end{cases}$

39. $\begin{cases} 2x + 3y - 13 = 0 \\ 3x - 2y = 0 \end{cases}$

40. $\begin{cases} 3x - 4y - 12 = 0 \\ 5x + 2y + 6 = 0 \end{cases}$

41. $\begin{cases} x + 2y - z = 6 \\ 2x - y + 3z = -13 \\ 3x - 2y + 3z = -16 \end{cases}$

42. $\begin{cases} x - y + z = 8 \\ 2x + 3y - z = -2 \\ 3x - y - 9z = 9 \end{cases}$

In Problems 43–52 solve each system of equations.

43. $\begin{cases} 2x + y + 3 = 0 \\ x^2 + y^2 = 5 \end{cases}$

44. $\begin{cases} x^2 + y^2 = 16 \\ 2x - y^2 = -8 \end{cases}$

45. $\begin{cases} 2xy + y^2 = 10 \\ 3y^2 - xy = 2 \end{cases}$

46. $\begin{cases} 3x^2 - y^2 = 1 \\ 7x^2 - 2y^2 - 5 = 0 \end{cases}$

47. $\begin{cases} x^2 + y^2 = 6y \\ x^2 = 3y \end{cases}$

48. $\begin{cases} 2x^2 + y^2 = 9 \\ x^2 + y^2 = 9 \end{cases}$

49. $\begin{cases} 3x^2 + 4xy + 5y^2 = 8 \\ x^2 + 3xy + 2y^2 = 0 \end{cases}$ **50.** $\begin{cases} 3x^2 + 2xy - 2y^2 = 6 \\ xy - 2y^2 + 4 = 0 \end{cases}$

51. $\begin{cases} x^2 - 3x + y^2 + y = -2 \\ \dfrac{x^2 - x}{y} + y + 1 = 0 \end{cases}$ **52.** $\begin{cases} x^2 + x + y^2 = y + 2 \\ x + 1 = \dfrac{2 - y}{x} \end{cases}$

In Problems 53–58 graph each system of inequalities. Tell whether the graph is bounded or unbounded and label the vertices.

53. $\begin{cases} 2x + y \le 4 \\ x + y \ge 2 \end{cases}$ **54.** $\begin{cases} x - 2y \le 6 \\ 2x + y \ge 2 \end{cases}$

55. $\begin{cases} x \ge 0 \\ y \ge 0 \\ x + y \le 4 \\ 2x + 3y \le 6 \end{cases}$ **56.** $\begin{cases} x \ge 0 \\ y \ge 0 \\ 3x + y \ge 6 \\ 2x + y \ge 2 \end{cases}$

57. $\begin{cases} x \ge 0 \\ y \ge 0 \\ 2x + y \le 8 \\ x + 2y \ge 2 \end{cases}$ **58.** $\begin{cases} x \ge 0 \\ y \ge 0 \\ 3x + y \le 9 \\ 2x + 3y \ge 6 \end{cases}$

59. Find A such that the system of equations below has infinitely many solutions.

$$\begin{cases} 2x + 5y = 5 \\ 4x + 10y = A \end{cases}$$

60. Find A such that the system in Problem 59 is inconsistent.

61. Find the quadratic function $y = ax^2 + bx + c$ that passes through the three points $(0, 1)$, $(1, 0)$, and $(-2, 1)$.

62. Find the general equation of the circle that passes through the three points $(0, 1)$, $(1, 0)$, and $(-2, 1)$. [*Hint:* The general equation of a circle is $x^2 + y^2 + Dx + Ey + F = 0$.]

63. Katy, Mike, Danny, and Colleen agreed to do yard work at home for $45 to be split among them. After they finished, their father determined that Mike deserves twice what Katy gets, Katy and Colleen deserve the same amount, and Danny deserves half of what Katy gets. How much does each one receive?

64. Rework Problem 63 if, after they finished, their father determined that Mike deserves $5 more than Katy, Katy deserves $5 more than Colleen, and Colleen deserves $5 more than Danny.

65. On a flight between Midway Airport in Chicago and Ft. Lauderdale, Florida, a Boeing 737 jet maintains an airspeed of 475 miles per hour. If the trip from Chicago to Ft. Lauderdale takes 2 hours, 30 minutes and the return flight takes 2 hours, 50 minutes, what is the speed of the jet stream? (Assume the jet-stream speed remains constant at the different altitudes of the plane.)

66. Rework Problem 65 if the airspeed of the jet is 500 miles per hour.

67. If Katy and Mike work together for 1 hour and 20 minutes, they will finish a certain job. If Mike and Danny work together for 1 hour and 36 minutes, the same job can be finished. If Danny and Katy work together, they can complete this job in 2 hours and 40 minutes. How long will it take each of them working alone to finish the job?

68. A chemistry laboratory has three containers of hydrochloric acid, HCl. One container holds a solution with a concentration of 10% HCl, the second holds 25% HCl, and the third holds 40% HCl. How many liters of each should be mixed to obtain 100 liters of a solution with a concentration of 30% HCl? Construct a table showing some of the possible combinations.

69. A small rectangular lot has a perimeter of 68 feet. If its diagonal is 26 feet, what are the dimensions of the lot?

70. The area of a rectangular window is 4 square feet. If the diagonal measures $2\sqrt{2}$ feet, what are the dimensions of the window?

71. A certain right triangle has a perimeter of 14 inches. If the hypotenuse is 6 inches long, what are the lengths of the legs?

72. A certain isosceles triangle has a perimeter of 18 inches. If the altitude is 6 inches long, what is the length of the base?

MISCELLANEOUS TOPICS

8.1 Matrix Algebra
8.2 Linear Programming
8.3 Partial Fraction Decomposition
8.4 Vectors
     Chapter Review

This chapter contains four topics. They are independent of each other and may be covered in any order.

The first section treats a matrix as an algebraic concept, introducing the ideas of addition and multiplication as well as citing some of the properties of matrix algebra.

The second section, linear programming, is a modern application of linear inequalities to certain types of problems. This topic is particularly useful for students interested in operations research.

The third section, partial fraction decomposition, provides an application of systems of equations. This particular application is one that is used later in the study of integral calculus.

Finally, the last section provides an introduction to the notion of a vector and some of its applications.

491

▰ 8.1 MATRIX ALGEBRA

LINEAR ALGEBRA

In Section 7.3, we defined a matrix as an array of real numbers and used a matrix to represent a system of linear equations. There is, however, a branch of mathematics, called **linear algebra**, that deals with matrices in such a way that an arithmetic, and hence an algebra, of matrices is permitted. In this section, we provide a survey of how this **matrix algebra** is developed.

MATRIX ALGEBRA

Before getting started, let's restate the definition of a matrix.

MATRIX

A **matrix** is defined as a rectangular array of numbers,

$$
\begin{array}{c}
\qquad\qquad\quad j\text{th column} \\
\begin{bmatrix}
a_{11} & a_{12} & \cdots & a_{1j} & \cdots & a_{1n} \\
a_{21} & a_{22} & \cdots & a_{2j} & \cdots & a_{2n} \\
\vdots & \vdots & & \vdots & & \vdots \\
a_{i1} & a_{i2} & \cdots & a_{ij} & \cdots & a_{in} \\
\vdots & \vdots & & \vdots & & \vdots \\
a_{m1} & a_{m2} & \cdots & a_{mj} & \cdots & a_{mn}
\end{bmatrix} \quad i\text{th row}
\end{array}
$$

ROW INDEX
COLUMN INDEX
ENTRIES

Each number a_{ij} of the matrix has two indices: the **row index** i and the **column index** j. The matrix shown above has m rows and n columns. The $m \cdot n$ numbers a_{ij} are usually referred to as the **entries** of the matrix.

Let's begin with an example that illustrates how matrices can be used to represent information compactly.

EXAMPLE 1

In a survey of 1000 people in June 1985, the following information was obtained:

200 males	Thought federal defense spending was too high
150 males	Thought federal defense spending was too low
45 males	Had no opinion
315 females	Thought federal defense spending was too high
125 females	Thought federal defense spending was too low
165 females	Had no opinion

We can compactly arrange the above data in a rectangular array as follows:

	TOO HIGH	TOO LOW	NO OPINION
Males	200	150	45
Females	315	125	165

or as

$$\begin{bmatrix} 200 & 150 & 45 \\ 315 & 125 & 165 \end{bmatrix}$$

This matrix has two rows (representing males and females) and three columns (representing "too high," "too low," and "no opinion"). ≡

The matrix we arrived at in Example 1 has two rows and three columns. In general, a matrix with m rows and n columns is called an **m by n matrix**. Thus, the matrix we developed in Example 1 is a 2 by 3 matrix. Notice that an m by n matrix will contain $m \cdot n$ entries.

m BY *n* MATRIX

If an m by n matrix has the same number of rows as columns, that is, if $m = n$, then the matrix is referred to as a **square matrix**.

SQUARE MATRIX

EXAMPLE 2   (a) $\begin{bmatrix} 5 & 0 \\ -6 & 1 \end{bmatrix}$ A 2 by 2 square matrix

(b) $\begin{bmatrix} 1 & 0 & 3 \end{bmatrix}$ A 1 by 3 matrix

(c) $\begin{bmatrix} 6 & -2 & 4 \\ 4 & 3 & 5 \\ 8 & 0 & 1 \end{bmatrix}$ A 3 by 3 square matrix   ≡

From now on, we shall represent matrices by capital letters, such as A, B, C, and so on. We are now ready to define an arithmetic for matrices.

Matrix Arithmetic

We begin our discussion of matrix arithmetic by defining what is meant by two matrices being equal, and then defining the operations of addition and subtraction. It is important to note that these definitions require each matrix to have the same number of rows *and* the same number of columns as a prerequisite for equality and for addition and subtraction.

EQUAL MATRICES

Two m by n matrices A and B are said to be **equal**, written as

$$A = B$$

provided each entry in A is equal to the corresponding entry in B.
For example,

$$\begin{bmatrix} 2 & 1 \\ 0.5 & -1 \end{bmatrix} = \begin{bmatrix} \sqrt{4} & 1 \\ \frac{1}{2} & -1 \end{bmatrix} \quad \text{and} \quad \begin{bmatrix} 3 & 2 & 1 \\ 0 & 1 & -2 \end{bmatrix} = \begin{bmatrix} \sqrt{9} & \sqrt{4} & 1 \\ 0 & 1 & \sqrt[3]{-8} \end{bmatrix}$$

$$\begin{bmatrix} 4 & 1 \\ 6 & 1 \end{bmatrix} \neq \begin{bmatrix} 4 & 0 \\ 6 & 1 \end{bmatrix}$$

Because the entries in row one, column two are not equal

$$\begin{bmatrix} 4 & 1 & 2 \\ 6 & 1 & 2 \end{bmatrix} \neq \begin{bmatrix} 4 & 1 & 2 & 3 \\ 6 & 1 & 2 & 4 \end{bmatrix}$$

Because the matrix on the left is 2 by 3 and the matrix on the right is 2 by 4

If each of A and B is an m by n matrix, and hence each contains $m \cdot n$ entries, the statement $A = B$ actually represents a system of $m \cdot n$ ordinary equations. We'll make use of this fact a little later.

Suppose A and B represent two m by n matrices. We define their

SUM $A + B$
OF TWO MATRICES
DIFFERENCE $A - B$
OF TWO MATRICES

sum $A + B$ to be the m by n matrix formed by adding the corresponding entries of A and B. The **difference $A - B$** is defined as the m by n matrix formed by subtracting the entries in B from the corresponding entries in A. Addition and subtraction of matrices is allowed only for matrices having the same number m of rows and the same number n of columns. Thus, for example, a 2 by 3 matrix and a 2 by 4 matrix cannot be added or subtracted.

EXAMPLE 3 Suppose

$$A = \begin{bmatrix} 2 & 4 & 8 & -3 \\ 0 & 1 & 2 & 3 \end{bmatrix} \qquad B = \begin{bmatrix} -3 & 4 & 0 & 1 \\ 6 & 8 & 2 & 0 \end{bmatrix}$$

Find:

(a) $A + B$ (b) $A - B$

Solution   (a) $A + B = \begin{bmatrix} 2 & 4 & 8 & -3 \\ 0 & 1 & 2 & 3 \end{bmatrix} + \begin{bmatrix} -3 & 4 & 0 & 1 \\ 6 & 8 & 2 & 0 \end{bmatrix}$

$$= \begin{bmatrix} 2 + (-3) & 4 + 4 & 8 + 0 & -3 + 1 \\ 0 + 6 & 1 + 8 & 2 + 2 & 3 + 0 \end{bmatrix}$$ Add corresponding entries.

$$= \begin{bmatrix} -1 & 8 & 8 & -2 \\ 6 & 9 & 4 & 3 \end{bmatrix}$$

(b) $A - B = \begin{bmatrix} 2 & 4 & 8 & -3 \\ 0 & 1 & 2 & 3 \end{bmatrix} - \begin{bmatrix} -3 & 4 & 0 & 1 \\ 6 & 8 & 2 & 0 \end{bmatrix}$

$$= \begin{bmatrix} 2 - (-3) & 4 - 4 & 8 - 0 & -3 - 1 \\ 0 - 6 & 1 - 8 & 2 - 2 & 3 - 0 \end{bmatrix}$$ Subtract corresponding entries.

$$= \begin{bmatrix} 5 & 0 & 8 & -4 \\ -6 & -7 & 0 & 3 \end{bmatrix}$$

Many of the algebraic properties of sums of real numbers are also true for sums of matrices. Suppose each of A, B, and C is an m by n

COMMUTATIVE PROPERTY   matrix. Then matrix addition is **commutative**. That is,

$$A + B = B + A$$

ASSOCIATIVE PROPERTY   Matrix addition is also **associative**. That is,

$$(A + B) + C = A + (B + C)$$

Although we shall not prove these results, the proofs, as the following example illustrates, are based on the commutative and associative laws for real numbers. (See page 6.)

EXAMPLE 4
$$\begin{bmatrix} 2 & 3 & -1 \\ 4 & 0 & 7 \end{bmatrix} + \begin{bmatrix} -1 & 2 & 1 \\ 5 & -3 & 4 \end{bmatrix} = \begin{bmatrix} 2 + (-1) & 3 + 2 & -1 + 1 \\ 4 + 5 & 0 + (-3) & 7 + 4 \end{bmatrix}$$

$$= \begin{bmatrix} -1 + 2 & 2 + 3 & 1 + (-1) \\ 5 + 4 & -3 + 0 & 4 + 7 \end{bmatrix}$$

$$= \begin{bmatrix} -1 & 2 & 1 \\ 5 & -3 & 4 \end{bmatrix} + \begin{bmatrix} 2 & 3 & -1 \\ 4 & 0 & 7 \end{bmatrix}$$

$\blacksquare$

ZERO MATRIX   A matrix whose entries are all equal to 0 is called a **zero matrix**. Each of the following matrices is a zero matrix:

$$\begin{bmatrix} 0 & 0 \\ 0 & 0 \end{bmatrix}$$ 2 by 2 square zero matrix

$$\begin{bmatrix} 0 & 0 & 0 \\ 0 & 0 & 0 \end{bmatrix}$$ 2 by 3 zero matrix

$$\begin{bmatrix} 0 & 0 & 0 \end{bmatrix}$$ 1 by 3 zero matrix

Zero matrices have properties similar to the real number 0. Thus, if A is an m by n matrix and 0 is an m by n zero matrix, then

$$A + 0 = A$$

In other words, the zero matrix is the additive identity in matrix algebra.

We also can multiply a matrix by a real number. If k is a real number
SCALAR MULTIPLE   and A is an m by n matrix, the matrix kA, called a **scalar multiple** of A, is the m by n matrix formed by multiplying each entry in A by k.
SCALAR   The number k is sometimes referred to as a **scalar**.

EXAMPLE 5   Suppose

$$A = \begin{bmatrix} 3 & 1 & 5 \\ -2 & 0 & 6 \end{bmatrix} \qquad B = \begin{bmatrix} 4 & 1 & 0 \\ 8 & 1 & -3 \end{bmatrix} \qquad C = \begin{bmatrix} 9 & 0 \\ -3 & 6 \end{bmatrix}.$$

Find:

(a) $4A$ (b) $\frac{1}{3}C$ (c) $3A - 2B$

Solution   (a) $4A = 4\begin{bmatrix} 3 & 1 & 5 \\ -2 & 0 & 6 \end{bmatrix} = \begin{bmatrix} 4\cdot 3 & 4\cdot 1 & 4\cdot 5 \\ 4(-2) & 4\cdot 0 & 4\cdot 6 \end{bmatrix}$

$$= \begin{bmatrix} 12 & 4 & 20 \\ -8 & 0 & 24 \end{bmatrix}$$

(b) $\frac{1}{3}C = \frac{1}{3}\begin{bmatrix} 9 & 0 \\ -3 & 6 \end{bmatrix} = \begin{bmatrix} \frac{1}{3}\cdot 9 & \frac{1}{3}\cdot 0 \\ \frac{1}{3}(-3) & \frac{1}{3}\cdot 6 \end{bmatrix} = \begin{bmatrix} 3 & 0 \\ -1 & 2 \end{bmatrix}$

(c) $3A - 2B = 3\begin{bmatrix} 3 & 1 & 5 \\ -2 & 0 & 6 \end{bmatrix} - 2\begin{bmatrix} 4 & 1 & 0 \\ 8 & 1 & -3 \end{bmatrix}$

$$= \begin{bmatrix} 3\cdot 3 & 3\cdot 1 & 3\cdot 5 \\ 3(-2) & 3\cdot 0 & 3\cdot 6 \end{bmatrix} - \begin{bmatrix} 2\cdot 4 & 2\cdot 1 & 2\cdot 0 \\ 2\cdot 8 & 2\cdot 1 & 2(-3) \end{bmatrix}$$

$$= \begin{bmatrix} 9 & 3 & 15 \\ -6 & 0 & 18 \end{bmatrix} - \begin{bmatrix} 8 & 2 & 0 \\ 16 & 2 & -6 \end{bmatrix}$$

$$= \begin{bmatrix} 9-8 & 3-2 & 15-0 \\ -6-16 & 0-2 & 18-(-6) \end{bmatrix}$$

$$= \begin{bmatrix} 1 & 1 & 15 \\ -22 & -2 & 24 \end{bmatrix} \qquad \blacksquare$$

We list below some of the algebraic properties of scalar multiplication.

PROPERTIES OF SCALAR   Let h and k be real numbers and let A and B be m by n matrices.
MULTIPLICATION   Then

$$k(hA) = (kh)A$$
$$(k + h)A = kA + hA$$
$$k(A + B) = kA + kB$$

The proofs of these properties are based on properties of real numbers. For example, if each of A and B is a 2 by 2 matrix, then

$$k(A + B) = k\left\{\begin{bmatrix} a_{11} & a_{12} \\ a_{21} & a_{22} \end{bmatrix} + \begin{bmatrix} b_{11} & b_{12} \\ b_{21} & b_{22} \end{bmatrix}\right\} = k\begin{bmatrix} a_{11} + b_{11} & a_{12} + b_{12} \\ a_{21} + b_{21} & a_{22} + b_{22} \end{bmatrix}$$

$$= \begin{bmatrix} k(a_{11} + b_{11}) & k(a_{12} + b_{12}) \\ k(a_{21} + b_{21}) & k(a_{22} + b_{22}) \end{bmatrix} = \begin{bmatrix} ka_{11} + kb_{11} & ka_{12} + kb_{12} \\ ka_{21} + kb_{21} & ka_{22} + kb_{22} \end{bmatrix}$$

$$= \begin{bmatrix} ka_{11} & ka_{12} \\ ka_{21} & ka_{22} \end{bmatrix} + \begin{bmatrix} kb_{11} & kb_{12} \\ kb_{21} & kb_{22} \end{bmatrix}$$

$$= k \begin{bmatrix} a_{11} & a_{12} \\ a_{21} & a_{22} \end{bmatrix} + k \begin{bmatrix} b_{11} & b_{12} \\ b_{21} & b_{22} \end{bmatrix} = kA + kB$$

Multiplication of Matrices

Unlike the straightforward rule for adding two matrices, the rule for multiplying two matrices is not what we might expect. Because of this, we'll need some preliminary ideas.

ROW VECTOR   A **row vector** R is a 1 by n matrix

$$R = [r_1 \quad r_2 \quad \cdots \quad r_n]$$

COLUMN VECTOR   A **column vector** C is an n by 1 matrix

$$C = \begin{bmatrix} c_1 \\ c_2 \\ \vdots \\ c_n \end{bmatrix}$$

PRODUCT   The **product** RC of R times C is defined as the number

$$RC = [r_1 \quad r_2 \quad \cdots \quad r_n] \begin{bmatrix} c_1 \\ c_2 \\ \vdots \\ c_n \end{bmatrix} = r_1 c_1 + r_2 c_2 + \cdots + r_n c_n$$

EXAMPLE 6   If $R = [3 \quad -5 \quad 2]$ and $C = \begin{bmatrix} 3 \\ 4 \\ -5 \end{bmatrix}$, then

$$RC = [3 \quad -5 \quad 2] \begin{bmatrix} 3 \\ 4 \\ -5 \end{bmatrix} = 3 \cdot 3 + (-5)4 + 2(-5)$$

$$= 9 - 20 - 10 = -21 \qquad \blacksquare$$

Notice that a row vector and a column vector can be multiplied only if they contain the same number of entries. Let's look at a practical use of the product of a row vector by a column vector.

EXAMPLE 7　A clothing store sells men's shirts for $25, silk ties for $8, and wool suits for $300. Last month the store had sales consisting of 100 shirts, 200 ties, and 50 suits. What was the total revenue due to these sales?

Solution　We set up a row vector R to represent the price of each item and a column vector C to represent the corresponding number of items sold. Then

$$
\underset{\text{Price}}{\underset{\text{Shirts \ Ties \ Suits}}{R = [\ 25 \quad 8 \quad 300\ ]}}
\qquad
\underset{\text{Number sold}}{C = \begin{bmatrix} 100 \\ 200 \\ 50 \end{bmatrix} \begin{matrix} \text{Shirts} \\ \text{Ties} \\ \text{Suits} \end{matrix}}
$$

The revenue obtained is

$$
\text{Total revenue} = RC = [25 \quad 8 \quad 300] \begin{bmatrix} 100 \\ 200 \\ 50 \end{bmatrix}
$$

$$
= \underset{\text{Shirt revenue}}{25 \cdot 100} + \underset{\text{Tie revenue}}{8 \cdot 200} + \underset{\text{Suit revenue}}{300 \cdot 50} = \underset{\text{Total revenue}}{\$19{,}100}
$$

The definition for multiplying two matrices is based on the idea of a row vector times a column vector.

PRODUCT　Let A denote an m by r matrix and let B denote an r by n matrix. The **product** AB is defined as the m by n matrix whose entry in row i, column j is the product of the ith row of A and the jth column of B.

An example will help clarify the definition.

EXAMPLE 8　Find the product AB if

$$
A = \begin{bmatrix} 2 & 4 & -1 \\ 5 & 8 & 0 \end{bmatrix} \quad \text{and} \quad B = \begin{bmatrix} 2 & 5 & 1 & 4 \\ 4 & 8 & 0 & 6 \\ -3 & 1 & -2 & -1 \end{bmatrix}
$$

Solution　First we note that A is 2 by 3 and B is 3 by 4 so that the product AB will be a 2 by 4 matrix. Suppose we want the entry in row two, column three of AB. To find it, we find the product of the row vector from row two of A and the column vector from column three of B, namely

$$
\begin{array}{c}
\text{Column three} \\
\text{of } B
\end{array}
$$

$$
\begin{array}{c}
\text{Row two} \\
\text{of } A
\end{array}
\begin{bmatrix} 5 & 8 & 0 \end{bmatrix}
\begin{bmatrix} 1 \\ 0 \\ -2 \end{bmatrix} = 5 \cdot 1 + 8 \cdot 0 + 0(-2) = 5
$$

So far, we have

$$
\text{Column three}
$$
$$
\downarrow
$$
$$
AB = \begin{bmatrix} - & - & - & - \\ - & - & 5 & - \end{bmatrix} \leftarrow \text{Row two}
$$

To find, say, the entry in row one, column four of AB, we find the product using row one of A and column four of B:

$$
\begin{array}{c}
\text{Column four} \\
\text{of } B
\end{array}
$$

$$
\begin{array}{c}
\text{Row one} \\
\text{of } A
\end{array}
\begin{bmatrix} 2 & 4 & -1 \end{bmatrix}
\begin{bmatrix} 4 \\ 6 \\ -1 \end{bmatrix} = 2 \cdot 4 + 4 \cdot 6 + (-1)(-1) = 33
$$

Continuing in this fashion, we find AB:

$$
AB = \begin{bmatrix} 2 & 4 & -1 \\ 5 & 8 & 0 \end{bmatrix}
\begin{bmatrix} 2 & 5 & 1 & 4 \\ 4 & 8 & 0 & 6 \\ -3 & 1 & -2 & -1 \end{bmatrix}
$$

$$
= \begin{bmatrix}
\begin{array}{l}\text{Row one of } A \\ \text{times column} \\ \text{one of } B\end{array} &
\begin{array}{l}\text{Row one of } A \\ \text{times column} \\ \text{two of } B\end{array} &
\begin{array}{l}\text{Row one of } A \\ \text{times column} \\ \text{three of } B\end{array} &
\begin{array}{l}\text{Row one of } A \\ \text{times column} \\ \text{four of } B\end{array} \\
\\
\begin{array}{l}\text{Row two of } A \\ \text{times column} \\ \text{one of } B\end{array} &
\begin{array}{l}\text{Row two of } A \\ \text{times column} \\ \text{two of } B\end{array} &
\begin{array}{l}\text{Row two of } A \\ \text{times column} \\ \text{three of } B\end{array} &
\begin{array}{l}\text{Row two of } A \\ \text{times column} \\ \text{four of } B\end{array}
\end{bmatrix}
$$

$$
= \begin{bmatrix}
2\cdot2+4\cdot4+(-1)(-3) & 2\cdot5+4\cdot8+(-1)1 & 2\cdot1+4\cdot0+(-1)(-2) & 33 \\
5\cdot2+8\cdot4+0(-3) & 5\cdot5+8\cdot8+0\cdot1 & 5 & 5\cdot4+8\cdot6+0(-1)
\end{bmatrix}
$$

$$
= \begin{bmatrix} 23 & 41 & 4 & 33 \\ 42 & 89 & 5 & 68 \end{bmatrix}
$$

The definition of the product AB of two matrices A and B requires that the number of columns of A equal the number of rows of B. Otherwise, no product is defined. See Figure 1.

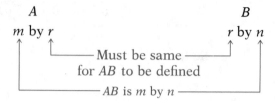

FIGURE 1

For example, for the matrices given in Example 8, the product AB is defined but the product BA is not. Another unusual result that can occur when multiplying two matrices is illustrated in the next example.

EXAMPLE 9   If

$$A = \begin{bmatrix} 2 & 1 & 3 \\ 1 & -1 & 0 \end{bmatrix} \text{ and } B = \begin{bmatrix} 1 & 0 \\ 2 & 1 \\ 3 & 2 \end{bmatrix}$$

find:

(a) AB (b) BA

Solution   (a) $AB = \begin{bmatrix} 2 & 1 & 3 \\ 1 & -1 & 0 \end{bmatrix}\begin{bmatrix} 1 & 0 \\ 2 & 1 \\ 3 & 2 \end{bmatrix} = \begin{bmatrix} 13 & 7 \\ -1 & -1 \end{bmatrix}$

(b) $BA = \begin{bmatrix} 1 & 0 \\ 2 & 1 \\ 3 & 2 \end{bmatrix}\begin{bmatrix} 2 & 1 & 3 \\ 1 & -1 & 0 \end{bmatrix} = \begin{bmatrix} 2 & 1 & 3 \\ 5 & 1 & 6 \\ 8 & 1 & 9 \end{bmatrix}$

$\equiv$

Notice in Example 9 that AB is 2 by 2 and BA is 3 by 3. Thus, it is possible for both AB and BA to be defined, yet be different. In fact, even if each of A and B is an n by n matrix, so that AB and BA are each defined and n by n, it may happen that AB and BA are not equal.

EXAMPLE 10   If

$$A = \begin{bmatrix} 2 & 1 \\ 0 & 4 \end{bmatrix} \text{ and } B = \begin{bmatrix} -3 & 1 \\ 1 & 2 \end{bmatrix}$$

find:

(a) AB (b) BA

Solution (a) $AB = \begin{bmatrix} 2 & 1 \\ 0 & 4 \end{bmatrix} \begin{bmatrix} -3 & 1 \\ 1 & 2 \end{bmatrix}$

$= \begin{bmatrix} 2(-3) + 1 \cdot 1 & 2 \cdot 1 + 1 \cdot 2 \\ 0(-3) + 4 \cdot 1 & 0 \cdot 1 + 4 \cdot 2 \end{bmatrix} = \begin{bmatrix} -5 & 4 \\ 4 & 8 \end{bmatrix}$

(b) $BA = \begin{bmatrix} -3 & 1 \\ 1 & 2 \end{bmatrix} \begin{bmatrix} 2 & 1 \\ 0 & 4 \end{bmatrix}$

$= \begin{bmatrix} (-3)2 + 1 \cdot 0 & (-3)1 + 1 \cdot 4 \\ 1 \cdot 2 + 2 \cdot 0 & 1 \cdot 1 + 2 \cdot 4 \end{bmatrix} = \begin{bmatrix} -6 & 1 \\ 2 & 9 \end{bmatrix}$ ≡

The preceding examples demonstrate that an important property of real numbers, the commutative law of multiplication, is not shared by matrices.

Matrix multiplication is not commutative

Below we list two of the properties of real numbers that are shared by matrices. Assuming each product and sum is defined, we have

ASSOCIATIVE PROPERTY

DISTRIBUTIVE PROPERTY

$$A(BC) = (AB)C$$
$$A(B + C) = AB + AC$$

The Identity Matrix

DIAGONAL ENTRIES

IDENITY MATRIX I_n

In an n by n square matrix, the entries located in row i, column i, $1 \le i \le n$, are called the **diagonal entries**. An n by n square matrix whose diagonal entries are 1's, while all other entries are 0's, is called the **identity matrix I_n**. For example,

$$I_2 = \begin{bmatrix} 1 & 0 \\ 0 & 1 \end{bmatrix} \qquad I_3 = \begin{bmatrix} 1 & 0 & 0 \\ 0 & 1 & 0 \\ 0 & 0 & 1 \end{bmatrix}$$

and so on. The identity matrix has properties analogous to the number 1.

EXAMPLE 11 Let

$$A = \begin{bmatrix} -1 & 2 & 0 \\ 0 & 1 & 3 \end{bmatrix} \quad \text{and} \quad B = \begin{bmatrix} 3 & 2 \\ 4 & 6 \\ 5 & 2 \end{bmatrix}$$

Find:

(a) AI_3 (b) I_2A (c) BI_2

Solution (a) $AI_3 = \begin{bmatrix} -1 & 2 & 0 \\ 0 & 1 & 3 \end{bmatrix} \begin{bmatrix} 1 & 0 & 0 \\ 0 & 1 & 0 \\ 0 & 0 & 1 \end{bmatrix} = \begin{bmatrix} -1 & 2 & 0 \\ 0 & 1 & 3 \end{bmatrix} = A$

(b) $I_2A = \begin{bmatrix} 1 & 0 \\ 0 & 1 \end{bmatrix} \begin{bmatrix} -1 & 2 & 0 \\ 0 & 1 & 3 \end{bmatrix} = \begin{bmatrix} -1 & 2 & 0 \\ 0 & 1 & 3 \end{bmatrix} = A$

(c) $BI_2 = \begin{bmatrix} 3 & 2 \\ 4 & 6 \\ 5 & 2 \end{bmatrix} \begin{bmatrix} 1 & 0 \\ 0 & 1 \end{bmatrix} = \begin{bmatrix} 3 & 2 \\ 4 & 6 \\ 5 & 2 \end{bmatrix} = B$ ≣

Example 11 demonstrates the following property:

If A is an m by n matrix, then

$$I_mA = A \qquad \text{and} \qquad AI_n = A$$

Thus, the identity matrix has properties analogous to the real number 1. In other words, the identity matrix is a multiplicative identity in matrix algebra.

The Inverse of a Matrix

Let A be a square n by n matrix. If there exists an n by n matrix A^{-1} for which

$$AA^{-1} = I_n$$

INVERSE MATRIX   then A^{-1} is called the **inverse** of the matrix A.

It can be shown that if A^{-1} is the inverse of A, then

$$A^{-1}A = I_n$$

As we shall soon see, not every square matrix has an inverse. When NONSINGULAR MATRIX   a matrix A does have an inverse A^{-1}, then A is said to be **nonsingular**.

EXAMPLE 12   Show that the inverse of

$$A = \begin{bmatrix} 3 & 1 \\ 2 & 1 \end{bmatrix} \quad \text{is} \quad A^{-1} = \begin{bmatrix} 1 & -1 \\ -2 & 3 \end{bmatrix}$$

Solution   We need to show that $AA^{-1} = I_2$.

$$AA^{-1} = \begin{bmatrix} 3 & 1 \\ 2 & 1 \end{bmatrix}\begin{bmatrix} 1 & -1 \\ -2 & 3 \end{bmatrix} = \begin{bmatrix} 3 \cdot 1 + 1(-2) & 3(-1) + 1 \cdot 3 \\ 2 \cdot 1 + 1(-2) & 2(-1) + 1 \cdot 3 \end{bmatrix}$$

$$= \begin{bmatrix} 1 & 0 \\ 0 & 1 \end{bmatrix} = I_2 \qquad\qquad \equiv$$

We now show one way to find the inverse of

$$A = \begin{bmatrix} 3 & 1 \\ 2 & 1 \end{bmatrix}$$

Suppose A^{-1} is given by

$$A^{-1} = \begin{bmatrix} x & y \\ z & w \end{bmatrix} \tag{1}$$

where x, y, z, and w are four unknowns. Based on the definition of an inverse, we have

$$AA^{-1} = I_2$$

$$\begin{bmatrix} 3 & 1 \\ 2 & 1 \end{bmatrix}\begin{bmatrix} x & y \\ z & w \end{bmatrix} = \begin{bmatrix} 1 & 0 \\ 0 & 1 \end{bmatrix}$$

$$\begin{bmatrix} 3x + z & 3y + w \\ 2x + z & 2y + w \end{bmatrix} = \begin{bmatrix} 1 & 0 \\ 0 & 1 \end{bmatrix}$$

Because corresponding entries must be equal, it follows that this matrix equation is equivalent to four ordinary equations:

$$\begin{cases} 3x + z = 1 \\ 2x + z = 0 \end{cases} \qquad \begin{cases} 3y + w = 0 \\ 2y + w = 1 \end{cases}$$

The augmented matrix of each system is

$$\left[\begin{array}{cc|c} 3 & 1 & 1 \\ 2 & 1 & 0 \end{array}\right] \qquad \left[\begin{array}{cc|c} 3 & 1 & 0 \\ 2 & 1 & 1 \end{array}\right] \tag{2}$$

The usual procedure would be to place each augmented matrix in reduced echelon form. However, rather than do this separately for each augmented matrix, we find it more efficient to combine the two augmented matrices in Display (2) as a single matrix and place it in reduced echelon form:

$$\left[\begin{array}{cc|cc} 3 & 1 & 1 & 0 \\ 2 & 1 & 0 & 1 \end{array}\right]$$

Now,

$$\begin{bmatrix} 3 & 1 & | & 1 & 0 \\ 2 & 1 & | & 0 & 1 \end{bmatrix} \rightarrow \begin{bmatrix} 1 & 0 & | & 1 & -1 \\ 2 & 1 & | & 0 & 1 \end{bmatrix}$$
$$\uparrow$$
$$R_1 = -1r_2 + r_1$$

$$\rightarrow \begin{bmatrix} 1 & 0 & | & 1 & -1 \\ 0 & 1 & | & -2 & 3 \end{bmatrix} \qquad (3)$$
$$\uparrow$$
$$R_2 = -2r_1 + r_2$$

The matrix in Display (3) is in reduced echelon form. Now we reverse the earlier step of combining the two augmented matrices in Display (2), and write the single matrix in Display (3) as two augmented matrices:

$$\begin{bmatrix} 1 & 0 & | & 1 \\ 0 & 1 & | & -2 \end{bmatrix} \text{ and } \begin{bmatrix} 1 & 0 & | & -1 \\ 0 & 1 & | & 3 \end{bmatrix}$$

We conclude from these that $x = 1$, $z = -2$, and $y = -1$, $w = 3$. From Equation (1) we find

$$A^{-1} = \begin{bmatrix} 1 & -1 \\ -2 & 3 \end{bmatrix}$$

Notice in Display (3) that the 2 by 2 matrix appearing to the right of the vertical bar is, in fact, precisely the inverse of A. Also notice that the identity matrix I_2 is the matrix that appears to the left of the vertical bar in Display (3). These observations and the procedures followed above are summarized below.

PROCEDURE FOR FINDING THE INVERSE OF A NONSINGULAR MATRIX

To find the inverse of an n by n nonsingular matrix A, proceed as follows:

Step 1: Form the matrix $[A \mid I_n]$.

Step 2: Place the matrix $[A \mid I_n]$ in reduced echelon form.

Step 3: The reduced echelon form of $[A \mid I_n]$ will contain the identity matrix I_n on the left of the vertical bar; the n by n matrix on the right of the vertical bar is the inverse of A.

In other words, we begin with the matrix $[A \mid I_n]$ and, after placing it in reduced echelon form, we end up with the matrix $[I_n \mid A^{-1}]$.

Let's do another example.

EXAMPLE 13   The matrix

$$A = \begin{bmatrix} 1 & 1 & 0 \\ -1 & 3 & 4 \\ 0 & 4 & 3 \end{bmatrix}$$

is nonsingular. Find its inverse.

Solution   First, we form the matrix

$$[A \mid I_3] = \begin{bmatrix} 1 & 1 & 0 & \mid & 1 & 0 & 0 \\ -1 & 3 & 4 & \mid & 0 & 1 & 0 \\ 0 & 4 & 3 & \mid & 0 & 0 & 1 \end{bmatrix}$$

Next, we use row operations to place $[A \mid I_3]$ in reduced echelon form:

$$\begin{bmatrix} 1 & 1 & 0 & \mid & 1 & 0 & 0 \\ -1 & 3 & 4 & \mid & 0 & 1 & 0 \\ 0 & 4 & 3 & \mid & 0 & 0 & 1 \end{bmatrix} \rightarrow \begin{bmatrix} 1 & 1 & 0 & \mid & 1 & 0 & 0 \\ 0 & 4 & 4 & \mid & 1 & 1 & 0 \\ 0 & 4 & 3 & \mid & 0 & 0 & 1 \end{bmatrix}$$

$$\uparrow$$
$$R_2 = r_2 + r_1$$

$$\rightarrow \begin{bmatrix} 1 & 1 & 0 & \mid & 1 & 0 & 0 \\ 0 & 1 & 1 & \mid & \frac{1}{4} & \frac{1}{4} & 0 \\ 0 & 4 & 3 & \mid & 0 & 0 & 1 \end{bmatrix}$$

$$\uparrow$$
$$R_2 = \frac{1}{4} r_2$$

$$\rightarrow \begin{bmatrix} 1 & 0 & -1 & \mid & \frac{3}{4} & -\frac{1}{4} & 0 \\ 0 & 1 & 1 & \mid & \frac{1}{4} & \frac{1}{4} & 0 \\ 0 & 0 & -1 & \mid & -1 & -1 & 1 \end{bmatrix}$$

$$\uparrow$$
$$R_1 = -1r_2 + r_1$$
$$R_3 = -4r_2 + r_3$$

$$\rightarrow \begin{bmatrix} 1 & 0 & -1 & \mid & \frac{3}{4} & -\frac{1}{4} & 0 \\ 0 & 1 & 1 & \mid & \frac{1}{4} & \frac{1}{4} & 0 \\ 0 & 0 & 1 & \mid & 1 & 1 & -1 \end{bmatrix}$$

$$\uparrow$$
$$R_3 = -1r_3$$

$$\rightarrow \begin{bmatrix} 1 & 0 & 0 & \mid & \frac{7}{4} & \frac{3}{4} & -1 \\ 0 & 1 & 0 & \mid & -\frac{3}{4} & -\frac{3}{4} & 1 \\ 0 & 0 & 1 & \mid & 1 & 1 & -1 \end{bmatrix}$$

$$\uparrow$$
$$R_1 = r_1 + r_3$$
$$R_2 = -1r_3 + r_2$$

The matrix $[A \mid I_3]$ is now in reduced echelon form, and the identity matrix I_3 is on the left of the vertical bar. Hence, the inverse of A is

$$A^{-1} = \begin{bmatrix} \frac{7}{4} & \frac{3}{4} & -1 \\ -\frac{3}{4} & -\frac{3}{4} & 1 \\ 1 & 1 & -1 \end{bmatrix}$$

You can (and should) verify that this is the correct inverse by showing that $AA^{-1} = I_3$. ∎

If placing the matrix $[A \mid I_n]$ in reduced echelon form does not result in the identity matrix I_n to the left of the vertical bar, then A has no inverse. The next example demonstrates such a matrix.

EXAMPLE 14   Show that the matrix

$$A = \begin{bmatrix} 4 & 6 \\ 2 & 3 \end{bmatrix}$$

has no inverse.

Solution   Proceeding as in Example 13, we form the matrix

$$[A \mid I_2] = \begin{bmatrix} 4 & 6 & | & 1 & 0 \\ 2 & 3 & | & 0 & 1 \end{bmatrix}$$

Now we use row operations to place $[A \mid I_2]$ in reduced echelon form:

$$[A \mid I_2] = \begin{bmatrix} 4 & 6 & | & 1 & 0 \\ 2 & 3 & | & 0 & 1 \end{bmatrix}$$

$$\rightarrow \begin{bmatrix} 1 & \frac{3}{2} & | & \frac{1}{4} & 0 \\ 2 & 3 & | & 0 & 1 \end{bmatrix}$$
$$\uparrow$$
$$R_1 = \tfrac{1}{4} r_1$$

$$\rightarrow \begin{bmatrix} 1 & \frac{3}{2} & | & \frac{1}{4} & 0 \\ 0 & 0 & | & -\frac{1}{2} & 1 \end{bmatrix}$$
$$\uparrow$$
$$R_2 = -2r_1 + r_2$$

The matrix $[A \mid I_2]$ is now in reduced echelon form. However, the identity matrix does not appear to the left of the vertical bar. We conclude that A has no inverse. ∎

Solving Systems of Equations

Inverse matrices can be used to solve systems of equations in which the number of equations is the same as the number of variables.

EXAMPLE 15   Solve the system of equations

$$\begin{cases} x + y = 3 \\ -x + 3y + 4z = -3 \\ 4y + 3z = 2 \end{cases}$$

Solution   If we let

$$A = \begin{bmatrix} 1 & 1 & 0 \\ -1 & 3 & 4 \\ 0 & 4 & 3 \end{bmatrix}, \quad X = \begin{bmatrix} x \\ y \\ z \end{bmatrix}, \quad \text{and} \quad B = \begin{bmatrix} 3 \\ -3 \\ 2 \end{bmatrix}$$

then the original system of equations can be written compactly as the matrix equation

$$AX = B \tag{4}$$

Now the matrix A has the inverse A^{-1} (see Example 13). We will multiply each side of Equation (4) by A^{-1}:

$$AX = B$$

$$A^{-1}(AX) = A^{-1}B$$
$$(A^{-1}A)X = A^{-1}B \quad \text{Associative law for multiplication}$$
$$I_3 X = A^{-1}B \quad \text{Definition of inverse matrix}$$

$$X = A^{-1}B \quad \text{Property of identity matrix}$$

Now, we use this to find $X = \begin{bmatrix} x \\ y \\ z \end{bmatrix}$:

$$X = \begin{bmatrix} x \\ y \\ z \end{bmatrix} = A^{-1}B = \begin{bmatrix} \frac{7}{4} & \frac{3}{4} & -1 \\ -\frac{3}{4} & -\frac{3}{4} & 1 \\ 1 & 1 & -1 \end{bmatrix} \begin{bmatrix} 3 \\ -3 \\ 2 \end{bmatrix} \quad \text{Example 13}$$

$$= \begin{bmatrix} 1 \\ 2 \\ -2 \end{bmatrix}$$

Thus, $x = 1$, $y = 2$, $z = -2$.   ◼

The method used in Example 15 to solve a system of equations is particularly useful when it is necessary to solve several systems of equations in which the constants appearing to the right of the equal signs are changing, while the coefficients of the variables on the left side remain the same. See Problems 31–50 for some illustrations.

≡ HISTORICAL
COMMENT Matrices were invented in 1857 by Arthur Cayley (1821–1895) as a way of efficiently computing the result of substituting one linear system into another (see Historical Problem 2). The resulting system had really incredible richness, in the sense that a very wide variety of mathematical systems could be mimicked by the matrices. Cayley and his friend J. J. Sylvester (1814–1897) spent much of the rest of their lives elaborating the theory. The torch was then passed to G. Frobenius (1848–1917), whose deep investigations established a central place for matrices in modern mathematics. In 1924, rather to the surprise of physicists, it was found that matrices (with *complex* numbers in them) were exactly the right tool for describing the behavior of atomic systems. Today matrices are used in a wide variety of applications.

≡ HISTORICAL
PROBLEMS 1. *Matrices and complex numbers* Frobenius emphasized in his research how matrices could be used to mimic other mathematical systems. Here, we mimic the behavior of complex numbers using matrices. Mathematicians call this an *isomorphism*.

$$\text{Complex number} \longleftrightarrow \text{Matrix}$$

$$a + bi \qquad \longleftrightarrow \qquad \begin{bmatrix} a & b \\ -b & a \end{bmatrix}$$

Note that the complex number can be read off the top line of the matrix. Thus,

$$2 + 3i \longleftrightarrow \begin{bmatrix} 2 & 3 \\ -3 & 2 \end{bmatrix} \quad \text{and} \quad \begin{bmatrix} 4 & -2 \\ 2 & 4 \end{bmatrix} \longleftrightarrow 4 - 2i$$

(a) Find the matrices corresponding to $2 - 5i$ and $1 + 3i$.
(b) Multiply the two matrices.
(c) Find the corresponding complex number for the matrix found in part (b).
(d) Multiply $2 - 5i$ by $1 + 3i$. The result should be the same as that found in part (c).

The process also works for addition and subtraction. Try it for yourself.

2. *Cayley's definition of matrix multiplication* Cayley in 1857 invented matrix multiplication to simplify the following problem:

$$\begin{cases} u = ar + bs \\ v = cr + ds \end{cases} \qquad \begin{cases} x = ku + lv \\ y = mu + nv \end{cases}$$

(a) Find x and y in terms of r and s by substituting u and v from the first system of equations into the second system of equations.
(b) Use the result of part (a) to find the 2 by 2 matrix A in

$$\begin{bmatrix} x \\ y \end{bmatrix} = A \begin{bmatrix} r \\ s \end{bmatrix}$$

(c) Now look at the following way to do it. Write the equations in matrix form,

$$\begin{bmatrix} u \\ v \end{bmatrix} = \begin{bmatrix} a & b \\ c & d \end{bmatrix} \begin{bmatrix} r \\ s \end{bmatrix} \qquad \begin{bmatrix} x \\ y \end{bmatrix} = \begin{bmatrix} k & l \\ m & n \end{bmatrix} \begin{bmatrix} u \\ v \end{bmatrix}$$

so

$$\begin{bmatrix} x \\ y \end{bmatrix} = \begin{bmatrix} k & l \\ m & n \end{bmatrix} \begin{bmatrix} a & b \\ c & d \end{bmatrix} \begin{bmatrix} r \\ s \end{bmatrix}$$

Do you see how Cayley defined matrix multiplication?

EXERCISE 8.1   *In Problems 1–16 use the matrices*

$$A = \begin{bmatrix} 0 & 3 & -5 \\ 1 & 2 & 6 \end{bmatrix} \qquad B = \begin{bmatrix} 4 & 1 & 0 \\ -2 & 3 & -2 \end{bmatrix} \qquad C = \begin{bmatrix} 4 & 1 \\ 6 & 2 \\ -2 & 3 \end{bmatrix}$$

to compute the given expression.

1. $A + B$ 2. $A - B$ 3. $4A$ 4. $-3B$
5. $3A - 2B$ 6. $2A + 4B$ 7. AC 8. BC
9. CA 10. CB 11. $C(A + B)$ 12. $(A + B)C$
13. $AC - 3I_2$ 14. $CA + 5I_3$ 15. $CA - CB$ 16. $AC + BC$

In Problems 17–20 compute each product.

17. $\begin{bmatrix} 2 & -2 \\ 1 & 0 \end{bmatrix} \begin{bmatrix} 2 & 1 & 4 & 6 \\ 3 & -1 & 3 & 2 \end{bmatrix}$ 18. $\begin{bmatrix} 4 & 1 \\ 2 & 1 \end{bmatrix} \begin{bmatrix} -6 & 6 & 1 & 0 \\ 2 & 5 & 4 & -1 \end{bmatrix}$

19. $\begin{bmatrix} 1 & 0 & 1 \\ 2 & 4 & 1 \\ 3 & 6 & 1 \end{bmatrix} \begin{bmatrix} 1 & 3 \\ 6 & 2 \\ 8 & -1 \end{bmatrix}$ 20. $\begin{bmatrix} 4 & -2 & 3 \\ 0 & 1 & 2 \\ -1 & 0 & 1 \end{bmatrix} \begin{bmatrix} 2 & 6 \\ 1 & -1 \\ 0 & 2 \end{bmatrix}$

In Problems 21–30 each matrix is nonsingular. Find the inverse of each matrix. Be sure to check your answer.

21. $\begin{bmatrix} 2 & 1 \\ 1 & 1 \end{bmatrix}$ 22. $\begin{bmatrix} 3 & -1 \\ -2 & 1 \end{bmatrix}$ 23. $\begin{bmatrix} 6 & 5 \\ 2 & 2 \end{bmatrix}$

24. $\begin{bmatrix} -4 & 1 \\ 6 & -2 \end{bmatrix}$ 25. $\begin{bmatrix} 2 & 1 \\ a & a \end{bmatrix}, \ a \neq 0$ 26. $\begin{bmatrix} b & 3 \\ b & 2 \end{bmatrix}, \ b \neq 0$

27. $\begin{bmatrix} 1 & -1 & 1 \\ 0 & -2 & 1 \\ -2 & -3 & 0 \end{bmatrix}$ 28. $\begin{bmatrix} 1 & 0 & 2 \\ -1 & 2 & 3 \\ 1 & -1 & 0 \end{bmatrix}$ 29. $\begin{bmatrix} 1 & 1 & 1 \\ 3 & 2 & -1 \\ 3 & 1 & 2 \end{bmatrix}$

30. $\begin{bmatrix} 3 & 3 & 1 \\ 1 & 2 & 1 \\ 2 & -1 & 1 \end{bmatrix}$

In Problems 31–50 use the inverses found in Problems 21–30 to solve each system of equations.

31. $\begin{cases} 2x + y = 8 \\ x + y = 5 \end{cases}$

32. $\begin{cases} 3x - y = 8 \\ -2x + y = 4 \end{cases}$

33. $\begin{cases} 2x + y = 0 \\ x + y = 5 \end{cases}$

34. $\begin{cases} 3x - y = 4 \\ -2x + y = 5 \end{cases}$

35. $\begin{cases} 6x + 5y = 7 \\ 2x + 2y = 2 \end{cases}$

36. $\begin{cases} -4x + y = 0 \\ 6x - 2y = 14 \end{cases}$

37. $\begin{cases} 6x + 5y = 13 \\ 2x + 2y = 5 \end{cases}$

38. $\begin{cases} -4x + y = 5 \\ 6x - 2y = -9 \end{cases}$

39. $\begin{cases} 2x + y = -3 \\ ax + ay = -a \end{cases}, \ a \neq 0$

40. $\begin{cases} bx + 3y = 2b + 3 \\ bx + 2y = 2b + 2 \end{cases}, \ b \neq 0$

41. $\begin{cases} 2x + y = \dfrac{7}{a} \\ ax + ay = 5 \end{cases}, \ a \neq 0$

42. $\begin{cases} bx + 3y = 14 \\ bx + 2y = 10 \end{cases}, \ b \neq 0$

43. $\begin{cases} x - y + z = 0 \\ -2y + z = -1 \\ -2x - 3y = -5 \end{cases}$

44. $\begin{cases} x + 2z = 6 \\ -x + 2y + 3z = -5 \\ x - y = 6 \end{cases}$

45. $\begin{cases} x - y + z = 2 \\ -2y + z = 2 \\ -2x - 3y = \frac{1}{2} \end{cases}$

46. $\begin{cases} x + 2z = 2 \\ -x + 2y + 3z = -\frac{3}{2} \\ x - y = 2 \end{cases}$

47. $\begin{cases} x + y + z = 9 \\ 3x + 2y - z = 8 \\ 3x + y + 2z = 1 \end{cases}$

48. $\begin{cases} 3x + 3y + z = 8 \\ x + 2y + z = 5 \\ 2x - y + z = 4 \end{cases}$

49. $\begin{cases} x + y + z = 2 \\ 3x + 2y - z = \frac{7}{3} \\ 3x + y + 2z = \frac{10}{3} \end{cases}$

50. $\begin{cases} 3x + 3y + z = 1 \\ x + 2y + z = 0 \\ 2x - y + z = 4 \end{cases}$

In Problems 51–56 show that each matrix has no inverse.

51. $\begin{bmatrix} 4 & 2 \\ 2 & 1 \end{bmatrix}$

52. $\begin{bmatrix} -3 & \frac{1}{2} \\ 6 & -1 \end{bmatrix}$

53. $\begin{bmatrix} 15 & 3 \\ 10 & 2 \end{bmatrix}$

54. $\begin{bmatrix} -3 & 0 \\ 4 & 0 \end{bmatrix}$

55. $\begin{bmatrix} -3 & 1 & -1 \\ 1 & -4 & -7 \\ 1 & 2 & 5 \end{bmatrix}$

56. $\begin{bmatrix} 1 & 1 & -3 \\ 2 & -4 & 1 \\ -5 & 7 & 1 \end{bmatrix}$

57. The Acme Steel Company is a producer of stainless steel and aluminum containers. On a certain day, 500 10-gallon stainless steel containers, 350 5-gallon stainless steel containers, and 400 1-gallon stainless steel containers were manufactured. On the same day, 700 10-gallon aluminum containers, 500 5-gallon aluminum containers, and 850 1-gallon aluminum containers were made.

(a) Find a 2 by 3 matrix representing the above data. Could a 3 by 2 matrix also have been formed?

(b) If the amount of material used in the 10-gallon container is 15 pounds, the amount used in the 5-gallon container is 8 pounds, and the amount used in the 1-gallon container is 3 pounds, find a 3 by 1 matrix representing the amount of material.

(c) Multiply the matrices found in parts (a) and (b) to get a 2 by 1 matrix showing the day's usage of material.

(d) If stainless steel costs Acme $0.10 per pound and aluminum costs $0.05 per pound, find a 1 by 2 matrix representing cost.

(e) Multiply the matrices found in parts (c) and (d) to determine what the total cost of the day's production was.

58. A car dealership has two locations, one in a city, the other in the suburbs. In January, the city location sold 400 subcompacts, 250 intermediate-size cars, and 50 station wagons; in February, it sold 350 subcompacts, 100 intermediates, and 30 station wagons. At the suburban location in January, 450 subcompacts, 200 intermediates, and 140 station wagons were sold. In February, the suburban location sold 350 subcompacts, 300 intermediates, and 100 station wagons.

(a) Find 2 by 3 matrices that summarize the sales data for each location for January and February (one matrix for each month).

(b) Use matrix addition to obtain total sales for the 2 month period.

(c) The profit on each kind of car is: $100 per subcompact, $150 per intermediate, and $200 per station wagon. Find a 3 by 1 matrix representing this profit.

(d) Multiply the matrices found in parts (b) and (c) to get a 2 by 1 matrix showing the profit at each location.

59. Consider the 2 by 2 square matrix

$$A = \begin{bmatrix} a & b \\ c & d \end{bmatrix}$$

If $\Delta = ad - bc \neq 0$, show that A is nonsingular and that

$$A^{-1} = \frac{1}{\Delta} \begin{bmatrix} d & -b \\ -c & a \end{bmatrix}$$

▤ 8.2 LINEAR PROGRAMMING

Many problems are best viewed as a system of working parts consisting of capital, raw materials, a labor force, and so on, that are to be allocated according to prescribed limitations and needs, with the ultimate goal in mind of achieving a certain objective, such as maximizing profit or minimizing costs. If such a system can be quantified—that is, represented by mathematical equations and/or inequalities—it may be possible to devise a computational procedure for identifying the *best* way of achieving the goal. Such procedures are usually referred to as *mathematical programs*.

LINEAR PROGRAM

Indeed, if the system can be represented by a system of linear inequalities and if the goal can be expressed as that of minimizing or maximizing a linear expression, then the procedure for solving the problem is called a **linear program**. Historically, linear programming evolved as a technique for solving problems involving resource allocation of goods and materials for the U.S. Air Force during World War II. Today linear programming techniques are used to solve a wide variety of problems such as optimizing airline scheduling, establishing telephone lines, and many others.

OBJECTIVE FUNCTION

CONSTRAINTS

Every linear programming problem requires that a certain linear expression, called the **objective function**, be maximized (or minimized). However, such problems further require that this maximization (or minimization) occur under certain conditions, or **constraints**, that can be expressed as linear inequalities. Although most practical linear programming problems involve systems of several hundred linear inequalities containing several hundred variables, we shall limit our discussion to problems containing only two variables because we can solve such problems using graphing techniques.*

LINEAR PROGRAMMING
PROBLEM

A **linear programming problem** in two variables x and y consists of maximizing (or minimizing) a linear objective function

$$z = Ax + By \qquad A \text{ and } B \text{ are real numbers}$$

subject to certain conditions or constraints expressible as linear inequalities in x and y.

FEASIBLE SOLUTION

In order to maximize (or minimize) the quantity $z = Ax + By$, we need to identify points (x, y) that make the expression for z the largest (or smallest) possible. But not all points (x, y) are eligible; only those that also satisfy each linear inequality (constraint) can be used. We refer to each point (x, y) that obeys the system of linear inequalities (the constraints) as a **feasible solution**. Thus, in a linear programming problem, we seek the feasible solution that maximizes (or minimizes) the objective function.

*The *simplex method* is a way to solve linear programming problems involving many inequalities and variables. This method was developed by George Dantzig in 1946 and is particularly well-suited for computerization. In 1984, Narendra Karmarkar of Bell Laboratories discovered a way of solving large linear programming problems that improves on the simplex method.

EXAMPLE 1 Consider the linear programming problem

$$\text{Maximize} \quad z = x + 3y$$

subject to the constraints

$$x \geq 0, \quad y \geq 0, \quad x + y \leq 6, \quad x \leq 4$$

(a) Graph the constraints.

(b) Graph the objective function for $z = 0, 9, 18, 21$.

Solution (a) The constraints are the system of linear inequalities

$$\begin{cases} x \geq 0 \\ y \geq 0 \\ x + y \leq 6 \\ x \leq 4 \end{cases}$$

The graph of this system is the darker shaded region shown in Figure 2(a).

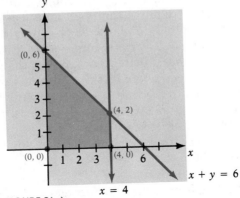

FIGURE 2(a)

(b) For $z = 0$, the objective function is the line $0 = x + 3y$, or $y = -\frac{1}{3}x$. For $z = 9$, the objective function is the line $9 = x + 3y$, or $y = -\frac{1}{3}x + 3$. For $z = 18$, the objective function is the line $18 = x + 3y$, or $y = -\frac{1}{3}x + 6$. For $z = 21$, the objective function is the line $21 = x + 3y$, or $y = -\frac{1}{3}x + 7$. Figure 2(b) on the next page shows the graphs. ≡

SOLUTION A **solution** to a linear programming problem consists of the feasible solutions that maximize (or minimize) the objective function, together with the corresponding values of the objective function.

If none of the feasible solutions maximizes (or minimizes) the objective function or if there are no feasible solutions, then the linear programming problem has no solution.

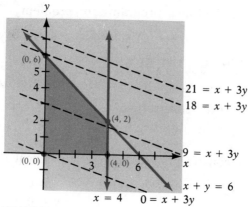

FIGURE 2(b)

Consider the linear programming problem stated in Example 1, and look again at Figure 2(b). The feasible solutions are the points that lie inside the darker shaded region. For example, (2, 2) is a feasible solution, as is (4, 2), (0, 6), (0, 0), etc. To find the solution of the problem requires that we find a feasible solution that makes z, where $z = x + 3y$, as large as possible. Notice that as z increases in value from $z = 0$ to $z = 9$ to $z = 18$ to $z = 21$, we obtain a collection of parallel lines. Furthermore, notice that the largest value of z that can be obtained while feasible solutions are present is $z = 18$, which corresponds to the line $18 = x + 3y$. Any larger value of z results in a line that does not pass through any feasible solutions. Finally, notice that the feasible solution that yields $z = 18$ is the point $(0, 6)$, a vertex. These observations form the basis of the following result, which we state without proof.

THEOREM   If a linear programming problem has a solution,* it is located at a vertex of the graph of the feasible solutions.

If a linear programming problem has multiple solutions, at least one of them is located at a vertex of the graph of the feasible solutions.

In either case, the corresponding value of the objective function is unique.   ≡

We shall not consider here linear programming problems that have no solution. As a result, we can outline the procedure for solving a linear programming problem as follows:

---

*One condition for a linear programming problem in two variables to have a solution is that the graph of the feasible solutions be bounded.

Step 1: Write an expression for the quantity to be maximized (or minimized). This expression is the objective function.

Step 2: Write all the constraints as a system of linear inequalities and graph it.

Step 3: List the vertices of this graph.

Step 4: List the corresponding value of the objective function at each vertex. The largest (or smallest) of these is the solution.

EXAMPLE 2   Minimize the expression

$$z = 2x + 3y$$

subject to the constraints

$$y \leq 5, \quad x \leq 6, \quad x + y \geq 2, \quad x \geq 0, \quad y \geq 0$$

Solution   The objective function is $z = 2x + 3y$. We seek the smallest value of z that can occur if x and y are solutions of the system of linear inequalities

$$\begin{cases} y \leq 5 \\ x \leq 6 \\ x + y \geq 2 \\ x \geq 0 \\ y \geq 0 \end{cases}$$

The graph of this system (the feasible solutions) is shown as the darker shaded region in Figure 3. There, we have also plotted the vertices.

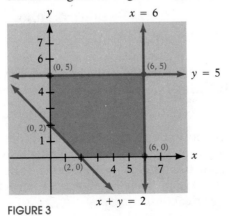

FIGURE 3

Table 1 lists the vertices and the corresponding values of the objective function. From the table we can see that the minimum value of z is 4, and it occurs at the point $(2, 0)$.

TABLE 1

VERTEX	VALUE OF THE OBJECTIVE FUNCTION
(x, y)	$z = 2x + 3y$
$(0, 2)$	$z = 2(0) + 3(2) = 6$
$(0, 5)$	$z = 2(0) + 3(5) = 15$
$(6, 5)$	$z = 2(6) + 3(5) = 27$
$(6, 0)$	$z = 2(6) + 3(0) = 12$
$(2, 0)$	$z = 2(2) + 3(0) = 4$

EXAMPLE 3 At the end of every month, after filling orders for its regular customers, a coffee company has some pure Colombian coffee and some special-blend coffee remaining. The practice of the company has been to package a mixture of the two coffees into 1-pound packages as follows: a low-grade mixture containing 4 ounces of Colombian coffee and 12 ounces of special-blend coffee and a high-grade mixture containing 8 ounces of Colombian and 8 ounces of special-blend coffee. A profit of $0.30 per package is made on the low-grade mixture, whereas a profit of $0.40 per package is made on the high-grade mixture. This month, 120 pounds of special-blend coffee and 100 pounds of pure Colombian coffee remain. How many packages of each mixture should be prepared to achieve a maximum profit?

Solution We begin by assigning symbols for the two variables:

x = Number of packages of the low-grade mixture

y = Number of packages of the high-grade mixture

If P denotes the profit, then

$$P = \$0.30x + \$0.40y$$

This expression is the objective function. We seek to maximize P subject to certain constraints on x and y. Because x and y represent numbers of packages, the only meaningful values for x and y are nonnegative. Thus, we have the two constraints

$$x \geq 0, \quad y \geq 0 \quad \text{Nonnegative constraints}$$

We also have only so much of each type of coffee available. For example, the total amount of Colombian coffee used in the two mixtures cannot exceed 100 pounds, or 1600 ounces. Because we use 4 ounces in each low-grade package and 8 ounces in each high-grade package, we are led to the constraint

$$4x + 8y \leq 1600 \quad \text{Colombian coffee constraint}$$

Similarly, the supply of 120 pounds, or 1920 ounces, of special-blend coffee leads to the constraint

$$12x + 8y \leq 1920 \quad \text{Special-blend coffee constraint}$$

The linear programming problem may be stated as:

$$\text{Maximize} \quad P = 0.3x + 0.4y$$

subject to the constraints

$$x \geq 0, \quad y \geq 0, \quad 4x + 8y \leq 1600, \quad 12x + 8y \leq 1920$$

The graph of the constraints (the feasible solutions) is illustrated in Figure 4. We list the vertices and evaluate the objective function at each vertex. In Table 2 we can see that the maximum profit, $84, is achieved with 40 packages of the low-grade mixture and 180 packages of the high-grade mixture.

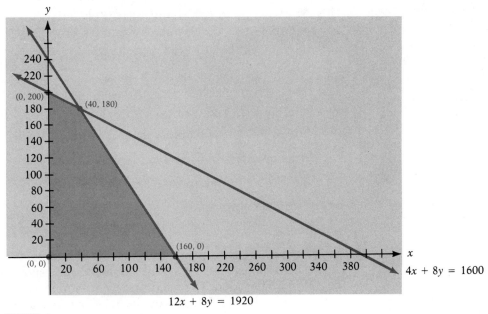

FIGURE 4

TABLE 2

VERTEX	VALUE OF PROFIT
(x, y)	$P = 0.3x + 0.4y$
$(0, 0)$	$P = 0$
$(0, 200)$	$P = 0.3(0) + 0.4(200) = \$80$
$(40, 180)$	$P = 0.3(40) + 0.4(180) = \$84$
$(160, 0)$	$P = 0.3(160) + 0.4(0) = \$48$

EXAMPLE 4 A retired couple has up to $50,000 to place in fixed-income securities. Their banker suggests two securities to them: one is a AAA bond that yields 12% per annum, the other is a Certificate of Deposit that yields 9%. After careful consideration of the alternatives, the couple decides to place at most $20,000 in the AAA bond and at least $15,000 in the Certificate of Deposit. They also instruct the banker to place at least as much in the Certificate of Deposit as in the AAA bond. How should the banker proceed to maximize the return on their investment?

Solution The variables are named as

$$x = \text{Amount invested in the AAA bond}$$
$$y = \text{Amount invested in the Certificate of Deposit}$$

The return R on investment is given by

$$R = 0.12x + 0.09y$$

This expression is to be maximized (the objective function). The conditions imposed on the variables x and y are:

$x \geq 0, \quad y \geq 0$ Nonnegative constraints

$x + y \leq 50,000$ Up to $50,000 to invest

$x \leq 20,000$ Place at most $20,000 in the AAA bond

$y \geq 15,000$ Place at least $15,000 in the Certificate of Deposit

$y \geq x$ Place at least as much in the certificate as in the bond

The linear programming problem may be stated as:

$$\text{Maximize} \quad R = 0.12x + 0.09y$$

subject to the contraints

$$x \geq 0, \quad y \geq 0, \quad x + y \leq 50,000, \quad x \leq 20,000, \quad y \geq 150,000, \quad y \geq x$$

The graph of the constraints (the feasible solutions) is illustrated in Figure 5. Table 3 lists the vertices and the corresponding values of the objective function. The maximum return on investment is $5,100, achieved by placing $20,000 in the AAA bonds and $30,000 in the Certificate of Deposit.

TABLE 3

VERTEX	RETURN ON INVESTMENT
(x, y)	$R = 0.12x + 0.09y$
$(0, 15)$	$R = 0.12(0) + 0.09(15) = 1.35$ thousand dollars
$(15, 15)$	$R = 0.12(15) + 0.09(15) = 3.15$ thousand dollars
$(20, 20)$	$R = 0.12(20) + 0.09(20) = 4.2$ thousand dollars
$(20, 30)$	$R = 0.12(20) + 0.09(30) = 5.1$ thousand dollars
$(0, 50)$	$R = 0.12(0) + 0.09(50) = 4.5$ thousand dollars

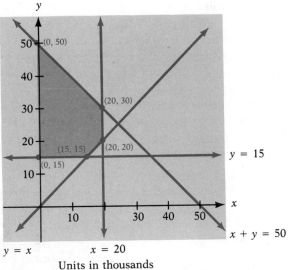

FIGURE 5

EXERCISE 8.2 *The figure illustrates the graph of the feasible solutions of a linear programming problem. In Problems 1–6 find the maximum and minimum value of each of the following objective functions.*

1. $z = x + y$
2. $z = 2x + 3y$
3. $z = x + 10y$
4. $z = 10x + y$
5. $z = 5x + 7y$
6. $z = 7x + 5y$

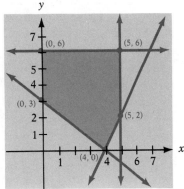

In Problems 7–16 solve each linear programming problem.

7. Maximize $z = 2x + y$
   subject to $x \geq 0, \quad y \geq 0, \quad x + y \leq 6, \quad x + y \geq 1$

8. Maximize $z = x + 3y$
   subject to $x \geq 0, \quad y \geq 0, \quad x + y \geq 3, \quad x \leq 5, \quad y \leq 7$

9. Minimize $z = 2x + 5y$
   subject to $x \geq 0, \quad y \geq 0, \quad x + y \geq 2, \quad x \leq 5, \quad y \leq 3$

10. Minimize $z = 3x + 4y$
    subject to $x \geq 0, \quad y \geq 0, \quad 2x + 3y \geq 6, \quad x + y \leq 8$

11. Maximize $z = 3x + 5y$
    subject to $x \geq 0, \quad y \geq 0, \quad x + y \geq 2, \quad 2x + 3y \leq 12, \quad 3x + 2y \leq 12$

12. Maximize $z = 5x + 3y$
    subject to $x \geq 0$, $y \geq 0$, $x + y \geq 2$, $x + y \leq 8$, $2x + y \leq 10$

13. Minimize $z = 5x + 4y$
    subject to $x \geq 0$, $y \geq 0$, $x + y \geq 2$, $2x + 3y \leq 12$, $3x + y \leq 12$

14. Minimize $z = 2x + 3y$
    subject to $x \geq 0$, $y \geq 0$, $x + y \geq 3$, $x + y \leq 9$, $2x + 3y \leq 6$

15. Maximize $z = 5x + 2y$
    subject to $x \geq 0$, $y \geq 0$, $x + y \leq 10$, $2x + y \geq 10$, $x + 2y \geq 10$

16. Maximize $z = 2x + 4y$
    subject to $x \geq 0$, $y \geq 0$, $2x + y \geq 4$, $x + y \leq 9$

17. Rework Example 3 if the profit on the low-grade mixture is $0.40 per package and the profit on the high-grade mixture is $0.30 per package.

18. Rework Example 4 if the retired couple no longer requires that at most $20,000 be placed in the AAA bond.

19. A manufacturer of skis produces two types: downhill and cross-country. Use the following table to determine how many of each kind of ski should be produced to achieve a maximum profit. What is the maximum profit?

	DOWNHILL	CROSS-COUNTRY	MAXIMUM TIME AVAILABLE
Manufacturing time per ski	2 hours	1 hour	40 hours
Finishing time per ski	1 hour	1 hour	32 hours
Profit per ski	$70	$50	

20. Rework Problem 19 if the manufacturing unit has 48 hours available.

21. A farmer has 70 acres of land available for planting either soybeans or wheat. The cost of preparing the soil, the workdays required, and the expected profit per acre planted for each type of crop are given in the following table:

	SOYBEANS	WHEAT
Preparation cost per acre	$60	$30
Workdays required per acre	3	4
Profit per acre	$180	$100

The farmer cannot spend more than $1800 in preparation costs nor more than a total of 120 workdays. How many acres of each crop should be planted in order to maximize the profit? What is the maximum profit?

22. Rework Problem 21 if the farmer is willing to spend no more than $2400 on preparation.

23. An investment broker is instructed by her client to invest up to $20,000, some in a Certificate of Deposit yielding 9% per annum and some in Treasury bills yielding 7% per annum. The client wants to invest at least $8000 in Treasury bills and no more than $12,000 in the Certificate of Deposit. The client also insists that the amount invested in Treasury bills must equal or exceed the amount placed in the certificate. How

much should the broker recommend the client place in each type of investment if the objective is to maximize return on investment?

24. Rework Problem 23 if the client insists instead that the amount invested in Treasury bills must not exceed the amount placed in certificates.

25. A factory manufactures two kinds of ice skates: racing skates and figure skates. The racing skate requires 6 work-hours in the fabrication department, whereas the figure skates require 4 work-hours there. The racing skates require 1 work-hour in the finishing department, whereas the figure skates require 2 work-hours there. The fabricating department has available at most 120 work-hours per day, and the finishing department has no more than 40 work-hours per day available. If the profit on each racing skate is $10 and the profit on each figure skate is $12, how many of each should be manufactured each day to maximize profit? (Assume all skates made are sold.)

26. A factory manufactures two kinds of ceramic figurines: a dancing girl and a mermaid, each requiring three processes—molding, painting, and glazing. The daily labor available for molding is no more than 90 work-hours; labor available for painting does not exceed 120 work-hours; and labor available for glazing is no more than 60 work-hours. The dancing girl requires 3 work-hours for molding, 6 work-hours for painting, and 2 work-hours for glazing. The mermaid requires 3 work-hours for molding, 4 work-hours for painting, and 3 work-hours for glazing. If the profit on each figurine is $25 for dancing girls and $30 for mermaids, how many of each should be produced each day to maximize profit? If management decides to produce the number of each figurine that maximizes profit, determine which of these processes has excess work-hours assigned to it.

27. An airline has two classes of service: first class and coach. Management's experience has been that each aircraft should have at least 8 but not more than 16 first-class seats and at least 80 but not more than 120 coach seats. Management has further decided the ratio of first class to coach should never exceed $\frac{1}{12}$. With how many of each type of seat should an aircraft be configured to maximize revenue? [*Hint:* Assume the airline charges $C for a coach seat and $C + $F for a first-class seat, $C > 0$, $F > 0$.]

28. Rework Problem 27 if management decides instead that the ratio of first class to coach should never exceed $\frac{1}{8}$.

29. A farm that specializes in raising frying chickens supplements the regular chicken feed with four vitamins. The owner wants the supplemental food to contain at least 50 units of Vitamin I, 90 units of Vitamin II, 60 units of Vitamin III, and 100 units of Vitamin IV per 100 ounces of feed. Two supplements are available: Supplement A, which per ounce contains 5 units of Vitamin I, 25 units of Vitamin II, 10 units of Vitamin III, and 35 units of Vitamin IV; and Supplement B, which per ounce contains 25 units of Vitamin I, 10 units of Vitamin II, 10 units of Vitamin III, and 20 units of Vitamin IV. If Supplement A costs $0.06 per ounce and Supplement B costs $0.08 per ounce, how much of each supplement should the manager of the farm buy to add to each 100 ounces of feed in order to keep the total cost at a minimum, while still meeting the owner's vitamin specifications?

■ 8.3 PARTIAL FRACTION DECOMPOSITION

Consider the problem of adding the two fractions $3/(x + 4)$ and $2/(x - 3)$. The result is

$$\frac{3}{x + 4} + \frac{2}{x - 3} = \frac{3(x - 3) + 2(x + 4)}{(x + 4)(x - 3)} = \frac{5x - 1}{x^2 + x - 12}$$

PARTIAL FRACTION
DECOMPOSITION

PARTIAL FRACTIONS

The reverse procedure, of starting with the rational expression $(5x - 1)/(x^2 + x - 12)$ and writing it as the sum (or difference) of the two simpler fractions $3/(x + 4)$ and $2/(x - 3)$, is referred to as **partial fraction decomposition**, and the two simpler fractions are called **partial fractions**. Decomposing a rational expression into a sum of partial fractions is important in solving certain types of calculus problems. This section presents a systematic way to decompose rational expressions.

We begin by recalling that a rational expression is the ratio of two polynomials, say P and $Q \neq 0$, that have no common factors. Recall also that a rational expression P/Q is called **proper** if the degree of the polynomial in the numerator is less than the degree of the polynomial in the denominator. Otherwise the rational expression is termed **improper**.

Because any improper rational expression can be reduced by division to a mixed form consisting of the sum of a polynomial and a proper rational expression, we shall restrict the discussion that follows to proper rational expressions.

The partial fraction decomposition of the rational expression P/Q depends on the factors of the denominator Q. Recall (from Section 5.4) that any polynomial whose coefficients are real numbers can be factored (over the real numbers) into products of linear and irreducible quadratic factors. Thus, the denominator Q of the rational expression P/Q will contain only factors of one or both of the following types:

1. *Linear factors* of the form $x - a$, where a is a real number.
2. *Irreducible quadratic factors* of the form $ax^2 + bx + c$, where a, b, and c are real numbers and $ax^2 + bx + c$ cannot be written as the product of two linear factors with real coefficients.

As it turns out, there are four cases to be examined. We begin with the case for which Q has only nonrepeated linear factors.

Case 1: Q has only nonrepeated linear factors.

Under the assumption that Q has only nonrepeated linear factors, the polynomial Q has the form

$$Q(x) = (x - a_1)(x - a_2) \cdot \cdots \cdot (x - a_n)$$

where none of the numbers $a_1, a_2, \ldots, a_n$ are equal. In this case, the partial fraction decomposition of P/Q is of the form

$$\frac{P(x)}{Q(x)} = \frac{A_1}{x - a_1} + \frac{A_2}{x - a_2} + \cdots + \frac{A_n}{x - a_n} \qquad (1)$$

where the numbers $A_1, A_2, \ldots, A_n$ are to be found.

We show how to find these numbers in the example that follows.

EXAMPLE 1   Write the partial fraction decomposition of

$$\frac{x}{x^2 - 5x + 6}$$

Solution   First we factor the denominator,

$$x^2 - 5x + 6 = (x - 2)(x - 3)$$

and conclude that the denominator contains only nonrepeated linear factors. Then we decompose according to Equation (1):

$$\frac{x}{x^2 - 5x + 6} = \frac{A}{x - 2} + \frac{B}{x - 3} \qquad (2)$$

where A and B are to be found. Now clear the fractions by multiplying each side by $(x - 2)(x - 3) = x^2 - 5x + 6$. The result is

$$x = A(x - 3) + B(x - 2)$$

or

$$x = (A + B)x + (-3A - 2B)$$

This equation is an identity in x. Thus, we may equate the coefficients of like powers of x to get

$$\begin{cases} 1 = A + B \\ 0 = -3A - 2B \end{cases}$$
Equate coefficients of x: $1x = (A + B)x$
Equate coefficients of x^0, the constants: $0x^0 = (-3A - 2B)x^0$

This system of two equations containing two variables A and B can be solved using whatever method you like. Solving it, we get

$$A = -2 \qquad B = 3$$

Thus, from Equation (2), the partial fraction decomposition is

$$\frac{x}{x^2 - 5x + 6} = \frac{-2}{x - 2} + \frac{3}{x - 3}$$

The decomposition in Example 1 can be checked by adding the fractions:

$$\frac{-2}{x-2} + \frac{3}{x-3} = \frac{-2(x-3) + 3(x-2)}{(x-2)(x-3)} = \frac{x}{(x-2)(x-3)}$$

$$= \frac{x}{x^2 - 5x + 6}$$

Case 2: Q has repeated linear factors.

If the polynomial Q has a repeated factor—say $(x-a)^n$, $n \geq 2$ an integer—then, in the partial fraction decomposition of P/Q, we allow for the terms

$$\frac{A_1}{x-a} + \frac{A_2}{(x-a)^2} + \cdots + \frac{A_n}{(x-a)^n}$$

where the numbers $A_1, A_2, \ldots, A_n$ are to be found.

EXAMPLE 2   Write the partial fraction decomposition of

$$\frac{x+2}{x^3 - 2x^2 + x}$$

Solution   First we factor the denominator,

$$x^3 - 2x^2 + x = x(x^2 - 2x + 1) = x(x-1)^2$$

and find that the denominator has the nonrepeated linear factor x and the twice-repeated linear factor $x-1$. By Case 1, we must allow for the term A/x in the decomposition; and, by Case 2, we must allow for the terms $B/(x-1) + C/(x-1)^2$ in the decomposition. Thus, we write

$$\frac{x+2}{x^3 - 2x^2 + x} = \frac{A}{x} + \frac{B}{x-1} + \frac{C}{(x-1)^2} \qquad (3)$$

Again we clear fractions by multiplying each side by $x^3 - 2x^2 + x = x(x-1)^2$. The result is the identity

$$x + 2 = A(x-1)^2 + Bx(x-1) + Cx$$
$$= (A+B)x^2 + (-2A - B + C)x + A$$

Equating coefficients of like powers of x, we get the system

$$\begin{cases} A + B = 0 & \text{The coefficient of } x^2 \text{ on the left is 0.} \\ -2A - B + C = 1 \\ A = 2 \end{cases}$$

The solution of this system is $A = 2, B = -2, C = 3$. From Equation (3), the partial fraction decomposition is

$$\frac{x + 2}{x^3 - 2x^2 + x} = \frac{2}{x} + \frac{-2}{x - 1} + \frac{3}{(x - 1)^2}$$

The numbers to be found in the partial fraction decomposition can sometimes be found more readily by using suitable choices for x (which may include complex numbers) in the identity obtained after fractions have been cleared. In Example 2, the identity obtained after clearing fractions is

$$x + 2 = A(x - 1)^2 + Bx(x - 1) + Cx \qquad (4)$$

If we let $x = 0$ in this expression, the terms containing B and C drop out, leaving $2 = A(-1)^2$, or $A = 2$. Similarly, if we let $x = 1$, the terms containing A and B drop out, leaving $3 = C$. Thus, Equation (4) becomes

$$x + 2 = 2(x - 1)^2 + Bx(x - 1) + 3x$$

Now, let $x = 2$ (any choice other than 0 or 1 will work as well). The result is

$$4 = 2(1)^2 + B(2)(1) + 3(2)$$
$$2B = 4 - 2 - 6 = -4$$
$$B = -2$$

We use this method in the next example.

EXAMPLE 3   Write the partial fraction decomposition of

$$\frac{x^3 - 8}{x^2(x - 1)^3}$$

Solution   The denominator contains the twice-repeated linear factor x and the linear factor $x - 1$ repeated three times. Thus, the partial fraction decomposition takes the form

$$\frac{x^3 - 8}{x^2(x - 1)^3} = \frac{A}{x} + \frac{B}{x^2} + \frac{C}{x - 1} + \frac{D}{(x - 1)^2} + \frac{E}{(x - 1)^3} \qquad (5)$$

As before, we clear fractions and get the identity

$$x^3 - 8 = Ax(x - 1)^3 + B(x - 1)^3 + Cx^2(x - 1)^2 + Dx^2(x - 1) + Ex^2 \qquad (6)$$

Let $x = 0$. (Do you see why this choice was made?) Then,

$$-8 = B(-1)$$
$$B = 8$$

Now let $x = 1$ in Equation (6). Then,

$$-7 = E$$

Use $B = 8$ and $E = -7$ in Equation (6) and collect like terms:

$$x^3 - 8 = Ax(x - 1)^3 + 8(x - 1)^3$$
$$+ Cx^2(x - 1)^2 + Dx^2(x - 1) - 7x^2$$
$$x^3 - 8 - 8(x^3 - 3x^2 + 3x - 1) + 7x^2 = Ax(x - 1)^3 + Cx^2(x - 1)^2 + Dx^2(x - 1)$$
$$-7x^3 + 31x^2 - 24x = x(x - 1)[A(x - 1)^2 + Cx(x - 1) + Dx]$$
$$x(x - 1)(-7x + 24) = x(x - 1)[A(x - 1)^2 + Cx(x - 1) + Dx]$$
$$-7x + 24 = A(x - 1)^2 + Cx(x - 1) + Dx \qquad (7)$$

We now work with Equation (7). Let $x = 0$. Then,

$$24 = A$$

Now, let $x = 1$ in Equation (7). Then,

$$17 = D$$

Use $A = 24$ and $D = 17$ in Equation (7) and collect like terms:

$$-7x + 24 = 24(x - 1)^2 + Cx(x - 1) + 17x$$
$$-24x^2 + 48x - 24 - 17x - 7x + 24 = Cx(x - 1)$$
$$-24x^2 + 24x = Cx(x - 1)$$
$$-24x(x - 1) = Cx(x - 1)$$
$$-24 = C$$

We know all the numbers A, B, C, D, and E so that, from Equation (5), we get the decomposition

$$\frac{x^3 - 8}{x^2(x - 1)^3} = \frac{24}{x} + \frac{8}{x^2} + \frac{-24}{x - 1} + \frac{17}{(x - 1)^2} + \frac{-7}{(x - 1)^3} \qquad \blacksquare$$

The method employed in Example 3, although somewhat tedious, is preferable to solving the system of five equations containing five unknowns that the expansion of Equation (5) leads to.

The final two cases involve irreducible quadratic factors. A quadratic factor is irreducible if it cannot be factored into linear factors with real coefficients. A quadratic expression $ax^2 + bx + c$ is irreducible whenever $b^2 - 4ac < 0$. For example, $x^2 + x + 1$ and $x^2 + 4$ are irreducible.

Case 3: Q contains a nonrepeated irreducible quadratic factor.

If Q contains a nonrepeated irreducible quadratic factor $ax^2 + bx + c$, then, in the partial fraction decomposition of P/Q, allow for the term

$$\frac{Ax + B}{ax^2 + bx + c}$$

where the numbers A and B are to be found.

EXAMPLE 4 Write the partial fraction decomposition of

$$\frac{3x - 5}{x^3 - 1}$$

Solution We factor the denominator,

$$x^3 - 1 = (x - 1)(x^2 + x + 1)$$

and find that it has a nonrepeated linear factor $x - 1$ and a nonrepeated irreducible quadratic factor $x^2 + x + 1$. Thus, by Case 1, we allow for the term $A/(x - 1)$ and, by Case 3, we allow for the term $(Bx + C)/(x^2 + x + 1)$. Hence, we write

$$\frac{3x - 5}{x^3 - 1} = \frac{A}{x - 1} + \frac{Bx + C}{x^2 + x + 1} \tag{8}$$

We clear fractions by multiplying each side of Equation (8) by $x^3 - 1 = (x - 1)(x^2 + x + 1)$ to get

$$3x - 5 = A(x^2 + x + 1) + (Bx + C)(x - 1) \tag{9}$$

Collecting like terms, we get

$$3x - 5 = (A + B)x^2 + (A - B + C)x + (A - C)$$

Equating coefficients, we get the system

$$\begin{cases} A + B = 0 \\ A - B + C = 3 \\ A - C = -5 \end{cases}$$

The solution is $A = -\frac{2}{3}$, $B = \frac{2}{3}$, $C = \frac{13}{3}$. Thus, from Equation (8), we see that

$$\frac{3x - 5}{x^3 - 1} = \frac{-\frac{2}{3}}{x - 1} + \frac{\frac{2}{3}x + \frac{13}{3}}{x^2 + x + 1} \qquad \blacksquare$$

An alternative to solving the system of three equations containing three variables in Example 4 would be to let $x = 1$ in Equation (9) to get $-\frac{2}{3} = A$. Using this value in Equation (9), collecting terms, and then equating coefficients is a little faster. Try it for yourself.

Case 4: Q contains repeated irreducible quadratic factors.

If the polynomial Q contains a repeated irreducible quadratic factor $(ax^2 + bx + c)^n$, $n \geq 2$, n an integer, then, in the partial fraction decomposition of P/Q, allow for the terms

$$\frac{A_1x + B_1}{ax^2 + bx + c} + \frac{A_2x + B_2}{(ax^2 + bx + c)^2} + \cdots + \frac{A_nx + B_n}{(ax^2 + bx + c)^n}$$

where the numbers $A_1, B_1, A_2, B_2, \ldots, A_n, B_n$ are to be found.

EXAMPLE 5   Write the partial fraction decomposition of

$$\frac{x^3 + x^2}{(x^2 + 4)^2}$$

Solution   The denominator contains the twice-repeated irreducible quadratic factor $(x^2 + 4)^2$, so we write

$$\frac{x^3 + x^2}{(x^2 + 4)^2} = \frac{Ax + B}{x^2 + 4} + \frac{Cx + D}{(x^2 + 4)^2} \qquad (10)$$

We clear fractions to get

$$x^3 + x^2 = (Ax + B)(x^2 + 4) + Cx + D$$

Collecting like terms yields

$$x^3 + x^2 = Ax^3 + Bx^2 + (4A + C)x + D + 4B$$

Equating coefficients, we get the system

$$\begin{cases} A = 1 \\ B = 1 \\ 4A + C = 0 \\ D + 4B = 0 \end{cases}$$

The solution is $A = 1, B = 1, C = -4, D = -4$. Hence, from Equation (10),

$$\frac{x^3 + x^2}{(x^2 + 4)^2} = \frac{x + 1}{x^2 + 4} + \frac{-4x - 4}{(x^2 + 4)^2}$$ ≡

EXERCISE 8.3   *In Problems 1–8 tell whether the given rational expression is proper or improper. If improper, rewrite it as the sum of a polynomial and a proper rational expression.*

1. $\dfrac{x}{x^2 - 1}$

2. $\dfrac{5x + 2}{x^3 - 1}$

3. $\dfrac{x^2 + 5}{x^2 - 4}$

4. $\dfrac{3x^2 - 2}{x^2 - 1}$ **5.** $\dfrac{5x^3 + 2x - 1}{x^2 - 4}$ **6.** $\dfrac{3x^4 + x^2 - 2}{x^3 + 8}$

7. $\dfrac{x(x - 1)}{(x + 4)(x - 3)}$ **8.** $\dfrac{2x(x^2 + 4)}{x^2 + 1}$

In Problems 9–38 write the partial fraction decomposition of each rational expression.

9. $\dfrac{4}{x(x - 1)}$ **10.** $\dfrac{3x}{(x + 2)(x - 1)}$

11. $\dfrac{1}{x(x^2 + 1)}$ **12.** $\dfrac{1}{(x + 1)(x^2 + 4)}$

13. $\dfrac{x}{(x - 1)(x - 2)}$ **14.** $\dfrac{3x}{(x + 2)(x - 4)}$

15. $\dfrac{x^2}{(x - 1)^2(x + 1)}$ **16.** $\dfrac{x + 1}{x^2(x - 2)}$

17. $\dfrac{1}{x^3 - 8}$ **18.** $\dfrac{2x + 4}{x^3 - 1}$

19. $\dfrac{x^2}{(x - 1)^2(x + 1)^2}$ **20.** $\dfrac{x + 1}{x^2(x - 2)^2}$

21. $\dfrac{x - 3}{(x + 2)(x + 1)^2}$ **22.** $\dfrac{x^2 + x}{(x + 2)(x - 1)^2}$

23. $\dfrac{x + 4}{x^2(x^2 + 4)}$ **24.** $\dfrac{10x^2 + 2x}{(x - 1)^2(x^2 + 2)}$

25. $\dfrac{x^2 + 2x + 3}{(x + 1)(x^2 + 2x + 4)}$ **26.** $\dfrac{x^2 - 11x - 18}{x(x^2 + 3x + 3)}$

27. $\dfrac{x}{(3x - 2)(2x + 1)}$ **28.** $\dfrac{1}{(2x + 3)(4x - 1)}$

29. $\dfrac{x}{x^2 + 2x - 3}$ **30.** $\dfrac{x^2 - x - 8}{(x + 1)(x^2 + 5x + 6)}$

31. $\dfrac{x^2 + 2x + 3}{(x^2 + 4)^2}$ **32.** $\dfrac{2x + 1}{(x^2 + 16)^2}$

33. $\dfrac{7x + 3}{x^3 - 2x^2 - 3x}$ **34.** $\dfrac{x^5 + 1}{x^6 - x^4}$

35. $\dfrac{x^2}{x^3 - 4x^2 + 5x - 2}$ **36.** $\dfrac{x^2 + 1}{x^3 + x^2 - 5x + 3}$

37. $\dfrac{x^3}{(x^2 + 16)^3}$ **38.** $\dfrac{x^2}{(x^2 + 4)^3}$

▤ 8.4 VECTORS

In simple terms, a vector (derived from the Latin *vehere*, meaning "to carry") is a quantity that has both magnitude and direction. For a vector in the plane, which is the only type we shall discuss, it is convenient to represent a vector by using an arrow. The arrow's length MAGNITUDE represents the **magnitude** of the vector, and the arrow's head indicates DIRECTION the **direction** of the vector.

Many quantities in physics are vectors. For example, the velocity of an aircraft can be represented by an arrow that points in the direction of movement; the length of the arrow represents speed. Thus, if the aircraft speeds up, we lengthen the arrow; if the aircraft changes direction, we point the arrow in the new direction. See Figure 6. Based on this representation, it is not surprising that vectors and directed line segments are somehow related.

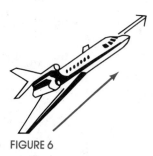

FIGURE 6

Directed Line Segments

If P and Q are two distinct points in the xy-plane, there is exactly one line containing both P and Q. The points on that part of the line that LINE SEGMENT joins P to Q, including P and Q, form what is called the **line segment** $\overline{PQ}$. If we order the points so that they proceed from P to Q, we have DIRECTED LINE SEGMENT a **directed line segment** from P to Q, which we denote by $\overrightarrow{PQ}$. In a INITIAL POINT directed line segment $\overrightarrow{PQ}$, we call P the **initial point** and Q the **ter-** TERMINAL POINT **minal point**. See Figure 7.

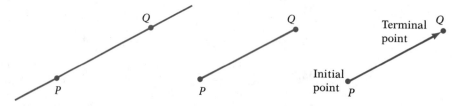

(a) Line containing P and Q (b) Line segment $\overline{PQ}$ (c) Directed line segment $\overrightarrow{PQ}$

FIGURE 7

If P and Q are the same point, then the directed line segment $\overrightarrow{PQ}$, DEGENERATE DIRECTED LINE which consists of a single point, is said to be **degenerate**. SEGMENT Directed line segments that are not degenerate have both magnitude and direction. The magnitude of the directed line segment $\overrightarrow{PQ}$ is the distance from the point P to the point Q. The direction of $\overrightarrow{PQ}$

is from P to Q. The directed line segment $\overrightarrow{QP}$, on the other hand, is equal in magnitude to $\overrightarrow{PQ}$ but has the opposite direction, because it is directed from Q to P. See Figure 8.

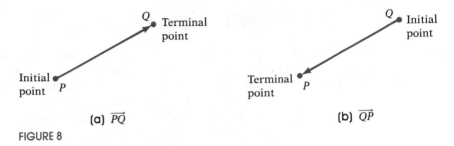

(a) $\overrightarrow{PQ}$ (b) $\overrightarrow{QP}$

FIGURE 8

Equivalent directed line segments

Two nondegenerate directed line segments are called **equivalent** if they have the same magnitude and the same direction. In Figure 9 the directed line segments $\overrightarrow{PQ}$, $\overrightarrow{RS}$, and $\overrightarrow{TU}$ are all equivalent. For completeness, we agree that all degenerate directed line segments are equivalent. A vector may be defined in terms of equivalent directed line segments.

Vector   A **vector** is defined as a collection of equivalent directed line segments.

What this definition implies is that we may represent a vector by any one of the equivalent directed line segments in the collection. Thus, if $\mathbf{v}$* is the vector defined by the collection of equivalent directed line segments $\overrightarrow{PQ}$, $\overrightarrow{RS}$, and $\overrightarrow{TU}$, then $\mathbf{v}$ can be represented by $\overrightarrow{PQ}$ or by $\overrightarrow{RS}$ or by $\overrightarrow{TU}$. It is precisely this flexibility of representation that makes vectors so useful in many applications. Thus, in our illustration (Figure 9), we may select any of the directed line segments from the collection defined by the vector and use it to represent the vector.

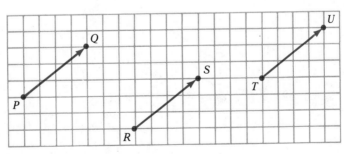

FIGURE 9

*Boldface letters will be used to denote vectors, in order to distinguish them from numbers. For handwritten work, an arrow is placed over the letter to signify a vector.

ZERO VECTOR 0
The vector defined by the collection of all degenerate directed line segments is called the **zero vector 0**.

Adding Vectors

EQUAL VECTORS
Two vectors **v** and **w** are said to be **equal**, written as $\mathbf{v} = \mathbf{w}$, if they have the same magnitude and the same direction.

SUM v + w OF VECTORS
The **sum v + w** of two vectors is defined as follows: We position the vectors **v** and **w** so that the terminal point of **v** coincides with the initial point of **w**. See Figure 10. The vector **v** + **w** is then the unique vector whose initial point coincides with the initial point of **v** and whose terminal point coincides with the terminal point of **w**.

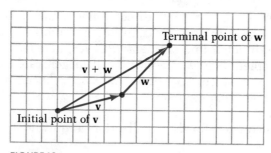

FIGURE 10

PARALLELOGRAM LAW OF
ADDITION
Based on the representation in Figure 11, we refer to this way of adding vectors as the **parallelogram law of addition**. Because opposite sides of a parallelogram are equal and parallel, Figure 11 also shows that vector addition is **commutative**. That is,

COMMUTATIVE LAW
$$\mathbf{v} + \mathbf{w} = \mathbf{w} + \mathbf{v}$$

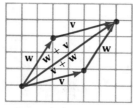

FIGURE 11

Vector addition is also **associative**. That is, if **u**, **v**, and **w** are vectors, then

ASSOCIATIVE LAW
$$\mathbf{u} + (\mathbf{v} + \mathbf{w}) = (\mathbf{u} + \mathbf{v}) + \mathbf{w}$$

Figure 12 illustrates the associative law for vectors.

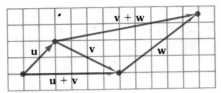

FIGURE 12
(u + v) + w = u + (v + w)

If **v** is a vector, then −**v** is defined as the vector having the same magnitude as **v** but whose direction is opposite to **v**. See Figure 13.

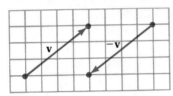

FIGURE 13

DIFFERENCE v − w
OF VECTORS

We define the **difference v − w** of the two vectors **v** and **w** as

$$v - w = v + (-w)$$

As the definition indicates, to subtract **w** from **v**, that is, to find **v − w**, we find the vector that, when added to **w**, results in **v**. See Figure 14.

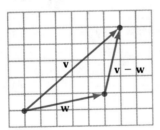

FIGURE 14

The zero vector **0** has the property that

$$v + 0 = v$$

IDENTITY

for any vector **v**. That is, the zero vector **0** is the **identity** under addition. Further, for any vector **v**,

$$v + (-v) = 0$$

ADDITIVE INVERSE

That is, −**v** is the **additive inverse** of **v**.

Multiplying Vectors by Numbers

SCALARS   In dealing with vectors, we refer to real numbers as **scalars**. Scalars, therefore, are quantities having only magnitude. Examples from physics of scalar quantities are mass, temperature, speed, and time. It is possible to multiply a scalar by a vector.

SCALAR PRODUCT   If α is a scalar and if **v** is a vector, the **scalar product** α**v** is defined as:

1. The zero vector **Ø** if $\alpha = 0$.
2. The vector whose magnitude is $|\alpha|$ times the magnitude of **v** and whose direction is the same as **v** if $\alpha > 0$.
3. The vector whose magnitude is $|\alpha|$ times the magnitude of **v** and whose direction is opposite that of **v** if $\alpha < 0$.

See Figure 15 for some illustrations.

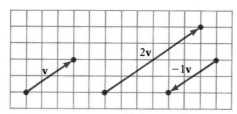

FIGURE 15

Scalar products have the following properties:

PROPERTIES OF SCALAR
PRODUCTS

$$0\mathbf{v} = \mathbf{Ø} \qquad 1\mathbf{v} = \mathbf{v} \qquad -1\mathbf{v} = -\mathbf{v}$$
$$(\alpha + \beta)\mathbf{v} = \alpha\mathbf{v} + \beta\mathbf{v} \qquad \alpha(\mathbf{v} + \mathbf{w}) = \alpha\mathbf{v} + \alpha\mathbf{w}$$
$$\alpha(\beta\mathbf{v}) = (\alpha\beta)\mathbf{v}$$

EXAMPLE 1   Use the vectors illustrated in Figure 16 to graph each expression.

(a) **v** − **w**   (b) **2v** + **3w**   (c) **2v** − **w** + **3u**

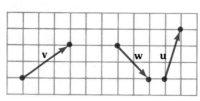

FIGURE 16

Solution   Figure 17 illustrates the solutions.

(a) See Figure 17(a).   (b) See Figure 17(b).

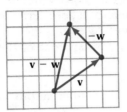

FIGURE 17(a)

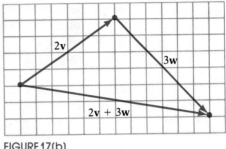

FIGURE 17(b)

(c) See Figure 17(c).

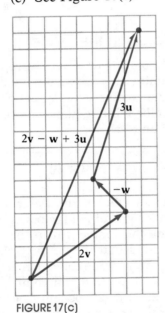

FIGURE 17(c)

Magnitudes of Vectors

MAGNITUDE ‖v‖ OF A VECTOR   If **v** is a vector, we use the symbol $\|\mathbf{v}\|$ to represent the **magnitude** of **v**. Since $\|\mathbf{v}\|$ equals the length of a directed line segment from the collection that represents **v**, it follows that $\|\mathbf{v}\|$ has the following properties:

THEOREM   If **v** is a vector and if α is a scalar, then

PROPERTIES OF ‖v‖

(a) $\|\mathbf{v}\| \geq 0$

(b) $\|\mathbf{v}\| = 0$ if and only if $\mathbf{v} = \mathbf{0}$

(c) $\|-\mathbf{v}\| = \|\mathbf{v}\|$

(d) $\|\alpha\mathbf{v}\| = |\alpha|\|\mathbf{v}\|$

Property (a) is a consequence of the fact that distance is a non-negative number. Property (b) follows because the length of the directed length segment $\overrightarrow{PQ}$ is positive unless P and Q are the same point in which case the length is zero. Property (c) follows because the length of the line segment $\overline{PQ}$ equals the length of the line segment $\overline{QP}$. Property (d) is a direct consequence of the definition of a scalar product.

UNIT VECTOR   A vector $\mathbf{v}$ for which $\|\mathbf{v}\| = 1$ is called a **unit vector**.

To compute the magnitude of a vector, we need to have an algebraic way of representing vectors.

Representing Vectors in the Plane

We use a rectangular coordinate system to represent vectors in the plane. Let $\mathbf{i}$ denote a unit vector whose direction is along the positive x-axis; let $\mathbf{j}$ denote a unit vector whose direction is along the positive y-axis. Then every vector $\mathbf{v}$ in the plane can be written uniquely in terms of the vectors $\mathbf{i}$ and $\mathbf{j}$ as

$$\mathbf{v} = a\mathbf{i} + b\mathbf{j}$$

COMPONENTS OF A VECTOR   for some choice of scalars a and b. The scalars a and b are called the **components** of the vector $\mathbf{v} = a\mathbf{i} + b\mathbf{j}$, a being the component in the direction $\mathbf{i}$ and b being the component in the direction $\mathbf{j}$. This fact may be justified by noting that the vector $\mathbf{v}$ can be represented by the directed line segment from the collection defining $\mathbf{v}$ whose initial point is at the origin O. If the terminal point of this line segment is the point $P = (a, b)$, then the component of $\mathbf{v}$ in the direction $\mathbf{i}$ is a and the component of the vector $\mathbf{v}$ in the direction $\mathbf{j}$ is b so that $\mathbf{v} = a\mathbf{i} + b\mathbf{j}$. See Figure 18 for an illustration.

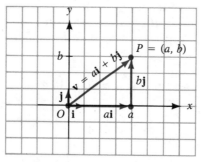

FIGURE 18

It is not necessary to represent **v** by the directed line segment $\overrightarrow{OP}$ from the collection defining **v**. Any line segment will do. Suppose the directed line segment $\overrightarrow{P_1P_2}$ is in the collection that defines **v** and suppose the coordinates P_1 and P_2 are

$$P_1 = (x_1, y_1) \qquad P_2 = (x_2, y_2)$$

Then, as Figure 19 illustrates, triangle OPA and triangle P_1P_2Q are congruent. Do you see why? The line segments have the same magnitude, so $d(O, P) = d(P_1, P_2)$; and they have the same direction, so $\angle POA = \angle P_2P_1Q$. Since the triangles are right triangles, we have Angle–Side–Angle. Thus, it follows that corresponding sides are equal. As a result, $x_2 - x_1 = a$ and $y_2 - y_1 = b$, so that **v** may be written as

$$\mathbf{v} = (x_2 - x_1)\mathbf{i} + (y_2 - y_1)\mathbf{j} \quad \text{or as} \quad \mathbf{v} = a\mathbf{i} + b\mathbf{j}$$

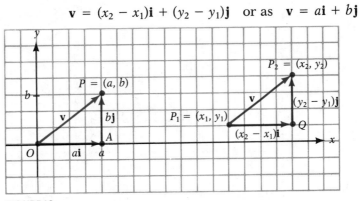

FIGURE 19
$\mathbf{v} = (x_2 - x_1)\mathbf{i} + (y_2 - y_1)\mathbf{j} = a\mathbf{i} + b\mathbf{j}$

EXAMPLE 2 Consider the directed line segment with initial point $(-1, 2)$ and terminal point $(4, 4)$. An equivalent directed line segment is one with initial point $(3, 0)$ and terminal point $(8, 2)$. Each of these is represented by the vector

$$\mathbf{v} = 5\mathbf{i} + 2\mathbf{j}$$

See Figure 20.

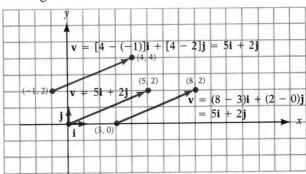

FIGURE 20

This discussion leads us to the following result:

THEOREM
EQUALITY OF VECTORS

Two vectors $\mathbf{v}$ and $\mathbf{w}$ are equal if and only if their corresponding components are equal. That is:

> If $\mathbf{v} = a_1\mathbf{i} + b_1\mathbf{j}$ and $\mathbf{w} = a_2\mathbf{i} + b_2\mathbf{j}$
> then $\mathbf{v} = \mathbf{w}$ if and only if $a_1 = a_2$ and $b_1 = b_2$

As Figure 21 illustrates, to add two vectors, merely add corresponding components. Thus, if $\mathbf{v} = a_1\mathbf{i} + b_1\mathbf{j}$ and $\mathbf{w} = a_2\mathbf{i} + b_2\mathbf{j}$, then

$$\mathbf{v} + \mathbf{w} = (a_1 + a_2)\mathbf{i} + (b_1 + b_2)\mathbf{j}$$

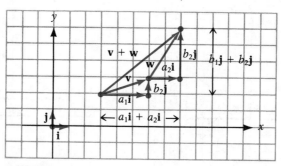

FIGURE 21

Similarly, to subtract two vectors, merely subtract corresponding components:

$$\mathbf{v} - \mathbf{w} = (a_1 - a_2)\mathbf{i} + (b_1 - b_2)\mathbf{j}$$

EXAMPLE 3   If $\mathbf{v} = 2\mathbf{i} + 3\mathbf{j}$ and $\mathbf{w} = 3\mathbf{i} - 4\mathbf{j}$, find:

(a) $\mathbf{v} + \mathbf{w}$ (b) $\mathbf{v} - \mathbf{w}$

Solution   (a) $\mathbf{v} + \mathbf{w} = (2\mathbf{i} + 3\mathbf{j}) + (3\mathbf{i} - 4\mathbf{j}) = (2 + 3)\mathbf{i} + (3 - 4)\mathbf{j}$
$$= 5\mathbf{i} - \mathbf{j}$$
(b) $\mathbf{v} - \mathbf{w} = (2\mathbf{i} + 3\mathbf{j}) - (3\mathbf{i} - 4\mathbf{j}) = (2 - 3)\mathbf{i} + [3 - (-4)]\mathbf{j}$
$$= -\mathbf{i} + 7\mathbf{j}$$

If $\mathbf{v} = a\mathbf{i} + b\mathbf{j}$ is a vector, its magnitude $\|\mathbf{v}\|$ is given by the formula

$$\|\mathbf{v}\| = \sqrt{a^2 + b^2}$$

To see why, look at Figure 22 and apply the Pythagorean Theorem.

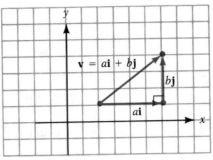

FIGURE 22

If $\mathbf{v} = a\mathbf{i} + b\mathbf{j}$ is a vector and if α is a scalar, then

$$\alpha\mathbf{v} = (\alpha a)\mathbf{i} + (\alpha b)\mathbf{j}$$

To see why for the case $\alpha > 0$, look at Figure 23. The triangles are similar.

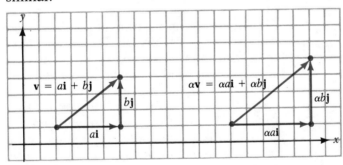

FIGURE 23

EXAMPLE 4 If $\mathbf{v} = 2\mathbf{i} + 3\mathbf{j}$ and $\mathbf{w} = 3\mathbf{i} - 4\mathbf{j}$, find:

(a) $3\mathbf{v}$ (b) $2\mathbf{v} - 3\mathbf{w}$ (c) $\|\mathbf{v}\|$

Solution (a) $3\mathbf{v} = 3(2\mathbf{i} + 3\mathbf{j}) = 6\mathbf{i} + 9\mathbf{j}$

(b) $2\mathbf{v} - 3\mathbf{w} = 2(2\mathbf{i} + 3\mathbf{j}) - 3(3\mathbf{i} - 4\mathbf{j}) = 4\mathbf{i} + 6\mathbf{j} - 9\mathbf{i} + 12\mathbf{j}$
$$= -5\mathbf{i} + 18\mathbf{j}$$

(c) $\|\mathbf{v}\| = \|2\mathbf{i} + 3\mathbf{j}\| = \sqrt{2^2 + 3^2} = \sqrt{13}$

≡

Sometimes it is necessary to find a unit vector having a given direction. If $\mathbf{v}$ is a vector, a unit vector having the same direction as $\mathbf{v}$ is the vector $\mathbf{v}/\|\mathbf{v}\| = (1/\|\mathbf{v}\|)\mathbf{v}$. The reason is

$$\left\|\frac{\mathbf{v}}{\|\mathbf{v}\|}\right\| = \left\|\frac{1}{\|\mathbf{v}\|}\mathbf{v}\right\| = \underset{\underset{\text{Property (d)}}{\uparrow}}{\frac{1}{\|\mathbf{v}\|}\|\mathbf{v}\|} = 1$$

EXAMPLE 5   Find a unit vector in the same direction as $\mathbf{v} = 4\mathbf{i} - 3\mathbf{j}$.

Solution   We find $\|\mathbf{v}\|$ first:

$$\|\mathbf{v}\| = \|4\mathbf{i} - 3\mathbf{j}\| = \sqrt{16 + 9} = 5$$

Now we multiply $\mathbf{v}$ by the scalar $1/\|\mathbf{v}\| = \frac{1}{5}$. The result is

$$\frac{\mathbf{v}}{\|\mathbf{v}\|} = \frac{4\mathbf{i} - 3\mathbf{j}}{5} = \frac{4}{5}\mathbf{i} - \frac{3}{5}\mathbf{j}$$

This vector is, in fact, a unit vector because $(\frac{4}{5})^2 + (-\frac{3}{5})^2 = \frac{16}{25} + \frac{9}{25} = \frac{25}{25} = 1$.   ≡

Forces provide an example of physical quantities that are vectors; two forces "combine" the way vectors "add." How do we know this? Well, physicists tell us they do, and laboratory experiments bear it out. Thus, if $\mathbf{F}_1$ and $\mathbf{F}_2$ are two forces that act on an object, the vector sum $\mathbf{F}_1 + \mathbf{F}_2$ is equal to the force that produces the same effect on the object as that obtained when the forces $\mathbf{F}_1$ and $\mathbf{F}_2$ act on the RESULTANT object. The force $\mathbf{F}_1 + \mathbf{F}_2$ is sometimes called the **resultant** of $\mathbf{F}_1$ and $\mathbf{F}_2$.

An important application of the parallelogram law of addition of vectors occurs with aircraft flying in the presence of a wind and with boats cruising across a river with a current. For example, consider the velocity of wind acting on the velocity of an airplane (see Figure 24). Suppose $\mathbf{w}$ is a vector describing the velocity of the wind; that is, $\mathbf{w}$ represents the direction and speed of the wind. If $\mathbf{v}$ is the velocity of the airplane in the absence of wind (called its velocity relative to the air), then $\mathbf{v} + \mathbf{w}$ is the vector equal to the actual velocity of the airplane (called its velocity relative to the earth).

(a) Velocity $\mathbf{w}$ of wind relative to earth

(b) Velocity $\mathbf{v}$ of airplane relative to air

(c) Resultant $\mathbf{w} + \mathbf{v}$ equals velocity of airplane relative to earth

FIGURE 24

Our next example illustrates this use of vectors in navigation.

EXAMPLE 6 A Boeing 737 aircraft maintains a constant airspeed of 500 miles per hour in the direction due south. The velocity of the jet stream is 80 miles per hour in a northeasterly direction.

(a) Find a unit vector having northeast as direction.

(b) Find a vector 80 units in magnitude having the same direction as the unit vector found in part (a).

C (c) Find the actual speed of the aircraft relative to the ground.

Solution We set up a coordinate system in which north (N) is along the positive y-axis. See Figure 25. The scale used on each axis is 1 unit = 100 miles per hour. Let

$$\mathbf{v}_a = \text{Velocity of aircraft in the air} = -500\mathbf{j}$$
$$\mathbf{v}_g = \text{Velocity of aircraft relative to ground}$$
$$\mathbf{v}_w = \text{Velocity of jet stream}$$

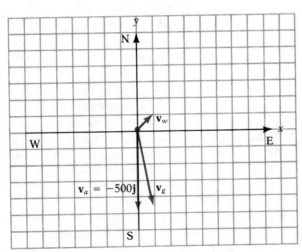

FIGURE 25

(a) A vector having northeast as direction is $\mathbf{i} + \mathbf{j}$. The unit vector in this direction is

$$\frac{\mathbf{i} + \mathbf{j}}{\|\mathbf{i} + \mathbf{j}\|} = \frac{\mathbf{i} + \mathbf{j}}{\sqrt{1 + 1}} = \frac{1}{\sqrt{2}}(\mathbf{i} + \mathbf{j})$$

(b) The velocity $\mathbf{v}_w$ of the jet stream is a vector with magnitude 80 in the direction of $(1/\sqrt{2})(\mathbf{i} + \mathbf{j})$. Thus,

$$\mathbf{v}_w = \frac{80}{\sqrt{2}}(\mathbf{i} + \mathbf{j}) = 40\sqrt{2}(\mathbf{i} + \mathbf{j})$$

(c) The velocity $\mathbf{v}_g$ of the aircraft relative to the ground is the resultant of the vectors $\mathbf{v}_a$ and $\mathbf{v}_w$. Thus,

$$\mathbf{v}_g = \mathbf{v}_a + \mathbf{v}_w$$
$$= -500\mathbf{j} + 40\sqrt{2}(\mathbf{i} + \mathbf{j})$$
$$= 40\sqrt{2}\mathbf{i} + (40\sqrt{2} - 500)\mathbf{j}$$

The actual speed of the aircraft is

$$\|\mathbf{v}_g\| = \sqrt{(40\sqrt{2})^2 + (40\sqrt{2} - 500)^2}$$
$$\approx 447 \text{ miles per hour}$$

≡

≡ **HISTORICAL**
   COMMENT
The history of vectors is surprisingly complicated for such a natural concept. In the xy-plane, complex numbers do a good job of imitating vectors. About 1840, mathematicians became interested in finding a system that would do for three dimensions what the complex numbers do for two dimensions. Hermann Grassmann (1809–1877) in Germany and William Rowan Hamilton (1805–1865) in Ireland both came up with solutions.

Hamilton's system was the *quaternions*, which are best thought of as a real number plus a vector. In this system the order of multiplication matters; that is, $\mathbf{ab} \neq \mathbf{ba}$. Hamilton spent the rest of his life working out quaternion theory and trying to get it accepted in applied mathematics, but he encountered fierce resistance due to the complicated nature of quaternion multiplication. In the work with quaternions, two products of vectors emerged, the scalar (or dot) and the vector (or cross) products.

Grassmann fared even worse than Hamilton; if people did not like Hamilton's work, at least they understood it. Grassmann's abstract style, though easily read today, was almost impenetrable during the previous century, and only a few of his ideas were appreciated. Among those few were the same scalar and vector products that Hamilton had found.

About 1880, the American physicist Josiah Willard Gibbs (1839–1903) worked out an algebra involving only the simplest concepts—the vectors and the two products. He then added some calculus and the resulting system was simple, flexible, and well adapted to expressing a large number of physical laws. This system remains in use essentially unchanged. Hamilton's and Grassmann's more extensive systems each gave birth to much interesting mathematics, but little of this mathematics is seen at elementary levels.   ≡

EXERCISE 8.4

In Problems 1–8 use the vectors in the figure in the margin to graph each expression.

1. $\mathbf{v} + \mathbf{w}$ 2. $\mathbf{u} + \mathbf{v}$ 3. $3\mathbf{v}$ 4. $4\mathbf{w}$

5. $\mathbf{v} - \mathbf{w}$ 6. $\mathbf{u} - \mathbf{v}$ 7. $3\mathbf{v} + \mathbf{u} - 2\mathbf{w}$ 8. $2\mathbf{u} - 3\mathbf{v} + \mathbf{w}$

In Problems 9–16 use the figure below to find each vector.

9. $\mathbf{x}$, if $\mathbf{x} + \mathbf{B} = \mathbf{F}$

10. $\mathbf{x}$, if $\mathbf{x} + \mathbf{D} = \mathbf{E}$

11. $\mathbf{C}$ in terms of $\mathbf{E}$, $\mathbf{D}$, and $\mathbf{F}$

12. $\mathbf{G}$ in terms of $\mathbf{C}$, $\mathbf{D}$, $\mathbf{E}$, and $\mathbf{K}$

13. $\mathbf{E}$ in terms of $\mathbf{G}$, $\mathbf{H}$, and $\mathbf{D}$

14. $\mathbf{E}$ in terms of $\mathbf{A}$, $\mathbf{B}$, $\mathbf{C}$, and $\mathbf{D}$

15. $\mathbf{x}$, if $\mathbf{x} = \mathbf{A} + \mathbf{B} + \mathbf{K} + \mathbf{G}$

16. $\mathbf{x}$, if $\mathbf{x} = \mathbf{A} + \mathbf{B} + \mathbf{C} + \mathbf{H} + \mathbf{G}$

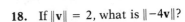

17. If $\|\mathbf{v}\| = 4$, what is $\|3\mathbf{v}\|$?

18. If $\|\mathbf{v}\| = 2$, what is $\|-4\mathbf{v}\|$?

In Problems 19–26 the vector $\mathbf{v}$ is represented by the directed line segment $\overrightarrow{PQ}$. Write $\mathbf{v}$ in the form $a\mathbf{i} + b\mathbf{j}$.

19. $P = (0, 0)$, $Q = (3, 4)$ 20. $P = (0, 0)$, $Q = (-3, -5)$

21. $P = (3, 2)$, $Q = (5, 6)$ 22. $P = (-3, 2)$, $Q = (6, 5)$

23. $P = (-2, -1)$, $Q = (6, -2)$ 24. $P = (-1, 4)$, $Q = (6, 2)$

25. $P = (1, 0)$, $Q = (0, 1)$ 26. $P = (1, 1)$, $Q = (2, 2)$

In Problems 27–32 find $\|\mathbf{v}\|$.

27. $\mathbf{v} = 3\mathbf{i} - 4\mathbf{j}$ 28. $\mathbf{v} = -5\mathbf{i} + 12\mathbf{j}$

29. $\mathbf{v} = \mathbf{i} - \mathbf{j}$ 30. $\mathbf{v} = -\mathbf{i} - \mathbf{j}$

31. $\mathbf{v} = -2\mathbf{i} + 3\mathbf{j}$ 32. $\mathbf{v} = 6\mathbf{i} + 2\mathbf{j}$

In Problems 33–38 find each quantity if $\mathbf{v} = 3\mathbf{i} - 5\mathbf{j}$ and $\mathbf{w} = -2\mathbf{i} + 3\mathbf{j}$.

33. $2\mathbf{v} + 3\mathbf{w}$ 34. $3\mathbf{v} - 2\mathbf{w}$

35. $\|\mathbf{v} - \mathbf{w}\|$ 36. $\|\mathbf{v} + \mathbf{w}\|$

37. $\|\mathbf{v}\| - \|\mathbf{w}\|$ 38. $\|\mathbf{v}\| + \|\mathbf{w}\|$

In Problems 39–44 find the unit vector having the same direction as $\mathbf{v}$.

39. $\mathbf{v} = 5\mathbf{i}$ 40. $\mathbf{v} = -3\mathbf{j}$

41. $\mathbf{v} = 3\mathbf{i} - 4\mathbf{j}$ 42. $\mathbf{v} = -5\mathbf{i} + 12\mathbf{j}$

43. $\mathbf{v} = \mathbf{i} - \mathbf{j}$ 44. $\mathbf{v} = 2\mathbf{i} - \mathbf{j}$

45. Find a vector $\mathbf{v}$ whose magnitude is 4 and whose component in the $\mathbf{i}$ direction is twice the component in the $\mathbf{j}$ direction.

46. Find a vector $\mathbf{v}$ whose magnitude is 3 and whose component in the $\mathbf{i}$ direction is equal to its component in the $\mathbf{j}$ direction.

47. If $\mathbf{v} = 2\mathbf{i} - \mathbf{j}$ and $\mathbf{w} = x\mathbf{i} + 3\mathbf{j}$, find all numbers x for which $\|\mathbf{v} + \mathbf{w}\| = 5$.

48. Find all numbers x such that the vector represented by $\overrightarrow{PQ}$ has length 5, if $P = (-3, 1)$ and $Q = (x, 4)$.

C **49.** An airplane has an airspeed of 500 kilometers per hour in an easterly direction. If the wind velocity is 60 kilometers per hour in a northwesterly direction, find the speed of the airplane relative to the ground.

C **50.** An airplane, after 1 hour in the air, arrives at a point 200 miles due south of its departure point. If there was a steady wind of 30 miles per hour from the northwest during the entire flight, what was the airplane's average airspeed?

C **51.** An airplane travels in a northwesterly direction at a constant ground speed of 250 miles per hour, due to an easterly wind of 50 miles per hour. How fast would the plane have gone if there had been no wind?

52. While a woman is jogging east at an average speed of 8 miles per hour, the wind appears to be coming from the north. Upon doubling her speed, the wind appears to be coming from the northeast. What is the velocity of the wind?

53. Show on the graph below the force needed to prevent an object at P from moving.

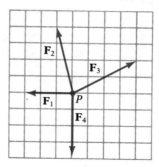

C·H·A·P·T·E·R   R·E·V·I·E·W

VOCABULARY

linear algebra	sum of two matrices	column vector
matrix algebra	difference of two	product
matrix	matrices	associative property
row index	commutative property	distributive property
column index	associative property	diagonal entries
entries	zero matrix	identity matrix I_n
m by n matrix	scalar multiple	inverse matrix
square matrix	scalar	nonsingular matrix
equal matrices	row vector	linear program

objective function
constraints
linear programming
  problem
feasible solutions
solution
partial fraction
  decomposition
partial fractions
magnitude
direction
line segment
directed line segment

initial point
terminal point
degenerate directed
  line segment
equivalent directed
  line segment
vector
zero vector $\mathbf{0}$
equal vectors
sum of vectors
parallelogram law of
  addition
commutative law

associative law
difference of vectors
identity
additive inverse
scalars
scalar product
magnitude $\|\mathbf{v}\|$ of a
  vector
properties of $\|\mathbf{v}\|$
unit vector
components of a
  vector
resultant

FILL-IN-THE-BLANK QUESTIONS

1. A matrix B for which $AB = I_n$, the identity matrix, is called the _____ of A.

2. *True or false* The statement that matrix multiplication is commutative is _____ .

3. In the algebra of matrices, the matrix that has properties similar to the number 1 is called the _____ matrix.

4. A linear programming problem requires that a linear expression, called the _____ _____ , be maximized or minimized.

5. Each point that obeys the constraints of a linear programming problem is called a(n) _____ _____ .

6. A rational function is called _____ if the degree of its numerator is less than the degree of its denominator.

7. A vector whose magnitude is 1 is called a(n) _____ vector.

REVIEW EXERCISES

In Problems 1–8 use the matrices

$$A = \begin{bmatrix} 1 & 0 \\ 2 & 4 \\ -1 & 2 \end{bmatrix} \quad B = \begin{bmatrix} 4 & -3 & 0 \\ 1 & 1 & -2 \end{bmatrix} \quad C = \begin{bmatrix} 3 & -4 \\ 1 & 5 \\ 5 & -2 \end{bmatrix}$$

to compute each expression.

1. $A + C$
2. $A - C$
3. $6A$
4. $-4B$
5. AB
6. BA
7. CB
8. BC

In Problems 9–14 find the inverse of each matrix, if there is one. If there is not an inverse, say the matrix is singular.

9. $\begin{bmatrix} 4 & 6 \\ 1 & 3 \end{bmatrix}$

10. $\begin{bmatrix} -3 & 2 \\ 1 & -2 \end{bmatrix}$

11. $\begin{bmatrix} 1 & 3 & 3 \\ 1 & 2 & 1 \\ 1 & -1 & 2 \end{bmatrix}$

12. $\begin{bmatrix} 3 & 1 & 2 \\ 3 & 2 & -1 \\ 1 & 1 & 1 \end{bmatrix}$ **13.** $\begin{bmatrix} 4 & -8 \\ -1 & 2 \end{bmatrix}$ **14.** $\begin{bmatrix} -3 & 1 \\ -6 & 2 \end{bmatrix}$

In Problems 15–20 solve each linear programming problem.

15. Maximize $z = 3x + 4y$
subject to $x \geq 0, \ y \geq 0, \ 3x + 2y \geq 6, \ x + y \leq 8$

16. Maximize $z = 2x + 4y$
subject to $x \geq 0, \ y \geq 0, \ x + y \leq 6, \ x \geq 2$

17. Minimize $z = 3x + 5y$
subject to $x \geq 0, \ y \geq 0, \ x + y \geq 1, \ 3x + 2y \leq 12, \ x + 3y \leq 12$

18. Minimize $z = 3x + y$
subject to $x \geq 0, \ y \geq 0, \ x \leq 8, \ y \leq 6, \ 2x + y \geq 4$

19. Maximize $z = 5x + 4y$
subject to $x \geq 0, \ y \geq 0, \ x + 2y \geq 2, \ 3x + 4y \leq 12, \ y \geq x$

20. Maximize $z = 4x + 5y$
subject to $x \geq 0, \ y \geq 0, \ 2x + 3y \geq 6, \ x \geq y, \ 2x + y \leq 12$

In Problems 21–30 write the partial fraction decomposition of each rational expression.

21. $\dfrac{6}{x(x - 4)}$

22. $\dfrac{x}{(x + 2)(x - 3)}$

23. $\dfrac{x - 4}{x^2(x - 1)}$

24. $\dfrac{2x - 6}{(x - 2)^2(x - 1)}$

25. $\dfrac{x}{(x^2 + 9)(x + 1)}$

26. $\dfrac{3x}{(x - 2)(x^2 + 1)}$

27. $\dfrac{x^3}{(x^2 + 4)^2}$

28. $\dfrac{x}{(x^2 + 16)^2}$

29. $\dfrac{x^2}{(x^2 + 1)(x^2 - 1)}$

30. $\dfrac{4}{(x^2 + 4)(x^2 - 1)}$

*In Problems 31–34 the vector **v** is represented by the directed line segment $\overrightarrow{PQ}$. Write **v** in the form $a\mathbf{i} + b\mathbf{j}$ and find $\|\mathbf{v}\|$.*

31. $P = (1, -2), \quad Q = (3, -6)$ **32.** $P = (-3, 1), \quad Q = (4, -2)$
33. $P = (0, -2), \quad Q = (-1, 1)$ **34.** $P = (3, -4), \quad Q = (-2, 0)$

In Problems 35–42 use the vectors $\mathbf{v} = -2\mathbf{i} + \mathbf{j}$ and $\mathbf{w} = 4\mathbf{i} - 3\mathbf{j}$.

35. Find $4\mathbf{v} - 3\mathbf{w}$.   **36.** Find $-\mathbf{v} + 2\mathbf{w}$.

37. Find $\|\mathbf{v}\|$.   **38.** Find $\|\mathbf{v} + \mathbf{w}\|$.

39. Find $\|\mathbf{v}\| + \|\mathbf{w}\|$.   **40.** Find $\|2\mathbf{v}\| - 3\|\mathbf{w}\|$.

41. Find a unit vector having the same direction as **v**.

42. Find a unit vector having the opposite direction of **w**.

C **43.** A swimmer can maintain a constant speed of 5 miles per hour. If the swimmer heads directly across a river that has a current moving at the rate of 2 miles per hour, what is the actual speed of the swimmer?

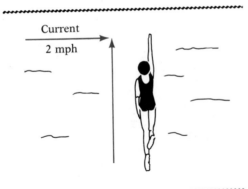

44. A small motorboat is moving at a true speed of 11 miles per hour in a southerly direction. The current is known to be from the northeast at 3 miles per hour. What is the speed of the motorboat relative to the water?

45. A factory produces gasoline engines and diesel engines. Each week the factory is obligated to deliver at least 20 gasoline engines and at least 15 diesel engines. Due to physical limitations, however, the factory cannot make more than 60 gasoline engines nor more than 40 diesel engines. Finally, to prevent layoffs, a total of at least 50 engines must be produced. If gasoline engines cost $450 each to produce and diesel engines cost $550 each to produce, how many of each should be produced per week to minimize the cost? What is the excess capacity of the factory; that is, how many of each kind of engine are being produced in excess of the number the factory is obligated to deliver?

46. Rework Problem 45 if the factory is obligated to deliver at least 25 gasoline engines and at least 20 diesel engines.

INDUCTION; SEQUENCES

9.1 Mathematical Induction
9.2 The Binomial Theorem
9.3 Sequences
9.4 Arithmetic and Geometric
Sequences
Chapter Review

Each topic in this chapter is related to the set of natural numbers. Mathematical induction is a technique for proving theorems involving the natural numbers. It is also the basis for recursively defined functions and recursive procedures used in computer programming. The binomial theorem is a formula for the expansion of $(x + a)^n$, where n is any natural number. The last two sections deal with sequences, which are functions whose domain is the set of natural numbers.

This chapter is intended to be introductory. Those of you who pursue mathematics further will use and encounter these topics in greater detail in more advanced courses. If you are interested in computer science, you will see these topics in a discrete mathematics course.

9.1 MATHEMATICAL INDUCTION

Mathematical induction is a method for proving that statements involving natural numbers are true for all natural numbers.* For example, the statement that "$2n$ is always an even integer" can be proven true for all natural numbers by using mathematical induction. The statement that "the sum of the first n positive odd integers equals n^2"—that is, the statement that

$$1 + 3 + 5 + \cdots + (2n - 1) = n^2 \tag{1}$$

can be proven for all natural numbers n by using mathematical induction.

Before stating the method of mathematical induction, let's try to gain a sense of the power of the method. We shall use the statement in Equation (1) for this purpose by restating it for various values of $n = 1, 2, 3, \ldots$:

$n = 1$ The sum of the first positive odd integer is 1^2; $1 = 1^2$.

$n = 2$ The sum of the first 2 positive odd integers is 2^2; $1 + 3 = 4 = 2^2$.

$n = 3$ The sum of the first 3 positive odd integers is 3^2; $1 + 3 + 5 = 9 = 3^2$.

$n = 4$ The sum of the first 4 positive odd integers is 4^2; $1 + 3 + 5 + 7 = 16 = 4^2$.

Although from this pattern we might conjecture that the statement "the sum of the first n positive odd integers is n^2" is true for any choice of n, can we really be sure that it doesn't fail for some choice of n? The method of proof by mathematical induction will, in fact, prove that the statement is true for all n.

THEOREM

THE PRINCIPLE OF MATHEMATICAL INDUCTION

Suppose the following two conditions are satisfied with regard to a statement about natural numbers:

Condition I.   The statement is true for the natural number 1.

Condition II.   If the statement is true for some natural number k, it is also true for the next natural number $k + 1$.

Then the statement is true for all natural numbers.   ≡

We shall not prove this principle. However, we can provide a physical interpretation that will help us see why the principle works.

---

*Recall from Chapter 1 that the natural numbers are the numbers 1, 2, 3, 4, ... In other words, the terms *natural numbers* and *positive integers* are synonymous.

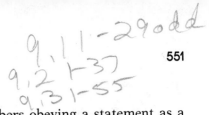

Think of a collection of natural numbers obeying a statement as a collection of infinitely many dominoes. See Figure 1 for an illustration. Suppose now we are told two things:

1. The first domino is pushed over.
2. If one of the dominoes falls over—say the kth one—then so will the next one—the $k + 1$st—fall over.

FIGURE 1

Is it safe to conclude that *all* the dominoes fall over? The answer is yes because, if the first one falls (by Condition I), then the second one does also (by Condition II); and if the second one falls, then so does the third (by Condition II); and so on.

Now let's prove some statements about natural numbers using mathematical induction.

EXAMPLE 1   Show that

$$1 + 3 + 5 + \cdots + (2n - 1) = n^2 \qquad (2)$$

is true for all natural numbers n.

Solution   We need to show first that Formula (2) holds for $n = 1$. Because $1 = 1^2$, Formula (2) is true for $n = 1$. Thus, Condition I holds.

Next, we need to show that Condition II holds. Suppose we know for some k that

$$1 + 3 + \cdots + (2k - 1) = k^2 \qquad (3)$$

We wish to show that, based on Equation (3), the formula holds for $k + 1$. Thus, we look at the sum of the first $(k + 1)$ positive odd integers to see whether this sum equals $(k + 1)^2$:

$$1 + 3 + \cdots + (2k - 1) + (2k + 1) = \underbrace{[1 + 3 + \cdots + (2k - 1)]}_{= k^2 \text{ by Equation (3)}} + (2k + 1)$$
$$= k^2 + (2k + 1)$$
$$= k^2 + 2k + 1$$
$$= (k + 1)^2$$

Conditions I and II are satisfied; thus, by the principle of mathematical induction, Formula (2) is true for all natural numbers. ∎

EXAMPLE 2   Show that

$$2^n > n$$

is true for all natural numbers n.

Solution   First, we show that the statement $2^n > n$ holds when $n = 1$. Because $2^1 = 2 > 1$, the inequality is true for $n = 1$. Thus, Condition I holds.

Next, we assume, for some natural number k, that $2^k > k$. We wish to show that the formula holds for $k + 1$; that is, we wish to show that $2^{k+1} > k + 1$. Now,

$$2^{k+1} = 2 \cdot 2^k > 2 \cdot k = k + k \geq k + 1$$

$$\uparrow \qquad\qquad\qquad \uparrow$$

We know that $\qquad\qquad k \geq 1$
$2^k > k$.

Thus, if $2^k > k$, then $2^{k+1} > k + 1$, so Condition II of the principle of mathematical induction is satisfied. Hence, the formula $2^n > n$ is true for all natural numbers n.   ■

EXAMPLE 3   Show that the formula

$$1 + 2 + 3 + \cdots + n = \frac{n(n + 1)}{2} \tag{4}$$

is true for all natural numbers.

Solution   First, we show that the formula is true when $n = 1$. Because $1(1 + 1)/2 = 1(2)/2 = 1$, Condition I of the principle of mathematical induction holds.

Next, we assume that the formula holds for some k and we see whether the formula then holds for $k + 1$. Thus, we assume

$$1 + 2 + 3 + \cdots + k = \frac{k(k + 1)}{2} \qquad \text{for some } k \tag{5}$$

Now, we need to show that

$$1 + 2 + 3 + \cdots + k + (k + 1) = \frac{(k + 1)(k + 1 + 1)}{2} = \frac{(k + 1)(k + 2)}{2}$$

We do this as follows:

$$1 + 2 + 3 + \cdots + k + (k + 1) = \underbrace{[1 + 2 + 3 + \cdots + k]}_{= \frac{k(k + 1)}{2} \text{ by Equation (5)}} + (k + 1)$$

$$= \frac{k(k + 1)}{2} + (k + 1)$$

$$= \frac{k^2 + k + 2k + 2}{2}$$

$$= \frac{k^2 + 3k + 2}{2}$$

$$= \frac{(k + 1)(k + 2)}{2}$$

Thus, Condition II also holds. As a result, Formula (4) is true for all natural numbers. ▀

EXAMPLE 4   Show that $3^n - 1$ is divisible by 2 for all natural numbers n.

Solution   First, we show that the statement is true when $n = 1$. Because $3^1 - 1 = 3 - 1 = 2$ is divisible by 2, the statement is true when $n = 1$. Thus, Condition I is satisfied. Next, we assume that the statement holds for some k and we see whether the statement then holds for $k + 1$. Thus, we assume that $3^k - 1$ is divisible by 2 for some k. We need to show that $3^{k+1} - 1$ is divisible by 2. Now,

$$3^{k+1} - 1 = 3^{k+1} - 3^k + 3^k - 1 = 3^k(3 - 1) + (3^k - 1) = 3^k \cdot 2 + (3^k - 1)$$

Because $3^k \cdot 2$ is divisible by 2 and because $3^k - 1$ is divisible by 2, it follows that $3^k \cdot 2 + (3^k - 1) = 3^{k+1} - 1$ is divisible by 2. Thus, Condition II is also satisfied. As a result, the statement "$3^n - 1$ is divisible by 2" is true for all natural numbers n. ▀

Warning: The conclusion that a statement involving natural numbers is true for all natural numbers is made only after *both* Conditions I and II of the principle of mathematical induction have been satisfied. Problem 27 demonstrates a statement for which only Condition I holds that is not true for all natural numbers. Problem 28 demonstrates a statement for which only Condition II holds that is true for *no* natural number.

EXERCISE 9.1   *In Problems 1–26 use the principle of mathematical induction to show that the given statement is true for all natural numbers.*

1. $2 + 4 + 6 + \cdots + 2n = n(n + 1)$
2. $1 + 5 + 9 + \cdots + (4n - 3) = n(2n - 1)$
3. $3 + 4 + 5 + \cdots + (n + 2) = \frac{1}{2}n(n + 5)$
4. $3 + 5 + 7 + \cdots + (2n + 1) = n(n + 2)$
5. $2 + 5 + 8 + \cdots + (3n - 1) = \frac{1}{2}n(3n + 1)$

6. $1 + 4 + 7 + \cdots + (3n - 2) = \frac{1}{2}n(3n - 1)$

7. $1 + 2 + 2^2 + \cdots + 2^{n-1} = 2^n - 1$

8. $1 + 3 + 3^2 + \cdots + 3^{n-1} = \frac{1}{2}(3^n - 1)$

9. $1 + 4 + 4^2 + \cdots + 4^{n-1} = \frac{1}{3}(4^n - 1)$

10. $1 + 5 + 5^2 + \cdots + 5^{n-1} = \frac{1}{4}(5^n - 1)$

11. $\dfrac{1}{1 \cdot 2} + \dfrac{1}{2 \cdot 3} + \dfrac{1}{3 \cdot 4} + \cdots + \dfrac{1}{n(n + 1)} = \dfrac{n}{n + 1}$

12. $\dfrac{1}{1 \cdot 3} + \dfrac{1}{3 \cdot 5} + \dfrac{1}{5 \cdot 7} + \cdots + \dfrac{1}{(2n - 1)(2n + 1)} = \dfrac{n}{2n + 1}$

13. $1^2 + 2^2 + 3^2 + \cdots + n^2 = \frac{1}{6}n(n + 1)(2n + 1)$

14. $1^3 + 2^3 + 3^3 + \cdots + n^3 = \frac{1}{4}n^2(n + 1)^2$

15. $4 + 3 + 2 + \cdots + (5 - n) = \frac{1}{2}n(9 - n)$

16. $-2 - 3 - 4 - \cdots - (n + 1) = -\frac{1}{2}n(n + 3)$

17. $1 \cdot 2 + 2 \cdot 3 + 3 \cdot 4 + \cdots + n(n + 1) = \frac{1}{3}n(n + 1)(n + 2)$

18. $1 \cdot 2 + 3 \cdot 4 + 5 \cdot 6 + \cdots + (2n - 1)(2n) = \frac{1}{3}n(n + 1)(4n - 1)$

19. $n^2 + n$ is divisible by 2.

20. $n^3 + 2n$ is divisible by 3.

21. $n^2 - n + 2$ is divisible by 2.

22. $n(n + 1)(n + 2)$ is divisible by 6.

23. If $x > 1$, then $x^n > 1$.

24. If $0 < x < 1$, then $0 < x^n < 1$.

25. $a - b$ is a factor of $a^n - b^n$.
    [*Hint:* $a^{k+1} - b^{k+1} = a(a^k - b^k) + b^k(a - b)$]

26. $a + b$ is a factor of $a^{2n+1} + b^{2n+1}$.

27. Show that the statement "$n^2 - n + 41$ is a prime number" is true for $n = 1$ but is not true for $n = 41$.

28. Show that the formula

$$2 + 4 + 6 + \cdots + 2n = n^2 + n + 2$$

obeys Condition II of the principle of mathematical induction. That is, show that, if the formula is true for some k, it is also true for $k + 1$. Then show that the formula is false for $n = 1$ (or for any other choice of n).

29. Prove that, if $r \neq 1$,

$$a + ar + ar^2 + \cdots + ar^{n-1} = a\frac{1 - r^n}{1 - r}$$

30. Prove that

$$a + (a + d) + (a + 2d) + \cdots + [a + (n - 1)d] = na + d\frac{n(n - 1)}{2}$$

☰ 9.2 THE BINOMIAL THEOREM

In Chapter 1 we listed some special products. Among these were formulas for expanding $(x + a)^n$ for $n = 2$ and $n = 3$. The binomial theorem* is a formula for the expansion of $(x + a)^n$ for n any positive integer. If $n = 1, 2,$ and 3, the expansion of $(x + a)^n$ is straightforward.

$(x + a)^1 = x + a$	2 terms, beginning with x^1 and ending with a^1
$(x + a)^2 = x^2 + 2ax + a^2$	3 terms, beginning with x^2 and ending with a^2
$(x + a)^3 = x^3 + 3ax^2 + 3a^2x + a^3$	4 terms, beginning with x^3 and ending with a^3

Notice that each expansion of $(x + a)^n$ begins with x^n and ends with a^n. Also, the number of terms that appear equals $n + 1$. Notice, too, that if we treat both x and a as variables, then the degree of each monomial in the expansion equals n. For example, in the expansion of $(x + a)^3$, each monomial (x^3, ax^2, a^2x, a^3) is of degree three. As a result, we might conjecture that the expansion of $(x + a)^n$ would look like this:

$$(x + a)^n = x^n + \_ax^{n-1} + \_a^2x^{n-2} + \cdots + \_a^{n-1}x + a^n$$

where the blanks are numbers to be found. That is, in fact, the case, as we shall shortly see.

First we need to introduce some symbols.

The Factorial Symbol

FACTORIAL SYMBOL $n!$ If $n \geq 0$ is an integer, the **factorial symbol** $n!$ is defined as

$$0! = 1 \qquad 1! = 1 \qquad \text{and}$$
$$n! = n(n - 1) \cdot \cdots \cdot 3 \cdot 2 \cdot 1 \quad \text{if } n \geq 2$$

For example, $2! = 2 \cdot 1 = 2$, $3! = 3 \cdot 2 \cdot 1 = 6$, $4! = 4 \cdot 3 \cdot 2 \cdot 1 = 24$, and so on. Table 1 lists the values of $n!$ for $0 \leq n \leq 6$.

TABLE 1

n	0	1	2	3	4	5	6
$n!$	1	1	2	6	24	120	720

*The name *binomial* derives from the fact that $x + a$ is a binomial—that is, contains two terms.

Because

$$n! = n(n-1)(n-2) \cdot \cdots \cdot 3 \cdot 2 \cdot 1$$
$$\underbrace{}_{(n-1)!}$$

we can use the formula

$$n! = n(n-1)!$$

to find successive factorials. For example, because $6! = 720$, we have

$$7! = 7 \cdot 6! = 7(720) = 5040$$

and

$$8! = 8 \cdot 7! = 8(5040) = 40{,}320$$

As you can see, factorials increase very rapidly. In fact, $70!$ is larger than 10^{100} (a *googol*), the largest number most calculators can display.

The Symbol $\binom{n}{j}$

In addition to the factorial symbol, we need to define the symbol $\binom{n}{j}$.

SYMBOL $\binom{n}{j}$ If j and n are integers with $0 \le j \le n$, the **symbol** $\binom{n}{j}$ is defined as

$$\binom{n}{j} = \frac{n!}{j!(n-j)!}$$

EXAMPLE 1   Find:

(a) $\binom{3}{1}$ (b) $\binom{4}{2}$ (c) $\binom{8}{7}$

Solution   (a) $\binom{3}{1} = \dfrac{3!}{1!(3-1)!} = \dfrac{3!}{1!2!} = \dfrac{3 \cdot 2 \cdot 1}{1(2 \cdot 1)} = \dfrac{6}{2} = 3$

(b) $\binom{4}{2} = \dfrac{4!}{2!(4-2)!} = \dfrac{4!}{2!2!} = \dfrac{4 \cdot 3 \cdot 2 \cdot 1}{(2 \cdot 1)(2 \cdot 1)} = \dfrac{24}{4} = 6$

(c) $\binom{8}{7} = \dfrac{8!}{7!(8-7)!} = \dfrac{8!}{7!1!} = \dfrac{8 \cdot 7!}{7! \cdot 1!} = \dfrac{8}{1} = 8$

$$\uparrow$$
$$8! = 8 \cdot 7!$$

Two useful formulas involving the symbol $\binom{n}{j}$ are

$$\binom{n}{0} = 1 \qquad \binom{n}{n} = 1$$

For example,

$$\binom{n}{0} = \frac{n!}{0!(n-0)!} = \frac{n!}{0!n!} = \frac{1}{1} = 1$$

You are asked to show that $\binom{n}{n} = 1$ in Problem 35 at the end of this section.

Suppose we arrange the various values of the symbol $\binom{n}{j}$ in a triangular display, as shown below and in Figure 2:

$$\binom{0}{0}$$

$$\binom{1}{0} \quad \binom{1}{1}$$

$$\binom{2}{0} \quad \binom{2}{1} \quad \binom{2}{2}$$

$$\binom{3}{0} \quad \binom{3}{1} \quad \binom{3}{2} \quad \binom{3}{3}$$

$$\binom{4}{0} \quad \binom{4}{1} \quad \binom{4}{2} \quad \binom{4}{3} \quad \binom{4}{4}$$

$$\binom{5}{0} \quad \binom{5}{1} \quad \binom{5}{2} \quad \binom{5}{3} \quad \binom{5}{4} \quad \binom{5}{5}$$

PASCAL TRIANGLE This display is called the **Pascal triangle**, named after Blaise Pascal, a French mathematician.

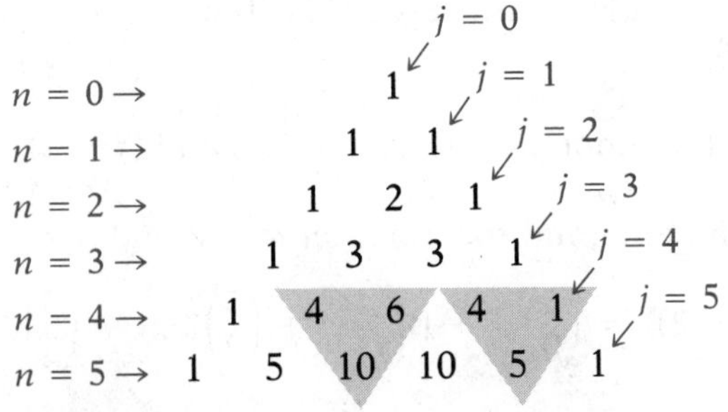

FIGURE 2
Pascal triangle

The Pascal triangle has 1's down the sides. To get any other entry, merely add the two nearest entries in the row above it. The shaded triangles in Figure 2 serve to illustrate this feature of the Pascal triangle. Based on this feature, the row corresponding to $n = 6$ is

$$
\begin{array}{ccccccc}
n = 5 \rightarrow & 1 & 5 & 10 & 10 & 5 & 1 \\
n = 6 \rightarrow & 1 & 6 & 15 & 20 & 15 & 6 & 1
\end{array}
$$

We shall later prove that this addition always works (see Example 6 near the end of this section).

Although the Pascal triangle provides an interesting and organized display of the symbol $\binom{n}{j}$, in practice it is not all that helpful. For example, if you wanted to know the value of $\binom{12}{5}$, you would need to produce twelve rows of the triangle before seeing the answer. It is much faster instead to use the definition stated earlier.

The Binomial Theorem

Now we are ready to state the binomial theorem. A proof is given at the end of this section.

THEOREM
BINOMIAL THEOREM

Let x and a be real numbers. For any positive integer n, we have

$$
\begin{aligned}
(x + a)^n &= \binom{n}{0}x^n + \binom{n}{1}ax^{n-1} + \cdots + \binom{n}{j}a^j x^{n-j} \\
&\quad + \cdots + \binom{n}{n}a^n
\end{aligned}
\tag{1}
$$

≡

Now you know why we needed to introduce the symbol $\binom{n}{j}$—these symbols are the numerical coefficients that appear in the expansion of $(x + a)^n$. Because of this, the symbol $\binom{n}{j}$ is called a **binomial coefficient**.

BINOMIAL COEFFICIENT

EXAMPLE 2   Use the binomial theorem to expand $(x + 2)^5$.

Solution   In the binomial theorem, let $a = 2$ and $n = 5$. Then

$$
(x + 2)^5 = \binom{5}{0}x^5 + \binom{5}{1}2x^4 + \binom{5}{2}2^2x^3 + \binom{5}{3}2^3x^2 + \binom{5}{4}2^4x + \binom{5}{5}2^5
$$

↑
Use Equation (1).

$$
= 1 \cdot x^5 + 5 \cdot 2x^4 + 10 \cdot 4x^3 + 10 \cdot 8x^2 + 5 \cdot 16x + 1 \cdot 32
$$

↑
Use row $n = 5$ of the Pascal triangle.

$$
= x^5 + 10x^4 + 40x^3 + 80x^2 + 80x + 32
$$

≡

EXAMPLE 3 Expand $(2y - 3)^4$ using the binomial theorem.

Solution We use the binomial theorem with $n = 4$, $a = -3$, and $x = 2y$. Then

$$(2y - 3)^4 = \binom{4}{0}(2y)^4 + \binom{4}{1}(-3)(2y)^3 + \binom{4}{2}(-3)^2(2y)^2 + \binom{4}{3}(-3)^3 2y + \binom{4}{4}(-3)^4$$

$$= 1 \cdot 16y^4 + 4(-3)8y^3 + 6 \cdot 9 \cdot 4y^2 + 4(-27)2y + 1 \cdot 81$$

↑
Use row $n = 4$ of the Pascal triangle.

$$= 16y^4 - 96y^3 + 216y^2 - 216y + 81 \qquad \blacksquare$$

The binomial theorem can be used to "pick out" a particular term in an expansion without writing the entire expansion. Based on the expansion of $(x + a)^n$, the term containing x^j is

$$\binom{n}{n - j} a^{n-j} x^j \qquad (2)$$

EXAMPLE 4 Find the coefficient of y^8 in the expansion of $(2y + 3)^{10}$.

Solution We shall use Formula (2) with $n = 10$, $a = 3$, $x = 2y$, and $j = 8$. Then the term containing y^8 is

$$\binom{10}{10 - 8} 3^{10-8}(2y)^8 = \binom{10}{2} \cdot 3^2 \cdot 2^8 \cdot y^8$$

$$= \frac{10!}{2!8!} \cdot 9 \cdot 2^8 y^8$$

$$= \frac{10 \cdot 9 \cdot 8!}{2!8!} \cdot 9 \cdot 2^8 y^8$$

$$= 90 \cdot 9 \cdot 2^7 y^8$$

$$= 103{,}680 y^8 \qquad \blacksquare$$

EXAMPLE 5 Find the sixth term in the expansion of $(x + 2)^9$.

Solution The sixth term in the expansion of $(x + 2)^9$, which has 10 terms total, contains x^4. (Do you see why?) Thus, the sixth term is

$$\binom{9}{9 - 4} 2^{9-4} x^4 = \binom{9}{5} 2^5 x^4 = \frac{9!}{5!4!} 32 x^4 = 4032 x^4 \qquad \blacksquare$$

Next, we show that the "triangular addition" feature of the Pascal triangle always works.

EXAMPLE 6   If n and j are integers with $0 \leq j \leq n$, show that

$$\binom{n}{j-1} + \binom{n}{j} = \binom{n+1}{j}$$

Solution

$$\binom{n}{j-1} + \binom{n}{j} = \frac{n!}{(j-1)![n-(j-1)]!} + \frac{n!}{j!(n-j)!}$$

$$= \frac{n!}{(j-1)!(n-j+1)!} + \frac{n!}{j!(n-j)!}$$

$$= \frac{jn!}{j(j-1)!(n-j+1)!} + \frac{(n-j+1)n!}{j!(n-j+1)(n-j)!}$$

$$= \frac{jn!}{j!(n-j+1)!} + \frac{(n-j+1)n!}{j!(n-j+1)!}$$

$$= \frac{jn! + (n-j+1)n!}{j!(n-j+1)!}$$

$$= \frac{n!(j+n-j+1)}{j!(n-j+1)!}$$

$$= \frac{n!(n+1)}{j!(n-j+1)!} = \frac{(n+1)!}{j![(n+1)-j]!} = \binom{n+1}{j} \quad \blacksquare$$

Proof of the binomial theorem   We use mathematical induction to prove the binomial theorem. First, we show that Formula (1) is true for $n = 1$:

$$(x + a)^1 = x + a = \binom{1}{0}x^1 + \binom{1}{1}a^1$$

Next, we suppose that Formula (1) is true for some k. That is, we assume

$$(x + a)^k = \binom{k}{0}x^k + \binom{k}{1}ax^{k-1} + \cdots + \binom{k}{j-1}a^{j-1}x^{k-j+1} + \binom{k}{j}a^jx^{k-j} + \cdots + \binom{k}{k}a^k \quad (3)$$

Now we calculate $(x + a)^{k+1}$:

$$(x + a)^{k+1} = (x + a)(x + a)^k$$

$$= x(x+a)^k + a(x+a)^k$$

$$= x\left[\binom{k}{0}x^k + \binom{k}{1}ax^{k-1} + \cdots + \binom{k}{j-1}a^{j-1}x^{k-j+1} + \binom{k}{j}a^jx^{k-j} + \cdots + \binom{k}{k}a^k\right]$$

↑
Use Equation (3).

$$+ a\left[\binom{k}{0}x^k + \binom{k}{1}ax^{k-1} + \cdots + \binom{k}{j-1}a^{j-1}x^{k-j+1} + \binom{k}{j}a^jx^{k-j} + \cdots + \binom{k}{k}a^k\right]$$

$$= \binom{k}{0}x^{k+1} + \binom{k}{1}ax^k + \cdots + \binom{k}{j}a^jx^{k-j+1} + \cdots + \binom{k}{k}a^kx + \binom{k}{0}ax^k$$

$$+ \cdots + \binom{k}{j-1}a^jx^{k-j+1} + \cdots + \binom{k}{k-1}a^kx + \binom{k}{k}a^{k+1}$$

$$= \binom{k}{0}x^{k+1} + \left[\binom{k}{1} + \binom{k}{0}\right]ax^k + \cdots + \left[\binom{k}{j} + \binom{k}{j-1}\right]a^jx^{k-j+1}$$

$$+ \cdots + \left[\binom{k}{k} + \binom{k}{k-1}\right]a^kx + \binom{k}{k}a^{k+1}$$

Because

$$\binom{k}{0} = 1 = \binom{k+1}{0}, \quad \binom{k}{1} + \binom{k}{0} = \binom{k+1}{1}, \quad \cdots, \quad \binom{k}{j} + \binom{k}{j-1}$$

↑
Example 6

$$= \binom{k+1}{j}, \quad \cdots, \quad \binom{k}{k} = 1 = \binom{k+1}{k+1}$$

↑
Example 6

we have

$$(x + a)^{k+1} = \binom{k+1}{0}x^{k+1} + \binom{k+1}{1}ax^k + \cdots + \binom{k+1}{j}a^jx^{k-j+1} + \cdots + \binom{k+1}{k+1}a^{k+1}$$

Thus, the conditions of the principle of mathematical induction are satisfied, and Formula (1) is therefore true for all n. ≡

≡ HISTORICAL COMMENT The case $n = 2$ of the binomial theorem, $(a + b)^2$, was known to Euclid in 300 B.C., but the general law seems to have been discovered by the Persian mathematician and astronomer Omar Khayyám (1044?–1123?), who is also well known as the author of the *Rubaiyat*, a collection of four-line poems making observations on the human condition. Omar Khayyám did not state the binomial theorem explicitly, but he claimed to have a method for extracting third, fourth, fifth roots, and so on. A little study shows that one must know the binomial theorem to create such a method.

The heart of the binomial theorem is the formula for the numerical coefficients and, as we saw, they can be written out in a symmetric triangular form. The Pascal triangle appears first in the books of Yang Hui (about 1270) and Chu Shih-chie (1303). Pascal's name is attached to the triangle because of the many applications he made of it, especially to counting and probability. In establishing these results, he was one of the earliest users of mathematical induction.

Many people worked on the proof of the binomial theorem, which was finally completed for all n (including complex numbers) by Niels Abel (1802–1829) about 1825. ≡

EXERCISE 9.2   *In Problems 1–12 evaluate each expression.*

1. $9!$ **2.** $10!$ **3.** $\dfrac{8!5!}{6!6!}$ **4.** $\dfrac{7!6!}{9!8!}$

5. $\dbinom{5}{3}$ **6.** $\dbinom{7}{3}$ **7.** $\dbinom{7}{5}$ **8.** $\dbinom{9}{7}$

9. $\dbinom{50}{49}$ **10.** $\dbinom{100}{98}$ **11.** $\dbinom{1000}{1000}$ **12.** $\dbinom{1000}{0}$

In Problems 13–24 expand each expression using the binomial theorem.

13. $(x + 1)^5$ **14.** $(x - 1)^5$ **15.** $(x - 2)^6$

16. $(x + 3)^4$ **17.** $(3x + 1)^4$ **18.** $(2x + 3)^5$

19. $(x^2 + y^2)^5$ **20.** $(x^2 - y^2)^6$ **21.** $(\sqrt{x} + \sqrt{2})^6$

22. $(\sqrt{x} - \sqrt{3})^4$ **23.** $(ax + by)^5$ **24.** $(ax - by)^4$

In Problems 25–34 use the binomial theorem to find the following.

25. The coefficient of x^6 in the expansion of $(x + 3)^{10}$

26. The coefficient of x^3 in the expansion of $(x - 3)^{10}$

27. The coefficient of x^8 in the expansion of $(2x - 1)^{12}$

28. The coefficient of x^2 in the expansion of $(2x + 1)^{12}$

29. The coefficient of x^7 in the expansion of $(2x + 3)^9$

30. The coefficient of x^2 in the expansion of $(2x - 3)^9$

31. The fifth term in the expansion of $(x + 3)^7$

32. The third term in the expansion of $(x - 3)^7$

C **33.** The third term in the expansion of $(3x - 2)^9$

C **34.** The sixth term in the expansion of $(3x + 2)^8$

35. Show that $\dbinom{n}{n} = 1$.

36. Show that, if n and j are integers with $0 \le j \le n$, then

$$\binom{n}{j} = \binom{n}{n - j}$$

Thus conclude that the Pascal triangle is symmetric with respect to a vertical line drawn from the topmost entry.

37. If n is a positive integer, show that

$$\binom{n}{0} + \binom{n}{1} + \cdots + \binom{n}{n} = 2^n$$

[*Hint:* $2^n = (1 + 1)^n$; now use the binomial theorem.]

38. If n is a positive integer, show that

$$\binom{n}{0} - \binom{n}{1} + \binom{n}{2} - \cdots + (-1)^n\binom{n}{n} = 0$$

[C] **39.** Stirling's formula for approximating $n!$ when n is large is given by

$$n! \approx \sqrt{2n\pi}\left(\frac{n}{e}\right)^n\left(1 + \frac{1}{12n-1}\right)$$

Calculate 12!, 20!, and 25!. Then use Stirling's formula to approximate 12!, 20!, and 25!.

▤ 9.3 SEQUENCES

SEQUENCE A **sequence** is a function whose domain is the set of positive integers.

Because a sequence is a function, it will have a graph. In Figure 3(a) you will recognize the graph of the function $f(x) = 1/x$, $x > 0$. If all the points on this graph were removed except those whose x-coordinates are positive integers—that is, if all points were removed except $(1, 1)$, $(2, \frac{1}{2})$, $(3, \frac{1}{3})$, and so on—the remaining points would be the graph of a sequence. See Figure 3(b).

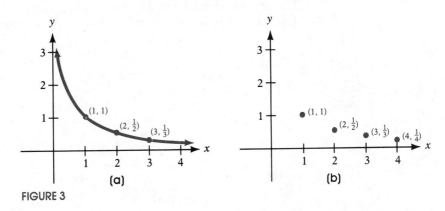

FIGURE 3

A sequence is sometimes represented by listing its values in order. For example, the sequence whose graph is given in Figure 3(b) might be represented as

$$f(1), f(2), f(3), f(4), \ldots \quad \text{or} \quad 1, \frac{1}{2}, \frac{1}{3}, \frac{1}{4}, \ldots$$

TERMS OF A SEQUENCE The list never ends, as the ellipsis dots indicate. The numbers in this list are called the **terms** of the sequence.

In dealing with sequences, we usually use subscripted letters, such as a_1 to represent the first term, a_2 for the second term, a_3 for the third term, and so on. Thus, for the sequence $f(n) = 1/n$, we write

$$a_1 = f(1) = 1$$

$$a_2 = f(2) = \frac{1}{2}$$

$$a_3 = f(3) = \frac{1}{3}$$

$$a_4 = f(4) = \frac{1}{4}$$

$$\vdots$$

$$a_n = f(n) = \frac{1}{n}$$

In other words, we usually do not use the traditional function notation $f(n)$ for sequences. For this particular sequence, we have a rule for the nth term, namely $a_n = 1/n$, so it is easy to find any other term of the sequence.

When the nth term of a sequence is known, rather than write each term of the sequence, we shall represent the entire sequence by placing braces around the nth term. For example, the sequence whose nth term is $b_n = (\frac{1}{2})^n$ may be represented compactly as

$$\{b_n\} = \left\{\left(\frac{1}{2}\right)^n\right\}$$

or by

$$b_1 = \frac{1}{2}$$

$$b_2 = \frac{1}{4}$$

$$b_3 = \frac{1}{8}$$

$$\vdots$$

$$b_n = \left(\frac{1}{2}\right)^n$$

$$\vdots$$

EXAMPLE 1   Write down the first six terms of each sequence and graph each one.

(a) $\{a_n\} = \left\{\dfrac{n-1}{n}\right\}$ (b) $\{b_n\} = \left\{(-1)^{n-1}\dfrac{2}{n}\right\}$

(c) $\{c_n\} = \begin{cases} n & \text{if } n \text{ is even} \\ 1/n & \text{if } n \text{ is odd} \end{cases}$

Solution (a) $a_1 = 0$

$a_2 = \frac{1}{2}$

$a_3 = \frac{2}{3}$

$a_4 = \frac{3}{4}$

$a_5 = \frac{4}{5}$

$a_6 = \frac{5}{6}$

See Figure 4(a).

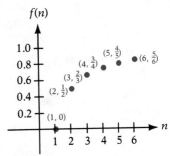

FIGURE 4(a)

$\{a_n\} = \left\{ \dfrac{n-1}{n} \right\}$

(b) $b_1 = 2$

$b_2 = -1$

$b_3 = \frac{2}{3}$

$b_4 = -\frac{1}{2}$

$b_5 = \frac{2}{5}$

$b_6 = -\frac{1}{3}$

See Figure 4(b).

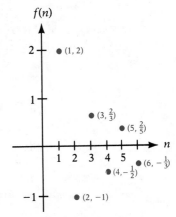

FIGURE 4(b)

$\{b_n\} = \left\{ (-1)^{n-1} \dfrac{2}{n} \right\}$

(c) $c_1 = 1$

$c_2 = 2$

$c_3 = \frac{1}{3}$

$c_4 = 4$

$c_5 = \frac{1}{5}$

$c_6 = 6$

See Figure 4(c).

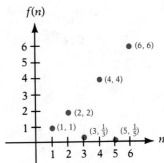

FIGURE 4(c)

$\{c_n\} = \begin{cases} n & \text{if } n \text{ is even} \\ 1/n & \text{if } n \text{ is odd} \end{cases}$

Sometimes a sequence is indicated by an observed pattern in the first few terms that makes it possible to infer the makeup of the nth term. In the example that follows, a sufficient number of terms of the sequence is given so that a natural choice for the nth term is suggested.

EXAMPLE 2   (a) $e, \dfrac{e^2}{2}, \dfrac{e^3}{3}, \dfrac{e^4}{4}, \ldots$ $a_n = \dfrac{e^n}{n}$

(b) $1, \dfrac{1}{3}, \dfrac{1}{9}, \dfrac{1}{27}, \ldots$ $b_n = \dfrac{1}{3^{n-1}}$

(c) $1, 3, 5, 7, \ldots$ $c_n = 2n - 1$

(d) $1, 4, 9, 16, 25, \ldots$ $d_n = n^2$

(e) $1, -\dfrac{1}{2}, \dfrac{1}{3}, -\dfrac{1}{4}, \dfrac{1}{5}, \ldots$ $e_n = (-1)^{n+1}\dfrac{1}{n}$ ≡

ALTERNATING SIGNS   Notice that in the sequence $\{e_n\}$ in Example 2(e) the signs of the terms **alternate**. When this occurs, we use factors such as $(-1)^{n+1}$, which equals 1 if n is odd and -1 if n is even, or $(-1)^n$, which equals -1 if n is odd and 1 if n is even.

Recursion Formulas

RECURSIVELY
RECURSIVE FORMULA   A second way of defining a sequence is to specify the first (or the first few) terms along with a rule or formula for the nth term in terms of one or more of the terms preceding it. Sequences defined this way are said to be defined **recursively**, and the rule or formula is called a **recursive formula**.

EXAMPLE 3   Write down the first five terms of each recursively defined sequence.

(a) $s_1 = 1,\quad s_n = 4s_{n-1}$ (b) $u_1 = 1,\quad u_2 = 1,\quad u_{n+2} = u_n + u_{n+1}$

(c) $f_1 = 1,\quad f_{n+1} = (n + 1)f_n$

Solution   (a) The first term is given as $s_1 = 1$. To get the second term, we use $n = 2$ in the formula to get $s_2 = 4s_1 = 4 \cdot 1 = 4$. To get the third term, we use $n = 3$ in the formula to get $s_3 = 4s_2 = 4 \cdot 4 = 16$. To get a new term requires that we know the value of the preceding one. The first five terms are

$$s_1 = 1$$
$$s_2 = 4 \cdot 1 = 4$$
$$s_3 = 4 \cdot 4 = 16$$
$$s_4 = 4 \cdot 16 = 64$$
$$s_5 = 4 \cdot 64 = 256$$

(b) We are given the first two terms. To get the third term requires that we know each of the previous two terms. Thus,

$$u_1 = 1$$
$$u_2 = 1$$
$$u_3 = u_1 + u_2 = 2$$
$$u_4 = u_3 + u_2 = 2 + 1 = 3$$
$$u_5 = u_4 + u_3 = 3 + 2 = 5$$

(c) Here,

$$f_1 = 1$$
$$f_2 = 2f_1 = 2 \cdot 1 = 2$$
$$f_3 = 3f_2 = 3 \cdot 2 = 6$$
$$f_4 = 4f_3 = 4 \cdot 6 = 24$$
$$f_5 = 5f_4 = 5 \cdot 24 = 120$$

≡

FIBONACCI SEQUENCE

FIBONACCI NUMBERS

The sequence defined in Example 3(b) is called a **Fibonacci sequence**, and the terms of this sequence are the **Fibonacci numbers**, which appear in a wide variety of applications (see Problems 55 and 56 at the end of this section). You should recognize the nth term of the sequence in Example 3(c) as the factorial of n.

Adding the First n Terms of a Sequence; Summation Notation

It is often important to be able to find the sum of the first n terms of a sequence $\{a_n\}$, namely,

$$a_1 + a_2 + a_3 + \cdots + a_n \tag{1}$$

SUMMATION NOTATION

Rather than write down all these terms, we introduce a more concise way to express the sum, called **summation notation**. Using summation notation we would write the sum (1) as

$$a_1 + a_2 + a_3 + \cdots + a_n = \sum_{k=1}^{n} a_k$$

INDEX

The symbol Σ (a stylized version of the Greek letter sigma, which is an S in our alphabet) is simply an instruction to sum or add up. The integer k is called the **index** of the sum; it tells you where to start the sum and where to end it. Therefore, the expression $\sum_{k=1}^{n} a_k$ is an instruction to add the terms a_k of the sequence $\{a_n\}$ from $k = 1$ through $k = n$. We read $\sum_{k=1}^{n} a_k$ as "the sum of a_k from $k = 1$ through $k = n$."

EXAMPLE 4   Write out each sum.

$$\text{(a) } \sum_{k=1}^{n} \frac{1}{k} \qquad \text{(b) } \sum_{k=0}^{n} k!$$

Solution   (a) $\displaystyle\sum_{k=1}^{n} \frac{1}{k} = \frac{1}{1} + \frac{1}{2} + \frac{1}{3} + \cdots + \frac{1}{n}$

(b) $\displaystyle\sum_{k=0}^{n} k! = 0! + 1! + 2! + \cdots + n!$ ≡

EXAMPLE 5   Express each sum using summation notation.

(a) $1^2 + 2^2 + 3^2 + \cdots + n^2$

(b) $1 + \dfrac{1}{2} + \dfrac{1}{4} + \dfrac{1}{8} + \cdots + \dfrac{1}{2^{n-1}}$

Solution   (a) The sum $1^2 + 2^2 + 3^2 + \cdots + n^2$ has n terms, each of the form k^2, and starts at $k = 1$ and ends at $k = n$. Thus,

$$1^2 + 2^2 + 3^2 + \cdots + n^2 = \sum_{k=1}^{n} k^2$$

(b) The sum

$$1 + \frac{1}{2} + \frac{1}{4} + \frac{1}{8} + \cdots + \frac{1}{2^{n-1}}$$

has n terms, each of the form $1/2^k$, and starts at $k = 0$ and ends at $k = n - 1$. Thus,

$$1 + \frac{1}{2} + \frac{1}{4} + \frac{1}{8} + \cdots + \frac{1}{2^{n-1}} = \sum_{k=0}^{n-1} \frac{1}{2^k}$$ ≡

Letters other than k can be used as the index. For example,

$$\sum_{j=0}^{n} j! \quad \text{and} \quad \sum_{i=0}^{n} i!$$

each represent the same sum as the one given in Example 4(b).
Next, we list some properties of summation:

If $\{a_n\}$ and $\{b_n\}$ are two sequences and c is a real number, then

1. $\displaystyle\sum_{k=1}^{n} ca_k = c\sum_{k=1}^{n} a_k$

2. $\displaystyle\sum_{k=1}^{n} (a_k + b_k) = \sum_{k=1}^{n} a_k + \sum_{k=1}^{n} b_k$

3. $\displaystyle\sum_{k=1}^{n} (a_k - b_k) = \sum_{k=1}^{n} a_k - \sum_{k=1}^{n} b_k$

4. $\displaystyle\sum_{k=1}^{n} a_k = \sum_{k=1}^{j} a_k + \sum_{k=j+1}^{n} a_k,$ when $1 < j < n$

Although we shall not prove these properties, the proofs are based on properties of real numbers.

EXERCISE 9.3   *In Problems 1–12 write down the first five terms of each sequence.*

1. $\{n\}$ 2. $\{n^2 + 1\}$

3. $\left\{\dfrac{n}{n + 1}\right\}$ 4. $\left\{\dfrac{2n + 3}{2n - 1}\right\}$

5. $\{(-1)^{n+1} n^2\}$ 6. $\left\{(-1)^{n-1}\dfrac{n}{2n - 1}\right\}$

7. $\left\{\dfrac{2^n}{3^n + 1}\right\}$ 8. $\left\{\left(\dfrac{4}{3}\right)^n\right\}$

9. $\left\{\dfrac{(-1)^n}{(n + 1)(n + 2)}\right\}$ 10. $\left\{\dfrac{3^n}{n}\right\}$

11. $\left\{\dfrac{n}{e^n}\right\}$ 12. $\left\{\dfrac{n^2}{2^n}\right\}$

In Problems 13–20 the pattern given continues. Write down the nth term of each sequence suggested by the pattern.

13. $\dfrac{1}{2}, \dfrac{2}{3}, \dfrac{3}{4}, \dfrac{4}{5}, \cdots$ 14. $\dfrac{1}{1 \cdot 2}, \dfrac{1}{2 \cdot 3}, \dfrac{1}{3 \cdot 4}, \dfrac{1}{4 \cdot 5}, \cdots$

15. $1, \dfrac{1}{2}, \dfrac{1}{4}, \dfrac{1}{8}, \cdots$ 16. $\dfrac{2}{3}, \dfrac{4}{9}, \dfrac{8}{27}, \dfrac{16}{81}, \cdots$

17. $1, -1, 1, -1, 1, -1, \ldots$ 18. $1, \dfrac{1}{2}, 1, \dfrac{1}{3}, 1, \dfrac{1}{4}, 1, \dfrac{1}{5}, \ldots$

19. $1, -2, 3, -4, 5, -6, \ldots$ 20. $2, -4, 6, -8, 10, \ldots$

In Problems 21–34 a sequence is defined recursively. Write down the first five terms.

21. $a_1 = 1,\quad a_{n+1} = 2 + a_n$ 22. $a_1 = 3,\quad a_{n+1} = 5 - a_n$

23. $a_1 = -2,\quad a_{n+1} = n + a_n$ 24. $a_1 = 1,\quad a_{n+1} = n - a_n$

25. $a_1 = 5,\quad a_{n+1} = 2a_n$ 26. $a_1 = 2,\quad a_{n+1} = -a_n$

27. $a_1 = 3,\quad a_{n+1} = \dfrac{a_n}{n}$ 28. $a_1 = -2,\quad a_{n+1} = n + 3a_n$

29. $a_1 = 1,\quad a_2 = 2,\quad a_{n+2} = a_n a_{n+1}$

30. $a_1 = -1,\quad a_2 = 1,\quad a_{n+2} = a_{n+1} + na_n$

31. $a_1 = A,\quad a_{n+1} = a_n + d$ 32. $a_1 = A,\quad a_{n+1} = ra_n,\quad r \neq 0$

33. $a_1 = \sqrt{2},\quad a_{n+1} = \sqrt{2 + a_n}$ 34. $a_1 = \sqrt{2},\quad a_{n+1} = \sqrt{\dfrac{a_n}{2}}$

In Problems 35–44 write out each sum.

35. $\displaystyle\sum_{k=1}^{n} (k + 1)$ **36.** $\displaystyle\sum_{k=1}^{n} (2k - 1)$ **37.** $\displaystyle\sum_{k=1}^{n} \frac{k^2}{2}$

38. $\displaystyle\sum_{k=1}^{n} (k + 1)^2$ **39.** $\displaystyle\sum_{k=0}^{n} \frac{1}{3^k}$ **40.** $\displaystyle\sum_{k=0}^{n} \left(\frac{3}{2}\right)^k$

41. $\displaystyle\sum_{k=0}^{n-1} \frac{1}{3^{k+1}}$ **42.** $\displaystyle\sum_{k=0}^{n-1} (2k + 1)$ **43.** $\displaystyle\sum_{k=2}^{n} (-1)^k \ln k$

44. $\displaystyle\sum_{k=3}^{n} (-1)^{k+1} 2^k$

In Problems 45–54 express each sum using summation notation.

45. $1 + 2 + 3 + \cdots + n$ **46.** $1^3 + 2^3 + 3^3 + \cdots + n^3$

47. $\dfrac{1}{2} + \dfrac{2}{3} + \dfrac{3}{4} + \cdots + \dfrac{n}{n + 1}$

48. $1 + 3 + 5 + 7 + \cdots + (2n - 1)$

49. $1 - \dfrac{1}{3} + \dfrac{1}{9} - \dfrac{1}{27} + \cdots + (-1)^n \dfrac{1}{3^n}$

50. $\dfrac{2}{3} - \dfrac{4}{9} + \dfrac{8}{27} - \cdots + (-1)^{n+1} \left(\dfrac{2}{3}\right)^n$

51. $3 + \dfrac{3^2}{2} + \dfrac{3^3}{3} + \cdots + \dfrac{3^n}{n}$ **52.** $\dfrac{1}{e} + \dfrac{2}{e^2} + \dfrac{3}{e^3} + \cdots + \dfrac{n}{e^n}$

53. $a + (a + d) + (a + 2d) + \cdots + (a + nd)$

54. $a + ar + ar^2 + \cdots + ar^{n-1}$

55. A colony of rabbits begins with one pair of mature rabbits, which will produce a pair of offspring (one male, one female) each month. Assume that all rabbits mature in 1 month and produce a pair of offspring (one male, one female) after 2 months. If no rabbits ever die, how many pairs of mature rabbits are there after 7 months? [*Hint:* The Fibonacci sequence models this colony. Do you see why?]

56. Let

$$u_n = \frac{(1 + \sqrt{5})^n - (1 - \sqrt{5})^n}{2^n \sqrt{5}}$$

define the *n*th term of a sequence.
(a) Show that $u_1 = 1$ and $u_2 = 1$.   (b) Show that $u_{n+2} = u_{n+1} + u_n$.
(c) Draw the conclusion that $\{u_n\}$ is the Fibonacci sequence.

In Problems 57–58 we use the fact that in some programming languages* it is possible to have a function subroutine include a call to itself.

57. *Programming exercise* Write a program that accepts a positive integer as input and prints the number and its factorial. Use a recursively defined function; that is, use a function subroutine that calls itself.

*Pascal, PL/1, ALGOL, and Logo to name a few.

58. *Programming exercise* Write a program that accepts a positive integer N as input and outputs the Nth Fibonacci number. Use a recursively defined subroutine.

■ 9.4 ARITHMETIC AND GEOMETRIC SEQUENCES

Arithmetic Sequences

ARITHMETIC SEQUENCE
ARITHMETIC PROGRESSION

When the difference between successive terms of a sequence is always the same number, the sequence is called **arithmetic**. Thus, an arithmetic sequence, sometimes called an **arithmetic progression**, may be defined recursively as

$$a_{n+1} - a_n = d \quad \text{or} \quad a_{n+1} = a_n + d \qquad (1)$$

COMMON DIFFERENCE

where the number d is called the **common difference**.

EXAMPLE 1 Show that each of the following sequences is arithmetic and find the common difference:

(a) $\{s_n\} = \{3n + 5\}$ (b) $\{t_n\} = \{4 - n\}$

Solution (a) The $(n + 1)$st and nth terms of the sequence $\{s_n\}$ are

$$s_{n+1} = 3(n + 1) + 5 = 3n + 8 \quad \text{and} \quad s_n = 3n + 5$$

Their difference is

$$s_{n+1} - s_n = (3n + 8) - (3n + 5) = 8 - 5 = 3$$

Thus, successive terms will always differ by the common difference 3, and so the sequence is arithmetic.

(b) Here,

$$t_{n+1} = 4 - (n + 1) = 3 - n \quad \text{and} \quad t_n = 4 - n$$

Their difference is

$$t_{n+1} - t_n = (3 - n) - (4 - n) = 3 - 4 = -1$$

The difference of two successive terms does not depend on n; it always equals the same number, -1. Hence, $\{t_n\}$ is an arithmetic sequence whose common difference is -1. ■

Suppose a_1 is the first term of an arithmetic sequence whose common difference is d. We seek a formula for the nth term a_n. To see the pattern, we write down the first few terms:

$$a_1 = a_1 + 0 \cdot d$$
$$a_2 = a_1 + d = a_1 + 1 \cdot d$$
$$a_3 = a_2 + d = (a_1 + d) + d = a_1 + 2 \cdot d$$
$$a_4 = a_3 + d = (a_1 + 2 \cdot d) + d = a_1 + 3 \cdot d$$
$$a_5 = a_4 + d = (a_1 + 3 \cdot d) + d = a_1 + 4 \cdot d$$
$$\vdots$$
$$a_n = a_{n-1} + d = [a_1 + (n - 2)d] + d = a_1 + (n - 1)d$$

We are led to the following result:

THEOREM For an arithmetic sequence $\{a_n\}$ whose first term is a_1 and whose common difference is d, the nth term obeys the formula

$$a_n = a_1 + (n - 1)d \qquad (2)$$

Proof A proof of this result requires mathematical induction. If $n = 1$, we have

$$a_1 = a_1 + (1 - 1)d = a_1 + 0 \cdot d = a_1$$

Thus, Condition I of the principle of mathematical induction is satisfied. Suppose for some k that

$$a_k = a_1 + (k - 1)d \qquad (3)$$

We want to show that

$$a_{k+1} = a_1 + [(k + 1) - 1]d = a_1 + kd$$

We do this as follows:

$$a_{k+1} = a_k + d = [a_1 + (k - 1)d] + d = [a_1 + kd - d] + d = a_1 + kd$$

Equation (3)

Definition of arithmetic sequence,
Equation (1)

Thus, Condition II of the principle of mathematical induction is also satisfied. The formula $a_n = a_1 + (n - 1)d$ therefore is true for all n.

EXAMPLE 2 Find the 13th term of the arithmetic sequence 2, 6, 10, 14, 18, ...

Solution The first term of this arithmetic sequence is $a_1 = 2$, and the common difference is 4. By Formula (2), the nth term is

$$a_n = 2 + (n - 1)4$$

Hence, the 13th term is

$$a_{13} = 2 + 12 \cdot 4 = 50$$

EXAMPLE 3 The 8th term of an arithmetic sequence is 75, and the 20th term is 39. Find the first term and the common difference. Give a recursive formula for the sequence.

Solution By Equation (2), we know that

$$\begin{cases} a_8 = a_1 + 7d = 75 \\ a_{20} = a_1 + 19d = 39 \end{cases}$$

This is a system of two equations containing two variables, which we shall solve by elimination. Thus, subtracting the second equation from the first equation, we get

$$-12d = 36$$
$$d = -3$$

With $d = -3$, we find $a_1 = 75 - 7d = 75 - 7(-3) = 96$. A recursive formula for this sequence is

$$a_{n+1} = a_n - 3, \qquad a_1 = 96$$

Based on Formula (2), a general formula for the sequence $\{a_n\}$ in Example 3 is

$$a_n = a_1 + (n - 1)d = 96 + (n - 1)(-3) = 99 - 3n$$

Adding the First n Terms of an Arithmetic Sequence

The next result gives a formula for finding the sum of the first n terms of an arithmetic sequence.

THEOREM For an arithmetic sequence $\{a_n\}$, the sum S_n of the first n terms is

$$S_n = \sum_{k=1}^{n} a_k = a_1 + a_2 + \cdots + a_n = na_1 + d\frac{n(n-1)}{2} \qquad (4)$$

Proof

$$S_n = \sum_{k=1}^{n} a_k = a_1 + a_2 + \cdots + a_n$$

$$= a_1 + (a_1 + d) + \cdots + [a_1 + (n - 1)d] \qquad \text{Equation (2)}$$

$$= \underbrace{(a_1 + a_1 + \cdots + a_1)}_{n \text{ terms}} + [d + 2d + \cdots + (n - 1)d]$$

$$= na_1 + d[1 + 2 + \cdots + (n - 1)]$$

$$= na_1 + d\frac{n(n - 1)}{2} \qquad \text{From Example 3,} \atop \text{page 552} \quad \blacksquare$$

This proof also can be done using the principle of mathematical induction. (Look back at Problem 30 in Exercise 9.1.)

EXAMPLE 4   Find the sum S_n of the first n terms of the sequence $\{3n + 5\}$; that is, find

$$8 + 11 + 14 + \cdots + (3n + 5)$$

Solution   The sequence $\{3n + 5\}$ is an arithmetic sequence with $a_1 = 8$ and $d = 3$ [see Example 1(a)]. The sum S_n we seek contains the first n terms of this sequence, so we use Formula (4) to get

$$S_n = \sum_{k=1}^{n} (3k + 5) = 8 + 11 + 14 + \cdots + (3n + 5)$$

$$= n \cdot 8 + 3\frac{n(n - 1)}{2} = \frac{1}{2}n(3n + 13) \qquad \blacksquare$$

Geometric Sequences

GEOMETRIC SEQUENCE
GEOMETRIC PROGRESSION

When the ratio of successive terms of a sequence is always the same nonzero number, the sequence is called **geometric**. Thus, a geometric sequence, sometimes called a **geometric progression**, may be defined recursively as

$$\frac{a_{n+1}}{a_n} = r \qquad \text{or} \qquad a_{n+1} = ra_n \qquad (5)$$

COMMON RATIO   where the fixed nonzero number r is called the **common ratio**.

EXAMPLE 5   Show that each of the following sequences is geometric, and find the common ratio:

(a) $\{s_n\} = 2^{-n}$ (b) $\{t_n\} = 4^n$

Solution (a) The $(n + 1)$st and nth terms of the sequence $\{s_n\}$ are

$$s_{n+1} = 2^{-(n+1)} \quad \text{and} \quad s_n = 2^{-n}$$

Their ratio is

$$\frac{s_{n+1}}{s_n} = \frac{2^{-(n+1)}}{2^{-n}} = 2^{-n-1+n} = 2^{-1} = \frac{1}{2}$$

$$\frac{a^m}{a^u} = a^{m-u}$$

Because the ratio of successive terms is a nonzero number independent of n, the sequence $\{s_n\}$ is geometric with common ratio $\frac{1}{2}$.

(b) Here,

$$t_{n+1} = 4^{n+1} \quad \text{and} \quad t_n = 4^n$$

The common ratio is

$$\frac{t_{n+1}}{t_n} = \frac{4^{n+1}}{4^n} = 4$$

Thus, $\{t_n\}$ is a geometric sequence with common ratio 4. ∎

Suppose a_1 is the first term of a geometric sequence with common ratio $r \neq 0$. We seek a formula for the nth term a_n. To see the pattern, we write down the first few terms:

$$a_1 = 1 \cdot a_1 = r^0 a_1$$
$$a_2 = ra_1 = r^1 a_1$$
$$a_3 = ra_2 = r(ra_1) = r^2 a_1$$
$$a_4 = ra_3 = r(r^2 a_1) = r^3 a_1$$
$$a_5 = ra_4 = r(r^3 a_1) = r^4 a_1$$
$$\vdots$$
$$a_n = ra_{n-1} = r(r^{n-2} a_1) = r^{n-1} a_1$$

We are led to the following result:

THEOREM For a geometric sequence $\{a_n\}$ whose first term is a_1 and whose common ratio is r, the nth term obeys the formula

$$a_n = r^{n-1} a_1 \qquad (6)$$

∎

A proof of this result requires mathematical induction. See Problem 86 at the end of this section.

EXAMPLE 6　Find the 9th term of the geometric sequence $2, \frac{2}{3}, \frac{2}{9}, \frac{2}{27}, \ldots$.

Solution　The first term of this geometric sequence is $a_1 = 2$, and the common ratio is $\frac{1}{3}$. (Use $\frac{2}{3}/2 = \frac{1}{3}$, or $\frac{2}{9}/\frac{2}{3} = \frac{1}{3}$, or any two successive terms.) By Formula (6), the nth term is

$$a_n = \left(\frac{1}{3}\right)^{n-1} 2$$

Hence, the 9th term is

$$a_9 = \left(\frac{1}{3}\right)^8 2 = \frac{2}{3^8} = \frac{2}{6561}$$

$\blacksquare$

Adding the First n Terms of a Geometric Sequence

The next result gives us a formula for finding the sum of the first n terms of a geometric sequence.

THEOREM　For a geometric sequence $\{a_n\}$, the sum S_n of the first n terms is

$$S_n = \sum_{k=1}^{n} a_k = a_1 + a_2 + \cdots + a_n = a_1 \frac{1 - r^n}{1 - r} \qquad r \neq 1 \quad (7)$$

Proof

$$S_n = \sum_{k=1}^{n} a_k = a_1 + a_2 + \cdots + a_n = a_1 + ra_1 + \cdots + r^{n-1}a_1 \quad \text{Equation (6)}$$

$$= a_1(1 + r + \cdots + r^{n-1})$$

$$= a_1(1 + r + \cdots + r^{n-1})\frac{1 - r}{1 - r}$$

$$= a_1 \frac{1 - r^n}{1 - r}$$

$\blacksquare$

This proof also can be done using the principle of mathematical induction. (Look back at Problem 29 in Exercise 9.1.)

EXAMPLE 7　Find the sum S_n of the first n terms of the sequence $\{(\frac{1}{2})^n\}$; that is, find

$$\frac{1}{2} + \frac{1}{4} + \frac{1}{8} + \cdots + \left(\frac{1}{2}\right)^n$$

Solution   The sequence $\{(\frac{1}{2})^n\}$ is a geometric sequence with $a_1 = \frac{1}{2}$ and $r = \frac{1}{2}$. The sum S_n we seek is the sum of the first n terms of the sequence, so we use Formula (7) to get

$$S_n = \sum_{k=1}^{n} \left(\frac{1}{2}\right)^k = \frac{1}{2} + \frac{1}{4} + \frac{1}{8} + \cdots + \left(\frac{1}{2}\right)^n = \frac{1}{2}\left[\frac{1 - (\frac{1}{2})^n}{1 - \frac{1}{2}}\right]$$

$$= \frac{1}{2}\left[\frac{1 - (\frac{1}{2})^n}{\frac{1}{2}}\right] = 1 - \frac{1}{2^n}$$

≡ HISTORICAL COMMENT

Sequences are among the oldest objects of mathematical investigation, having been studied for over 3500 years. After the initial steps, however, little progress was made until about 1600.

Arithmetic and geometric sequences were present in the Rhind papyrus, a mathematical text containing 85 problems copied around 1650 B.C. by the Egyptian scribe Ahmes from an earlier work. (See the Historical Problems following this section.) Fibonacci (A.D. 1220) wrote about problems similar to those found in the Rhind papyrus, leading one to suspect that Fibonacci may have had material available that is now lost. This material would have been in the non-Euclidean Greek tradition of Heron (about A.D. 75) and Diophantus (about A.D. 250). One problem, again modified slightly, is still with us in the puzzle rhyme "As I was going to St. Ives . . ." (see Historical Problem 2).

The Rhind papyrus indicates the Egyptians knew how to add up the terms of an arithmetic or geometric sequence, as did the Babylonians. The rule for summing up a geometric sequence is found in Euclid's *Elements* (book IX, 35, 36) where, like all of Euclid's algebra, it is presented in a geometric form.

Investigations of other kinds of sequences began in the 1500s, when algebra became sufficiently developed to handle the more complicated problems. The development of calculus in the 1600s added a powerful new tool, especially for finding the sum of infinite sequences, and the subject continues to flourish today.

≡ HISTORICAL PROBLEMS

1. *Arithmetic sequence problem from the Rhind papyrus* (1650 B.C.); *statement modified slightly for clarity* One hundred loaves of bread are to be divided among five people so that the amounts they receive form an arithmetic sequence. The first two together receive one-seventh of what the last three receive. How many does each receive? [*Partial answer:* First person receives $1\frac{2}{3}$ loaves.]

2. The following old English children's rhyme resembles one of the Rhind papyrus problems:

> As I was going to St. Ives
> I met a man with seven wives
> Each wife had seven sacks
> Each sack had seven cats
> Each cat had seven kits [kittens]
> Kits, cats, sacks, wives,
> How many were going to St. Ives?

(a) Assuming that the speaker and the cat fanciers met by traveling in opposite directions, what is the answer?
(b) How many kittens are being transported?
(c) Kits, cats, sacks, wives; how many? [*Hint:* It is easier to include the man, find the sum with the formula, and then subtract 1 for the man.] ≡

EXERCISE 9.4   *In Problems 1–10 an arithmetic sequence is given. Find the common difference and write out the first four terms.*

1. $\{n + 5\}$

2. $\{n - 3\}$

3. $\{2n - 5\}$

4. $\{3n + 1\}$

5. $\{6 - 2n\}$

6. $\{4 - 2n\}$

7. $\left\{\dfrac{1}{2} - \dfrac{1}{3}n\right\}$

8. $\left\{\dfrac{2}{3} + \dfrac{n}{4}\right\}$

9. $\{\ln 3^n\}$

10. $\{e^{\ln n}\}$

In Problems 11–20 a geometric sequence is given. Find the common ratio and write out the first four terms.

11. $\{2^n\}$

12. $\{(-4)^n\}$

13. $\left\{-3\left(\dfrac{1}{2}\right)^n\right\}$

14. $\left\{\left(\dfrac{5}{2}\right)^n\right\}$

15. $\left\{\dfrac{2^{n-1}}{4}\right\}$

16. $\left\{\dfrac{3^n}{9}\right\}$

17. $\{2^{n/3}\}$

18. $\{3^{2n}\}$

19. $\left\{\dfrac{3^{n-1}}{2^n}\right\}$

20. $\left\{\dfrac{2^n}{3^{n-1}}\right\}$

In Problems 21–34 determine whether the given sequence is arithmetic, geometric, or neither. If the sequence is arithmetic, find the common difference; if it is geometric, find the common ratio.

21. $\{n + 2\}$

22. $\{2n - 5\}$

23. $\{2n^2\}$

24. $\{3n^2 + 1\}$

25. $\left\{3 - \dfrac{2}{3}n\right\}$

26. $\left\{8 - \dfrac{3}{4}n\right\}$

27. 1, 3, 6, 10, . . .

28. 2, 4, 6, 8, . . .

29. $\left\{\left(\dfrac{2}{3}\right)^n\right\}$

30. $\left\{\left(\dfrac{5}{4}\right)^n\right\}$

31. $-1, -2, -4, -8, \ldots$

32. 1, 1, 2, 3, 5, 8, . . .

33. $\{3^{n/2}\}$

34. $\{(-1)^n\}$

In Problems 35–42 find the fifth term and the nth term of the arithmetic sequence whose initial term a_1 and common difference d are given.

35. $a_1 = 1; \quad d = 2$

36. $a_1 = -1; \quad d = 3$

37. $a_1 = 5; \quad d = -3$

38. $a_1 = 6; \quad d = -2$

39. $a_1 = 0; \quad d = \dfrac{1}{2}$

40. $a_1 = 1; \quad d = -\dfrac{1}{3}$

41. $a_1 = \sqrt{2}; \quad d = \sqrt{2}$

42. $a_1 = 0; \quad d = \pi$

In Problems 43–50 find the fifth term and the nth term of the geometric sequence whose initial term a_1 and common ratio r are given.

43. $a_1 = 1; \quad r = 2$

44. $a_1 = -1; \quad r = 3$

45. $a_1 = 5; \quad r = -1$

46. $a_1 = 6; \quad r = -2$

47. $a_1 = 0; \quad r = \dfrac{1}{2}$

48. $a_1 = 1; \quad r = -\dfrac{1}{3}$

49. $a_1 = \sqrt{2}; \quad r = \sqrt{2}$

50. $a_1 = 0; \quad r = 1/\pi$

In Problems 51–56 find the indicated term in each arithmetic sequence.

51. 12th term of 2, 4, 6, . . .

52. 8th term of $-1, 1, 3, \ldots$

53. 10th term of $1, -2, -5, \ldots$

54. 9th term of $5, 0, -5, \ldots$

55. 8th term of $a, a + b, a + 2b, \ldots$

56. 7th term of $2\sqrt{5}, 4\sqrt{5}, 6\sqrt{5}, \ldots$

In Problems 57–62 find the indicated term in each geometric sequence.

57. 7th term of $1, \frac{1}{2}, \frac{1}{4}, \ldots$

58. 8th term of 1, 3, 9, . . .

59. 9th term of $1, -1, 1, \ldots$

60. 10th term of $-1, 2, -4, \ldots$

61. 8th term of 0.4, 0.04, 0.004, . . .

62. 7th term of 0.1, 1.0, 10.0, . . .

In Problems 63–70 find the first term and the common difference of the arithmetic sequence described.

63. 8th term is 8; 20th term is 44

64. 4th term is 3; 20th term is 35

65. 9th term is -5; 15th term is 31

66. 8th term is 4; 18th term is -96

67. 15th term is 0; 40th term is -50

68. 5th term is −2;   13th term is 30

69. 14th term is −1;   18th term is −9

70. 12th term is 4;   18th term is 28

In Problems 71–80 find the sum.

71. $6 + 11 + 16 + \cdots + (1 + 5n)$

72. $-2 + 1 + 4 + \cdots + (3n - 5)$

73. $\dfrac{1}{4} + \dfrac{2}{4} + \dfrac{2^2}{4} + \dfrac{2^3}{4} + \cdots + \dfrac{2^{n-1}}{4}$

74. $\dfrac{3}{9} + \dfrac{3^2}{9} + \dfrac{3^3}{9} + \cdots + \dfrac{3^n}{9}$

75. $\dfrac{2}{3} + \left(\dfrac{2}{3}\right)^2 + \left(\dfrac{2}{3}\right)^3 + \cdots + \left(\dfrac{2}{3}\right)^n$

76. $4 + 4 \cdot 3 + 4 \cdot 3^2 + \cdots + 4 \cdot 3^{n-1}$

77. $2 + 4 + 6 + \cdots + 2n$

78. $1 + 3 + 5 + \cdots + (2n - 1)$

79. $-1 - 2 - 4 - 8 - \cdots - (2^{n-1})$

80. $2 + 0 - 2 - 4 - \cdots - (2n - 4)$

C **81.** A ball is dropped from a height of 20 feet. Each time it strikes the ground, it bounces up to $\frac{3}{4}$ of the previous height. What height will the ball bounce up to after it strikes the ground for the third time? What is its height after it strikes the ground for the nth time?

C **82.** Your friend has just been hired at an annual salary of $20,000. If she expects to receive annual increases of 4%, what will her salary be as she begins her fifth year?

C **83.** Compute the distance the ball in Problem 81 has traveled when it strikes the ground for the fifth time.

C **84.** A rich man promises to give you $1000 on September 1, 1987. Each day thereafter he will give you $\frac{9}{10}$ of what he gave you the previous day. What is the first date on which the amount you receive is less than 1¢? How much have you received when this happens?

C **85.** In an old fable, a commoner, who had just saved the king's life, was told he could ask the king for any just reward. Being a shrewd man, the commoner said, "A simple wish, sire. Place one grain of wheat on the first square of a chessboard, two grains on the second square, four grains on the third square, continuing until you have filled the board. This is all I seek." Compute the total number of grains needed to do this to see why the request, seemingly simple, could not be granted. [A chessboard (or checkerboard) consists of $8 \times 8 = 64$ squares.]

86. Use mathematical induction to prove that, for a geometric sequence $\{a_n\}$ with common ratio r, the nth term a_n obeys the formula

$$a_n = r^{n-1}a_1$$

C H A P T E R   R E V I E W

VOCABULARY

principle of
  mathematical
  induction
 factorial symbol $n!$
 symbol $\binom{n}{r}$
 Pascal triangle
 binomial theorem
 binomial coefficient
	sequence
	 terms of a sequence
	 alternating signs
	 recursively
	 recursive formula
	 Fibonacci sequence
	 Fibonacci numbers
	 summation notation
		index
		 arithmetic sequence
		 arithmetic
		  progression
		 common difference
		 geometric sequence
		 geometric progression
		 common ratio

**FILL-IN-THE-BLANK
QUESTIONS**

1. The _____ _____ is a triangular display of the binomial coefficients.

2. $\binom{6}{2}$ = _____

3. A(n) _____ is a function whose domain is the set of positive integers.

4. In a(n) _____ sequence, the difference between successive terms is always the same number.

5. In a(n) _____ sequence, the ratio of successive terms is always the same number.

**REVIEW
EXERCISES**

In Problems 1–6 use the principle of mathematical induction to show that the given statement is true for all natural numbers.

1. $3 + 6 + 9 + \cdots + 3n = \dfrac{3n}{2}(n + 1)$

2. $2 + 6 + 10 + \cdots + (4n - 2) = 2n^2$

3. $2 + 6 + 18 + \cdots + 2 \cdot 3^{n-1} = 3^n - 1$

4. $3 + 6 + 12 + \cdots + 3 \cdot 2^{n-1} = 3(2^n - 1)$

5. $1^2 + 4^2 + 7^2 + \cdots + (3n - 2)^2 = \dfrac{1}{2}n(6n^2 - 3n - 1)$

6. $1 \cdot 3 + 2 \cdot 4 + 3 \cdot 5 + \cdots + n(n + 2) = \dfrac{n}{6}(n + 1)(2n + 7)$

In Problems 7–10 evaluate each expression.

7. $5!$ 8. $6!$ 9. $\binom{5}{2}$ 10. $\binom{8}{6}$

In Problems 11–14 expand each expression using the binomial theorem.

11. $(x + 2)^4$ 12. $(x - 3)^5$ 13. $(2x + 3)^5$ 14. $(3x - 4)^4$

15. Find the coefficient of x^7 in the expansion of $(x + 2)^9$.

16. Find the coefficient of x^3 in the expansion of $(x - 3)^8$.

17. Find the coefficient of x^2 in the expansion of $(2x + 1)^7$.

18. Find the coefficient of x^6 in the expansion of $(2x + 1)^8$.

In Problems 19–26 write down the first five terms of each sequence.

19. $\left\{(-1)^n \dfrac{n + 1}{n + 2}\right\}$

20. $\{(-1)^{n+1}(2n - 3)\}$

21. $\left\{\dfrac{2^n}{n^2}\right\}$

22. $\left\{\dfrac{e^n}{n}\right\}$

23. $a_1 = 3, \quad a_{n+1} = \frac{2}{3}a_n$

24. $a_1 = 4, \quad a_{n+1} = -\frac{1}{4}a_n$

25. $a_1 = 2, \quad a_{n+1} = 2 - a_n$

26. $a_1 = -3, \quad a_{n+1} = 4 + a_n$

In Problems 27–38 determine whether the given sequence is arithmetic, geometric, or neither. If the sequence is arithmetic, find the common difference and the sum of the first n terms. If the sequence is geometric, find the common ratio and the sum of the first n terms.

27. $\{n + 3\}$

28. $\{4n + 1\}$

29. $\{2n^3\}$

30. $\{2n^2 - 1\}$

31. $\{2^{3n}\}$

32. $\{3^{2n}\}$

33. $0, 4, 8, 12, \ldots$

34. $1, -3, -7, -11, \ldots$

35. $3, \frac{3}{2}, \frac{3}{4}, \frac{3}{8}, \frac{3}{16}, \ldots$

36. $5, -\frac{5}{3}, \frac{5}{9}, -\frac{5}{27}, \frac{5}{81}, \ldots$

37. $\frac{2}{3}, \frac{3}{4}, \frac{4}{5}, \frac{5}{6}, \ldots$

38. $\frac{3}{2}, \frac{5}{4}, \frac{7}{6}, \frac{9}{8}, \frac{11}{10}, \ldots$

In Problems 39–44 find the indicated term in each sequence.

39. 7th term of $3, 7, 11, 15, \ldots$

40. 7th term of $1, -1, -3, -5, \ldots$

41. 11th term of $1, \frac{1}{10}, \frac{1}{100}, \ldots$

42. 11th term of $1, 2, 4, 8, \ldots$

43. 9th term of $\sqrt{2}, 2\sqrt{2}, 3\sqrt{2}, \ldots$

44. 9th term of $\sqrt{2}, 2, 2^{3/2}, \ldots$

In Problems 45–48 find a general formula for each arithmetic sequence.

45. 7th term is 31;   20th term is 96

46. 8th term is -20;   17th term is -47

47. 10th term is 0;   18th term is 8

48. 12th term is 30;   22nd term is 50

SETS; COUNTING; PROBABILITY

10.1   Sets and Counting
10.2   Permutations
10.3   Combinations
10.4   Probability
        Chapter Review

The first three sections of this chapter deal with techniques and formulas for counting the number of objects in a set, a part of the branch of mathematics called *combinatorics*. These formulas are used in computer science to analyze algorithms and recursive functions and to study stacks and queues. They are also used to determine probabilities, the likelihood that a certain outcome of a random experiment will occur, which is the subject of the fourth section in this chapter.

Applications of the topics in this chapter can be found in fields including engineering, business and economics, the social sciences, and the biological sciences. You will see these topics again in more detail if you take a course in finite mathematics or discrete mathematics.

≡ 10.1   SETS AND COUNTING

Sets

SET
ELEMENTS
WELL-DEFINED
EMPTY SET ∅
NULL SET ∅

A **set** is a well-defined collection of distinct objects. The objects of a set are called its **elements**. By well-defined we mean there is a rule that enables us to tell whether or not a given object is an element of the set. If a set has no elements, it is called the **empty set**, or **null set**, and is denoted by the symbol ∅.

Because the elements of a set are distinct, we never repeat elements. Thus, we would never write {1, 2, 3, 2}; the correct listing would be {1, 2, 3}. Furthermore, because a set is a collection, the order in which the elements are listed is immaterial. Thus, {1, 2, 3}, {1, 3, 2}, {2, 1, 3}, and so on, all represent the same set.

EXAMPLE 1   Write the set consisting of the possible outcomes from tossing a coin twice. Use H for "heads" and T for "tails."

Solution   In tossing a coin twice, we can get heads each time, HH; or heads the first time and tails the second, HT; or tails the first time and heads the second, TH; or tails each time, TT. Because no other possibilities exist, the set of outcomes is

$$\{HH, HT, TH, TT\}$$ ≡

If two sets A and B have precisely the same elements, then we say that A and B are **equal** and write $A = B$.

EQUAL SETS

If each element of a set A is also an element of a set B, then we say that A is a **subset** of B and write $A \subseteq B$.

SUBSET
PROPER SUBSET

If $A \subseteq B$ and $A \neq B$, then we say A is a **proper subset** of B and write $A \subset B$.

Thus, if $A \subseteq B$, every element in set A is also in set B, but B may or may not have additional elements. If $A \subset B$, every element in A is also in B, and B does have elements not found in A. Finally, we agree that the empty set is a subset of every set; that is,

$$\emptyset \subseteq A \qquad \text{for any set } A$$

EXAMPLE 2   Consider the sets

$$A = \{1, 2, 3, 4, 5, 6, 7, 8, 9, 10\}$$
$$B = \{1, 3, 5, 7, 9\}$$
$$C = \{3, 5, 7\}$$

Some possible relationships involving these sets are:

1. $B \subseteq A$, because every element in B is also in A.
2. $\varnothing \subseteq C$, because the empty set is a subset of any set.
3. $C \subseteq B$, because every element in C is also in B.
4. $C \subset B$, because every element in C is also in B, and B has elements not found in C. ≡

EXAMPLE 3   Write down all the subsets of the set {a, b, c}.

Solution   To organize our work, we write down all the subsets with no elements, then those with one element, then those with two elements, and finally those with three elements. These will give us all the subsets—do you see why?

0 ELEMENTS	1 ELEMENT	2 ELEMENTS	3 ELEMENTS
$\varnothing$	{a}, {b}, {c}	{a, b}, {b, c}, {a, c}	{a, b, c}

≡

Intersection, Union, and Complements

INTERSECTION   If A and B are sets, the **intersection** of A with B, denoted by $A \cap B$, is the set consisting of elements that belong to *both A and B*. The

UNION   **union** of A with B, denoted by $A \cup B$, is the set consisting of elements that belong to *either A or B*.

EXAMPLE 4   Let $A = \{1, 3, 5, 8\}$, $B = \{3, 5, 7\}$, and $C = \{2, 4, 6, 8\}$. Find:

(a) $A \cap B$ (b) $A \cup B$ (c) $B \cap (A \cup C)$

Solution   (a) $A \cap B = \{1, 3, 5, 8\} \cap \{3, 5, 7\} = \{3, 5\}$

(b) $A \cup B = \{1, 3, 5, 8\} \cup \{3, 5, 7\} = \{1, 3, 5, 7, 8\}$

(c) $B \cap (A \cup C) = \{3, 5, 7\} \cap [\{1, 3, 5, 8\} \cup \{2, 4, 6, 8\}]$
$$= \{3, 5, 7\} \cap \{1, 2, 3, 4, 5, 6, 8\}$$
$$= \{3, 5\}$$

≡

UNIVERSAL SET   Usually in working with sets, we designate a **universal set**—namely, the set consisting of all the elements we wish to consider. Once a universal set has been designated, we can consider elements not found in a given set.

COMPLEMENT OF A SET   If A is a set, the **complement** of A, denoted by A', is the set consisting of all the elements of the universal set that are not in A.

EXAMPLE 5   If the universal set is $U = \{1, 2, 3, 4, 5, 6, 7, 8, 9\}$, and if $A = \{1, 3, 5, 7, 9\}$, then $A' = \{2, 4, 6, 8\}$. ≡

Venn Diagrams

It is often helpful to draw pictures of sets. Such pictures, called **Venn diagrams**, represent sets as overlapping circles enclosed in a large rectangle, which represents the universal set. See Figure 1.

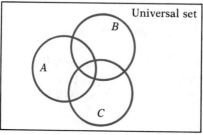

FIGURE 1

If we know in advance that $A \subseteq B$, we would use the Venn diagram in Figure 2(a). If we know that A and B have no elements in common, that is, if $A \cap B = \varnothing$, we would use the Venn diagram in Figure 2(b).

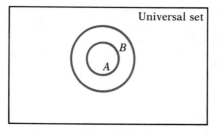
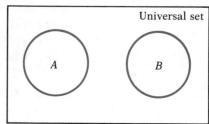

(a) $A \subseteq B$ **(b)** A and B have no elements in common, $A \cap B = \varnothing$

FIGURE 2

Figures 3(a), 3(b), and 3(c) use Venn diagrams to illustrate the definitions of intersection, union, and complement, respectively.

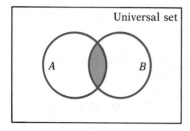
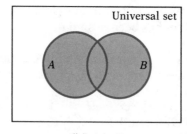
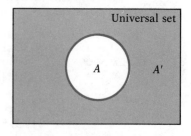

(a) $A \cap B$ **(b)** $A \cup B$ **(c)** A'

FIGURE 3

Counting

As you count the number of students in a classroom or the number of pennies in your pocket, what you are really doing is matching each object to be counted exactly once with the counting numbers 1, 2, 3, ... until no objects remain. If a set A to be counted matched up in this fashion to the set $\{1, 2, \ldots, 25\}$, you would conclude there were 25 elements in the set A. We use the notation $n(A) = 25$ to indicate there are 25 elements in the set A.

Because the empty set has no elements, we write

$$n(\varnothing) = 0$$

FINITE SET
INFINITE SET

If the number of elements in a set is a nonnegative integer, we say the set is **finite**. Otherwise, it is **infinite**. We shall concern ourselves only with finite sets.

From Example 3, we can see that a set with three elements has $2^3 = 8$ subsets. In fact, it can be shown that a set with n elements has exactly 2^n subsets. This fact has an important application to computers, which we take up at the end of this section.

EXAMPLE 6   In a survey of 100 college students, 35 were registered in College Algebra, 52 were registered in Introduction to Computer Science, and 18 were in both courses. How many were registered in neither course?

Solution   First, we let

$A =$ Set of students in College Algebra

$B =$ Set of students in Introduction to Computer Science

Then the information tells us that

$$n(A) = 35 \qquad n(B) = 52 \qquad n(A \cap B) = 18$$

Refer to Figure 4. Do you see how the numerical entries were determined? Based on the diagram, we conclude that $17 + 18 + 34 = 69$ students were registered in either course. Since 100 students were surveyed, it follows that $100 - 69 = 31$ were registered in neither course.

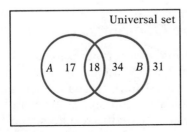

FIGURE 4

The conclusions drawn in Example 6 lead us to formulate a general counting formula. If we count the elements in each of two sets A and B, we necessarily count twice any elements that are in both A and B, that is, those elements in $A \cap B$. Thus, to count correctly the elements that are in A or B, that is, to find $n(A \cup B)$, we need to subtract those in $A \cap B$ from $n(A) + n(B)$.

THEOREM   If A and B are finite sets, then

COUNTING FORMULA

$$n(A \cup B) = n(A) + n(B) - n(A \cap B) \qquad (1)$$

≡

A special case of the counting formula, Formula (1), occurs if A and B have no elements in common. In this case, $A \cap B = \varnothing$ so that $n(A \cap B) = 0$.

THEOREM   If two sets A and B have no elements in common, then

ADDITION PRINCIPLE OF
COUNTING

$$n(A \cup B) = n(A) + n(B) \qquad (2)$$

≡

EXAMPLE 7   A certain code is to consist of either a letter of the alphabet or a digit. How many codes are possible?

Solution   Let the sets A and B be defined as

$$A = \text{Set of letters in the alphabet}$$
$$B = \text{Set of digits } \{0, 1, 2, \ldots, 9\}$$

Then,

$$n(A) = 26 \qquad n(B) = 10$$

Because letters and digits are different, $A \cap B = \varnothing$. The number of ways either a letter or a digit can be chosen is, therefore,

$$n(A \cup B) = n(A) + n(B) = 26 + 10 = 36 \qquad ≡$$

Application to Computers

Information is stored in a computer using a series of switches, which may be on or off. These switches are denoted by either the number 0 (off) or the number 1 (on); these numbers are the binary digits, or **bits**. A **register** holds a certain fixed number of bits. For example, the Z-80 microprocessor has 8-bit registers; the PDP-11 minicomputer has 16-bit registers; and the IBM-370 computer has 32-bit registers.

BITS

REGISTER

Thus, a Z-80 register might hold an entry that looks like this: 01111001 (8 bits). We wish to find out how many different representations are possible in a given register.

We proceed in steps, first looking at a hypothetical 3-bit register. Look again at the solution to Example 3 and arrange all the subsets of {a, b, c} as shown in Table 1. As the table illustrates, the number of subsets of a set with 3 elements equals the number of different representations in a 3-bit register. A set with n elements has 2^n subsets; thus, an n-bit register has 2^n representations. So an 8-bit register can hold $2^8 = 256$ different symbols.

TABLE 1

a	b	c	SUBSET
0	0	0	$\varnothing$
1	0	0	{a}
0	1	0	{b}
0	0	1	{c}
1	1	0	{a, b}
0	1	1	{b, c}
1	0	1	{a, c}
1	1	1	{a, b, c}

EXERCISE 10.1   *In Problems 1–10 use* $A = \{1, 3, 5, 7, 9\}$, $B = \{1, 5, 6, 7\}$, *and* $C = \{1, 2, 4, 6, 8, 9\}$ *to find each set.*

1. $A \cup B$ **2.** $A \cup C$ **3.** $A \cap B$

4. $A \cap C$ **5.** $(A \cup B) \cap C$ **6.** $(A \cap C) \cup (B \cap C)$

7. $(A \cap B) \cup C$ **8.** $(A \cup B) \cup C$ **9.** $(A \cup C) \cap (B \cup C)$

10. $(A \cap B) \cap C$

In Problems 11–20 use $U = $ *Universal set* $= \{0, 1, 2, 3, 4, 5, 6, 7, 8, 9\}$, $A = \{1, 3, 4, 5, 9\}$, $B = \{2, 4, 6, 7, 8\}$, *and* $C = \{1, 3, 4, 6\}$ *to find each set.*

11. A' **12.** C' **13.** $(A \cap B)'$

14. $(B \cup C)'$ **15.** $A' \cup B'$ **16.** $B' \cap C'$

17. $(A \cap C')'$ **18.** $(B' \cup C)'$ **19.** $(A \cup B \cup C)'$

20. $(A \cap B \cap C)'$

21. Write down all the subsets of {a, b, c, d}.

22. Write down all the subsets of {a, b, c, d, e}.

23. If $n(A) = 15$, $n(B) = 20$, and $n(A \cap B) = 10$, find $n(A \cup B)$.

24. If $n(A) = 20$, $n(B) = 40$, and $n(A \cup B) = 35$, find $n(A \cap B)$.

25. If $n(A \cup B) = 50$, $n(A \cap B) = 10$, and $n(B) = 20$, find $n(A)$.

26. If $n(A \cup B) = 60$, $n(A \cap B) = 40$, and $n(A) = n(B)$, find $n(A)$.

In Problems 27–34 use the information given in the figure below.

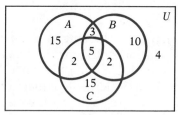

27. How many are in set A?
28. How many are in set B?
29. How many are in A or B?
30. How many are in A and B?
31. How many are in A but not C?
32. How many are not in A?
33. How many are in A and B and C?
34. How many are in A or B or C?
35. In a consumer survey of 500 people, 200 indicated they would be buying a major appliance within the next month; 150 indicated they would buy a car, and 25 said they would purchase both a major appliance and a car. How many will purchase neither? How many will purchase only a car?
36. In a student survey, 200 indicated they would attend Summer Session I and 150 indicated Summer Session II. If 75 students plan to attend both summer sessions and 275 indicated they would attend neither session, how many students participated in the survey?
37. In a survey of 100 investors in the stock market,

    50 owned shares in IBM

    40 owned shares in AT&T

    45 owned shares in GE

    20 owned shares in both IBM and AT&T

    20 owned shares in both IBM and GE

    15 owned shares in both AT&T and GE

    5 owned shares in all three

    (a) How many of the investors surveyed did not have shares in any of the three stocks?
    (b) How many owned just IBM shares?
    (c) How many owned just GE shares?
    (d) How many owned neither IBM nor GE?
    (e) How many owned either IBM or AT&T but no GE?
38. Human blood is classified as either Rh+ or Rh−. Blood is also classified by type: A, if it contains an A antigen; B, if it contains a B antigen; AB, if it contains both A and B antigens; and O, if it contains neither antigen. Draw a Venn diagram illustrating the various blood types. Based on this classification, how many different kinds of blood are there?

▚ 10.2 PERMUTATIONS

Counting plays a major role in many diverse areas, such as probability, statistics, and computer science. In this section and the next, we shall look at special types of counting problems and develop general formulas for solving them.

We begin with an example that will demonstrate a general counting principle.

EXAMPLE 1 The fixed-price dinner at a restaurant provides the following choices:

Appetizer: soup, salad, or cole slaw

Entree: baked chicken, broiled beef patty, baby beef liver, or roast beef au jus

Dessert: ice cream or cheese cake

How many different meals can be ordered?

Solution Ordering such a meal requires three separate decisions:

CHOOSE AN APPETIZER　　　CHOOSE AN ENTREE　　　CHOOSE A DESSERT

3 choices　　　　　　　　4 choices　　　　　　　2 choices

TREE DIAGRAM Look at the **tree diagram** in Figure 5 on the next page. We see that, for each choice of appetizer, there are 4 choices of entrees. And for each of these $3 \cdot 4 = 12$ choices, there are 2 choices for dessert. Thus, there are a total of

$$3 \cdot 4 \cdot 2 = 24$$

different meals that can be ordered. ▬

Example 1 illustrates a general counting principle.

THEOREM If a task consists of a sequence of choices in which there are p selections for the first choice, q selections for the second choice, r selections for the third choice, and so on, then the task of making these selections can be done in

MULTIPLICATION PRINCIPLE OF COUNTING

$$p \cdot q \cdot r \cdot \cdot \cdot \cdot$$

different ways. ▬

EXAMPLE 2 The International Airline Transportation Association (IATA) assigns 3-letter codes to represent airport locations. For example, JFK represents Kennedy International in New York. How many different airport codes are possible?

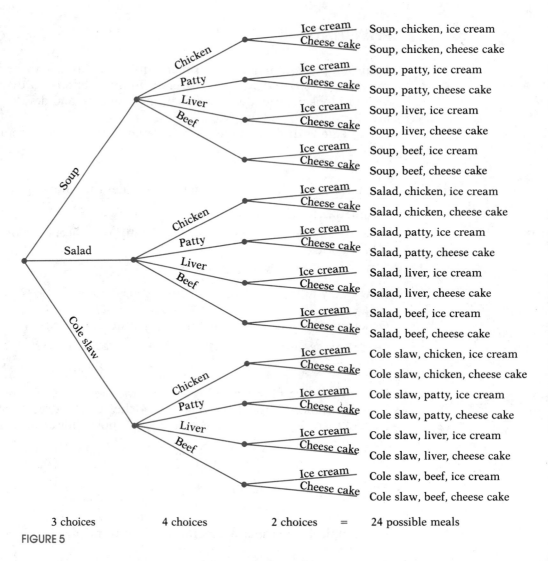

			Ice cream	Soup, chicken, ice cream
	Chicken		Cheese cake	Soup, chicken, cheese cake
	Patty		Ice cream	Soup, patty, ice cream
	Liver		Cheese cake	Soup, patty, cheese cake
Soup	Beef		Ice cream	Soup, liver, ice cream

3 choices   4 choices   2 choices   =   24 possible meals

FIGURE 5

Solution   The task consists of making three selections. Each selection requires choosing a letter of the alphabet (26 choices). Thus, by the multiplication principle, there are

$$26 \cdot 26 \cdot 26 = 17{,}576$$

different airport codes.   ≡

In Example 2, we were allowed to repeat a letter. For example, a valid airport code is FLL (Ft. Lauderdale International Airport), in which the letter L appears twice. In the next example, such repetition is not allowed.

EXAMPLE 3   Suppose we wish to establish a 3-letter code using any of the 26 letters of the alphabet, but we require that no letter be used more than once. How many different 3-letter codes are there?

Solution   The task consists of making three selections. The first selection requires choosing from 26 letters. Because no letter can be used more than once, the second selection requires choosing from 25 letters. The third selection requires choosing from 24 letters. (Do you see why?) By the multiplication principle, there are

$$26 \cdot 25 \cdot 24 = 15,600$$

different 3-letter codes.   ▬

Permutations

Example 3 illustrates a type of counting problem referred to as a *permutation*.

PERMUTATION   A **permutation** is an ordered arrangement of n distinct objects. The symbol $P(n, r)$ represents the number of ordered arrangements of n objects, using $r \leq n$ of the objects, without using any of them more than once.*

For example, the question posed in Example 3 asks for the number of ways the 26 letters of the alphabet can be arranged using 3 non-repeated letters. The answer is

$$P(26, 3) = 26 \cdot 25 \cdot 24 = 15,600$$

To arrive at a formula for $P(n, r)$, we note that the task of obtaining an ordered arrangement of n objects in which only $r \leq n$ of them are used, without repeating any of them, requires making r selections. For the first selection, there are n choices; for the second selection, there are $n - 1$ choices; for the third, there are $n - 2$ choices; ...; for the rth selection, there are $n - (r - 1)$ choices. By the multiplication principle, we have

$$
\begin{array}{c}
\quad \text{1st} \quad\ \text{2nd} \quad\ \ \text{3rd} \qquad\qquad\quad\ \text{rth} \\
P(n, r) = \ n \ \cdot (n - 1) \cdot (n - 2) \cdot \cdots \cdot n - (r - 1)
\end{array}
$$

*The symbol $P(n, r)$ is read as "the number of permutations of n objects taken r at a time."

This formula for $P(n, r)$ can be compactly written by using factorial notation (introduced in Section 9.2):

$$P(n, r) = n \cdot (n - 1) \cdot (n - 2) \cdot \cdots \cdot (n - r + 1)$$

$$= n \cdot (n - 1) \cdot (n - 2) \cdot \cdots \cdot (n - r + 1) \cdot \frac{(n - r) \cdot \cdots \cdot 3 \cdot 2 \cdot 1}{(n - r) \cdot \cdots \cdot 3 \cdot 2 \cdot 1}$$

$$= \frac{n!}{(n - r)!}$$

THEOREM   The number of different arrangements of n objects using $r \leq n$ of them in which

1. the n objects are distinct
2. once an object is used it cannot be repeated
3. order is important

is given by the formula

$$P(n, r) = \frac{n!}{(n - r)!} \tag{1}$$

≡

EXAMPLE 4   Evaluate:

(a) $P(7, 3)$

(b) $P(6, 1)$

(c) $P(4, 4)$

Solution   We shall work each problem in two ways.

(a) $P(7, 3) = \underbrace{7 \cdot 6 \cdot 5}_{3 \text{ factors}} = 210$

$$P(7, 3) = \frac{7!}{(7 - 3)!} = \frac{7!}{4!} = \frac{7 \cdot 6 \cdot 5 \cdot \cancel{4!}}{\cancel{4!}} = 210$$

(b) $P(6, 1) = \underbrace{6}_{1 \text{ factor}} = 6$

$$P(6, 1) = \frac{6!}{(6 - 1)!} = \frac{6!}{5!} = \frac{6 \cdot \cancel{5!}}{\cancel{5!}} = 6$$

(c) $P(4, 4) = \underbrace{4 \cdot 3 \cdot 2 \cdot 1}_{4 \text{ factors}} = 24$

$$P(4, 4) = \frac{4!}{(4 - 4)!} = \frac{4!}{0!} = \frac{4 \cdot 3 \cdot 2 \cdot 1}{1} = 24$$

≡

EXAMPLE 5   In how many ways can 5 people be lined up?

Solution   The 5 people are obviously distinct; once a person is in line, that person will not be repeated elsewhere in the line; and, in lining up people, order is important. Thus, we have a permutation of 5 objects taken 5 at a time. We can line up the 5 people in

$$P(5, 5) = 5! = 120 \text{ ways}$$

EXERCISE 10.2

1. A man has 5 shirts and 3 ties. How many different shirt and tie combinations can he make?

2. A woman has 3 blouses and 5 skirts. How many different outfits can she wear?

3. How many 2-letter codes can be formed using the letters A, B, C, D? Repeated letters are allowed.

4. How many 2-letter codes can be formed using the letters A, B, C, D, E? Repeated letters are allowed.

5. How many 3-digit numbers can be formed using the digits 0 and 1? Repeated digits are allowed.

6. How many 3-digit numbers can be formed using the digits 0, 1, 2, 3, 4, 5, 6, 7, 8, 9? Repeated digits are allowed.

 7. How many arrangements of answers are possible for a true/false test with 10 questions?

8. How many arrangements of answers are possible in a multiple-choice test with 5 questions, each of which has 4 possible answers?

9. How many 4-digit numbers can be formed using the digits 0, 1, 2, 3, 4, 5, 6, 7, 8, 9 if the first digit cannot be 0? Repeated digits are allowed.

10. How many 5-digit numbers can be formed using the digits 0, 1, 2, 3, 4, 5, 6, 7, 8, 9 if the first digit cannot be 0 or 1? Repeated digits are allowed.

In Problems 11–18 find the value of each permutation.

11. $P(6, 2)$	**12.** $P(7, 2)$	**13.** $P(5, 5)$	**14.** $P(4, 4)$
15. $P(8, 0)$	**16.** $P(9, 0)$	**17.** $P(8, 3)$	**18.** $P(8, 5)$

19. In how many ways can 4 people be lined up?

20. In how many ways can 5 books be arranged on a shelf?

21. How many different 3-letter codes are there if only the letters A, B, C, D, E can be used and no letter can be used more than once?

22. How many different 4-letter codes are there if only the letters A, B, C, D, E, F can be used and no letter can be used more than once?

23. How many arrangements are there of the letters in the word MONEY?

24. How many arrangements are there of the digits in the number 51,342?

C **25.** In the American Baseball League, a designated hitter is used. How many batting orders are possible for a manager to use? (There are nine players on a team.)

C **26.** In the National Baseball League, the pitcher usually bats ninth. If this is the case, how many batting orders are possible for a manager to use?

C **27.** How many different license plates can be issued if every plate consists of two letters, without repetition, followed by any 4-digit number that does not start with a 0?

C **28.** Answer Problem 27 if repetition of the letters is allowed.

≡ 10.3 COMBINATIONS

In a permutation, order is important; for example, the arrangements ABC, CAB, BAC, . . . are considered different arrangements of the letters A, B, and C. In many situations, though, order is unimportant. For example, in the card game of poker, the order in which the cards are received does not matter—it is the *combination* of the cards that matters.

The symbol $C(n, r)$ is used to represent the number of arrangements of n distinct objects, using $r \leq n$ of the objects, without regard to order and without using any of them more than once. Such a collec-

COMBINATION   tion of objects is called a **combination** of n objects taken $r \leq n$ at a time.

EXAMPLE 1   List all the combinations of the four objects a, b, c, d taken two at a time. What is $C(4, 2)$?

Solution   One combination of a, b, c, d taken two at a time is

$$ab$$

The object ba is excluded, because order is not important in a combination. The list of all such combinations (convince yourself of this) is

$$ab, \quad ac, \quad ad, \quad bc, \quad bd, \quad cd$$

Thus,

$$C(4, 2) = 6 \qquad\qquad ≡$$

We can find a formula for $C(n, r)$ by noting the difference between a permutation and a combination. Consider n distinct objects. Then $P(n, r)$ is the number of permutations of these n objects taken $r \leq n$ at a time. This number can be obtained in a different way. List the combinations of these n objects taken $r \leq n$ at a time. [There are

$C(n, r)$ of them.] This list represents the unordered arrangements of the r objects. But there are $r!$ ways to order each of these arrangements. By the multiplication principle, there are $C(n, r) \cdot r!$ ways to list n objects taken r at a time with order being important. Thus,

$$P(n, r) = C(n, r) \cdot r!$$

Now solve for $C(n, r)$ and use Formula (1) in Section 10.2:

$$C(n, r) = \frac{P(n, r)}{r!} = \frac{n!/(n - r)!}{r!} = \frac{n!}{(n - r)!r!}$$

$\uparrow$
Use Formula (1), Section 10.2

We have proved the following result.

THEOREM   The number of different arrangements of n objects using $r \leq n$ of them in which

1. the n objects are distinct
2. once an object is used, it cannot be repeated
3. order is not important

is given by the formula

$$C(n, r) = \frac{n!}{(n - r)!r!} \qquad (1)$$

≡

Based on Formula (1) above, we discover that the symbol $C(n, r)$ and the symbol $\binom{n}{r}$ for the binomial coefficients are in fact the same. Thus, the Pascal triangle (see Section 9.2) could be used to find the value of $C(n, r)$; however, we shall use Formula (1) instead, for convenience.

EXAMPLE 2   Use Formula (1) to find the value of each expression:

(a) $C(3, 1)$ (b) $C(6, 3)$ (c) $C(n, n)$ (d) $C(n, 0)$

Solution   (a) $C(3, 1) = \dfrac{3!}{(3 - 1)!1!} = \dfrac{3!}{2!1!} = \dfrac{3 \cdot 2 \cdot 1}{2 \cdot 1 \cdot 1} = 3$

(b) $C(6, 3) = \dfrac{6!}{(6 - 3)!3!} = \dfrac{6 \cdot 5 \cdot 4 \cdot 3!}{3! \cdot 3!} = \dfrac{6 \cdot 5 \cdot 4}{6} = 20$

(c) $C(n, n) = \dfrac{n!}{(n - n)!n!} = \dfrac{n!}{0!n!} = \dfrac{1}{1} = 1$

(d) $C(n, 0) = \dfrac{n!}{(n - 0)!0!} = \dfrac{n!}{n!0!} = \dfrac{1}{1} = 1$

≡

EXAMPLE 3   How many different 3-person committees can be formed from a pool of 7 people?

Solution   The 7 people are, of course, distinct. More important, though, is the observation that the order of being selected on a committee is not significant. Thus, the problem asks for the number of combinations of 7 objects taken 3 at a time, namely,

$$C(7, 3) = \frac{7!}{4!3!} = \frac{7 \cdot 6 \cdot 5 \cdot \cancel{4!}}{\cancel{4!}3!} = \frac{7 \cdot \cancel{6} \cdot 5}{\cancel{6}} = 35$$ ≡

C EXAMPLE 4   Using a regular deck of 52 cards, how many different 5-card hands can be dealt?

Solution   As mentioned earlier, it is the combination, not the order, of the cards that is important. Thus, we seek the number

$$C(52, 5) = \frac{52!}{47!5!} = \frac{52 \cdot 51 \cdot 50 \cdot 49 \cdot 48 \cdot \cancel{47!}}{\cancel{47!}5!}$$

$$= \frac{52 \cdot 51 \cdot 50 \cdot 49 \cdot 48}{5 \cdot 4 \cdot 3 \cdot 2 \cdot 1} = 2,598,960$$ ≡

EXAMPLE 5   In how many ways can a committee consisting of 2 faculty members and 3 students be formed if there are 6 faculty members and 10 students eligible to serve on the committee?

Solution   The problem can be separated into two parts: the number of ways the faculty members can be chosen—namely, $C(6, 2)$—and the number of ways the student members can be chosen—namely, $C(10, 3)$. By the multiplication principle, the committee can be formed in

$$C(6, 2) \cdot C(10, 3) = \frac{6!}{4!2!} \cdot \frac{10!}{7!3!} = \frac{6 \cdot 5 \cdot \cancel{4!}}{\cancel{4!}2!} \cdot \frac{10 \cdot 9 \cdot 8 \cdot \cancel{7!}}{\cancel{7!}3!}$$

$$= \frac{30}{2} \cdot \frac{720}{6} = 1800 \text{ ways}$$ ≡

EXERCISE 10.3

1. List all the combinations of the 5 objects a, b, c, d, e taken 3 at a time. What is $C(5, 3)$?

2. List all the combinations of the 5 objects a, b, c, d, e taken 2 at a time. What is $C(5, 2)$?

3. List all the combinations of the 4 objects 1, 2, 3, 4 taken 3 at a time. What is $C(4, 3)$?

4. List all the combinations of the 6 objects 1, 2, 3, 4, 5, 6 taken 3 at a time. What is $C(6, 3)$?

In Problems 5–12 use Formula (1) to find the value of each expression.

5. $C(8, 2)$ 6. $C(8, 6)$ 7. $C(6, 4)$

8. $C(6, 2)$ **9.** $C(15, 15)$ **10.** $C(18, 1)$

C **11.** $C(26, 13)$ C **12.** $C(18, 9)$

13. In how many ways can a committee of 4 students be formed from a pool of 7 students?

14. In how many ways can a committee of 3 professors be formed from a department having 8 professors?

C **15.** How many different bridge hands are possible? (A bridge hand consists of 13 cards dealt from a regular deck of 52 cards.)

C **16.** How many different 7-card-stud hands are possible? (A 7-card-stud hand consists of 7 cards dealt from a regular deck of 52 cards.)

17. A student dance committee is to be formed consisting of 2 boys and 3 girls. If the membership is to be chosen from 4 boys and 8 girls, how many different committees are possible?

18. A baseball team has 15 members. Four of the players are pitchers, and the remaining 11 members can play any position. How many different teams of 9 players can be formed?

C **19.** The student relations committee of a college consists of 2 administrators, 3 faculty members, and 5 students. There are 4 administrators, 8 faculty members, and 20 students eligible to serve. How many different committees are possible?

C **20.** A defensive football squad consists of 25 players. Of these, 10 are linemen, 10 are linebackers, and 5 are safeties. How many different teams of 5 linemen, 3 linebackers, and 3 safeties can be formed?

21. A jar contains 6 red balls and 4 green balls. In how many ways can 4 balls be chosen such that exactly 2 are green?

22. Answer Problem 21 if the 4 balls are to be chosen such that exactly 1 is green.

23. *Programming exercise* When both n and r are large, finding $C(n, r)$ on a computer may lead to integers too large to compute. To avoid this, we can approximate the values of $C(n, r)$. One way to do this is the following:

$$C(40, 20) = \frac{40!}{20!20!} = \frac{40 \cdot 39 \cdot 38 \cdots \cdots 21}{20 \cdot 19 \cdot 18 \cdots \cdots 1}$$

$$= \frac{40}{20} \cdot \frac{39}{19} \cdot \frac{38}{18} \cdots \cdots \frac{21}{1}$$

$$\approx 2.000 \cdot 2.053 \cdot 2.111 \cdots \cdots 21.000$$

$$= 1.3784652 \times 10^{11}$$

(a) Write a program that inputs two integers N and R and computes $C(N, R)$ using Formula (1).

(b) Use the program to determine where overflow occurs on your computer.

(c) Write a program that inputs two integers N and R and computes $C(N, R)$ by the approximation technique shown above.

(d) Compare the answers found in parts (a) and (c).

▓ 10.4 PROBABILITY

Probability is an area of mathematics that deals with experiments that yield random results yet admit a certain regularity. Such experiments do not always produce the same result or outcome, so the result of any one observation is not predictable. However, the results of the experiment over a long period do produce regular patterns that enable us to predict with remarkable accuracy.

EXAMPLE 1 In tossing a fair coin, we know that the outcome is either a head or a tail. On any particular throw, we cannot predict what will happen, but, if we toss the coin many times, we observe that the number of times a head comes up is approximately equal to the number of times we get a tail. It seems reasonable, therefore, to assign a probability of $\frac{1}{2}$ that a head comes up and a probability of $\frac{1}{2}$ that a tail comes up.

▓

Probabilistic Models

PROBABILISTIC MODEL   The discussion in Example 1 constitutes the construction of a **probabilistic model** for the experiment of tossing a fair coin once. A probabilistic model has two components: a sample space and an

SAMPLE SPACE S assignment of probabilities. A **sample space S** is a set whose elements represent all the logical possibilities that can occur as a result of the

OUTCOME   experiment. Each element of S is called an **outcome**. To each outcome,

PROBABILITY   we assign a number, the **probability** of that outcome, which has two properties:

1. Each outcome is assigned a probability that is nonnegative.
2. The sum of all the probabilities equals 1.

Thus, if a probabilistic model has the sample space

$$S = \{e_1, e_2, \ldots, e_n\}$$

where $e_1, e_2, \ldots, e_n$ are the outcomes, and if $P(e_1), P(e_2), \ldots, P(e_n)$ denote the probabilities assigned to each outcome, then

$$P(e_1) \geq 0, P(e_2) \geq 0, \ldots, P(e_n) \geq 0 \qquad (1)$$
$$P(e_1) + P(e_2) + \cdots + P(e_n) = 1 \qquad (2)$$

Let's look at an example.

EXAMPLE 2 An experiment consists of rolling a fair die once.* Construct a probabilistic model for this experiment.

*A die is a cube with each face having either 1, 2, 3, 4, 5, or 6 spots on it.

Solution   A sample space S consists of all the logical possibilities that can occur. Because rolling the die will result in one of six faces showing, the sample space S consists of

$$S = \{1, 2, 3, 4, 5, 6\}$$

Because the die is fair, one face is no more likely to occur than another. As a result, our assignment of probabilities is

$$P(1) = \tfrac{1}{6} \qquad P(2) = \tfrac{1}{6} \qquad P(3) = \tfrac{1}{6}$$
$$P(4) = \tfrac{1}{6} \qquad P(5) = \tfrac{1}{6} \qquad P(6) = \tfrac{1}{6}$$

■

EXAMPLE 3   An experiment consists of tossing a fair die and then a fair coin. Construct a probabilistic model for this experiment.

Solution   A tree diagram is helpful in listing all the possible outcomes. See Figure 6. The sample space consists of the outcomes

$$S = \{1H, 1T, 2H, 2T, 3H, 3T, 4H, 4T, 5H, 5T, 6H, 6T\}$$

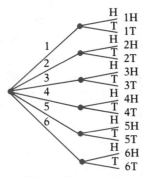

FIGURE 6

The die and the coin are fair; thus, no one outcome is more likely to occur than another. As a result, we assign the probability $\tfrac{1}{12}$ to each of the 12 outcomes.   ■

EVENT   In work with probabilistic models, the term **event** is used to describe a set of possible outcomes of the experiment. Thus, an event

PROBABILITY OF AN EVENT E is some subset of the sample space S. The **probability of an event** E, denoted by $P(E)$, is defined to be equal to the sum of the probabilities of the outcomes in E.

EXAMPLE 4   For the experiment described in Example 3, what is the probability that an even number followed by a head occurs?

Solution   The event E, an even number followed by a head, consists of

$$E = \{2H, 4H, 6H\}$$

The probability of E is

$$P(E) = P(2H) + P(4H) + P(6H) = \tfrac{1}{12} + \tfrac{1}{12} + \tfrac{1}{12} = \tfrac{1}{4}$$ ▀

Equally Likely Outcomes

EQUALLY LIKELY OUTCOMES   When the same probability is assigned to each outcome of the sample space, the experiment is said to have **equally likely outcomes**.

THEOREM   If an experiment has n equally likely outcomes, and if the event E occurs m times, then the probability of E is

$$P(E) = \frac{\text{Number of ways that } E \text{ can occur}}{\text{Number of all logical possibilities}} = \frac{m}{n}$$

Thus, if S is the sample space of this experiment, then

$$P(E) = \frac{n(E)}{n(S)}$$

▀

Based on this result, the solution of Example 4 can now be given as

$$P(E) = \frac{n(E)}{n(S)} = \frac{3}{12} = \frac{1}{4}$$

EXAMPLE 5   What is the probability of selecting a ten, jack, queen, or king from a well-shuffled regular deck of 52 cards?

Solution   Because no one card is more likely to be selected than another, the outcomes are equally likely. The sample space S consists of the 52 cards. The event E consists of a ten (4 possibilities), a jack (4 possibilities), a queen (4 possibilities), or a king (4 possibilities). Thus, $n(S) = 52$ and, by the addition principle, $n(E) = 16$. As a result, the probability of E is

$$P(E) = \frac{n(E)}{n(S)} = \frac{16}{52} = \frac{4}{13} \approx 0.3077$$ ▀

EXAMPLE 6   In the game of "craps," two fair dice are rolled. If the total of the faces equals 7 or 11, you win. If the totals are 2, 3, or 12, you have thrown craps and you lose. In all other cases, you throw again.

(a) What is the probability you will win?
(b) What is the probability you will lose?
(c) What is the probability you will need to throw again?

Solution   We begin by constructing a probabilistic model for the experiment. A tree diagram will help us see all the logical probabilities. See Figure 7. Because the dice are fair, no one of the 36 possible outcomes in the sample space S is more likely to occur than any other. Thus, we have equally likely outcomes with $n(S) = 36$.

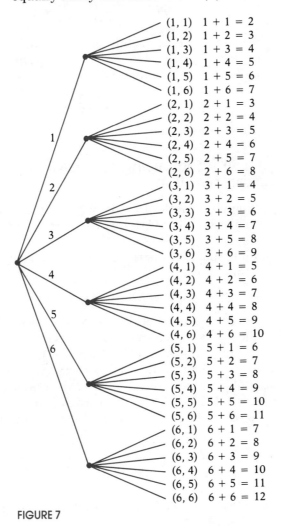

FIGURE 7

(a) The event E, "the dice total 7 or 11," consists of the outcomes

$$E = \{(1, 6), (2, 5), (3, 4), (4, 3), (5, 2), (6, 1), (5, 6), (6, 5)\}$$

Because $n(E) = 8$, we have

$$P(E) = \frac{n(E)}{n(S)} = \frac{8}{36} = \frac{2}{9} \approx 0.222$$

(b) The event F, "the dice total 2, 3, or 12," consists of the outcomes
$$F = \{(1, 1), (1, 2), (2, 1), (6, 6)\}$$
Because $n(F) = 4$,
$$P(F) = \frac{n(F)}{n(S)} = \frac{4}{36} = \frac{1}{9} \approx 0.111$$

(c) The number of possibilities that lead to throwing again is $36 - n(E) - n(F) = 36 - 8 - 4 = 24$. Thus, the probability that another throw is required is
$$\frac{24}{36} = \frac{2}{3} \approx 0.667$$ ≡

Applications Involving Permutations and Combinations

C   EXAMPLE 7   After shuffling a deck of cards, what is the probability that, when they are all dealt, the first 4 cards dealt are aces?

Solution   The sample space S consists of all the arrangements of 52 cards taken 52 at a time. Thus,
$$n(S) = P(52, 52) = 52!$$

The event E requires that the first 4 cards dealt be aces. Because any one of the 4 aces could be dealt first, leaving 3 aces to be dealt second, and so on, it follows that there are $4 \cdot 3 \cdot 2 \cdot 1 = 24$ ways for the 4 aces to be dealt as the first 4 cards. We now need to count the number of possible arrangements of the remaining 48 cards, namely $P(48, 48) = 48!$. By the multiplication principle, the number of elements in E is $n(E) = 24 \cdot 48!$. Because all the outcomes are equally likely, we have
$$P(E) = \frac{n(E)}{n(S)} = \frac{24 \cdot 48!}{52!} = \frac{24}{52 \cdot 51 \cdot 50 \cdot 49} \approx 0.000003693$$ ≡

C   EXAMPLE 8   In the poker game of 5-card stud, what is the probability of being dealt a:

(a) Royal flush (ace, king, queen, jack, ten of the same suit)?
(b) Flush (5 cards of the same suit)?

(In 5-card stud, 5 cards are dealt—no replacements allowed—from a regular deck of 52 cards.)

Solution   (a) Because a regular deck contains 4 suits (clubs, diamonds, hearts, spades), there are 4 ways to be dealt a royal flush. All outcomes are equally likely; thus we can use the result of Example 4 in Section 10.3 to find

$$\text{Probability(royal flush)} = \frac{4}{C(52, 5)} = \frac{4}{2,598,960} \approx 0.0000015$$

(b) The number of ways a flush can be obtained in a given suit equals the number of combinations of 13 cards taken 5 at a time, namely $C(13, 5)$. Because there are 4 possible suits, we use the multiplication principle and compute the total number of flushes as

$$4 \cdot C(13, 5) = 4 \cdot \frac{13!}{8!5!} = \frac{4 \cdot 13 \cdot 12 \cdot 11 \cdot 10 \cdot 9 \cdot 8!}{8! \cdot 5 \cdot 4 \cdot 3 \cdot 2 \cdot 1} = 5148$$

From this number, though, we need to subtract the royal flushes (4) and the straight flushes (36).* Thus, the number of flushes is $5148 - 4 - 36 = 5108$. Because all outcomes are equally likely,

$$\text{Probability(flush)} = \frac{5108}{C(52, 5)} = \frac{5108}{2,598,960} \approx 0.0020 \qquad \blacksquare$$

≡ HISTORICAL COMMENTS

Set theory, counting, and probability first took form as a systematic theory in the exchange of letters (1654) between Fermat (1601–1665) and Pascal (1623–1662). They discussed the problem of how to divide the stakes in a game that is interrupted before completion, knowing how many points each player needs to win. Fermat solved the problem by listing all possibilities and counting the favorable ones, whereas Pascal made use of the triangle that now bears his name. As mentioned in the text, the entries in Pascal's triangle are equivalent to $C(n, r)$. This recognition of the role of $C(n, r)$ in counting is the foundation of all further developments.

The first book on probability, the work of Christian Huygens (1629–1695), appeared in 1657. In it, the notion of mathematical expectations is explored. This allows the calculation of the profit or loss a gambler may expect, knowing the probabilities involved in the game (see the Historical Problems that follow).

It is interesting to note that Girolamo Cardano (1501–1576) wrote a treatise on probability, but it was not published until 1663 in Cardano's collected works, and this was too late to have any effect on the development of the theory.

In 1713 the posthumously published *Ars Conjectandi* of Jacob Bernoulli gave the theory the form it would have until 1900. In the current century both combinatorics (counting) and probability have undergone rapid development.

---

*A straight flush consists of 5 consecutive cards of the same suit. Besides royal flushes, there are 9 ways {A, 1, 2, 3, 4}, . . . , {9, 10, J, Q, K}, in each of 4 suits, for a total of 36 ways, to get a straight flush. Royal flushes and straight flushes are better poker hands than regular flushes since they are more difficult to obtain.

A final comment about notation. The notations $C(n, r)$ and $P(n, r)$ are variants of a form of notation developed in England after 1830. The notation $\binom{n}{r}$ for $C(n, r)$ goes back to Euler (1707–1783), but is now losing ground because it has no clearly related symbolism of the same type for permutations. The set symbols $\cup$ and $\cap$ were introduced by Giuseppe Peano (1858–1932) in 1888 in a slightly different context. The inclusion symbol $\subset$ was introduced by E. Schroeder (1841–1902) about 1890. The treatment of set theory in the text is due to George Boole (1815–1864), who wrote $A + B$ for $A \cup B$ and AB for $A \cap B$ (statisticians still use AB for $A \cap B$).

≡ HISTORICAL
PROBLEMS

1. *The problem discussed by Fermat and Pascal* A game between two equally skilled players, A and B, is interrupted when A needs 2 points to win and B needs 3 points. In what proportion should the stakes be divided? (Clearly, if each play results in 1 point for either player, at most four more plays will decide the game.)
    (a) *Fermat's solution* List all possible outcomes that will end the game to form the sample space (for example, $ABAA$, $ABBB$, etc.). The probabilities for A to win and B to win then determine how the stakes should be divided.
    (b) *Pascal's solution* Use combinations to determine the number of ways the 2 points needed for A to win could occur in four plays. Then use combinations to determine the number of ways the 3 points needed for B to win could occur. This is trickier than it looks, since A can win with 2 points in either two plays, three plays, or four plays. Compute the probabilities and compare with the results in part (a).

2. *Huygens' mathematical expectation* In a game with n possible outcomes with probabilities $p_1, p_2, \ldots, p_n$, suppose that the *net* winnings (the amount won less the amount paid to play) are $w_1, w_2, \ldots, w_n$, respectively. Then the mathematical expectation is

$$E = p_1 w_1 + p_2 w_2 + \cdots + p_n w_n$$

The number E represents the profit or loss per game in the long run. The following problems are a modification of those of Huygens.
    (a) A fair die is tossed. A gambler wins \$3 if he throws a 6 and \$6 if he throws a 5. What is his expectation?
       [*Note:* $w_1 = w_2 = w_3 = w_4 = 0$]
    (b) A gambler plays the same game as in part (a), but now the gambler must pay \$1 to play. This means $w_5 = \$5$, $w_6 = \$2$, and $w_1 = w_2 = w_3 = w_4 = -\$1$. What is the expectation?   ≡

EXERCISE 10.4   *In Problems 1–6 construct a probabilistic model for each experiment.*

1. Tossing a fair coin twice

2. Tossing two fair coins once

3. Tossing two fair coins, then a fair die

4. Tossing a fair coin, a fair die, and then a fair coin

5. Tossing three fair coins once

6. Tossing one fair coin three times

In Problems 7–12 use the spinners pictured below and construct a probabilistic model for each experiment.

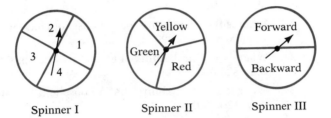

Spinner I Spinner II Spinner III

7. Spin Spinner I, then Spinner II. What is the probability of getting a 2 or a 4, followed by Red?

8. Spin Spinner III, then Spinner II. What is the probability of getting Forward, followed by Yellow or Green?

9. Spin Spinner I, then II, then III. What is the probability of getting a 1, followed by Red or Green, followed by Backward?

10. Spin Spinner II, then I, then III. What is the probability of getting Yellow, followed by a 2 or a 4, followed by Forward?

11. Spin Spinner I twice, then Spinner II. What is the probability of getting a 2, followed by a 2 or a 4, followed by Red or Green?

12. Spin Spinner III, then Spinner I twice. What is the probability of getting Forward, followed by a 1 or a 3, followed by a 2 or a 4?

13. What is the probability of selecting a face card (jack, queen, or king) from a well-shuffled deck of 52 cards?

14. What is the probability of selecting a red 3 or a black face card from a well-shuffled deck of 52 cards?

15. What is the probability of throwing a 6 or an 8 in a game of craps?

16. What is the probability of throwing a 5 or a 9 in a game of craps?

In Problems 17–20 a golf ball is selected at random from a container. If the container has 9 white balls, 8 green ones, and 3 orange ones, find the probability of each event.

17. The golf ball is white.

18. The golf ball is green.

19. The golf ball is white or green.

20. The golf ball is not white.

Problems 21–24 are based on a consumer survey of annual incomes in 100 households. The following table gives the data:

Income	$0–9999	$10,000–$19,999	$20,000–$29,999	$30,000–$39,999	$40,000 or More
Number of households	5	35	30	20	10

21. What is the probability that a household has an annual income in excess of $30,000?

22. What is the probability that a household has an annual income between $10,000 and $29,999?

23. What is the probability that a household has an annual income less than $20,000?

24. What is the probability that a household has an annual income more than $20,000?

[C] **25.** In a certain lottery, there are ten balls, numbered 0, 1, 2, 3, 4, 5, 6, 7, 8, 9. Of these, five are drawn. If you pick five numbers that match those drawn, in the correct order, you win $1,000,000. What is the probability of winning such a lottery?

[C] **26.** Rework Problem 25 if only four balls are drawn.

[C] **27.** A committee of 6 people is to be chosen at random from a group of 14 people consisting of 2 supervisors, 5 skilled laborers, and 7 unskilled laborers. What is the probability that the committee chosen consists of 2 skilled and 4 unskilled laborers?

[C] **28.** Rework Problem 27 if the committee chosen consists only of unskilled laborers.

[C] **29.** What is the probability of being dealt a straight in a game of 5-card stud? (A straight consists of 5 consecutive cards, such as 3, 4, 5, 6, 7, of any suit. The ace may be used as a 1 or as a card following the king.)

[C] **30.** What is the probability of being dealt four of a kind in a game of 5-card stud?

C H A P T E R   R E V I E W

VOCABULARY

set
elements
well-defined
empty, or null, set ∅
equal sets
subset
proper subset
finite set
infinite set
intersection
union

universal set
complement of a set
Venn diagram
counting formula
addition principle of
   counting
bits
register
tree diagram
multiplication
   principle of counting

permutation
combination
probabilistic model
sample space S
outcome
probability
event
probability of an
   event
equally likely
   outcomes

FILL-IN-THE-BLANK QUESTIONS

1. The _____ of A with B consists of all elements either in A or in B; the _____ of A with B consists of all elements both in A and in B.

2. $P(5, 2)$ = _____; $C(5, 2)$ = _____.

3. A(n) _____ is an ordered arrangement of n distinct objects.

4. A(n) _____ is an arrangement of n distinct objects without regard to order.

5. When the same probability is assigned to each outcome of a sample space, the experiment is said to have _____ _____ outcomes.

REVIEW EXERCISES

In Problems 1–8 use U = Universal set = {1, 2, 3, 4, 5, 6, 7, 8, 9}, A = {1, 3, 5, 7}, B = {3, 5, 6, 7, 8}, and C = {2, 3, 7, 8, 9} to find each set.

1. $A \cup B$	2. $B \cup C$	3. $A \cap C$	4. $A \cap B$
5. $A' \cup B'$	6. $B' \cap C'$	7. $(B \cap C)'$	8. $(A \cup B)'$

9. If $n(A) = 8$, $n(B) = 12$, and $n(A \cap B) = 3$, find $n(A \cup B)$.

10. If $n(A) = 12$, $n(A \cup B) = 30$, and $n(A \cap B) = 6$, find $n(B)$.

In Problems 11–16 use the information supplied in the following figure:

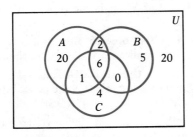

11. How many are in A?   12. How many are in A or B?

13. How many are in A and C?   14. How many are not in B?

15. How many are neither in A nor in C?

16. How many are in B but not in C?

In Problems 17–20 find the value of each expression.

17. $P(8, 3)$ 18. $P(7, 3)$ 19. $C(8, 3)$ 20. $C(7, 3)$

21. A clothing store sells pure wool and polyester/wool suits. Each suit comes in 3 colors and 10 sizes. How many suits are required for a complete assortment?

22. In connecting a certain electrical device, 5 wires are to be connected to 5 different terminals. How many different wirings are possible if 1 wire is connected to each terminal?

23. On a given day, the American Baseball League schedules 7 games. How many different outcomes are possible assuming each game is played to completion?

24. On a given day, the National Baseball League will schedule 6 games. How many different outcomes are possible assuming each game is played to completion?

25. If 4 people enter a bus having 9 vacant seats, in how many ways can they be seated?

26. How many different arrangements are there of the letters in the word ROSE?

27. In how many ways can a squad of 4 relay runners be chosen from a track team of 8 runners?

28. A professor has 10 similar problems to put on a 3-item test. How many different tests can she design?

29. In how many different ways can the 14 baseball teams in the American League be paired without regard to which team is at home?

30. Rework Problem 29 for the National League, which has 12 teams.

31. Rework Problem 29 if the pairings are with regard to which team is at home.

32. Rework Problem 30 if the pairings are with regard to which team is at home.

33. From a box containing three 40-watt bulbs, six 60-watt bulbs, and eleven 75-watt bulbs, a bulb is drawn at random. What is the probability that the bulb is 40 watts? What is the probability that it is not a 75-watt bulb?

34. You have four $1 bills, three $5 bills, and two $10 bills in your wallet. If you pick a bill at random, what is the probability it will be a $1 bill?

35. Each of the letters in the word ROSE is written on an index card and the cards are then shuffled. What is the probability that, when the cards are dealt out, they spell the word ROSE?

36. Each of the numbers 1, 2, . . . , 100 is written on an index card and the cards are then shuffled. If a card is selected at random, what is the probability that the number on the card is divisible by 5? What is the probability that the card selected either is a 1 or names a prime number?

C H A P T E R   11

THE CONICS

11.1 Preliminaries

11.2 The Parabola

11.3 The Ellipse

11.4 The Hyperbola
     Chapter Review

Historically, Apollonius (200 B.C.) was among the first to study conics and discover some of their interesting properties. Today, conics are still studied because of their many uses. Paraboloids of revolution (parabolas rotated about their axes of symmetry) are used as signal collectors (the satellite dishes used with radar and cable TV, for example), as solar energy collectors, and as reflectors (telescopes, light projection, and so on). The planets circle the sun in approximately elliptical orbits. Elliptical surfaces can be used to reflect signals such as light and sound from one place to another. Positions of ships at sea can be found using hyperbolas.

The Greeks used the methods of Euclidean geometry to study conics. We shall use the more powerful methods of analytic geometry, bringing to bear both algebra and geometry, for our study of conics. Thus, we shall give a geometric description of each conic and then, using rectangular coordinates and the distance formula, we shall find equations that represent conics. We used this same development, you may recall, when we first defined a circle in Section 3.2.

▤ 11.1 PRELIMINARIES

The word *conic* derives from the word *cone*, which is a geometric figure that can be constructed in the following way: Let *a* and *g* be two distinct lines that intersect at a point *V*. Keep the line *a* fixed. Now rotate the line *g* about *a* while maintaining the same angle between *a* and *g*. The collection of points swept out (generated) by the line *g* is called a (right circular) **cone**. See Figure 1. The fixed line *a* is called the **axis** of the cone; the point *V* is called its **vertex**; the lines that pass through *V* and make the same angle with *a* as *g* are called **generators** of the cone. Thus, each generator is a line that lies entirely in the cone.

CONE
AXIS OF A CONE
VERTEX OF A CONE
GENERATORS OF A CONE

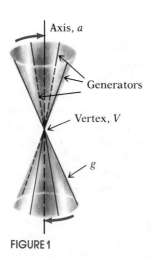

FIGURE 1

CONICS
CONIC SECTIONS

Conics, an abbreviation for **conic sections**, are curves that result from the intersection of a (right circular) cone and a plane.

The conics we shall study arise when the plane does not contain the vertex. See Figure 2. These conics are parabolas (when the plane is parallel to one and only one generator); hyperbolas (when the plane is parallel to two generators); and ellipses (when the plane is parallel to no generator, in which case it intersects each generator). A special case of the ellipse is a circle (which results if the plane intersects each generator and is also perpendicular to the axis of the cone). If the plane does contain the vertex, the intersection of the plane and the cone is a point, a line, or a pair of intersecting lines. These are usually called **degenerate conics**.

DEGENERATE CONICS

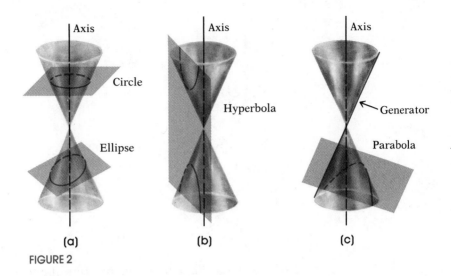

(a)   (b)   (c)

FIGURE 2

■ 11.2 THE PARABOLA

We stated earlier (Section 5.1) that the graph of a quadratic function is a parabola. In this section we begin with a geometric definition of parabola.

PARABOLA   A **parabola** is defined as the collection of points P in a plane such that the distance from each point to a fixed point F equals its distance to
FOCUS   a fixed line D. The point F is called the **focus** of the parabola and the
DIRECTRIX   line D is its **directrix**. As a result, a parabola is the set of points P for which

$$d(F, P) = d(P, D) \qquad (1)$$

Figure 3 (on the next page) shows a parabola. The line through the
AXIS OF SYMMETRY OF A   focus F and perpendicular to the directrix D is called the **axis of sym-**
PARABOLA   **metry** of the parabola. The point of intersection of the parabola with
VERTEX OF A PARABOLA   its axis of symmetry is called the **vertex** V. Because the vertex V lies on the parabola, it must satisfy Equation (1), namely $d(F, V) = d(V, D)$. Thus, the vertex is midway between the focus and the directrix. We shall let a equal the distance $d(F, V)$ from F to V.

Now we are ready to derive an equation for a parabola. To do this, we use a rectangular system of coordinates, positioned so that the vertex V, focus F, and directrix D of the parabola are conveniently located. If we choose to locate the vertex V at the origin, $(0, 0)$, then we can position the focus F on either the x-axis or the y-axis.

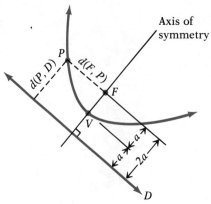

FIGURE 3

Let's see what happens if we position the focus F on the positive x-axis. See Figure 4. Because the distance from F to V is a, the coordinates of F will be $(a, 0)$ with $a > 0$. Similarly, because the distance from V to the directrix D is also a and because D must be perpendicular to the x-axis (do you see why?), the equation of the directrix D must be $x = -a$. Now, if $P = (x, y)$ is any point on the parabola, then P must obey Equation (1), namely,

$$d(F, P) = d(P, D)$$

$$\sqrt{(x - a)^2 + y^2} = |x + a| \qquad \text{Use the distance formula.}$$

$$(x - a)^2 + y^2 = (x + a)^2 \qquad \text{Square both sides.}$$

$$x^2 - 2ax + a^2 + y^2 = x^2 + 2ax + a^2$$

$$y^2 = 4ax$$

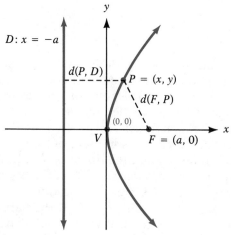

FIGURE 4

$y^2 = 4ax$

THEOREM

The equation of a parabola with vertex at $(0, 0)$, focus at $(a, 0)$, and directrix $x = -a$, $a > 0$, is

EQUATION OF A PARABOLA;
VERTEX AT $(0, 0)$,
FOCUS AT $(a, 0)$, $a > 0$

$$y^2 = 4ax \qquad (2)$$

By reversing the steps, it follows that the graph of an equation of the form of Equation (2) is a parabola; its vertex is at $(0, 0)$, its focus is at $(a, 0)$, its directrix is the line $x = -a$, and its axis of symmetry is the x-axis.

EXAMPLE 1

(a) Find an equation of the parabola with vertex at $(0, 0)$ and focus at $(3, 0)$. Graph the equation.

(b) Discuss and graph the equation $y^2 = 8x$.

Solution

(a) The distance from the vertex $(0, 0)$ to the focus $(3, 0)$ is $a = 3$. Based on Equation (2), the equation of this parabola is

$$y^2 = 4ax$$
$$y^2 = 12x \quad a = 3$$

See Figure 5(a).

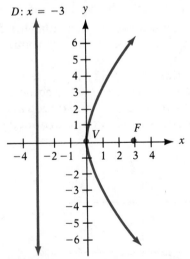

FIGURE 5(a)
$y^2 = 12x$

(b) The equation $y^2 = 8x$ is of the form $y^2 = 4ax$, where $4a = 8$, so that $a = 2$. Consequently, the graph of the equation is a parabola with vertex at $(0, 0)$ and focus on the positive x-axis at $(2, 0)$. The directrix is the vertical line $x = -2$. See Figure 5(b).

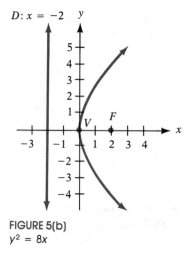

FIGURE 5(b)
$y^2 = 8x$

Recall that we arrived at Equation (2) after placing the focus on the positive x-axis. If the focus is placed on the negative x-axis, positive y-axis, or negative y-axis, a different form of the equation for the parabola results. The four forms of the equation of a parabola with vertex at $(0, 0)$ and focus on a coordinate axis a distance a from $(0, 0)$ are given in Table 1. Notice that each graph is symmetric with respect to its axis of symmetry.

EXAMPLE 2   (a) Discuss and graph the equation $x^2 = -12y$.

(b) Find the equation of the parabola with focus at $(0, 4)$ and directrix the line $y = -4$. Graph the equation.

Solution   (a) The equation $x^2 = -12y$ is of the form $x^2 = -4ay$, with $a = 3$. Consequently, the graph of the equation is a parabola with vertex at $(0, 0)$, focus at $(0, -3)$, and directrix the line $y = 3$. The parabola opens down, and its axis of symmetry is the y-axis. See Figure 6(a).

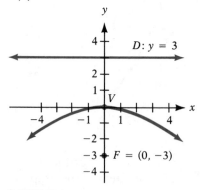

FIGURE 6(a)
$x^2 = -12y$

TABLE 1   Equations of a Parabola: Vertex at (0, 0); Focus on an Axis; $a > 0$

VERTEX	FOCUS	DIRECTRIX	EQUATION	DESCRIPTION	GRAPH
$(0, 0)$	$(a, 0)$	$x = -a$	$y^2 = 4ax$	Parabola, axis of symmetry is the x-axis, opens to right	
$(0, 0)$	$(-a, 0)$	$x = a$	$y^2 = -4ax$	Parabola, axis of symmetry is the x-axis, opens to left	
$(0, 0)$	$(0, a)$	$y = -a$	$x^2 = 4ay$	Parabola, axis of symmetry is the y-axis, opens up	
$(0, 0)$	$(0, -a)$	$y = a$	$x^2 = -4ay$	Parabola, axis of symmetry is the y-axis, opens down	

(b) A parabola whose focus is at (0, 4) and whose directrix is the horizontal line $y = -4$ will have its vertex at (0, 0). (Do you see why? The vertex is midway between the focus and the directrix.) Thus, the equation is of the form

$$x^2 = 4ay$$
$$x^2 = 16y \quad a = 4$$

Figure 6(b) shows the graph.

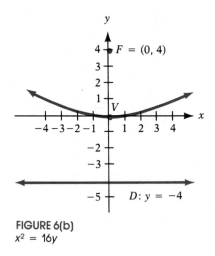

FIGURE 6(b)
$x^2 = 16y$

EXAMPLE 3   Find the equation of a parabola with vertex at (0, 0) if its axis of symmetry is the x-axis and its graph contains the point $(-2, 4)$. Find its focus and directrix and graph the equation.

Solution   Because the vertex is at the origin and the axis of symmetry is the x-axis, we see from Table 1 that the form of the equation is either

$$y^2 = 4ax \quad \text{or} \quad y^2 = -4ax \qquad a > 0 \qquad (3)$$

Because the point $(-2, 4)$ is on the parabola, the coordinates $x = -2$, $y = 4$ must satisfy the equation. Putting $x = -2$ and $y = 4$ into each equation, we find

$$16 = 4a(-2) \quad \text{or} \quad 16 = -4a(-2)$$
$$16 = -8a \qquad \text{or} \quad 16 = 8a$$
$$a = -2 \qquad \text{or} \quad a = 2$$

Because $a > 0$, we discard the left equation in Display (3) and use $a = 2$ in the right equation of Display (3). Thus, the equation of the parabola is

$$y^2 = -8x$$

The focus is at $(-2, 0)$ and the directrix is the line $x = 2$. See Figure 7.

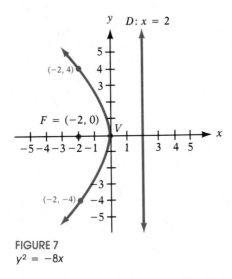

FIGURE 7
$y^2 = -8x$

Vertex at (h, k)

If a parabola with vertex at the origin and axis of symmetry along a coordinate axis is shifted horizontally h units and then vertically k units, the result is a parabola with vertex at (h, k) and axis of symmetry parallel to a coordinate axis. Table 2 gives the forms of the equations of such parabolas. Figure 8 (on the next page) illustrates the graphs for $h > 0, k > 0$.

TABLE 2 Parabolas with Vertex at (h, k) and Axis of Symmetry Parallel to a Coordinate Axis

VERTEX	FOCUS	DIRECTRIX	EQUATION	DESCRIPTION
(h, k)	$(h + a, k)$	$x = -a + h$	$(y - k)^2 = 4a(x - h)$	Parabola, axis of symmetry parallel to x-axis, opens to right
(h, k)	$(h - a, k)$	$x = a + h$	$(y - k)^2 = -4a(x - h)$	Parabola, axis of symmetry parallel to x-axis, opens to left
(h, k)	$(h, k + a)$	$y = -a + k$	$(x - h)^2 = 4a(y - k)$	Parabola, axis of symmetry parallel to y-axis, opens up
(h, k)	$(h, k - a)$	$y = a + k$	$(x - h)^2 = -4a(y - k)$	Parabola, axis of symmetry parallel to y-axis, opens down

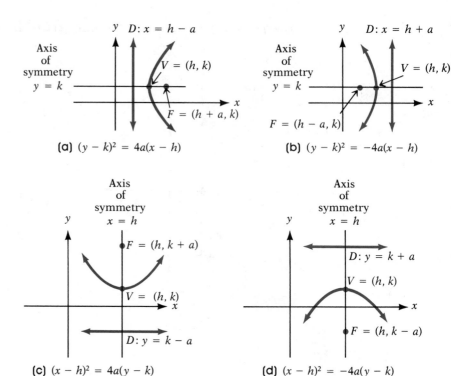

(a) $(y - k)^2 = 4a(x - h)$

(b) $(y - k)^2 = -4a(x - h)$

(c) $(x - h)^2 = 4a(y - k)$

(d) $(x - h)^2 = -4a(y - k)$

FIGURE 8

EXAMPLE 4   (a) Find an equation of the parabola with vertex at $(-2, 3)$ and focus at $(0, 3)$. Graph the equation.

(b) Discuss and graph the equation

$$x^2 + 4x - 4y = 0$$

Solution   (a) The vertex $(-2, 3)$ and focus $(0, 3)$ both lie on the horizontal line $y = 3$ (the axis of symmetry). The distance a from $(-2, 3)$ to $(0, 3)$ is $a = 2$. Also, because the focus lies to the right of the vertex, we know the parabola opens to the right. Consequently, the form of the equation is

$$(y - k)^2 = 4a(x - h)$$

where $(h, k) = (-2, 3)$ and $a = 2$. The equation is, therefore,

$$(y - 3)^2 = 4 \cdot 2[x - (-2)]$$
$$(y - 3)^2 = 8(x + 2)$$

Figure 9(a) shows the graph.

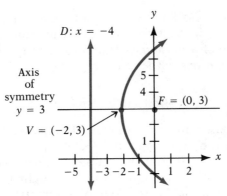

FIGURE 9(a)
$(y - 3)^2 = 8(x + 2)$

(b) To discuss the equation, we complete the square involving the variable x. Thus,

$$x^2 + 4x - 4y = 0$$
$$x^2 + 4x = 4y \qquad \text{Isolate the squared variable.}$$
$$x^2 + 4x + 4 = 4y + 4 \quad \text{Complete the square.}$$
$$(x + 2)^2 = 4(y + 1)$$

The equation is of the form

$$(x - h)^2 = 4a(y - k)$$

with $h = -2$, $k = -1$, and $a = 1$. Its graph is a parabola with vertex at $(h, k) = (-2, -1)$ that opens up. The focus is at $(-2, 0)$ and the directrix is the line $y = -2$. See Figure 9(b).

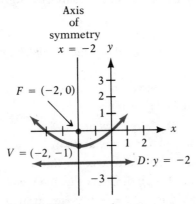

FIGURE 9(b)
$x^2 + 4x - 4y = 0$

Parabolas find their way into many applications. For example, we learned in Section 5.1 that suspension bridges have cables in the shape of a parabola. Another property of parabolas that is used in applications is the reflecting property.

Reflecting Property

PARABOLOID OF REVOLUTION   Suppose a mirror is shaped like a **paraboloid of revolution**, a surface formed by rotating a parabola about its axis of symmetry. If a light (or any other emitting source) is placed at the focus of the parabola, all the rays emanating from the light will reflect off the mirror in lines parallel to the axis of symmetry. See Figure 10. This is the principle behind the design of flashlights, automobile headlights, and other such devices.

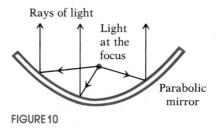

Rays of light

Light at the focus

Parabolic mirror

FIGURE 10

Conversely, suppose rays of light (or other signals) emanate from a distant source, so that they are essentially parallel. When these rays strike the surface of a parabolic mirror whose axis of symmetry is parallel to these rays, they will be reflected onto a single point, the focus. This is the principle behind some solar energy devices and the cable-TV dishes that receive satellite signals.

EXERCISE 11.2   *In Problems 1–16 find the equation of the parabola described. Graph the equation.*

1. Focus at $(2, 0)$; vertex at $(0, 0)$
2. Focus at $(0, 1)$; vertex at $(0, 0)$
3. Focus at $(0, -3)$; vertex at $(0, 0)$
4. Focus at $(-4, 0)$; vertex at $(0, 0)$
5. Focus at $(-2, 0)$; directrix the line $x = 2$
6. Focus at $(0, -1)$; directrix the line $y = 1$
7. Directrix the line $y = -\frac{1}{2}$; vertex at $(0, 0)$
8. Directrix the line $x = -\frac{1}{2}$; vertex at $(0, 0)$
9. Vertex at $(2, -3)$; focus at $(2, -5)$

10. Vertex at $(4, -2)$; focus at $(6, -2)$

11. Vertex at $(0, 0)$; axis of symmetry is the y-axis; containing the point $(2, 3)$

12. Vertex at $(0, 0)$; axis of symmetry is the x-axis; containing the point $(2, 3)$

13. Focus at $(-3, 4)$; directrix the line $y = 2$

14. Focus at $(2, 4)$; directrix the line $x = -4$

15. Focus at $(-3, -2)$; directrix the line $x = 1$

16. Focus at $(-4, 4)$; directrix the line $y = -2$

In Problems 17–28 find the vertex, focus, and directrix of each parabola. Graph the equation.

17. $x^2 = 8y$ **18.** $y^2 = 4x$

19. $y^2 = -16x$ **20.** $x^2 = -4y$

21. $(y - 2)^2 = 8(x + 1)$ **22.** $(x + 4)^2 = 16(y + 2)$

23. $(x - 3)^2 = -(y + 1)$ **24.** $(y + 1)^2 = -4(x - 2)$

25. $y^2 + 2y - x = 0$ **26.** $x^2 - 4x = 2y$

27. $x^2 - 4x = y + 4$ **28.** $y^2 + 12y = -x + 1$

29. A cable-TV receiving dish is in the shape of a paraboloid of revolution. Find the location of the receiver, which is placed at the focus, if the dish is 10 feet across at its opening and is 3 feet deep.

30. The reflector of a flashlight is in the shape of a paraboloid of revolution. Its diameter is 4 inches and its depth is 1 inch. How far from the vertex should the light bulb be placed so that the rays will be reflected parallel to the axis?

11.3 THE ELLIPSE

ELLIPSE
FOCI

An **ellipse** is the collection of all points in the plane the sum of whose distances from two fixed points, called the **foci**, is a constant.

The definition actually contains within it a physical means for drawing an ellipse. Find a piece of string (the length of this string is the constant referred to in the definition). Then take two thumbtacks (the foci) and stick them on a piece of paper so that the distance between them is less than the length of the string. Now attach the ends of the string to the thumbtacks and, using the point of a pencil, pull the string taut. Keeping the string taut, rotate the pencil around the two thumbtacks. The pencil mark traces out an ellipse. See Figure 11.

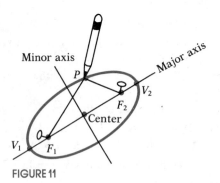

FIGURE 11

In Figure 11, the foci are labeled F_1 and F_2. The line containing the
MAJOR AXIS   foci is called the **major axis**. The midpoint of the foci is called the
CENTER   **center** of the ellipse. The line through the center and perpendicular
MINOR AXIS   to the major axis is called the **minor axis**.

The two points of intersection of the ellipse and the major axis are
VERTICES   called the **vertices**, V_1 and V_2, of the ellipse. The distance from one
LENGTH OF THE MAJOR AXIS   vertex to the other is called the **length of the major axis**. The ellipse
is symmetric with respect to its major axis and with respect to its
minor axis.

With these ideas in mind, we are now ready to place the ellipse on
a system of rectangular coordinates. First, we place the center of the
ellipse at the origin. Second, we position the ellipse so that its major
axis coincides with a coordinate axis.

Suppose the major axis coincides with the x-axis. See Figure 12. If
c is the distance from the center to a focus, then one focus will be at
$F_1 = (-c, 0)$ and the other one at $F_2 = (c, 0)$. Now we let $2a$ denote
the constant distance referred to in the definition. Thus, if $P = (x, y)$
is any point on the ellipse, we have

$$d(F_1, P) + d(F_2, P) = 2a$$ Sum of the distances of P from the foci equals a constant.

$$\sqrt{(x + c)^2 + y^2} + \sqrt{(x - c)^2 + y^2} = 2a$$ Use the distance formula.

$$\sqrt{(x + c)^2 + y^2} = 2a - \sqrt{(x - c)^2 + y^2}$$ Isolate one radical.

$$(x + c)^2 + y^2 = 4a^2 - 4a\sqrt{(x - c)^2 + y^2} + (x - c)^2 + y^2$$ Square both sides.

$$x^2 + 2cx + c^2 + y^2 = 4a^2 - 4a\sqrt{(x - c)^2 + y^2} + x^2 - 2cx + c^2 + y^2$$

$$4cx - 4a^2 = -4a\sqrt{(x - c)^2 + y^2}$$ Isolate the radical.

$$cx - a^2 = -a\sqrt{(x - c)^2 + y^2}$$ Divide each side by 4.

$$c^2x^2 - 2a^2cx + a^4 = a^2[(x - c)^2 + y^2]$$ Square both sides again.

$$c^2x^2 - 2a^2cx + a^4 = a^2(x^2 - 2cx + c^2 + y^2)$$

$$(c^2 - a^2)x^2 - a^2y^2 = a^2c^2 - a^4$$

$$(a^2 - c^2)x^2 + a^2y^2 = a^2(a^2 - c^2)$$ Multiply each side by -1;   **(1)**
factor a^2 on the right side.

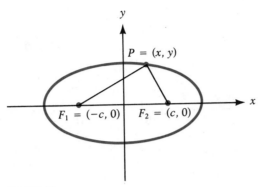

FIGURE 12
$d(F_1, P) + d(F_2, P) = 2a$

To get points on the ellipse off the x-axis, it must be that $a > c$. Do you see why? Look again at Figure 12: $d(F_1, P) + d(F_2, P) = 2a$, $d(F_1, F_2) = 2c$, and $d(F_1, P) + d(F_2, P) > d(F_1, F_2)$. Thus, $2a > 2c$ and $a > c$. Since $a > c$, then $a^2 > c^2$, so $a^2 - c^2 > 0$. Let $b^2 = a^2 - c^2$, $b > 0$. Then $a > b$ and Equation (1) can be written as

$$b^2x^2 + a^2y^2 = a^2b^2$$

$$\frac{x^2}{a^2} + \frac{y^2}{b^2} = 1 \qquad b^2 = a^2 - c^2 \quad \text{Divide each side by } a^2b^2.$$

THEOREM   An equation of the ellipse with center at $(0, 0)$ and foci at $(-c, 0)$ and $(c, 0)$ is

EQUATION OF AN ELLIPSE;
CENTER AT (0, 0);
FOCI AT ($\pm c$, 0);
MAJOR AXIS ALONG
THE x-AXIS

$$\frac{x^2}{a^2} + \frac{y^2}{b^2} = 1 \qquad \text{where } a > b \text{ and } b^2 = a^2 - c^2 \qquad (2)$$

The major axis is the x-axis.

Conversely, an equation of the form of Equation (2), with $a > b$, is the equation of an ellipse with center at the origin, foci on the x-axis at $(-c, 0)$ and $(c, 0)$, where $c^2 = a^2 - b^2$, and major axis along the x-axis.

As you can verify from Equation (2), the ellipse defined by Equation (2) is symmetric with respect to the x-axis, y-axis, and origin.

To find the vertices of the ellipse defined by Equation (2), let $y = 0$. The vertices obey the equation $x^2/a^2 = 1$, the solutions of which are $x = \pm a$. Consequently, the vertices of the ellipse given by Equation (2) are $V_1 = (-a, 0)$ and $V_2 = (a, 0)$. The y-intercepts of the ellipse, found by letting $x = 0$, have coordinates $(0, -b)$ and $(0, b)$. These four intercepts, $(a, 0)$, $(-a, 0)$, $(0, b)$, and $(0, -b)$, are used to graph the ellipse. See Figure 13.

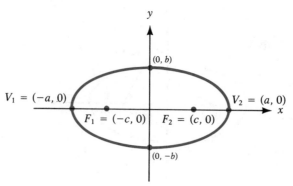

FIGURE 13

$\dfrac{x^2}{a^2} + \dfrac{y^2}{b^2} = 1$, $a > b$ and $b^2 = a^2 - c^2$

EXAMPLE 1   (a) Find an equation of the ellipse with center at the origin, one focus at $(3, 0)$, and a vertex at $(-4, 0)$. Graph the equation.

(b) Discuss and graph the equation

$$\frac{x^2}{25} + \frac{y^2}{9} = 1$$

Solution   (a) The ellipse has its center at the origin, and the major axis coincides with the x-axis. One focus is at $(c, 0) = (3, 0)$, so $c = 3$. One vertex is at $(-a, 0) = (-4, 0)$, so $a = 4$. From Equation (2), it follows that

$$b^2 = a^2 - c^2 = 16 - 9 = 7$$

so an equation of the ellipse is

$$\frac{x^2}{16} + \frac{y^2}{7} = 1$$

Figure 14(a) shows the graph.

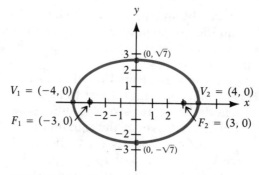

FIGURE 14(a)

$\dfrac{x^2}{16} + \dfrac{y^2}{7} = 1$

(b) The given equation

$$\frac{x^2}{25} + \frac{y^2}{9} = 1$$

is of the form of Equation (2), with $a^2 = 25$ and $b^2 = 9$. The equation is that of an ellipse with center at $(0, 0)$ and major axis along the x-axis. The vertices are at $(\pm a, 0) = (\pm 5, 0)$. Because $b^2 = a^2 - c^2$, we find

$$c^2 = a^2 - b^2 = 25 - 9 = 16$$

The foci are at $(\pm c, 0) = (\pm 4, 0)$. Figure 14(b) shows the graph.

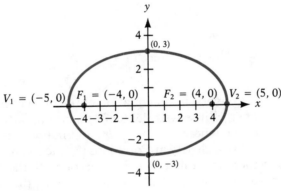

FIGURE 14(b)
$\frac{x^2}{25} + \frac{y^2}{9} = 1$

You should have observed in the graphs given in Figures 13 and 14 that we plotted the intercepts. This makes it much easier to get an accurate graph of the ellipse.

If the major axis of an ellipse with center at $(0, 0)$ coincides with the y-axis, then the foci are at $(0, -c)$ and $(0, c)$. Using the same steps as before, the definition of an ellipse leads to the following result:

THEOREM   An equation of the ellipse with center at $(0, 0)$ and foci at $(0, -c)$ and $(0, c)$ is

EQUATION OF AN ELLIPSE;
CENTER AT (0, 0);
FOCI AT (0, ±c);
MAJOR AXIS ALONG
THE y-AXIS

$$\frac{x^2}{b^2} + \frac{y^2}{a^2} = 1 \qquad \text{where } a > b \text{ and } b^2 = a^2 - c^2 \qquad (3)$$

The major axis is the y-axis; the vertices are at $(0, -a)$ and $(0, a)$.

Figure 15 illustrates the graph of such an ellipse.

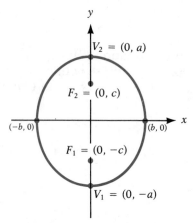

FIGURE 15
$$\frac{x^2}{b^2} + \frac{y^2}{a^2} = 1, a > b \text{ and } b^2 = a^2 - c^2$$

Conversely, an equation of the form of Equation (3) is the equation of an ellipse with center at the origin, foci on the y-axis at $(0, -c)$ and $(0, c)$, where $c^2 = a^2 - b^2$, and major axis along the y-axis.

Look closely at Equations (2) and (3). Although they may look alike, there is a difference. In Equation (2), the larger number, a^2, is in the denominator of the x^2 term so that the major axis of the ellipse is along the x-axis. In Equation (3), the larger number, a^2, is in the denominator of the y^2 term so that the major axis is along the y-axis.

EXAMPLE 2   Discuss and graph the equation

$$9x^2 + y^2 = 9$$

Solution   To get the equation in the proper form, we need to divide each side by 9. The result is

$$x^2 + \frac{y^2}{9} = 1$$

The larger number, 9, is in the denominator of the y^2 term so that, based on Equation (3), this equation is of an ellipse with center at the origin and major axis along the y-axis. Also, we conclude that $a^2 = 9$, $b^2 = 1$, and $c^2 = a^2 - b^2 = 9 - 1 = 8$. The vertices are at $(0, \pm a) = (0, \pm 3)$, and the foci are at $(0, \pm c) = (0, \pm 2\sqrt{2})$. The graph is given in Figure 16.

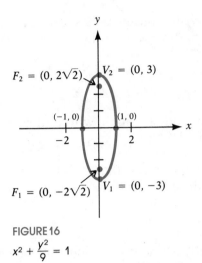

FIGURE 16

$$x^2 + \frac{y^2}{9} = 1$$

EXAMPLE 3 Find an equation of the ellipse having one focus at $(0, 2)$ and vertices at $(0, -3)$ and $(0, 3)$. Graph the equation.

Solution Because the vertices are at $(0, -3)$ and $(0, 3)$, the center of this ellipse is at the origin. Also, its major axis coincides with the y-axis. The given information also reveals that $c = 2$ and $a = 3$, so $b^2 = a^2 - c^2 = 9 - 4 = 5$. The form of the equation of this ellipse is given by Equation (3):

$$\frac{x^2}{b^2} + \frac{y^2}{a^2} = 1$$

$$\frac{x^2}{5} + \frac{y^2}{9} = 1$$

Figure 17 shows the graph.

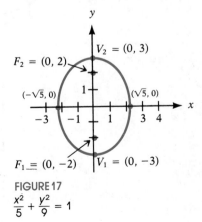

FIGURE 17

$$\frac{x^2}{5} + \frac{y^2}{9} = 1$$

The circle may be considered a special kind of ellipse. To see how, let $a = b$ in Equation (2) or in Equation (3). Then

$$\frac{x^2}{a^2} + \frac{y^2}{a^2} = 1$$

$$x^2 + y^2 = a^2$$

This is the equation of a circle with center at the origin and radius a. The value of c is

$$c^2 = a^2 - b^2 = 0$$

We interpret this to mean that the closer the two foci of an ellipse are, the more the ellipse will look like a circle.

Center at (h, k)

If an ellipse with center at the origin and major axis coinciding with a coordinate axis is shifted horizontally h units and then vertically k units, the result is an ellipse whose center is at (h, k) and whose major axis is parallel to a coordinate axis. Table 3 gives the forms of the equations of such an ellipse.

TABLE 3   Ellipses with Center at (h, k) and Major Axis Parallel to a Coordinate Axis

CENTER	MAJOR AXIS	FOCI	VERTICES	EQUATION	GRAPH
(h, k)	Parallel to x-axis	$(h \pm c, k)$	$(h \pm a, k)$	$\dfrac{(x - h)^2}{a^2} + \dfrac{(y - k)^2}{b^2} = 1,$ $a > b$ and $b^2 = a^2 - c^2$	

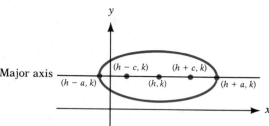

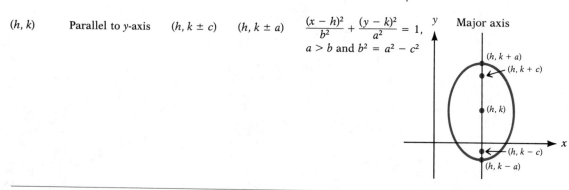

| (h, k) | Parallel to y-axis | $(h, k \pm c)$ | $(h, k \pm a)$ | $\dfrac{(x - h)^2}{b^2} + \dfrac{(y - k)^2}{a^2} = 1,$ $a > b$ and $b^2 = a^2 - c^2$ | |

EXAMPLE 4   (a) Find an equation for the ellipse with center at $(2, -3)$, one focus at $(3, -3)$, and one vertex at $(5, -3)$. Graph the equation.

     (b) Discuss and graph the equation $4x^2 + y^2 - 8x + 4y + 4 = 0$.

Solution   (a) The center is at $(h, k) = (2, -3)$, so $h = 2$ and $k = -3$. The major axis is parallel to the x-axis. The distance from the center $(2, -3)$ to a focus $(3, -3)$ is $c = 1$; the distance from the center $(2, -3)$ to a vertex $(5, -3)$ is $a = 3$. Thus, $b^2 = a^2 - c^2 = 9 - 1 = 8$. The form of the equation is

$$\frac{(x - h)^2}{a^2} + \frac{(y - k)^2}{b^2} = 1 \qquad \text{where } h = 2, k = -3, a = 3, b = 2\sqrt{2}$$

$$\frac{(x - 2)^2}{9} + \frac{(y + 3)^2}{8} = 1$$

Figure 18(a) shows the graph.

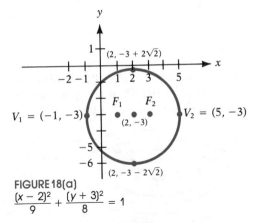

FIGURE 18(a)
$$\frac{(x - 2)^2}{9} + \frac{(y + 3)^2}{8} = 1$$

(b) We proceed to complete the square in x and in y:

$$4x^2 + y^2 - 8x + 4y + 4 = 0$$
$$4x^2 - 8x + y^2 + 4y = -4$$
$$4(x^2 - 2x) + (y^2 + 4y) = -4$$
$$4(x^2 - 2x + 1) + (y^2 + 4y + 4) = -4 + 4 + 4 \qquad \text{Complete each square.}$$
$$4(x - 1)^2 + (y + 2)^2 = 4$$
$$(x - 1)^2 + \frac{(y + 2)^2}{4} = 1 \qquad\qquad \text{Divide each side by 4.}$$

This is the equation of an ellipse with center at $(1, -2)$ and major axis parallel to the y-axis. Since $a^2 = 4$ and $b^2 = 1$, we have $c^2 = a^2 - b^2 = 4 - 1 = 3$. The vertices are at $(h, k \pm a) = (1, -2 \pm 2) = (1, 0)$ and $(1, -4)$. The foci are at $(h, k \pm c) = (1, -2 \pm \sqrt{3}) = (1, -2 - \sqrt{3})$ and $(1, -2 + \sqrt{3})$. Figure 18(b) shows the graph.

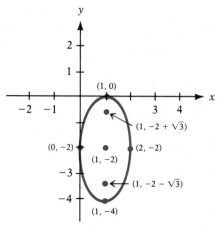

FIGURE 18(b)

$$(x - 1)^2 + \frac{(y + 2)^2}{4} = 1$$

Applications

Ellipses are found in many applications in science and engineering. For example, the orbits of the planets around the sun are elliptical, with the sun's position at a focus. See Figure 19.

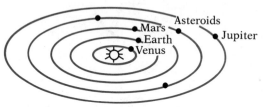

FIGURE 19

Ellipses also have an interesting reflection property. If a source of light (or sound) is placed at one focus, the waves transmitted by the source will reflect off the ellipse and concentrate at the other focus. This is the principle behind "whispering galleries," which are rooms designed with elliptical ceilings. A person standing at one focus of the ellipse can whisper and be heard by a person standing at the other focus, because all the sound waves reflect off the ceiling to the other person. See Figure 20.

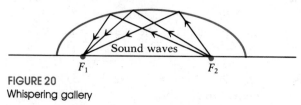

FIGURE 20
Whispering gallery

Stone and concrete bridges are often shaped as semielliptical arches. Elliptical gears are used in machinery when a variable rate of motion is required.

EXERCISE 11.3

In Problems 1–10 find the vertices and foci of each ellipse. Also graph each equation.

1. $\dfrac{x^2}{9} + \dfrac{y^2}{4} = 1$

2. $\dfrac{x^2}{16} + \dfrac{y^2}{4} = 1$

3. $\dfrac{x^2}{9} + \dfrac{y^2}{25} = 1$

4. $\dfrac{x^2}{4} + \dfrac{y^2}{16} = 1$

5. $4x^2 + y^2 = 16$

6. $x^2 + 9y^2 = 18$

7. $4y^2 + x^2 = 8$

8. $4y^2 + 9x^2 = 36$

9. $x^2 + y^2 = 16$

10. $x^2 + y^2 = 4$

In Problems 11–20 find an equation for each ellipse. Graph the equation.

11. Center at (0, 0); focus at (3, 0); vertex at (6, 0)

12. Center at (0, 0); focus at (−2, 0); vertex at (3, 0)

13. Center at (0, 0); focus at (0, −4); vertex at (0, 5)

14. Center at (0, 0); focus at (0, 1); vertex at (0, −2)

15. Foci at (±2, 0); length of the major axis is 6

16. Focus at (0, −4); vertices at (0, ±8)

17. Foci at (0, ±3); the *x*-intercepts are ±2

18. Foci at (0, ±2); the length of the major axis is 8

19. Center at (0, 0); vertex at (0, 4); $b = 1$

20. Vertices at (±5, 0); $c = 2$

In Problems 21–28 find the center, foci, and vertices of each ellipse. Also graph each equation.

21. $\dfrac{(x - 2)^2}{4} + \dfrac{(y + 1)^2}{9} = 1$

22. $\dfrac{(x + 4)^2}{9} + \dfrac{(y + 1)^2}{4} = 1$

23. $(x + 5)^2 + 4(y - 4)^2 = 16$

24. $9(x - 3)^2 + (y + 2)^2 = 18$

25. $x^2 + 4x + 4y^2 - 8y + 4 = 0$

26. $x^2 + 3y^2 - 12y + 9 = 0$

27. $2x^2 + 3y^2 - 8x + 6y + 5 = 0$

28. $4x^2 + 3y^2 + 8x - 6y = 5$

In Problems 29–34 find an equation for each ellipse. Graph the equation.

29. Center at (2, −2); vertex at (5, −2); focus at (4, −2)

30. Center at (−3, 1); vertex at (−3, 4); focus at (−3, 0)

31. Vertices at (4, 3) and (4, 9); focus at (4, 8)

32. Foci at (1, 2) and (−3, 2); vertex at (−4, 2)

33. Foci at (5, 1) and (−1, 1); the length of the major axis is 8

34. Vertices at (2, 5) and (2, −1); $c = 2$

35. An arch in the shape of the upper half of an ellipse is used to support a bridge that is to span a river 20 meters wide. The center of the arch is 6 meters above the center of the river. Consult the figure. Write an equation for the ellipse in which the *x*-axis coincides with the water level and the *y*-axis passes through the center of the arch.

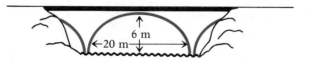

36. The arch of a bridge is a semiellipse with a horizontal major axis. The span is 30 feet, and the top of the arch is 10 feet above the major axis. The roadway is horizontal and is 2 feet above the top of the arch. Find the vertical distance from the roadway to the arch at 5-foot intervals along the roadway.

C **37.** An arch in the form of half an ellipse is 40 feet wide and 15 feet high at the center. Find the height of the arch at intervals of 10 feet along its width.

ECCENTRICITY OF
AN ELLIPSE

38. The **eccentricity** *e* of an ellipse is defined as the number *c/a*. Because $a > c$, it follows that $e < 1$. Describe the general shape of an ellipse whose eccentricity is:
(a) Close to 0   (b) Equal to $\frac{1}{2}$ (c) Close to 1

C **39.** The orbit of the earth is an ellipse with the sun at one focus. If the length of the **semimajor axis** (half the length of the major axis) is approximately 92 million miles and the eccentricity is $\frac{1}{60}$, find the greatest and least distances of the earth from the sun.

SEMIMAJOR AXIS

■ 11.4   THE HYPERBOLA

HYPERBOLA

FOCI

A **hyperbola** is the collection of all points in the plane the difference of whose distances from two fixed points, called the **foci**, is a constant.

Figure 21 illustrates a hyperbola with foci F_1 and F_2. The line containing the foci is called the **transverse axis**. The midpoint of the foci is called the **center** of the hyperbola. The line through the center and perpendicular to the transverse axis is called the **conjugate axis**. The hyperbola consists of two separate curves, called **branches**, that are symmetric with respect to the transverse axis, conjugate axis, and center. The two points of intersection of the hyperbola and the transverse axis are called the **vertices** V_1 and V_2 of the hyperbola.

TRANSVERSE AXIS

CENTER

CONJUGATE AXIS

BRANCHES

VERTICES

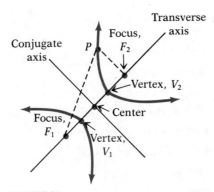

FIGURE 21

With these ideas in mind, we are now ready to place the hyperbola on a system of rectangular coordinates. First, we place the center at the origin. Next, we position the hyperbola so that its transverse axis coincides with a coordinate axis.

Suppose the transverse axis coincides with the x-axis. See Figure 22. If c is the distance from the center to a focus, then one focus will be at $F_1 = (-c, 0)$ and the other one at $F_2 = (c, 0)$. Now we let the constant difference of the distances from any point $P = (x, y)$ on the hyperbola to the foci F_1 and F_2 be denoted by $\pm 2a$. (If P is on the right branch, the $+$ sign is used; if P is on the left branch, the $-$ sign is used.) The coordinates of P must obey the equation

$$d(F_1, P) - d(F_2, P) = \pm 2a$$
Difference of the distances from the foci equals $\pm 2a$.

$$\sqrt{(x + c)^2 + y^2} - \sqrt{(x - c)^2 + y^2} = \pm 2a$$
Use the distance formula.

$$\sqrt{(x + c)^2 + y^2} = \pm 2a + \sqrt{(x - c)^2 + y^2}$$
Isolate one radical.

$$(x + c)^2 + y^2 = 4a^2 \pm 4a\sqrt{(x - c)^2 + y^2} \\ + (x - c)^2 + y^2$$
Square both sides.

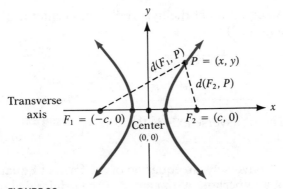

FIGURE 22

$$\frac{x^2}{a^2} - \frac{y^2}{b^2} = 1, \ b^2 = c^2 - a^2$$

Next, we remove the parentheses:

$$x^2 + 2cx + c^2 + y^2 = 4a^2 \pm 4a\sqrt{(x-c)^2 + y^2} + x^2 - 2cx + c^2 + y^2$$

$$4cx - 4a^2 = \pm 4a\sqrt{(x-c)^2 + y^2} \qquad \text{Isolate the radical.}$$

$$cx - a^2 = \pm a\sqrt{(x-c)^2 + y^2} \qquad \text{Divide each side by 4.}$$

$$(cx - a^2)^2 = a^2[(x-c)^2 + y^2] \qquad \text{Square both sides.}$$

$$c^2x^2 - 2ca^2x + a^4 = a^2(x^2 - 2cx + c^2 + y^2)$$

$$c^2x^2 + a^4 = a^2x^2 + a^2c^2 + a^2y^2$$

$$(c^2 - a^2)x^2 - a^2y^2 = a^2c^2 - a^4$$

$$(c^2 - a^2)x^2 - a^2y^2 = a^2(c^2 - a^2) \qquad (1)$$

To get points on the hyperbola off the x-axis, it must be that $a < c$. Do you see why? Look again at Figure 22. The point P is on the right branch, so $d(F_1, P) - d(F_2, P) = 2a$. Now,

$$d(F_1, P) < d(F_2, P) + d(F_1, F_2) \quad \text{Use triangle } F_1PF_2.$$

$$d(F_1, P) - d(F_2, P) < d(F_1, F_2)$$

$$2a < 2c$$

$$a < c$$

Since $a < c$, then $a^2 < c^2$, so $c^2 - a^2 > 0$. Let $b^2 = c^2 - a^2$, $b > 0$. Then Equation (1) can be written as

$$b^2x^2 - a^2y^2 = a^2b^2$$

$$\frac{x^2}{a^2} - \frac{y^2}{b^2} = 1$$

THEOREM   An equation of the hyperbola with center at $(0, 0)$ and foci at $(-c, 0)$ and $(c, 0)$ is

EQUATION OF A HYPERBOLA;
•   CENTER AT (0, 0);
FOCI AT (±c, 0);
TRANSVERSE AXIS ALONG
x-AXIS

$$\frac{x^2}{a^2} - \frac{y^2}{b^2} = 1 \qquad \text{where } b^2 = c^2 - a^2 \qquad (2)$$

The transverse axis is the x-axis.   ≡

Conversely, an equation of the form of Equation (2) is the equation of a hyperbola with center at the origin, foci on the x-axis at $(-c, 0)$ and $(c, 0)$, where $c^2 = a^2 + b^2$, and transverse axis along the x-axis.

As you can verify from Equation (2), the hyperbola defined by Equation (2) is symmetric with respect to the x-axis, y-axis, and origin.

To find the vertices of the hyperbola defined by Equation (2), let $y = 0$. The vertices obey the equation $x^2/a^2 = 1$, the solutions of which are $x = \pm a$. Consequently, the vertices of the hyperbola given by Equation (2) are $V_1 = (-a, 0)$ and $V_2 = (a, 0)$. To find the y-intercepts, if any, let $x = 0$. This results in the equation $y^2/b^2 = -1$, which has no solution. We conclude that the hyperbola defined by Equation (2) has no y-intercepts. See Figure 23.

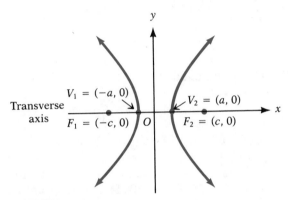

FIGURE 23

EXAMPLE 1   (a) Find an equation of the hyperbola with center at the origin, one focus at $(3, 0)$, and one vertex at $(-2, 0)$. Graph the equation.

   (b) Discuss and graph the equation

$$\frac{x^2}{16} - \frac{y^2}{4} = 1$$

Solution   (a) The hyperbola has its center at the origin, and the transverse axis coincides with the x-axis. One focus is at $(c, 0) = (3, 0)$, so $c = 3$. One vertex is at $(-a, 0) = (-2, 0)$, so $a = 2$. From Equation (2), it follows that $b^2 = c^2 - a^2 = 9 - 4 = 5$, so an equation of the hyperbola is

$$\frac{x^2}{4} - \frac{y^2}{5} = 1$$

See Figure 24(a).

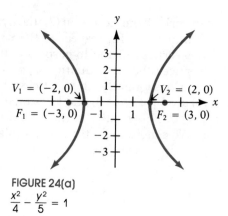

FIGURE 24(a)
$$\frac{x^2}{4} - \frac{y^2}{5} = 1$$

(b) The given equation,

$$\frac{x^2}{16} - \frac{y^2}{4} = 1$$

is of the form of Equation (2), with $a^2 = 16$ and $b^2 = 4$. Thus, the graph of the equation is a hyperbola with center at $(0, 0)$ and transverse axis along the x-axis. Also, we know that $c^2 = a^2 + b^2 = 16 + 4 = 20$. The vertices are at $(\pm a, 0) = (\pm 4, 0)$, and the foci are at $(\pm c, 0) = (\pm 2\sqrt{5}, 0)$. Figure 24(b) shows the graph.

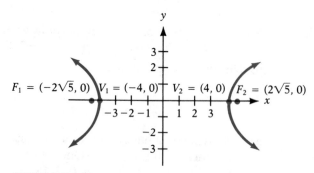

FIGURE 24(b)
$$\frac{x^2}{16} - \frac{y^2}{4} = 1$$

The next result gives the form of the equation of a hyperbola with center at the origin and transverse axis along the y-axis.

THEOREM

An equation of the hyperbola with center at $(0, 0)$ and foci at $(0, -c)$ and $(0, c)$ is

EQUATION OF A HYPERBOLA;
CENTER AT (0, 0);
FOCI AT (0, ±c);
TRANSVERSE AXIS ALONG
y-AXIS

$$\frac{y^2}{a^2} - \frac{x^2}{b^2} = 1 \qquad \text{where } b^2 = c^2 - a^2 \qquad (3)$$

The transverse axis is the y-axis; the vertices are at $(0, -a)$ and $(0, a)$. ∎

Figure 25 shows the graph of the hyperbola defined in Equation (3).

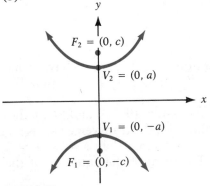

FIGURE 25
$\frac{y^2}{a^2} - \frac{x^2}{b^2} = 1, b^2 = c^2 - a^2$

Notice the difference in the form of Equations (2) and (3). When the y^2-term is subtracted from the x^2-term, the transverse axis is the x-axis. When the x^2-term is subtracted from the y^2-term, the transverse axis is the y-axis.

EXAMPLE 2

Discuss and graph the equation

$$y^2 - 4x^2 = 4$$

Solution

To get the equation in the proper form, we need to divide each side by 4. The result is

$$\frac{y^2}{4} - x^2 = 1$$

Because the x^2-term is subtracted from the y^2-term, the equation is that of a hyperbola with center at the origin and transverse axis along the y-axis. Also, comparing the above equation to Equation (3), we find $a^2 = 4$, $b^2 = 1$, and $c^2 = a^2 + b^2 = 5$. The vertices are at $(0, \pm a) = (0, \pm 2)$; the foci are at $(0, \pm c) = (0, \pm\sqrt{5})$. The graph is given in Figure 26.

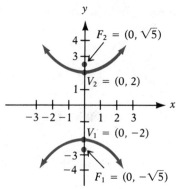

FIGURE 26

$$\frac{y^2}{4} - x^2 = 1$$

EXAMPLE 3   Find an equation of the hyperbola having one vertex at $(0, 2)$ and foci at $(0, -3)$ and $(0, 3)$. Graph the equation.

Solution   Because the foci are at $(0, -3)$ and $(0, 3)$, the center of the hyperbola is at the origin. Also, the transverse axis is along the y-axis. The given information also reveals that $c = 3$, $a = 2$, and $b^2 = c^2 - a^2 = 9 - 4 = 5$. The form of the equation of the hyperbola is given by Equation (3):

$$\frac{y^2}{a^2} - \frac{x^2}{b^2} = 1$$

$$\frac{y^2}{4} - \frac{x^2}{5} = 1$$

See Figure 27.

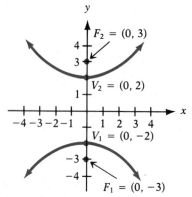

FIGURE 27

$$\frac{y^2}{4} - \frac{x^2}{5} = 1$$

Look at the equations of the hyperbolas in Examples 2 and 3. For the hyperbola in Example 2, $a^2 = 4$ and $b^2 = 1$, so $a > b$; for the hyperbola in Example 3, $a^2 = 4$ and $b^2 = 5$, so $a < b$. We conclude that, for hyperbolas, there are no requirements involving the relative size of a and b. This is in contrast, you should recall, to the case of an ellipse, in which the relative sizes of a and b dictate which axis is the major axis. Hyperbolas have another feature to distinguish them from ellipses and parabolas: Hyperbolas have asymptotes.

Asymptotes

Recall from Section 5.6 that an asymptote of a graph is a line with the property that the distance from the line to points on the graph gets closer to 0 as $x \to -\infty$ or as $x \to +\infty$.

THEOREM

ASYMPTOTES OF A HYPERBOLA

The hyperbola

$$\frac{x^2}{a^2} - \frac{y^2}{b^2} = 1$$

has the two asymptotes

$$y = \frac{b}{a}x \quad \text{and} \quad y = -\frac{b}{a}x$$

Proof   We begin by solving for y in the equation of the hyperbola:

$$\frac{x^2}{a^2} - \frac{y^2}{b^2} = 1$$

$$\frac{y^2}{b^2} = \frac{x^2}{a^2} - 1$$

$$y^2 = b^2\left(\frac{x^2}{a^2} - 1\right)$$

If $x \neq 0$, we can rearrange the right side in the form

$$y^2 = \frac{b^2 x^2}{a^2}\left(1 - \frac{a^2}{x^2}\right)$$

$$y = \pm\frac{bx}{a}\sqrt{1 - \frac{a^2}{x^2}}$$

Now, as $x \to -\infty$ or as $x \to +\infty$, the term a^2/x^2 gets closer to 0, so the expression under the radical gets closer to 1. Thus, as $x \to -\infty$ or as $x \to +\infty$, the value of y gets closer to $\pm bx/a$; that is, the graph of the hyperbola gets closer to the lines

$$y = -\frac{b}{a}x \quad \text{and} \quad y = \frac{b}{a}x$$

Thus, these lines are asymptotes of the hyperbola. ≡

The asymptotes of a hyperbola are not part of the hyperbola, but they do serve as a guide for graphing a hyperbola. For example, suppose we want to graph the equation

$$\frac{x^2}{a^2} - \frac{y^2}{b^2} = 1$$

We begin by plotting the vertices $(-a, 0)$ and $(a, 0)$. Then we plot the points $(0, -b)$ and $(0, b)$. Use these four points to construct a rectangle. The diagonals of this rectangle have slopes b/a and $-b/a$, and their extensions are the asymptotes $y = (b/a)x$ and $y = -(b/a)x$ of the hyperbola. See Figure 28.

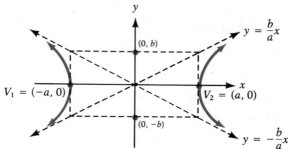

FIGURE 28
$$\frac{x^2}{a^2} - \frac{y^2}{b^2} = 1$$

THEOREM

ASYMPTOTES OF A HYPERBOLA

The hyperbola

$$\frac{y^2}{a^2} - \frac{x^2}{b^2} = 1$$

has the two asymptotes

$$y = \frac{a}{b}x \quad \text{and} \quad y = -\frac{a}{b}x$$

You are asked to prove this result in Problem 38 at the end of this section.

EXAMPLE 4   Discuss the equation

$$9x^2 - 4y^2 = 36$$

Solution   First we divide each side by 36 to get the equation in the proper form. The result is

$$\frac{x^2}{4} - \frac{y^2}{9} = 1$$

This is the equation of a hyperbola with center at the origin and transverse axis along the x-axis. Using $a^2 = 4$ and $b^2 = 9$, we find $c^2 = a^2 + b^2 = 13$. The vertices are at $(\pm a, 0) = (\pm 2, 0)$; the foci are at $(\pm c, 0) = (\pm\sqrt{13}, 0)$; the asymptotes have the equations

$$y = \frac{3}{2}x \quad \text{and} \quad y = -\frac{3}{2}x$$

Now form the rectangle containing the points $(\pm a, 0)$ and $(0, \pm b)$, namely $(-2, 0), (2, 0), (0, -3)$, and $(0, 3)$. The extension of the diagonals of this rectangle are the asymptotes. See Figure 29 for the graph.

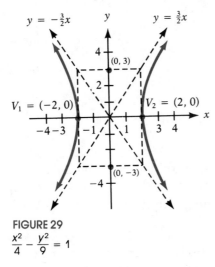

FIGURE 29
$$\frac{x^2}{4} - \frac{y^2}{9} = 1$$

Center at (h, k)

If a hyperbola with center at the origin and transverse axis coinciding with a coordinate axis is shifted horizontally h units and then vertically k units, the result is a hyperbola with center at (h, k) and transverse axis parallel to a coordinate axis. Table 4 gives the form of the equations of such hyperbolas. See Figure 30 (on the next page) for the graphs.

TABLE 4   Hyperbolas with Center at (h, k) and Transverse Axis Parallel to a Coordinate Axis

CENTER	TRANSVERSE AXIS	FOCI	VERTICES	EQUATION	ASYMPTOTES
(h, k)	Parallel to x-axis	$(h \pm c, k)$	$(h \pm a, k)$	$\dfrac{(x-h)^2}{a^2} - \dfrac{(y-k)^2}{b^2} = 1,$ $\quad b^2 = c^2 - a^2$	$y - k = \pm\dfrac{b}{a}(x - h)$
(h, k)	Parallel to y-axis	$(h, k \pm c)$	$(h, k \pm a)$	$\dfrac{(y-k)^2}{a^2} - \dfrac{(x-h)^2}{b^2} = 1,$ $\quad b^2 = c^2 - a^2$	$y - k = \pm\dfrac{a}{b}(x - h)$

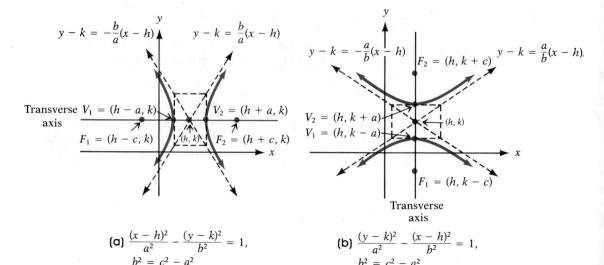

(a) $\dfrac{(x - h)^2}{a^2} - \dfrac{(y - k)^2}{b^2} = 1,$

$b^2 = c^2 - a^2$

(b) $\dfrac{(y - k)^2}{a^2} - \dfrac{(x - h)^2}{b^2} = 1,$

$b^2 = c^2 - a^2$

FIGURE 30

EXAMPLE 5   (a) Find an equation for the hyperbola with center at $(1, -2)$, one focus at $(4, -2)$, and one vertex at $(3, -2)$. Graph the equation.

(b) Discuss and graph the equation $-x^2 + 4y^2 - 2x - 16y + 11 = 0$.

Solution   (a) The center is at $(h, k) = (1, -2)$, so $h = 1$ and $k = -2$. The transverse axis is parallel to the x-axis. The distance from the center $(1, -2)$ to the focus $(4, -2)$ is $c = 3$; the distance from the center $(1, -2)$ to the vertex $(3, -2)$ is $a = 2$. Thus, $b^2 = c^2 - a^2 = 9 - 4 = 5$. The equation is

$$\frac{(x - h)^2}{a^2} - \frac{(y - k)^2}{b^2} = 1$$

$$\frac{(x - 1)^2}{4} - \frac{(y + 2)^2}{5} = 1$$

See Figure 31(a).

(b) We complete the squares in x and in y:

$$-x^2 + 4y^2 - 2x - 16y + 11 = 0$$

$$-(x^2 + 2x) + 4(y^2 - 4y) = -11 \qquad \text{Group the terms.}$$

$$-(x^2 + 2x + 1) + 4(y^2 - 4y + 4) = -1 + 16 - 11 \quad \text{Complete each square.}$$

$$-(x + 1)^2 + 4(y - 2)^2 = 4$$

$$(y - 2)^2 - \frac{(x + 1)^2}{4} = 1 \qquad \text{Divide by 4.}$$

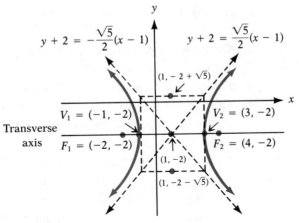

FIGURE 31(a)

$$\frac{(x-1)^2}{4} - \frac{(y+2)^2}{5} = 1$$

This is the equation of a hyperbola with center at $(-1, 2)$ and transverse axis parallel to the y-axis. Also, $a^2 = 1$ and $b^2 = 4$, so $c^2 = a^2 + b^2 = 5$. The vertices are at $(h, k \pm a) = (-1, 2 \pm 1)$, or $(-1, 1)$ and $(-1, 3)$. The foci are at $(h, k \pm c) = (-1, 2 \pm \sqrt{5})$. The asymptotes are $y - 2 = \frac{1}{2}(x + 1)$ and $y - 2 = -\frac{1}{2}(x + 1)$. Figure 31(b) shows the graph.

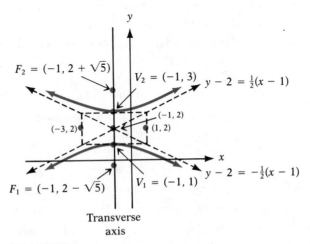

FIGURE 31(b)

$$(y-2)^2 - \frac{(x+1)^2}{4} = 1$$

EXERCISE 11.4   *In Problems 1–10 find an equation for the hyperbola described. Graph the equation.*

1. Center at (0, 0); focus at (4, 0); vertex at (1, 0)
2. Center at (0, 0); focus at (0, 5); vertex at (0, 4)
3. Center at (0, 0); focus at (0, −6); vertex at (0, 4)
4. Center at (0, 0); focus at (−3, 0); vertex at (2, 0)
5. Foci at (−5, 0) and (5, 0); vertex at (3, 0)
6. Focus at (0, 6); vertices at (0, −2) and (0, 2)
7. Vertices at (0, −6) and (0, 6); asymptote the line $y = 2x$
8. Vertices at (−4, 0) and (4, 0); asymptote the line $y = 2x$
9. Foci at (−4, 0) and (4, 0); asymptote the line $y = -x$
10. Foci at (0, −2) and (0, 2); asymptote the line $y = -x$

In Problems 11–18 discuss and graph each equation.

11. $\dfrac{x^2}{9} - \dfrac{y^2}{4} = 1$

12. $\dfrac{y^2}{9} - \dfrac{x^2}{4} = 1$

13. $4x^2 - y^2 = 16$

14. $y^2 - 4x^2 = 16$

15. $y^2 - 9x^2 = 9$

16. $x^2 - y^2 = 4$

17. $y^2 - x^2 = 25$

18. $2x^2 - y^2 = 4$

In Problems 19–26 find an equation for the hyperbola described. Graph the equation.

19. Center at (4, −1); focus at (7, −1); vertex at (6, −1)
20. Center at (−3, 1); focus at (−3, 6); vertex at (−3, 4)
21. Center at (−3, −4); focus at (−3, −8); vertex at (−3, −2)
22. Center at (1, 4); focus at (−2, 4); vertex at (0, 4)
23. Foci at (3, 7) and (7, 7); vertex at (6, 7)
24. Focus at (−4, 0); vertices at (−4, 4) and (−4, 2)
25. Vertices at (−1, −1) and (3, −1); asymptote the line $(x - 1)/2 = (y + 1)/3$
26. Vertices at (1, −3) and (1, 1); asymptote the line $(x - 1)/2 = (y + 1)/3$

In Problems 27–34 discuss and graph each equation.

27. $\dfrac{(x - 3)^2}{4} - \dfrac{(y + 2)^2}{9} = 1$

28. $\dfrac{(y + 4)^2}{4} - \dfrac{(x - 1)^2}{9} = 1$

29. $(y - 2)^2 - 4(x + 2)^2 = 4$

30. $(x + 4)^2 - 9(y - 3)^2 = 9$

31. $4x^2 - y^2 - 24x - 4y + 16 = 0$

32. $2y^2 - x^2 + 2x + 8y + 3 = 0$

33. $y^2 - 4x^2 - 16x - 2y - 19 = 0$

34. $x^2 - 3y^2 + 8x - 6y + 4 = 0$

ECCENTRICITY OF A HYPERBOLA   35. The **eccentricity** e of a hyperbola is defined as the number c/a. Because $c > a$, it follows that $e > 1$. Describe the general shape of a hyperbola whose eccentricity is close to 1. What is the shape if e is very large?

EQUALATERAL HYPERBOLA   36. A hyperbola for which $a = b$ is called an **equilateral hyperbola**. Find the eccentricity e of an equilateral hyperbola.

CONJUGATE HYPERBOLAS **37.** Two hyperbolas that have the same set of asymptotes are called **conjugate**. Show that the hyperbolas

$$\frac{x^2}{4} - y^2 = 1 \quad \text{and} \quad y^2 - \frac{x^2}{4} = 1$$

are conjugate. Graph each hyperbola.

38. Prove that the hyperbola

$$\frac{y^2}{a^2} - \frac{x^2}{b^2} = 1$$

has the two asymptotes

$$y = \frac{a}{b}x \quad \text{and} \quad y = -\frac{a}{b}x$$

C H A P T E R R E V I E W

VOCABULARY

cone
axis of a cone
vertex of a cone
generator of a cone
conics
conic sections
degenerate conics
parabola
focus
directrix
axis of symmetry of a
  parabola

vertex of a parabola
equation of a parabola
paraboloid of
  revolution
ellipse
foci
major axis
center
minor axis
vertices
length of the major
  axis

equation of an ellipse
hyperbola
foci
transverse axis
center
conjugate axis
branches
vertices
equation of a
  hyperbola
asymptotes

FILL-IN-THE-BLANK QUESTIONS

1. A(n) _____ is the collection of all points in a plane such that the distance from each point to a fixed point equals its distance to a fixed line.

2. A(n) _____ is the collection of all points in a plane the sum of whose distances from two fixed points is a constant.

3. A(n) _____ is the collection of all points in a plane the difference of whose distances from two fixed points is a constant.

4. For an ellipse, the foci lie on the _____ axis; for a hyperbola, the foci lie on the _____ axis.

5. For the ellipse $x^2/9 + y^2/16 = 1$, the major axis is along the

   _____.

6. The equations of the asymptotes of the hyperbola $y^2/9 - x^2/4 = 1$ are

   _____ and _____.

REVIEW EXERCISES

In Problems 1–20 identify each equation. If it is a parabola, give its vertex, focus, and directrix; if it is an ellipse, give its center, vertices, and foci; if it is a hyperbola, give its center, vertices, foci, and asymptotes.

1. $y^2 = -16x$
2. $16x^2 = y$
3. $\dfrac{x^2}{4} - y^2 = 1$

4. $\dfrac{x^2}{9} - y^2 = 1$
5. $\dfrac{y^2}{25} + \dfrac{x^2}{16} = 1$
6. $\dfrac{x^2}{9} + \dfrac{y^2}{16} = 1$

7. $x^2 + 4y = 4$
8. $3y^2 - x^2 = 9$

9. $4x^2 - y^2 = 8$
10. $9x^2 + 4y^2 = 36$

11. $x^2 - 4x = 2y$
12. $2y^2 - 4y = x - 2$

13. $y^2 - 4y - 4x^2 + 8x = 4$
14. $4x^2 + y^2 + 8x - 4y + 4 = 0$

15. $4x^2 + 9y^2 - 16x - 18y = 11$
16. $4x^2 + 9y^2 - 16x + 18y = 11$

17. $4x^2 - 16x + 16y + 32 = 0$
18. $4y^2 + 3x - 16y + 19 = 0$

19. $9x^2 + 4y^2 - 18x + 8y = 23$
20. $x^2 - y^2 - 2x - 2y = 1$

In Problems 21–36 obtain an equation of the conic described. Graph the equation.

21. A parabola; focus at $(-2, 0)$; directrix the line $x = 2$
22. An ellipse; center at $(0, 0)$; focus at $(0, 3)$; vertex at $(0, 5)$
23. A hyperbola; center at $(0, 0)$; focus at $(0, 4)$; vertex at $(0, -2)$
24. A parabola; vertex at $(0, 0)$; directrix the line $y = -3$
25. An ellipse; foci at $(-3, 0)$ and $(3, 0)$; vertex at $(4, 0)$
26. A hyperbola; vertices at $(-2, 0)$ and $(2, 0)$; focus at $(4, 0)$
27. A parabola; vertex at $(2, -3)$; focus at $(2, -4)$
28. An ellipse; center at $(-1, 2)$; focus at $(0, 2)$; vertex at $(2, 2)$
29. A hyperbola; center at $(-2, -3)$; focus at $(-4, -3)$; vertex at $(-3, -3)$
30. A parabola; focus at $(3, 6)$; directrix the line $y = 8$
31. An ellipse; foci at $(-4, 2)$ and $(-4, 8)$; vertex at $(-4, 10)$
32. A hyperbola; vertices at $(-3, 3)$ and $(5, 3)$; focus at $(7, 3)$
33. Center at $(-1, 2)$; $a = 3$; $c = 4$; transverse axis parallel to the x-axis
34. Center at $(4, -2)$; $a = 1$; $c = 4$; transverse axis parallel to the y-axis
35. Vertices at $(0, 1)$ and $(6, 1)$; asymptote the line $3y + 2x - 9 = 0$
36. Vertices at $(4, 0)$ and $(4, 4)$; asymptote the line $y + 2x - 10 = 0$
37. Find an equation of the hyperbola whose foci are the vertices of the ellipse $4x^2 + 9y^2 = 36$ and whose vertices are the foci of this ellipse.
38. Find an equation of the ellipse whose foci are the vertices of the hyperbola $x^2 - 4y^2 = 16$ and whose vertices are the foci of this hyperbola.
39. Describe the collection of points in a plane so that the distance from each point to the point $(3, 0)$ is three-fourths of its distance from the line $x = \frac{16}{3}$.
40. Describe the collection of points in a plane so that the distance from each point to the point $(5, 0)$ is five-fourths of its distance from the line $x = \frac{16}{5}$.

TABLE I Exponential Functions

x	e^x	e^{-x}	x	e^x	e^{-x}	x	e^x	e^{-x}
0.00	1.0000	1.0000	0.50	1.6487	0.6065	3.0	20.086	0.0498
0.01	1.0101	0.9901	0.55	1.7333	0.5769	3.1	22.198	0.0450
0.02	1.0202	0.9802	0.60	1.8221	0.5488	3.2	24.533	0.0408
0.03	1.0305	0.9705	0.65	1.9155	0.5220	3.3	27.113	0.0369
0.04	1.0408	0.9608	0.70	2.0138	0.4966	3.4	29.964	0.0334
0.05	1.0513	0.9512	0.75	2.1170	0.4724	3.5	33.115	0.0302
0.06	1.0618	0.9418	0.80	2.2255	0.4493	3.6	36.598	0.0273
0.07	1.0725	0.9324	0.85	2.3396	0.4274	3.7	40.447	0.0247
0.08	1.0833	0.9331	0.90	2.4596	0.4066	3.8	44.701	0.0224
0.09	1.0942	0.9139	0.95	2.5857	0.3867	3.9	49.402	0.0202
0.10	1.1052	0.9048	1.0	2.7183	0.3679	4.0	54.598	0.0183
0.11	1.1163	0.8958	1.1	3.0042	0.3329	4.1	60.340	0.0166
0.12	1.1275	0.8869	1.2	3.3201	0.3012	4.2	66.686	0.0150
0.13	1.1388	0.8781	1.3	3.6693	0.2725	4.3	73.700	0.0136
0.14	1.1503	0.8694	1.4	4.0552	0.2466	4.4	81.451	0.0123
0.15	1.1618	0.8607	1.5	4.4817	0.2231	4.5	90.017	0.0111
0.16	1.1735	0.8521	1.6	4.9530	0.2019	4.6	99.484	0.0101
0.17	1.1853	0.8437	1.7	5.4739	0.1827	4.7	109.95	0.0091
0.18	1.1972	0.8353	1.8	6.0496	0.1653	4.8	121.51	0.0082
0.19	1.2092	0.8270	1.9	6.6859	0.1496	4.9	134.29	0.0074
0.20	1.2214	0.8187	2.0	7.3891	0.1353	5.0	148.41	0.0067
0.21	1.2337	0.8106	2.1	8.1662	0.1225	5.5	244.69	0.0041
0.22	1.2461	0.8025	2.2	9.0250	0.1108	6.0	403.43	0.0025
0.23	1.2586	0.7945	2.3	9.9742	0.1003	6.5	665.14	0.0015
0.24	1.2712	0.7866	2.4	11.023	0.0907	7.0	1096.6	0.0009
0.25	1.2840	0.7788	2.5	12.182	0.0821	7.5	1808.0	0.0006
0.30	1.3499	0.7408	2.6	13.464	0.0743	8.0	2981.0	0.0003
0.35	1.4191	0.7047	2.7	14.880	0.0672	8.5	4914.8	0.0002
0.40	1.4918	0.6703	2.8	16.445	0.0608	9.0	8103.1	0.0001
0.45	1.5683	0.6376	2.9	18.174	0.0550	10.0	22026	0.00005

TABLE II Common Logarithms

x	0	1	2	3	4	5	6	7	8	9
1.0	.0000	.0043	.0086	.0128	.0170	.0212	.0253	.0294	.0334	.0374
1.1	.0414	.0453	.0492	.0531	.0569	.0607	.0645	.0682	.0719	.0755
1.2	.0792	.0828	.0864	.0899	.0934	.0969	.1004	.1038	.1072	.1106
1.3	.1139	.1173	.1206	.1239	.1271	.1303	.1335	.1367	.1399	.1430
1.4	.1461	.1492	.1523	.1553	.1584	.1614	.1644	.1673	.1703	.1732
1.5	.1761	.1790	.1818	.1847	.1875	.1903	.1931	.1959	.1987	.2014
1.6	.2041	.2068	.2095	.2122	.2148	.2175	.2201	.2227	.2253	.2279
1.7	.2304	.2330	.2355	.2380	.2405	.2430	.2455	.2480	.2504	.2529
1.8	.2553	.2577	.2601	.2625	.2648	.2672	.2695	.2718	.2742	.2765
1.9	.2788	.2810	.2833	.2856	.2878	.2900	.2923	.2945	.2967	.2989
2.0	.3010	.3032	.3054	.3075	.3096	.3118	.3139	.3160	.3181	.3201
2.1	.3222	.3243	.3263	.3284	.3304	.3324	.3345	.3365	.3385	.3404
2.2	.3424	.3444	.3464	.3483	.3502	.3522	.3541	.3560	.3579	.3598
2.3	.3617	.3636	.3655	.3674	.3692	.3711	.3729	.3747	.3766	.3784
2.4	.3802	.3820	.3838	.3856	.3874	.3892	.3909	.3927	.3945	.3962
2.5	.3979	.3997	.4014	.4031	.4048	.4065	.4082	.4099	.4116	.4133
2.6	.4150	.4166	.4183	.4200	.4216	.4232	.4249	.4265	.4281	.4298
2.7	.4314	.4330	.4346	.4362	.4378	.4393	.4409	.4425	.4440	.4456
2.8	.4472	.4487	.4502	.4518	.4533	.4548	.4564	.4579	.4594	.4609
2.9	.4624	.4639	.4654	.4669	.4683	.4698	.4713	.4728	.4742	.4757
3.0	.4771	.4786	.4800	.4814	.4829	.4843	.4857	.4871	.4886	.4900
3.1	.4914	.4928	.4942	.4955	.4969	.4983	.4997	.5011	.5024	.5038
3.2	.5051	.5065	.5079	.5092	.5105	.5119	.5132	.5145	.5159	.5172
3.3	.5185	.5198	.5211	.5224	.5237	.5250	.5263	.5276	.5289	.5302
3.4	.5315	.5328	.5340	.5353	.5366	.5378	.5391	.5403	.5416	.5428
3.5	.5441	.5453	.5465	.5478	.5490	.5502	.5514	.5527	.5539	.5551
3.6	.5563	.5575	.5587	.5599	.5611	.5623	.5635	.5647	.5658	.5670
3.7	.5682	.5694	.5705	.5717	.5729	.5740	.5752	.5763	.5775	.5786
3.8	.5798	.5809	.5821	.5832	.5843	.5855	.5866	.5877	.5888	.5899
3.9	.5911	.5922	.5933	.5944	.5955	.5966	.5977	.5988	.5999	.6010
4.0	.6021	.6031	.6042	.6053	.6064	.6075	.6085	.6096	.6107	.6117
4.1	.6128	.6138	.6149	.6160	.6170	.6180	.6191	.6201	.6212	.6222
4.2	.6232	.6243	.6253	.6263	.6274	.6284	.6294	.6304	.6314	.6325
4.3	.6335	.6345	.6355	.6365	.6375	.6385	.6395	.6405	.6415	.6425
4.4	.6435	.6444	.6454	.6464	.6474	.6484	.6493	.6503	.6513	.6522
4.5	.6532	.6542	.6551	.6561	.6571	.6580	.6590	.6599	.6609	.6618
4.6	.6628	.6637	.6646	.6656	.6665	.6675	.6684	.6693	.6702	.6712
4.7	.6721	.6730	.6739	.6749	.6758	.6767	.6776	.6785	.6794	.6803
4.8	.6812	.6821	.6830	.6839	.6848	.6857	.6866	.6875	.6884	.6893
4.9	.6902	.6911	.6920	.6928	.6937	.6946	.6955	.6964	.6972	.6981
5.0	.6990	.6998	.7007	.7016	.7024	.7033	.7042	.7050	.7059	.7067
5.1	.7076	.7084	.7093	.7101	.7110	.7118	.7126	.7135	.7143	.7152
5.2	.7160	.7168	.7177	.7185	.7193	.7202	.7210	.7218	.7226	.7235
5.3	.7243	.7251	.7259	.7267	.7275	.7284	.7292	.7300	.7308	.7316
5.4	.7324	.7332	.7340	.7348	.7356	.7364	.7372	.7380	.7388	.7396

x	0	1	2	3	4	5	6	7	8	9
5.5	.7404	.7412	.7419	.7427	.7435	.7443	.7451	.7459	.7466	.7474
5.6	.7482	.7490	.7497	.7505	.7513	.7520	.7528	.7536	.7543	.7551
5.7	.7559	.7566	.7574	.7582	.7589	.7597	.7604	.7612	.7619	.7627
5.8	.7634	.7642	.7649	.7657	.7664	.7672	.7679	.7686	.7694	.7701
5.9	.7709	.7716	.7723	.7731	.7738	.7745	.7752	.7760	.7767	.7774
6.0	.7782	.7789	.7796	.7803	.7810	.7818	.7825	.7832	.7839	.7846
6.1	.7853	.7860	.7868	.7875	.7882	.7889	.7896	.7903	.7910	.7917
6.2	.7924	.7931	.7938	.7945	.7952	.7959	.7966	.7973	.7980	.7987
6.3	.7993	.8000	.8007	.8014	.8021	.8028	.8035	.8041	.8048	.8055
6.4	.8062	.8069	.8075	.8082	.8089	.8096	.8102	.8109	.8116	.8122
6.5	.8129	.8136	.8142	.8149	.8156	.8162	.8169	.8176	.8182	.8189
6.6	.8195	.8202	.8209	.8215	.8222	.8228	.8235	.8241	.8248	.8254
6.7	.8261	.8267	.8274	.8280	.8287	.8293	.8299	.8306	.8312	.8319
6.8	.8325	.8331	.8338	.8344	.8351	.8357	.8363	.8370	.8376	.8382
6.9	.8388	.8395	.8401	.8407	.8414	.8420	.8426	.8432	.8439	.8445
7.0	.8451	.8457	.8463	.8470	.8476	.8482	.8488	.8494	.8500	.8506
7.1	.8513	.8519	.8525	.8531	.8537	.8543	.8549	.8555	.8561	.8567
7.2	.8573	.8579	.8585	.8591	.8597	.8603	.8609	.8615	.8621	.8627
7.3	.8633	.8639	.8645	.8651	.8657	.8663	.8669	.8675	.8681	.8686
7.4	.8692	.8698	.8704	.8710	.8716	.8722	.8727	.8733	.8739	.8745
7.5	.8751	.8756	.8762	.8768	.8774	.8779	.8785	.8791	.8797	.8802
7.6	.8808	.8814	.8820	.8825	.8831	.8837	.8842	.8848	.8854	.8859
7.7	.8865	.8871	.8876	.8882	.8887	.8893	.8899	.8904	.8910	.8915
7.8	.8921	.8927	.8932	.8938	.8943	.8949	.8954	.8960	.8965	.8971
7.9	.8976	.8982	.8987	.8993	.8998	.9004	.9009	.9015	.9020	.9025
8.0	.9031	.9036	.9042	.9047	.9053	.9058	.9063	.9069	.9074	.9079
8.1	.9085	.9090	.9096	.9101	.9106	.9112	.9117	.9122	.9128	.9133
8.2	.9138	.9143	.9149	.9154	.9159	.9165	.9170	.9175	.9180	.9186
8.3	.9191	.9196	.9201	.9206	.9212	.9217	.9222	.9227	.9232	.9238
8.4	.9243	.9248	.9253	.9258	.9263	.9269	.9274	.9279	.9284	.9289
8.5	.9294	.9299	.9304	.9309	.9315	.9320	.9325	.9330	.9335	.9340
8.6	.9345	.9350	.9355	.9360	.9365	.9370	.9375	.9380	.9385	.9390
8.7	.9395	.9400	.9405	.9410	.9415	.9420	.9425	.9430	.9435	.9440
8.8	.9445	.9450	.9455	.9460	.9465	.9469	.9474	.9479	.9484	.9489
8.9	.9494	.9499	.9504	.9509	.9513	.9518	.9523	.9528	.9533	.9538
9.0	.9542	.9547	.9552	.9557	.9562	.9566	.9571	.9576	.9581	.9586
9.1	.9590	.9595	.9600	.9605	.9609	.9614	.9619	.9624	.9628	.9633
9.2	.9638	.9643	.9647	.9652	.9657	.9661	.9666	.9671	.9675	.9680
9.3	.9685	.9689	.9694	.9699	.9703	.9708	.9713	.9717	.9722	.9727
9.4	.9731	.9736	.9741	.9745	.9750	.9754	.9759	.9763	.9768	.9773
9.5	.9777	.9782	.9786	.9791	.9795	.9800	.9805	.9809	.9814	.9818
9.6	.9823	.9827	.9832	.9836	.9841	.9845	.9850	.9854	.9859	.9863
9.7	.9868	.9872	.9877	.9881	.9886	.9890	.9894	.9899	.9903	.9908
9.8	.9912	.9917	.9921	.9926	.9930	.9934	.9939	.9943	.9948	.9952
9.9	.9956	.9961	.9965	.9969	.9974	.9978	.9983	.9987	.9991	.9996

TABLE III Natural Logarithms

x	$\ln x$	x	$\ln x$	x	$\ln x$
		4.5	1.5041	9.0	2.1972
0.1	−2.3026	4.6	1.5261	9.1	2.2083
0.2	−1.6094	4.7	1.5476	9.2	2.2192
0.3	−1.2040	4.8	1.5686	9.3	2.2300
0.4	−0.9163	4.9	1.5892	9.4	2.2407
0.5	−0.6931	5.0	1.6094	9.5	2.2513
0.6	−0.5108	5.1	1.6292	9.6	2.2618
0.7	−0.3567	5.2	1.6487	9.7	2.2721
0.8	−0.2231	5.3	1.6677	9.8	2.2824
0.9	−0.1054	5.4	1.6864	9.9	2.2925
1.0	0.0000	5.5	1.7047	10	2.3026
1.1	0.0953	5.6	1.7228	11	2.3979
1.2	0.1823	5.7	1.7405	12	2.4849
1.3	0.2624	5.8	1.7579	13	2.5649
1.4	0.3365	5.9	1.7750	14	2.6391
1.5	0.4055	6.0	1.7918	15	2.7081
1.6	0.4700	6.1	1.8083	16	2.7726
1.7	0.5306	6.2	1.8245	17	2.8332
1.8	0.5878	6.3	1.8405	18	2.8904
1.9	0.6419	6.4	1.8563	19	2.9444
2.0	0.6931	6.5	1.8718	20	2.9957
2.1	0.7419	6.6	1.8871	25	3.2189
2.2	0.7885	6.7	1.9021	30	3.4012
2.3	0.8329	6.8	1.9169	35	3.5553
2.4	0.8755	6.9	1.9315	40	3.6889
2.5	0.9163	7.0	1.9459	45	3.8067
2.6	0.9555	7.1	1.9601	50	3.9120
2.7	0.9933	7.2	1.9741	55	4.0073
2.8	1.0296	7.3	1.9879	60	4.0943
2.9	1.0647	7.4	2.0015	65	4.1744
3.0	1.0986	7.5	2.0149	70	4.2485
3.1	1.1314	7.6	2.0281	75	4.3175
3.2	1.1632	7.7	2.0412	80	4.3820
3.3	1.1939	7.8	2.0541	85	4.4427
3.4	1.2238	7.9	2.0669	90	4.4998
3.5	1.2528	8.0	2.0794	100	4.6052
3.6	1.2809	8.1	2.0919	110	4.7005
3.7	1.3083	8.2	2.1041	120	4.7875
3.8	1.3350	8.3	2.1163	130	4.8676
3.9	1.3610	8.4	2.1282	140	4.9416
4.0	1.3863	8.5	2.1401	150	5.0106
4.1	1.4110	8.6	2.1518	160	5.0752
4.2	1.4351	8.7	2.1633	170	5.1358
4.3	1.4586	8.8	2.1748	180	5.1930
4.4	1.4816	8.9	2.1861	190	5.2470

CHAPTER 1 SECTION 1.1

1. $A = lw$ **3.** $C = \pi d$ **5.** $A = \dfrac{\sqrt{3}}{4}x^2$ **7.** $V = \frac{4}{3}\pi R^3$ **9.** $V = x^3$ **11.** $0.666\ldots$ **13.** 0.125 **15.** -1.6

17. $0.111\ldots$ **19.** 0.16 **21.** $-0.4285714285714\ldots$ **23.** 2 **25.** -1 **27.** $\frac{13}{3}$ **29.** -11 **31.** 11 **33.** -4 **35.** $\frac{7}{2}$

37. $\frac{52}{15}$ **39.** $\frac{81}{10}$ **41.** $3x + 12$ **43.** $x^2 - 3x$ **45.** $x^2 + 6x + 8$ **47.** $x^2 - x - 2$ **49.** $x^2 - 10x + 16$ **51.** $x^2 - 4$

53. No; $2 - 3 \neq 3 - 2$ **55.** No; $\frac{2}{3} \neq \frac{3}{2}$ **57.** Symmetric property

SECTION 1.2

1. $>$ **3.** $>$ **5.** $>$ **7.** $=$ **9.** $<$ **11.**

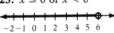

13. $x > 0$ **15.** $x < 2$ **17.** $x \leq 1$ **19.** $2 < x < 5$

21. $x \geq -1$

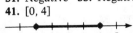

23. $x \geq 4$ and $x < 6$

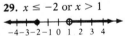

25. $x \leq 0$ or $x < 6$

27. There are no numbers x for which $x \leq -2$ and $x > 1$. **29.** $x \leq -2$ or $x > 1$

31. Negative **33.** Negative **35.** Positive **37.** Positive **39.** Positive

41. $[0, 4]$

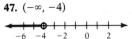

43. $[4, 6)$

45. $[2, +\infty)$

47. $(-\infty, -4)$

49. $2 \leq x \leq 5$

51. $-3 < x < -2$

53. $x \geq 4$

55. $x < -1$

57. 1 **59.** 5 **61.** 1 **63.** 23 **65.** 7

67. $a \leq b, c > 0$ $a - b \leq 0$
$$(a - b)c \leq 0(c)$$
$$ac - bc \leq 0$$
$$ac \leq bc$$

69. $\dfrac{a + b}{2} - a = \dfrac{a + b - 2a}{2} = \dfrac{b - a}{2} > 0$; therefore, $a < \dfrac{a + b}{2}$

$b - \dfrac{a + b}{2} = \dfrac{2b - a - b}{2} = \dfrac{b - a}{2} > 0$; therefore, $b > \dfrac{a + b}{2}$

SECTION 1.3

1. 1 **3.** $\frac{1}{16}$ **5.** $\frac{4}{9}$ **7.** $\frac{1}{8}$ **9.** $\frac{1}{4}$ **11.** $\frac{1}{9}$ **13.** 8 **15.** $\frac{9}{4}$ **17.** 324 **19.** $\frac{4}{81}$ **21.** y^2 **23.** $\frac{y}{x^2}$ **25.** $\frac{1}{x^3y}$ **27.** $\frac{1}{xy}$ **29.** $\frac{y}{x}$

31. $\frac{25y^2}{16x^2}$ **33.** $\frac{1}{x} + \frac{1}{y^2}$ **35.** $\frac{1}{x^3y^3z^2}$ **37.** $\frac{-8x^3}{9yz^2}$ **39.** $\frac{y^2}{x^2 + y^2}$ **41.** $\frac{y^3}{x^8}$ **43.** $\frac{16x^2}{9y^2}$ **45.** $\frac{y}{x^3}$ **47.** x^4y **49.** $\frac{y^2}{x^2}$ **51.** 1, −11

SECTION 1.4

1. Yes; 2 **3.** Yes; 0 **5.** No **7.** Yes; 3 **9.** Yes; 3 **11.** $x^2 + 4x + 3$ **13.** $x^3 - 5x^2 + 6x + 9$ **15.** $6x^5 + 5x^4 + 3x^2 + x$
17. $9x^2 - 7x - 5$ **19.** $-2x^3 + 18x^2 - 18$ **21.** $2x^2 - 4x + 6$ **23.** $15y^2 - 18y + 3$ **25.** $2ax + a^2$
27. $3ax^2 + 3a^2x + a^3$ **29.** $x^2 + x - 12$ **31.** $2x^2 + 17x + 8$ **33.** $-3x^2 - 11x + 4$ **35.** $12x^2 - 11x + 2$ **37.** $x^3 - 1$
39. $2x^3 + 3x^2 - 4x - 6$ **41.** $6x^2 - x - 15$ **43.** $2x^3 + 7x^2 + 7x + 5$ **45.** $6x^3 - 11x^2 + 9x - 2$ **47.** $x^4 - x^2 + 2x - 1$
49. $6x^4 - 11x^3 + 23x^2 - 16x + 16$ **51.** $2x^5 - 6x^4 - x^3 + 13x^2 - 5$ **53.** $x^3 + 3x^2 - 4$
55. $2x^3 - x^2 - 4x + 3$ **57.** $x^3 - 6x^2 + 11x - 6$ **59.** $6x^2 + 2$ **61.** $x^4 - 2x^2 + 1$
63. $x^6 - 4x^5 + 6x^4 - 6x^3 + 4x^2 - 2x + 1$ **65.** $x^2 - xy - 2y^2$ **67.** $x^2 - 2y^2 + 3x + y$ **69.** $-4xy$
71. $2x^2 - 2xy + 13y^2$ **73.** $2x^2 + 2y^2 + 2z^2 + 2xy + 2yz$ **75.** $x^3 - y^3$ **77.** $6x^4 + x^3y + 6x^2y^2 + xy^3 + 2y^4$
79. $x^2 - y^2 - z^2 - 2yz$
81. The degree of the product equals the degree of the product of the leading terms:

$$(a_nx^n + a_{n-1}x^{n-1} + \cdots + a_1x + a_0)(b_mx^m + b_{m-1}x^{m-1} + \cdots + b_1x + b_0)$$
$$= a_nb_mx^{n+m} + (a_nb_{m-1} + a_{n-1}b_m)x^{n+m-1} + \cdots$$

83. (2) $(x - a)(x + a) = x(x + a) - a(x + a) = x^2 + ax - ax - a^2 = x^2 - a^2$
(3a) $(x + a)^2 = (x + a)(x + a) = x(x + a) + a(x + a) = x^2 + 2ax + a^2$
(3b) $(x - a)^2 = (x - a)(x - a) = x(x - a) - a(x - a) = x^2 - 2ax + a^2$
(6) $(x - a)(x^2 + ax + a^2) = x(x^2 + ax + a^2) - a(x^2 + ax + a^2)$
$= x^3 + ax^2 + a^2x - ax^2 - a^2x - a^3 = x^3 - a^3$
85. $(x + a)^4 = (x + a)^2(x + a)^2 = (x^2 + 2ax + a^2)(x^2 + 2ax + a^2) = x^4 + 4ax^3 + 6a^2x^2 + 4a^3x + a^4$

SECTION 1.5

1. $3(x + 2)$ **3.** $a(x^2 + 1)$ **5.** $x(x^2 + x + 1)$ **7.** $2(x^2 + x + 1)$ **9.** $3xy(x - 2y + 4)$ **11.** $(x - 1)(x + 1)$
13. $(2x - 1)(2x + 1)$ **15.** $(x + 2)(x + 5)$ **17.** $(x - 7)(x - 3)$ **19.** $(x - 8)(x + 1)$ **21.** $(x + 1)^2$ **23.** $(x + 2)^2$
25. $(x - 5)(x + 3)$ **27.** $3(x + 2)(x - 6)$ **29.** $y^2(y + 5)(y + 6)$ **31.** $(4x + 1)^2$ **33.** $(2x + 3)^2$ **35.** $a(x - 9a)(x + 5a)$
37. $(x - 3)(x^2 + 3x + 9)$ **39.** $(2x + 3)(4x^2 - 6x + 9)$ **41.** $(3x + 1)(x + 1)$ **43.** $(x^2 + 9)(x + 3)(x - 3)$
45. $(x - 1)^2(x^2 + x + 1)^2$ **47.** $x^5(x - 1)(x + 1)$ **49.** $(2z + 3)(z + 1)$ **51.** $(4x - 3)^2$ **53.** $(4x - 5)(4x + 1)$
55. $(2y + 5)(2y + 3)$ **57.** $9(2x - 3)(x + 1)$ **59.** $2(4x^2 + x + 3)$ **61.** Prime **63.** $2x(2x + 1)(x - 3)$
65. $(x^2 - x + 1)(x^2 + x + 1)$ **67.** $(3x - 1)(3x + 1)(x^2 + 1)$ **69.** $(x + 3)(x - 6)$ **71.** $(x + 2)(x - 3)$
73. $(x + 5)(3x + 13)$ **75.** $(x - 1)(x + 1)(x + 2)$ **77.** $(x - 1)(x + 1)(x^2 - x + 1)$ **79.** $(x^2 + 1)(x + 2)(x^2 - 2x + 4)$

SECTION 1.6

1. $\frac{5}{x - 2}$ **3.** $\frac{x}{3}$ **5.** $\frac{4}{2x - 1}$ **7.** $\frac{y + 5}{2}$ **9.** $x + 5$ **11.** $\frac{x - 2}{x + 3}$ **13.** $-(x + 7)$ **15.** $\frac{2x - 5}{x - 4}$ **17.** $\frac{x - 7}{x + 7}$ **19.** $\frac{6(x + 1)}{x^2 + x + 6}$

21. $\frac{3(x - 3)}{5x}$ **23.** $\frac{x(2x - 1)}{x + 4}$ **25.** $\frac{8}{3x}$ **27.** $\frac{x - 3}{x + 7}$ **29.** $\frac{(x - 4)(x + 3)}{(x - 1)(2x + 1)}$ **31.** $\frac{3x - 2}{3x + 2}$ **33.** $\frac{x^2(5x - 1)^2}{(x - 1)(3x + 2)^2}$ **35.** $\frac{2x - 3}{x - 3}$

37. $\frac{x + 9}{2x - 1}$ **39.** $\frac{7x - 2}{(x + 1)(x - 2)}$ **41.** $\frac{3(x + 3)}{(x - 1)(x + 2)}$ **43.** $\frac{3x^2 - 2x - 3}{(x + 1)(x - 1)}$ **45.** $\frac{-8x}{(x + 2)(x - 2)}$ **47.** $\frac{2(x^2 - 2)}{x(x - 2)(x + 2)}$

49. $\frac{2x^3 - 2x^2 + 2x - 1}{x(x - 1)^2}$ **51.** $\frac{x^3 - 7x^2 + 3x + 5}{(x + 1)(x - 1)(x - 2)}$ **53.** $\frac{x^2 - 2x - 1}{x(x - 1)(x + 1)}$ **55.** $(x - 2)(x + 2)(x + 1)$

57. $x(x - 1)(x + 1)$ **59.** $x^3(2x - 1)^2$ **61.** $x(x - 1)^2(x + 1)(x^2 + x + 1)$ **63.** $\frac{5x}{(x - 6)(x - 1)(x + 4)}$

65. $\dfrac{2(x + 4)}{(x - 2)(x + 2)(x + 3)}$ **67.** $\dfrac{5x + 1}{(x - 1)^2(x + 1)^2}$ **69.** $\dfrac{-x^2 + 3x + 13}{(x - 2)(x + 1)(x + 4)}$ **71.** $\dfrac{x^3 - 2x^2 + 4x + 3}{x^2(x + 1)(x - 1)}$ **73.** $\dfrac{-1}{x(x + h)}$

75. $\dfrac{(x - 1)(x + 1)}{x^2 + 1}$ **77.** $\dfrac{(x - 1)(-x^2 + 3x + 3)}{x^2 + x - 1}$ **79.** $\dfrac{2(5x - 1)}{(x - 2)(x + 1)^2}$ **81.** $\dfrac{-2x(x^2 - 2)}{(x + 2)(x^2 - x - 3)}$ **83.** $\dfrac{-1}{x - 1}$

85. $\dfrac{4}{(x + h + 2)(x + 2)}$

SECTION 1.7

1. 8 **3.** 3 **5.** -1 **7.** $\frac{1}{2}$ **9.** $3x$ **11.** $2(1 + x)$ **13.** $2\sqrt{2}$ **15.** $2x\sqrt[3]{2x}$ **17.** x **19.** $\frac{4}{3}x\sqrt{2}$ **21.** x^3y^2 **23.** x^2y

25. $6\sqrt{x}$ **27.** $6x\sqrt{x}$ **29.** 1 **31.** $15\sqrt[3]{3}$ **33.** $\dfrac{1}{x(2x + 3)}$ **35.** $\sqrt[24]{x}$ **37.** 1 **39.** $4\sqrt{2}$ **41.** $4\sqrt{2}$ **43.** $\sqrt[3]{2}$

45. $(x + 5)(x - 3)\sqrt{2x}$ **47.** $(-x - 5y)\sqrt[3]{2xy}$ **49.** $12\sqrt{3}$ **51.** $2\sqrt{3}$ **53.** $12(4 + \sqrt{7})$ **55.** $x - 2\sqrt{x} + 1$

57. $x - 3\sqrt[3]{x^2} + 3\sqrt[3]{x} - 1$ **59.** $4x + 4\sqrt{xy} - 15y$ **61.** $\dfrac{-x^2}{\sqrt{1 - x^2}}$ **63.** $\dfrac{\sqrt{2}}{2}$ **65.** $\dfrac{-\sqrt{15}}{5}$ **67.** $\dfrac{\sqrt{3}(5 + \sqrt{2})}{23}$

69. $\dfrac{-19 + 8\sqrt{5}}{41}$ **71.** $5 - 2\sqrt{6}$ **73.** $\dfrac{2x + h - 2\sqrt{x(x + h)}}{h}$ **75.** $\dfrac{\sqrt[3]{x^2} - \sqrt[3]{x} + 1}{x + 1}$ **77.** $\dfrac{1}{\sqrt{x + h} + \sqrt{x}}$

SECTION 1.8

1. 4 **3.** -3 **5.** 64 **7.** $\frac{1}{27}$ **9.** $\dfrac{27\sqrt{2}}{32}$ **11.** $\dfrac{27\sqrt{2}}{32}$ **13.** 8 **15.** 8 **17.** 27 **19.** $\frac{1}{2}$ **21.** $x^{7/12}$ **23.** xy^2 **25.** $x^{4/3}y^{5/3}$

27. $\dfrac{8x^{3/2}}{y^{1/4}}$ **29.** $\dfrac{x^{11}}{y^{13}}$ **31.** $\dfrac{3x + 2}{(1 + x)^{1/2}}$ **33.** $\dfrac{x(3x^2 + 2)}{(x^2 + 1)^{1/2}}$ **35.** $\dfrac{22x + 5}{10\sqrt{x} - 5\sqrt{4x + 3}}$ **37.** $\dfrac{2 + x}{2(1 + x)^{3/2}}$ **39.** $\dfrac{4 - x}{(x + 4)^{3/2}}$

41. $\dfrac{1}{x^2(x^2 - 1)^{1/2}}$ **43.** $\dfrac{1 - 3x^2}{2\sqrt{x}(1 + x^2)^2}$ **45.** $\frac{1}{2}(5x + 2)(x + 1)^{1/2}$ **47.** $2x^{1/2}(3x - 4)(x + 1)$ **49.** $(x^2 + 4)^{1/3}(11x^2 + 12)$

51. $(3x + 5)^{1/3}(2x + 3)^{1/2}(17x + 27)$

SECTION 1.9

1. $8 + 5i$ **3.** $1 - 2i$ **5.** $-6 - 11i$ **7.** $6 - 18i$ **9.** $6 + 4i$ **11.** $10 - 5i$ **13.** 37 **15.** $\frac{6}{5} + \frac{8}{5}i$ **17.** $\frac{1}{5} + \frac{2}{5}i$ **19.** $\frac{5}{2} - \frac{7}{2}i$

21. $-\frac{1}{2} + (\sqrt{3}/2)i$ **23.** $2i$ **25.** $\frac{2}{5} - \frac{11}{5}i$ **27.** $\frac{7}{50} - \frac{1}{50}i$ **29.** $-i$ **31.** i **33.** -6 **35.** $-10i$ **37.** $-2 + 2i$ **39.** 0

41. $\frac{1}{2} - \frac{1}{2}i$ **43.** 0 **45.** 6 **47.** 25 **49.** $36 + 23i$ **51.** $11 - 7i$

53. $z + \bar{z} = (a + bi) + (a - bi) = 2a; \; z - \bar{z} = (a + bi) - (a - bi) = 2bi$

55. $\overline{z + w} = \overline{(a + bi) + (c + di)} = \overline{(a + c) + (b + d)i} = (a + c) - (b + d)i = (a - bi) + (c - di) = \bar{z} + \bar{w}$

SECTION 1.10

1. 4.542×10^2 **3.** 1.3×10^{-2} **5.** 3.2155×10^4 **7.** 4.23×10^{-4} **9.** 21,500 **11.** 0.001214 **13.** 110,000,000

15. 0.081 **17.** 3 **19.** 3 **21.** 4 **23.** 2 **25.** (a) 6.79 (b) 6.78 **27.** (a) 189 (b) 189

29. (a) 0.000126 (b) 0.000125 **31.** (a) 1,000,000 (b) 999,000 **33.** 18.953 **35.** 28.653 **37.** 0.063 **39.** 0.429

41. 1.141 **43.** 4.796

SECTION 1.11

1. 13 **3.** 26 **5.** 25 **7.** Yes, 5 **9.** Not a right triangle **11.** Yes, 25 **13.** Not a right triangle **15.** 46.7 mi

17. 3 mi

FILL-IN-THE-BLANK QUESTIONS

1. symmetric **3.** distributive **5.** base, exponent **7.** index, radicand

REVIEW EXERCISES

1. -16 **3.** $\frac{1}{6}$ **5.** $\frac{87}{8}$ **7.** -2 **9.** $\frac{1}{128}$ **11.** 16 **13.** $\frac{3\sqrt[3]{3}}{4}$ **15.** $\frac{1}{2}$

17. $x > -4$ and $x < 4$ **19.** $x < -4$ or $x > 0$

21. $\frac{y^2}{x^2}$ **23.** $\frac{1}{x^5 y}$ **25.** $\frac{125}{x^2 y}$ **27.** $\frac{x^2}{16y^3}$ **29.** $2x^3 - 13x^2 + 19x - 6$ **31.** $9x^3 - 20x^2 + 6x + 13$ **33.** $6x^2 - 11x - 10$

35. $x^3 + 6x^2 + 11x + 6$ **37.** $x^2 + 2xy - 8y^2$ **39.** $(x + 7)(x - 2)$ **41.** $(3x + 2)(2x - 3)$ **43.** $(3x + 2)(x - 7)$

45. $(2x + 1)(4x^2 - 2x + 1)$ **47.** $(2x + 3)(x - 1)(x + 1)$ **49.** $(5x - 2)(5x + 2)$ **51.** Prime **53.** $\frac{2x + 7}{x - 2}$

55. $\frac{3(3x - 1)}{(x + 3)(3x + 1)}$ **57.** $\frac{4x}{(x + 1)(x - 1)}$ **59.** $\frac{x^2 + 17x + 2}{(x - 2)(x + 2)^2}$ **61.** $\frac{2x + 1}{x + 1}$ **63.** $\frac{2\sqrt{3}}{3}$ **65.** $-2(1 + \sqrt{2})$

67. $\frac{-(3 + \sqrt{5})}{2}$ **69.** $\frac{2(1 + x^2)}{(2 + x^2)^{1/2}}$ **71.** $\frac{x(3x + 16)}{2(x + 4)^{3/2}}$ **73.** $4 - 7i$ **75.** $-3 + 2i$ **77.** $\frac{9}{10} - \frac{3}{10}i$ **79.** 1 **81.** $-46 + 9i$

83.
$$\sqrt{1 + \left[\frac{1}{2}\left(x^3 - \frac{1}{x^3}\right)\right]^2} = \sqrt{1 + \frac{1}{4}\left(x^6 - 2 + \frac{1}{x^6}\right)}$$
$$= \sqrt{1 + \frac{1}{4}\left(\frac{x^{12} - 2x^6 + 1}{x^6}\right)}$$
$$= \sqrt{\frac{4x^6 + x^{12} - 2x^6 + 1}{4x^6}}$$
$$= \frac{1}{2}\sqrt{\frac{x^{12} + 2x^6 + 1}{x^6}}$$
$$= \frac{1}{2}\sqrt{x^6 + 2 + \frac{1}{x^6}}$$
$$= \frac{1}{2}\sqrt{\left(x^3 + \frac{1}{x^3}\right)^2}$$
$$= \frac{1}{2}\left(x^3 + \frac{1}{x^3}\right)$$

85.
$$1 + \left[\frac{1}{2}\left(x^2 - \frac{1}{x^2}\right)\right]^2 = 1 + \frac{1}{4}\left(x^4 - 2 + \frac{1}{x^4}\right)$$
$$= 1 + \frac{x^8 - 2x^4 + 1}{4x^4}$$
$$= \frac{4x^4 + x^8 - 2x^4 + 1}{4x^4}$$
$$= \frac{1}{4}\left(\frac{x^8 + 2x^4 + 1}{x^4}\right)$$
$$= \frac{1}{4}\left(x^4 + 2 + \frac{1}{x^4}\right)$$
$$= \frac{1}{4}\left(x^2 + \frac{1}{x^2}\right)^2$$

CHAPTER 2   SECTION 2.1

1. 6 **3.** -2 **5.** 4 **7.** $\frac{5}{4}$ **9.** -4 **11.** 3 **13.** -1 **15.** $-\frac{4}{3}$ **17.** -18 **19.** -3 **21.** $-\frac{3}{4}$ **23.** -16 **25.** 2 **27.** 0.5

29. $\frac{46}{5}$ **31.** 2 **33.** 2 **35.** -1 **37.** 3 **39.** 2 **41.** 21 **43.** $\{-2, 2\}$ **45.** $-\frac{20}{39}$ **47.** 6 **49.** $-\frac{11}{6}$ **51.** $x = \frac{b + c}{a}$

53. $x = \frac{abc}{a + b}$ **55.** $x = a^2$ **57.** $a = 3$ **59.** $R = \frac{R_1 R_2}{R_1 + R_2}$ **61.** $R = \frac{mv^2}{F}$ **63.** $r = \frac{S - a}{S}$

65. In Step 7 we divided by $x - 2$. Since $x = 2$ (from Step 1), we actually divided by 0.

SECTION 2.2

1. $A = \pi R^2$; R = radius, A = area **3.** $A = s^2$; A = area, s = length of a side

5. $F = ma$; F = force, m = mass, a = acceleration **7.** $W = Fd$; W = work, F = force, d = distance **9.** 41, 42

11. 13, 14 **13.** 65, 25 **15.** \$8.50 per hr **17.** 5 **19.** \$31,250 in B-bonds, \$18,750 in a certificate **21.** \$8000

23. 40 lb **25.** 30 cc of 15% HCl, 70 cc of 5% HCl **27.** 3260 adults **29.** \$110,000; \$16,500 **31.** 12 min **33.** 85

35. At tight end's 45-yard line **37.** $\frac{2}{3}$ gal **39.** 2.14 mph **41.** 9:45 A.M. **43.** 1 hr

SECTION 2.3

1. $\{0, 4\}$ **3.** $\{-3, 3\}$ **5.** $\{-4, 3\}$ **7.** $\{-\frac{1}{2}, 3\}$ **9.** $\{-\frac{1}{2}, \frac{1}{3}\}$ **11.** $\{3, 4\}$ **13.** $\frac{3}{2}$ **15.** $\{-\frac{2}{3}, \frac{3}{2}\}$ **17.** $\{-\frac{2}{3}, \frac{3}{2}\}$ **19.** $\{-\frac{3}{4}, 2\}$
21. $\{2 - \sqrt{2}, 2 + \sqrt{2}\}$ **23.** $\left\{\dfrac{5 - \sqrt{29}}{2}, \dfrac{5 + \sqrt{29}}{2}\right\}$ **25.** $\{1, \frac{3}{2}\}$ **27.** No real solution **29.** $\left\{\dfrac{-1 - \sqrt{5}}{2}, \dfrac{-1 + \sqrt{5}}{2}\right\}$
31. $\{0, \frac{9}{4}\}$ **33.** $\frac{1}{3}$ **35.** $\{1 - \sqrt{2}, 1 + \sqrt{2}\}$ **37.** $\left\{\dfrac{1 - \sqrt{33}}{8}, \dfrac{1 + \sqrt{33}}{8}\right\}$ **39.** No real solution **41.** $\{0.59, 3.41\}$
43. $\{-2.80, 1.07\}$ **45.** $\{-0.85, 1.17\}$ **47.** No real solution **49.** $\{-\sqrt{6}, \sqrt{6}\}$ **51.** $\frac{1}{4}$ **53.** $\{-\frac{3}{5}, \frac{5}{2}\}$ **55.** $\{-\frac{1}{2}, \frac{2}{3}\}$
57. $\left\{\dfrac{-\sqrt{2} + 2}{2}, \dfrac{-\sqrt{2} - 2}{2}\right\}$ **59.** $\left\{\dfrac{-1 - \sqrt{17}}{2}, \dfrac{-1 + \sqrt{17}}{2}\right\}$ **61.** No real solution **63.** Repeated real solution
65. Two unequal real solutions **67.** 4 **69.** $\frac{1}{16}$ **71.** $\frac{1}{9}$ **73.** $\{-7, 3\}$ **75.** $\{-\frac{1}{4}, \frac{3}{4}\}$ **77.** $\left\{\dfrac{-1 - \sqrt{7}}{6}, \dfrac{-1 + \sqrt{7}}{6}\right\}$
79. 11, 13 **81.** 4 ft by 4 ft **83.** (a) 6 sec (b) 5 sec **85.** 10 days **87.** 175 boxes **89.** 2.56 ft
91. Length = 11.55 cm, width = 6.55 cm, thickness = 3 cm (as given) **93.** 2.71 ft **95.** 5 mph
97. $\dfrac{-b + \sqrt{b^2 - 4ac}}{2a} + \dfrac{-b - \sqrt{b^2 - 4ac}}{2a} = \dfrac{-2b}{2a} = \dfrac{-b}{a}$

SECTION 2.4

1. $2i$ **3.** $5i$ **5.** $5i$ **7.** $-15 + 8i$ **9.** $12 + 5i$ **11.** $\{-2i, 2i\}$ **13.** $\{-4, 4\}$ **15.** $\{3 - 2i, 3 + 2i\}$ **17.** $\{3 - i, 3 + i\}$
19. $\{\frac{1}{4} - \frac{1}{4}i, \frac{1}{4} + \frac{1}{4}i\}$ **21.** $\{-\frac{1}{5} - \frac{2}{5}i, -\frac{1}{5} + \frac{2}{5}i\}$ **23.** $\left\{-\dfrac{1}{2} - \dfrac{\sqrt{3}}{2}i, -\dfrac{1}{2} + \dfrac{\sqrt{3}}{2}i\right\}$ **25.** Two complex solutions
27. Two unequal real solutions **29.** A repeated real solution **31.** $2 - 3i$

SECTION 2.5

1. 3 **3.** $\{-2, 2\}$ **5.** No real solution **7.** 9 **9.** -8 **11.** 4 **13.** No real solution **15.** $\{-\frac{1}{8}, \frac{1}{8}\}$ **17.** $\frac{1}{2}$ **19.** $\{-\frac{1}{2}, \frac{1}{2}\}$
21. $\frac{1}{32}$ **23.** No real solution **25.** No real solution **27.** 1 **29.** No real solution **31.** -13 **33.** 3 **35.** 2 **37.** $-\frac{8}{5}$
39. 2 **41.** $\{-1, 3\}$ **43.** $\{1, 5\}$ **45.** $\{0, 16\}$ **47.** $\{-64, 27\}$ **49.** $\{-\frac{3}{2}, 2\}$ **51.** $\{-2, -1, 1, 2\}$ **53.** 1
55. $\left\{\left(\dfrac{9 - \sqrt{17}}{8}\right)^4, \left(\dfrac{9 + \sqrt{17}}{8}\right)^4\right\}$ **57.** $\{\sqrt{2}, \sqrt{3}\}$ **59.** $\{-4, 1\}$ **61.** $\{-2, -\frac{1}{2}\}$ **63.** $\{-2, -\frac{4}{5}\}$ **65.** $\{0, 64\}$ **67.** $\{0, 2\}$
69. $\{-4, 0, 3\}$ **71.** -1 **73.** $\{-2, 2, 3\}$ **75.** $\{-1, 1\}$ **77.** $\{0.34, 11.66\}$ **79.** $\{-1.03, 1.03\}$ **81.** $\{-1.85, 0.17\}$
83. 229.94 ft

SECTION 2.6

1. (a) $0 < 2$ (b) $-2 < 0$ (c) $9 < 15$ (d) $-6 > -10$
3. (a) $2x - 2 < -1$ (b) $2x - 4 < -3$ (c) $6x + 3 < 6$ (d) $-4x - 2 > -4$
5. $x < 4$

7. $x \geq -1$

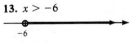

9. $x > 3$

11. $x \geq 2$

13. $x > -6$

15. $x \leq \frac{2}{3}$

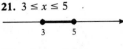

17. $x < -20$

19. $x \geq \frac{4}{3}$

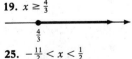

21. $3 \leq x \leq 5$

23. $-\frac{1}{3} \leq x \leq \frac{7}{3}$

25. $-\frac{11}{2} < x < \frac{1}{2}$

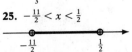

27. $-6 < x < 0$

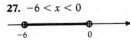

29. $x < -5$

31. $x \geq -1$

33. $\frac{1}{2} \leq x < \frac{5}{4}$

35. From 1600 to 2400 cc, inclusive **37.** From \$45,000 to \$95,000; from 5% to 8.6%, inclusive
39. From \$78.16 to \$90.66, inclusive **41.** From 675.46 kwhr to 2500.82 kwhr, inclusive
43. From \$7457.63 to \$7857.14, inclusive **45.** Fifth test score ≥ 74 **47.** From 12 to 20 gal, inclusive

SECTION 2.7

1. $-1 < x < 3$ **3.** $x < 0$ or $x > 4$ **5.** $-3 < x < 3$ **7.** $x < -4$ or $x > 3$ **9.** $-\frac{1}{2} < x < 3$ **11.** $x < 3$ or $x > 4$
13. No real solution **15.** $x < -\frac{2}{3}$ or $x > \frac{3}{2}$ **17.** $x > 1$ **19.** $x < 1$ or $2 < x < 3$ **21.** $-2 < x < 0$ or $x > 4$
23. $-1 < x < 0$ or $x > 1$ **25.** $x > 1$ **27.** $x < -1$ or $x > 1$ **29.** $x < -1$ or $x > 1$ **31.** $x < -1$ or $0 < x < 1$
33. $-1 < x < 1$ or $x \geq 2$ **35.** $x < -\frac{2}{3}$ or $0 < x < \frac{3}{2}$ **37.** $x < 2$ **39.** $x < -3$ or $-1 < x < 1$ or $x > 2$
41. $x < -5$ or $-4 < x < -3$ or $x > 1$ **43.** Numbers larger than 4 **45.** $2 < t < 3$ **47.** $10 \leq x \leq 50$
49. $x^2 - a < 0$; $(x - \sqrt{a})(x + \sqrt{a}) < 0$; therefore, $-\sqrt{a} < x < \sqrt{a}$ **51.** $-1 < x < 1$ **53.** $x \geq 3$ or $x \leq -3$
55. $-4 \leq x \leq 4$ **57.** $x > 2$ or $x < -2$

SECTION 2.8

1. $\{-3, 3\}$ **3.** $\{-4, 1\}$ **5.** $\{-1, \frac{3}{2}\}$ **7.** $\{-4, 4\}$ **9.** 2 **11.** $\{-12, 12\}$ **13.** $\{-\frac{36}{5}, \frac{24}{5}\}$ **15.** No real solution **17.** $\{-2, 2\}$
19. $\{-1, 3\}$ **21.** $\{-2, -1, 0, 1\}$ **23.** $-3 < x < 3$ **25.** $x < -4$ or $x > 4$ **27.** $1 < x < 3$ **29.** $-\frac{2}{3} \leq t \leq 2$
31. $x \leq -1$ or $x \geq 3$ **33.** $-1 < x < \frac{3}{2}$ **35.** $x < -1$ or $x > 2$ **37.** $x < -2$ or $x > -2$ **39.** All real numbers
41. $0.49 < x < 0.51$ **43.** $\left|\dfrac{x}{y}\right| = \sqrt{\left(\dfrac{x}{y}\right)^2} = \sqrt{\dfrac{x^2}{y^2}} = \dfrac{\sqrt{x^2}}{\sqrt{y^2}} = \dfrac{|x|}{|y|}$
45. $(x + y)^2 = x^2 + 2xy + y^2 \leq |x|^2 + 2|x||y| + |y|^2 = [|x| + |y|]^2$; therefore,
$$\sqrt{(x + y)^2} \leq \sqrt{(|x| + |y|)^2} \text{ or } |x + y| \leq |x| + |y|$$
47. $|x - 2| < \frac{1}{2}, \frac{3}{2} < x < \frac{5}{2}$ **49.** $|x + 3| > 2, x < -5$ or $x > -1$ **51.** $|x - 98.6| \geq 1.5, x \leq 97.1° \text{F}$ or $x \geq 100.1° \text{F}$

FILL-IN-THE-BLANK QUESTIONS

1. equivalent **3.** add, $\frac{25}{4}$ **5.** $-a$ **7.** extraneous

REVIEW EXERCISES

1. -9 **3.** 6 **5.** $\frac{1}{5}$ **7.** 5 **9.** No real solution **11.** $\frac{11}{8}$ **13.** $\{-2, \frac{3}{2}\}$ **15.** $\left\{\dfrac{1 - \sqrt{13}}{4}, \dfrac{1 + \sqrt{13}}{4}\right\}$ **17.** $\{-3, 3\}$

19. No real solution **21.** $\{-2, -1, 1, 2\}$ **23.** 2 **25.** 0 **27.** $\dfrac{\sqrt{5}}{2}$ **29.** 1 **31.** $\{-1, \frac{1}{2}\}$

33. $\left\{\dfrac{m}{1 - n}, \dfrac{m}{1 + n}\right\}$ **35.** $\left\{\dfrac{-9b}{5a}, \dfrac{2b}{a}\right\}$ **37.** $\{-\frac{9}{5}, 3\}$ **39.** $x \geq 4$ **41.** $-\frac{31}{2} \leq x \leq \frac{33}{2}$ **43.** $-23 < x < -7$
45. $-4 < x < \frac{3}{2}$ **47.** $-2 < x \leq 4$ **49.** $x < 1$ or $x > \frac{5}{4}$ **51.** $1 < x < 2$ or $x > 3$ **53.** $x < -4$ or $2 < x < 4$ or $x > 6$
55. $-\frac{3}{2} < x < -\frac{7}{6}$ **57.** $x \leq -1$ or $x \geq 6$ **59.** $\left\{\dfrac{-1 - \sqrt{3}i}{2}, \dfrac{-1 + \sqrt{3}i}{2}\right\}$ **61.** $\left\{\dfrac{-1 - \sqrt{17}}{4}, \dfrac{-1 + \sqrt{17}}{4}\right\}$
63. $\left\{\dfrac{1 - \sqrt{11}i}{2}, \dfrac{1 + \sqrt{11}i}{2}\right\}$ **65.** $\left\{\dfrac{1 - \sqrt{23}i}{2}, \dfrac{1 + \sqrt{23}i}{2}\right\}$ **67.** $\{-\sqrt{2}, \sqrt{2}, -2i, 2i\}$ **69.** $k = \frac{1}{2}, k = -\frac{1}{2}$
71. $-2 < k < 2$ **73.** Equivalent **75.** Not equivalent; the resulting equation may have more solutions.
77. $ax^2 + bx + c = 0, x = \dfrac{-b \pm \sqrt{b^2 - 4ac}}{2a}; ax^2 - bx + c = 0, x = \dfrac{b \pm \sqrt{(-b)^2 - 4ac}}{2a}$ **79.** 36 **81.** 3300 ft
83. 616 mi **85.** A little less than 1 hr, 35 min **87.** 190.67 ft **89.** 36; \$13.40

CHAPTER 3   SECTION 3.1

1. (a) Quadrant II
(b) Positive x-axis
(c) Quadrant III
(d) Quadrant I
(e) Negative y-axis
(f) Quadrant IV

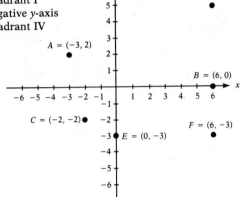

3. The points will be on a vertical line that is two units to the right of the y-axis.

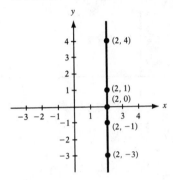

5. 5   **7.** $\sqrt{85}$ **9.** $2\sqrt{5}$ **11.** 2.625   **13.** $\sqrt{a^2 + b^2}$
15. $d(A, B) = \sqrt{13}$
$d(B, C) = \sqrt{13}$
$d(A, C) = \sqrt{26}$
$(\sqrt{13})^2 + (\sqrt{13})^2 = (\sqrt{26})^2$
Area $= \frac{13}{2}$ square units

17. $d(A, B) = \sqrt{130}$
$d(B, C) = \sqrt{26}$
$d(A, C) = \sqrt{104}$
$(\sqrt{26})^2 + (\sqrt{104})^2 = (\sqrt{130})^2$
Area $= 26$ square units

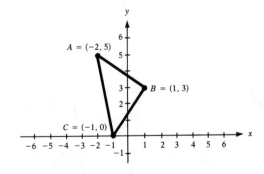

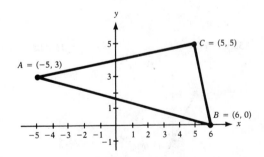

19. $(-2, 0), (6, 0)$ **21.** $(3, -\frac{3}{2})$ **23.** $(\frac{3}{2}, 1)$ **25.** $(5, -1)$ **27.** $(1.05, 0.7)$ **29.** $(a/2, b/2)$ **31.** $\sqrt{73}, 2\sqrt{13}, 5$
33. $d(P_1, P_2) = 6; d(P_2, P_3) = 4; d(P_1, P_3) = 2\sqrt{13}$; right triangle
35. $d(P_1, P_2) = \sqrt{68}; d(P_2, P_3) = \sqrt{34}; d(P_1, P_3) = \sqrt{34}$; isosceles right triangle
37. $\dfrac{x - x_1}{x_2 - x_1} = r$ and $\dfrac{y - y_1}{y_2 - y_1} = r; x = (1 - r)x_1 + rx_2$ and $y = (1 - r)y_1 + ry_2$ **39.** P_2 **41.** $(9, 8)$
43. $90\sqrt{2} \approx 127.28$ ft   **45.** $M_1 = (s/2, s/2), M_2 = (s/2, s/2)$ **47.** $d = 50t$

SECTION 3.2

1.

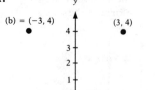

3.

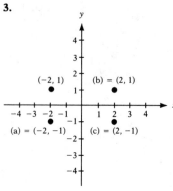

5.

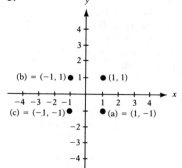

7.

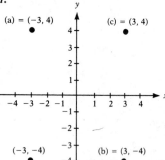

9.

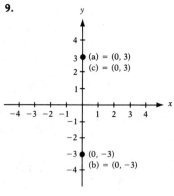

11. $y = 3x + 2$

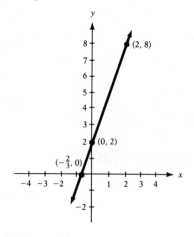

13. $y = -2x + 1$

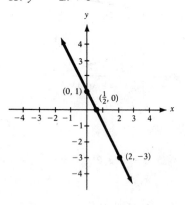

15. $3x - 2y + 6 = 0$

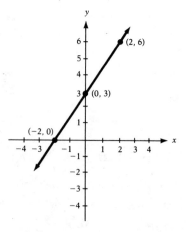

17. $y = 2x^2$

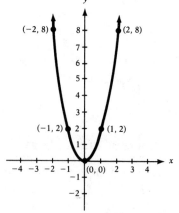

19. $y = -x^2$

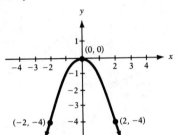

21. $y = x^2 + 3$

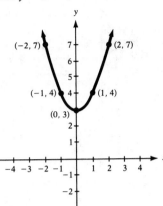

23. $y = x^3 - 1$

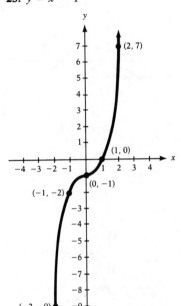

25. $x^2 = 4y$

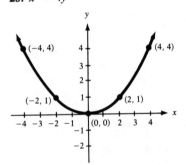

27. $x^2 = y + 1$

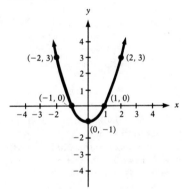

29. $y = \sqrt{x}$

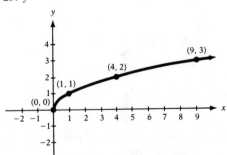

31. $(x - 1)^2 + (y + 1)^2 = 1$; $x^2 + y^2 - 2x + 2y + 1 = 0$ **33.** $x^2 + (y - 2)^2 = 4$; $x^2 + y^2 - 4y = 0$
35. $(x - 4)^2 + (y + 3)^2 = 25$; $x^2 + y^2 - 8x + 6y = 0$ **37.** $x^2 + y^2 = 4$; $x^2 + y^2 - 4 = 0$ **39.** $R = 2$; $(h, k) = (0, 0)$
41. $R = 2$; $(h, k) = (3, 0)$ **43.** $R = 3$; $(h, k) = (-2, 2)$ **45.** $R = \frac{1}{2}$; $(h, k) = (\frac{1}{2}, -1)$ **47.** $R = 5$; $(h, k) = (3, -2)$

49.

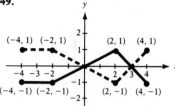

51.

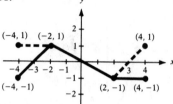

53.

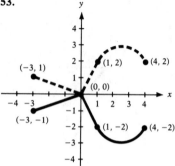

55.

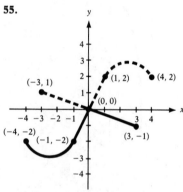

57. $(0, 0)$; symmetric with respect to the y-axis   **59.** $(0, 0)$; symmetric with respect to the origin
61. $(0, 9), (3, 0), (-3, 0)$; symmetric with respect to the y-axis
63. $(0, 2), (0, -2), (3, 0), (-3, 0)$; symmetric with respect to the x-axis, y-axis, and origin
65. $(0, -27), (3, 0)$; no symmetry   **67.** $x^2 + y^2 - 13 = 0$ **69.** $x^2 + y^2 - 4x - 6y + 4 = 0$
71. $x^2 + y^2 + 2x - 6y + 5 = 0$ **73.** $x^2 + y^2 + 2x + 4y - 4168.16 = 0$

SECTION 3.3

1. Slope $= 3$

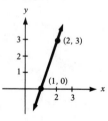

3. Slope $= -\frac{1}{2}$

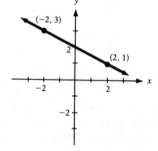

5. Slope $= 0$

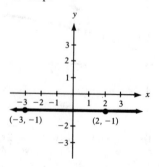

7. Slope undefined

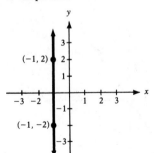

9. Slope $= \dfrac{\sqrt{3} - 3}{1 - \sqrt{2}} \approx 3.06$

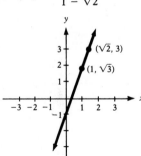

11.

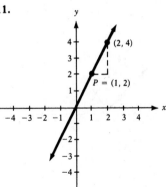

13.

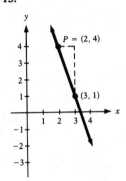

15.

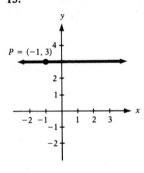

17.

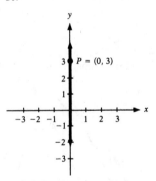

19. $2x - y + 7 = 0$ **21.** $2x + 3y + 1 = 0$ **23.** $x - 2y + 5 = 0$ **25.** $3x + y - 3 = 0$ **27.** $x - 2y - 2 = 0$
29. $x - 1 = 0$ **31.** $3x - y + 5 = 0$ **33.** $2x - y = 0$ **35.** $x - 4 = 0$ **37.** $2x + y = 0$ **39.** $x - 2y + 3 = 0$
41. $y - 4 = 0$
43. Slope $= 2$; y-intercept $= 3$ **45.** Slope $= 2$; y-intercept $= -2$ **47.** Slope $= \frac{2}{3}$; y-intercept $= -2$

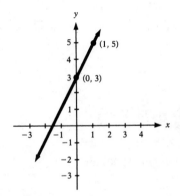

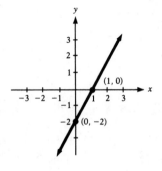

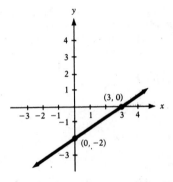

49. Slope $= -1$; y-intercept $= 1$ **51.** Slope undefined; no y-intercept   **53.** Slope $= \frac{3}{2}$; y-intercept $= 0$

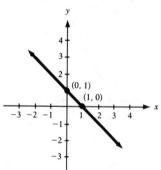   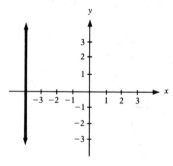   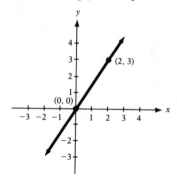

55. $x - 2y = 0$ **57.** $x + y - 2 = 0$ **59.** $y = 0$
61. $P_1 = (-2, 5)$, $P_2 = (1, 3)$, $m_1 = -\frac{2}{3}$; $P_2 = (1, 3)$, $P_3 = (-1, 0)$, $m_2 = \frac{3}{2}$; because $m_1 m_2 = -1$, the lines are perpendicular; the points P_1, P_2, and P_3 form a right triangle.
63. $P_1 = (-1, 0)$, $P_2 = (2, 3)$, $m_1 = 1$; $P_3 = (1, -2)$, $P_4 = (4, 1)$, $m_2 = 1$; $P_1 = (-1, 0)$, $P_3 = (1, -2)$, $m_3 = -1$; $P_2 = (2, 3)$, $P_4 = (4, 1)$, $m_4 = -1$; opposite sides are parallel; adjacent sides are perpendicular; the points form a rectangle
65. $°C = \frac{5}{9}(°F - 32)$; approx. $21° C$
67. All have same slope; the lines are parallel.   **69.** $P = 0.5x - 100$

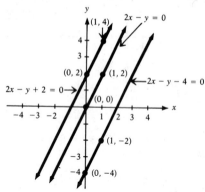   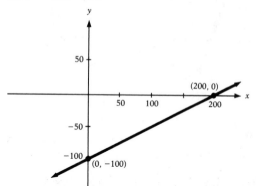

71. 12 square units   **73.** 4 square units   **75.** 5 square units
77. Refer to Figure 43. If $m_1 m_2 = -1$, then $d(A, B) = \sqrt{(m_2 - m_1)^2}$; $d(O, A) = \sqrt{1 + m_2^2}$; $d(O, B) = \sqrt{1 + m_1^2}$. Now show that $d(O, B)^2 + d(O, A)^2 = d(A, B)^2$.

SECTION 3.4

1. $y = \frac{1}{2}x$ **3.** $A = \pi x^2$ **5.** $F = \dfrac{250}{d^2}$ **7.** $z = \frac{1}{5}(x^2 + y^2)$ **9.** $M = \dfrac{9d^2}{2\sqrt{x}}$ **11.** $T^2 = 64\dfrac{a^3}{d^2}$ **13.** $V = \dfrac{4\pi}{3}R^3$

15. $A = \frac{1}{2}bh$ **17.** $V = \pi R^2 h$ **19.** $F = 6.67 \times 10^{-11}\left(\dfrac{mM}{d^2}\right)$ **21.** 144 ft; 2 sec   **23.** 2.25   **25.** 45.45 lb

27. $3\sqrt[3]{\frac{3}{4}} \approx 2.73$ in.   **29.** 900 ft-lb   **31.** 384 psi   **33.** $\frac{720}{49} \approx 14.69$ ohms   **35.** $V = \sqrt{gR}$ **37.** 18,001 mph

39. Approx. 505 mi   **41.** $F = \dfrac{mv^2}{R}$ **43.** By 21%   **45.** 9 times

FILL-IN-THE-BLANK QUESTIONS

1. x-coordinate, y-coordinate   **3.** y-axis   **5.** undefined, 0   **7.** $\dfrac{kx^2y^3}{\sqrt{t}}$

REVIEW EXERCISES

1. $2x + y - 3 = 0$ **3.** $x + 3 = 0$ **5.** $x + 5y + 10 = 0$ **7.** $2x - 3y + 19 = 0$ **9.** $-x + y + 4 = 0$

11.

13.

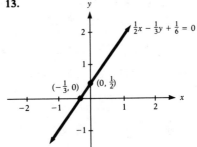

15.

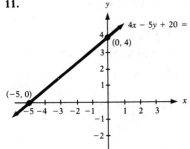

17. Center $(1, -2)$, radius $= 3$ **19.** Center $(1, -2)$, radius $= \sqrt{5}$

21. Intercept: $(0, 0)$; symmetric with respect to the x-axis

23. Intercepts: $(0, -1)$, $(0, 1)$, $(\tfrac{1}{2}, 0)$, $(-\tfrac{1}{2}, 0)$; symmetric with respect to the x-axis, the y-axis, and the origin

25. Intercept: $(0, 1)$; symmetric with respect to the y-axis   **27.** Intercepts: $(0, 0)$, $(0, -2)$, $(-1, 0)$; no symmetry

29. $A = ks^2 = \dfrac{\sqrt{3}}{4}s^2$; $s = \dfrac{8}{\sqrt[4]{3}} \approx 6.08$ cm   **31.** $a \approx 36$ million miles

33. $M = \left(\dfrac{a}{2}, \dfrac{b}{2}\right)$; $d(O, M) = \tfrac{1}{2}\sqrt{a^2 + b^2}$; $d(A, M) = \tfrac{1}{2}\sqrt{a^2 + b^2}$; $d(B, M) = \tfrac{1}{2}\sqrt{a^2 + b^2}$

35. $d(A, B) = \sqrt{13}$; $d(B, C) = \sqrt{13}$

37. Slope using A and B is -1; slope using B and C is -1.

39. Center $(1, -2)$; radius $= 4\sqrt{2}$; $x^2 + y^2 - 2x + 4y - 27 = 0$

CHAPTER 4   SECTION 4.1

1. (a) -4 (b) -5 (c) -9 (d) -12 **3.** (a) 0 (b) $\tfrac{1}{2}$ (c) $-\tfrac{1}{2}$ (d) $\tfrac{2}{5}$ **5.** (a) 4 (b) 5 (c) 5 (d) 6

7. (a) $-\tfrac{1}{5}$ (b) $-\tfrac{3}{2}$ (c) $\tfrac{1}{8}$ (d) 5 **9.** (a) $-2x + 3$ (b) $-2x - 3$ (c) $4x + 3$ (d) $2x - 3$ (e) $\dfrac{3x + 2}{x}$

11. (a) $2x^2 - 4$ (b) $-2x^2 + 4$ (c) $8x^2 - 4$ (d) $2x^2 - 12x + 14$ (e) $\dfrac{2 - 4x^2}{x^2}$

13. (a) $-x^3 + 3x$ (b) $-x^3 + 3x$ (c) $8x^3 - 6x$ (d) $x^3 - 9x^2 + 24x - 18$ (e) $\dfrac{1}{x^3} - \dfrac{3}{x}$

15. (a) $-\dfrac{x}{x^2 + 1}$ (b) $-\dfrac{x}{x^2 + 1}$ (c) $\dfrac{2x}{4x^2 + 1}$ (d) $\dfrac{x - 3}{x^2 - 6x + 10}$ (e) $\dfrac{x}{x^2 + 1}$

17. (a) $|x|$ (b) $-|x|$ (c) $2|x|$ (d) $|x - 3|$ (e) $\dfrac{1}{|x|}$ **19.** All real numbers   **21.** All real numbers

23. All real numbers except 1 and -1 **25.** All real numbers except 0   **27.** All real numbers x except $-3 < x < 3$

29. All real numbers x except $1 \le x < 2$ **31.** 0   **33.** -3 **35.** $6x + 3h - 2$ **37.** $3x^2 + 3xh + h^2 - 1$

39. $-\dfrac{1}{x(x + h)}$ **41.** $\dfrac{\sqrt{x + h} - \sqrt{x}}{h} \cdot \dfrac{\sqrt{x + h} + \sqrt{x}}{\sqrt{x + h} + \sqrt{x}} = \dfrac{h}{h(\sqrt{x + h} + \sqrt{x})} = \dfrac{1}{\sqrt{x + h} + \sqrt{x}}$ **43.** $A = -4$

45. $A = -4$ **47.** $A = 8$; at 3   **49.** $f\left(\dfrac{1}{x}\right) = \dfrac{2x + 3}{2 - 3x}, x \ne \dfrac{2}{3}$ **51.** (a) 15.1 m, 14.07 m, 12.94 m, 11.72 m   (b) 2.02 sec

53. $A(x) = \frac{1}{2}x^2$ **55.** $G(x) = 5x$ **57.** $A(x) = (7 - 2x)(11 - 2x), 0 \le x \le \frac{7}{2}, 0 \le A \le 77$

59. 1.11 sec; 0.46 sec

SECTION 4.2

1. $f(0) = 3; f(2) = 4$ **3.** Positive   **5.** $-3, 6,$ and 10   **7.** $-6 \le x \le 11$ **9.** $-6 \le x \le 2, 8 \le x \le 11$ **11.** 3 times

13. No   **15.** Yes   **17.** Not a function

19. Function   (a) Domain: $-\pi \le x \le \pi$; range: $-1 \le y \le 1$ (b) Increasing on the interval $[-\pi, 0]$; decreasing on the interval $[0, \pi]$ (c) Even   (d) Intercepts: $\left(-\dfrac{\pi}{2}, 0\right), \left(\dfrac{\pi}{2}, 0\right), (0, 1)$

21. Not a function

23. Function   (a) Domain: $x > 0$; range: all real numbers   (b) Increasing on its domain
(c) Neither   (d) Intercept: $(1, 0)$

25. Function   (a) Domain: all real numbers; range: $y \le 2$ (b) Increasing on the interval $(-\infty, -1]$; decreasing on the interval $[1, +\infty)$; stationary on the interval $[-1, 1]$ (c) Even   (d) Intercepts: $(-3, 0), (3, 0), (0, 2)$

27. Function   (a) Domain: $-4 \le x < 4$; range: $y = -2, 0, 2, 3$ (b) Stationary on $[-4, -2), [-2, 0), [0, 2), [2, 4)$
(c) Neither   (d) Intercepts: $(x, 0)$ for $0 \le x < 2$

29. Function   (a) Domain: $-4 \le x < 4$; range: $-3 \le y \le -2, -1 \le y \le 0, 1 \le y \le 2$ (b) Increasing on the interval $[-4, -1)$; decreasing on the interval $[1, 4)$; stationary on the interval $[-1, 1)$ (c) Neither
(d) Intercepts: $(2, 0), (0, 1)$

31. Odd   **33.** Even   **35.** Odd   **37.** Neither   **39.** Even   **41.** Odd   **43.** At most one

45. $f(x) = 3x - 3$ **47.** $g(x) = x^2 - 4$ **49.** $F(x) = 2x^3$

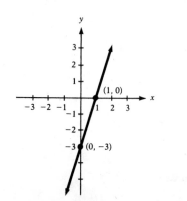

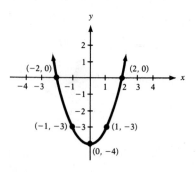

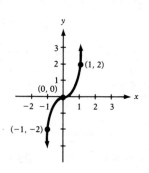

51. $f(x) = |x - 2|$

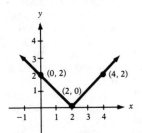

53. $f(x) = \begin{cases} 1 & \text{if } x \text{ is an integer} \\ -1 & \text{if } x \text{ is not an integer} \end{cases}$

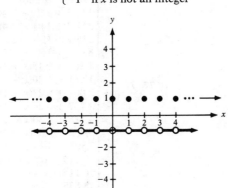

55. $f(x) = \begin{cases} 2x & \text{if } x \neq 0 \\ 0 & \text{if } x = 0 \end{cases}$

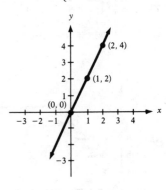

57. $f(x) = \begin{cases} 3x + 2 & \text{if } x \neq -1 \\ 1 & \text{if } x = -1 \end{cases}$

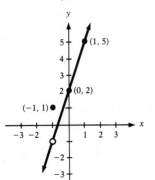

59. $f(x) = \begin{cases} -2x - 3 & \text{if } x < 0 \\ x - 3 & \text{if } 0 \leq x < 5 \end{cases}$

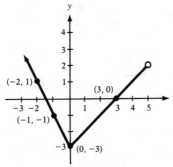

61. $f(x) = \begin{cases} x + 3 & \text{if } -2 \leq x < 0 \\ 1 & \text{if } x = 0 \\ \frac{1}{2}x^2 & \text{if } x \geq 0 \end{cases}$

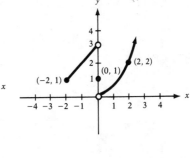

63. $f(x) = [\![x - \tfrac{1}{2}]\!]$

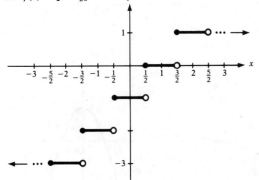

65. Function **67.** Function **69.** Not a function **71.** Function

73. (a) $E(-x) = \frac{1}{2}[f(-x) + f(x)] = E(x)$ (b) $O(-x) = \frac{1}{2}[f(-x) - f(x)] = -\frac{1}{2}[f(x) - f(-x)] = -O(x)$
    (c) $E(x) + O(x) = \frac{1}{2}[f(x) + f(-x)] + \frac{1}{2}[f(x) - f(-x)] = f(x)$ (d) Combine the results of parts (a), (b), and (c).

75. $C = \begin{cases} 95 & \text{if } x = 7 \\ 119 & \text{if } 7 < x \le 8 \\ 143 & \text{if } 8 < x \le 9 \\ 167 & \text{if } 9 < x \le 10 \\ 190 & \text{if } 10 < x \le 14 \end{cases}$

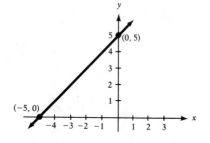

77. $T(x) = \begin{cases} 0 & \text{if } 0 < x \le 2{,}300 \\ 0.11(x - 2{,}300) & \text{if } 2{,}300 < x \le 3{,}400 \\ 121 + 0.12(x - 3{,}400) & \text{if } 3{,}400 < x \le 4{,}400 \\ 241 + 0.14(x - 4{,}400) & \text{if } 4{,}400 < x \le 6{,}500 \\ 535 + 0.15(x - 6{,}500) & \text{if } 6{,}500 < x \le 8{,}500 \\ 835 + 0.16(x - 8{,}500) & \text{if } 8{,}500 < x \le 10{,}800 \\ 1{,}203 + 0.18(x - 10{,}800) & \text{if } 10{,}800 < x \le 12{,}900 \\ 1{,}581 + 0.20(x - 12{,}900) & \text{if } 12{,}900 < x \le 15{,}000 \\ 2{,}001 + 0.23(x - 15{,}000) & \text{if } 15{,}000 < x \le 18{,}200 \\ 2{,}737 + 0.26(x - 18{,}200) & \text{if } 18{,}200 < x \le 23{,}500 \\ 4{,}115 + 0.30(x - 23{,}500) & \text{if } 23{,}500 < x \le 28{,}800 \\ 5{,}705 + 0.34(x - 28{,}800) & \text{if } 28{,}800 < x \le 34{,}100 \\ 7{,}507 + 0.38(x - 34{,}100) & \text{if } 34{,}100 < x \le 41{,}500 \\ 10{,}319 + 0.42(x - 41{,}500) & \text{if } 41{,}500 < x \le 55{,}300 \\ 16{,}115 + 0.48(x - 55{,}300) & \text{if } 55{,}300 < x \le 81{,}800 \\ 28{,}835 + 0.50(x - 81{,}800) & \text{if } x > 81{,}800 \end{cases}$

SECTION 4.3

1. $y = x + 5$

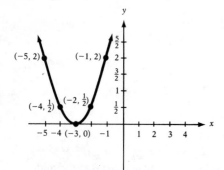

3. $y = x^2 - 1$

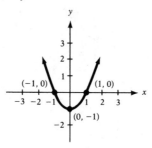

5. $y = |x + 1|$

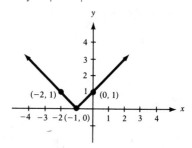

7. $y = \frac{1}{2}(x + 3)^2$

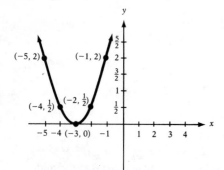

9. $y = \sqrt{x - 1} - 2$

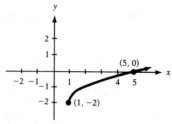

11. $y = -x^2 + 2$

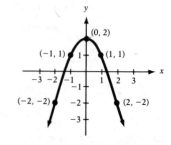

13. $y = -\frac{1}{2}x^2 + 2$

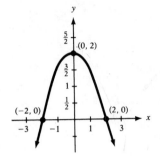

15. $y = 2\sqrt{x - 3}$

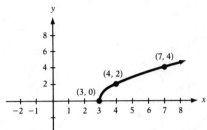

17. $y = [\![x - 1]\!]$

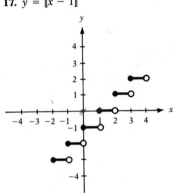

19. $y = -\frac{1}{2}(x + 1)^3 + 1$

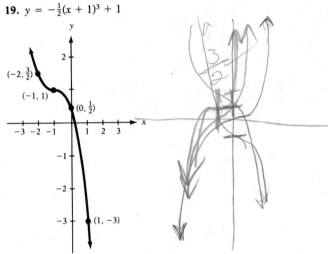

21. (a) $F(x) = f(x) + 3$

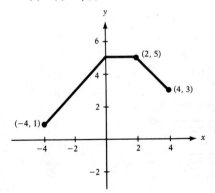

(b) $G(x) = f(x + 2)$

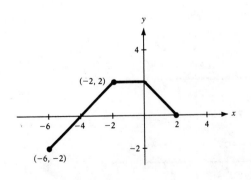

(c) $P(x) = -f(x)$

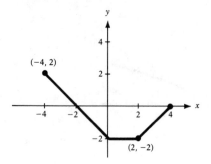

(d) $Q(x) = \frac{1}{2}f(x)$

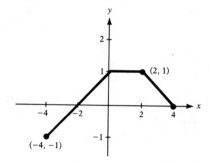

(e) $g(x) = f(-x)$

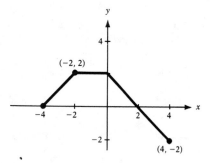

(f) $h(x) = 3f(x)$

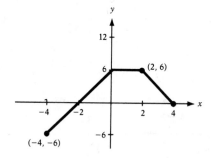

(g) $r(x) = f(\frac{1}{2}x)$

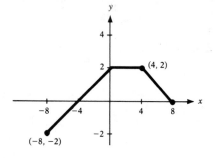

(h) $H(x) = 3f(2x)$

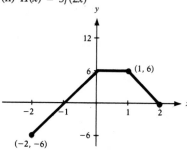

23. (a) $F(x) = f(x) + 3$

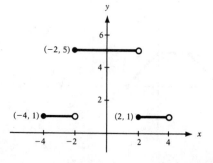

(b) $G(x) = f(x + 2)$

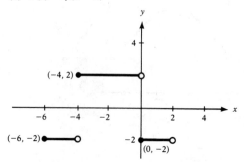

(c) $P(x) = -f(x)$

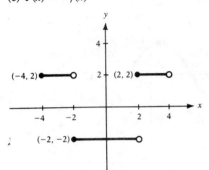

(d) $Q(x) = \frac{1}{2}f(x)$

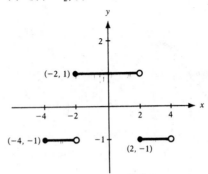

(e) $g(x) = f(-x)$

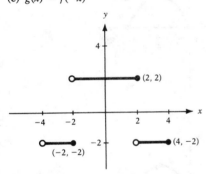

(f) $h(x) = 3f(x)$

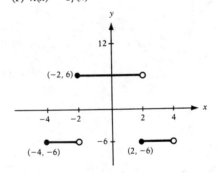

(g) $r(x) = f(\frac{1}{2}x)$

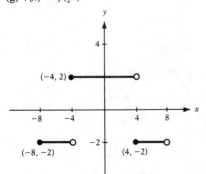

(h) $H(x) = 3f(2x)$

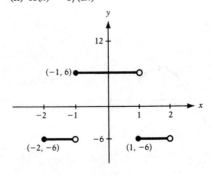

25. (a) $F(x) = f(x) + 3$

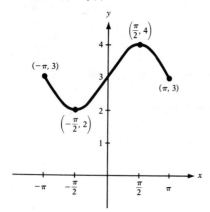

(b) $G(x) = f(x + 2)$

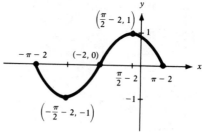

(c) $P(x) = -f(x)$

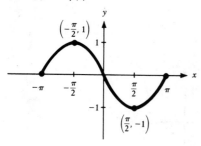

(d) $Q(x) = \frac{1}{2}f(x)$

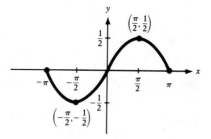

(e) $g(x) = f(-x)$

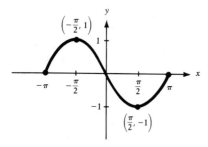

(f) $h(x) = 3f(x)$

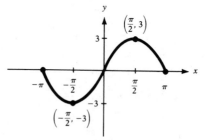

(g) $r(x) = f(\frac{1}{2}x)$

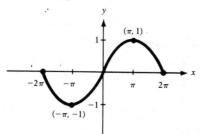

(h) $H(x) = 3f(2x)$

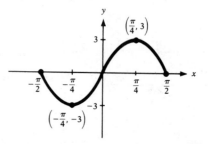

27. $f(x) = (x + 1)^2 - 1$

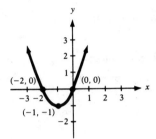

29. $f(x) = (x - 4)^2 - 15$

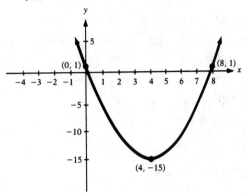

31. $f(x) = (x + \frac{1}{2})^2 + \frac{3}{4}$

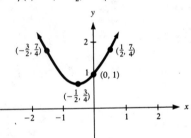

33.

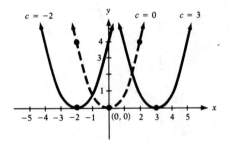

35.

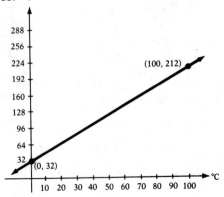

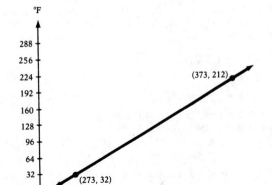

SECTION 4.4

1. (a) $(f + g)(x) = 5x - 1$, all real numbers (b) $(f - g)(x) = x - 7$, all real numbers (c) $(f \cdot g)(x) =$ $6x^2 + x - 12$, all real numbers (d) $\left(\dfrac{f}{g}\right)(x) = \dfrac{3x - 4}{2x + 3}$, all real numbers x except $x = -\dfrac{3}{2}$

3. (a) $(f + g)(x) = 2x^2 + x - 1$, all real numbers (b) $(f - g)(x) = -2x^2 + x - 1$, all real numbers (c) $(f \cdot g)(x) = 2x^3 - 2x^2$, all real numbers (d) $\left(\dfrac{f}{g}\right)(x) = \dfrac{x - 1}{2x^2}$, all real numbers x except $x = 0$

5. (a) $(f + g)(x) = \sqrt{x} + 3x - 5, x \geq 0$ (b) $(f - g)(x) = \sqrt{x} - 3x + 5, x \geq 0$ (c) $(f \cdot g)(x) = 3x\sqrt{x} - 5\sqrt{x}, x \geq 0$ (d) $\left(\dfrac{f}{g}\right)(x) = \dfrac{\sqrt{x}}{3x - 5}, x \geq 0$ except $x = \dfrac{5}{3}$

7. (a) $(f + g)(x) = 1 + \dfrac{2}{x}, x \neq 0$ (b) $(f - g)(x) = 1, x \neq 0$ (c) $(f \cdot g)(x) = \dfrac{1}{x} + \dfrac{1}{x^2}, x \neq 0$ (d) $\left(\dfrac{f}{g}\right)(x) = x + 1, x \neq 0$

9. (a) $(f + g)(x) = \dfrac{3x + 3}{3x - 2}, x \neq \dfrac{2}{3}$ (b) $(f - g)(x) = \dfrac{x + 3}{3x - 2}, x \neq \dfrac{2}{3}$ (c) $(f \cdot g)(x) = \dfrac{2x^2 + 3x}{(3x - 2)^2}, x \neq \dfrac{2}{3}$ (d) $\left(\dfrac{f}{g}\right)(x) = \dfrac{2x + 3}{x}, x \neq \dfrac{2}{3}, x \neq 0$

11. $g(x) = 5 - \dfrac{7}{2}x$ 13.

15.

17. (a) 98 (b) 49 (c) 4 (d) 4 19. (a) 97 (b) $-\dfrac{163}{2}$ (c) 1 (d) $-\dfrac{3}{2}$ 21. (a) $2\sqrt{2}$ (b) $2\sqrt{2}$ (c) 1 (d) 0
23. (a) $\dfrac{1}{17}$ (b) $\dfrac{1}{5}$ (c) 1 (d) $\dfrac{1}{2}$ 25. (a) $\dfrac{3}{5}$ (b) $\sqrt{15}/5$ (c) $\dfrac{12}{13}$ (d) 0
27. (a) $(f \circ g)(x) = 6x + 1$ (b) $(g \circ f)(x) = 6x + 3$ (c) $(f \circ f)(x) = 4x + 3$ (d) $(g \circ g)(x) = 9x$
29. (a) $(f \circ g)(x) = 3x^2 + 1$ (b) $(g \circ f)(x) = 9x^2 + 6x + 1$ (c) $(f \circ f)(x) = 9x + 4$ (d) $(g \circ g)(x) = x^4$
31. (a) $(f \circ g)(x) = \sqrt{x^2 - 1}$ (b) $(g \circ f)(x) = x - 1$ (c) $(f \circ f)(x) = \sqrt[4]{x}$ (d) $(g \circ g)(x) = x^4 - 2x^2$
33. (a) $(f \circ g)(x) = \dfrac{1 - x}{1 + x}$ (b) $(g \circ f)(x) = \dfrac{x + 1}{x - 1}$ (c) $(f \circ f)(x) = -\dfrac{1}{x}$ (d) $(g \circ g)(x) = x$
35. (a) $(f \circ g)(x) = x$ (b) $(g \circ f)(x) = |x|$ (c) $(f \circ f)(x) = x^4$ (d) $(g \circ g)(x) = \sqrt[4]{x}$
37. (a) $(f \circ g)(x) = \dfrac{1}{4x + 9}$ (b) $(g \circ f)(x) = \dfrac{2}{2x + 3} + 3 = \dfrac{6x + 11}{2x + 3}$ (c) $(f \circ f)(x) = \dfrac{2x + 3}{6x + 11}$ (d) $(g \circ g)(x) = 4x + 9$
39. (a) $(f \circ g)(x) = acx + ad + b$ (b) $(g \circ f)(x) = acx + bc + d$ (c) $(f \circ f)(x) = a^2x + ab + b$ (d) $(g \circ g)(x) = c^2x + cd + d$
41. $g(x) = \dfrac{1}{3}x$ 43. $g(x) = \sqrt[3]{x}$ 45. $g(x) = \dfrac{1}{2}x + 3$ 47. $g(x) = \dfrac{1}{a}(x - b), a \neq 0$ 49. $(f \circ g)(x) = 11, (g \circ f)(x) = 2$
51. $f \circ (g \circ h) = 5 - 3x + 4\sqrt{1 - 3x}$ 53. $(f + g) \circ h = 9x^2 - 6x + 3 + \sqrt{1 - 3x}$ 55. $F = f \circ g$ 57. $H = h \circ f$
59. $g = f \circ h$ 61. $P = f \circ f$ 63. $f(x) = x^3, g(x) = 2x + 5$ 65. $f(x) = \sqrt{x}, g(x) = x^2 + x + 1$
67. $f(x) = x^2, g(x) = 1 - \dfrac{1}{x^2}$ 69. $f(x) = [\![x]\!], g(x) = x^2 + 1$ 71. $S(R(t)) = \dfrac{16}{9}\pi t^6$
73. $C(N(t)) = 5{,}000 + 600{,}000t - 30{,}000t^2$

SECTION 4.5

1. One-to-one **3.** Not one-to-one **5.** One-to-one
7.

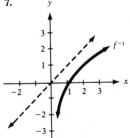

9.

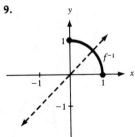

11.

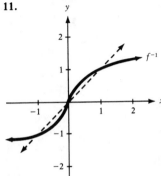

13. $f(g(x)) = 3[\frac{1}{3}(x + 4)] - 4 = (x + 4) - 4 = x$ **15.** $f(g(x)) = 4\left(\frac{x}{4} + 2\right) - 8 = (x + 8) - 8 = x$

17. $f(g(x)) = (\sqrt[3]{x + 8})^3 - 8 = (x + 8) - 8 = x$ **19.** $f(g(x)) = \dfrac{1}{1/x} = x$

21. $f(g(x)) = \dfrac{2\left(\dfrac{4x - 3}{2 - x}\right) + 3}{\dfrac{4x - 3}{2 - x} + 4} = \dfrac{\dfrac{2(4x - 3) + 3(2 - x)}{2 - x}}{\dfrac{4x - 3 + 4(2 - x)}{2 - x}} = \dfrac{5x}{2 - x} \cdot \dfrac{2 - x}{5} = x$

23. $f^{-1}(x) = \frac{x}{4} - \frac{1}{2}$; $f(f^{-1}(x)) = 4\left(\frac{x}{4} - \frac{1}{2}\right) + 2 = x$ **25.** $f^{-1}(x) = \sqrt[3]{x + 1}$; $f(f^{-1}(x)) = (\sqrt[3]{x + 1})^3 - 1 = x$

27. $f^{-1}(x) = \frac{4}{x}$; $f(f^{-1}(x)) = \frac{4}{4/x} = x$ **29.** $f^{-1}(x) = \frac{x}{x - 1}$; $f(f^{-1}(x)) = \dfrac{\dfrac{x}{x - 1}}{\dfrac{x}{x - 1} - 1} = \dfrac{x}{x - 1} \cdot \dfrac{x - 1}{1} = x$

31. $f^{-1}(x) = \sqrt{x - 4}, x \geq 4$; $f(f^{-1}(x)) = (\sqrt{x - 4})^2 + 4 = x - 4 + 4 = x$
33. $f^{-1}(x) = -\sqrt{x - 9}, x \geq 9$; $f(f^{-1}(x)) = (-\sqrt{x - 9})^2 + 9 = x$ **35.** $f^{-1}(x) = \sqrt[4]{x}, x \geq 0$; $f(f^{-1}(x)) = (\sqrt[4]{x})^4 = x$

37. $f^{-1}(x) = \frac{1}{m}(x - b), m \neq 0$

39. No. Whenever x and $-x$ are in the domain of f, two equal y values, $f(x)$ and $f(-x)$, are present.

41. First quadrant **43.** $f(g(x)) = \frac{9}{5}[\frac{5}{9}(x - 32)] + 32 = (x - 32) + 32 = x$ **45.** $l(T) = \dfrac{gT^2}{4\pi^2}$

FILL-IN-THE-BLANK QUESTIONS

1. independent, dependent **3.** even, odd **5.** $g(f(x)) = (g \circ f)(x)$ **7.** $y = x$

REVIEW EXERCISES

1. $f(x) = -2x + 6$ **3.** $A = 11$ **5.** All real numbers except $x = 2$ **7.** $x > 0$ **9.** $x \geq -1$

11. $f(x) = \sqrt{x + 2}$

13. $f(x) = 1 - x^2$

15. $F(x) = \begin{cases} x^2 + 4 & \text{if } x < 0 \\ 4 - x^2 & \text{if } x \geq 0 \end{cases}$

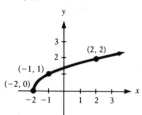

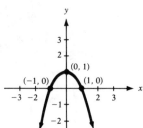

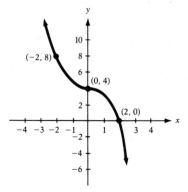

17. (a) B, C, D   (b) D   **19.** Odd   **21.** Even   **23.** Neither

25. $f^{-1}(x) = \dfrac{3 + 2x}{x - 2}$; $f(f^{-1}(x)) = \dfrac{2\left(\dfrac{3 + 2x}{x - 2}\right) + 3}{\dfrac{3 + 2x}{x - 2} - 2} = \dfrac{7x}{7} = x$ **27.** $f^{-1}(x) = \dfrac{1 + x}{x}$; $f(f^{-1}(x)) = \dfrac{1}{\dfrac{1 + x}{x} - 1} = x$

29. $f^{-1}(x) = \dfrac{27}{x^3}$; $f(f^{-1}(x)) = \dfrac{3}{\left(\dfrac{27}{x^3}\right)^{1/3}} = \dfrac{3}{\dfrac{3}{x}} = x$ **31.** (a) -26 (b) -241 (c) 16 (d) -1

33. (a) $\sqrt{11}$ (b) 1 (c) $\sqrt{\sqrt{6} + 2}$ (d) 19 **35.** $\frac{1}{20}$ (b) $-\frac{13}{8}$ (c) $\frac{400}{1601}$ (d) -17

37. $f \circ g = \dfrac{-3x}{3x + 2}$, $g \circ f = \dfrac{6 - x}{x}$, $f \circ f = \dfrac{3x - 2}{2 - x}$, $g \circ g = 9x + 8$

39. $(f \circ g)(x) = 27x^2 + 3|x| + 1$, $(g \circ f)(x) = 3|3x^2 + x + 1|$, $(f \circ f)(x) = 3(3x^2 + x + 1)^2 + 3x^2 + x + 2$, $(g \circ g)(x) = 9|x|$

41. $f \circ g = \dfrac{1 + x}{1 - x}$, $g \circ f = \dfrac{x - 1}{x + 1}$, $f \circ f = x$, $g \circ g = x$

43. (a)

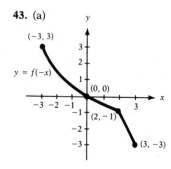

(b)

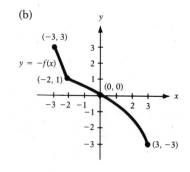

(c)

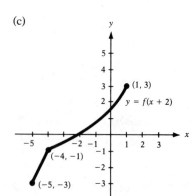

(d)

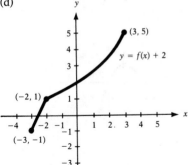

(e)

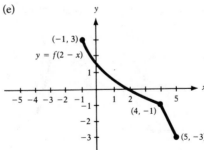

(f)

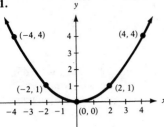

45. $T(h) = -0.0025h + 30$

CHAPTER 5   SECTION 5.1

1.

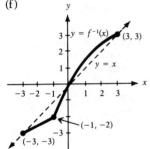

3.

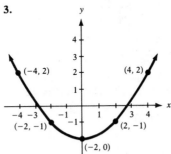

5.

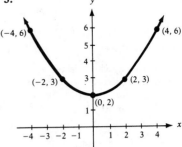

7.

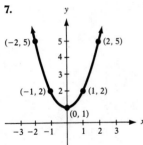

9.

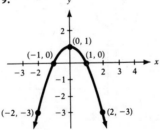

11.

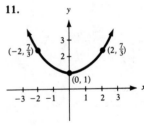

13. $f(x) = (x + 2)^2 - 2$

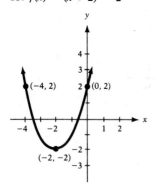

15. $f(x) = 2(x - 1)^2 - 1$

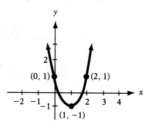

17. $f(x) = -(x + 1)^2 + 1$

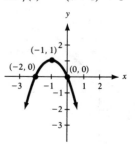

19. $f(x) = \frac{1}{2}(x + 1)^2 - \frac{3}{2}$

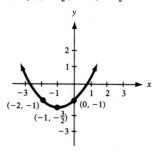

21. Opens upward; vertex at $(-1, -4)$; axis of symmetry $x = -1$; x-intercepts $(1, 0)$, $(-3, 0)$; y-intercept $(0, -3)$

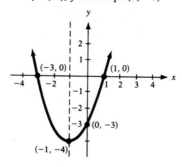

23. Opens downward; vertex at $(-\frac{3}{2}, \frac{25}{4})$; axis of symmetry $x = -\frac{3}{2}$; x-intercepts $(-4, 0)$, $(1, 0)$; y-intercept $(0, 4)$

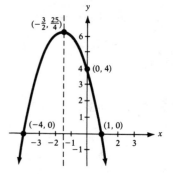

25. Opens upward; vertex at $(-1, 0)$; axis of symmetry $x = -1$; x-intercept $(-1, 0)$; y-intercept $(0, 1)$

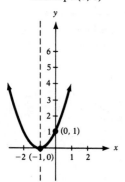

27. Opens upward; vertex at $(\frac{1}{4}, \frac{15}{8})$; axis of symmetry $x = \frac{1}{4}$; no x-intercept; y-intercept $(0, 2)$

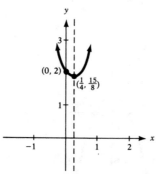

29. Opens downward; vertex at $(\frac{1}{2}, -\frac{5}{2})$; axis of symmetry $x = \frac{1}{2}$; no x-intercept; y-intercept $(0, -3)$

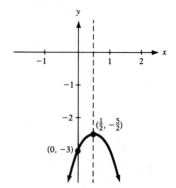

31. Opens upward; vertex at $(1, -1)$; axis of symmetry $x = 1$; x-intercepts $(1.58, 0)$, $(0.42, 0)$; y-intercept $(0, 2)$

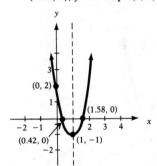

33. Opens downward; vertex at $(-\frac{3}{4}, \frac{17}{4})$; axis of symmetry at $x = -\frac{3}{4}$; x-intercepts $(-1.78, 0)$, $(0.28, 0)$; y-intercept $(0, 2)$

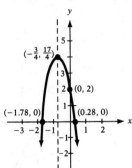

35. Minimum value; -9

37. Maximum value; 21 **39.** Maximum value; 13
41. Opens upward; vertex at $(-1, f(-1))$; axis of symmetry $x = -1$

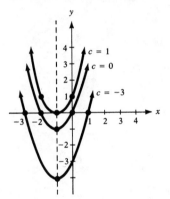

43. 15, 15 **45.** Price: \$500; maximum revenue: \$1,000,000 **47.** 8 P.M. **49.** 2,000,000 m² **51.** 37,500,000 m²
53. 80 members **55.** $\frac{64}{3}$ m

SECTION 5.2

1. Yes; degree three **3.** Yes; degree two **5.** No; x is raised to the -1 power **7.** No; x is raised to the $\frac{3}{2}$ power
9. Yes; degree four
11.

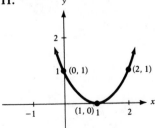

13.

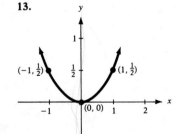

15.

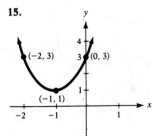

17.

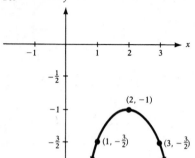

19. x-intercepts: 0, 4
y-intercept: 0
Above x-axis: $x < 0, x > 4$
Below x-axis: $0 < x < 4$

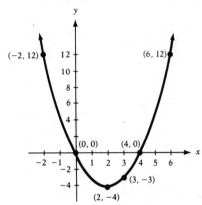

21. x-intercepts: $-1, 0, 3$
y-intercept: 0
Above x-axis:
$-1 < x < 0, x > 3$
Below x-axis:
$x < -1, 0 < x < 3$

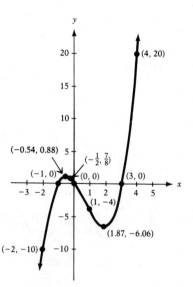

23. x-intercepts: $-2, 0, 2$
y-intercept: 0
Above x-axis:
$-2 < x < 0; x > 2$
Below x-axis:
$x < -2, 0 < x < 2$

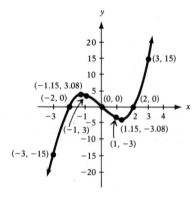

25. x-intercepts: $-1, 0, 3$
y-intercept: 0
Above x-axis: $x < -1, x > 3$
Below x-axis: $-1 < x < 0$,
$0 < x < 3$

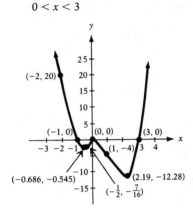

27. x-intercepts: 0, 2
y-intercept: 0
Above x-axis: $x > 2$
Below x-axis: $x < 0, 0 < x < 2$

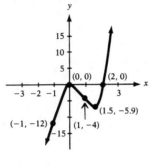

29. x-intercepts: $-1, 3$
y-intercept: -3
Above x-axis: $x > 3$
Below x-axis: $x < -1, -1 < x < 3$

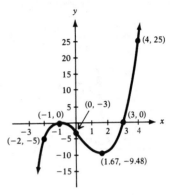

SECTION 5.3

1. $q(x) = 4x - 6; r(x) = x + 7$ **3.** $q(x) = -2x^3 + 7x^2 - 24x + 72; r(x) = -214$
5. $q(x) = 3x^2 - 5x; r(x) = -6x^2 + 12x - 4$ **7.** $q(x) = -x^3 - x; r(x) = -x + 1$
9. $q(x) = x^3 + cx^2 + c^2x + c^3; r(x) = 0$ **11.** $q(x) = \frac{5}{3}x; r(x) = \frac{1}{3}x + 3$ **13.** $q(x) = 3x^2 + 11x + 32; R = 99$
15. $q(x) = x^4 - 3x^3 + 5x^2 - 15x + 46; R = -138$ **17.** $q(x) = 4x^5 + 4x^4 + x^3 + x^2 + 2x + 2; R = 7$
19. $q(x) = 0.1x^2 - 0.11x + 0.321; R = -0.3531$ **21.** $q(x) = x^4 + x^3 + x^2 + x + 1; R = 0$ **23.** No; $f(3) = 93$
25. Yes; $f(2) = 0$ **27.** No; $f(-3) = 4428$ **29.** No; $f(-4) = 1$ **31.** Yes; $f(\frac{1}{2}) = 0$ **33.** 41 **35.** -4 **37.** 15
39. $[(3x + 2)x - 3]x + 3$ **41.** $[(3x - 6)x \cdot x \cdot x - 5]x + 10$ **43.** $(3x \cdot x \cdot x - 82)x \cdot x \cdot x + 27$
45. $[(4x \cdot x - 64)x \cdot x + 1]x \cdot x - 15$ **47.** $[(2x - 1)x \cdot x + 2]x - 1$ **49.** 7.464 **51.** -0.1472 **53.** -105.738
55. -134.327 **57.** 3.8192 **59.** $k = 5$ **61.** -7 **63.** If $f(x) = x^n - c^n$, then $f(c) = c^n - c^n = 0$.
65. (a) 201,498 nanoseconds (b) 101,499 nanoseconds

SECTION 5.4

1. 4, multiplicity one; -5, multiplicity two **3.** 2, multiplicity three **5.** $-\frac{1}{2}$, multiplicity two
7. 5, multiplicity three; -4, multiplicity two **9.** No zeros **11.** 7 **13.** 6 **15.** 2 or 0 positive; 1 negative
17. 2 or 0 positive; 2 or 0 negative **19.** 0 positive; 3 or 1 negative **21.** 1 positive; 1 negative **23.** ± 1
25. $\pm 1, \pm 3$ **27.** $\pm\frac{1}{2}, \pm 1$ **29.** $\pm\frac{1}{3}, \pm\frac{2}{3}, \pm 1, \pm 2$ **31.** $\pm\frac{1}{2}, \pm 1, \pm 2, \pm 4$ **33.** $\pm\frac{1}{6}, \pm\frac{1}{3}, \pm\frac{1}{2}, \pm\frac{2}{3}, \pm 1, \pm 2$
35. $-3, -1, 2; f(x) = (x + 3)(x + 1)(x - 2)$ **37.** $\frac{1}{2}; f(x) = 2(x - \frac{1}{2})(x^2 + 1)$ **39.** $-1, 1; f(x) = (x + 1)(x - 1)(x^2 + 2)$
41. $-\frac{1}{2}, \frac{1}{2}; f(x) = 4(x + \frac{1}{2})(x - \frac{1}{2})(x^2 + 2)$ **43.** $-2, -1, 1, 1; f(x) = (x + 2)(x + 1)(x - 1)^2$
45. $-\sqrt{2}/2, \sqrt{2}/2, 2; f(x) = 4(x + \sqrt{2}/2)(x - \sqrt{2}/2)(x - 2)(x^2 + \frac{1}{2})$ **47.** $-1, 2; f(x) = (x + 1)(x - 2)(x^2 + 4)$
49. $\frac{2}{3}, -1 + \sqrt{2}, -1 - \sqrt{2}; f(x) = 3(x - \frac{2}{3})(x + 1 - \sqrt{2})(x + 1 + \sqrt{2})$
51. $\frac{1}{3}, \sqrt{5}, -\sqrt{5}; f(x) = 3(x - \frac{1}{3})(x - \sqrt{5})(x + \sqrt{5})$ **53.** $-3, -2; f(x) = (x + 3)(x + 2)(x^2 - x + 1)$
55. $-\frac{1}{3}; f(x) = (x + \frac{1}{3})(x^2 - x + 3)$

57. $f(0) = -1$
$x < \frac{1}{2}, f(0) = 1$, below x-axis
$x > \frac{1}{2}, f(1) = 2$, above x-axis

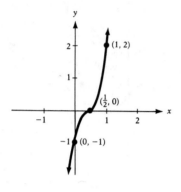

59. $f(0) = -2$
$x < -1, f(-2) = 18$, above x-axis
$-1 < x < 1, f(0) = -2$, below x-axis
$x > 1, f(2) = 18$, above x-axis
(symmetric with respect to the y-axis)

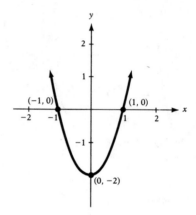

61. $f(0) = -2$
$x < -\frac{1}{2}, f(-1) = 9$, above x-axis
$-\frac{1}{2} < x < \frac{1}{2}, f(0) = -2$, below x-axis
$x > \frac{1}{2}, f(1) = 9$, above x-axis
(symmetric with respect to the y-axis)

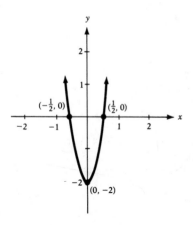

63. $f(0) = 2$
$x < -2, f(-3) = 32$, above x-axis
$-2 < x < -1, f(-\frac{3}{2}) = -\frac{25}{16}$, below x-axis
$-1 < x < 1, f(0) = 2$, above x-axis
$x > 1, f(2) = 12$, above x-axis

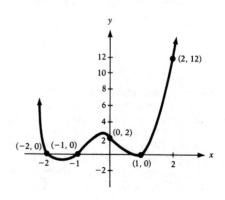

65. $f(0) = 2$
$x < -\sqrt{2}/2, f(-1) = -9$, below x-axis
$-\sqrt{2}/2 < x < \sqrt{2}/2, f(0) = 2$, above x-axis
$\sqrt{2}/2 < x < 2, f(1) = -3$, below x-axis
$x > 2, f(3) = 323$, above x-axis

67. 7 in.

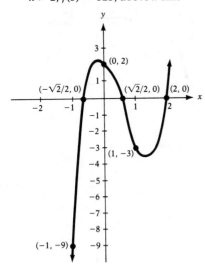

69. All the potential rational zeros are integers. Hence, r either is an integer or is not a rational root (and is therefore irrational).

SECTION 5.5

1. $f(0) = -1, f(1) = 10$ **3.** $f(-5) = -58, f(-4) = 2$ **5.** $f(1.4) = -0.17536, f(1.5) = 1.40625$ **7.** -1 and 1
9. -5 and 1 **11.** -5 and 2 **13.** Between 1.15 and 1.16 **15.** Between 2.53 and 2.54

SECTION 5.6

1. All real numbers except 2 **3.** All real numbers except 2 and -1 **5.** All real numbers except $-\frac{1}{2}$ and 3
7. All real numbers except 2 **9.** All real numbers

11.

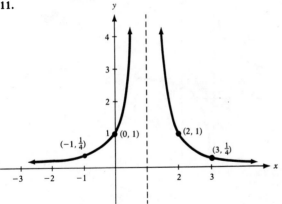

13.

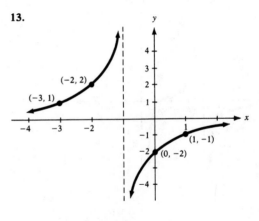

15.

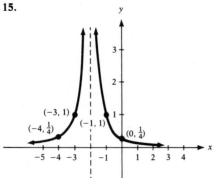

17.

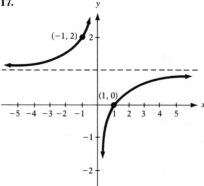

19.

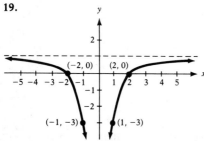

21. Horizontal asymptote: $y = 1$; vertical asymptote: $x = -1$ **23.** No asymptotes
25. Horizontal asymptote: $y = 0$; vertical asymptotes: $x = 1, x = -1$
27. Horizontal asymptote: $y = 0$; vertical asymptote: $x = 0$
29. Oblique asymptote: $y = 3x$; vertical asymptote: $x = 0$
31. (a) x-intercept $(-1, 0)$; no y-intercept
   (b) No symmetry
   (c) Vertical asymptotes: $x = 0, x = -4$
   (d) Horizontal asymptote: $y = 0$
   (e) $x < -4$, below x-axis
      $-4 < x < -1$, above x-axis
      $-1 < x < 0$, below x-axis
      $x > 0$, above x-axis
   (f)

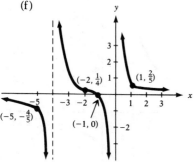

33. (a) x-intercept $(-1, 0)$;
      y-intercept $(0, \frac{3}{4})$
   (b) No symmetry
   (c) Vertical asymptote: $x = -2$
   (d) Horizontal asymptote: $y = \frac{3}{2}$
   (e) $x < -2$, above x-axis
      $-2 < x < -1$, below x-axis
      $x > -1$, above x-axis
   (f)

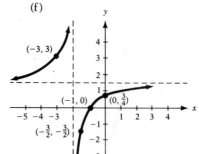

35. (a) No x-intercept; y-intercept $(0, -\frac{3}{4})$
   (b) Symmetric with respect to y-axis
   (c) Vertical asymptotes: $x = 2, x = -2$
   (d) Horizontal asymptote: $y = 0$
   (e) $x < -2$, above x-axis
      $-2 < x < 2$, below x-axis
      $x > 2$, above x-axis
   (f)

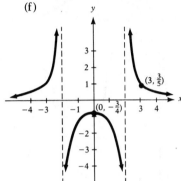

37. (a) No x-intercept; y-intercept $(0, -1)$
   (b) Symmetric with respect to y-axis
   (c) Vertical asymptotes: $x = -1, x = 1$
   (d) No horizontal or oblique asymptotes
   (e) $x < -1$, above x-axis
      $-1 < x < 1$, below x-axis
      $x > 1$, above x-axis
   (f)

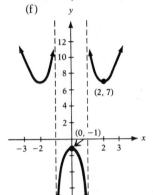

39. (a) x-intercept $(1, 0)$; y-intercept $(0, \frac{1}{9})$
   (b) No symmetry
   (c) Vertical asymptotes: $x = 3, x = -3$
   (d) Oblique asymptote: $y = x$
   (e) $x < -3$, below x-axis
      $-3 < x < 1$, above x-axis
      $1 < x < 3$, below x-axis
      $x > 3$, above x-axis
   (f)

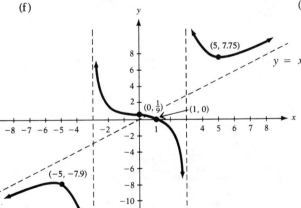

41. (a) Intercept $(0, 0)$
   (b) No symmetry
   (c) Vertical asymptotes: $x = 2, x = -3$
   (d) Horizontal asymptote: $y = 1$
   (e) $x < -3$, above x-axis
      $-3 < x < 0$, below x-axis
      $0 < x < 2$, below x-axis
      $x > 2$, above x-axis
   (f)

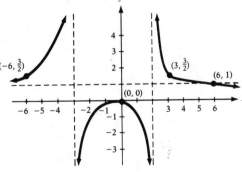

43. (a) Intercept $(0, 0)$
   (b) Symmetry with respect to origin
   (c) Vertical asymptotes: $x = -2, x = 2$
   (d) Horizontal asymptote: $y = 0$
   (e) $x < -2$, below x-axis
      $-2 < x < 0$, above x-axis
      $0 < x < 2$, below x-axis
      $x > 2$, above x-axis

(f)

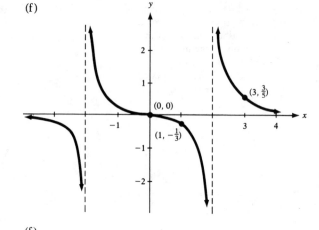

45. (a) No x-intercept; y-intercept: $(0, \frac{3}{4})$
   (b) No symmetry
   (c) Vertical asymptotes: $x = -2, x = 1$,
      $x = 2$
   (d) Horizontal asymptote: $y = 0$
   (e) $x < -2$, below x-axis
      $-2 < x < 1$, above x-axis
      $1 < x < 2$, below x-axis
      $x > 2$, above x-axis

(f)

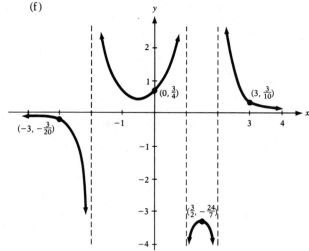

47. (a) x-intercepts $(-1, 0)$, $(1, 0)$; y-intercept $(0, \frac{1}{4})$
   (b) Symmetric with respect to y-axis
   (c) Vertical asymptotes: $x = -2, x = 2$
   (d) Horizontal asymptote: $y = 0$
   (e) $x < -2$, above x-axis
      $-2 < x < -1$, below x-axis
      $-1 < x < 1$, above x-axis
      $1 < x < 2$, below x-axis
      $x > 2$, above x-axis

(f)

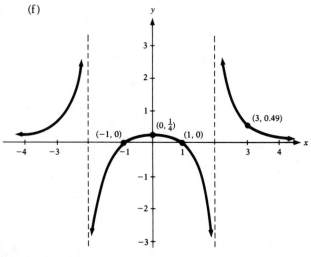

49. (a) x-intercepts $(-1, 0)$, $(4, 0)$;
    y-intercept $(0, -2)$
  (b) No symmetry
  (c) Vertical asymptote: $x = -2$
  (d) Oblique asymptote: $y = x - 5$
  (e) $x < -2$, below x-axis
    $-2 < x < -1$, above x-axis
    $-1 < x < 4$, below x-axis
    $x > 4$, above x-axis

(f)

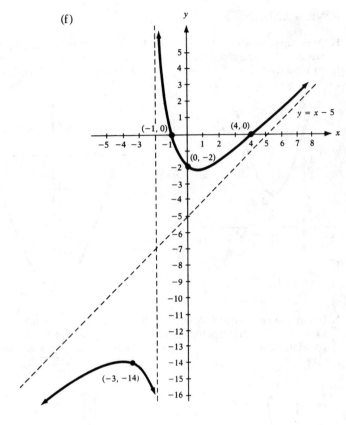

51. 4 must be a zero of the denominator; hence, $x - 4$ must be a factor.

SECTION 5.7

1. $-3 + i$ **3.** $-1 + 5i$ **5.** $-4 + 4i$ **7.** $-18 - 16i$ **9.** $16 - 18i$ **11.** $38 + 31i$
13. $z^3 + (-7 - 2i)z^2 + (15 + 12i)z - 9 - 18i$ **15.** $z^3 - 3z^2 + (3 - i)z - 2 + 2i$
17. $z^4 + (2i - 6)z^3 + (8 - 12i)z^2 + (6 + 18i)z - 9$ **19.** $1 + i$; $f(z) = z^3 - 4z^2 + 6z - 4$
21. $-i, 1 - i$; $f(z) = z^4 - 2z^3 + 3z^2 - 2z + 2$ **23.** $-i, -2i$; $f(z) = z^5 - z^4 + 5z^3 - 5z^2 + 4z - 4$
25. Zeros that are complex numbers must occur in conjugate pairs; or a polynomial with real coefficients of odd degree must have at least one real zero.
27. If the remaining zero were a complex number, then its conjugate would also be a zero.
29. $1, -\dfrac{1}{2} + \dfrac{\sqrt{3}}{2}i, -\dfrac{1}{2} - \dfrac{\sqrt{3}}{2}i$

FILL-IN-THE-BLANK QUESTIONS

1. parabola, vertex   **3.** $f(c)$ **5.** zero   **7.** $\pm 1, \pm \frac{1}{2}$ **9.** $x = -1$

REVIEW EXERCISES

1. Opens upward; vertex (2, 2); axis $x = 2$; y-intercept (0, 6); no x-intercept

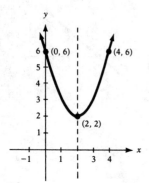

3. Opens upward; vertex (0, −16); axis $x = 0$; y-intercept (0, −16); x-intercepts (−8, 0), (8, 0)

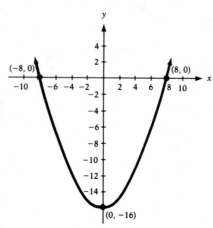

5. Opens downward; vertex $(\frac{1}{2}, 1)$; axis $x = \frac{1}{2}$; y-intercept (0, 0); x-intercepts (0, 0), (1, 0)

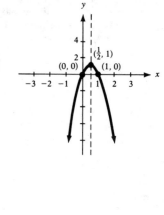

7. Opens upward; vertex $(-\frac{1}{3}, \frac{1}{2})$; axis $x = -\frac{1}{3}$; y-intercept (0, 1); no x-intercept

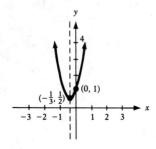

9. Opens upward; vertex $(-\frac{2}{3}, -\frac{7}{3})$; axis $x = -\frac{2}{3}$; y-intercept (0, −1); x-intercepts (−1.55, 0), (0.22, 0)

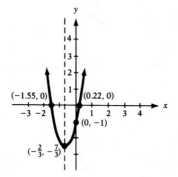

11.

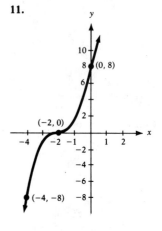

13.

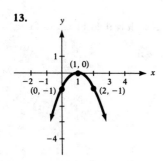

15.

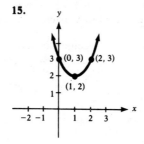

17. $q(x) = 8x^2 + 6x + 7; R = 3$ **19.** $q(x) = x^3 - 4x^2 + 8x - 15; R = 29$

21. $f(x) = [(8x - 2)x + 1]x - 4; f(1.5) = 20$ **23.** $f(x) = [(x - 2)xx + 1]x - 1; f(1.5) = -1.1875$

25. $-2, 1, 4; f(x) = (x + 2)(x - 1)(x - 4)$ **27.** $\frac{1}{2}$, multiplicity two; $-2; f(x) = 4(x - \frac{1}{2})^2(x + 2)$

29. 2, multiplicity two; $f(x) = (x - 2)^2(x^2 + 5)$ **31.** $-3, 2; f(x) = 2(x + 3)(x - 2)(x^2 + \frac{1}{2})$

33. $-3, -1, -\frac{1}{2}, 1; f(x) = 2(x + 3)(x + 1)(x + \frac{1}{2})(x - 1)$

35. x-intercepts: $-2, 1, 4$
    y-intercept: 8
    Above x-axis: $-2 < x < 1, x > 4$
    Below x-axis: $x < -2, 1 < x < 4$

37. x-intercepts: $-2, \frac{1}{2}$
    y-intercept: 2
    Above x-axis: $x > -2$
    Below x-axis: $x < -2$

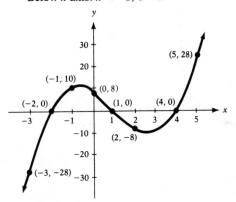

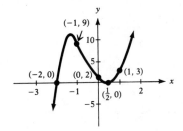

39. x-intercept: 2
    y-intercept: 20
    Above or on x-axis: all x

41. x-intercepts: $-3, 2$
    y-intercept: -6
    Above x-axis: $x < -3, x > 2$
    Below x-axis: $-3 < x < 2$

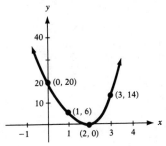

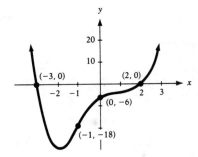

43. x-intercepts: $-3, -1, -\frac{1}{2}, 1$
    y-intercept: -3
    Above x-axis: $x < -3, -1 < x < -\frac{1}{2}, x > 1$
    Below x-axis: $-3 < x < -1, -\frac{1}{2} < x < 1$

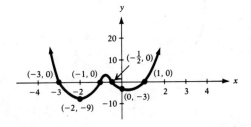

45. $f(0) = -1; f(1) = 1$　**47.** $f(0) = -1; f(1) = 1$　**49.** -2 and 2　**51.** -3 and 5　**53.** Between 1.52 and 1.53
55. Between 0.93 and 0.94

57. (a) x-intercept $(3, 0)$; no y-intercept
　(b) No symmetry
　(c) Vertical asymptote: $x = 0$
　(d) Horizontal asymptote: $y = 2$
　(e) $x < 0$, above x-axis
　　 $0 < x < 3$, below x-axis
　　 $x > 3$; above x-axis
　(f)

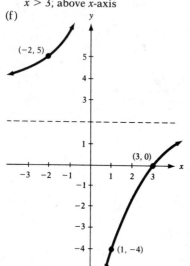

59. (a) x-intercept $(-2, 0)$; no y-intercept
　(b) No symmetry
　(c) Vertical asymptotes: $x = 0, x = 2$
　(d) Horizontal asymptote: $y = 0$
　(e) $x < -2$, below x-axis
　　 $-2 < x < 0$, above x-axis
　　 $0 < x < 2$, below x-axis
　　 $x > 2$, above x-axis
　(f)

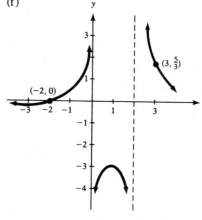

61. (a) Intercept $(0, 0)$
　(b) No symmetry
　(c) Vertical asymptote: $x = 1$
　(d) Horizontal asymptote: $y = 1$
　(e) $x < 0$, above x-axis
　　 $0 < x < 1$, above x-axis
　　 $x > 1$, above x-axis
　(f)

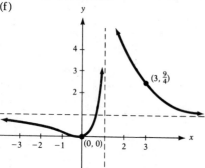

63. (a) Intercept $(0, 0)$
　(b) Symmetric with respect to the origin
　(c) Vertical asymptotes: $x = -2, x = 2$
　(d) Oblique asymptote: $y = x$
　(e) $x < -2$, below x-axis
　　 $-2 < x < 0$, above x-axis
　　 $0 < x < 2$, below x-axis
　　 $x > 2$, above x-axis
　(f)

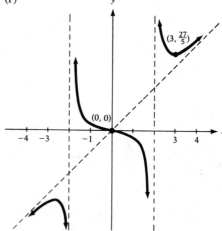

65. (a) Intercept $(0, 0)$
(b) No symmetry
(c) Vertical asymptote: $x = 1$
(d) No oblique or horizontal asymptote
(e) $x < 0$, above x-axis
$0 < x < 1$, above x-axis
$x > 1$, above x-axis
(f)

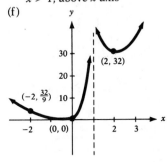

67. $f(z) = z^4 - (4 + i)z^3 + (5 + 4i)z^2 - (2 + 5i)z + 2i$ **69.** $f(z) = z^3 - (6 + i)z^2 + (11 + 5i)z - 6 - 6i$
71. $1 - i$; $f(z) = z^3 - 3z^2 + 4z - 2$ **73.** $-i, 1 - i$; $f(z) = z^4 - 2z^3 + 3z^2 - 2z + 2$ **75.** $q(x) = x^2 + 5x + 6$; $R = 0$
77. $x = -3, 2$ **79.** $x = \frac{1}{3}, 1, -i, i$ **81.** $f(4) = 47,105$ **83.** No (use the Rational Zeros Theorem)
85. No (use the Rational Zeros Theorem) **87.** 5, 3, or 1 positive zeros; 1 negative zero
89. $\pm\frac{1}{12}, \pm\frac{1}{6}, \pm\frac{1}{4}, \pm\frac{1}{3}, \pm\frac{1}{2}, \pm\frac{3}{4}, \pm1, \pm\frac{3}{2}, \pm3$ **91.** 1 is an upper bound; -1 is a lower bound **93.** $(2, 2)$ **95.** 3.6 ft

CHAPTER 6   SECTION 6.1

1. (a) 11.211578   (b) 11.587251   (c) 11.663882   (d) 11.664753
3. (a) 8.8152409   (b) 8.8213533   (c) 8.8244111   (d) 8.8249778
5. (a) 21.216638   (b) 22.21669   (c) 22.440403   (d) 22.45916

7.

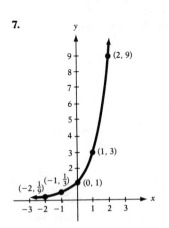

9.

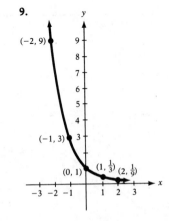

11.

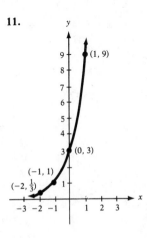

x	-2	-1	0	1	2
$f(x)$	$\frac{1}{9}$	$\frac{1}{3}$	1	3	9

13.

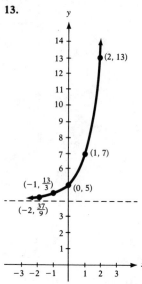

15.

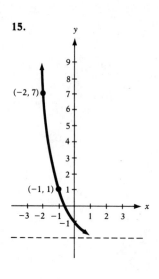

17.

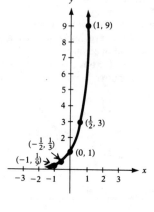

19. $\frac{1}{2}$ **21.** $0, -\sqrt{2}, \sqrt{2}$ **23.** $1 + \dfrac{\sqrt{6}}{3}, 1 - \dfrac{\sqrt{6}}{3}$ **25.** 0 **27.** 2 **29.** 0

31. (a) 56.47%, 68.17% (b) (c) 70%

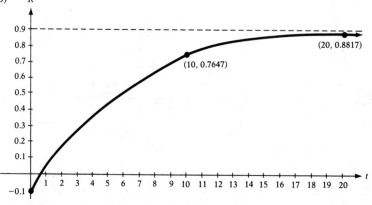

33. (a) 76.47%, 88.17% (b) (c) 90%

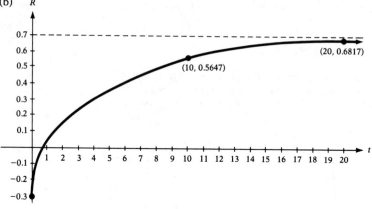

35. (a) 0.2376, 7.5854   (b)   (c) 12 amp

37. (a) $98,125, $99,941.41   (b)   (c) $100,000

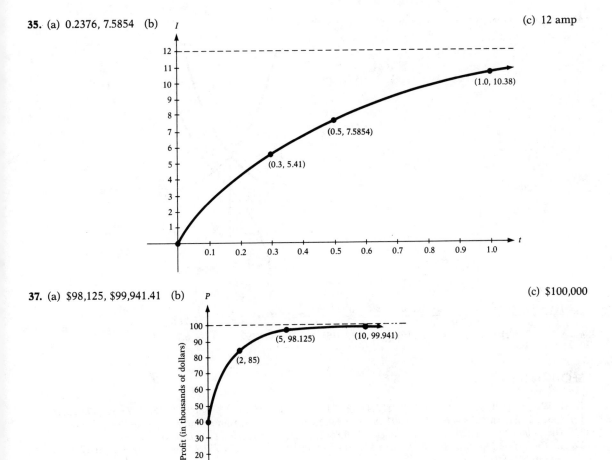

39. 12.696481   **41.** 2.5937425, 2.7048138, 2.7169239, 2.7181459, 2.7182682 Each result is less than e.

43. (a) $\sinh(-x) = \frac{1}{2}(e^{-x} - e^x) = -\frac{1}{2}(e^x - e^{-x}) = -\sinh x$

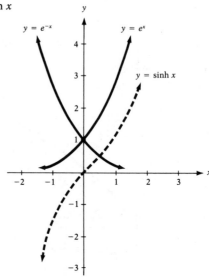

45. $\dfrac{f(x + h) - f(x)}{h} = \dfrac{a^{x+h} - a^x}{h} = \dfrac{a^x a^h - a^x}{h} = \dfrac{a^x(a^h - 1)}{h}$ **47.** $f(-x) = a^{-x} = \dfrac{1}{a^x} = \dfrac{1}{f(x)}$

49. $f(1) = 5, f(2) = 17, f(3) = 257, f(4) = 65,537, f(5) = 4,294,967,297 = 641 \times 6,700,417$

SECTION 6.2

1. \$117.29   **3.** \$640.04   **5.** \$697.09   **7.** \$12.46   **9.** \$125.23   **11.** \$85.26   **13.** \$860.72   **15.** \$473.65   **17.** \$59.71
19. \$361.93   **21.** (a) \$1364.62   (b) \$907.18   **23.** You have \$11,632.73. Your friend has \$10,947.89.
25. The \$1,000.00 invested at 10% becomes \$1349.86. It is better to receive \$1000.00 now.
27. (a) Interest is \$30,000.   (b) Interest is \$38,613.59.   (c) Interest is \$37,752.73. Simple interest at 12% per annum is best.
29. Quarterly compounding   **31.** $A = Ve^{rt}, V = \dfrac{A}{e^{rt}}, V = Ae^{-rt}$

SECTION 6.3

1. $2 = \log_3 9$ **3.** $2 = \log_a 1.6$ **5.** $2 = \log_{1.1} M$ **7.** $x = \log_2 7.2$ **9.** $\sqrt{2} = \log_x \pi$ **11.** $2^3 = 8$ **13.** $a^6 = 3$
15. $3^x = 2$ **17.** $2^{1.3} = M$ **19.** $(\sqrt{2})^x = \pi$ **21.** 0   **23.** 2   **25.** -4 **27.** $\frac{1}{2}$ **29.** 4   **31.** $\log_5 u^3 v^4$ **33.** $-\frac{5}{2}\log_{1/2} x$
35. $-2\ln(x - 1)$ **37.** $\log_2[x(3x - 2)^4]$ **39.** $\log_a \dfrac{25x^6}{(2x + 3)^{1/2}}$ **41.** $2\ln x + \frac{1}{2}\ln(1 - x)$ **43.** $3\log_2 x - \log_2(x - 3)$
45. $\log x + \log(x + 2) - 2\log(x + 3)$ **47.** $\frac{1}{3}\ln(x - 2) + \frac{1}{3}\ln(x + 1) - \frac{2}{3}\ln(x + 4)$
49. $\ln 5 + \ln x + \frac{1}{2}\ln(1 - 3x) - \ln 3 - 3\ln(x - 4) - \frac{1}{2}\ln(3 - 4x)$ **51.** 81   **53.** 3   **55.** 5   **57.** 2   **59.** $-2, 4$ **61.** 21
63. $\frac{9}{2}$ **65.** 2   **67.** $\frac{11}{3}, -\frac{7}{3}$ **69.** -1 **71.** 1.7917   **73.** 0.4055   **75.** 1.6931   **77.** 2.4848   **79.** 0.5781   **81.** 1.5851
83. $\log_a(x + \sqrt{x^2 - 1}) + \log_a(x - \sqrt{x^2 - 1}) = \log_a[(x + \sqrt{x^2 - 1})(x - \sqrt{x^2 - 1})] = \log_a[x^2 - (x^2 - 1)] = \log_a 1 = 0$
85. 2.7712437   **87.** -3.880058 **89.** 5.6147098   **91.** 0.8735685   **93.** 3.3219281   **95.** -0.0876781 **97.** 0.3065736
99. 1.3559551   **101.** 0   **103.** 0.5337408   **105.** 3   **107.** 1   **109.** 104.32 months, 103.97 months
111. 61.02 months, 60.82 months
113. If $A = \log_a M$ and $B = \log_a N$, then $a^A = M$ and $a^B = N$; then $\log_a(M/N) = \log_a(a^A/a^B) = \log_a a^{A-B} = A - B = \log_a M - \log_a N$.

SECTION 6.4

1.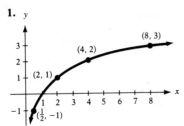

$f(x)$	-1	0	1	1.585	2	2.3219	2.585	2.8074	3
x	0.5	1	2	3	4	5	6	7	8

3.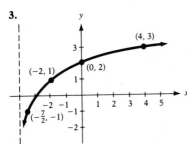

5.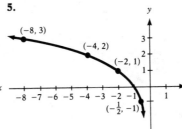

7.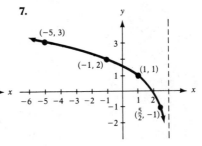

9.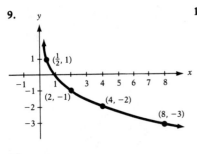

11.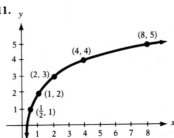

13.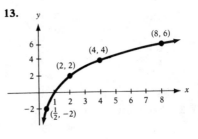

15.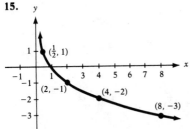

17. $x < 3$ **19.** All real numbers except 0 **21.** $x < 0$ **23.** $x > 0, x \neq 1$ **25.** $\sqrt{2}$
27. Domain of f: all real numbers except 0; domain of g: all positive real numbers; the equality holds only for $x > 0$.
29. $y = f(x) = \log_a x,\ a^y = x,\ \left(\dfrac{1}{a}\right)^{-y} = x,\ -y = \log_{1/a} x,\ -f(x) = \log_{1/a} x$
31. $f(AB) = \log_a AB = \log_a A + \log_a B = f(A) + f(B)$ **33.** $y = Cx$ **35.** $y = Cx(x + 1)$ **37.** $y = Ce^{3x}$

39. $y = Ce^{-4x} + 3$ **41.** $y = \dfrac{\sqrt[3]{C}(2x+1)^{1/6}}{(x+4)^{1/9}}$ **43.** 70 decibels **45.** 111.76 decibels **47.** 10 watts per m²

49. 4.0 on the Richter scale

51. 70,794.58 mm, the San Francisco earthquake was 11.22 times as intense as the one in Mexico City.

53. 3229.54 m above sea level

SECTION 6.5

1. 34.7 days, 69.3 days **3.** 28.4 years **5.** 94.4 years **7.** 5832, 3.9 days **9.** 25,198.4 **11.** 9.797 g

13. 9727 years ago **15.** 5:18 P.M. **17.** 18.63° C, 25.1° C **19.** 7.34 kg, 76.6 hr **21.** 28.15%, 0.24, 0.15

23. 0.2695 sec, 0.8959 sec

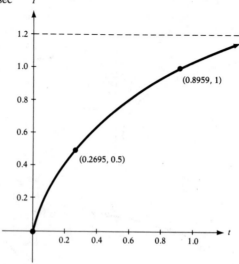

25. $k \approx 0.02107$, 38 words, 54 words, 109 min

FILL-IN-THE-BLANK QUESTIONS

1. (0, 1) **3.** 4 **5.** 1 **7.** $x > 0$ **9.** 1

REVIEW EXERCISES

1. -3 **3.** $\sqrt{2}$ **5.** 0.4 **7.** $\log_4 x^{25/4}$ **9.** $\ln \dfrac{1}{(x+1)^2} = \ln(x+1)^{-2}$ **11.** $\log \dfrac{4x^3}{[(x+3)(x-2)]^{1/2}}$ **13.** $y = Ce^{2x^2}$

15. $y = (Ce^{3x^2})^2$ **17.** $y = \sqrt{e^{x+C} + 9}$ **19.** $y = \ln(x^2 + 4) - C$

21.

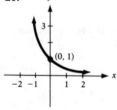

23.

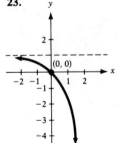

25.

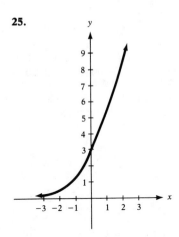

27.

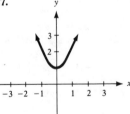

29.

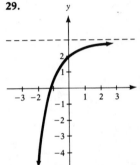

31. $\frac{1}{4}$ **33.** $\frac{-1-\sqrt{3}}{2}, \frac{-1+\sqrt{3}}{2}$ **35.** $\frac{1}{4}$ **37.** 5 **39.** 4.301 **41.** $\frac{12}{5}$ **43.** $-3, \frac{1}{2}$ **45.** -1 **47.** -0.6094 **49.** -9.3273

51. 10.436% **53.** The first earthquake was approximately 1.58 times as intense as the second.

55. 24,203 years ago

CHAPTER 7   SECTION 7.1

1. $2(2) - (-1) = 5$ and $5(2) + 2(-1) = 8$ **3.** $3(2) - 4(\frac{1}{2}) = 4$ and $2 - 3(\frac{1}{2}) = \frac{1}{2}$ **5.** $2^2 - 1^2 = 3$ and $(2)(1) = 2$

7. $\frac{0}{1+0} + 3(2) = 6$ and $0 + 9(2)^2 = 36$ **9.** $3(1) + 3(-1) + 2(2) = 4$, $1 - (-1) - 2 = 0$, and $2(-1) - 3(2) = -8$

11. $x = 6, y = 2$ **13.** $x = 3, y = 2$ **15.** $x = 8, y = -4$ **17.** $x = 4, y = -2$ **19.** Inconsistent **21.** $x = \frac{1}{2}, y = \frac{3}{4}$

23. $x = 4 - 2y, y$ is any real number **25.** $x = \frac{1}{10}, y = \frac{2}{5}$ **27.** $x = \frac{3}{2}, y = 1$ **29.** $y = \frac{5}{3} - \frac{2}{3}x, x$ is any real number

31. $x = \frac{4}{3}, y = \frac{1}{5}$ **33.** $x = 8, y = 2, z = 0$ **35.** $x = 2, y = -1, z = 1$ **37.** Inconsistent

39. $x = 5z - 2, y = 4z - 3$ or $x = \frac{5}{4}y + \frac{7}{4}, z = \frac{1}{4}y + \frac{3}{4}$ or $y = \frac{4}{5}x - \frac{7}{5}, z = \frac{1}{5}x + \frac{2}{5}$ **41.** Inconsistent

43. $x = 1, y = 3, z = -2$ **45.** $x = -3, y = \frac{1}{2}, z = 1$ **47.** $x = \frac{1}{5}, y = \frac{1}{3}$ **49.** 20 and 61 **51.** 15 ft by 30 ft

53. Cheeseburger $1.55, shake $0.85 **55.** Average wind speed 25 mph, average airspeed 175 mph **57.** 22.5 lb

59. $b = -\frac{1}{2}, c = \frac{3}{2}$ **61.** $a = \frac{4}{3}, b = -\frac{5}{3}, c = 1$ **63.** $x = \frac{b_1 - b_2}{m_2 - m_1}, y = \frac{m_2 b_1 - m_1 b_2}{m_2 - m_1}$

65. $y = mx + b, x$ is any real number

SECTION 7.2

1. $x = 6, y = 2$ **3.** $x = 3, y = 2$ **5.** $x = 8, y = -4$ **7.** $x = 4, y = -2$ **9.** Inconsistent **11.** $x = \frac{1}{2}, y = \frac{3}{4}$
13. $x = 4 - 2y$, y is any real number **15.** $x = \frac{1}{10}, y = \frac{2}{5}$ **17.** $x = \frac{3}{2}, y = 1$ **19.** $y = \frac{5}{3} - \frac{2}{3}x$, x is any real number
21. $x = \frac{4}{3}, y = \frac{1}{5}$ **23.** $x = 8, y = 2, z = 0$ **25.** $x = 2, y = -1, z = 1$ **27.** Inconsistent
29. $x = 5z - 2, y = 4z - 3$ or $x = \frac{5}{4}y + \frac{7}{4}, z = \frac{1}{4}y + \frac{3}{4}$ or $y = \frac{4}{5}x - \frac{7}{5}, z = \frac{1}{5}x + \frac{2}{5}$ **31.** Inconsistent
33. $x = 1, y = 3, z = -2$ **35.** $x = -3, y = \frac{1}{2}, z = 1$ **37.** 10 two-student and 6 three-student stations **39.** \$5.56
41. 12, 15, 21 **43.** 100 orchestra, 210 main, and 190 balcony seats

SECTION 7.3

1. $\begin{bmatrix} 1 & -3 & | & 5 \\ 4 & 1 & | & 6 \end{bmatrix}$ **3.** $\begin{bmatrix} 2 & 3 & | & 6 \\ 4 & -6 & | & -2 \end{bmatrix}$ **5.** $\begin{bmatrix} 0.01 & -0.03 & | & 0.06 \\ 0.13 & 0.10 & | & 0.20 \end{bmatrix}$ **7.** $\begin{bmatrix} 1 & -1 & 1 & | & 10 \\ 3 & 2 & 0 & | & 5 \\ 1 & 1 & 2 & | & 2 \end{bmatrix}$

9. $\begin{bmatrix} 1 & 1 & -1 & | & 2 \\ 3 & -2 & 0 & | & 2 \end{bmatrix}$ **11.** $R_1 = -1r_2 + r_1$ **13.** $R_1 = -4r_2 + r_1$ **15.** $\begin{bmatrix} 1 & 2 & 3 & | & 0 \\ 0 & 0 & -3 & | & 3 \\ -3 & 2 & 1 & | & -2 \end{bmatrix}$

17. $\begin{bmatrix} 1 & 2 & 3 & | & 0 \\ 8 & 0 & 1 & | & 7 \\ -3 & 2 & 1 & | & -2 \end{bmatrix}$ **19.** $\begin{bmatrix} 1 & 2 & 3 & | & 0 \\ 2 & 4 & 3 & | & 3 \\ -1 & 6 & 4 & | & 1 \end{bmatrix}$ **21.** $x = 6, y = 2$ **23.** $x = 3, y = 2$ **25.** $x = 8, y = -4$

27. $x = 4, y = -2$ **29.** Inconsistent **31.** $x = \frac{1}{2}, y = \frac{3}{4}$ **33.** $x = 4 - 2y$, y is any real number **35.** $x = \frac{1}{10}, y = \frac{2}{5}$
37. $x = \frac{3}{2}, y = 1$ **39.** $x = \frac{5}{2} - \frac{3}{2}y$, y is any real number **41.** $x = \frac{4}{3}, y = \frac{1}{5}$ **43.** $x = 8, y = 2, z = 0$
45. $x = 2, y = -1, z = 1$ **47.** Inconsistent
49. $x = 5z - 2, y = 4z - 3$ or $x = \frac{5}{4}y + \frac{7}{4}, z = \frac{1}{4}y + \frac{3}{4}$ or $y = \frac{4}{5}x - \frac{7}{5}, z = \frac{1}{5}x + \frac{2}{5}$ **51.** Inconsistent
53. $x = 1, y = 3, z = -2$ **55.** $x = -3, y = \frac{1}{2}, z = 1$ **57.** $x = \frac{1}{3}, y = \frac{2}{3}, z = 1$ **59.** $x = 1, y = 2, z = 0, w = 1$
61. $y = 0, z = 1 - x$, x is any real number **63.** $x = 2, y = z - 3$, z is any real number **65.** $x = \frac{13}{9}, y = \frac{7}{18}, z = \frac{19}{18}$
67. $\begin{cases} x = \frac{7}{5} - \frac{3}{5}z - \frac{2}{5}w \\ y = -\frac{8}{5} + \frac{7}{5}z + \frac{13}{5}w \end{cases}$, where z and w are any real numbers **69.** $y = -2x^2 + x + 3$ **71.** $f(x) = 3x^3 - 4x^2 + 5$
73. $x = $ liters of 15% H_2SO_4, $y = $ liters of 25% H_2SO_4, $z = $ liters of 50% H_2SO_4

$$\begin{cases} x = \frac{5}{2}z - 150 \\ y = 250 - \frac{7}{2}z \end{cases}$$

15%	25%	50%	40%
0	40	60	100
10	26	64	100
20	12	68	100

75. If $a_1 \neq 0$,

$$\begin{bmatrix} a_1 & b_1 & | & c_1 \\ a_2 & b_2 & | & c_2 \end{bmatrix} \rightarrow \begin{bmatrix} 1 & \dfrac{b_1}{a_1} & | & \dfrac{c_1}{a_1} \\ a_2 & b_2 & | & c_2 \end{bmatrix} \rightarrow \begin{bmatrix} 1 & \dfrac{b_1}{a_1} & | & \dfrac{c_1}{a_1} \\ 0 & \dfrac{-a_2 b_1}{a_1} + b_2 & | & \dfrac{-a_2 c_1}{a_1} + c_2 \end{bmatrix} \rightarrow \begin{bmatrix} 1 & \dfrac{b_1}{a_1} & | & \dfrac{c_1}{a_1} \\ 0 & \dfrac{-a_2 b_1 + b_2 a_1}{a_1} & | & \dfrac{-a_2 c_1 + c_2 a_1}{a_1} \end{bmatrix}$$

$$\rightarrow \begin{bmatrix} 1 & \dfrac{b_1}{a_1} & | & \dfrac{c_1}{a_1} \\ 0 & 1 & | & \dfrac{-a_2 c_1 + c_2 a_1}{a_1} \cdot \dfrac{a_1}{-a_2 b_1 + b_2 a_1} \end{bmatrix} \rightarrow \begin{bmatrix} 1 & \dfrac{b_1}{a_1} & | & \dfrac{c_1}{a_1} \\ 0 & 1 & | & \dfrac{-a_2 c_1 + c_2 a_1}{-a_2 b_1 + b_2 a_1} \end{bmatrix}$$

$$\rightarrow \begin{bmatrix} 1 & 0 & | & \dfrac{-b_1 c_2 + b_2 c_1}{-a_2 b_1 + b_2 a_1} \\ 0 & 1 & | & \dfrac{-a_2 c_1 + c_2 a_1}{-a_2 b_1 + b_2 a_1} \end{bmatrix}$$

$$x = \frac{1}{a_1 b_2 - a_2 b_1}(c_1 b_2 - c_2 b_1) = \frac{1}{D}(c_1 b_2 - c_2 b_1)$$

$$y = \frac{1}{a_1 b_2 - a_2 b_1}(a_1 c_2 - a_2 c_1) = \frac{1}{D}(a_1 c_2 - a_2 c_1)$$

If $a_1 = 0$, then $a_2 \neq 0$, $b_1 \neq 0$, and

$$\begin{bmatrix} 0 & b_1 & | & c_1 \\ a_2 & b_2 & | & c_2 \end{bmatrix} \rightarrow \begin{bmatrix} a_2 & b_2 & | & c_2 \\ 0 & b_1 & | & c_1 \end{bmatrix} \rightarrow \begin{bmatrix} 1 & \dfrac{b_2}{a_2} & | & \dfrac{c_2}{a_2} \\ 0 & b_1 & | & c_1 \end{bmatrix} \rightarrow \begin{bmatrix} 1 & \dfrac{b_2}{a_2} & | & \dfrac{c_2}{a_2} \\ 0 & 1 & | & \dfrac{c_1}{b_1} \end{bmatrix} \rightarrow \begin{bmatrix} 1 & 0 & | & \dfrac{c_2}{a_2} - \dfrac{b_2 c_1}{a_2 b_1} = \dfrac{c_1 b_2 - c_2 b_1}{-a_2 b_1} \\ 0 & 1 & | & \dfrac{c_1}{b_1} = \dfrac{-a_2 c_1}{-a_2 b_1} \end{bmatrix}$$

SECTION 7.4

1. 2 **3.** 22 **5.** −2 **7.** 10 **9.** −26 **11.** $x = 6, y = 2$ **13.** $x = 3, y = 2$ **15.** $x = 8, y = -4$ **17.** $x = 4, y = -2$
19. Not applicable **21.** $x = \frac{1}{2}, y = \frac{3}{4}$ **23.** $x = \frac{1}{10}, y = \frac{2}{5}$ **25.** $x = \frac{3}{2}, y = 1$ **27.** $x = \frac{4}{3}, y = \frac{1}{5}$
29. $x = 1, y = 3, z = -2$ **31.** $x = -3, y = \frac{1}{2}, z = 1$ **33.** Not applicable **35.** $x = 0, y = 0, z = 0$
37. Not applicable **39.** $x = \frac{1}{5}, y = \frac{1}{3}$ **41.** −5 **43.** $\frac{13}{11}$ **45.** 0 or −9 **47.** 0 **49.** 12 **51.** 8 **53.** 8

55.
$$\begin{vmatrix} x^2 & x & 1 \\ y^2 & y & 1 \\ z^2 & z & 1 \end{vmatrix} = x^2 \begin{vmatrix} y & 1 \\ z & 1 \end{vmatrix} - x \begin{vmatrix} y^2 & 1 \\ z^2 & 1 \end{vmatrix} + \begin{vmatrix} y^2 & y \\ z^2 & z \end{vmatrix} = x^2(y - z) - x(y^2 - z^2) + yz(y - z)$$

$$= (y - z)[x^2 - x(y + z) + yz] = (y - z)[(x^2 - xy) - (xz - yz)]$$
$$= (y - z)[x(x - y) - z(x - y)] = (y - z)(x - y)(x - z)$$

57.
$$\begin{vmatrix} a_{13} & a_{12} & a_{11} \\ a_{23} & a_{22} & a_{21} \\ a_{33} & a_{32} & a_{31} \end{vmatrix} = a_{13}(a_{22}a_{31} - a_{32}a_{21}) - a_{12}(a_{23}a_{31} - a_{33}a_{21}) + a_{11}(a_{23}a_{32} - a_{33}a_{22})$$

$$= -[a_{11}(a_{22}a_{33} - a_{32}a_{23}) - a_{12}(a_{21}a_{33} - a_{31}a_{23}) + a_{13}(a_{21}a_{32} - a_{31}a_{22})]$$

$$= - \begin{vmatrix} a_{11} & a_{12} & a_{13} \\ a_{21} & a_{22} & a_{23} \\ a_{31} & a_{32} & a_{33} \end{vmatrix}$$

59.
$$\begin{vmatrix} a_{11} & a_{12} & a_{11} \\ a_{21} & a_{22} & a_{21} \\ a_{31} & a_{32} & a_{31} \end{vmatrix} = a_{11}(a_{22}a_{31} - a_{32}a_{21}) - a_{12}(a_{21}a_{31} - a_{31}a_{21}) + a_{11}(a_{21}a_{32} - a_{31}a_{22})$$

$$= a_{11}a_{22}a_{31} - a_{11}a_{32}a_{21} - a_{12}(0) + a_{11}a_{21}a_{32} - a_{11}a_{31}a_{22} = 0$$

SECTION 7.5

1. $x = 1, y = -2; x = -\frac{11}{5}, y = -\frac{2}{5}$

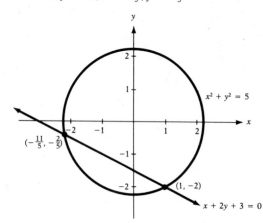

3. $x = 0, y = 2; x = 0, y = -2;$
  $x = -1, y = \sqrt{3}; x = -1, y = -\sqrt{3}$

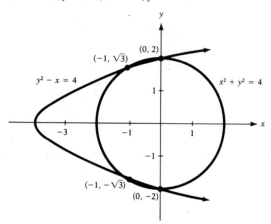

5. $x = 4 - \sqrt{2}, y = 4 + \sqrt{2}; x = 4 + \sqrt{2}, y = 4 - \sqrt{2}$ **7.** $x = 2, y = 2; x = -2, y = -2$

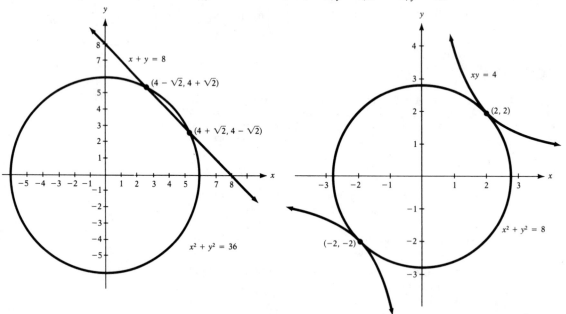

9. $x = 1, y = 4; x = -1, y = -4; x = 2\sqrt{2}, y = \sqrt{2}; x = -2\sqrt{2}, y = -\sqrt{2}$ **11.** $x = 1, y = 2; x = -\frac{13}{37}, y = -\frac{76}{37}$
13. $x = 4, y = -5; x = -\frac{3}{2}, y = \frac{1}{2}$ **15.** $x = 2, y = \frac{1}{3}; x = \frac{1}{2}, y = \frac{4}{3}$
17. $x = 3, y = 2; x = 3, y = -2; x = -3, y = 2; x = -3, y = -2$
19. $x = \frac{1}{2}, y = \frac{3}{2}; x = \frac{1}{2}, y = -\frac{3}{2}; x = -\frac{1}{2}, y = \frac{3}{2}; x = -\frac{1}{2}, y = -\frac{3}{2}$ **21.** $x = \sqrt{2}, y = 2\sqrt{2}; x = -\sqrt{2}, y = -2\sqrt{2}$
23. $x = \frac{8}{3}, y = \frac{2\sqrt{10}}{3}; x = -\frac{8}{3}, y = \frac{2\sqrt{10}}{3}; x = \frac{8}{3}, y = -\frac{2\sqrt{10}}{3}; x = -\frac{8}{3}, y = -\frac{2\sqrt{10}}{3}$

25. $x = 1, y = \frac{1}{2}; x = -1, y = \frac{1}{2}; x = 1, y = -\frac{1}{2}; x = -1, y = -\frac{1}{2}$ **27.** No solution; system is inconsistent

29. $x = 2, y = 1; x = -2, y = -1; x = \sqrt{3}, y = \sqrt{3}; x = -\sqrt{3}, y = -\sqrt{3}$

31. $x = 3, y = 2; x = -3, y = -2; x = 2, y = \frac{1}{2}; x = -2, y = -\frac{1}{2}$ **33.** $x = 3, y = 1; x = -1, y = -3$

35. $x = 0, y = -2; x = 0, y = 1; x = 2, y = -1$ **37.** $4 + \sqrt{2}$ and $4 - \sqrt{2}$ **39.** 1 and 7; -1 and -7 **41.** $\frac{1}{2}$ and $\frac{1}{3}$

43. $12 \text{ cm} \times 18 \text{ cm}$ **45.** 8 cm   **47.** $y = -\frac{3}{4}x + \frac{25}{4}$ **49.** $y = 2x + 1$ **51.** $y = -\frac{1}{3}x + \frac{7}{3}$ **53.** $y = 2x - 3$

55. $l = \dfrac{p + \sqrt{p^2 - 16A}}{4}, w = \dfrac{p - \sqrt{p^2 - 16A}}{4}$

57.

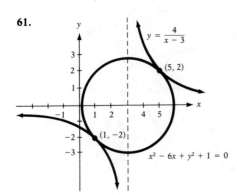

59. Solutions: $(0, -\sqrt{3} - 2), (0, \sqrt{3} - 2), (1, 0), (1, -4)$

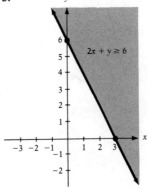

61.

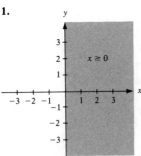

63. $r_1 = \dfrac{-b + \sqrt{b^2 - 4ac}}{2a}, r_2 = \dfrac{-b - \sqrt{b^2 - 4ac}}{2a}$

SECTION 7.6

1.

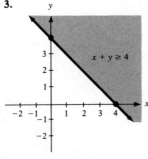

3.

5.

7.

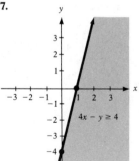

$4x - y \geq 4$

9.

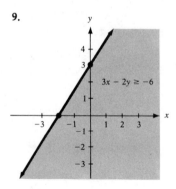

$3x - 2y \geq -6$

11.

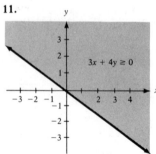

$3x + 4y \geq 0$

13.

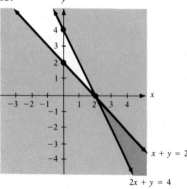

$x + y = 2$

$2x + y = 4$

15.

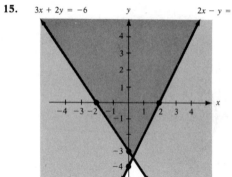

$3x + 2y = -6$ $2x - y = 4$

17.

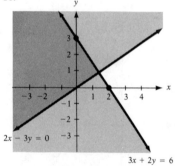

$2x - 3y = 0$

$3x + 2y = 6$

19.

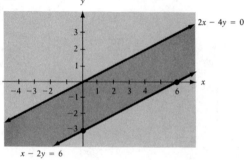

$2x - 4y = 0$

$x - 2y = 6$

21.

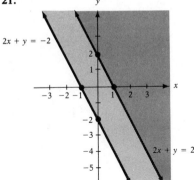

$2x + y = -2$

$2x + y = 2$

23. No solution

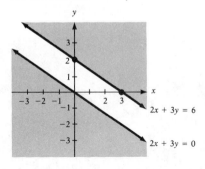

$2x + 3y = 6$

$2x + 3y = 0$

25. Bounded; vertices
$(0, 0), (3, 0), (2, 2), (0, 3)$

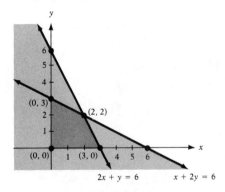

$(0, 3)$ $(2, 2)$

$(0, 0)$ $(3, 0)$

$2x + y = 6$ $x + 2y = 6$

27. Unbounded; vertices $(2, 0), (0, 4)$

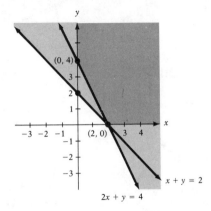

$(0, 4)$

$(2, 0)$

$x + y = 2$

$2x + y = 4$

29. Bounded; vertices $(2, 0), (4, 0),$
$(\frac{24}{7}, \frac{12}{7}), (0, 4), (0, 2)$

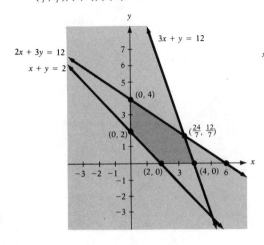

$2x + 3y = 12$ $3x + y = 12$

$x + y = 2$

$(0, 4)$

$(0, 2)$ $(\frac{24}{7}, \frac{12}{7})$

$(2, 0)$ $(4, 0)$

31. Bounded; vertices $(2, 0), (5, 0),$
$(2, 6), (0, 8), (0, 2)$

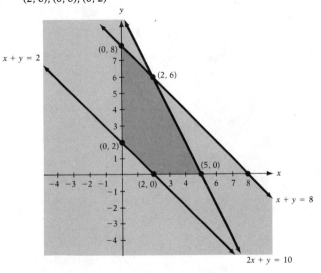

$x + y = 2$

$(0, 8)$ $(2, 6)$

$(0, 2)$

$(2, 0)$ $(5, 0)$

$x + y = 8$

$2x + y = 10$

33. Bounded; vertices $(1, 0)$, $(10, 0)$, $(0, 5)$, $(0, \frac{1}{2})$

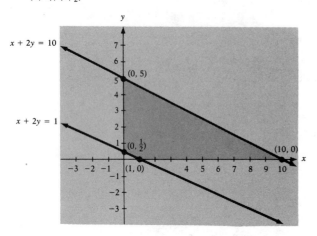

FILL-IN-THE-BLANK QUESTIONS

1. inconsistent **3.** determinants **5.** half-plane

REVIEW EXERCISES

1. $x = 2, y = -1$ **3.** $x = 2, y = \frac{1}{2}$ **5.** $x = 2, y = -1$ **7.** $x = \frac{11}{5}, y = -\frac{3}{5}$ **9.** $x = -\frac{8}{5}, y = \frac{12}{5}$ **11.** $x = 6, y = -1$
13. $x = -4, y = 3$ **15.** $x = 2, y = 3$ **17.** Inconsistent **19.** $x = -1, y = 2, z = -3$ **21.** $x = \frac{2}{5}, y = \frac{1}{10}$
23. $x = \frac{1}{2}, y = \frac{2}{3}, z = \frac{1}{6}$ **25.** $x = -\frac{1}{2}, y = -\frac{2}{3}, z = -\frac{3}{4}$ **27.** $z = -1, x = y + 1, y$ any number
29. $x = 1, y = 2, z = -3, t = 1$ **31.** 5 **33.** 108 **35.** -100 **37.** $x = 2, y = -1$ **39.** $x = 2, y = 3$
41. $x = -1, y = 2, z = -3$ **43.** $x = -\frac{2}{5}, y = -\frac{11}{5}; x = -2, y = 1$ **45.** $x = 2\sqrt{2}, y = \sqrt{2}; x = -2\sqrt{2}, y = -\sqrt{2}$
47. $x = 0, y = 0; x = 3, y = 3; x = -3, y = 3$
49. $x = \sqrt{2}, y = -\sqrt{2}; x = -\sqrt{2}, y = \sqrt{2}; x = \frac{4}{3}\sqrt{2}, y = -\frac{4}{3}\sqrt{2}; x = -\frac{4}{3}\sqrt{2}, y = \frac{4}{3}\sqrt{2}$
51. $x = 1, y = -1$
53. Unbounded; vertex $(2, 0)$

55. Bounded; vertices $(0, 0)$, $(0, 2)$, $(3, 0)$

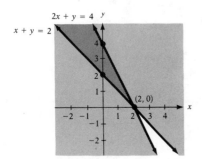

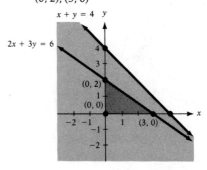

57. Bounded; vertices $(0, 1)$, $(0, 8)$, $(4, 0)$, $(2, 0)$

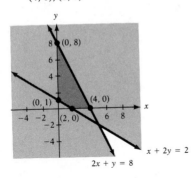

59. 10 **61.** $y = -\frac{1}{3}x^2 - \frac{2}{3}x + 1$ **63.** Katy gets \$10, Mike gets \$20, Danny gets \$5, Colleen gets \$10.
65. 29.69 mph **67.** Katy: 4 hr, Mike: 2 hr, Danny: 8 hr **69.** 24 ft by 10 ft **71.** $4 + \sqrt{2}$ in. and $4 - \sqrt{2}$ in.

CHAPTER 8   SECTION 8.1

1. $\begin{bmatrix} 4 & 4 & -5 \\ -1 & 5 & 4 \end{bmatrix}$ **3.** $\begin{bmatrix} 0 & 12 & -20 \\ 4 & 8 & 24 \end{bmatrix}$ **5.** $\begin{bmatrix} -8 & 7 & -15 \\ 7 & 0 & 22 \end{bmatrix}$ **7.** $\begin{bmatrix} 28 & -9 \\ 4 & 23 \end{bmatrix}$ **9.** $\begin{bmatrix} 1 & 14 & -14 \\ 2 & 22 & -18 \\ 3 & 0 & 28 \end{bmatrix}$

11. $\begin{bmatrix} 15 & 21 & -16 \\ 22 & 34 & -22 \\ -11 & 7 & 22 \end{bmatrix}$ **13.** $\begin{bmatrix} 25 & -9 \\ 4 & 20 \end{bmatrix}$ **15.** $\begin{bmatrix} -13 & 7 & -12 \\ -18 & 10 & -14 \\ 17 & -7 & 34 \end{bmatrix}$ **17.** $\begin{bmatrix} -2 & 4 & 2 & 8 \\ 2 & 1 & 4 & 6 \end{bmatrix}$ **19.** $\begin{bmatrix} 9 & 2 \\ 34 & 13 \\ 47 & 20 \end{bmatrix}$

21. $\begin{bmatrix} 1 & -1 \\ -1 & 2 \end{bmatrix}$ **23.** $\begin{bmatrix} 1 & -\frac{5}{2} \\ -1 & 3 \end{bmatrix}$ **25.** $\begin{bmatrix} 1 & -\frac{1}{a} \\ -1 & \frac{2}{a} \end{bmatrix}$ **27.** $\begin{bmatrix} 3 & -3 & 1 \\ -2 & 2 & -1 \\ -4 & 5 & -2 \end{bmatrix}$ **29.** $\begin{bmatrix} -\frac{5}{7} & \frac{1}{7} & \frac{3}{7} \\ \frac{9}{7} & \frac{1}{7} & -\frac{4}{7} \\ \frac{3}{7} & -\frac{2}{7} & \frac{1}{7} \end{bmatrix}$ **31.** $x = 3, y = 2$

33. $x = -5, y = 10$ **35.** $x = 2, y = -1$ **37.** $x = \frac{1}{2}, y = 2$ **39.** $x = -2, y = 1$ **41.** $x = \frac{2}{a}, y = \frac{3}{a}$

43. $x = -2, y = 3, z = 5$ **45.** $x = \frac{1}{2}, y = -\frac{1}{2}, z = 1$ **47.** $x = -\frac{34}{7}, y = \frac{85}{7}, z = \frac{12}{7}$ **49.** $x = \frac{1}{3}, y = 1, z = \frac{2}{3}$

51. $\begin{bmatrix} 4 & 2 & | & 1 & 0 \\ 2 & 1 & | & 0 & 1 \end{bmatrix} \rightarrow \begin{bmatrix} 1 & \frac{1}{2} & | & \frac{1}{4} & 0 \\ 2 & 1 & | & 0 & 1 \end{bmatrix} \rightarrow \begin{bmatrix} 1 & \frac{1}{2} & | & \frac{1}{4} & 0 \\ 0 & 0 & | & -\frac{1}{2} & 1 \end{bmatrix}$

53. $\begin{bmatrix} 15 & 3 & | & 1 & 0 \\ 10 & 2 & | & 0 & 1 \end{bmatrix} \rightarrow \begin{bmatrix} 1 & \frac{1}{5} & | & \frac{1}{15} & 0 \\ 10 & 2 & | & 0 & 1 \end{bmatrix} \rightarrow \begin{bmatrix} 1 & \frac{1}{5} & | & \frac{1}{15} & 0 \\ 0 & 0 & | & -\frac{2}{3} & 1 \end{bmatrix}$

55. $\begin{bmatrix} -3 & 1 & -1 & | & 1 & 0 & 0 \\ 1 & -4 & -7 & | & 0 & 1 & 0 \\ 1 & 2 & 5 & | & 0 & 0 & 1 \end{bmatrix} \rightarrow \begin{bmatrix} 1 & 2 & 5 & | & 0 & 0 & 1 \\ 1 & -4 & -7 & | & 0 & 1 & 0 \\ -3 & 1 & -1 & | & 1 & 0 & 0 \end{bmatrix} \rightarrow \begin{bmatrix} 1 & 2 & 5 & | & 0 & 0 & 1 \\ 0 & -6 & -12 & | & 0 & 1 & -1 \\ 0 & 7 & 14 & | & 1 & 0 & 3 \end{bmatrix}$

$\rightarrow \begin{bmatrix} 1 & 2 & 5 & | & 0 & 0 & 1 \\ 0 & 1 & 2 & | & 0 & -\frac{1}{6} & \frac{1}{6} \\ 0 & 1 & 2 & | & \frac{1}{7} & 0 & \frac{3}{7} \end{bmatrix} \rightarrow \begin{bmatrix} 1 & 2 & 5 & | & 0 & 0 & 1 \\ 0 & 1 & 2 & | & 0 & -\frac{1}{6} & \frac{1}{6} \\ 0 & 0 & 0 & | & \frac{1}{7} & \frac{1}{6} & \frac{11}{42} \end{bmatrix}$

57. (a) $\begin{bmatrix} 500 & 350 & 400 \\ 700 & 500 & 850 \end{bmatrix}$; yes, $\begin{bmatrix} 500 & 700 \\ 350 & 500 \\ 400 & 850 \end{bmatrix}$ (b) $\begin{bmatrix} 15 \\ 8 \\ 3 \end{bmatrix}$ (c) $\begin{bmatrix} 11,500 \\ 17,050 \end{bmatrix}$ (d) $[0.10 \quad 0.05]$ (e) \$2002.50

59. If $a \neq 0$,
$$\begin{bmatrix} a & b \\ c & d \end{bmatrix} \begin{array}{|cc} 1 & 0 \\ 0 & 1 \end{array} \rightarrow \begin{bmatrix} 1 & \dfrac{b}{a} \\ c & d \end{bmatrix} \begin{array}{|cc} \dfrac{1}{a} & 0 \\ 0 & 1 \end{array} \rightarrow \begin{bmatrix} 1 & \dfrac{b}{a} \\ 0 & \dfrac{-cb + da}{a} \end{bmatrix} \begin{array}{|cc} \dfrac{1}{a} & 0 \\ -\dfrac{c}{a} & 1 \end{array}$$

$$\rightarrow \begin{bmatrix} 1 & \dfrac{b}{a} \\ 0 & 1 \end{bmatrix} \begin{array}{|cc} \dfrac{1}{a} & 0 \\ \dfrac{-c}{-cb + da} & \dfrac{a}{-cb + da} \end{array} \rightarrow \begin{bmatrix} 1 & 0 \\ 0 & 1 \end{bmatrix} \begin{array}{|cc} \dfrac{d}{-cb + da} & \dfrac{-b}{-cb + da} \\ \dfrac{-c}{-cb + da} & \dfrac{a}{-cb + da} \end{array}.$$

Therefore, $A^{-1} = \dfrac{1}{\Delta} \begin{bmatrix} d & -b \\ -c & a \end{bmatrix}.$

If $a = 0$,
$$\begin{bmatrix} 0 & b \\ c & d \end{bmatrix} \begin{array}{|cc} 1 & 0 \\ 0 & 1 \end{array} \rightarrow \begin{bmatrix} c & d \\ 0 & b \end{bmatrix} \begin{array}{|cc} 0 & 1 \\ 1 & 0 \end{array} \rightarrow \begin{bmatrix} 1 & \dfrac{d}{c} \\ 0 & b \end{bmatrix} \begin{array}{|cc} 0 & \dfrac{1}{c} \\ 1 & 0 \end{array}$$

$$\rightarrow \begin{bmatrix} 1 & 0 \\ 0 & 1 \end{bmatrix} \begin{array}{|cc} \dfrac{-d}{cb} & \dfrac{1}{c} \\ \dfrac{1}{b} & 0 \end{array} \rightarrow \begin{bmatrix} 1 & 0 \\ 0 & 1 \end{bmatrix} \begin{array}{|cc} \dfrac{d}{-bc} & \dfrac{-b}{-bc} \\ \dfrac{-c}{-bc} & 0 \end{array}.$$

SECTION 8.2

1. Maximum value is 11, minimum value is 3.   **3.** Maximum value is 65, minimum value is 4.
5. Maximum value is 67, minimum value is 20.   **7.** The maximum value of z is 12, and it occurs at the point $(6, 0)$.
9. The minimum value of z is 4, and it occurs at the point $(2, 0)$.
11. The maximum value of z is 20, and it occurs at the point $(0, 4)$.
13. The minimum value of z is 8, and it occurs at the point $(0, 2)$.
15. The maximum value of z is 50, and it occurs at the point $(10, 0)$.
17. The maximum profit is \$70. It is achieved with 40 packages of the low-grade mixture and 180 packages of the high-grade mixture.
19. 8 downhill, 24 cross-country; \$1760   **21.** 24 acres of soybeans, 12 acres of wheat; \$5520
23. \$10,000 in the Certificate of Deposit, \$10,000 in Treasury bills   **25.** 10 racing skates, 15 figure skates
27. 10 first-class, 120 coach   **29.** 5 oz of Supplement A, 1 oz of Supplement B

SECTION 8.3

1. Proper   **3.** Improper, $1 + \dfrac{9}{x^2 - 4}$ **5.** Improper, $5x + \dfrac{22x - 1}{x^2 - 4}$ **7.** Improper, $1 + \dfrac{-2(x - 6)}{(x + 4)(x - 3)}$ **9.** $\dfrac{-4}{x} + \dfrac{4}{x - 1}$

11. $\dfrac{1}{x} + \dfrac{-x}{x^2 + 1}$ **13.** $\dfrac{-1}{x - 1} + \dfrac{2}{x - 2}$ **15.** $\dfrac{\frac{1}{4}}{x + 1} + \dfrac{\frac{3}{4}}{x - 1} + \dfrac{\frac{1}{2}}{(x - 1)^2}$ **17.** $\dfrac{\frac{1}{12}}{x - 2} + \dfrac{-\frac{1}{12}(x + 4)}{x^2 + 2x + 4}$

19. $\dfrac{\frac{1}{4}}{x - 1} + \dfrac{\frac{1}{4}}{(x - 1)^2} - \dfrac{\frac{1}{4}}{x + 1} + \dfrac{\frac{1}{4}}{(x + 1)^2}$ **21.** $\dfrac{-5}{x + 2} + \dfrac{5}{x + 1} + \dfrac{-4}{(x + 1)^2}$ **23.** $\dfrac{\frac{1}{4}}{x} + \dfrac{1}{x^2} - \dfrac{\frac{1}{4}(x + 4)}{x^2 + 4}$

25. $\dfrac{\frac{2}{3}}{x + 1} + \dfrac{\frac{1}{3}(x + 1)}{x^2 + 2x + 4}$ **27.** $\dfrac{\frac{2}{7}}{3x - 2} + \dfrac{\frac{1}{7}}{2x + 1}$ **29.** $\dfrac{\frac{3}{4}}{x + 3} + \dfrac{\frac{1}{4}}{x - 1}$ **31.** $\dfrac{1}{x^2 + 4} + \dfrac{2x - 1}{(x^2 + 4)^2}$ **33.** $\dfrac{-1}{x} + \dfrac{2}{x - 3} + \dfrac{-1}{x + 1}$

35. $\dfrac{4}{x - 2} + \dfrac{-3}{x - 1} + \dfrac{-1}{(x - 1)^2}$ **37.** $\dfrac{x}{(x^2 + 16)^2} + \dfrac{-16x}{(x^2 + 16)^3}$

SECTION 8.4

1.

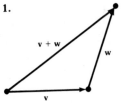

3.

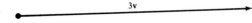

5.

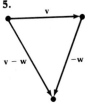

7.

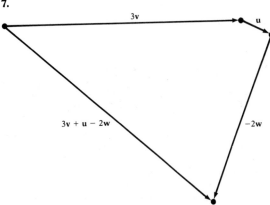

9. $X = A$ **11.** $C = D - E + F$ **13.** $E = -G - H + D$ **15.** $X = \emptyset$ **17.** 12   **19.** $v = 3i + 4j$ **21.** $v = 2i + 4j$
23. $v = 8i - j$ **25.** $v = -i + j$ **27.** 5   **29.** $\sqrt{2}$ **31.** $\sqrt{13}$ **33.** $-j$ **35.** $\sqrt{89}$ **37.** $\sqrt{34} - \sqrt{13}$ **39.** i   **41.** $\frac{3}{5}i - \frac{4}{5}j$
43. $\frac{\sqrt{2}}{2}i - \frac{\sqrt{2}}{2}j$ **45.** $v = \frac{8}{\sqrt{5}}i + \frac{4}{\sqrt{5}}j$ or $v = -\frac{8}{\sqrt{5}}i - \frac{4}{\sqrt{5}}j$ **47.** $-2 + \sqrt{21}, -2 - \sqrt{21}$ **49.** 460 mph
51. 218 mph   **53.**

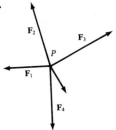

FILL-IN-THE-BLANK QUESTIONS

1. inverse   **3.** identity   **5.** feasible solution   **7.** unit

REVIEW EXERCISES

1. $\begin{bmatrix} 4 & -4 \\ 3 & 9 \\ 4 & 0 \end{bmatrix}$ **3.** $\begin{bmatrix} 6 & 0 \\ 12 & 24 \\ -6 & 12 \end{bmatrix}$ **5.** $\begin{bmatrix} 4 & -3 & 0 \\ 12 & -2 & -8 \\ -2 & 5 & -4 \end{bmatrix}$ **7.** $\begin{bmatrix} 8 & -13 & 8 \\ 9 & 2 & -10 \\ 18 & -17 & 4 \end{bmatrix}$ **9.** $\begin{bmatrix} \frac{1}{2} & -1 \\ -\frac{1}{6} & \frac{2}{3} \end{bmatrix}$ **11.** $\begin{bmatrix} -\frac{5}{7} & \frac{9}{7} & \frac{3}{7} \\ \frac{1}{7} & \frac{1}{7} & -\frac{2}{7} \\ \frac{3}{7} & -\frac{4}{7} & \frac{1}{7} \end{bmatrix}$

13. Singular **15.** The maximum value is 32, when $x = 0$ and $y = 8$.
17. The minimum value is 3, when $x = 1$ and $y = 0$. **19.** The maximum value is $\frac{108}{7}$, when $x = \frac{12}{7}$ and $y = \frac{12}{7}$.
21. $\dfrac{-\frac{3}{2}}{x} + \dfrac{\frac{3}{2}}{x - 4}$ **23.** $\dfrac{-3}{x - 1} + \dfrac{3}{x} + \dfrac{4}{x^2}$ **25.** $\dfrac{-\frac{1}{10}}{x + 1} + \dfrac{\frac{1}{10}x + \frac{9}{10}}{x^2 + 9}$ **27.** $\dfrac{x}{x^2 + 4} - \dfrac{4x}{(x^2 + 4)^2}$ **29.** $\dfrac{\frac{1}{2}}{x^2 + 1} + \dfrac{\frac{1}{4}}{x - 1} - \dfrac{\frac{1}{4}}{x + 1}$
31. $\mathbf{v} = 2\mathbf{i} - 4\mathbf{j}, \|\mathbf{v}\| = 2\sqrt{5}$ **33.** $\mathbf{v} = -\mathbf{i} + 3\mathbf{j}, \|\mathbf{v}\| = \sqrt{10}$ **35.** $-20\mathbf{i} + 13\mathbf{j}$ **37.** $\sqrt{5}$ **39.** $\sqrt{5} + 5 \approx 7.24$
41. $\dfrac{-2\sqrt{5}}{5}\mathbf{i} + \dfrac{\sqrt{5}}{5}\mathbf{j}$ **43.** $\sqrt{29} \approx 5.39$ mpg
45. 35 gasoline engines, 15 diesel engines; 15 gasoline engines, 0 diesel engines

CHAPTER 9   SECTION 9.1

1. (I) $n = 1$: $2 \cdot 1 = 2$ and $1(1 + 1) = 2$
  (II) If $2 + 4 + 6 + \cdots + 2k = k(k + 1)$, then $2 + 4 + 6 + \cdots + 2k + 2(k + 1)$
    $= [2 + 4 + 6 + \cdots + 2k] + 2(k + 1) = k(k + 1) + 2(k + 1) = k^2 + 3k + 2 = (k + 1)(k + 2)$.
3. (I) $n = 1$: $1 + 2 = 3$ and $\frac{1}{2}(1)(1 + 5) = \frac{1}{2}(6) = 3$
  (II) If $3 + 4 + 5 + \cdots + (k + 2) = \frac{1}{2}k(k + 5)$, then $3 + 4 + 5 + \cdots + (k + 2) + [(k + 1) + 2]$
    $= [3 + 4 + 5 + \cdots + (k + 2)] + (k + 3) = \frac{1}{2}k(k + 5) + k + 3 = \frac{1}{2}(k^2 + 7k + 6) = \frac{1}{2}(k + 1)(k + 6)$.
5. (I) $n = 1$: $3 \cdot 1 - 1 = 2$ and $\frac{1}{2}(1)[3(1) + 1] = \frac{1}{2}(4) = 2$
  (II) If $2 + 5 + 8 + \cdots + (3k - 1) = \frac{1}{2}k(3k + 1)$, then $2 + 5 + 8 + \cdots + (3k - 1) + [3(k + 1) - 1]$
    $= [2 + 5 + 8 + \cdots + (3k - 1)] + 3k + 2 = \frac{1}{2}k(3k + 1) + (3k + 2) = \frac{1}{2}(3k^2 + 7k + 4) = \frac{1}{2}(k + 1)(3k + 4)$.
7. (I) $n = 1$: $2^{1-1} = 1$ and $2^1 - 1 = 1$
  (II) If $1 + 2 + 2^2 + \cdots + 2^{k-1} = 2^k - 1$, then $1 + 2 + 2^2 + \cdots + 2^{k-1} + 2^{(k+1)-1}$
    $= (1 + 2 + 2^2 + \cdots + 2^{k-1}) + 2^k = 2^k - 1 + 2^k = 2(2^k) - 1 = 2^{k+1} - 1$.
9. (I) $n = 1$: $4^{1-1} = 1$ and $\frac{1}{3}(4^1 - 1) = \frac{1}{3}(3) = 1$
  (II) If $1 + 4 + 4^2 + \cdots + 4^{k-1} = \frac{1}{3}(4^k - 1)$, then $1 + 4 + 4^2 + \cdots + 4^{k-1} + 4^{(k+1)-1}$
    $= (1 + 4 + 4^2 + \cdots + 4^{k-1}) + 4^k = \frac{1}{3}(4^k - 1) + 4^k = \frac{1}{3}[4^k - 1 + 3(4^k)] = \frac{1}{3}[4(4^k) - 1] = \frac{1}{3}(4^{k+1} - 1)$.
11. (I) $n = 1$: $\dfrac{1}{1 \cdot 2} = \dfrac{1}{2}$ and $\dfrac{1}{1 + 1} = \dfrac{1}{2}$
  (II) If $\dfrac{1}{1 \cdot 2} + \dfrac{1}{2 \cdot 3} + \dfrac{1}{3 \cdot 4} + \cdots + \dfrac{1}{k(k + 1)} = \dfrac{k}{k + 1}$,
    then $\dfrac{1}{1 \cdot 2} + \dfrac{1}{2 \cdot 3} + \dfrac{1}{3 \cdot 4} + \cdots + \dfrac{1}{k(k + 1)} + \dfrac{1}{(k + 1)[(k + 1) + 1]}$
    $= \left[\dfrac{1}{1 \cdot 2} + \dfrac{1}{2 \cdot 3} + \dfrac{1}{3 \cdot 4} + \cdots + \dfrac{1}{k(k + 1)}\right] + \dfrac{1}{(k + 1)(k + 2)} = \dfrac{k}{k + 1} + \dfrac{1}{(k + 1)(k + 2)} = \dfrac{k + 1}{k + 2}$.
13. (I) $n = 1$: $1^2 = 1$ and $\frac{1}{6} \cdot 1 \cdot 2 \cdot 3 = 1$
  (II) If $1^2 + 2^2 + 3^2 + \cdots + k^2 = \frac{1}{6}k(k + 1)(2k + 1)$, then $1^2 + 2^2 + 3^2 + \cdots + k^2 + (k + 1)^2$
    $= (1^2 + 2^2 + 3^2 + \cdots + k^2) + (k + 1)^2 = \frac{1}{6}k(k + 1)(2k + 1) + (k + 1)^2 = \frac{1}{6}(2k^3 + 9k^2 + 13k + 6)$
    $= \frac{1}{6}(k + 1)(k + 2)(2k + 3)$.
15. (I) $n = 1$: $5 - 1 = 4$ and $\frac{1}{2}(9 - 1) = \frac{1}{2} \cdot 8 = 4$
  (II) If $4 + 3 + 2 + \cdots + (5 - k) = \frac{1}{2}k(9 - k)$, then $4 + 3 + 2 + \cdots + (5 - k) + 5 - (k + 1)$
    $= [4 + 3 + 2 + \cdots + (5 - k)] + 5 - (k + 1) = \frac{1}{2}k(9 - k) + 4 - k = \frac{1}{2}(-k^2 + 7k + 8) = \frac{1}{2}(8 - k)(k + 1)$
    $= \frac{1}{2}(k + 1)[9 - (k + 1)]$.
17. (I) $n = 1$: $1 \cdot (1 + 1) = 2$ and $\frac{1}{3} \cdot 1 \cdot 2 \cdot 3 = 2$
  (II) If $1 \cdot 2 + 2 \cdot 3 + 3 \cdot 4 + \cdots + k(k + 1) = \frac{1}{3}k(k + 1)(k + 2)$,
    then $1 \cdot 2 + 2 \cdot 3 + 3 \cdot 4 + \cdots + k(k + 1) + (k + 1)(k + 2)$
    $= [1 \cdot 2 + 2 \cdot 3 + 3 \cdot 4 + \cdots + k(k + 1)] + (k + 1)(k + 2) = \frac{1}{3}k(k + 1)(k + 2) + (k + 1)(k + 2)$
    $= \frac{1}{3}(k + 1)(k + 2)(k + 3)$.
19. (I) $n = 1$: $1^2 + 1 = 2$ is divisible by 2.
  (II) If $k^2 + k$ is divisible by 2, then $(k + 1)^2 + (k + 1) = k^2 + 2k + 1 + k + 1 = (k^2 + k) + 2k + 2$.
    Since $k^2 + k$ is divisible by 2 and $2k + 2$ is divisible by 2, therefore, $(k + 1)^2 + k + 1$ is divisible by 2.

21. (I) $n = 1$: $1^2 - 1 + 2 = 2$ is divisible by 2.
   (II) If $k^2 - k + 2$ is divisible by 2, then $(k + 1)^2 - (k + 1) + 2 = k^2 + 2k + 1 - k - 1 + 2$
   $= (k^2 - k + 2) + 2k$. Since $k^2 - k + 2$ is divisible by 2 and $2k$ is divisible by 2, therefore,
   $(k + 1)^2 - (k + 1) + 2$ is divisible by 2.

23. (I) $n = 1$: If $x > 1$, then $x^1 = x > 1$.
   (II) Assume, for any natural number k, that, if $x > 1$, then $x^k > 1$. Show that, if $x > 1$, then $x^{k+1} > 1$:
   $$x^{k+1} = x^k \cdot x^1 > 1 \cdot x = x > 1$$
   $$\underset{\uparrow}{}$$
   $$x^k > 1$$

25. (I) $n = 1$: $a - b$ is a factor of $a^1 - b^1 = a - b$.
   (II) If $a - b$ is a factor of $a^k - b^k$, show that $a - b$ is a factor of $a^{k+1} - b^{k+1}$:
   $a^{k+1} - b^{k+1} = a(a^k - b^k) + b^k(a - b)$. Since $a - b$ is a factor of $a^k - b^k$ and $a - b$ is a factor of $a - b$,
   therefore, $a - b$ is a factor of $a^{k+1} - b^{k+1}$.

27. $n = 1$: $1^2 - 1 + 41 = 41$ is a prime number.
   $n = 41$: $41^2 - 41 + 41 = 1681 = 41^2$ is not prime.

29. (I) $n = 1$: $ar^{1-1} = a \cdot 1 = a$ and $a \cdot \dfrac{1 - r^1}{1 - r} = a$ because $r \neq 1$.

   (II) If $a + ar + ar^2 + \cdots + ar^{k-1} = a\dfrac{(1 - r^k)}{1 - r}$, then $a + ar + ar^2 + \cdots + ar^{k-1} + ar^{(k+1)-1}$

   $= (a + ar + ar^2 + \cdots + ar^{k-1}) + ar^k = a\dfrac{(1 - r^k)}{1 - r} + ar^k = \dfrac{a(1 - r^k) + ar^k(1 - r)}{1 - r}$

   $= \dfrac{a - ar^k + ar^k - ar^{k+1}}{1 - r} = a\dfrac{(1 - r^{k+1})}{1 - r}$.

SECTION 9.2

1. 362,880 **3.** $\frac{28}{3}$ **5.** 10 **7.** 21 **9.** 50 **11.** 1 **13.** $x^5 + 5x^4 + 10x^3 + 10x^2 + 5x + 1$
15. $x^6 - 12x^5 + 60x^4 - 160x^3 + 240x^2 - 192x + 64$ **17.** $81x^4 + 108x^3 + 54x^2 + 12x + 1$
19. $x^{10} + 5y^2x^8 + 10y^4x^6 + 10y^6x^4 + 5y^8x^2 + y^{10}$ **21.** $x^3 + 6\sqrt{2}x^{5/2} + 30x^2 + 40\sqrt{2}x^{3/2} + 60x + 24\sqrt{2}x^{1/2} + 8$
23. $(ax)^5 + 5by(ax)^4 + 10(by)^2(ax)^3 + 10(by)^3(ax)^2 + 5(by)^4(ax) + (by)^5$ **25.** 17,010 **27.** 126,720 **29.** 41,472
31. $2835x^3$ **33.** $314,928x^7$ **35.** $\dbinom{n}{n} = \dfrac{n!}{n!(n - n)!} = \dfrac{n!}{n!0!} = \dfrac{n!}{n!} = 1$

37. $2^n = (1 + 1)^n = \dbinom{n}{0}1^n + \dbinom{n}{1}(1)(1)^{n-1} + \cdots + \dbinom{n}{n}1^n = \dbinom{n}{0} + \dbinom{n}{1} + \cdots + \dbinom{n}{n}$

39. $12! = 4.790016 \times 10^8$, $20! = 2.432302 \times 10^{18}$, $25! = 1.551121 \times 10^{25}$; $12! \approx 4.790139724 \times 10^8$,
   $20! \approx 2.432924 \times 10^{18}$, $25! \approx 1.5511299 \times 10^{25}$

SECTION 9.3

1. 1, 2, 3, 4, 5 **3.** $\frac{1}{2}, \frac{2}{3}, \frac{3}{4}, \frac{4}{5}, \frac{5}{6}$ **5.** 1, −4, 9, −16, 25 **7.** $\frac{1}{2}, \frac{2}{5}, \frac{2}{7}, \frac{8}{41}, \frac{8}{61}$ **9.** $-\frac{1}{6}, \frac{1}{12}, -\frac{1}{20}, \frac{1}{30}, -\frac{1}{42}$
11. $1/e, 2/e^2, 3/e^3, 4/e^4, 5/e^5$ **13.** $n/(n + 1)$ **15.** $1/2^{n-1}$ **17.** $(-1)^{n+1}$ **19.** $(-1)^{n+1}n$
21. $a_1 = 1, a_2 = 3, a_3 = 5, a_4 = 7, a_5 = 9$ **23.** $a_1 = -2, a_2 = -1, a_3 = 1, a_4 = 4, a_5 = 8$
25. $a_1 = 5, a_2 = 10, a_3 = 20, a_4 = 40, a_5 = 80$ **27.** $a_1 = 3, a_2 = 3, a_3 = \frac{3}{2}, a_4 = \frac{1}{2}, a_5 = \frac{1}{8}$
29. $a_1 = 1, a_2 = 2, a_3 = 2, a_4 = 4, a_5 = 8$ **31.** $a_1 = A, a_2 = A + d, a_3 = A + 2d, a_4 = A + 3d, a_5 = A + 4d$
33. $a_1 = \sqrt{2}, a_2 = \sqrt{2 + \sqrt{2}}, a_3 = \sqrt{2 + \sqrt{2 + \sqrt{2}}}, a_4 = \sqrt{2 + \sqrt{2 + \sqrt{2 + \sqrt{2}}}}$,

   $a_5 = \sqrt{2 + \sqrt{2 + \sqrt{2 + \sqrt{2 + \sqrt{2}}}}}$

35. $2 + 3 + 4 + \cdots + (n + 1)$ **37.** $\dfrac{1}{2} + 2 + \dfrac{9}{2} + \cdots + \dfrac{n^2}{2}$ **39.** $1 + \dfrac{1}{3} + \dfrac{1}{9} + \cdots + \dfrac{1}{3^n}$ **41.** $\dfrac{1}{3} + \dfrac{1}{9} + \cdots + \dfrac{1}{3^n}$

43. $\ln 2 - \ln 3 + \ln 4 - \cdots + (-1)^n \ln n$ **45.** $\displaystyle\sum_{k=1}^{n} k$ **47.** $\displaystyle\sum_{k=1}^{n} \dfrac{k}{k + 1}$ **49.** $\displaystyle\sum_{k=0}^{n} (-1)^k \dfrac{1}{3^k}$ **51.** $\displaystyle\sum_{k=1}^{n} \dfrac{3^k}{k}$ **53.** $\displaystyle\sum_{k=0}^{n} (a + kd)$

55. 21

SECTION 9.4

1. $d = 1; 6, 7, 8, 9$ **3.** $d = 2; -3, -1, 1, 3$ **5.** $d = -2; 4, 2, 0, -2$ **7.** $d = -\frac{1}{3}; \frac{1}{6}, -\frac{1}{6}, -\frac{1}{2}, -\frac{5}{6}$
9. $d = \ln 3; \ln 3, 2 \ln 3, 3 \ln 3, 4 \ln 3$ **11.** $r = 2; 2, 4, 8, 16$ **13.** $r = \frac{1}{2}; -\frac{3}{2}, -\frac{3}{4}, -\frac{3}{8}, -\frac{3}{16}$ **15.** $r = 2; \frac{1}{4}, \frac{1}{2}, 1, 2$
17. $r = 2^{1/3}; 2^{1/3}, 2^{2/3}, 2, 2^{4/3}$ **19.** $r = \frac{3}{2}; \frac{1}{2}, \frac{3}{4}, \frac{9}{8}, \frac{27}{16}$ **21.** Arithmetic, $d = 1$ **23.** Neither **25.** Arithmetic, $d = -\frac{2}{3}$
27. Neither **29.** Geometric, $r = \frac{2}{3}$ **31.** Geometric, $r = 2$ **33.** Geometric, $r = 3^{1/2}$ **35.** $a_5 = 9, a_n = 2n - 1$
37. $a_5 = -7, a_n = 8 - 3n$ **39.** $a_5 = 2, a_n = \frac{1}{2}(n-1)$ **41.** $a_5 = 5\sqrt{2}, a_n = \sqrt{2}n$ **43.** $a_5 = 16, a_n = 2^{n-1}$
45. $a_5 = 5, a_n = (-1)^{n-1}(5)$ **47.** $a_5 = 0, a_n = 0$ **49.** $a_5 = 4\sqrt{2}, a_n = (\sqrt{2})^n$ **51.** $a_{12} = 24$ **53.** $a_{10} = -26$
55. $a_8 = a + 7b$ **57.** $a_7 = \frac{1}{64}$ **59.** $a_9 = 1$ **61.** $a_8 = 0.00000004$ **63.** $a_1 = -13, d = 3$ **65.** $a_1 = -53, d = 6$
67. $a_1 = 28, d = -2$ **69.** $a_1 = 25, d = -2$ **71.** $\frac{1}{2}n(5n + 7)$ **73.** $-\frac{1}{4}(1 - 2^n)$ **75.** $2[1 - (\frac{2}{3})^n]$ **77.** $n(1 + n)$
79. $1 - 2^n$ **81.** $(\frac{3}{4})^3 \cdot 20 = \frac{135}{16}$ ft; $(\frac{3}{4})^n \cdot 20$ ft **83.** About 102 ft **85.** 1.845×10^{19}

FILL-IN-THE-BLANK QUESTIONS

1. Pascal triangle **3.** sequence **5.** geometric

REVIEW EXERCISES

1. (I) $n = 1$: $3 \cdot 1 = 3$ and $\dfrac{3 \cdot 1}{2}(2) = 3$

(II) If $3 + 6 + 9 + \cdots + 3k = \dfrac{3k}{2}(k + 1)$, then $3 + 6 + 9 + \cdots + 3k + 3(k + 1)$

$= (3 + 6 + 9 + \cdots + 3k) + (3k + 3) = \dfrac{3k}{2}(k + 1) + (3k + 3) = \dfrac{3k^2}{2} + \dfrac{9k}{2} + \dfrac{6}{2} = \dfrac{3}{2}(k^2 + 3k + 2)$

$= \dfrac{3}{2}(k + 1)(k + 2)$.

3. (I) $n = 1$: $2 \cdot 3^{1-1} = 2$ and $3^1 - 1 = 2$
(II) If $2 + 6 + 18 + \cdots + 2 \cdot 3^{k-1} = 3^k - 1$, then $2 + 6 + 18 + \cdots + 2 \cdot 3^{k-1} + 2 \cdot 3^{(k+1)-1}$
$= (2 + 6 + 18 + \cdots + 2 \cdot 3^{k-1}) + 2 \cdot 3^k = 3^k - 1 + 2 \cdot 3^k = 3 \cdot 3^k - 1 = 3^{k+1} - 1$.

5. (I) $n = 1$: $1^2 = 1$ and $\frac{1}{2}(6 - 3 - 1) = \frac{1}{2}(2) = 1$
(II) If $1^2 + 4^2 + 7^2 + \cdots + (3k - 2)^2 = \frac{1}{2}k(6k^2 - 3k - 1)$, then $1^2 + 4^2 + 7^2 + \cdots + (3k - 2)^2 + [3(k + 1) - 2]^2$
$= [1^2 + 4^2 + 7^2 + \cdots + (3k - 2)^2] + (3k + 1)^2 = \frac{1}{2}k(6k^2 - 3k - 1) + (3k + 1)^2$
$= \frac{1}{2}(6k^3 + 15k^2 + 11k + 2) = \frac{1}{2}(k + 1)(6k^2 + 9k + 2)$
$= \frac{1}{2}(k + 1)[6(k + 1)^2 - 3(k + 1) - 1]$.

7. 120 **9.** 10 **11.** $x^4 + 8x^3 + 24x^2 + 32x + 16$ **13.** $32x^5 + 240x^4 + 720x^3 + 1080x^2 + 810x + 243$ **15.** 144 **17.** 84
19. $-\frac{2}{3}, \frac{3}{4}, -\frac{4}{5}, \frac{5}{6}, -\frac{6}{7}$ **21.** $2, 1, \frac{8}{9}, 1, \frac{32}{25}$ **23.** $3, 2, \frac{4}{3}, \frac{8}{9}, \frac{16}{27}$ **25.** $2, 0, 2, 0, 2$ **27.** Arithmetic; $d = 1; \frac{n}{2}(7 + n)$
29. Neither **31.** Geometric; $r = 8; \frac{8}{7}(8^n - 1)$ **33.** Arithmetic; $d = 4; 2n(n - 1)$
35. Geometric; $r = \frac{1}{2}; 6[1 - (\frac{1}{2})]^n$ **37.** Neither **39.** 27 **41.** $(\frac{1}{10})^{10}$ **43.** $9\sqrt{2}$ **45.** $5n - 4$ **47.** $n - 10$

CHAPTER 10   SECTION 10.1

1. $\{1, 3, 5, 6, 7, 9\}$ **3.** $\{1, 5, 7\}$ **5.** $\{1, 6, 9\}$ **7.** $\{1, 2, 4, 5, 6, 7, 8, 9\}$ **9.** $\{1, 2, 4, 5, 6, 7, 8, 9\}$ **11.** $\{0, 2, 6, 7, 8\}$
13. $\{0, 1, 2, 3, 5, 6, 7, 8, 9\}$ **15.** $\{0, 1, 2, 3, 5, 6, 7, 8, 9\}$ **17.** $\{0, 1, 2, 3, 4, 6, 7, 8\}$ **19.** $\{0\}$
21. $\varnothing, \{a\}, \{b\}, \{c\}, \{d\}, \{a, b\}, \{a, c\}, \{a, d\}, \{b, c\}, \{b, d\}, \{c, d\}, \{a, b, c\}, \{b, c, d\}, \{a, c, d\}, \{a, b, d\}, \{a, b, c, d\}$ **23.** 25
25. 40 **27.** 25 **29.** 37 **31.** 18 **33.** 5 **35.** 175; 125 **37.** (a) 15 (b) 15 (c) 15 (d) 25 (e) 40

SECTION 10.2

1. 15 **3.** 16 **5.** 8 **7.** 1024 **9.** 9000 **11.** 30 **13.** 120 **15.** 1 **17.** 336 **19.** 24 **21.** 60 **23.** 120 **25.** 362,880
27. 5,850,000

SECTION 10.3

1. (a, b, c), (a, b, d), (a, b, e), (a, c, d), (a, c, e), (a, d, e), (b, c, d), (b, c, e), (b, d, e), (c, d, e); 10
3. (1, 2, 3), (2, 3, 4), (1, 2, 4), (1, 3, 4); 4 **5.** 28 **7.** 15 **9.** 1 **11.** 10,400,600 **13.** 35 **15.** 6.3501355×10^{11}
17. 336 **19.** 5,209,344 **21.** 90

SECTION 10.4

1. $S = \{HH, HT, TH, TT\}$; $P(HH) = \frac{1}{4}$, $P(HT) = \frac{1}{4}$, $P(TH) = \frac{1}{4}$, $P(TT) = \frac{1}{4}$
3. $S = \{HH1, HH2, HH3, HH4, HH5, HH6, HT1, HT2, HT3, HT4, HT5, HT6, TH1, TH2, TH3, TH4, TH5, TH6,$
TT1, TT2, TT3, TT4, TT5, TT6\}; each outcome has the probability of $\frac{1}{24}$.
5. $S = \{HHH, HHT, HTH, HTT, THH, THT, TTH, TTT\}$; each outcome has the probability of $\frac{1}{8}$.
7. $S = \{1$ Yellow, 1 Red, 1 Green, 2 Yellow, 2 Red, 2 Green, 3 Yellow, 3 Red, 3 Green, 4 Yellow, 4 Red, 4 Green\};
each outcome has the probability of $\frac{1}{12}$; thus, $P(2 \text{ Red}) + P(4 \text{ Red}) = \frac{1}{12} + \frac{1}{12} = \frac{1}{6}$.
9. $S = \{1$ Yellow Forward, 1 Yellow Backward, 1 Red Forward, 1 Red Backward, 1 Green Forward, 1 Green
Backward, 2 Yellow Forward, 2 Yellow Backward, 2 Red Forward, 2 Red Backward, 2 Green Forward, 2
Green Backward, 3 Yellow Forward, 3 Yellow Backward, 3 Red Forward, 3 Red Backward, 3 Green Forward,
3 Green Backward, 4 Yellow Forward, 4 Yellow Backward, 4 Red Forward, 4 Red Backward, 4 Green
Forward, 4 Green Backward\}; each outcome has the probability of $\frac{1}{24}$; thus, $P(1 \text{ Red Backward}) + P(1 \text{ Green}$
Backward) $= \frac{1}{24} + \frac{1}{24} = \frac{1}{12}$.
11. $S = \{11$ Red, 11 Yellow, 11 Green, 12 Red, 12 Yellow, 12 Green, 13 Red, 13 Yellow, 13 Green, 14 Red, 14
Yellow, 14 Green, 21 Red, 21 Yellow, 21 Green, 22 Red, 22 Yellow, 22 Green, 23 Red, 23 Yellow, 23 Green, 24
Red, 24 Yellow, 24 Green, 31 Red, 31 Yellow, 31 Green, 32 Red, 32 Yellow, 32 Green, 33 Red, 33 Yellow, 33
Green, 34 Red, 34 Yellow, 34 Green, 41 Red, 41 Yellow, 41 Green, 42 Red, 42 Yellow, 42 Green, 43 Red, 43
Yellow, 43 Green, 44 Red, 44 Yellow, 44 Green\}; each outcome has the probability of $\frac{1}{48}$; thus,
$E = \{22$ Red, 22 Green, 24 Red, 24 Green\}, $P(E) = n(E)/n(S) = \frac{4}{48} = \frac{1}{12}$.
13. $\frac{3}{13}$ **15.** $\frac{5}{18}$ **17.** $\frac{9}{20}$ **19.** $\frac{17}{20}$ **21.** $\frac{3}{10}$ **23.** $\frac{2}{5}$ **25.** 0.000033068 **27.** ≈ 0.11655 **29.** ≈ 0.0039

FILL-IN-THE-BLANK QUESTIONS

1. union, intersection **3.** permutation **5.** equally likely

REVIEW EXERCISES

1. $\{1, 3, 5, 6, 7, 8\}$ **3.** $\{3, 7\}$ **5.** $\{1, 2, 4, 6, 8, 9\}$ **7.** $\{1, 2, 4, 5, 6, 9\}$ **9.** 17 **11.** 29 **13.** 7 **15.** 25 **17.** 336 **19.** 56
21. 60 **23.** 128 **25.** 3024 **27.** 70 **29.** 91 **31.** 182 **33.** $\frac{3}{20}, \frac{9}{20}$ **35.** $\frac{1}{24}$

CHAPTER 11   SECTION 11.2

1. $y^2 = 8x$

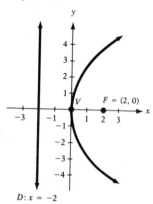

$D: x = -2$

3. $x^2 = -12y$

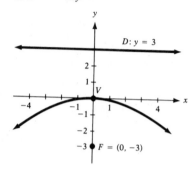

5. $y^2 = -8x$

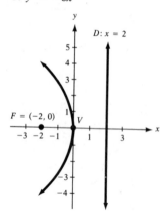

7. $x^2 = 2y$

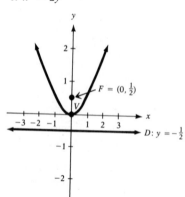

9. $(x - 2)^2 = -8(y + 3)$

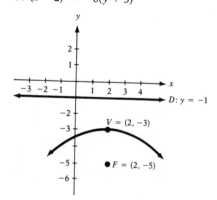

11. $x^2 = \frac{4}{3}y$

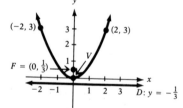

13. $(x + 3)^2 = 4(y - 3)$

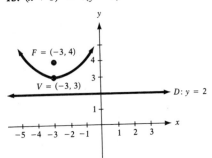

15. $(y + 2)^2 = -8(x + 1)$

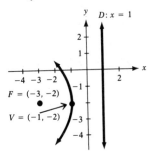

17.

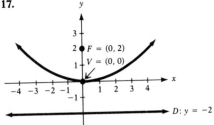

19.

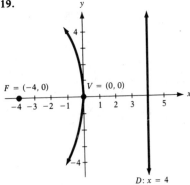

21.

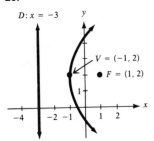

23.

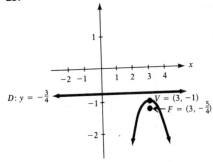

25.

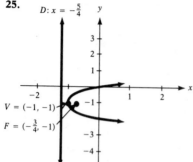

$D: x = -\frac{5}{4}$

$V = (-1, -1)$

$F = (-\frac{3}{4}, -1)$

27.

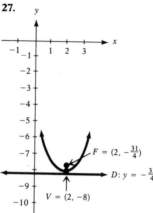

$F = (2, -\frac{31}{4})$

$D: y = -\frac{33}{4}$

$V = (2, -8)$

29. 2.083 ft from the vertex

SECTION 11.3

1. Vertices (3, 0), (−3, 0); foci ($\sqrt{5}$, 0), (−$\sqrt{5}$, 0)

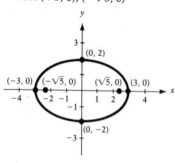

3. Vertices (0, 5), (0, −5); foci (0, 4), (0, −4)

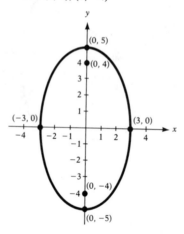

5. Vertices (0, 4), (0, −4); foci (0, $2\sqrt{3}$), (0, $-2\sqrt{3}$)
$$\frac{x^2}{4} + \frac{y^2}{16} = 1$$

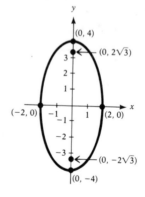

7. Vertices ($2\sqrt{2}$, 0), (−$2\sqrt{2}$, 0); foci ($\sqrt{6}$, 0), (−$\sqrt{6}$, 0)
$$\frac{x^2}{8} + \frac{y^2}{2} = 1$$

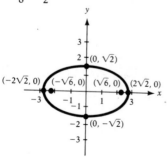

9. The graph is a circle, center at (0, 0), radius 4.

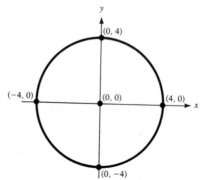

11. $\dfrac{x^2}{36} + \dfrac{y^2}{27} = 1$

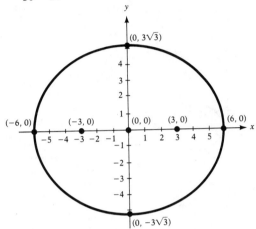

13. $\dfrac{x^2}{9} + \dfrac{y^2}{25} = 1$

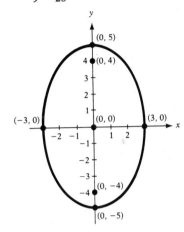

15. $\dfrac{x^2}{9} + \dfrac{y^2}{5} = 1$

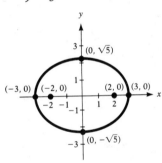

17. $\dfrac{x^2}{4} + \dfrac{y^2}{13} = 1$

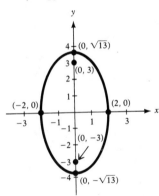

19. $x^2 + \dfrac{y^2}{16} = 1$

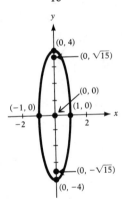

21. Center $(2, -1)$;
foci $(2, -1 + \sqrt{5})$, $(2, -1 - \sqrt{5})$;
vertices $(2, 2)$, $(2, -4)$

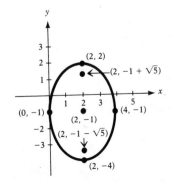

23. Center $(-5, 4)$;
foci $(-5 + 2\sqrt{3}, 4)$, $(-5 - 2\sqrt{3}, 4)$;
vertices $(-1, 4)$, $(-9, 4)$
$$\dfrac{(x + 5)^2}{16} + \dfrac{(y - 4)^2}{4} = 1$$

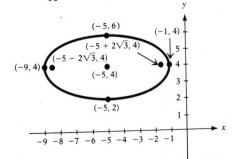

25. Center $(-2, 1)$;
foci $(-2 - \sqrt{3}, 1), (-2 + \sqrt{3}, 1)$;
vertices $(-4, 1), (0, 1)$
$$\frac{(x + 2)^2}{4} + (y - 1)^2 = 1$$

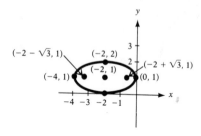

27. Center $(2, -1)$;
foci $(1, -1), (3, -1)$;
vertices $(2 + \sqrt{3}, -1), (2 - \sqrt{3}, -1)$
$$\frac{(x - 2)^2}{3} + \frac{(y + 1)^2}{2} = 1$$

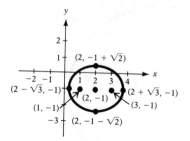

29. $\dfrac{(x - 2)^2}{9} + \dfrac{(y + 2)^2}{5} = 1$

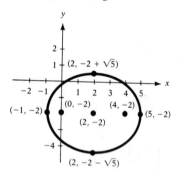

31. $\dfrac{(x - 4)^2}{5} + \dfrac{(y - 6)^2}{9} = 1$

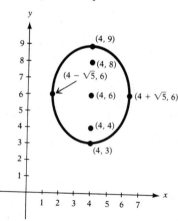

33. $\dfrac{(x - 2)^2}{16} + \dfrac{(y - 1)^2}{7} = 1$

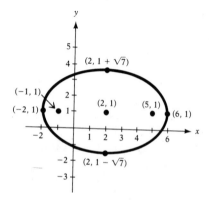

35. $\dfrac{x^2}{100} + \dfrac{y^2}{36} = 1$

37. $0, \approx 12.99, 15, \approx 12.99, 0$
39. Greatest distance $\approx 93{,}533{,}333$ mi, least distance $\approx 90{,}466{,}667$ mi

SECTION 11.4

1. $x^2 - \dfrac{y^2}{15} = 1$

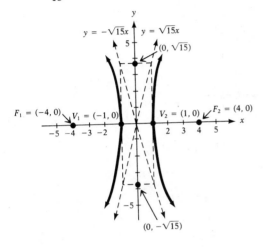

3. $\dfrac{y^2}{16} - \dfrac{x^2}{20} = 1$

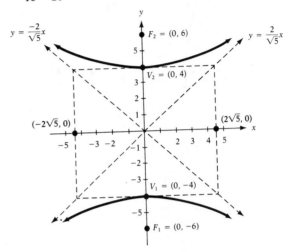

5. $\dfrac{x^2}{9} - \dfrac{y^2}{16} = 1$

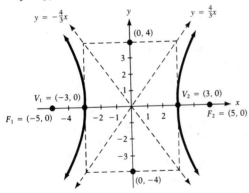

7. $\dfrac{y^2}{36} - \dfrac{x^2}{9} = 1$

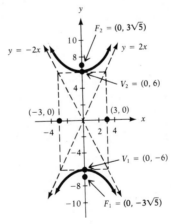

9. $\dfrac{x^2}{8} - \dfrac{y^2}{8} = 1$

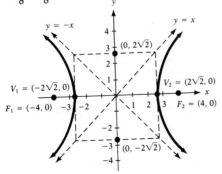

11.

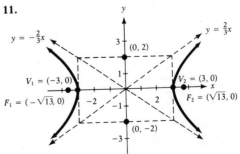

13. $\dfrac{x^2}{4} - \dfrac{y^2}{16} = 1$

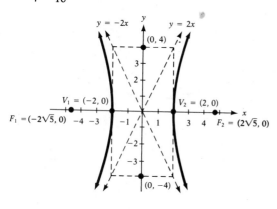

15. $\dfrac{y^2}{9} - x^2 = 1$

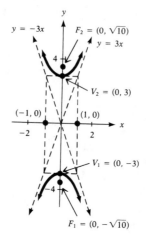

17. $\dfrac{y^2}{25} - \dfrac{x^2}{25} = 1$

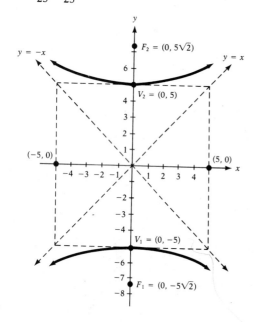

19. $\dfrac{(x-4)^2}{4} - \dfrac{(y+1)^2}{5} = 1$

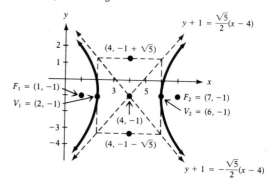

21. $\dfrac{(y + 4)^2}{4} - \dfrac{(x + 3)^2}{12} = 1$

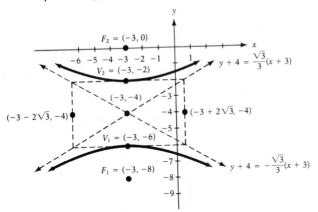

23. $(x - 5)^2 - \dfrac{(y - 7)^2}{3} = 1$

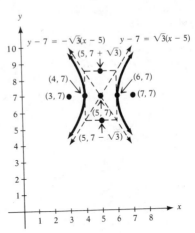

25. $\dfrac{(x - 1)^2}{4} - \dfrac{(y + 1)^2}{9} = 1$

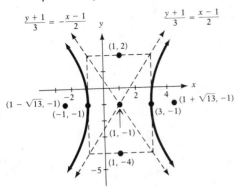

27. Transverse axis is parallel to the x-axis;
center $(3, -2)$;
foci $(3 \pm \sqrt{13}, -2)$;
vertices $(3 \pm 2, -2)$;
asymptotes: $y + 2 = \pm\frac{3}{2}(x - 3)$

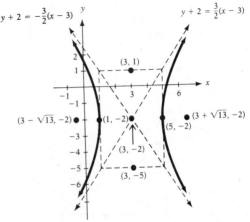

29. $\dfrac{(y-2)^2}{4} - (x+2)^2 = 1$

Transverse axis is parallel to the y-axis;
center $(-2, 2)$;
foci $(-2, 2 \pm \sqrt{5})$;
vertices $(-2, 2 \pm 2)$;
asymptotes: $y - 2 = \pm 2(x + 2)$

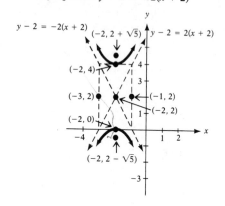

31. $\dfrac{(x-3)^2}{4} - \dfrac{(y+2)^2}{16} = 1$

Transverse axis is parallel to the x-axis;
center $(3, -2)$;
foci $(3 \pm 2\sqrt{5}, -2)$;
vertices $(3 \pm 2, -2)$;
asymptotes: $y + 2 = \pm 2(x - 3)$

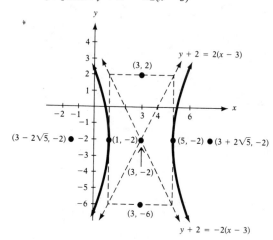

33. $\dfrac{(y-1)^2}{4} - (x+2)^2 = 1$

Transverse axis is parallel to the y-axis;
center $(-2, 1)$;
foci $(-2, 1 \pm \sqrt{5})$;
vertices $(-2, 1 \pm 2)$;
asymptotes: $y - 1 = \pm 2(x + 2)$

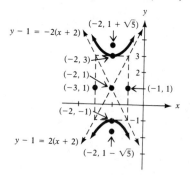

35. If e is close to 1, narrow hyperbola; if e is very large, wide hyperbola

37. $\frac{x^2}{4} - y^2 = 1$, asymptotes $y = \pm\frac{1}{2}x$

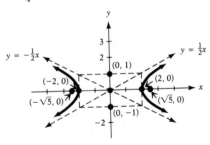

$y^2 - \frac{x^2}{4} = 1$, asymptotes $y = \pm\frac{1}{2}x$

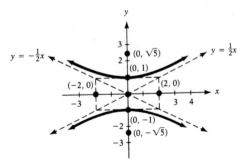

FILL-IN-THE-BLANK QUESTIONS

1. parabola **3.** hyperbola **5.** y-axis

REVIEW EXERCISES

1. Parabola, vertex $(0, 0)$, focus $(-4, 0)$, directrix $x = 4$
3. Hyperbola, center $(0, 0)$, vertices $(2, 0)$ and $(-2, 0)$, foci $(\sqrt{5}, 0)$ and $(-\sqrt{5}, 0)$, asymptotes $y = \frac{1}{2}x$ and $y = -\frac{1}{2}x$
5. Ellipse, center $(0, 0)$, vertices $(0, 5)$ and $(0, -5)$, foci $(0, 3)$ and $(0, -3)$
7. $x^2 = -4(y - 1)$: Parabola, vertex $(0, 1)$, focus $(0, 0)$, directrix $y = 2$
9. $x^2/2 - y^2/8 = 1$: Hyperbola, center $(0, 0)$, vertices $(\sqrt{2}, 0)$ and $(-\sqrt{2}, 0)$, foci $(\sqrt{10}, 0)$ and $(-\sqrt{10}, 0)$, asymptotes $y = 2x$ and $y = -2x$
11. $(x - 2)^2 = 2(y + 2)$: Parabola, vertex $(2, -2)$, focus $(2, -\frac{3}{2})$, directrix $y = -\frac{5}{2}$
13. $\frac{(y - 2)^2}{4} - (x - 1)^2 = 1$: Hyperbola, center $(1, 2)$, vertices $(1, 4)$ and $(1, 0)$, foci $(1, 2 + \sqrt{5})$ and $(1, 2 - \sqrt{5})$, asymptotes $y - 2 = \pm2(x - 1)$
15. $\frac{(x - 2)^2}{9} + \frac{(y - 1)^2}{4} = 1$: Ellipse, center $(2, 1)$, vertices $(5, 1)$ and $(-1, 1)$, foci $(2 + \sqrt{5}, 1)$ and $(2 - \sqrt{5}, 1)$
17. $(x - 2)^2 = -4(y + 1)$: Parabola, vertex $(2, -1)$, focus $(2, -2)$, directrix $y = 0$
19. $\frac{(x - 1)^2}{4} + \frac{(y + 1)^2}{9} = 1$: Ellipse, center $(1, -1)$, vertices $(1, 2)$ and $(1, -4)$, foci $(1, -1 + \sqrt{5})$ and $(1, -1 - \sqrt{5})$

21. $y^2 = -8x$

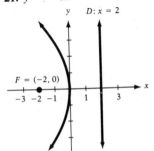

23. $\frac{y^2}{4} - \frac{x^2}{12} = 1$

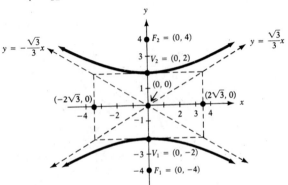

25. $\dfrac{x^2}{16} + \dfrac{y^2}{7} = 1$

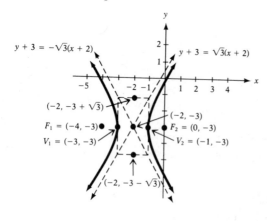

27. $(x - 2)^2 = -4(y + 3)$

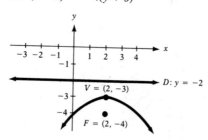

29. $(x + 2)^2 - \dfrac{(y + 3)^2}{3} = 1$

31. $\dfrac{(x + 4)^2}{16} + \dfrac{(y - 5)^2}{25} = 1$

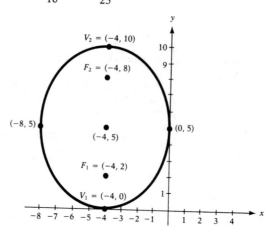

33. $\dfrac{(x + 1)^2}{9} - \dfrac{(y - 2)^2}{7} = 1$

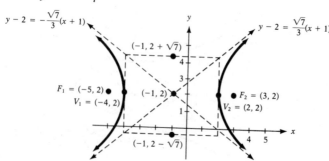

35. $\dfrac{(x-3)^2}{9} - \dfrac{(y-1)^2}{4} = 1$

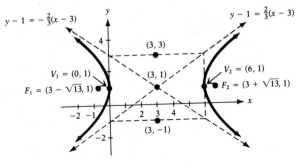

37. $\dfrac{x^2}{5} - \dfrac{y^2}{4} = 1$ **39.** The ellipse $\dfrac{x^2}{16} + \dfrac{y^2}{7} = 1$

Abel, N., 329, 562
Abscissa, 163
Absolute value, 22
   equations and inequalities
      involving, 151
   function, 237
   properties of, 151
Adding y-coordinates, 261
Addition principle of counting,
   588
Additive inverse, 7
   of a vector, 533
Algebra, 1
   fundamental theorem of, 355
   linear, 492
   matrix, 492
Algebraic function, 365
Algorithm, 309
   division, 309
al-jabr, 1
Alternating signs, 566
Apollonius, 161, 611
Argument of a function, 224
Arithmetic mean, 24
Arithmetic progression, 571
Arithmetic sequence, 571
   common difference of, 571
   nth term of, 572
   sum of, 573
Arrow notation of a function, 226
Ars Conjectandi, 605
Ars Magna, 328
Associative laws
   for matrices, 495, 501
   for real numbers, 7
   for vectors, 532

Asymptotes
   horizontal, 340
   oblique, 341
   of a hyperbola, 641, 642
   vertical, 340
Augmented matrix, 440
Axis
   conjugate, 634
   coordinate, 162
   major, minor, 624
   of a cone, 612
   of symmetry, of a parabola, 286,
      613
   semimajor, 634
   transverse, 634
   x-, 162
   y-, 162

Back-substitute, 433
Base, 25
Bell, A., 400
Bernoulli, J., 605
Bezoit, E., 475
Bilgere, Merry, xii, 1954
Binomial, 31
   coefficients, 558
   theorem, 558
Bits, 588
Boole, G., 606
Boundary of a linear inequality,
   479
Boundary points, 144
Bounded graph, 484
Bounds on zeros
   lower, 332
   theorem, 332

Bounds on zeros (*continued*)
   upper, 332
Bracket x, 238
Braden, T., xii, 1932
Briggs, H., 394
Bürgi, J., 394

Calculators, 78, 226, 373
   algebraic system for, 78
   arithmetic, 78
   reverse Polish, 78
   scientific, 78
Cancellation laws, 9
Cantor, G., 11
Carbon dating, 408
Cardano, G., 328, 605
Cauchy, A., 141
Cayley, A., 417, 508
Center
   of a circle, 182
   of an ellipse, 624, 630
   of a hyperbola, 634
Chain, 19
Change
   in horizontal scale, 253
   in vertical scale, 251
Change-of-base formula, 392
Circle, 182
   center of, 182
   general form of equation, 184
   radius of, 182
   standard form of equation, 182
   unit, 184
Closed interval, 20
Coefficient
   binomial, 558

Coefficient (*continued*)
  leading, 31, 354
  of a monomial, 30
  of a polynomial, 31, 35
Coefficient matrix, 440
Coincident lines, 422
Colleen, xii, 1972
Column index, 439, 492
Column vector, 497
Combinations, 596
  formula for, 597
Combinatorics, 583
Commutative laws
  for matrices, 495
  for real numbers, 6
  for vectors, 532
Complement of a set, 586
Completing the square, 112
Complex function, 354
  domain of, 354
  zero of, 354
Complex numbers, 67, 68
  conjugate of, 70
  imaginary part, 68
  real part, 68
Complex polynomial function, 354
  leading coefficient of, 354
Complex ratio, 52
Complex variable, 354
Components of a vector, 536
Composite function, 262
Composition, 262
Compound interest, 378
  formula for, 380
Compounded continuously, 381
Cone, 612
  axis of, 612
  generator of, 612
  vertex of, 612
Conic, 612
  degenerate, 612
Conic section, 612
Conjugate complex number, 70
Conjugate hyperbola, 647
Conjugate pairs theorem, 357
Consistent system, 419
Constant, 2
  of proportionality, 208
Constant function, 235
Constant rate jobs, 106
Constraints, 512
Coordinate, 13, 163
  axes, 162
  Cartesian, 162
  rectangular, 162

Coordinate (*continued*)
  x-, 163
  y-, 163
Counting
  addition principle of, 588
  multiplication principle of, 591
Counting formula, 588
Counting numbers, 3
Cramer, G., 417
Cramer's Rule, 457, 459, 462
Cube root, 56
Cubing functions, 236

Danny, xii, 1970
Decay
  law of uninhibited, 407
  radioactive, 408
Decibel, 400
Decimal, 4
  repeating, 4
  terminating, 4
Decreasing function, 232
Dedekind, R., 11
Degenerate conic, 612
Degenerate line segment, 530
Degree
  of a monomial, 30
  of a polynomial, 30, 35, 297
Dellen, D., xii, 1940
Denominator, 4, 46
  rationalizing, 59
Dependent variable, 223, 354
Depressed equation, 323
Descartes, R., 161, 219
Descartes' Rule of Signs, 321
Determinants
  2 by 2, 457
  3 by 3, 460
  properties of, 464, 465
  value of, 457, 461
Diagonal entries of a matrix, 501
Dibuadius, C., 61
Difference, 8
  of two cubes, 34
  of two matrices, 494
  of two squares, 34
  of two vectors, 533
Difference function, 260
Difference quotient, 225
Digit, significant, 76
Directed line segment, 530
  equivalent, 531
  initial point of, 530
  terminal point of, 530

Direction of a vector, 530
Directrix, 613
Dirichlet, L., 219
Discriminant, 115
Disguised quadratic equation, 132
Distance formula, 164
  proof of, 165
Distributive law, 7
  for matrices, 501
Dividend, 309
Division
  long, 308
  synthetic, 311
Division algorithm, 309
Divisor, 309
Double root, 111

e, 372
Earthquake
  intensity of, 403
  magnitude of, 403
  zero-level, 403
Eccentricity
  of an ellipse, 634
  of a hyperbola, 646
Echelon form of a matrix, 446
  reduced, 447
Effective rate of interest, 381
Elements of a set, 3, 584
Ellipse
  center of, 624
  definition of, 623
  equations of, 625, 627, 630
  foci of, 623
  major axis of, 624
  minor axis of, 624
  reflecting property, 632
  vertices of, 624
Ellipsis, 3
Empty set, 584
Entries of a matrix, 439, 492
  diagonal, 501
Equal matrices, 493
Equal sets, 3
Equal sign, 2
Equal vectors, 532
Equality
  of vectors, 538
  properties of, 2
Equally likely outcomes, 602
Equation(s), 2, 88
  depressed, 323
  disguised quadratic, 132
  equivalent, 88

Equation(s) (*continued*)
  exponential, 375
  first-degree, 92
  general form for a circle, 184
  general form for a line, 198
  graph of, 172
  intercept form for a line, 207
  linear, 92, 198
  of an ellipse, 625, 627, 630
  of a horizontal line, 197
  of a hyperbola, 636, 639, 643
  of a parabola, 615, 617, 619
  of a vertical line, 195
  point–slope form of a line, 195
  quadratic, 109
  radical, 131
  root of, 88
  sides of, 88
  slope–intercept form of a line,
    199
  solution of, 88
  standard form for a circle, 182
  systems of, 418
Equilateral hyperbola, 646
Equilateral triangle, 170
Equivalent directed line
    segments, 531
Equivalent equations, 88
Equivalent inequalities, 137
Equivalent systems of equations,
    432
Euclid, 121, 577
Euler, L., 219, 606
Even function, 232
Event, 601
  probability of, 601
Exponent, 25
  laws of, 29, 64, 367
  rational, 63
Exponential equations, 375
Exponential function, 367
Exponential law, 407
Exponential tables, 649
Extraneous solution, 91

Factor(s), 37
Factor theorem, 310
Factored completely, 38
Factorial symbol, 550, 555
Factoring, 37
  by grouping, 43
  over the integers, 37
  over the rational numbers, 37
  over the real numbers, 37

Feasible solution, 512
Fermat, 161, 605, 606
Ferrari, L., 328
Fibonacci, 577
  numbers, 567
  sequence, 567
Finite set, 585
First-degree equation, 92
Focus
  of an ellipse, 623
  of a hyperbola, 634
  of a parabola, 613
Formulas
  change-of-base, 390
  compound interest, 380
  counting, 588
  distance, 164
  midpoint, 169
  present value, 382
  quadratic, 115, 126
  recursive, 566
Fritz, xii, 1974
Frobenius, G., 508
Function, 220
  absolute value, 237
  algebraic, 365
  approximating the zeros of, 335
  argument of, 224
  arrow notation for, 226
  asymptote of, 340
  complex, 354
  composite, 262
  constant, 235
  cubing, 236
  decreasing, 232
  difference, 260
  difference quotient of, 225
  domain of, 220, 354
  even, 233
  exponential, 367
  graph of, 230
  greatest integer, 238
  hyperbolic cosine, 378
  hyperbolic sine, 377
  identity, 235
  increasing, 232
  inverse, 270
  linear, 235
  logarithmic, 397
  multiple zeros of, 320
  natural logarithm, 398
  objective, 512
  odd, 233
  one-to-one, 268
  ordered pair notation for, 231

Function (*continued*)
  piecewise-defined, 239
  polynomial, 297
  power, 299
  product, 260
  quadratic, 284
  quotient, 260
  range of, 220
  rational, 337
  square root, 237
  squaring, 236
  stationary, 232
  step, 239
  sum, 260
  transcendental, 365
  value of, 221
  vertical line test for, 230
  zero of, 320, 354
Function notation, 225
Fundamental Theorem of Algebra,
    355

Galois, 329
Gauss, K., 355, 417
General form of the equation of a
    circle, 184
General form of the equation of a
    line, 198
Generator of a cone, 612
Geometric progression, 574
Geometric sequence, 574
  common ratio, 574
  *n*th term of, 575
  sum of, 576
Gibbs, J., 542
Graph
  bounded, 484
  of an equation, 172
  of a function, 230
  of a system of linear
    inequalities, 480
  unbounded, 484
  vertices of, 484
Grassmann, H., 542
Greater than, 14
Greater than or equal to, 15
Greatest integer function, 238
Growth, law of uninhibited, 407

Half-closed interval, 20
Half-life, 408
Half-open interval, 20
Half-plane, 479
Hamilton, W., 542

Harriot, 121
Horizontal asymptote, 340
Horizontal line, 192
  equation of, 197
  test, 268
Horizontal scale, change in, 253
Horizontal shift, 249
Hui, Y., 561
Huygens, C., 605
Hyperbola, 634
  asymptotes of, 641, 642
  branches of, 634
  center of, 634
  conjugate, 647
  conjugate axis of, 634
  eccentricity of, 646
  equations of, 636, 639, 643
  equilateral, 646
  foci of, 634
  transverse axis of, 634
  vertices of, 634
Hyperbolic cosine, 378
Hyperbolic sine, 377
Hypotenuse, 79

i, 68
  powers of, 73
Identity, 88
Identity function, 235
Identity laws, 7
Identity matrix, 501
Image, 220
Imaginary part, 68
  pure, 68
  unit, 68
Improper rational function, 343, 522
Inconsistent system, 419
Increasing function, 232
Independent variable, 223, 354
Index, 56
  column, 439
  of a sum, 567
  row, 439
Inequalities, 136
  equivalent, 137
  graphing, 16, 478, 480
  linear, 137, 478
  nonstrict, 15, 478
  properties of, 15, 16
  solutions of, 136
  strict, 15, 478
  to solve, 136
Inequality, triangle, 151

Inequality signs, 15
Infinite set, 585
Infinity
  approaches, 338, 340
  minus, 21
  plus, 21
Intensity
  of an earthquake, 403
  of sound, 400
Intercept(s), 176
  finding, 177
  x-, 176
  y-, 176
Intercept form of equation of a line, 207
Interest, 102
  compound, 378
  compounded continuously, 381
  effective rate of, 381
  rate of, 102
  simple, 102
Intermediate Value Theorem, 334
Intersection of sets, 585
Intervals, 20
  closed, 20
  half-closed, 20
  half-open, 20
  left endpoint of, 20
  open, 20
  right endpoint of, 20
Inverse function, 270
  graph of, 273
  procedure for finding, 272
Inverse laws, 7
Inverse matrix, 502
Irrational number, 4
Irreducible quadratic, 327
Isosceles triangle, 171

Jones, R., xii, 1948
Jordan, C., 417

Katy, xii, 1965
Khayyám, O., 561
Kōwa, S., 417

Law of cooling, 409
Law of uninhibited decay, 407
Law of uninhibited growth, 407
Laws of exponents, 29, 64, 367
Leading coefficient, 31, 354
Least common multiple, 49
Left endpoint, 20

Leibniz, G., 417
Less than, 14
Less than or equal to, 15
Line(s)
  coincident, 422
  family of, 207
  horizontal, 192
  intercept form of the equation, 207
  parallel, 202
  perpendicular, 203
  real number, 13
  slope of, 189
  slope–intercept form of the equation, 199
  vertical, 189
Line segment, 530
Linear algebra, 492
Linear equation, 92, 198
  procedure to solve, 96
  systems of (*see* Systems of linear equations)
Linear function, 235
Linear inequality, 137, 478
  graph of, 16, 478, 480
  systems of, 478
Linear program, 512
Linear programming problem, 512
  feasible solution, 512
  solution of, 512
Logarithm, 385
  change-of-base formula for, 392
  common, 391
  natural, 392
  properties of, 386, 387, 390
  table of values, 650, 652
Logarithm to the base a, 385
Logarithmic equations, 390
Logarithmic function, 397
Loudness of sound, 400
Lowest terms, 10, 46

Magnitude of an earthquake, 403
Magnitude of a vector, 530, 535
Major axis, 624
  length of, 624
Mathematical induction, 550
Matrix, 439, 492
  augmented, 440
  coefficient, 440
  diagonal entries of, 501
  difference of, 494
  echelon form of, 446

Matrix (*continued*)
  entries of, 439, 492
  equal, 493
  identity, 501
  index of, 439, 492
  inverse, 502
  m by n, 493
  method for solving systems of
    equations, 445
  nonsingular, 502
  product of, 498
  properties of, 495, 496, 501
  reduced echelon form, 447
  row operations on, 443
  scalar multiple of, 495
  square, 493
  sum of, 494
  zero, 495
Matrix algebra, 492
Maximum value of a quadratic
  function, 292
Mean, arithmetic, 24
Medians of a triangle, 170
Midpoint of a line segment, 168
  formula, 169
Minimum value of a quadratic
  function, 292
Minor axis, 624
Minus infinity, 21
  approaches, 340
Mitosis, 407
Mixture problem, 104
Monomial, 30
  coefficient, 30
  degree, 30
  like terms, 30
Motion, uniform, 105
Multiple, least common, 49
Multiple zero, 320
Multiplication principle of
  counting, 591
Multiplicative inverse, 7
Multiplicity of zeros, 320, 356

Napier, J., 394
Natural logarithm function, 398
  table of, 652
Natural numbers, 3
Negative real number, 14
Nested form of a polynomial, 315
Newton, I., 409
  law of cooling, 409
Nicolo of Brescia, 328
Niklas, P., xii

Nolan, Clancy, xii
Nonsingular matrix, 502
  procedure for finding the
    inverse, 504
Nonstrict inequalities, 15, 478
Null set, 584
Numbers
  complex, 67, 68
  counting, 3
  e, 372
  i, 68
  irrational, 4
  natural, 3
  pure imaginary, 68
  rational, 4
  real, 4
Numerator, 4, 46

Objective function, 512
Oblique asymptote, 341
Odd function, 233
One-to-one functions, 268
Open interval, 20
Ordered pair, 163, 231
Ordinate, 163
Origin, 13, 162
  symmetry with respect to, 179
Outcome, 600
  equally likely, 602
  probability of, 600

Parabola, 174, 285, 613
  axis of symmetry of, 286, 613
  directrix of, 613
  equations of, 615, 617, 619
  focus of, 613
  opens downward, 285
  opens upward, 285
  vertex of, 285, 613
Paraboloid of revolution, 611, 622
Parallel lines, 202
Parallelogram law of addition,
  532
Partial fraction(s), 522
Partial fraction decomposition,
  522
Pascal, B., 557, 561, 605, 606
Pascal triangle, 557
Peano, G., 606
Pendulum, simple, 213
  period of, 213
Perfect cube, 34
Perfect root, 56

Perfect square, 34
Period of a pendulum, 213
Permutations, 593
  formula for, 594
Piecewise-defined function, 239
Plane, xy-, 162
Plot, 163
Plus infinity, 21
  approaches, 338
Polynomial, 31
  coefficients of, 31, 35
  complex, 354
  degree of, 31, 35, 297
  division algorithm for, 309
  function, 297
  in three variables, 35
  in two variables, 35
  leading coefficient of, 31
  nested form of, 315
  number of zeros, 320
  prime, 38
  terms of, 31
  zero, 31
Positive real number, 14
Power, 25
Power function, 299
Present value, 382
  formula, 382
Prime polynomial, 38
Principal, 102
Principal of substitution, 2
Principal qth root, 56
Principal square root, 125, 126
Probabilistic model, 600
Probability, 600
  of an event, 601
Product function, 260
Product law, 9
Product of matrices, 498
Product of row and column
  vectors, 497
Progression
  arithmetic, 571
  geometric, 574
Proper rational function, 343, 522
Proper subset, 584
Properties
  of absolute value, 151
  of determinants, 464, 465
  of equality, 2
  of logarithms, 386, 387, 390
  of matrices, 495, 496, 501
  of real numbers, 6, 7
  of vectors, 532–535

Proportional
  directly, 208
  inversely, 210
Proportionality, constant of, 208
Pythagorean brotherhood, 10
Pythagorean Theorem, 79
  converse of, 80
  proof of, 81

Quadrant, 163
Quadratic, irreducible, 327
Quadratic formula, 115, 126
Quadratic equation, 109
  discriminant of, 115
  disguised, 132
  procedure to solve, 118
  standard form of, 109
Quadratic function, 284
  maximum value of, 292
  minimum value of, 292
  vertex of, 285
Quotient, 309
  difference, 225
Quotient function, 260

Radical, 56
  equation, 131
  properties of, 57
Radicand, 56
Radius, 182
Rate of interest, 102
  effective, 381
Ratio, 8
  arithmetic of, 9
  complex, 52
  in lowest terms, 10
Rational exponent, 67
Rational expression, 46
  improper, 522
  partial fraction decomposition, 522
  proper, 522
Rational function, 337
  improper, 343, 522
  proper, 343, 522
Rational number, 4
Rational Zeros Theorem, 322
Rationalizing the denominator, 59
Real number, 4
  line, 13
  negative, 14
  positive, 14
Reciprocal, 7

Recursive formula, 566
Recursively defined sequences, 566
Reduced echelon form of a matrix, 447
Reflecting property
  of an ellipse, 632
  of a parabola, 622
Reflection
  about x-axis, 255
  about y-axis, 255
Reflexive property, 2
Register, 588
Remainder, 309
  theorem, 310
Repeated solution, 111
  zero, 320
Repeating decimal, 4
Resultant, 540
Richter, C. F., 403
  scale, 403
Right endpoint, 20
Right triangle, 79
Rise, 189
Root
  cube, 56
  double, 111
  of equation, 88
  of multiplicity two, 111
  perfect, 56
  principal qth, 56
  square, 56
  (see also Zero)
Rounding, 77
Row index, 439, 492
Row vector, 497
Row operations on a matrix, 443
Ruffini, 329
Rule of signs, Descartes', 321
Rules
  for division, 8
  of signs, 8
Run, 189

Sample space, 600
  outcome of, 600
Satisfies, 88
Scalar, 495, 534
  multiple of a matrix, 495
  product of a vector, 534
Scale, 13
  Richter, 403
Scientific notation, 75
Second-degree equation, 109

Sequence, 563
  alternating signs of, 566
  arithmetic, 571
  defined recursively, 566
  Fibonacci, 567
  geometric, 574
  terms of, 563
Sets, 3, 584
  complement of, 586
  elements of, 3, 584
  empty, 584
  equal, 3, 584
  finite, 585
  infinite, 585
  intersection of, 585
  null, 584
  solution, 88
  subset of, 584
  union of, 585
  universal, 586
  well-defined, 584
Shanks, Peter, xii, 1949
Shift
  horizontal, 249
  vertical, 247
Shih-chie, C., 561
Sigma symbol, 567
Significant digits, 76
Simple interest, 102
Slope, 189
  undefined, 189
Solution
  extraneous, 91
  feasible, 512
  of an equation, 88
  of an inequality, 136
  of a linear programming problem, 513
  repeated, 111
  set, 88
Sound
  intensity of, 400
  loudness of, 400
Square, completing the, 112
Square root, 56
  of a negative number, 125
Square root function, 236
Squaring function, 236
Standard form
  of a polynomial, 31
  of a quadratic equation, 109
  of the equation of a circle, 182
Stationary function, 232
Step function, 239
Stevin, S., 11

Strict inequalities, 15, 478
Subscript, 31
Subset, 3, 584
    proper, 584
Sum function, 260
Sum
    index of, 567
    of an arithmetic sequence, 573
    of a geometric sequence, 576
    of two cubes, 34
    of two matrices, 494
    of two vectors, 532
Summation notation, 567
Sylvester, J., 508
Symmetric property of equality, 2
Symmetry, 178
    tests for, 179
    with respect to the origin, 179
    with respect to the x-axis, 178
    with respect to the y-axis, 178
Synthetic division, 311
Systems
    consistent, 419
    equivalent, 432
    inconsistent, 419
    of equations, 418
    of linear equations, 419
    of linear inequalities, 478
    of nonlinear equations, 469
    solution of, 418
    to solve, 418
Systems of linear equations
        (solutions), 419
    Cramer's Rule, 459, 462
    matrix method, 445
    method of determinants, 457
    method of elimination, 433
    substitution method, 420
    using the inverse of a matrix,
        507

Tables
    common logarithms, 650
    exponential functions, 649
    natural logarithms, 652
Tartaglia, 328
Terminating decimal, 4
Terms of a sequence, 563
Test number, 145
Transcendental function, 365

Transitive property, 15
Tree diagram, 591
Triangle
    equilateral, 170
    isosceles, 171
    legs, 79
    medians of, 170
    Pascal, 557
    right, 79
Triangle inequality, 151
Trichotomy property, 16
Trinomial, 31
Truncation, 77
Tucker, Kevin, xii, 1956

Unbounded
    in the negative direction, 340
    in the positive direction, 338
    graph, 484
Uniform motion, 105
Union of sets, 585
Unit circle, 184
Unit vector, 536
Universal set, 586

Value
    of a determinant, 457, 461
    of a function, 220, 221
Value, present, 382
Variable, 2
    complex, 354
    dependent, 223, 354
    independent, 223, 354
Variation, 208
    combined, 211
    direct, 208
    inverse, 210
Varies
    directly, 208
    inversely, 210
    jointly, 211
Vector, 531
    column, 497
    components of, 536
    difference of, 533
    equal, 532, 538
    magnitude of, 535
    properties of, 532, 533, 534, 535
    resultant, 540
    row, 497

Vector (continued)
    scalar product of, 534
    sum of, 532
    unit, 536
    zero, 532
Venn diagram, 586
Vertex
    of a cone, 612
    of an ellipse, 624
    of a hyperbola, 634
    of a parabola, 285, 613, 619
    of the graph of a system of
        linear inequalities, 484
Vertical asymptote, 340
    locating, 342
Vertical line, 189
    equation of, 195
    test, 230
Vertical scale, change in, 251
Vertical shift, 247
Vièta, 121
Vinculum, 61

Weierstrauss, 23, 141

x-axis, 162
    reflection about, 255
    symmetry with respect to, 178
x-coordinate, 163
x-intercept, 176

y-axis, 162
    reflection about, 255
    symmetry with respect to, 178
y-coordinate, 163
    adding, 261
y-intercept, 176

Zero, 14
    approximating, 335
    bounds on, 332
    matrix, 495
    multiple, 320
    of a function, 320, 354
    of multiplicity m, 320, 356
    polynomial, 31
    repeated, 320
    vector, 532
Zero-level earthquake, 403

Designer: Janet Bollow
Cover designer: John Williams
Technical artist: Ben Turner Graphics
Copy editor: Patricia Cain
Production coordinator: Phyllis Niklas
Typesetter: Jonathan Peck Typographers, Ltd.
Printer and binder: R. R. Donnelley & Sons Company